CYCLODEXTRINS IN PHARMACEUTICS, COSMETICS, AND BIOMEDICINE

CYCLODEXTRINS IN PHARMACEUTICS, COSMETICS, AND BIOMEDICINE

Current and Future Industrial Applications

Edited by

EREM BILENSOY
Hacettepe University
Ankara, Turkey

WILEY

A JOHN WILEY & SONS, INC., PUBLICATION

Published by John Wiley & Sons, Inc., Hoboken, New Jersey.
Published simultaneously in Canada.

For general information on our other products and services or for technical support, please contact our Customer Care Department within the United States at (800) 762-2974, outside the United States at (317) 572-3993 or fax (317) 572-4002.

Wiley also publishes its books in a variety of electronic formats. Some content that appears in print may not be available in electronic formats. For more information about Wiley products, visit our web site at www.wiley.com.

Library of Congress Cataloging-in-Publication Data:

Cyclodextrins in pharmaceutics, cosmetics, and biomedicine : current and future industrial applications / edited by Erem Bilensoy.
     p. ; cm.
  Includes bibliographical references and index.
  ISBN 978-0-470-47422-8 (cloth)
  1. Cyclodextrins. I. Bilensoy, Erem.
  [DNLM: 1. Cyclodextrins–chemistry. 2. Technology, Pharmaceutical. 3. Biomedical Technology. QU 83]
  TP248.65.C92C93 2011
  660.6'3–dc22

                                       2010036222

Printed in Singapore

oBook ISBN: 9780470926819
ePDF ISBN: 9780470926802
ePub ISBN: 9780470934616

10 9 8 7 6 5 4 3 2 1

CONTENTS

CONTRIBUTORS

Füsun Acartürk, Gazi University, Etiler-Ankara, Turkey

Alka Ahuja, Oman Medical College, Muscat, Oman

Javed Ali, Hamdard University, New Delhi, India

Carmen Alvarez-Lorenzo, Universidad de Santiago de Compostela, Santiago de Compostela, Spain

Catherine Amiel, Institut de Chimie et Matériaux Paris Est, Thiais, France

Hidetoshi Arima, Kumamoto University, Kumamoto, Japan

Sanjula Baboota, Hamdard University, New Delhi, India

Uttam C. Banerjee, National Institute of Pharmaceutical Education and Research, SAS Nagar, India

Erem Bilensoy, Hacettepe University, Ankara, Turkey

Amélie Bochot, Université Paris–Sud, Paris, France

Marcus E. Brewster, Johnson & Johnson Pharmaceutical Research and Development, Beerse, Belgium

Nevin Çelebi, Gazi University, Etiler-Ankara, Turkey

Angel Concheiro, Universidad de Santiago de Compostela, Santiago de Compostela, Spain

François Donati, Université de Montréal, Montréal, Québec, Canada

Justin M. Dreyfuss, California State University–Northridge, Northridge, California

Dominique Duchêne, Université Paris–Sud, Châtenay Malabry, France

Hakan Eroğlu, Hacettepe University, Ankara, Turkey

Ana Rita Figueiras, University of Coimbra, Coimbra, Portugal; University of Beira Interior, Covilhã, Portugal

Ruxandra Gref, Université Paris–Sud, Châtenay, Malabry, France

A. Atilla Hincal, Education, Consultancy Ltd. Co. and Hacettepe University, Ankara, Turkey

Tetsumi Irie, Kumamoto University, Kumamoto, Japan

Phatsawee Jansook, University of Iceland, Reykjavik, Iceland

Abhishek Kaler, National Institute of Pharmaceutical Education and Research, SAS Nagar, India

Thorsteinn Loftsson, University of Iceland, Reykjavik, Iceland

Antonino Mazzaglia, Università di Messina, Messina, Italy

Agnese Miro, University of Naples, Naples, Italy

Keiichi Motoyama, Kumamoto University, Kumamoto, Japan

Maria D. Moya-Ortega, University of Iceland, Reykjavik, Iceland

Gulam Mustafa, Hamdard University, New Delhi, India

Steven B. Oppenheimer, California State University–Northridge, Northridge, California

Rachit Patil, National Institute of Pharmaceutical Education and Research, SAS Nagar, India

Hélène Parrot-Lopez, Université de Lyon, Lyon, France

Florent Perret, Université de Lyon, Lyon, France

Géraldine Piel, University of Liège, Liège, Belgium

Fabiana Quaglia, University of Naples, Naples, Italy

Stefano Salmaso, University of Padua, Padua, Italy

Amit Singh, National Institute of Pharmaceutical Education and Research, SAS Nagar, India

Vachan Singh, National Institute of Pharmaceutical Education and Research, SAS Nagar, India

Fabio Sonvico, University of Parma, Parma, Italy

NilüferTarimci, Ankara University, Ankara, Turkey

Juan J. Torres-Labandeira, Universidad de Santiago de Compostela, Santiago de Compostela, Spain

Francesco Trotta, Università di Torino, Torino, Italy

Francesca Ungaro, University of Naples, Naples, Italy

Francisco Veiga, University of Coimbra, Coimbra, Portugal

Amelia Vieira, University of Coimbra, Coimbra, Portugal

Véronique Wintgens, Institut de Chimie et Matériaux Paris Est, Thiais, France

PREFACE

Discovered toward the end of the nineteenth century, cyclodextrins have attracted the interest of scientists and industries in a variety of sectors. The main reason for this growing interest is the unique structure of the natural cyclodextrins, which enables inclusion of guest molecules in their apolar cavity and masking of the physicochemical properties of the included molecule. The included molecules, mostly hydrophobic, enter a cyclodextrin cavity totally or partially, depending on the size and configuration of the molecule. This book is limited to applications of natural and chemically modified cyclodextrins in the pharmaceutical, biomedical, and cosmetic fields. However, cyclodextrins find use in textile, food, agricultural, and environmental technologies, owing to their unique inclusion complex–forming capability. The relatively low cost of cyclodextrins, being enzymatic degradation products of starch, contributes to their large-scale production as pharmaceutical and cosmetic excipients and resulted recently in the use of a cyclodextrin derivative as an active ingredient in a pharmaceutical product.

Although they were discovered more than a century ago, these "100-year-old spinsters," as Prof. Dominique Duchene had called them at the 1998 CRS Workshop on Cyclodextrins, have been characterized by an ever-increasing number of publications and patents in the literature, which suggests that cyclodextrins continue to offer new horizons to scientists, with a wide range of possible modifications for adding novel properties to the natural cyclodextrins.

When one reviews the literature on cyclodextrins, the major characteristic of these ying-yang molecules seems to be their solubility enhancement and stability improvement effects on hydrophobic and/or labile active therapeutic or cosmetic ingredients. This effect causes a significant bioavailability enhancement of drug molecules with reduced efficacy due to lower drug absorption and plasma profiles as a result of their low solubility and stability problems, arising from hydrolysis, pH, and photodegradation.

Cyclodextrins produced on a large scale as industrial excipients are used primarily for their solubilizing effect, incorporated in the formulation of analgesic or anesthetic drugs with expected rapid onset. On the other hand, new groups of cyclodextrins are introduced in the pharmaceutical and biomedical fields every day. These exhibit a wide range of properties, including self-assembly, polymerization/condensation, gene delivery, swelling and gelling properties, encapsulation of perfumes and ingredients, and nano- and microencapsulation, which allows cyclodextrins to be actively researched as promising excipients in the nanomedicine, drug delivery, cosmetics, and biomedical fields.

This book consists of two main sections. Part I focuses on the general physicochemical properties of cyclodextrins, such as complexation, as well as drug solubilization and stabilization, which made them come into use in the first place, followed by specific chapters dedicated to various routes of administration, such as oral, mucosal, and skin. This part also covers the most recent findings on the toxicological overview and safety profiles of cyclodextrin derivatives and the regulatory status of cyclodextrins as excipients in the pharmaceutical industry, including the views and applications of regulatory authorities in different parts of the world and corresponding to different markets. The effects of cyclodextrins on the drug release properties of polymeric systems of different types are also discussed in this section, with examples from current literature.

Part II consists of novel and specialized applications of cyclodextrins based on the diversity of modified cyclodextrins. A major group of novel cyclodextrin derivatives are amphiphilic cyclodextrins with different surface charges.

Anionic, nonionic, and cationic amphiphilic cyclodextrins have been reported by research groups, and the self-assembly properties of these new cyclodextrin derivatives give them the capability to form nanoparticles spontaneously in addition to complex-forming properties. Applications of cyclodextrin polymers in gene delivery, peptide and protein delivery, biotechnological applications of cyclodextrins, and novel targeted cyclodextrins destined to carry their load to tumor cells or specific sites such as the colon in complex or conjugated form are also reviewed extensively in this part. Cyclodextrins and their incorporation into polymeric nanoparticles forming new drug delivery systems, cyclodextrin hydrogels, cellular interactions of cyclodextrins, and their relevance in the pharmaceutical and medical fields are discussed as well as the development and marketing story of sugammadex, a pharmaceutical product containing a cyclodextrin derivative as an active molecule. The emergence of cyclodextrins as active molecules rather than smart excipients in therapeutic or cosmetic products seems to be the next step in the discovery and development of cyclodextrin technology.

The goal of this book is to introduce readers of academic or industrial backgrounds to the diverse properties of cyclodextrins, different natural and modified cyclodextrins, and their applications and trends in cyclodextrin research which may be applicable to a variety of industries, such as the pharmaceutical, cosmetic, textile, environmental, and food industries.

ACKNOWLEDGMENTS

I would like to express my sincere thanks to my Ph.D. thesis supervisors, Professor Atilla Hincal and Professor Dominique Duchêne, who have opened for me the gates of the ever-promising cyclodextrin world. They kindly contributed to this book with significant chapters explaining properties, applications, and the regulatory status of cyclodextrins. The valuable contributions of all chapter authors have made this book possible, and I would like to thank all authors for their effort, time, and support. I owe special thanks to Dr. Hakan Eroğlu for his assistance in the preparation of the book, and I would also like to thank my editor at Wiley, Jonathan Rose, for encouragement, brilliant ideas, and support throughout the preparation and publication process.

Last but not least, I am indebted to my family—my husband, Tamer, and my daughter, Deniz—for their love and support during realization of this book.

EREM BILENSOY

PART I

CYCLODEXTRINS: HISTORY, PROPERTIES, APPLICATIONS, AND CURRENT STATUS

1

CYCLODEXTRINS AND THEIR INCLUSION COMPLEXES

DOMINIQUE DUCHÊNE

UMR CNRS 8612, Physico-Chimie–Pharmacotechnie–Biopharmacie, Université Paris–Sud, Châtenay, Malabry, France

1. INTRODUCTION

Cyclodextrins (CDs) are molecules of natural origin discovered in 1891 by Villiers. Studied by Schardinger at the beginning of the twentieth century, they became the topic of prominent scientific interest only in the late 1970s, early 1980s [1]. The main value of these oligosaccharides resides in their ring structure and their consequent ability to include guest molecules inside their internal cavity. This is at the origin of many applications: modification of the physico-chemical properties of the included molecule (i.e., physical state, stability, solubility, and bioavailability), preparation of conjugates, and linking to various polymers. This results in the use of CDs in many industries, such as agro-food, cosmetology, pharmacy, and chemistry. Presently, the annual average number of articles, book chapters, lectures, and scientific contributions is between 1500 and 2000.

Presented briefly in this chapter are the main cyclodextrins available on the market, and their major characteristics, focusing on their ability to yield inclusion complexes. Also described is the manner in which complexes can be obtained and studied.

2. MAIN CDs AND THEIR ABILITY TO INCLUDE GUEST MOLECULES

2.1. Main CDs

2.1.1. Natural CDs
CDs result from starch degradation by cycloglycosyl transferase amylases (CGTases) produced by various bacilli, among them *Bacillus macerans* and *B. circulans* [2]. Depending on the exact reaction conditions, three main CDs can be obtained: α-, β-, and γ-cyclodextrin, comprising six, seven, or eight α(1,4)-linked D(+)-glucopyranose units, respectively [3]. CDs are ring molecules, but due to the lack of free rotation at the level of bonds between glucopyranose units, they are not cylindrical but, rather, toroidal or cone shaped [4]. The primary hydroxyl groups are located on the narrow side; the secondary groups, on the wider side (Fig. 1).

Due to steric factors and tensions in the ring, CDs with fewer than six glucopyranose units cannot exist. On the other hand, although cyclodextrins with 9, 10, 11, 12, or 13 glucopyranose units (δ-, ε-, ζ-, η-, or θ-CD, respectively) have been described, only δ-CD has been well characterized [4]. The largest CDs, those with a helicoidal conformation, are rapidly reduced to smaller products.

The aqueous solubility of CDs is much lower than that of similar acyclic saccharides. This is the consequence of strong binding of CD molecules inside the crystal lattice. Furthermore, for β-CD, with its odd number of glucopyranose units, intramolecular hydrogen bonds appear between hydroxyl groups, preventing hydrogen bond formation with surrounding water molecules and resulting in poor water solubility [4] (Table 1).

The central cavity of CDs, which is composed of glucose residues, is hydrophobic when the external part is hydrophilic because of the presence of hydroxyl groups. In aqueous solution, water molecules inside the CD cavity can easily be replaced by apolar molecules or apolar parts of molecules, leading (reversibly) to an inclusion host–guest complex [5] which can be isolated.

When compared with its free molecular state, the included guest molecule has (apparent) new physicochemical

Cyclodextrins in Pharmaceutics, Cosmetics, and Biomedicine: Current and Future Industrial Applications, First Edition. Edited by Erem Bilensoy.
© 2011 John Wiley & Sons, Inc. Published 2011 by John Wiley & Sons, Inc.

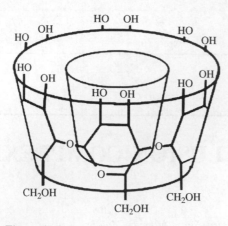

Figure 1. Schematic representation of α-CD.

Table 1. Main Natural CDs and Their Characteristics

Cyclodextrin	α	β	γ
Glucopyranose units	6	7	8
Molecular weight (Da)	972	1135	1297
Central cavity diameter (ext./int., Å)	5.3/4.7	6.5/6.0	8.3/7.5
Water solubility (at 25°C, g/100 mL)	14.5	1.85	23.2

properties, among which is higher apparent water solubility. This increase in water solubility depends on the CD water solubility, but this parameter is limited compared with linear oligosaccharides. This is one reason that highly water-soluble CD derivatives have been synthesized.

2.1.2. CD Derivatives   CDs' low aqueous solubility results from hydrogen bonds between hydroxyl groups. Any substitution on the hydroxyl groups, even by hydrophobic moieties, leads to a dramatic increase in water solubility [4]. The different CD derivatives still have the ability to include molecules inside their cavity, but with a different affinity than that of the parent CD. Among the water-soluble CD derivatives most often employed are three classes of modified CDs: methylated, hydroxypropylated (both neutral), and sulfobutylated (negatively charged).

Theoretically, methylation of CDs can occur on either two or three hydroxyl groups per glucopyranose unit. In the first case [dimethyl-cyclodextrins (DM-CDs)] the methylation takes place on all the primary hydroxyl groups (position C_6) and all the secondary hydroxyl groups in position C_2, the secondary hydroxyl groups in position C_3 remaining free. In the second case [trimethyl-cyclodextrins (TM-CDs)] all the hydroxyl groups are substituted, including those in C_3.

Most often, and in the case of β-CD, it is a randomly substituted CD that is used with an average substitution degree (number of substitutions per glucopyranose unit) of 1.8 (e.g., RAMEB, which is an amorphous product). There

also exists a very slightly substituted β-CD: Crysmeb, with a substitution degree of 0.5.

Hydroxypropylation occurs in a purely random manner on the primary or secondary hydroxyl groups, leading to an amorphous mixture. Most often, in the case of β-CD, it is 2-hydroxypropyl-β-cyclodextrin (HP-β-CD) that is used; this means that it is a 2-hydroxylpropyl moiety that is linked. Because of different producers, the substitution degree has to be mentioned.

There is only one sulfobutylated CD, the β-derivative, with 6.8 substituents per CD (SBE$_{7m}$-β-CD). It has about seven negative charges per CD, which are counterbalanced with sodium ions. Usually, a charged group reduces the CD complexation ability, but in the case of SBE$_{7m}$-β-CD, it shows high binding properties, due to the significant separation from the CD cavity of the charged sulfonate moieties [5].

2.2. Formation of Inclusion Compounds

2.2.1. Principle   The CD central cavity, composed of glucose residues, is lipophilic and in aqueous solutions can reversibly entrap suitably sized molecules (or parts of molecules) to form an inclusion complex [4]. Formation of an inclusion complex is the result of equilibrium between the free guest and CD molecules and the supramolecules of inclusion:

free CD + free guest ↔ CD/guest(inclusion complex)

Formation and dissociation of an inclusion complex is governed by a constant K, which may have different names: *affinity constant* (affinity of the guest molecule for the CD cavity), *stability constant* (stability of the inclusion complex in a nondissociated form), *association constant*, or *binding constant*. The higher the K value, the more stable the inclusion, and the less dissociation that occurs. The value of K depends on, among other factors, the size of the CD cavity and that of the guest molecule (or part of the molecule). It also depends on the more-or-less good fitting of the guest molecule inside the CD cavity. As a general rule, the complex is strong when there is size complementarity between the guest and the CD cavity [6]. Depending on their respective size, the guest molecule will enter the CD cavity at the narrow side (primary hydroxyl groups) or at the wide side (secondary hydroxyl groups) (Fig. 2).

2.2.2. Driving Force   The driving force for complex formation has been attributed to many factors, among them the extrusion of water from the cavity; hydrophobic, hydrogen bonding, and electrostatic interactions; induction forces; and London dispersion forces [7]. To better understand the inclusion mechanism, it is important to consider the thermodynamic parameters: the standard free-energy change

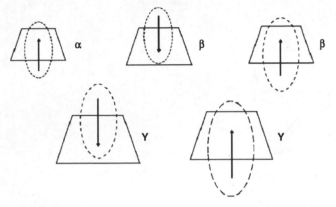

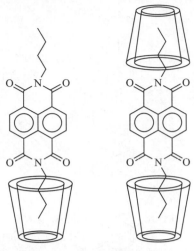

Figure 2. Influence of the guest and CD cavity size on the inclusion mechanism.

Figure 3. Example of type 1 : 1 (*left*) and 1 : 2 (*right*) inclusion complexes. (From [13], with permission.)

(ΔG), the standard enthalpy change (ΔH), and the standard entropy change (ΔS). Hydrophobic interactions are entropy driven (slightly positive ΔH and large positive ΔS). Van der Waals forces are characterized by negative ΔH and negative ΔS. Compensation (increasing enthalpy related to less negative entropy) is often correlated with water acting as the driving force. In this case, being unable to satisfy their hydrogen-bonding potentials, the enthalpy-rich water molecules from the cyclodextrin cavity are released from the cavity and replaced by guest molecules less polar than water, with a simultaneous decrease in the system energy [4].

2.2.3. Different Types of Complexes

When speaking of inclusion complexes, it is clear that an apolar molecule, or at least an apolar part of a molecule, is inside the CD cavity. But other complexes can be formed which are not inclusion complexes but in which the guest molecule is linked at the external part of the cyclodextrin [8]. Furthermore, depending on the respective size of the guest and host molecules, one guest molecule can interact with one or two (or more) CD (complexes 1 : 1 and 1 : 2) [9], or one or two guest molecules can interact with one CD (complexes 1 : 1 and 2 : 1). For example, Gabelica et al. [6] demonstrated that α-CD forms both inclusion and noninclusion complexes with dicarboxylic acids. The 1 : 1 acid/α-CD complex is mostly (but not totally) an inclusion complex, and the 2 : 1 complex results from the additional formation of a noninclusion complex by interaction of the acid with the 1 : 1 complex.

Loftsson et al. [10, 11] have shown that drug–CD complexes (such as CDs themselves) can self-associate to form aggregates or micelles in aqueous solutions, and that these aggregates can solubilize drugs inside their structures through noninclusion complexation. Furthermore, the less the CDs self-aggregate, the more likely it is that they are involved in interactions with guests [12].

Because of their conformation and size, some guest molecules can be included in one or two CDs (Fig. 3), and depending on the CD size, it is a different part of the guest

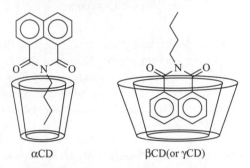

αCD   βCD(or γCD)

Figure 4. Influence of the CD size on the inclusion complex structure. (From [13], with permission.)

molecule that can be included (Fig. 4) [13]. Molecules with aliphatic chains fit better into the small α-CD cavity, whereas molecules containing phenyl groups fit better into the larger cavity of either β- or γ-CD [6]. Finally, in solution, multiple inclusion equilibria can coexist (Fig. 5) [14].

2.2.4. Influence of CD Characteristics

The nature of CDs has a tremendous influence on their complexation ability. Obviously, the size of a CD is important: It has to be large enough to allow guest entrance but not so large as to be unable to create guest–CD interactions by maintaining the guest molecule inside the cavity, thus preventing a too easy dissociation of the inclusion (low stability constant). The CD derivative substitutions also play a prominent role. In fact, they can either hinder the entrance of the guest or contribute to increasing guest–CD interactions, such as hydrogen bonds between the hydroxyl groups of hydroxylpropyl-CDs and the guest.

In the case of charged SBE_{7m}-β-CD, it is known that placing a charged group on or around a CD usually reduces its complexation ability. This is the consequence of a change in

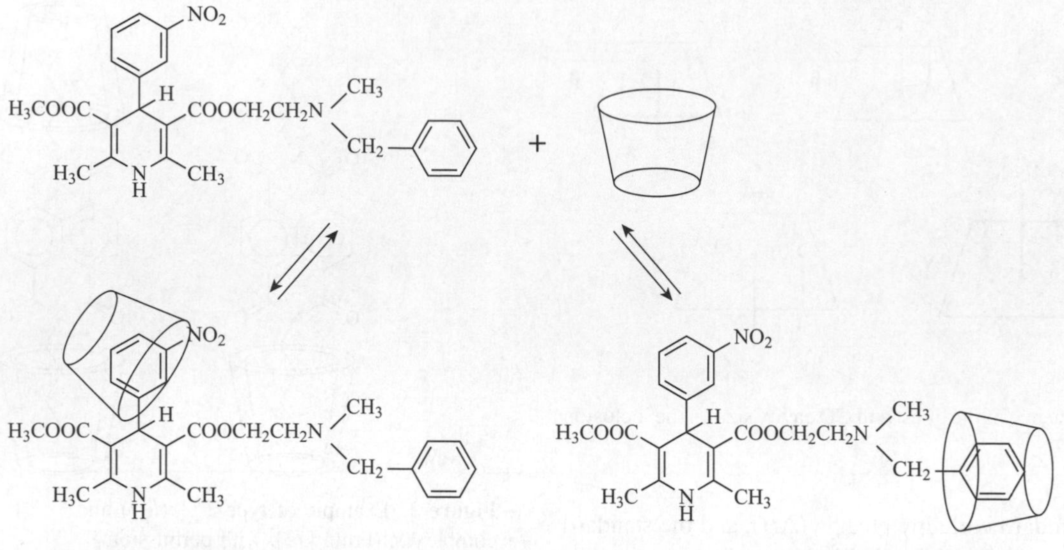

Figure 5. Example of coexistence of multiple inclusion equilibria in solution. (From [14], with permission of Elsevier.)

the CD cavity hydrophobicity and/or a change in the inclusion complexation geometry [15]. However, due to significant separation of the charged substrate moiety from the CD and the possible interaction of some substrates with portions of the butyl moiety, SBE_{7m}-β-CD often shows better binding than that of neutral CDs [5].

2.2.5. Influence of the Reaction Medium An inclusion complex between a guest and a CD can be obtained when both entities are in a molecular state. Thus, the complexation efficiency depends on the guest intrinsic solubility (S_0) and the complex affinity constant (K):

$$\text{complexation efficiency} = KS_0$$

An increase in the complexation efficiency can be obtained by an increase in either the guest intrinsic solubility or the complex affinity constant, or by a simultaneous increase in both parameters [16]. In fact, the problem relies most on the guest solubility, the CD solubility most often being much higher than that of the guest. Any substance capable of increasing the guest solubility could be considered to be favorable to the inclusion. But that is not always true.

Organic solvents such as ethanol can increase the guest's water solubility. However, it competes with the guest for space in the CD cavity, and most often the results are not those expected [17]. In the case of low-water-soluble basic guests, a better method consists of using acids as solubilizers. The enhancement in complexation ability results from both an increase in water solubility of the guest [17] and an increase in the affinity constant due to noncovalent multicomponent (or ion pair) association between the CD, the basic drug, and the acid [18–20]. If ionization of the guest increases its solubility, it can decrease the stability constant,

but the increase in solubility remains predominant. Water-soluble polymers can increase the complexation efficiency by an increase in the stability constant [21].

3. PREPARATION OF INCLUSION COMPLEXES

The method used to prepare an inclusion complex between a CD and a guest compound has a significant influence on the final product: yield, solubility, and stability of the complex. Most often the nature of the CD to choose depends on the future role of the inclusion. For example, in pharmacy the CD chosen depends on the drug administration route, the fact that the CD is registered in one (or more) of the main pharmacopeia, and the price of the CD. The preparation method has to be adapted to the production level (i.e., industry or laboratory scale) and the objective (i.e., increase in solubility, in stability, etc). Finally, the necessity to add a third or a fourth component for better product solubility has to be considered.

3.1. Preparation Methods

Many methods have been described for the preparation of inclusion complexes. It should, however, be kept in mind that except when the inclusion precipitates spontaneously from the preparation medium, the product obtained is a mixture of three compounds: inclusion complex, empty CD, and free guest. The proportion of inclusion compound is related to the affinity constant of the inclusion complex obtained.

To avoid this drawback, many years ago it was proposed that inclusion compounds be prepared by spontaneous precipitation of the complex from a solution or dispersion of the guest ingredient dispersed in an aqueous CD solution. The final product had to be washed by organic solvent to eliminate the excess of nonincluded guest. Of course, there are many disadvantages to this technique. To obtain an acceptable yield, it is usually necessary to use a cosolvent of the guest, such as an organic solvent, which unfortunately competes with the guest for inclusion in the CD; and in any case, the yield is very low and the method rather long. Also, and significantly, to be able to precipitate, the complex (and the parent CD) must have low solubility, so the method is restricted to β-CD. The only advantage is that the product obtained is the inclusion only, not a mixture. In fact, it has no industrial use.

Very often, the characteristics of the inclusion complex are compared with those of a physical mixture prepared using the same proportions of guest compound and CD. The hydrophilic character of CDs, comparable to that of saccharides, leads to an increase in the guest compound's solubility. Furthermore, one cannot exclude the progressive formation of an inclusion complex during a dissolution study. Anyhow, a physical mixture is just a blend and not an inclusion compound.

The co-grinding method [22], in which a physical mixture is submitted to ball-milling in a high-energy vibrational micromill, is interesting because it leads to an almost amorphous product that presents a high level of solubility with fast dissolution. Very similar results have been reported for many products co-ground with cellulose derivatives. In these cases the explanation was the polymer's role in facilitating amorphization. A similar explanation seems to be logical for co-grinding with CDs.

3.1.1. Co-Evaporation
Co-evaporation consists of mixing the guest ingredient in water with the CD (generally, in equimolar amounts) and other components when necessary. The mixing time can be some hours. The solvent can be removed, at a temperature compatible with the stability of the products, in hot air [22] or a vacuum oven [23], or better, to accelerate the process, by evaporation under vacuum in a rotary evaporator [24–26]. The product obtained is more or less crystalline, depending on the nature of the constituents and the exact drying method employed.

3.1.2. Spray-Drying and Freeze-Drying
Spray-drying [24, 27, 28] and freeze-drying [24, 25, 27–29] methods are derivatives of the co-evaporation method. To have solutions of good quality adapted to the drying process, they are stirred previously for one or two days or even sonicated. As shown by x-ray diffractograms, the products undergo amorphization during the drying process. Furthermore, the spray-dried product has the appearance of small spheres [24, 27], whereas the freeze-dried product is more amorphous but still has a few crystalline particles [24, 25, 27]. Because of the amorphization, dissolution of both

products is very rapid, freeze-drying leading to the fastest dissolution [27].

3.1.3. Kneading
Although many processes have been named kneading method, the use of kneading seems to be restricted to the preparation of CD inclusions. Briefly, the guest compound is kneaded together with CD and a small proportion of water or an aqueous solution of ethanol [22, 24, 29], acid [25], or base [27] is progressively added to obtain a slurry. The product may be set aside to equilibrate for 24 or 48 h [24, 27]; or it is kneaded to complete evaporation [22] or dried at 40°C [25] or under vacuum [29]. Amorphization results from the kneading process, but it is still possible to observe some crystals of the original CD and guest compound, as confirmed by x-ray diffractometry and differential scanning calorimetry [24, 27]. Generally, dissolution is better than that of the corresponding physical mixture but slower than that obtained by spray-drying or freeze-drying [27]. However, the results depend on the complex composition: the nature of the CD, the guest compound, and the additives.

3.1.4. Sealed-Heating
In the sealed-heating method, CD, guest compound, and additional products, at the desired molar ratio, are placed in a glass container with a very small amount of water. The container is then sealed and kept for 10 to 60 min, or more often 3 h, in an oven at a temperature of 75 to 90°C [22, 26, 30]. Despite the fact that the complex obtained retains some crystallinity [22, 26], its dissolution can be increased dramatically [22].

3.1.5. Supercritical Carbon Dioxide
Supercritical fluids are fluids used at temperatures and pressures above their critical value. They are good solvents for nonvolatile and thermolabile compounds. They have gaslike viscosities and diffusivities that promote mass transfer. Their density is similar to that of liquid solvents [31]. In supercritical fluids the diffusivity of dissolved entities is higher than in liquid solvents [32]. This characteristic is favorable to inclusion formation. Carbon dioxide has been widely used since it is safe, inexpensive, nonflammable, and usable in relatively mild processing (supercritical point: 73.8 bar, 31.1°C) [33].

Various types of equipment have been described for the preparation of inclusion complexes. In one type, a physical mixture of guest compound and CD is submitted to supercritical carbon dioxide in a static mode under the required pressure. Following depressurization at the end of the process, the product can be ground and homogenized [31, 34, 35]. Another type of equipment utilizes two main units: one for extraction and one for complex formation. The extraction cell consists of a high-pressure sight gauge packed with alternate layers of glass wool and guest compound. The complex formation unit is loaded with CD. The supercritical solution of guest compound passes through the CD, and the complex formation cell is isolated and left in static mode. Depressurization is complete within 10 min [36]. The products obtained are less crystalline than the corresponding physical mixtures [22, 26], the intensity of crystallinity depending on the exact preparation conditions (e.g., temperature, pressure). Dissolution is faster than that of the physical mixture [22, 35], and the bioavailability (liver and kidney tissues) is very good [35].

3.1.6. Microwave Treatment
Microwave treatment makes it possible to obtain rapidly high temperatures inside irradiated products. Applied to the preparation of inclusions, it reduces the reaction time significantly [9]. For the preparation of complex, a mixture of guest compound and CD with a minimum amount of solvent is subjected to microwave treatment, most often for 90 s at 60°C (150 W) [9, 37–39]. It also seems possible to subject the physical mixture itself to microwave treatment (500 or 750 W for 5 to 10 min) [30]. The products obtained using this technique show a practically unchanged solid state and are very stable under ambient conditions. Microwave seems more efficient than kneading and co-grinding with respect to dissolution [30].

3.1.7. Choosing a Preparation Method
The physicochemical and dissolution properties of inclusion compounds are influenced not only by the constituents—guest compound, CD, ternary or even quaternary systems (organic solvent, acid, base, polymer)—but also by the preparation method. The co-grinding and kneading techniques require only a short operating time and are both potentially industrializable. They could be of great interest for obtaining a limited increase in solubility and dissolution without the necessity of true inclusion [40]. Sealed heating achieves only small amounts of product and is not industrializable. Spray-drying is more expensive, and freeze-drying is both expensive and time consuming, although it seems to be efficient for obtaining true inclusion and amorphization, leading to a rather fast-dissolving product [40]. The use of supercritical carbon dioxide is still experimental. The microwave technique is extremely fast and leads to true inclusions when carried out on products in the presence of liquid [38]. When it is carried out on dried products the unchanged solid state suggests an absence of true inclusion [30]. The absence of true inclusion or a low proportion of true inclusion in the product obtained means that the guest molecule is not protected from the surrounding medium and that its stability will not be improved. However, its solubility can be increased enough for the objective looked at.

3.2. Additives

Various additives can be added either to increase the yield by increasing the affinity constant or to improve the solubility of the inclusion complex. As noted earlier, these additives are water-soluble polymers, and acids, or bases used to form ternary or quaternary systems.

3.2.1. Water-Soluble Polymers
The role of water-soluble polymers in the formation and/or solubility of inclusion complexes is multiple and probably depends on both the guest compound and the CD itself. It appears that the role of polymers is not only additive but also synergistic to that of CDs [41, 42]. Polymers can improve the water solubility of the guest compound, a factor favorable to inclusion in CD.

The thermodynamic role of poly(vinylpyrrolidone) (PVP) has been demonstrated. The addition of PVP to the complexation medium of a series of drugs by HP-β-CD results in an increased negative enthalpy change ($\Delta H°$), together with an increased negative entropy change ($\Delta S°$). Thus, the complexation is enhanced (the affinity constant K is increased) upon addition of PVP [41].

When comparing polymers, it has been shown that poly(ethylene glycol) (PEG) has little or no effect on the dissolution of guest/CD complex [42, 43]. This is due to the linearity of the polymer, which can form an inclusion with the CD itself, thus competing with the guest drug. On the other hand, "bulky" polymers such as PVP and hydroxypropyl methylcellulose (HPMC) can form hydrogen bonds with hydroxyl groups of CDs and, more especially, HP-β-CD, leading to a noninclusion complex in which guest molecules can be included in the form of a ternary complex [43]. A very interesting example is that of the inclusion of nabumetone in β-CD, in which the drug is wrapped at both ends by a β-CD molecule, the PVP polymer acting as a bridge between the two β-CD molecules [44].

Numerous polymers have been investigated for their ability to increase the affinity constant and solubility of guest/CD inclusion complexes. The most frequently studied have been PEG [42, 43], ploy(vinyl alcohol) (PVA) [45], PVP [41–44, 46–49], HPMC [42, 43, 46], carboxymethylcellulose (CMC) [46], and NaCMC [42, 49]. Unfortunately for future users of CDs and polymers, there is no general conclusion. Depending on the nature of the guest and the CD, the best products are PVP, HPMC, and NaCMC.

The exact preparation method may have some influence on the result. For example, heating the preparation medium (guest, CD, and polymer in aqueous solution) to 120 to 140°C for 20 to 40 min has been claimed to increase the affinity constant and the complexation efficiency of HP-β-CD [16]. Similar results have been obtained by heating at 70°C under sonication for 1 h [21, 43]. The mechanism of this phenomenon, called *polymer activation*, is not known. Some contradictory results have been published describing the absence of the effects of such treatment [42].

3.2.2. Acids and Bases
In the case of acidic or basic ionizable guest compounds, it seems appropriate to adjust

the pH to obtain the higher solubility of the guest, allowing easier inclusion formation and leading to better complexation efficiency [16]. This can be obtained by the addition of a base [23, 50], an acid [50], or by the use of phosphate buffers [43, 49]. Volatile bases (ammonia) or acids (acetic acid) can be removed from the complex during the drying process. A general increase in the guest compound's apparent solubility is obtained [16, 43, 49, 50]. In any case, the relative increment in solubility with respect to the guest alone obtained by cyclodextrin complexation at the optimal pH can be rather low because of a concomitant reduction in the stability of the complex formed with the ionized guest [49].

In dissolution experiments, precipitation of the guest compound can occur because of a thermodynamically unstable oversaturation of the solution [50]. The high-energy guest/CD complex obtained could lead to enhanced drug delivery through biological membranes and, consequently, enhanced drug bioavailability compared to conventional guest/CD complexes [50]. Tartaric acid has been proved to increase the water solubility and oral bioavailability of vinpocetine when included in either β-CD or SBE-β-CD [51].

A very interesting study has been carried out on the role of maleic, fumaric, and tartaric acids on the inclusion ability of miconazole in β-CD and HP-β-CD [52]. For the ternary complexes obtained, depending on their conformation and/or their structures, the acids can either stabilize or destabilize the complex. In β-CD, maleic acid presents the best conformation for forming a ternary complex. The inclusion yield with this acid is higher than with fumaric. Tartaric acid (L or D) does not affect the inclusion yield; in fact, it has affinity for the CD cavity and can extract miconazole. With HP-β-CD, L-tartaric acid stabilizes the complex, increases the interaction and complexation energies, and promotes miconazole inclusion. L-Tartaric acid does not interact with the imidazole ring of miconazole as maleic and fumaric acid do.

3.2.3. Other Additives

The role of various additives, known as *solubilizing agents*, on the solubility of guest/CD inclusion complex has been investigated.

Anionic organic salts such as sodium acetate and sodium benzoate increase the aqueous solubility of hydrocortisone/β-CD complex [8]. Normally, sodium salicylate forms inclusion complexes with β-CD and should compete with hydrocortisone, resulting in reduced complexation and CD solubilization of hydrocortisone. The favorable effect of sodium salicylate cannot be explained by simple inclusion formation. In the case of sodium acetate, the enhanced solubilization is partially due to increased β-CD and hydrocortisone/β-CD complex solubility. The acetate ions solubilize the hydrocortisone/β-CD microaggregates formed in aqueous solution [8].

For its part, the cationic organic salt benzalkonium chloride has only a limited effect on the hydrocortisone/β-CD complex solubility, possibly because of competing effects between benzalkonium and hydrocortisone for space in the CD cavity [8].

The role of phospholipids (egg phosphatidyl choline and phosphatidylglycerol) on ketoprofen/β-CD and ketoprofen/M-β-CD solubility has also been investigated [30, 39]. The ternary systems obtained have higher solubility, especially ketoprofen/phosphatidylcholine/β-CD. The synergistic effect between cyclodextrins and phospholipids in enhancing drug dissolution is attributed to a combination of the surfactant properties of phospholipids and the wetting and solubilizing power of CDs and/or the possible formation of a multicomponent complex [39].

3.2.4. Quaternary Systems

Considering the dissolution enhancements obtained with ternary systems in which are present the guest compound, the CD, and an additive such as polymer, acid, or anionic organic salts, it was logical to investigate the effect of several additives used in quaternary systems. For example, the association sodium acetate–HPMC in the preparation of a quaternary complex of hydrocortisone/β-CD appears to have a better solubilizing effect than that of one or the other additive used alone [8]. Similar results were obtained with the association tartaric acid–PVP added to vinpocetine/SBE-β-CD [51].

4. PHYSICAL STUDIES OF INCLUSION COMPLEXES

When preparing an inclusion complex, the objectives can be of different types: modification of the physical state (liquid or gas transformed into solid), masking of an unpleasant odor or taste, enhancement of solubility, enhancement of stability, and so on. It is not enough to control these effects; it is necessary to know exactly what complex has been obtained. Different physical studies can be carried out in order to check the existence of a true complex, evaluate its stoichiometry, calculate its stability constant, and discover its structure. Different types of studies can be carried out.

4.1. Characterization of the Complex

The objective is to know if there is a complex or just a mixture of guest, CD, and guest compound.

4.1.1. Scanning Electron Microscopy

Very often, study of an inclusion complex begins by observing it using scanning electron microscopy (SEM). For the observation, samples are fixed on a brass stub and made electrically conductive by coating with a thin layer of copper [27], gold [23, 24, 40], or gold–palladium alloy [53]. For a good comparison, samples should be observed at the same magnification. It is, however, difficult to conclude as to the formation of an inclusion because of the morphological change that occurs between the physical mixture and the product obtained. The preparation process generally has a great influence on the characteristics of the product. For example, spray-drying very often leads to small spheroids; co-evaporation or kneading, to large fragments; and freeze-drying, to thin and more or less crystalline particles [24, 25].

4.1.2. Ultraviolet Spectroscopy

Because it is a very simple method, the ultraviolet (UV) absorbance spectrum of complexes has been used to study inclusion complexes [23, 29, 54]. Cyclodextrins do not show any significant UV absorbance, so the increase in absorbance observed in a guest/CD solution results from perturbation of the chromophore electrons of the guest by its inclusion in the CD [29].

4.1.3. Circular Dichroism

Being symmetrical molecules, CDs have no dichroic activity, but they can modify that of guest molecules by perturbation of the microenvironment polarity resulting from the inclusion [54, 55].

4.1.4. Differential Scanning Calorimetry

Differential scanning calorimetry (DSC) analysis is very often carried out on raw materials and products resulting from the complex preparation process [22, 24–27, 29, 30, 40, 53, 55]. In fact, when guest molecules are included in the CD cavity, their melting, boiling, and sublimation points usually shift to a different temperature or disappear within the temperature range at which the CD is decomposed [53].

Generally, the samples are heated in a sealed pan from 25 to 250°C, 300°C, or even 450°C at a rate of 5 or 10°C/min, under nitrogen or air. An empty sealed pan is taken as a reference. CDs, in particular β-CD, contain water molecules inside their cavity. This water is released at the beginning of the temperature increase. In a sealed pan, the presence of vapor can perturb the observation of further thermal accidents, especially if the scanning rate is too fast. To prevent this drawback, some authors prefer to work with pierced pans [22, 30].

In a classical experiment, the melting peak of the guest will disappear or decrease (or shift) by its inclusion in a CD, depending on the proportion or true inclusion and free guest in the product under investigation. However, the disappearance of or decrease in the guest melting peak can result from its amorphization by the preparation process, in particular freeze-drying. Thus, the results have to be interpreted with great care.

4.1.5. Infrared Spectroscopy

Infrared (IR) analysis can give interesting information on the products obtained by association of a guest to a CD and is frequently employed [22, 24–27, 29, 30, 40, 53], either as classical IR spectroscopy or as Fourier-transformed infrared spectrometry (FTIR). IR analysis is carried out on the powders included in a KBr disk [29, 53]. FTIR analysis can be performed either on powder samples dispersed in Nujol [22, 26, 30, 40] or directly on the powder samples themselves [24, 27], which technique prevents any transformation of the products. At present the FTIR method is the one most commonly employed. A classical FTIR analysis is performed by application of 16 scans at a resolution of $4\,cm^{-1}$ over the range 4500–4000 to 600–400 cm^{-1}.

The physical mixture leads to a superposition of the two spectra (i.e., guest, CD) without any change. Normally, in a simple inclusion complexation no new bands should appear, which would be indicative of new chemical bonds in the product obtained corresponding to another type of interaction [27]. On the other hand, inclusion complexation leads to significant changes in the characteristic bands of the guest molecule. For example, the strong reduction or complete disappearance of the characteristic bands is indicative of strong guest–CD interactions and possibly inclusion complexation [24, 53]. A shift of a carbonyl stretching band to higher frequencies with concomitant broadening and decrease in intensity can be attributed to the dissociation of intermolecular hydrogen bonds associated with crystalline molecules and can be observed for complexes obtained by freeze-drying [29].

4.1.6. X-ray Diffractometry

Powder x-ray diffractometry (XRD) is used to measure the crystallinity of a product. Even if a change (lost) of crystallinity does not prove the inclusion, it is very frequently employed in the study of inclusion complexes [22, 24–27, 29, 30, 40, 53]. Most often the analysis is carried out with Cu [22, 26, 29, 30, 53] or Co [24, 25, 27, 40] Kα radiation with a voltage of 40 to 45 kV and a current of 35 to 40 mA over the 2θ range 2–5°C to 38–70°C.

In an XRD pattern the intensity of diffraction peaks is indicative of the crystalline character of the product. A hollow pattern is characteristic of amorphous products [27]. The relative degree of crystallinity can be calculated as the ratio of the peak height of the sample under investigation to that of the same angle for the reference with the highest intensity [24, 25, 27].

The different cyclodextrins do not exhibit the same crystallinity: β-CD and DM-β-CD are rather crystalline, whereas M-β-CD (Crysmeb), SBE-β-CD, and HP-β-CD exhibit an amorphous character.

Diffractograms of physical mixtures result from a combination of the diffractograms of the components analyzed separately. When a CD has an amorphous character, a decrease in peak intensity can be observed [27].

When studying a prepared inclusion complex, a decrease in crystallinity, shifts in and the disappearance of peaks, the appearance of new diffraction peaks, or a completely diffuse pattern might be related to possible guest amorphization and/or complexation [26]. A strong reduction in or the complete disappearance of the guest characteristic peaks can be indicative of strong guest–CD interactions and the possible inclusion complexation of the guest [24] or its molecular dispersion in the CDs [53]. Very often the amorphization observed for the inclusion complex is largely dependent on the preparation method. For example, and depending on the products, co-evaporation and sealed heating lead to rather crystalline profiles [26], whereas kneading or supercritical fluids have a variable effect [26]. Most often, freeze-drying leads to some amorphization [25, 53].

4.1.7. Electrospray Mass Photometry

Electrospray (or electrospray ionization) mass spectrophotometry (ES-MS) is a very powerful method of studying inclusion complexes [5, 54, 56–58]. It is a soft method of ionization for nonvolatile and thermolabile molecules which can hardly induce fragmentation [54]. It can provide evidence of complexation and stoichiometry on the basis of the molecular weights of all vaporized species [56–58]. However, there are still ambiguities in the spectra obtained for supramolecular assemblies: Do the species present in the mass spectra correspond to those present in solution, or do they result from processes that occur under high-vacuum conditions? [58].

4.1.8. Proton Nuclear Magnetic Resonance

Proton nuclear magnetic resonance (^{1}HNMR) spectroscopy is probably the technique that can give the most accurate information about inclusion formation [48, 53, 55, 56, 58, 59]. It can be used to prove the inclusion existence, and to determine its stoichiometry, affinity constant, and structure. Samples are dissolved in D_2O [48, 58], D_2O/CD_3OD [53, 55], or CD_3OOD/D_2O [56] and the experiment is performed at 300, 400, 500, or 600 MHz. Of course, the product that is investigated is a solution and not the solid complex. Insertion of a guest molecule into a CD cavity results in the modification of ^{1}H NMR frequencies. Major changes in the chemical shift values of the CD protons—more specifically, H3 and H5 located inside the cavity, or H6 on the cavity rim—indicate the formation of an inclusion complex [14, 48, 53, 59]. Guest protons interacting with the CD can be evidenced, and noninclusion complexes can be characterized.

4.2. Stoichiometry and Constant of the Complex

As already mentioned, guest/CD complexes can involve one or more guest molecules for one or more CDs and are characterized by their stability constant. If m guest molecules (G) associate with n CD molecules (CD),

$$_mG + {_n}CD \leftrightarrow G_m \cdot CD_n$$

$K_{m:n}$ is the stability constant of the guest–CD complex [60].

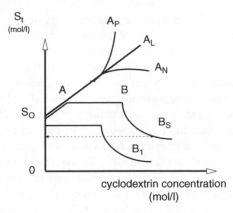

Figure 6. Higuchi phase solubility diagram.

4.2.1. Higuchi Phase Solubility Diagram

The phase solubility analysis described by Higuchi and Connors [61] is a very classical investigation carried out to better define the complex type [10, 62]. To obtain the corresponding diagram (Fig. 6), a fixed amount of guest compound is added to a series of CD solutions of increasing concentration with a constant volume. It is necessary to use an excess of guest compound in order to maintain the highest possible thermodynamic activity. These solutions are agitated for several hours (or days) up to equilibrium. After filtration the dissolved guest concentration is measured by an appropriate method. The value obtained corresponds to the guest really dissolved (the intrinsic solubility) plus the guest dissolved in inclusion form; it is the guest apparent solubility [62]. Two types of complexes can be obtained: A (a soluble inclusion complex is formed) and B (an inclusion complex with definite solubility is formed) [63].

In type A, the apparent solubility of the guest increases as a function of CD concentration. Three possible profiles exist: A_L, A_P, and A_N. A_L corresponds to a linear increase in solubility with an increase in CD concentration. A_P corresponds to a positive deviation from linearity (the CD is more effective at high concentrations), and A_N corresponds to a negative deviation (the CD is less effective) [62]. In the A_L case and assuming that the complex is of $1:1$ type, the stability constant of the complex can be calculated from the slope of the isotherm [63]:

$$K = \frac{\text{slope}}{S_0(1-\text{slope})}$$

In the A_P case, indicative of higher-order inclusion complexes, the K value can be calculated using the iteration method. It is difficult to analyze the diagram quantitatively because this system is associated with factors such as solute–solvent or solute–solute interactions [63]. In type B there is formation of complexes with limited water solubility, which are traditionally observed with β-CD. Two different possible profiles exist: B_S and B_I. B_S corresponds first to the formation of a soluble complex, which increases the total solubility of the guest, but at a particular point of this solubilization process, maximum solubility is achieved, corresponding to the guest intrinsic solubility plus the guest solubilized in inclusion complex form. Additional CD generates additional complex that precipitates, but as long as solid guest remains, dissolution and complexation can occur, maintaining a plateau. When the entire solid guest has been consumed, further addition of CD results in the formation of additional insoluble complex, and the final solubility observed in the system is that of the complex itself [62]. B_I is similar to B_S except that the complex is so insoluble that it does not increase the guest apparent solubility. The stability constant of B_S complexes can be calculated from the slope of the ascending part of the isotherm, using the same equation as for A_L complexes [63].

4.2.2. Permeation
When using artificial membranes permeable to the guest but impermeable to the larger CD, the permeation profile of the guest in the presence of CDs in the donor phase is related not only to its permeation rate but also to the complex stability constant [64, 65]. The relationship between the guest permeation rate and the stability constant of the complex is given by

$$[G_A] = \frac{[G_0]\{1 - \exp(-At)\}}{2 + K[CD_f]}$$

where $A = k(2 + K[CD_f])/(1 + K[CD_f])$, with $[G_A]$ being the guest concentration in the acceptor phase, $[G_0]$ the concentration in the donor phase at time 0, $[CD_f]$ the concentration of the free CD in the donor phase, k the guest permeation rate constant, K the complex stability constant, and t the time. Therefore, K and k can be calculated by analyzing the guest concentration data in the acceptor phase as a function of time using a nonlinear least-squares method [64, 65]. This method has been employed in the determination of the stability constant of hydrocortisone/HP-β-CD, a 1:1 A_L-type complex [11] and to that of flurbiprofen/M-β-CD and flurbiprofen/HP-β-CD [65]. It cannot be used for complexes with β-CD because its poor water solubility does not allow having a high enough CD concentration in the donor phase [65].

4.2.3. Nuclear Magnetic Resonance
The association constant can be derived from NMR data using the Benesi–Hildebrand method [66]. This is a graphical approach based on the observation of any parameter A (provided that it is affected by the interaction process considered) for one of the entities in the presence of a large but variable excess of the other entity, B. Chemical shift differences for a given proton can be used as variables [59, 67]. The equilibrium constant can be written

$$K([A]_t - [C])[B]_t = [C]$$

[C] is related to the chemical shift difference between the free molecule and the complex by

$$[C] = \Delta P_{obs}[A]_t \Delta \delta_c^{-1}$$

where $\Delta \delta_c$ is the chemical shift difference. This leads to

$$\Delta \delta_{obs} = K[B]_t \Delta \delta_c (1 + K[B]_t)^{-1}$$

The Benesi–Hildebrand graphical method allows to rewrite this equation in the form

$$(\Delta \delta_{obs})^{-1} = (K \Delta \delta_c)^{-1}([B]_t)^{-1} + (\Delta \delta_c)^{-1}$$

Plots of $(\Delta_{obs})^{-1}$ against $([B]_t)^{-1}$ are linear. The slope, abscissa, and ordinate intercepts are $(K \Delta \delta_c)^{-1}$, $-K$, and $(\Delta \delta_c)^{-1}$, respectively [67].

A 1:1 stoichiometry is assumed in the theoretical basis.

This method has a number of limitations: the entities should be soluble enough, there is a lack of sensitivity for low concentrations, and the effect of viscosity in the presence of a large excess of one of the entities must be accessed. The accuracy of the method drops rapidly as K increases [59].

Other methods have been described for obtaining the stability constant from NMR data. For example, the diffusion-ordered spectroscopy (DOSY) technique can be used [68]. The association constant K for a complex of n-molecule host (CD, H) and m-molecule guest (G),

$$n\text{CD} + m\text{G} \leftrightarrow \text{C}[H_n G_m]$$

could be reduced to

$$K = \frac{[C]}{[H]^n [G]^m} = \frac{[C]}{([H]_0 - n[C])^n ([G]_0 - m[C])^m}$$

where $[G]_0$ and $[H]_0$ are the total concentration of the guest and host, and [G], [H], and [C] are the equilibrium concentrations of the free host (H), free guest (G), and the complex (guest/CD, C). If the mole fraction χ_b of the bound entities is known, K is

$$K = \frac{\chi_b}{(1 - \chi_b)([H]_0 - \chi_b[G]_0)}$$

The diffusion coefficient observed (D_{obs}) in the NMR experiment (fast exchange conditions) is the weighted average of the diffusion coefficient of bound (D_{bound}) and free (D_{free}) guest:

$$D_{obs} = \chi D_{bound} + (1 - \chi) D_{free}$$

and the fraction of bound guest is

$$\chi_b = \frac{D_{\text{free}} - D_{\text{obs}}}{D_{\text{free}} - D_{\text{bound}}}$$

Accurate determination of the association constant implies that the stoichiometry is known unambiguously [68]. The association constant can also be obtained by using a nonlinear least-squares procedure, resorting to the Levenberg–Maquardt algorithm on the differences observed in the chemical shifts due to the presence of CD [69]. The values are calculated using the protons of the guest that lead to largest chemical shift variations in the presence of increased cyclodextrin concentrations [27].

NMR studies are also a tool for determination of the complex stoichiometry. The method is that proposed many years ago by Job [70]. It deals with fast exchange systems and can be applied to any technique provided that a given experimental parameter is different in the free and bound states [59]. This parameter is determined for a series of samples prepared by mixing, to constant total volumes, equimolar solutions of the two interacting entities, the total concentration being kept constant. The ratio between the concentrations of the two entities A and B is

$$r = \frac{[A]}{[A] + [B]}$$

this parameter varying from 0 (pure B) to 1 (pure A).

A parameter P being observed as a function of r, its measured value P_{obs} is given by

$$P_{\text{obs}}[A_t] = P_c[C] + P_f[A]$$

where P_c and P_f are the values of the parameter P observed in the complexed and free forms of A, $[A_t]$ and $[A]$ being the total and free concentrations of A. Plotting $\Delta P_{\text{obs}}[A_t]$ as a function of r gives a bell-shaped curve exhibiting a maximum for $r = 1/1 + n$, allowing direct determination of n. ΔP_{obs} being equal to $P_{\text{obs}} - P_f$, the general shape of the curve depends only on the difference between the value of the parameter observed in the free and complexed states, and not on K. When all plots show a maximum at $r = 0.5$, it indicates that the complex formed has a 1 : 1 stoichiometry; a 1 : 2 complex should provide a nonsymmetrical plot with a maximum at $r = 0.33$, a 2 : 1 complex corresponding to $r = 0.66$ [59]. This method is very often used for the determination of the complex stoichiometry [14, 48, 71–74]. Interestingly, this method also allows to evidence the simultaneous presence of two complexes of different stoichiometry. This is the case for β-CD/triclosan, for which the maximum of the curve is not at 0.5 (for 1 : 1 complex) or at 0.66 (for 2 : 1 complex) but at 0.6, indicating that complexes of both stoichiometries are present in solution simultaneously [75].

4.2.4. Ultraviolet Spectroscopy

UV spectroscopy can be used similarly to NMR for determination of either the complex stiochiometry or its association constant [76, 77]. In this case, it is the absorbance difference that is used. It was demonstrated that the relative error of the Benesi–Hildebrand method in measuring the association constant is often poorly reliable except for the K values $<1000^{-1}$ [78, 79]. When the complexation is strong, a nonlinear regression estimation of binding constants is chosen [76].

4.2.5. Fluorescence Spectrometry

Guest fluorescence variation can be the parameter used to calculate the complex stoichiometry [80] and the complex constant [37, 80] by the methods as described earlier.

4.2.6. Affinity Capillary Electrophoresis

Affinity capillary electrophoresis (ACE) can be used to determine the binding constant of an inclusion complex [65, 81]. When a charged solute (guest) is included in a CD cavity, the inclusion complex has a charge identical to that of the free solute but an increased molecular mass. Since the mass-to-charge ratio of the complex is greater than that of the free solute, the mobility of the solute–cyclodextrin (G-CD) complex is lower than that of the free solute. The electrophoretic mobility of a compound G (μ_{ep}) is a function of the proportion of the time that this compound is free and the proportion of the time that it is complexed:

$$\mu_{\text{ep}} = \left(\frac{[G]}{[G] + [G\text{-}CD]} \right) \mu_0 + \left(\frac{[G\text{-}CD]}{[G] + [G\text{-}CD]} \right) \mu_c$$

where μ_0 is the electrophoretic mobility of the free guest, μ_c the electrophoretic mobility of the G-CD complex, and [G] and [G-CD] the concentrations of the free guest and the inclusion complex, respectively. Given that

$$[G\text{-}CD] = K[G][CD]$$

then

$$\mu_{\text{ep}} = \frac{\mu_0 + \mu_c K[CD]}{1 + K[CD]}$$

where [CD] represents the concentration of CD in the buffer solution. The electrophoretic mobility of the guest, measured from an electropherogram, is

$$\mu_{\text{app}} = \mu_{\text{ep}} + \mu_{\text{eo}}$$

and

$$\mu_{\text{app}} = \frac{Ll}{V t_M}$$

where μ_{app} is the apparent mobility of the guest, L and l the total and effective length of the capillary, respectively, V the apparent voltage, and t_M the migration time of the guest. Furthermore, μ_{eo} is the mobility of the electrophoretic flow, calculated from the t_M of a neutral compound (t_{eo}). Experimentally, μ_{eo} is determined using the t_M of the peak corresponding to water and the equation

$$\mu_{eo} = \frac{Ll}{Vt_{eo}}$$

The determination of the K values is achieved by calculation of μ_{ep} of the guest in buffers containing increasing concentrations of CD. The data are analyzed by nonlinear regression to assess the agreement with the theoretical model and determine values for μ_0, μ_c, and K [81].

4.2.7. Isothermal Titration Calorimetry The formation of an inclusion complex is associated with changes in thermodynamic parameters [82]. Isothermal titration calorimetry, a powerful and versatile method for the study of molecular interactions [83], has been used to determine not only the thermodynamic parameters of guest/CD complexation [84], but also to calculate the affinity constant of complexes [85]. During an isothermal titration calorimetry (ITC) experiment, the heat generated or absorbed during a binding reaction is measured. For the experiment, a CD solution (titrant) is added to a guest solution (titrate) over time using one or more individual injections. The heat can be measured either as a change in temperature or as the change in power necessary to maintain the sample and the reference cell at the same temperature. The energy is converted into a binding enthalpy. Calculation of the enthalpy observed includes not only the heat of binding but also any additional heat sources associated with the reaction, including solvent effects, molecular reorganization and conformational changes, heats of dilution, and mechanical artifacts. Thus, careful preparation of solutions and measurement of appropriate background heats are required to obtain thermodynamic parameters that accurately reflect the event(s) of interest [83]. The titration can be either continuous or sequential. The heat produced during each injection is proportional to the amount of complex formed. The change in heat during the experiment allows evaluation of the stoichiometry of interaction, the affinity constant K, and the enthalpy (ΔH) of the interaction, from which the entropy (ΔS) and the Gibbs free energy of the progress (ΔG) can be derived [85].

4.3. Structure of the Complex

Most often the guest molecule is not totally included in the CD cavity; the part inside is hindered from any surrounding influence (i.e., humidity, oxidation, pH, etc.) when the part outside can be subjected to all these phenomenon. For this reason it is of prominent interest to know the exact structure of the complex.

4.3.1. NMR and ROESY Studies If classical NMR studies enable evidence for the existence of an inclusion, it does not give direct information on the inclusion structure. ROESY (Rotating frame Overhauser Effect SpectroscopY) experiments provide structural information and allow study of the complex geometry in aqueous solutions. It is a two-dimensional method in which a cross peak can be observed between the protons when the internuclear distance is smaller than about 3 to 4 Å [59]. The intensities of the cross peaks depend on the distance between the interacting nuclei, the intensity decreasing with the distance [86]. This method is now a reference for determination of an inclusion complex structure [11, 14, 23, 74, 86].

Cross peaks are displayed between the inner CD protons H3 and H5 and the interacting protons of the guest. For example, in the case of taginin/β-Cd complex [77] it has been shown that the internal H3 proton is correlated with the protons of the lactone part and the unsaturated ketone cycle, while the internal H5 proton is correlated with the protons of the ester part, which could suggest that the taginin is inserted deeply into the cavity by the largest rim, where secondary hydroxyl groups are present, with the ester and lactone parts oriented toward the primary hydroxyl groups of the CD.

4.3.2. Molecular Modeling Molecular modeling makes it possible to obtain the possible geometric structures of the inclusion complex with the docking energies. It constitutes a powerful method to use to predict or explain the inclusion mechanism and the complex structure [52, 65, 71, 77, 87, 88]. Molecular modeling enables geometrical representation of the most probable complex structure (Fig. 7). They are based on the search for a correlation between experimentally determined equilibrium constants of the complexes and some important theoretically evaluated parameters describing the inclusion process, such as the docking energy (gain of potential energy as a consequence of the inclusion), the host–guest contact surfaces (related to the hydrophobic interactions), and the intermolecular interaction fields (related to the hydrophilicity and lipohilicity of the interacting molecules) [87]. It must be emphasized that this method has only a predictive value, which could be useful in preformulation studies to select the best CD to use.

5. CONCLUSIONS

CDs are truly exceptional molecules. Not only have they an unusual shape; they are ring molecules, but because of this structure, they have unique properties. They can form inclusion complexes with molecules of size and charge adapted to their cavity. These inclusion complexes are real

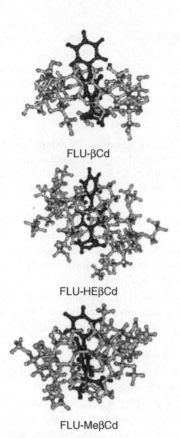

FLU-βCd

FLU-HEβCd

FLU-MeβCd

Figure 7. Molecular modeling of the most probable complex structures of flurbiprofen (FLU) included in β-cyclodextrin (β-CD), hydroxyethyl-β-cyclodextrin (HE-β-CD), and methyl-β-cyclodextrin (M-β-CD). (From [65], with permission of Elsevier.)

molecular encapsulation. Of course, depending on the CD used, the guest compound chosen, and the reaction medium, the inclusion yield can vary and the complex obtained can be a noninclusion complex. The preparation methods are numerous and not necessarily all adapted to the main purpose of the scientist preparing the inclusion. Similarly, the methods used to study the complex are also numerous, but they do not all have the same objective. The general conclusion could be very simple: When preparing and studying an inclusion complex, one must be clear as to his or her objective so as to make the right choice between the various tools proposed in the literature.

REFERENCES

1. Loftsson, T., Duchêne, D. (2007). Cyclodextrins and their pharmaceutical applications: historical perspectives. *Int. J. Pharm.*, *329*, 1–11.

2. Sicard, P.J., Saniez, M.-H. (1987). Biosynthesis of cycloglycosyl transferase and obtention of its enzymatic reaction products. In: Duchêne, D., Ed., *Cyclodextrins and Their Industrial Uses.* Editions de Santé, Paris, pp. 75–103.

3. Le Bas, G., Rysanek, N. (1987). Structural aspects of cyclodextrins. In: Duchêne, D. Ed., *Cyclodextrins and Their Industrial Uses.* Editions de Santé, Paris, pp. 105–130.

4. Loftsson, T., Brewster, M.E. (1996). Pharmaceutical applications of cyclodextrins: 1. Drug solubilisation and stabilization. *J. Pharm. Sci.*, *85*, 1017–1025.

5. Zia, V., Rajewski, R.A., Stella, V.J. (2001). Effect of cyclodextrin charge on complexation of neutral and charged substrates: comparison of $(SBE)_{7M}$-β-CD to HP-β-CD. *Pharm. Res.*, *18*, 667–673.

6. Gabelica, V., Galic, N., De Pauw, E. (2002). On the specificity of cyclodextrin complexes detected by electrospray mass spectroscopy. *J. Am. Soc. Mass Spectrom.*, *13*, 946–953.

7. Bikádi, Z., Iványi, R., Szente, L., Hazai, E. (2007). Cyclodextrin complexes: chiral recognition and complexation behavior. *Curr. Drug Discov. Technol.*, *4*, 282–294.

8. Loftsson, T., Matthíasson, K., Másson, M. (2003). The effects of organic salts on the cyclodextrin solubilization of drugs. *Int. J. Pharm.*, *262*, 101–107.

9. Zhao, D.G., Liao, K.I., Ma, X.Y., Yan, X.H. (2002). Study of the supramolecular inclusion of β-cyclodextrin with andrographolide. *J. Inclusion Phenom. Macrocyclic Chem.*, *43*, 259–264.

10. Loftsson, T., Magnúsdóttir, A., Másson, M., Sigurjónsdóttir, J.F. (2002). Self-association and cyclodextrin solubilization of drugs. *J. Pharm. Sci.*, *91*, 2307–2316.

11. Loftsson, T., Másson, M., Sigurdsson, H.H. (2002). Cyclodextrins and drug permeability through semi-permeable cellophane membranes. *Int. J. Pharm.*, *232*, 35–43.

12. Bikádi, Z., Kurdi, R., Balogh, S., Szemán, J., Hazai, E. (2006). Aggregation of cyclodextrins as an important factor to determine their complexation behavior. *Chem. Biodiversity*, *3*, 1266–1278.

13. Brochsztain, S., Politi, M.J. (1999). Solubilization of 1,4,5,8-naphthalenediimides and 1, 8-naphthalimides through the formation of novel host–guest complexes with α-cyclodextrin. *Langmuir*, *15*, 4486–4494.

14. Fernandes, C.M., Carvalho, R.A., Pereira da Costa, S., Veiga, F.J.B. (2003). Multimodal molecular encapsulation of nicardipine hydrochloride by β-cyclodextrin, hydroxylpropyl-β-cyclodextrin and triacetyl-β-cyclodextrin in solution: structural studies by ^{1}H NMR and ROESY experiments. *Eur. J. Pharm. Sci.*, *18*, 285–296.

15. Másson, M., Loftsson, T., Jónsdottir, S., Fridriksdóttir, H., Petersen, D.S. (1998). Stabilisation of ionic drugs through complexation with non-ionic and ionic cyclodextrins. *Int. J. Pharm.*, *164*, 45–55.

16. Loftsson, T., Másson, M., Sigurjónsdóttir, J.F. (1999). Methods to enhance the complexation efficiency of cyclodextrins. *STP Pharma Sci.*, *9*, 237–242.

17. Pitha, J., Hoshino, T. (1992). Effects of ethanol on the formation of inclusion complexes of hydroxypropylcyclodextrins with testosterone or with methyl orange. *Int. J. Pharm.*, *80*, 243–251.

18. Kim, Y., Oksanen, D.A., Massefski, W., Blake, J.F., Duffy, E. M., Chrunyk, B. (1998). Inclusion of ziprasidone mesylate with β-cyclodextrin sulfobutyl ether. *J. Pharm. Sci.*, *87*, 1560–1567.

19. Selva, A., Redenti, E., Ventura, P., Zanol, M., Casetta, B. (1998). Study of β-cyclodextrin–ketokonazole–tartaric acid multicomponent non-covalent association by positive and negative ionspray mass spectrometry. *J. Mass Spectrom.*, *33*, 729–734.

20. Csabai, K., Vikmon, M., Szejtli, J., Redenti, E., Poli, G., Ventura, P. (1998). Complexation of manidipine with cyclodextrins and their derivatives. *J. Inclusion Phenom. Mol. Recognit. Chem.*, *31*, 169–178.

21. Loftsson, T. (1998). Increasing the cyclodextrin complexation of drugs and drug bioavailability through addition of water-soluble polymers. *Pharmazie*, *53*, 733–740.

22. Al-Marzouqi, A.H., Jobe, B., Dowaidar, A., Maestrelli, F., Mura, P. (2007). Evaluation of supercritical fluid technology as preparative technique of benzocaine–cyclodextrin complexes; comparison with conventional methods. *J. Pharm. Biomed. Anal.*, *43*, 566–574.

23. Jambhekar, S., Casella, R., Maher, T. (2004). The physicochemical characteristics and bioavailability of indomethacin from β-cyclodextrin, hydroxyethyl-β-cyclodextrin, and hydroxypropyl-β-cyclodextrin complexes. *Int. J. Pharm.*, *270*, 149–166.

24. Ribeiro, A., Figueiras, A., Santos, D., Veiga, F. (2008). Preparation and solid state characterization on inclusion complexes formed between miconazole and methyl-β-cyclodextrin. *AAPS PharmSciTech*, *9*, 1102–1109.

25. Ribeiro, L.S.S., Ferreira, D.C., Veiga, F.J.B. (2003). Physicochemical investigation of the effects of water-soluble polymers on vinpocetine complexation with β-cyclodextrin and its sulfobutyl ether derivative in solution and solid state. *Eur. J. Pharm. Sci.*, *20*, 253–266.

26. Al-Marzouqi, A.H., Elwy, H.M., Shehadi, I., Adem, A. (2009). Physicochemical properties of antifungal drug–cyclodextrin complexes prepared by supercritical carbon dioxide and by conventional techniques. *J. Pharm. Biomed. Anal.*, *49*, 227–233.

27. Figueiras, A., Carvalho, R.A., Ribeiro, L., Torres-Labandeira, J.J., Veiga, F.J.B. (2007). Solid-state characterization and dissolution profiles of the inclusion complexes of omeprazole with native and chemically modified β-cyclodextrin. *Eur. J. Pharm. Biopharm.*, *67*, 531–539.

28. Salústio, P.J., Feio, G., Figueirinhas, J.L., Pinto, J.F., Cabral Marques, H.M. (2009). The influence of the preparation methods on the inclusion of model drugs in a β-cyclodextrin cavity. *Eur. J. Pharm. Biopharm.*, *71*, 377–386.

29. Badr-Eldin, S.M., Elkheshen, S.A., Ghorab, M.M. (2008). Inclusion complexes of tadalafil with natural and chemically modified β-cyclodextrins: I. Preparation and in-vitro evaluation. *Eur. J. Pharm. Biopharm.*, *70*, 819–827.

30. Cirri, M., Maestrelli, F., Mennini, N., Mura, P. (2009). Influence of the preparation method on the physical–chemical properties of ketoprofen–cyclodextrin–phosphatidylcholine ternary systems. *J. Pharm. Biomed. Anal.*, *50*, 690–694.

31. Van Hees, T., Piel, G., Evrard, B., Otte, X., Thunus, L., Delattre, L. (1999). Application of supercritical carbon dioxide for the preparation of a piroxicam-β-cyclodextrin inclusion compound. *Pharm. Res.*, *16*, 1864–1870.

32. York, P. (1999). Strategies for particle design using supercritical fluid technologies. *Pharm. Sci. Technol. Today*, *2*, 430–440.

33. Toropainen, T., Velaga, S., Heikkilä, T., Matilainen, L., Jarho, P., Carlfors, J., Lehto, V.P., Järvinen, T., Järvinen, K. (2006). Preparation of budesonide/γ-cyclodextrin complexes in supercritical fluids with a novel SEDS method. *J. Pharm. Sci.*, *95*, 2235–2245.

34. Bandi, N., Wei, W., Roberts, C.B., Kotra, L.P., Kompella, U.B. (2004). Preparation of budesonide– and indomethacin–hydroxypropyl-β-cyclodextrin (HPBCD) complexes using a single-step, organic-solvent-free supercritical fluid porcess. *Eur. J. Pharm. Sci.*, *23*, 159–168.

35. Hassan H.A., Al-Marzouqi, A.H., Jobe, B., Hamza, A.A., Ramadan, G.A. (2007). Enhancement of dissolution amount and in vivo bioavailability of itraconazole by complexation with β-cyclodextrin using supercritical carbon dioxide. *J. Pharm. Biomed. Anal.*, *45*, 243–250.

36. Charoenchaitrakool, M., Dehghani, F., Foster, N.R. (2002). Utilization of supercritical carbon dioxide for complex formation of ibuprofen and methyl-β-cyclodextrin. *Int. J. Pharm.*, *239*, 103–112.

37. Wen, X.H., Tan, F., Jing, Z.J., Liu, Z.Y. (2004). Preparation and study the 1 : 2 inclusion of carvedilol with β-cyclodextrin. *J. Pharm. Biomed. Anal.*, *34*, 517–523.

38. Nacsa, Á., Ambrus, R., Berkesi, O., Szabó-Révész, P., Aigner, Z. (2008). Water-soluble loratadine inclusion complex: analytical control of the preparation by microwave irradiation. *J. Pharm. Biomed. Anal.*, *48*, 1020–1023.

39. Cirri, M., Maestrelli, F., Mennini, N., Mura, P. (2009). Physical–chemical characterization of binary and ternary systems of ketoprofen with cyclodextins and phospholipids. *J. Pharm. Biomed. Anal.*, *50*, 683–689.

40. Mura, P., Fauci, M.T., Parrini, P.L., Furlanetto, S., Pinzauti, S. (1999). Influence of the preparation method on the physicochemical properties of ketoprofen–cyclodextrin binary systems. *Int. J. Pharm.*, *179*, 117–128.

41. Loftsson, T., Fridriksdóttir, H., Sigurdardóttir, A.M., Ueda, H. (1994). The effect of water-soluble polymers on drug–cyclodextrin complexation. *Int. J. Pharm.*, *110*, 169–177.

42. Faucci, M.T., Mura, P. (2001). Effect of water-soluble polymers on naproxen complexation with natural and chemically modified β-cyclodextrins. *Drug Dev. Ind. Pharm.*, *27*, 909–917.

43. Alexanian, C., Papademou, H., Vertzoni, M., Archontaki, H., Valsami, G. (2008). Effect of pH and water-soluble polymers on the aqueous solubility of nimesulide in the absence and presence of β-cyclodextrin derivatives. *J. Pharm. Pharmacol.*, *60*, 1433–1439.

44. Valero, M., Tejedor, J., Rodríguez, L.J. (2007). Encapsulation of nabumetone by means of drug:(β-cyclodextrin)$_2$:polyvinylpyrrolidone ternary complex formation. *J. Luminesc.*, *126*, 297–302.

45. Selvam, A.P., Geetha, D. (2008). Ultrasonic studies on lamivudine:β-cyclodextrin and polymer inclusion compexes. *Pak. J. Biol. Sci.*, *11*, 656–659.

46. Loftsson, T., Gudmundsdóttir, T.K., Fridriksdóttir, H. (1996). The influence of water-soluble polymers and pH on hydro-

xypropyl-β-cyclodextrin complexation of drugs. *Drug Dev. Ind. Pharm.*, *22*, 401–405.

47. Valero, M., Esteban, B., Pelaez, R., Rodríguez, L. (2004). Naproxen:hydroxypropyl-β-cyclodextr:polyvinylpyrrolidone ternary complex formation. *J. Inclusion Phenom. Macrocyclic Chem.*, *48*, 157–163.

48. Ribeiro, L., Carvalho, R.A., Ferreira, D.C., Veiga, F.J.B. (2005). Multicomponent complex formation between vinpocetine, cyclodextrins, tartaric acid and water-soluble polymers monitored by NMR and solubility studies. *Eur. J. Pharm. Sci.*, *24*, 1–13.

49. Cirri, M., Maestrelli, F., Corti, G., Furlanetto, S., Mura, P. (2006). Simultaneous effect of cyclodextrin complexation, pH, and hydrophilic polymers on naproxen solubilization. *J. Pharm. Biomed. Anal.*, *42*, 126–131.

50. Loftsson, T., Sigurdsson, H.H., Másson, M., Schipper, N. (2004). Preparation of solid drug/cyclodextrin complexes of acidic and basic drugs. *Pharmazie*, *59*, 25–29.

51. Ribeiro, L.S.S., Falcão, A.C., Patrício, J.A.B., Ferreira, D.C., Veiga, F.J.B. (2007). Cyclodextrin multicomponent complexation and controlled release delivery strategies to optimize the oral bioavailability of vinpocetine. *J. Pharm. Sci.*, *96*, 2018–2028.

52. Barillaro, V., Dive, G., Bertholet, P., Evrard, B., Delattre, L., Ziémons, E., Piel, G. (2007). Theoretical and experimental investigations on miconazole/cyclodextrin/acid complexes: molecular modeling studies. *Int. J. Pharm.*, *342*, 152–160.

53. Sinha, V.R., Anitha, R., Ghosh, S., Nanda, A., Kumria, R. (2005). Complexation of celecoxib with β-cyclodextrin: characterization of the interaction in solution and in solid state. *J. Pharm. Sci.*, *94*, 676–687.

54. Lemesle-Lamache, V., Wouessidjewe, D., Chéron, M., Duchêne, D. (1996). Study of β-cyclodextrin and ethylated β-cyclodextrin salbutamol complexes, in vitro evaluation of sustained-release behavior of salbutamol. *Int. J. Pharm.*, *141*, 117–124.

55. Ventura, C.A., Giannone, I., Musumeci, T., Pignatello, R., Ragni, L., Landolfi, C., Milanese, C., Paolino, D., Puglisi, G. (2006). Physico-chemical characterization of disoxaril–dimethyl-β-cyclodextrin inclusion complex and in vitro permeation studies. *Eur. J. Med. Chem.*, *41*, 233–240.

56. Dotsikas, Y., Loukas, Y.L. (2002). Kinetic degradation study of insulin complexed with methyl-beta cyclodextrin: confirmation of complexation with electrospray mass spectrometry and ^{1}H NMR. *J. Pharm. Biomed. Anal.*, *29*, 487–494.

57. Guernelli, S., Lagana, M.F., Mezzina, E., Ferroni, F., Siani, G. (2003). Supramolecular complex formation: a study of the interactions between β-cyclodextrin and some different classes of organic compounds by ESI-MS, surface tension measurements, and UV/Vis and ^{1}H NMR spectroscopy. *Eur. J. Org. Chem.*, *24*, 4765–4776.

58. Béni, S., Szakács, Z., Csernák, O., Barcza, L., Noszál, B. (2007). Cyclodextrin/imatinib complexation: binding mode and charge dependent stabilities. *Eur. J. Pharm. Sci.*, *30*, 167–174.

59. Djedaïni, F., Perly, B. (1991). Nuclear magnetic resonance of cyclodextrins, derivatives and inclusion compounds. In:

Duchêne, D., Ed., *New Trends in Cyclodextrins and Derivatives*. Editions de Santé, Paris, pp. 215–246.

60. Loftsson, T., Másson, M., Brewster, M.E. (2004). Self-association of cyclodextrins and cyclodextrin complexes. *J. Pharm. Sci.*, *93*, 1091–1099.

61. Higuchi, T., Connors, K.A. (1965). Phase-solubility techniques. *Adv. Anal. Chem. Instrum.*, *4*, 117–212.

62. Brewster, M.E., Loftsson, T. (2007). Cyclodextrins as pharmaceutical solubilizers. *Adv. Drug Deliv. Rev.*, *59*, 645–666.

63. Hirayama, F., Uekama, K. (1987). Methods of investigating and preparing inclusion compounds. In: Duchêne, D. Ed., *Cyclodextrins and Their Industrial Uses*. Editions de Santé, Paris, pp. 131–172.

64. Ono, N., Hirayama, F., Arima, H., Uekama, K. (1999). Determination of stability constant of β-cyclodextrin complexes using the membrane permeation technique and the permeation behaviour of drug–competing agent–β-cyclodextrin ternary systems. *Eur. J. Pharm. Sci.*, *8*, 133–139.

65. Cirri, M., Maestrelli, F., Orlandini S., Furlanetto, S., Pinzauti, S., Mura, P. (2005). Determination of stability constant values of flurbiprofen–cyclodextrin complexes using different techniques. *J. Pharm. Biomed. Anal.*, *37*, 995–1002.

66. Hildebrand, J.A., Benesi, H.A. (1949). A graphical method for the determination of binding constants. *J. Am. Chem. Soc.*, *71*, 2703–2710.

67. Djedaïni, F., Lin, S.Z., Perly, B., Wouessidjewe, D. (1990). High-field nuclear magnetic resonance techniques for the investigation of a β-cyclodextrin:indomethacin inclusion complex. *J. Pharm. Sci.*, *79*, 643–646.

68. Jullian, C., Miranda, S., Zapata-Torres, G., Mendizábal, F., Olea-Azar, C. (2007). Studies on inclusion complexes of natural and modified cyclodextrin with (+)catechin by NMR and molecular modeling. *Bioorg. Med. Chem.*, *15*, 3217–3224.

69. Tapia, M.J., Burrows, H.D., García, J.M., Pais, A.A.C.C. (2004). Lanthanide ion interaction with a crown ether methacrylic polymer, poly(1,4,7,10-tetraoxacyclododecan-2-methyl methacrylate), as seen by spectroscopic, calorimetric, and theoretical studies. *Macromolecules*, *37*, 856–862.

70. Job, P. (1928). Recherches sur la formation de complexes minéraux en solution et sur leur stabilité. *Ann. Chim.*, *9*, 113–125.

71. Veiga, F.J.B., Fernandes, C.M., Carvalho, R.A., Geraldes, C.F.G.C. (2001). Molecular modelling and ^{1}H-NMR: ultimate tools for the investigation of tolbutamide:β-cyclodextrin and tolbutamide:hydroxypropyl-β-cyclodextrin complexes. *Chem. Pharm. Bull.*, *49*, 1251–1256.

72. Figueiras, A., Sarraguça, J.M.G., Carvalho, R.A., Pais, A.A.C.C., Veiga, J.B. (2007). Interaction of omeprazole with a methylated derivative of β-cyclodextrin: phase solubility, NMR spectroscopy and molecular simulation. *Pharm. Res.*, *24*, 377–389.

73. Grillo, R., de Melo, N.F.S., Moraes, C.M., de Lima, R., Menezes, C.M.S., Ferreira, E.I., Rosa, A.H., Fraceto, L.F. (2008). Study of the interaction between hydroxymethylnitrofurazone and 2-hydroxypropyl-β-cyclodextrin. *J. Pharm. Biomed. Anal.*, *47*, 295–302.

74. Whang, H.S., Vendeix, F.A.P., Gracz, H.S., Gadsby, J., Tonelli, A. (2007). NMR studies of the inclusion complex of cloprostenol sodium salt with β-cyclodextrin in aqueous solution. *Pharm. Res.*, *25*, 1142–1149.

75. Paulidou, A., Maffeo, D., Yannakopoulou, K., Mavridis, I.M. (2008). Crystal structure of the inclusion complex of the antibacterial agent triclosan with cyclomaltoheptaose and NMR study of its molecular encapsulation in positively and negatively charged cyclomaltoheptaose derivatives. *Carbohydr. Res.*, *343*, 2634–2640.

76. Gibaud, S., Ben Zizar, S., Mutzenhardt, P., Fries, I., Astier, A. (2005). Melarsoprol–cyclodextrins inclusion complexes. *Int. J. Pharm.*, *306*, 107–121.

77. Ziémons, E., Dive, G., Debrus, B., Barillaro, V., Frederich, M., Lejeune, R., Angenot, L., Delattre, L., Thunus, L., Hubert, Ph. (2007). Study of the physicochemical properties in aqueous medium and molecular modeling of tagitinin C/cyclodextrin complexes. *J. Pharm. Biomed. Anal.*, *43*, 910–919.

78. Salvatierra, D., Diez, C., Jaime, C. (1997). Host/guest interactions and NMR spectroscopy: a computer program for association constant determination. *J. Inclusion Phenom. Mol. Recognit.*, *27*, 215–231.

79. Yang, C., Liu, L., Mu, T.W., Guo, Q.X. (2000). The performance of the Benesi–Hildebrand method in measuring the binding constants of the cyclodextrin complexation. *Anal. Sci.*, *16*, 537–539.

80. Song, L.S., Wang, H.M., Xu, P., Yang, Y., Zhang, Z.Q. (2008). Experimental and theoretical studies on the inclusion complexation of syringic acid with α-, β-, γ- and heptakis(2, 6-di-*O*-methyl)-β-cyclodextrin. *Chem. Pharm. Bull.*, *56*, 468–474.

81. Lemesle-Lamache, V., Taverna, M., Wouessidjewe, D., Duchêne, D., Ferrier, D. (1996). Determination of the binding constant of salbutamol to unmodified and ethylated cyclodextrins by affinity capillary electrophoresis. *J. Chromatogr. A*, *735*, 321–331.

82. Castronuovo, G., Niccoli, M. (2006). Thermodynamics of inclusion complexes of natural and modified cyclodextrins with propranolol in aqueous solution at 298 K. *Bioorg. Med. Chem.*, *14*, 3883–3887.

83. Salim, N.N., Feig, A.L. (2009). Isothermal titration calorimetry of RNA. *Methods*, *47*, 198–205.

84. Illapakurthy, A.C., Wyandt, C.M., Stodghill, S.P. (2005). Isothermal titration calorimetry method for determination of cyclodextrin complexation thermodynamic between artemisinin and naproxen under varying environmental conditions. *Eur. J. Pharm. Biopharm.*, *59*, 325–332.

85. Segura-Sanchez, F., Bouchemal, K., Le Bas, G., Vauthier, C., Santos-Magalhaes, N.S., Ponchel, G. (2009). Elucidation of the complexation mechanism between (+)-usnic acid and cyclodextrins studied by isothermal titration calorimetry and phase-solubility diagram experiments. *J. Mol. Recognit.*, *22*, 232–241.

86. Metha, S.K., Bhasin, K.K., Dham, S. (2008). Energetically favorable interactions between diclofenac sodium and cyclodextrin molecules in aqueous media. *J. Colloid Interface Sci.*, *326*, 374–381.

87. Faucci, M.T., Melani, F., Mura, P. (2002). Computer-aided molecular modeling techniques for predicting the stability of drug–cyclodextrin inclusion complexes in aqueous solutions. *Chem. Phys. Lett.*, *358*, 383–390.

88. Melani, F., Mura, P., Adamo, M., Maestrelli, F., Gratteri, P., Bonaccini, C. (2003). New docking CFF91 parameters specific for cyclodextrin inclusion complexes. *Chem. Phys. Lett.*, *370*, 280–292.

2

CYCLODEXTRINS AS POTENTIAL EXCIPIENTS IN PHARMACEUTICAL FORMULATIONS: SOLUBILIZING AND STABILIZING EFFECTS

ALKA AHUJA

Oman Medical College, Muscat, Oman

SANJULA BABOOTA, JAVED ALI, AND GULAM MUSTAFA

Department of Pharmaceutics, Faculty of Pharmacy, Hamdard University, New Delhi, India

1. INTRODUCTION

In this chapter we discuss and summarize some of the interesting applications of cyclodextrins (CDs) in various areas of drug delivery focusing on their well-known effects on drug solubility and stability. CDs are hollow, truncated-cone-shaped molecules made up of six or more glucose units linked together covalently by oxygen atoms and held in shape via hydrogen bonding between the secondary hydroxy groups on adjacent units at the wider rim of the cavity. CDs (also called *cycloamyloses*) make up a family of cyclic oligosaccharides. They are composed of five or more α-D-glucopyranoside units linked 1 to 4, as in amylose (a fragment of starch). The five-membered macrocycle is not natural. Recently, the largest well-characterized CD contains thirty-two 1,4-anhydroglucopyranoside units, while as a poorly characterized mixture, at least 150-membered cyclic oligosaccharides are also known. Typical CDs contain a number of glucose monomers, ranging from six to eight units in a ring, creating a cone shape thus denoting:

- *α-cyclodextrin:* six-membered sugar ring molecule
- *β-cyclodextrin:* seven-membered sugar ring molecule
- *γ-cyclodextrin:* eight-membered sugar ring molecule

These are shown in Figs. 1 and 2.

The characteristics of α-, β-, γ-, and δ-CDs are given in Table 1. The physical characteristics of CDs are given in Table 2. The CDs are manufactured by the enzymatic degradation of starch using specialized bacteria. However, Endo et al. [1] established an isolation and purification method for several types of large-ring CDs and also obtained a relatively large amount of δ-CD (cyclomaltonose) with nine glucose units [1–3]. The cavity size of α-CD is insufficient for many drugs, and γ-CD is expensive. In general, δ-CD has a weaker complex-forming ability than that of conventional CDs.

2. USES

Over the last few years CDs have found a wide range of applications in the food, pharmaceutical, and chemical industries as well as in agriculture and environmental engineering. CDs are used widely as host molecules in a number of areas of chemistry where molecular recognition is required, such as material sciences, supramolecular chemistry, molecular sensing, and artificial enzymes. Moreover, CDs are of great utility in the field of medicinal chemistry as solubilizing agents for lipophilic drugs [4–6], as photostabilizers of light-sensitive drugs, [7,8], and as sustained-release [4–6,9] and drug delivery systems [10–12] for various

Cyclodextrins in Pharmaceutics, Cosmetics, and Biomedicine: Current and Future Industrial Applications, First Edition. Edited by Erem Bilensoy.
© 2011 John Wiley & Sons, Inc. Published 2011 by John Wiley & Sons, Inc.

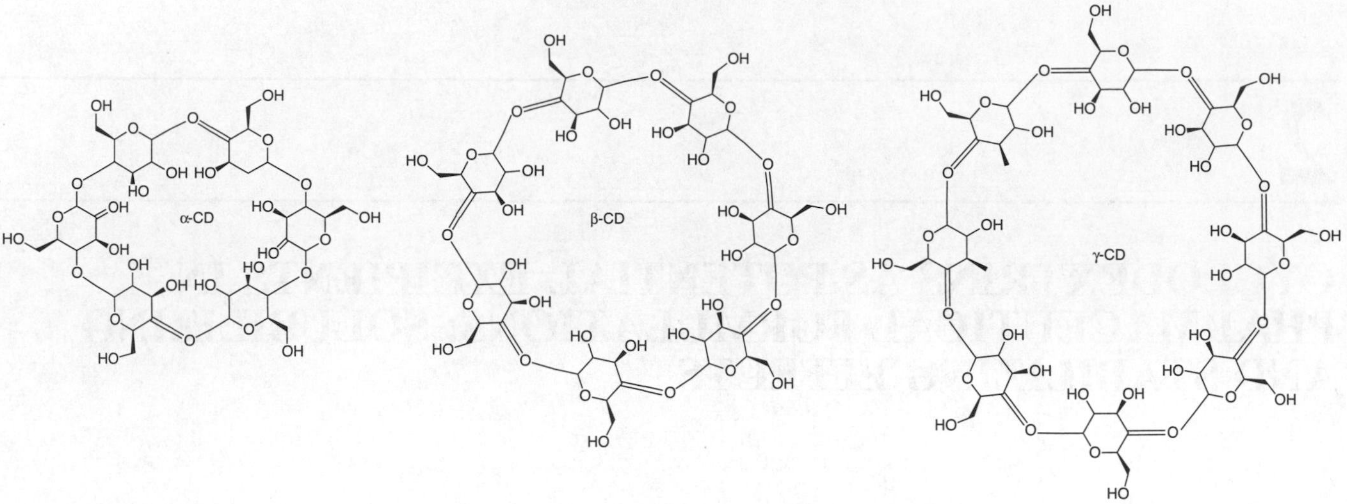

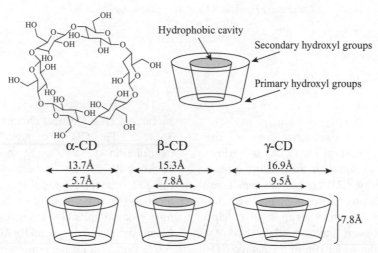

Figure 1. Comparative structures of CDs.

Figure 2. CD cavity sizes.

drugs. CDs and their derivatives play an important role in formulation development due to their effect on solubility, dissolution rate, chemical stability, and absorption of drugs. Although CDs have been investigated widely during the last two decades, their commercial application in pharmaceutical formulations began only in recent years with drugs such as piroxicam and nimesulide [13–16]. After this, various studies were done to investigate the possibility of improving the

solubility and dissolution rate of norfloxacin in the presence of solubilizing agents such as ascorbic acid and citric acid, which are incorporated into β-cyclodextrin complex (β-CD), α-cyclodextrin, and γ-cyclodextrin. Among them, β-CD has

Table 1. Characteristics of α-, β-, γ-, and δ-CDs

Type of CD	Cavity Diameter (Å)	Molecular Weight (Da)	Solubility (g/100 mL)
α-CD	4.7–5.3	972	14.5
β-CD	6.0–6.5	1135	1.85
γ-CD	7.5–8.3	1297	23.2
δ-CD	10.3–11.2	1459	8.19

Table 2. Physical Characteristics of CDs

	Cyclodextrin		
Characteristic	α-CD	β-CD	γ-CD
Cavity diameter (Å)	4.7–5.3	6.0–6.5	7.5–8.3
Height of torus (Å)	7.9	7.9	7.9
Diameter of periphery (Å)	14.6	15.4	17.5
Approximate volume of cavity (Å)	174	262	472
Per mole of CD (mL)	104	157	256
Per gram of CD (mL)	0.1	0.14	0.20

been used for a long time, as it has a bigger cavity size (7.5 Å) and is the least toxic of all the natural CDs [17].

Possible effects of complex formation between a guest molecule and α-, β-, or γ-CD include enhanced or reduced aqueous solubility, bioavailability-enhanced or reduced stability against hydrolysis, oxidation, ultraviolet (UV)-light-induced degradation, heat, and reduced volatility. Introduction of CDs into the pharmaceutical process can alter the solubility and stability of included medicines so as to be applied as drug carriers in some chemotherapies.

3. EFFECTS OF CDs ON IMPORTANT DRUG PROPERTIES IN FORMULATION

3.1. Effect on Drug Solubility and Dissolution

CD complexation has been widely used to improve the physicochemical properties of various drug molecules. CDs are of considerable practical and theoretical interest because of their ability, as host molecules, to form stable noncovalent inclusion complexes with numerous inorganic and organic guest molecules. In addition, CDs and their complexes form water-soluble aggregates in aqueous solutions. These aggregates are able to solubilize lipophillic water-insoluble drugs through noninclusion complexation or micellelike structures [18]. Such a drug–ligand complex has a rigid structure and a definite stoichiometry, usually 1 : 1 at low concentration. The formation of stable inclusion complexes in aqueous solutions may be attributed, in part, to the fact that CDs possess an apolar cavity with a well-defined geometry. However, the development of chemically modified CDs with favorable water solubility and lower hemolytic properties has circumvented this problem [19,20].

The hemolysis of erythrocytes by CDs has been attributed to two possible effects [21]: (1) the inducement of an osmotic hypotonic effect, and (2) the complexation of lipid components such as cholesterol or phospholipids. However, the use of CDs in pharmaceutical dosage forms is limited by their relatively high cost and due to problems of formulation, all principally related to the large amount necessary to obtain the desired drug-solubilizing effect [15]. Out of various commercially available CDs, methylated CDs with a relatively low molar substitution appear to be the most powerful solubilizers. Reduction in drug crystallinity on complexation or solid dispersion with CDs also contributes to the CD increased apparent drug solubility and dissolution rate [22,23]. As a result of their ability to form in situ inclusion complexes in a dissolution medium, CDs can enhance drug dissolution even when there is no complexation in the solid state [24]. Sulfobutylated CD (SBE-β-CD) was shown to be an excellent solubilizer for several drugs and was more effective than β-CD but not as effective as dimethyl CDs such as DM-β-CD [25]. CDs can also act as release

enhancers; for example, β-CD enhanced the release rate of poorly soluble naproxen and ketoprofen from inert acrylic resins and hydrophilic swellable [high-viscosity hydroxypropyl methylcellulose (HPMC)] tableted matrices. β-CD also enhanced the release of theophylline from a HPMC matrix by increasing the apparent solubility and dissolution rate of the drug [26,27]. With drugs such as digitoxin and spiranolactone, δ-CD showed a greater solubilizing effect than that of α-CD, but the effect of δ-CD was less than that of β- and γ-CDs. β-CD has been used widely in the early stages of pharmaceutical applications because of its ready availability and cavity size suitable for the widest range of drugs. But the low aqueous solubility and nephrotoxicity limited the use of β-CD, especially in parenteral drug delivery [28]. Chemically modified CD derivatives have been prepared with a view to extending the physicochemical properties and inclusion capacity of parent CDs. Several amorphous, noncrystallizable CD derivatives with enhanced aqueous solubility, physical and microbiological stability, and reduced parenteral toxicity have been developed by chemical modification of parenteral CDs [29,30].

To improve the solubility and stability of poorly soluble dihydroartemisinin (DHA), hydroxypropyl-β-cyclodextrin (HP-β-CD) inclusion complexes were prepared with recrystallized DHA to study its thermal stability also. The equilibrium solubility of DHA was enhanced as a function of HP-β-CD concentration. DHA/HP-β-CD complexes showed an 89-fold increase in solubility compared to DHA. Hydrogen bonding was found between DHA and HP-β-CD, and it was stronger in complexes prepared in water than in buffers. DHA/HP-β-CD complexes prepared using commercial (untreated) or recrystallized DHA showed a 40% increase in thermal stability and a 29-fold decrease in hydrolysis rates compared to DHA [31]. CD applications as solubilizing agents are summarized in Table 3.

3.2. Effect on Drug Stability

CDs can improve the stability of several labile drugs against dehydration, hydrolysis, oxidation, and photodecomposition and thus increase the shelf life of drugs [15]. Electrostatic, hydrophobic, and hydrogen-bonding interactions are the principal forces determining the stability of biological macromolecules. To improve the solubility and stability of poorly soluble dihydroartemisinin (DHA), HP-β-CD inclusion complexes were prepared with recrystallized DHA to study its thermal stability also. This is shown in Figure 3. The complexes were characterized by differential scanning calorimetery (DSC), Fourier transform infrared spectroscopy (FTIR), x-ray diffraction patterns (XRD), and thermal stability, phase, and equilibrium solubility studies. Pure DHA was crystalline in nature and remained crystalline after recrystallization, but its unit cell dimensions changed as exhibited by XRD. DHA/HP-β-CD complexes showed phase

Table 3. Effect of CDs on the Solubility of Various Drugs

Drug	Type of CD Used[a]	Effect on Solubility	Ref.
Irbesartan	B	The solubility and dissolution were improved. Among various methods, co-evaporation was best for increasing the solubility and dissolution rate of the drug.	[32]
Finasteride	D	Inclusion complexes improved the solubility.	[33]
Propolis ethanolic (PE) extracts	B	Encapsulation in β-CD was reported to increase the solubility of PE constituents in a manner related to their structure, while the amount of substance released was dependent on both their chemical properties and their relative abundance.	[34]
Caffeine	B	The solubility of caffeine molecules was enhanced throughout the inclusion interactions and prevented caffeine self-association.	[35]
Finasteride (FIN)	P	Equimolar FIN/HP-β-CD solid systems suggested that true binary and ternary inclusion complexes were formed.	[33]
Cortisone acetate	D	The solubility of cortisone acetate increased from 0.039 g/L to 7.382 g/L. HP-β-CD was far superior to dimethylformamide and ethanol.	[34]
Dihydroartemisinin	D	The solubility of dihydroartemisinin, a major metabolite of artemisinin and its derivatives, including arteether, artemether, and artesunate, was improved 89-fold.	[31]
Benzophenone and tamoxifen	X	The highest benzophenone loadings were obtained by solubilizing it in p-β-CD, and it showed solubility enhancement.	[36]
Glyburide	Y	In the presence of HB-β-CD, an almost 400-fold increase in glyburide aqueous solubility was observed.	[37]
Glyburide	P	A significant improvement in the drug dissolution profile was achieved from tablets containing drug–CD systems. 100% drug dissolution was never reached. Better results were obtained with ternary systems. In particular, poly(vinylpyrrolidone) (PVP) emerged as the most effective polymer, and tablets with drug–PVP–hydroxypropyl-β-CD co-evaporated products showed the best dissolution profiles, reaching 100% dissolved drug within 15 min.	[38]
Itraconazole	B	The efficacy and bioavailability of these drugs have been limited by their poor aqueous solubility and dissolution rate, which was improved by complexation.	[39]
Econazole	B	The efficacy and bioavailability of these drugs have been limited by their poor aqueous solubility and dissolution rate, which was improved by complexation.	[39]
Fluconazole	B	The efficacy and bioavailability of these drugs have been limited by their poor aqueous solubility and dissolution rate, which was improved by complexation.	[39]
Etoricoxib	P	Among all binary systems, a lyophilized product showed superior performance in enhancing solubility and hence dissolution of etoricoxib.	[40]
Irbesartan	B	Phase solubility studies revealed an increase in solubility of the drug upon CD addition, showing an A(L) type of graph with slope less than 1 indicating formation of a 1:1 stoichiometry inclusion complex.	[32]
Coumestrol	B	To improve its solubility in water, CDs were used.	[41]
Albendazole	X	The apparent solubility of albendazole was enhanced, especially with poly-α-CD.	[42]
Itraconazole hydrochloride	B	Aqueous solubility measurements showed that the solubility of the salt, its 1:1, 1:2, and 1:3 (w/w) physical mixtures with β-CD, were 6, 99, 236, and 388 times greater than that of itraconazole. More than 94% of itraconazole was dissolved out of the salt/β-CD 1:3 physical mixture after 60 min.	[43]
Glutathione	P and F	Solid-state FTIR, thermal analysis (DSC), and x-ray diffraction studies suggested that the nanoencapsulation process produced a marked decrease in the crystallinity of GSH, and thus solubility.	[44]
Omeprazole	B and F	Permeation studies indicated an 1.1- to 1.7-fold increase in drug permeation in a complexed form of β-CD and methyl-β-CD.	[45]
Flavonoid dioclein	B	The inclusion of dioclein in β-CD increased the water solubility to 44% compared to free dioclein. The in vitro vasodilator effect of dioclein was unchanged by its inclusion in β-CD, showing that the IC value does not change the interaction between dioclein and its cellular targets.	[46]

Table 3. (*Continued*)

Drug	Type of CD Used[a]	Effect on Solubility	Ref.
Glyburide	X	A significant improvement of the drug dissolution profile was achieved in tablets containing drug–CD systems, but 100% drug dissolution was never reached. Drug/PVP/hydroxypropyl-β-CD co-evaporated products showed the best dissolution profiles, showing 100% dissolved drug within only 15 min.	[38]
Caffeic acid	P	The solubility was enhanced.	[47]
Quercetin	B	Enhancement of the aqueous solubility of quercetin was 4.6- and 2.2-fold in the presence of 15 mM of β-CD using the spray-drying process and physical mixture method, respectively.	[48]
Cortisone acetate	P	The water solubility of cortisone acetate increased from 0.039 to 7.382 g/L at 32°C. The solubilization effect of HP-β-CD was far superior to that of dimethylformamide and ethanol.	[34]
Efavirenz (EFV)	I and P	The dissolution of EFV was substantially higher with HP-β-CD and RM-β-CD inclusion complexes prepared by the freeze-drying method.	[49]
Pyrimethamine (PYR)	A	A linear increase of PYR solubility was verified as a function of α-CD concentration.	[50]
Ibuprofen	P	Solubility enhancement was reported.	[51]
Albendazole	B	Solubility enhancement was reported.	[52]
Pb, Sr, Zn, and perchloroethylene	E	Cyclodextrins were capable of simultaneously enhancing the solubility.	[53]
Hexachlorobenzene	F	Solubility-enhanced electrokinetics were reported.	[54]
Formoterol	P and Q	An increase in the aqueous solubility of formoterol-complexed CDs led to the generation of aerosols with a particle size compatible with pulmonary deposition.	[55]
Nimesulide	B, P, and Q	β-CDs increased the aqueous solubility of nimesulide in the following order: methyl-β-CD > β-CD > hydroxypropyl-β-CD. β-CD, hydroxypropyl-β-CD, and methyl-β-CD were proposed as good solubilizing agents for nimesulide in the presence and absence of hydroxypropyl methylcellulose to enhance its oral bioavailability.	[56]
Itraconazole dihydrochloride	B	Aqueous solubility measurements showed that the solubility of the salt, its 1 : 1, 1 : 2, and 1 : 3 (w/w) physical mixtures with β-CD, was 6, 99, 236, and 388 times greater than that of itraconazole. 94% of itraconazole was dissolved out of the salt/β-CD 1 : 3 physical mixture after 60 min.	[43]
Fullerene	B	It was reported that the clustering effect of CD and PEG at the surface of the dendrimer might be crucial for the solubilization of fullerene.	[57]
Hesperidin	D	Hesperidin demonstrated poor, pH-independent aqueous solubility. Solubility improved dramatically in the presence of 2-hydroxypropyl-β-CD, and the results supported 1 : 1 complex formation.	[58]
Model drug	A, B, and C	It was reported that solubility of native α-, β-, γ-CDs in water rises with temperature. The opposite is true for their methylated derivatives (mCDs; per-dimethylated β-CD and per-trimethylated γ-CD). Results of the comparison indicated that (1) in solution, CDs and mCDs are in monomeric form; (2) van der Waals and solute-excluded volumes which can be related by introducing a shell of a thickness that correlates with the solute's structure and solute–water interactions, and (3) the SAXS curves calculated under the assumption of a uniform distribution of electron density in the solute molecules agree with experimental ones for CDs but not for mCDs.	[59]
Lamotrigine	B	The solubility profile of lamotrigine with β-CD was classified as A(L) type, indicating formation of a 1 : 1 stoichiometry inclusion complex with a stability constant of $369.96 \pm 2.26 \, M^{-1}$. The interaction was evaluated by powder x-ray diffractometry, Fourier transform infrared spectroscopy, and differential scanning calorimetry, confirming that it was no longer present in the crystalline state but was converted to an amorphous form.	[60]

(*continued*)

Table 3. (*Continued*)

Drug	Type of CD Used[a]	Effect on Solubility	Ref.
Thalidomide enantiomers	P	Complexes were obtained by using hydroxypropyl-β-CD, and the solubility of both thalidomide enantiomers was increased directly depending on the amount of hydroxylpropyl-β-CD.	[61]
Sodium dicloxacillin	A, B, C, and P	Phase solubility diagrams obtained were A(L) or B(S) type, depending on the CD used and the pH of the solution.	[62]
Bisphenol A (BPA)	A, B, and C	The results showed that β- and γ-CDs gave the satisfactory solubilization ability to BPA up to 7.2×10^3 mg/L and 9.0×10^3 mg/L, respectively.	[63]
Resveratrol	P and I	Both HP-β-CD and RM-β-CD enhanced the aqueous solubility of an intravenous dose of resveratrol; rapid elimination of resveratrol was observed at all doses tested regardless of the formulation types with nonlinear elimination.	[64]
Pseudo[1]rotaxane	X	Flexibility of the altro-α-CD cavity resulted in an induced fit [from (1)C(4) to (4)C (1)] to the arm moiety, and introducing a bulky end group allowed the stability of this pseudo[1]rotaxane to be enhanced.	[65]
Miconazole	D and F	The apparent stability constants [$K(S)$] calculated from the phase solubility diagram were 145.69 M^{-1} [$K(1:1)$] and 11.11 M^{-1} [$K(1:2)$] for M-β-CD and 126.94 M^{-1} [$K(1:1)$] and 2.20 M^{-1} [$K(1:2)$] for HP-β-CD.	[66]
Lamotrigine	B	Solubility enhancement was reported.	[60]
Cellulose acetate fibers	B	Solubility enhancement was reported.	[67]
Albendazole (ABZ)	B	Complexation of ABZ with β-cyclodextrin lead to the formation of an ABZ solution with potent antiproliferative effects.	[68]
Progesterone	P	The increased aqueous solubility of P by complexation with HP-β-CD was the main factor that increased the yield of 17-α-HP.	[69]
17-β-Estradiol	B and P	The complex was formed with E2, and its low aqueous solubility was improved. However, the hydrophilic E2/β-CD and E2/HP-β-CD complexes do not penetrate the membrane. Therefore, these CDs are able to suppress the hormone activity of E2.	[70]
Lamivudine	B	Binary and ternary mixtures prepared with inclusion complexes of lamivudine in β-CD and PVA lead to increase in solubility.	[71]
Ketoconazole (KET)	P	The highest levels of KET in aqueous humor (C_{max}, 2.67 μg/mL) were obtained after a 20-min application of KET–CD, 6.1 times greater than that corresponding to the KET–SP at 30 min. The KET concentrations in aqueous humor for post-120- and 180-min instillations of KET–CD were 57.9 and 34.5 times higher than that of KET–SP post-120 min, respectively.	[67]
Di(8-hydroxyquino-line)magnesium	B	The thermal stability and solubility of di(8-hydroxyquinoline)magnesium were improved when forming an inclusion complex.	[72]
Curcumin and curcuminoides	P	Hydroxypropyl-β-CD and propylene glycol alginate seemed to be the best choice with respect to curcumin solubility and release from the vehicle.	[73]
Genistein, an isoflavone	A, B, and C	Noncovalent inclusion complex with both β- and γ-CD was formed but did not form a stable complex with α-CD. A significantly improved aqueous solubility of genistein was reported.	[74]
Glycyrrhetic acid (GTA)	P	GTA/HP-β-CD inclusion complex was prepared by an ultrasonic-lyophilization technique. Solubility was increased 54.6-fold using an optimized method.	[75]
4-Methylbenzylidene camphor	A, B, C, and I	Among the CDs studied, random methyl-β-CD (RM-β-CD) had the greatest solubilizing activity.	[76]
Paclitaxel	I, K, P, and Q	Methylated β-CDs (randomly methylated and 2,6-dimethylated) showed the best ability to solubilize paclitaxel compared to sulfobutylether- and hydroxypropyl-β-CD.	[77]
Curcumin and curcuminoid	P, Q, and I′	Solubility and phase distribution studies showed that curcuminoids with side groups on the phenyl moiety have higher affinity for HP-γ-CD than for β-CDs.	[78]
Paclitaxel	A, B, C, and R	2,6-Dimethyl-β-CD was the most effective and its solubility was 2.3 mM in a 0.1 M 2,6-dimethyl-β-CD aqueous solution.	[79]
Progesterone	B, C′, C, P, J′, and U	Results showed that HP-β-CD and PM-β-CD were the most efficient among the four CD derivatives and two natural CDs for the solubilization of progesterone.	[80]

Table 3. (*Continued*)

Drug	Type of CD Used[a]	Effect on Solubility	Ref.
The herbicide diclo-fop-methyl (DM)	X	The solubility of DM was enhanced in the presence of β-CDs, the extent of which was dependent on the modification of β-CDs.	[81]
Glimepiride	B, P, and U	The dissolution rate of glimepiride/HP-β-CD–5% PEG 4000 was high compared to others.	[82]
Lycopene	B, F, and P	Solubility enhancement was reported.	[83]
Prazosin	B and P	Phase solubility diagrams and the solubility of PRH could be enhanced by 27.6% for β-CD and 226.4% for HP-β-CD, respectively.	[84]
Melarsoprol	B, I, and P	The solubility enhancement factor of melarsoprol (solubility in 250 mM of CD/solubility in water) was about 7.2×10^3 with both β-CD derivatives.	[85]
Diazepam	P, C′, U, and A′	The increase in solubility displayed a concentration dependency for the four CDs used. Diazepam's solubility was enhanced linearly as a function of each CD concentration. The highest improvements in solubility (dissolved concentration ca. 3.5 mg/mL in 40% CD) were found by adding HP-β-CD or SBE7-β-CD.	[86]
Benzocaine (BZC)	B	Threefold increase in BZC solubility could be reached upon complexation with β-CD.	[87]
Irisquinone	P	Solubility of irisquinone was enhanced markedly by inclusion with HP-β-CD when the host and guest stoichiometry was 2 : 1.	[88]
Azadirachtin	B, P, G, and K′	The water solubility of azadirachtin was obviously increased after the resulting inclusion complex with CDs. Typically, β-CD, DM-β-CD, TM-β-CD, and HP-β-CD were found to be able to solubilize azadirachtin to high levels up to 2.7, 1.3, 3.5, and 1.6 mg/mL.	[89]
Quercetin	B, P, and U	Solubility enhancements of quercetin obtained with the three β-CDs followed the rank order: SBE-β-CD > HP-β-CD > β-CD.	[90]
13-*cis*-Retinoic acid	A, B, and P	The effect of complexation of 13-*cis*-RA with α-CD and HP-β-CD on its phase solubility was studied. HP-β-CD was found to be more effective in increasing the aqueous solubility of 13-*cis*-RA compared to α-CD.	[91]
Propofol	H′	Aqueous solubility enhancement was reported.	[92]
Celecoxib	B	Solubility enhancement was reported.	[93]
Artelinic acid and artesunic acid	B	These compounds exhibited poor solubility in aqueous solution. NMR results are most consistent for artelinic acid and β-CD forming complexes in a ratio of 2 : 1.	[94]
Dexamethasone and digoxin	P	Digoxin and dexamethasone are insoluble in water. The inclusion of hydroxy acids in CD complexes in the necessary molar proportions led to a considerable increase in the solubility.	[95]
Pentachlorophenol	B, E, P, and Q	All CDs were found to form 1 : 1 inclusion complexes, which led to an increase in the aqueous solubility.	[96]
Celecoxib	B	Celecoxib has a very low water solubility, formation of a complex with β-CD both in an aqueous medium and in solid state, led to solubility and dissolution rate enhancement.	[97]
Triamterene	B	Improved the solubility and therefore dissolution and bioavailability.	[98]
Gliquidone	P	The molecular encapsulation and amorphization of gliquidone enhanced the oral bioavailability.	[99]
Ganciclovir	P	The aqueous solubility was enhanced.	[100]
Natamycin	B, C, and P	The increase in solubility of natamycin with added β-CD was observed to be linear. The water solubility of natamycin was increased 16-, 73-, and 152-fold with β-CD, γ-CD, and HP β-CD, respectively.	[101]
Nifedipine	B	Nifedipine is practically insoluble in water and aqueous fluids. Complexation of nifedipine with β-CD markedly enhanced the solubility and dissolution rate.	[102]
Acitretin	I and P	The solubility of acitretin was improved dramatically by formation of complexes and increased further by pH adjustment. Stability constants were much higher for acitretin complexed with RM-β-CD than with HP-β-CD.	[103]

(*continued*)

Table 3. (*Continued*)

Drug	Type of CD Used[a]	Effect on Solubility	Ref.
Vinpocetine (VP)	B and U	SBE-β-CD showed higher solubilizing efficacy toward VP than did the parent β-CD, due to its greater solubility and complexing abilities.	[104]
Rofecoxib	B	Formation of solid inclusion complexes of rofecoxib and cyclodextrin at different molar ratios exhibited a higher rate of solubilization and thus dissolution than did the physical mixture and the pure drug.	[105]
2,4,6-Trichlorophenol	B, P, and Q	The enhancement solubility method revealed that the stability of the complexes was dependent on the polarity of the compound and on the CD used. Solubilization efficiencies toward TCP could be ranked in the following order: methylated CD > HPCD > β-CD.	[106]
Fentanyl	P, U, and A′	The solubility of fentanyl increased linearly as a function of the CD concentration and with decreasing Ph.	[107]
Vinpocetine (VP)	U	The water-soluble polymers were shown to improve the complexation efficiency of SBE-β-CD, and thus less SBE-β-CD was needed to prepare solid VP-SBE-β-CD complexes in the presence of the polymers, but VP solubilization in water increased linearly with increasing SBE-β-CD concentration.	[108]
Ibuprofen (IBU)	B	A phase solubility study of IBU in an aqueous β-CD solution in the presence of the bile salt (sodium cholate) showed an increase in the solubility of IBU.	[109]
Insulin	B	A complex was prepared to improve the solubility of insulin. It was available as a dry powder, after encapsulation into poly(D,L-lactic-*co*-glycolic acid) microspheres.	[110]
Itraconazole	P	Solubility enhancement was reported.	[111]
Clarithromycin	X	The water solubility of clarithromycin was enhanced about 700-fold by complexation with CD.	[112]
The alkaloids harmane and harmine	X	Both 1:1 and 1:2 β-carboline alkaloid/CD complexes with different solubility properties were prepared using different CDs, such as β-CD, HP-β-CD, DM-β-CD, TM-β-CD. Association constants varied from 112 for harmine/DM-β-CD to 418 for harmane/HP-β-CD.	[113]
Ketoprofen	P	The solubility of the ionized complex of the drug was 2.5-fold greater than the nonionized complex. The flux of the ionized drug at 10% w/v HP-β-CD concentration was enhanced order of approximately eightfold compared to the intrinsic permeability of the nonionized drug.	[114]
Iodine	A, B, and P	The solubility order was KI < β-CD < α-CD < 2-HP-α-CD.	[115]
Nifedipine	B, P, and T	The inclusion complexation of nifedipine was shown to retard drug otodegradation with values dependent on light source and type of complexing agent. The effect was less for β-CD than for modified β-CD. It was noticed that the inclusion complexation of nifedipine offered much higher protection against the effect of a fluorescent lamp than of sunlight.	[116]
Naphthoquinone	P	A 38-fold solubility enhancement was possible by using a combined approach of pH adjustment and complexation with HP-β-CD.	[117]
Tenoxicam	B, C, P, C′, and Q	M-β-CD complexes yielded the best results: good solubility and the highest stability constant.	[118]
Camptothecin (CPT)	A, B, C, P, and Q	The results showed a linear increase in the solubility of CPT with an increasing concentration of CDs. The solubility of CPT was $228.45 \pm 8.45 \, \mu g/mL$, about 171-fold higher than that in 0.02 N HCl.	[119]
Nicardipine	B and P	The aqueous solubility was enhanced.	[120]
Naproxen	B, P, and Q	Water-soluble polymers increased the complexation efficacy of CDs toward naproxen, which resulted in enhanced drug solubility.	[121]
Naproxen (NAP)	P	The combined use of PVP and HP-β-CD resulted in a synergistic increasing effect of the aqueous solubility of NAP (120 times that of the pure drug).	[122]
Tolbutamide	B	Solubility enhancement was reported.	[123]
Sulfamethizole	B and P	The aqueous solubility of sulfamethizole was increased by complexation with β-CD or HP-β-CD.	[124]
Phosphatidylcholine	B	The aqueous solubility increased after complexation.	[125]

Table 3. (*Continued*)

Drug	Type of CD Used[a]	Effect on Solubility	Ref.
Phenytoin	C, P, and Q	Drug solubility in 0.05 M potassium phosphate buffer were notably improved by employing the β-CDs. A 45% w/v HP-β-CD or M-β-CD solution gave rise to an increase of dissolved drug by 420- and 578-fold.	[126]
Cyclosporin A	A and P	Cyclodextrins can increase the CsA solubility, but α-CD was more effective than HP-β-CD.	[127]
Hydrocortisone	P	The dissolution rate for HC increased after complexation, with 25 to 40% of the drug being released in the first hour compared with about 5% for pure HC.	[128]
Terpenes	B	The ability of GUG-β-CD was almost the same as that of G2-β-CD.	[129]
Loteprednol etabonate	A′, B′, C, D, and K	Among the five CDs used, DMCD showed the highest effects on the solubility (4.2 to 18.3 mg/mL in 10 to 50% DMCD) and stability ($t_{90} > 4$ years at 4°C).	[130]
all-trans-Retinoic acid (ATRA)	D	The aqueous solubility of ATRA could be greatly increased by the inclusion of ATRA in 2-hydroxypropyl-β-CD (HP-β-CD). Adjusting the pH value further improved the water solubility of ATRA.	[131]
Hydrocortisone	P	The drug and complex–polymer interactions in each system could be responsible for the solubility enhancement of the drug molecule.	[132]
Gliclazide	B	The solubility was enhanced.	[133]
Phenothiazine	B and R	It was established that the improvement in the solubility of phenothiazine was dependent on the type of CD.	[134]
Artemisinin	A, B, C, D, K, L, and V	The solubility order is β-CD < HP-β-CD < SBE7-β-CD < DM-β-CD.	[135]
Albendazole, mebendazole, ricobendazole	B, P, and Q	The solubility and bioavailability were enhanced.	[136]
Ursodeoxycholic acid (UDCA)	B	Both β-CD/CDC and β-CD increased the water solubility of UDCA, particularly β-CD/CDC.	[137]
Praziquantel	A, B, and C	The solubility (dissolution) of α-, β- and γ-CD complexes was 2.6-, 5-, and 8-fold greater, respectively, than that of the pure drug.	[24]
Cyclosporin A	S and T	The solubility increased 87-fold in the presence of 5.0×10^{-2} M DM-β-CD.	[138]
Rutin	B	The dissolubility of rutin in water was added to 643.19 mg/L (20°C).	[139]
Levemopamil	P	The solubility was enhanced.	[140]
Ziprasidone	U	The dissociation constants increased eightfold.	[141]
Praziquantel	B	The solubility data were linear for up to the highest concentration of β-CD added. The 30°C decrease in the melting point was reflected in increased solubility of the enantiomers.	[142]
Propofol	P	The aqueous solubility increased linearly as a function of HP-β-CD concentration. It is potentially useful for parenteral administration of the drug.	[143]
Melatonin (MT)	P	The solubility of MT in PG increased slowly until it reached 40% PG and then increased steeply. The solubility of MT increased linearly as the concentration of 2-HP-β-CD without PG increased. MT solubility in the mixtures of PG and 2-HP-β-CD also increased linearly but was less than the sum of its solubility in 2-HP-β-CD and PG individually.	[144]
Haloperidol	P and Q	The solubility was increased 20-fold in the presence of a 10-fold excess of methyl-β-CD (M-β-CD) and 12-fold in the presence of a 10-fold excess of 2-hydroxypropyl β-CD (HP-β-CD).	[145]
Ampicillin	B	A drug with β-CD was found to increase both the solubility and the dissolution rate of the drug.	[146]
Anandamide	P	The aqueous solubility increased 1000- to 30,000-fold, depending on the type of CD (10% solution) used.	[147]
Methylparaben	P	HP-β-CD increased significantly the solubility of methylparaben in water.	[148]
Paclitaxel	B, C, P, N, and T	These CDs increased paclitaxel solubility 2×10^3-fold or more and did not alter the cytostatic properties of paclitaxel in vitro.	[149]
Pilocarpine	P	The solubility of the prodrugs was shown to increase markedly in phase solubility studies.	[150]

(*continued*)

Table 3. (*Continued*)

Drug	Type of CD Used[a]	Effect on Solubility	Ref.
Psoralen	B, K, and L	The solubility and dissolution rate of the complexed forms improved, particularly for the DM-β-CD complex.	[151]
Thalidomide	P	Thalidomide, however, is sparingly soluble in aqueous solution (50 μg/mL) and unstable. Complexation with hydroxypropyl-β-CD has significantly improved the aqueous solubility and stability of thalidomide. Results obtained with HPLC and ^{1}H NMR spectrometry demonstrated that the solubility increased to 1.7 mg/mL, and the half-life of a diluted solution was extended from 2.1 h to 4.1 h.	[152]
Carbamazepine	B and C	The poor aqueous solubility of carbamazepine was dramatically increased via complexation with various chemically modified β- and γ-CDs.	[153]
Hydroflumethiazide, bendrofluazide, and cyclopenthiazide	B	The solubility was enhanced.	[154]
Nimodipine	B, D, M, N, and O	The order of increase was 2,3-dihydroxypropyl-β-CD < β-CD < 2-hydroxyethyl-β-CD < 3-hydroxypropyl-β-CD < 2-hydroxypropyl-β-CD.	[155]
Chlorambucil (CHL)	K	The aqueous solubility of CHL was increased more than 40-fold in the presence of 1.74×10^{-2} M DIMEB and with 1.74×10^{-2} M β-CD, there was a threefold increase in the aqueous solubility of CHL under similar conditions.	[156]

[a] A, α-Cyclodextrin; B, β-cyclodextrin (β-CD); C, γ-cyclodextrin; D, 2-hydroxypropyl-β-cyclodextrin (HP-β-CD); E, carboxymethyl-β-cyclodextrin (CMCD); F, methyl-β-CD; G, 2,3-di-*O*-methyl-β-cyclodextrin (DM-β-CD); H, 2,3,6-tri-*O*-methyl-β-cyclodextrin; I, randomly methylated β-CD (RM-β-CD); J, HE-β-CD; K, heptakis(2,6-di-*O*-methyl)-β-cyclodextrin (DM-β-CD; DMIEB); L, hepatakis (2,3,6-tri-*O*-methyl)-β-cyclodextrin (TM-β-CD); M, 2,3-dihydroxypropyl-β-CD; N, 2-hydroxyethyl-β-CD; O, 3-hydroxypropyl-β-CD; P, hydroxypropyl-β-CD; Q, methyl-β-CD; R, substituted-β-CD; S, DM-α-CD; T, DM-β-CD; U, β-cyclodextrin sulfobutyl ether (SBE-CD); V, (SBE)$_{7m}$-β-CD; W, α-L-Fuc-β-CD; X, cyclodextrin (CD); Y, hydroxybutenyl-β-cyclodextrin; Z, sulfobutyl ether-β-CD (SBE7-β-CD); A′, maltosyl-β-cyclodextrin (MBCD); B′, mixture of glucosyl/maltosyl-α-, β-, and γ-cyclodextrin (GMCD); C′; HP-γ-CD; D′, GUG-β-CD; E′, G2-β-CD; F′, β-CD-tetradecasulfate; G′, 6-(3′S-carboline-3′-carboxylaminoethylamino)-6-deoxy-β-cyclodextrin; H′, sulfobutyl ether 7-β-cyclodextrin (SBE-CD); I′, HP-γ-CD; J′, permethyl-β-cyclodextrin (PM-β-CD).

transitions toward amorphous in DSC thermograms, FTIR spectra, and XRD patterns. Hydrogen bonding was found between DHA and HP-β-CD, and it was stronger in complexes prepared in water than in buffers. DHA/HP-β-CD complexes prepared using commercial (untreated) or recrystallized DHA showed a 40% increase in thermal stability and

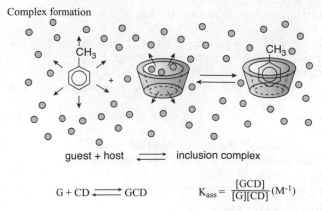

Complex formation

guest + host ⇌ inclusion complex

$$G + CD \rightleftharpoons GCD \qquad K_{ass} = \frac{[GCD]}{[G][CD]} \, (M^{-1})$$

Figure 3. Complex formation with CDs. (*See insert for color representation of the figure.*)

a 29-fold decrease in hydrolysis rates compared with DHA. The rank order of stability constants (*K*) was water, acetate buffer (pH 3.0), phosphate buffer (pH 3.0), and phosphate buffer (pH 7.4) [31].

The requisite values of the thermodynamic quantities at different temperatures have been obtained by isothermal titration calorimetry. The studies done with CDs to establish a relation to stability are given in Table 4. It was reported that CD-induced enhancement of drug stability may be a result of inhibition of drug interaction with vehicles and/or inhibition of drug bioconversion at the absorption site [30]. By providing a molecular shield, CD complexation encapsulates labile drug molecules at the molecular level and thus insulates them against various degradation processes. SBE-β-CD showed greater stability enhancement of many chemically unstable drugs than that of other CDs [25]. The stabilizing effect of CDs depends on the nature and effect of the included functional group on the drug stability and the nature of the vehicle. Both the catalyzing effect of the nitro group as well as the stabilizing effect of the halogen and cyanogen groups on photodegradation of 1,4-dihydropyrimidine derivatives were reduced by complexation with CDs [157]. HP-β-CD significantly reduced the

Table 4. Effect of CDs on the Stability of Various Drugs

Drug	Type of CD Used[a]	Effect on Stability	Ref.
Efavirenz	B, I, and P	Stability constants [$K(s)$] for EFV-β-CD, EFV-HPβ-CD, and EFV-RM-β-CD systems were 288, 469, and 1073 M^{-1}, respectively.	[49]
α-Tocopherol and quercetin	C	Natural antioxidant/CD inclusion complexes served as novel additives in controlled-release active packaging to extend the oxidative stability of foods.	[178]
α-Tocopherol and quercetin	B	Natural antioxidant/CD inclusion complexes served as novel additives in controlled-release active packaging to extend the oxidative stability of foods.	[178]
Tizanidine hydrochloride	B and P	Stability studies in natural saliva indicated that optimized formulation using CDs had good stability in human saliva.	[179]
Cortisone acetate	P	The enzymatic stability of Δ^1 dehydrogenase from *Arthrobacter simplex* TCCC 11037 was not influenced by the increasing concentrations of HP-β-CD.	[34]
Flavonoid dioclein	B	The mechanism underlying the increase in bioavailability was probably a consequence of a protective effect of β-CD against in vivo biodegradation by enzymes and possibly increased water solubility.	[46]
cis-Cyclooctene, *cis,cis*-1,3-cyclooctadiene, and *cis,cis*-1,5-cyclooctadiene	B	The trend of stability of the three inclusion complexes deduced from their calculated stabilization energies agreed well with the order of their association constants obtained from NMR experiments.	[180]
Omeprazole	B and Q	Results showed that complexation with CDs increased drug stability at neutral conditions; furthermore, L-arginine contributed to higher drug stability.	[45]
Glucagon	C	In the solid state, glucagon was degraded via oxidation and aggregation and in the presence of lactose via the Maillard reaction. The solid-state stability of glucagon/γ-CD powder was better than that of glucagon/lactose powder.	[181]
Itraconazole hydrochloride	B	Results obtained clearly indicated that the presence of β-CD improved the thermal stability of nutraceutical antioxidants present in *H. sabdariffa* L. extract, both in solution and in solid state.	[43]
Caffeine	B	The inclusion interactions prevented caffeine self-association.	[35]
Branched β-cyclodextrins	B and W	The order of binding affinity, as a function of the fucose-binding position, was 6(1), 6(4) → 6(1), 6(3) → 6(1), 6(2)-di-*O* (α-L-Fuc) β-CD > 6-*O* (α-L-Fuc) β-CD.	[182]
Irbesartan	B	Stability constant [$K(s)$] was found to be 104.39 M^{-1}.	[32]
Dihydroartemisinin (DHA)	P	Thermal stability of DHA complex was studied. A 40% increase in thermal stability (50°C) and a 29-fold decrease in hydrolysis rates was reported.	[31]
ClO_4^- ions	A	Results suggested that the stability of the ClO_4^- ion encapsulation involves not only the individual ion but also its first solvation shell.	[183]
Pyrimethamine	A	DSC measurements provided additional evidence of complexation, such as the absence of the endothermic peak assigned to the melting of the drug, indicating thermal stability.	[50]
Etoricoxib	P	The apparent stability constant [$K(c)$] of binary complex obtained at room temperature, $371.80 \pm 2.61\,M^{-1}$, was decreased with the addition of PVP and arginine, indicating no benefit of addition of auxiliary substances to promote higher complexation efficiency.	[40]
Itraconazole hydrochloride	B	The formulation in a complex form was an environmentally friendly, economical, and practical alternative to commercially available itraconazole capsules.	[43]
(R,S)-2-acetyl-1-(4'-chlorophenyl)-6,7-dimethoxy-1,2,3,4-tetrahydroisoquinoline	B	The binding constant [$K(R)^{-1}$] = 15,889 M^{-1}, $K(R,S)^{-1}$) = 1079 M^{-1}] and the complexation efficiency were increased.	[184]
Methylcyclopropene	A	Solid state was studied by means of TG and DSC. The apparent activation energy of dissociation [$E(D)$] decreased with increasing inclusion ratio, indicating higher complex stability at lower inclusion ratios, $E(D)$.	[185]

(continued)

Table 4. (*Continued*)

Drug	Type of CD Used[a]	Effect on Stability	Ref.
Anthocyanin extract	B	In the presence of β-CD, anthocyanins degraded at a decreased rate, evidently due to their complexation with β-CD, having the same activation energy. DSC revealed that the inclusion complex of *H. sabdariffa* L. extract with β-CD in the solid state was more stable against oxidation then was to the free extract, as the complex remained intact at temperatures of 100 to 250°C, where the free extract was oxidized.	[186]
Miconazole	D and Q	Phase solubility diagrams with Me-β-CD and HP-β-CD were classified as A(P) type, indicating the formation of 1 : 1 and 1 : 2 stoichiometric inclusion complexes.	[66]
Sodium dicloxacillin	A, B, C, and P	The highest stability constants of the inclusion complexes were obtained with γ-CD at pH 1 and 2 and HP-β-CD at pH 3. Both increased the stability of the drug, but the efficacy was higher with γ-CD.	[62]
Albendazole, mebendazole, thiabendazole	A, B, C, and P	Albendazole and mebendazole exhibited similar complexation behavior, whereas thiabendazole acted differently regarding the thermodynamic profile.	[187]
Di(8-hydroxyquinoline) magnesium	B	The thermal stability and solubility of di(8-hydroxyquinoline)magnesium were improved when forming inclusion complex.	[72]
Thalidomide enantiomers	P	The chemical stability of thalidomide enantiomers was clearly improved by hydroxypropyl-β-CD. No enantioselective degradation of thalidomide was observed in sodium chloride solution (0.9%) samples stored at 6°C for 9 days when hydroxypropyl-β-CD was used.	[61]
Itraconazole dihydrochloride	B	The stability studies indicated that the physical mixture remained stable for 24 months. Itraconazole dihydrochloride/β-CD (1 : 3) was an environmentally friendly, economical, and practical alternative to the commercially available itraconazole capsules (Sporanox).	[43]
Hesperidin	D	The solutions were stable under the pH and temperature (25, 40°C) conditions tested, except for samples stored at pH 9.	[58]
Anticancer agent EO-9	D	Chemical stability was enhanced.	[188]
Ketoprofen	B and Q	Advantages of the microwave technology for preparing ternary complexes of ketoprofen with β-CD or methylated β-CD were studied. The irradiation energy helped in obtaining totally dehydrated samples, which maintained unchanged solid-state characteristics and showed no susceptibility to ambient humidity after 2 years' storage at ambient temperature.	[189]
Meloxicam	P	Ternary system of M/HP-β-CD/L-arginine exhibited a stability constant 30.3 times higher than the binary system of M/HP-β-CD, while the ternary system of M/HP-β-CD/PVP increased the stability constant only 2.2-fold.	[190]
Quercetin	B, P, and U	The results showed that the inclusion antioxidant stability was in the order. SBE-β-CD > HP-β-CD > β-CD.	[191]
Paclitaxel	I, K, P, and Q	After 24 h of storage at room temperature or 2 h at HIPEC conditions (41.5°C), the 1 : 40 (mol/mol) ratio showed the highest stability at paclitaxel concentrations of 0.1 and 0.5 mg/mL. Hydroxypropyl methylcellulose also aided the stability significantly, offering the opportunity to reduce the amount of RAME-β-CDs in the formulation.	[77]
4-Methylbenzylidene camphor (MBC)	A, B, C, and I	The light-induced decomposition of 4-MBC in emulsion vehicles was decreased markedly by complexation with RM-β-CD.	[76]
Curcumin and curcuminoid	P, Q, and I'	Hydrolytic stability and photochemical stability, in the general order of the stabilizing effect was HP-β-CD > M-β-CD ≫ HP-γ-CD.	[78]
Salinosporamide A	U	Enhanced aqueous stabilization in aqueous solutions was noted.	[192]
Glimepiride	B, P, and U	Chemical stabilization was observed after complexation.	[82]
The herbicide diclofopmethyl	X	Complexation reduced the hydrolysis of DM and hence increased its stability. Small inconsistency in the power of β-CDs between hydrolysis retardation and solubilization suggested that hydrolysis was affected by the properties of β-CDs and the configuration of DM in the complexes.	[81]

Table 4. (*Continued*)

Drug	Type of CD Used[a]	Effect on Stability	Ref.
Melarsoprol	B, I, and P	RAME-β-CD had a pronounced effect on the drug hydrolysis.	[85]
Quercetin	B and D	Stability constants of QURC-β-CD (1 : 1) was 402 M^{-1} and QURC-HP-β-CD (1 : 1) was 532 M^{-1}.	[193]
Celecoxib	B	The apparent stability constants calculated by these techniques were increased to 881.5 and 341.5 M^{-1}, respectively, for aqueous solutions and in solid state.	[97]
Pentachlorophenol	B, E, P, and Q	Solubility enhancement experiments revealed that the stability of the complexes was increased simultaneously, which was dependent on the polarity of the compound, the ionic strength, and the CD type.	[96]
Dexamethasone and digoxin	P	Digoxin is sensitive to light and is subject to acidic hydrolysis. A digoxin–CD complex showed stability in a water medium, and the optimum molar ratio of digoxin/HP-β-CD was 1 : 6. The same results can be achieved through HP-β-CD, by including dexamethasone in a multicomponent composition containing HP-β-CD and citric acid in a molar ratio of 1 : 4 : 1.	[95]
Artelinic acid and artesunic acid	B	These compounds exhibited poor stability in aqueous solution. NMR results were most consistent for artelinic acid- and β-CD-forming complexes in a ratio of 2 : 1.	[94]
Celecoxib	B	The physical stability was enhanced.	[93]
Propofol	H'	Short-term stability studies of the liquid formulation showed that they were stable for a month at 4°C. Short-term stability studies of the freeze-dried cakes showed that the product was stable for over a month at 4, 37, and 50°C.	[161]
Insulin	B	To improve its hydrolytic stability, encapsulation into poly(D,L-lactic-*co*-glycolic acid) microspheres was done along with cyclodextrins.	[110]
Acitretin	I and P	Both cyclodextrins acted to decrease aqueous as well as photodegradation of acitretin in solution.	[103]
Ganciclovir (GCV)	P	Butyryl cholinesterase-mediated enzymatic hydrolysis of the GCV prodrugs was studied using various percentages of w/v HP-β-CD. Considerable improvement in chemical and enzymatic stability of the GCV prodrugs was observed in the presence of HP-β-CD.	[100]
Angelica sinensis	B	The results showed *Angelica sinensis* essential oil was more stable and well preserved with β-CD inclusion.	[194]
Camptothecin (CPT)	A, B. C, I, P, and T	Stability constants [$K(c)$] for the CPT complexes with α-CD, β-CD, γ-CD, HP-β-CD, RDM-β-CD, and RDM-γ-CD were 188, 266, 73, 160, 910, and 40.6 M^{-1}, respectively, suggesting that RDM-β-CD afforded the most stable complex. The increase in the half-life of CPT was from 58.7 to 587.3 min.	[119]
Melatonin	A, B, and C	The stability order was seen as β-CD > γ-CD > α-CD in water and β-CD > α-CD > γ-CD in protonated or alkali-cationated complexes.	[195]
Iodine	A, B, and D	The stability constant order was KI < β-CD < α-CD < 2-HP-α-CD.	[115]
Phosphatidylcholine	B	Oxidative stability was observed.	[125]
Tolbutamide	P	TBM and HP-β-CD formed 1 : 1 inclusion complexes in aqueous solution with an apparent stability constant of 63 M^{-1}. HP-β-CDs and TBM/HP-β-CD complexes were amorphous, whereas the freeze-dried and spray-dried TBMs were polymorphic forms II and I.	[196]
Naproxen (NAP)	P	The strongest complexation capacity of HP-β-CD toward NAP was reflected by an about 65% increase in the apparent stability constant of the NAP/HP-β-CD complex in the presence of only 0.1% w/v PVP.	[122]
Sulfamethizole	B and P	The stability constants calculated from the phase solubility method were in the order HP-β-CD < β-CD.	[124]
Rac-nicardipine	B and D	The solutions exposed to UVA and UVB radiation showed a photoprotective effect by β-CD; conversely, HP-β-CD proved to favor drug photodegradation.	[197]

(*continued*)

Table 4. (*Continued*)

Drug	Type of CD Used[a]	Effect on Stability	Ref.
Melphalan and carmustine	P and V	The chemical stability was enhanced.	[166]
Loteprednol etabonate	A′, B′, C, D, and K	Among the five CDs used, DMCD showed the highest effects on the solubility (4.2 to 18.3 mg/mL in 10 to 50% DMCD) and the stability was $t_{90} > 4$ years at 4°C.	[130]
Phenothiazine	B, C, and R	It was established that the improvement in stability of phenothiazine was dependent on the type of CD.	[160]
Terpenes	D′ and E′	The stabilizing ability of GUG-β-CD was superior to G2-β-CD. The difference was in the structure of the side chain: namely, the hydroxymethyl group in G2-β-CD and the carboxyl group in GUG-β-CD.	[129]
Artemisinin	A, B, C, D, K, L, and V	The stability order was α- < γ- < or = β-CD.	[135]
all-trans-Retinoic acid (ATRA)	D	The photostability of HP-β-CD-based formulation of ATRA was evaluated and it was found that the inclusion of ATRA in HP-β-CD did improve the photostability of ATRA.	[131]
Adriamycin, adriamycinol, adriamycinone, and daunomycin	A, B, and C	It was found that α-CD did not affect the degradation of tested compounds; β-CD caused a little effect and γ-CD resulted in pronounced stabilizing effect. The formation of complexes was monitored by fluorescence spectroscopy.	[198]
Nimesulide (N)	B	An increase of 25.6- and 38.7-fold in the dissolution rate was indicated with N/β-CD 1 : 1 and 1 : 2 kneaded complexes.	[199]
Praziquantel	A, B, and C	The β-complex had a stability constant in the optimum range for pharmaceutical use, suggesting that the preferred complex for further development would be a water-soluble β-CD derivative.	[24]
Melphalan	P and V	The shelflife of the reconstituted melphalan was greatly enhanced.	[200]
2-Ethylhexyl-*p*-dimethylaminobenzoate	A, B, and C	The photostability/degradation value was 25.5% for the complex compared to 54.6% for free compound.	[158]
Isradipine (IS)	Q	Inclusion complexes of IS with M-β-CD was proved to increase twice the photostability of the drug.	[201]
Phenytoin, fosphenytoin	V	From the solubility of phenytoin and the kinetic information, the fosphenytoin shelf life was as high as nine years at 25°C and pH 7.4 in the presence of 60 mM of SBE$_{7m}$-β-CD, while longer shelf lives were possible at pH 8.	[202]
Melatonin (MT)	D	The solubility of MT in PG solution increased slowly until reaching 40% PG and then increased steeply. The solubility of MT increased linearly as a concentration of 2-HP-β-CD without increased PG ($r = 0.993$). MT was unstable in a strongly acidic solution (HCl-NaCl buffer, pH 1.4) but relatively stable in other pH values of 4 to 10 at 70°C. In HCl-NaCl buffer, MT in 10% PG was degraded more quickly and then slowed down at a higher concentration. However, the degradation rate constant of MT in 2-HP-β-CD was not changed significantly compared to water. The curent studies can be applied to the dosage formulations for the purpose of enhancing the percutaneous absorption or bioavailability of MT.	[144]
Psoralen	B, K, and L	The stability constants were 663 M^{-1} for β-CD, 603 M^{-1} for DM-β-CD, and 69.6 M^{-1} for TM-β-CD.	[151]
Pilocarpine	P	The stability of prodrug increased as a function of HP-β-CD concentration over the temperature range studied. The shelf life ($t_{90\%}$, calculated by the Arrhenius equation) of the prodrug in 72.5 mM HP-β-CD solution increased 5.1- and 6.1-fold at 25 and 4°C, respectively. The degradation rate of prodrug in stability studies was shown to be slower in the 1 : 2 complex than in the 1 : 1-complex, and the relative amounts of complex species were found to be dependent on CD concentration.	[150]

Table 4. (*Continued*)

Drug	Type of CD Used[a]	Effect on Stability	Ref.
Nisoldipine, nimodipine, nitrendipine, and nicardipine	B	The complexation with β-CD improved their photostability by five- to 10-fold. The greatest decrease in the photodegradation rate was observed for the most photosensitive compound NS, whose photostability increased 100 times when in inclusion complex with β-CD.	[203]
Nitrazepam	K	Inclusion complexation of nitrazepam in DM-β-CD resulted in a relatively improved stability of the drug in solution at 30°C.	[204]
Thalidomide	P	Thalidomide, is sparingly soluble in aqueous solution (50 μg/mL) and is unstable. Complexation with hydroxypropyl-β-CD has improved the aqueous solubility and stability of thalidomide significantly. Results obtained with HPLC and ^{1}H NMR spectrometry have demonstrated that the half-life of a diluted solution was extended from 2.1 h to 4.1 h.	[152]
Prostaglandin E1	A'	Increased stability was reported.	[205]
Carbamazepine	B and C	A preparation of carbamazepine and 2-hydroxypropyl-β-CD was found to be stable to steam sterilization and storage under a variety of conditions.	[153]
S-Nitrosothiol	A, B, C, P, and F'	The usefulness of several CDs to stabilize (chemical stability) this polar compound in solution was studied. At CD concentrations of 12 mM, hydroxypropyl-β-CD was most effective at stabilizing SNAP ($t_{1/2} = 77$ h) compared to α-CD (41 h), β-CD (69 h), γ-CD (36 h), and β-CD-tetradecasulfate (38 h).	[161]
Nimodipine	B, M, N, O, and P	The order of stability was 2,3-dihydroxypropyl-β-CD < β-CD < 2-hydroxyethyl-β-CD < 3-hydroxypropyl-β-CD < 2-hydroxypropyl-β-CD.	[155]
Hydroflumethiazide, bendrofluazide, and cyclopenthiazide	B	The complexes were found to have 1:1 stoichiometric ratios, and stability constants were 165.4, 55.7, and 27.9 M^{-1} for cyclopenthiazide, bendrofluazide, and hydroflumethiazide, respectively.	[154]
Chlorambucil (CHL)	K	In the presence of 1.3×10^{-3} M DIMEB, there is a greater than 20-fold increase in the stability of chlorambucil at 37°C, pH 4.13. In the presence of 1.3×10^{-3} M β-CD, there is a fourfold increase in the stability, and with 1.74×10^{-2} M β-CD, there is a threefold increase in the aqueous solubility of CHL under similar conditions.	[156]

[a] See Table 1 for definitions of the abbreviations.

photodegradation of 2-ethylhexyl *p*-dimethylaminobenzoate in solution compared to photodegradation in an emulsion vehicle [158]. CDs improved the photostability of trimeprazine (when the solution pH is reduced) [159] and promethazine [160]. CDs also enhanced the solid-state stability and shelf life of drugs [161–163]. CDs were reported to enhance the physical stability of viral vectors for gene therapy, and formulations containing sucrose and CDs were stable for two years when stored at 20°C [164]. Since the hydrolysis of drugs encapsulated in CDs is slower than that of free drugs [14], the stability of the drug–CD complex (i.e., the magnitude of the complex stability constant) plays a significant role in determining the extent of protection [119,165–167].

Very low concentrations of HP-β-CD (1% or lower), due to formation of a more physically unstable complex, did not protect taxol as effectively as did higher CD concentrations. Following is a simple model representing the effect of complex stability constant on drug degradation [168].

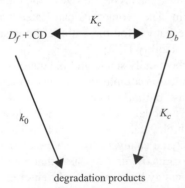

The effect of complexation on drug stability can be represented by

$$\frac{1}{k_0 - k_{obs}} = \frac{1}{K_c(k_0 - k_c)[CD]} + \frac{1}{(k_0 - k_c)} \quad (1)$$

where k_0 is the degradation rate constant of free drug, k_{obs} is the degradation rate constant observed in the presence of

CD, k_c is the degradation rate constant of the drug within CD, K_c is the stability constant for the complex, and [CD] is the concentration of CD [168]. Under specific conditions, CD complexation may accelerate drug degradation, depending on the type of the CD. CDs catalyzed deacetylation and degradation of spiranolactone. The effect was correlated qualitatively with the ionization state of hydroxyl groups on CDs that were lower in SBE-CDs [169]. Structural changes in drug molecules on CD complexation can also accelerate drug degradation [170]. β-CD did not improve the photostability of oflaxacin, as there was only partial inclusion of the methylpiperazinyl moiety in the CD [171].

The feverish level of research activity in the field of CDs continues unabated, and there have been many studies concerning the formation and stability of CD–incluscate complexes. Unfortunately, there is a considerable variation in the magnitude of the binding constants reported and a lack of agreement concerning the key factors that govern the formation and stability of CD–incluscate complexes [172].

As far as stability is concerned, much work has been done to study the interaction forces involved in the inclusion processes of drug molecules with several CDs and to assess the best CD for complexing. The behavior of the inclusion complexes of drug with α-, β-, and γ-CDs can be evaluated by ultraviolet/visible (UV/Vis) direct spectroscopy, proton nuclear magnetic resonance (^{1}H NMR), and molecular mechanics. Thermodynamic parameters for the binding processes were obtained from the temperature variations in binding. The thermodyanamic stability study of an anti-inflammatory drug (ketoprofen) complexed with cyclodextrin shows that binding constants of β- and γ-CDs form more stable 1 : 1 complexes with ketoprofen than does α-CDs [173]. Similarly, ^{1}H NMR spectra showed that the inclusion degree depends on the size of the internal diameter of cyclodextrin. The geometries calculated on the basis of molecular mechanics for these three-dimensional models indicate high stability [173].

They are chemically stable and in aqueous solution form host–guest inclusion complexes with molecules and ions that can fit at least partially into the cavity. They are used extensively as models to study noncovalent interactions important in biological processes such as molecular recognition and enzyme catalysis. The shape of α-CD, which is made up of six glucose units, is such that a benzene ring fits snugly in the cavity but cannot easily enter or leave via the narrow end [14]. Hence, it seems likely that directional forces pushing a substituted or 1,4-disubstituted benzene derivative into the cavity result in stronger binding, whereas forces pushing the derivative out of the wider end of the cavity result in weaker binding [15]. The situation is far more complicated for loosely fitting host–guest complexes, such as those involving β-CD (with seven glucose units) and benzene derivatives, where additional cross-interaction forces

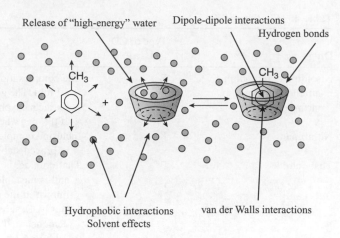

Figure 4. Driving forces for complex formation. (*See insert for color representation of the figure.*)

determine the optimum set of positions of the guest in the cavity [15]. The anisotropic nature of the interaction of α-CD and substituted or 1,4-disubstituted benzene derivatives is reflected in the following correlation equation [15], based on literature stability constants of about 50 inclusion complexes:

$$\log K_{1:1} = 1.28 \pm 0.11 + (1.38 \pm 0.16)\sigma_x - (2.35 \pm 0.33)\sigma_x\sigma_y$$
$$+ (0.120 \pm 0.013)(1 - \text{carb}^-)R_{mx} - (0.27 \pm 0.12)Y_{\text{sub}}^-$$
$$(2)$$

Here, $K_{1:1}$ is the stability constant of the 1 : 1 complex; σ_x and R_{mx} are, respectively, the Hammett σ_p substituent constant and the molar refractivity of the more electron-withdrawing substituent that we propose is located in the narrower, primary O_6H end of the α-CD cavity that represents the positive end of the CD dipole; and σ_y is the Hammett constant for the substituent that protrudes from the wider secondary O_2H and O_3H rim of the cavity; Y_{sub}^- is an identity variable for complexes involving negatively charged y-substituents; and carb$^-$ is an identity variable specifically for use when this substituent is a carboxylate. The correlation analysis leading to equation (2) covered values of log $K_{1:1}$ from about 0.5 to 3.5 and yielded a standard deviation of 0.34 and a correlation coefficient 0.92 [15]. There was only one outlier, the value of log $K_{1:1}$ for 1,4-diacetylbenzene was 1.01, whereas equation (2) predicted a value of 2.72. Major driving forces for complex formation include hydrophobic interactions, van der Waals interactions, and dipole-dipole interactions. These are shown in Fig. 4.

3.3. Characterization

The use of techniques such as NMR, thermodynamic methods, and optical spectrophotometry provides an opportunity to measure different physical properties associated with complex formation and a better means to achieve the objectives described above. For example, information about the inclusion mode, stoichiometry, and binding constant(s) of the host–guest complex can be derived from NMR [174]. Also, the relative contribution of solute–solute and solute–solvent interactions can be estimated from NMR chemical shift changes and dipolar couplings [175] of the host and guest nuclei.

Similarly, for addressing stability concerns, many studies were carried out to examine the interaction forces involved in the inclusion processes of drug molecules with several CDs and to assess the best CD for complexing. The behavior of the inclusion complexes of drug with α-, β-, and γ-CDs can be evaluated by UV/Vis direct spectroscopy, ^{1}H NMR, and molecular mechanics. Thermodynamic parameters for the binding processes were obtained from the temperature variations in binding. The thermodynamic stability study of an anti-inflammatory drug (ketoprofen) complexed with CD showed that binding constants of β- and γ-CDs form more stable 1 : 1 complexes with ketoprofen than does α-CD. Similarly, ^{1}H NMR spectra showed that the inclusion degree depends on the size of the internal diameter of the CD. The geometries calculated on the basis of molecular mechanics for these three-dimensional models indicate high stability [173]. Similarly, to improve the solubility and stability of poorly soluble dihydroartemisinin (DHA), HP-β-CD inclusion complexes were prepared with recrystallized DHA to study its thermal stability. The complexes were characterized by DSC, FTIR, XRD, and thermal stability, phase, and equilibrium solubility studies. Pure DHA was crystalline and remained crystalline after recrystallization, but its unit cell dimensions changed as exhibited by XRD. DHA/HP-β-CD complexes showed phase transitions toward amorphous in DSC thermograms, FTIR spectra, and XRD patterns. The equilibrium solubility of DHA was enhanced as a function of HP-β-CD concentration. DHA/HP-β-CD complexes showed an 89-fold increase in solubility compared to DHA. Hydrogen bonding was found between DHA and HP-β-CD, and it was stronger in complexes prepared in water than in buffers. DHA-HP-β-CD complexes prepared using commercial (untreated) or recrystallized DHA (no detectable impurity) showed a 40% increase in thermal stability (50°C) and a 29-fold decrease in hydrolysis rates compared with DHA. The rank order of stability constants (K) was water, acetate buffer (pH 3.0), phosphate buffer (pH 3.0), and phosphate buffer (pH 7.4) [31].

3.4. Evaluation of the Stability Profile of Complexes

The thermodynamic quantities from 15 to 45°C for binding to β-CD of the reference molecules phenethylamine and hydrocinnamate and their phenolic hydroxyl-substituted derivatives, which are capable of forming an additional hydrogen bond, are reported. Although the enthalpy change always makes a stabilizing contribution, the presence of the hydrogen bond may result in either a net stabilizing effect, as in 3-(4-hydroxyphenyl)propionate, or have reduced stability, as in 3-(2-hydroxyphenyl)propionate. In the latter case, the destabilization relative to hydrocinnamate arises from an enhanced unfavorable (negative) entropic contribution. Like most association reactions in water, the heat capacity change of these binding reactions is always negative.

The solid-state stability and the dissolution of glucagon/γ-CD and glucagon–lactose powders was evaluated. Freeze-dried powders were stored at an increased temperature and/or humidity for up to 39 weeks. Preweighed samples were withdrawn at predetermined intervals and analyzed with high-performance liquid chromatography (HPLC)–UV, HPLC–electrospray ionization mass spectrometry, size-exclusion chromatography, turbidity measurements, and solid-state FTIR. Dissolution of glucagon was evaluated at pH 2.5, 5.0, and 7.0. In addition, before storage, proton rotating-frame relaxation experiments of solid glucagon/γ-CD powder were conducted with ^{13}C cross-polarization magic-angle spinning (CPMAS) NMR spectroscopy. In the solid state, glucagon degraded via oxidation and aggregation and in the presence of lactose via the Maillard reaction. The solid-state stability of glucagon/γ-CD powder was better than that of glucagon–lactose powder. In addition, γ-CD improved the dissolution of glucagon at pH 5.0 and 7.0 and delayed the aggregation of glucagon after its dissolution at pH 2.5, 5.0, and 7.0. There was no marked difference between the proton rotating-frame relaxation times of pure glucagon and γ-CD, and thus the presence of inclusion complexes in the solid state could not be ascertained by CPMAS NMR. In conclusion, when compared to glucagon–lactose powder, glucagon/γ-CD powder exhibited better solid-state stability and more favorable dissolution properties.

3.5. Safety

Some CDs are reported to have significant renal toxicity [16]. CDs are starch derivatives and are used primarily in oral and parenteral pharmaceutical formulations. They are also used in topical and ophthalmic formulations [176]. CDs are also used in cosmetics and food products and are generally regarded as essentially nontoxic and nonirritant materials. However, when administered parenterally, β^2-CD is not metabolized but accumulates in the kidneys as insoluble cholesterol complexes, resulting in severe nephrotoxicity [6]. Other CDs, such as 2-hydroxypropyl-β-cyclodextrin, have been the subject of extensive toxicological studies. They are not associated with nephrotoxicity and are reported to be safe for use in parenteral formulations [15].

CD administered orally is metabolized by microflora in the colon, forming the metabolites maltodextrin, maltose, and glucose; which are themselves metabolized further before finally being excreted as carbon dioxide and water. Although a study published in 1957 suggested that orally administered CDs were highly toxic [177], more recent animal toxicity studies in rats and dogs have shown this not to be the case, and CDs are now approved for use in food products and orally administered pharmaceuticals in a number of countries. CDs are not an irritant to the skin and eyes, or upon inhalation. There is also no evidence to suggest that CDs are mutagenic or teratogenic.

4. CONCLUSIONS

Cyclodextrins have been playing a very important role in the formulation of poorly water-soluble drugs by improving apparent drug solubility and/or dissolution as well as the stability of several labile drugs against dehydration, hydrolysis, oxidation, and photodecomposition. As a result of their complexation ability and other versatile characteristics, CDs are continuing to have a variety of applications in different areas of drug delivery and the pharmaceutical industry. However, it is necessary to find any possible interaction between these agents and other formulation additives because the interaction can adversely affect the performance of both. It is also important to have knowledge of different factors that can influence complex formation for economical preparation of drug–CD complexes with desirable properties. Since CDs continue to find novel applications in drug delivery, we may expect these polymers to solve many problems associated with the delivery of different novel drugs via different routes.

REFERENCES

1. Endo, T., Nagase, H., Ueda, H., Kobayashi, S., Nagai, T. (1997). Isolation, purification, and characterization of cyclomaltodecaose (curly epsilon-cyclodextrin), cyclomaltoundecaose (zeta-cyclodextrin) and cyclomaltotridecaose (é-cyclodextrin). *Chem. Pharm. Bull. Tokyo*, *45*, 532–536.

2. Miyazawa, H., Ueda, H., Nagase, T., Endo, T., Kobayashi, S., Nagai, T. (1995). Physicochemical properties and inclusion complex formation of δ-cyclodextrin. *Eur. J. Pharm. Sci.*, *3*, 153–162.

3. Endo, T., Nagase, H., Ueda, H., Shigihara, A., Kobayashi, S., Nagai, T. (1998). Isolation, purification and characterization of cyclomaltooctadecaose (ν-cyclodextrin), cyclomaltononadecaose (xi-cyclodextrin), cyclomaltoeicosaose (o-cyclodextrin) and cyclomaltoheneicosaose (ã-cyclodextrin). *Chem. Pharm. Bull. (Tokyo)*, *46*, 1840–1843.

4. Muller, B.W., Brauns, U. (1986). Hydroxypropyl-β-cyclodextrin derivatives: influence of average degree of substitution on complexing ability and surface activity. *J. Pharm. Sci.*, *75*, 571–572.

5. Yoshida, A., Yamamoto, M., Irie, T., Hirayama, F., Uekama, K. (1989). Some pharmaceutical properties of 3-hydroxypropyl and 2,3-dihydroxypropyl-beta-cyclodextrins and their solubilizing and stabilizing abilities. *Chem. Pharm. Bull.*, *37*, 1059–1063.

6. Hirayama, F., Usami, M., Kimura, K., Uekama, K. (1997). Crystallization and polymorphic transition behavior of chloramphemicol palmitate in 2-hydroxypropyl-β-cyclodextrin matrix. *Eur. J. Pharm. Sci.*, *5*, 23–30.

7. Loukas, Y. L., Vraka, V., Gregoriadis, G. (1996). Use of a nonlinear least-squares model for the kinetic determination of the stability constant of cyclodextirn inclusion complexes. *Int. J. Pharm.*, *144*, 225–231.

8. Sortino, S., Scaiano, J.C., de Guidi, G., Monti, S. (1999). Effect of beta cyclodextrin complexation on the photochemical and photosensitizing properties of tolmetin: a steady state and time resolved study. *Photochem. Photobiol.*, *70*, 549–556.

9. Liu, Y., Han, B.H., Zhang, H.Y. (2004). Spectroscopic studies on molecular recognition of modified cyclodextrins, *Curr. Org. Chem.*, *8*, 35–46.

10. Uekama, K., Hirayama, F., Irie, T. (1998). Cyclodextrin drug carrier systems, *Chem. Rev.*, *98*, 2045–2076.

11. Irie, T., Uekama, K. (1999). Cyclodextrins in peptide and protein delivery. *Adv. Drug Deliv. Rev.*, *36*, 101–123.

12. Uekama, K. (2004). Design and evaluation of cyclodextrin-based drug formulation. *Chem. Pharm. Bull.*, *52*, 900–915.

13. Martin, A., Bustamante, P., Chun, A.H.C. (1994). *Physical Pharmacy: Physical Chemical Principles in the Pharmaceutical Sciences*. B. I. Waverly, New Delhi, India, p. 257.

14. Mishra, P.R., Mishra, M., Namdeo, Jain, N.K. (1999). Pharmaceutical potential of cyclodextrins. *Ind. J. Pharm. Sci.*, *61*(4), 193.

15. Loftsson, T., Brewester, M. (1996). Pharmaceutical applications of cyclodextrins: 1. Drug solubilization and stabilization. *J. Pharm. Sci.*, *85*, 1017–1025.

16. Rajewski, R.A., Stella, A.J. (1996). Pharmaceutical applications of cyclodextrins: in vivo drug delivery. *J. Pharm. Sci.*, *85*(11), 1142–1168.

17. Chowdary, K.P.R., Reddy, K. G., Kamlakar, G. (2002). Complexes of nifedipine with beta and hydroxypropyl beta cyclodextrins in the design of nifedipine SR tablets. *Int. J. Pharm. Exp.*, *2*, 12.

18. Loftsson, T., Másson, M., Brewster, M.E. (2004). Self-association of cyclodextrins and cyclodextrin complexes. *J. Pharm. Sci.*, *93*, 1091–1099.

19. Bost, M., Laine, V., Pilard, F., Gadelle, A., Defaye, J., Perly, B. (1997). The hemolytic properties of chemically modified cyclodextrins, *J. Inclusion Phenom. Mol. Recognit. Chem.*, *29*, 57.

20. Leray, E., Leroy-Lechat, F., Parrot-Lopez, H., Duchêne, D. (1995). Reduction of the haemolytic effect in a biologically recognizable β-cyclodextrin. *Supramol. Chem.*, *5*, 149.

21. Irie, T., Otagiri, M., Sunada, K., Uekama, Y., Ohtani, Y., Yamada, Y., Sugiyama, Y. (1982). Cyclodextrin-induced hemolysis and shape changes of human erythrocytes in vitro. *J. Pharm. Dyn.*, *5*, 741.

22. Londhe, V., Nagarsenker, M. (1999). Comparision between hydroxypropyl-β-cyclodextrin and polyvinyl pyrrolidine as carriers for carbamazepine solid dispersions. *Indian J. Pharm. Sci.*, *61*, 237–240.

23. Bettinetti, G., Gazzaniga, A., Mura, P., Giordano, F., Setti, M. (1992). Thermal behavior and dissolution properties of naproxen in combinations with chemically modified β-cyclodextrins. *Drug Dev. Ind. Pharm.*, *18*, 39–53.

24. Becket, G., Schep, L.J., Tan, M.Y. (1999). Improvement of the in vitro dissolution of praziquantel by complexation with α-, β- and γ-cyclodextrins. *Int. J. Pharm. 179*(1), 65–71.

25. Ueda, H., Ou, D., Endo, T., Nagase, H., Tomono, K., Nagai, T. (1998). Evaluation of a sulfobutyl ether β-cyclodextrin as a solubilizing/stabilizing agent for several drugs. *Drug Dev. Ind. Pharm.*, *24*, 863–867.

26. Sangalli, M.E., Zema, L., Moroni, A., Foppoli, A., Giordano, F., Gazzania, A. (2001). Influence of β-cyclodextrin on the release of poorly soluble drugs from inert and hydrophilic heterogeneous polymeric matrices. *Biomaterials*, *22*, 2647–2651.

27. Pina, M.E., Veiga, F. (2000). The influence of diluent on the release of theophylline from hydrophilic matrix tablets. *Drug Dev. Ind. Pharm.*, *26*, 1125–1128.

28. Szejtli, J. (1991). Cyclodextrin in drug formulations: I. *Pharm. Technol. Int.*, *3*, 15Y23.

29. Szente, L., Szejtli, J. (1999). Highly soluble cyclodextrin derivatives: chemistry, properties, and trends in development. *Adv. Drug Deliv. Rev.*, *36*, 17Y38.

30. Matsuda, H., Arima, H. (1999). Cyclodextrins in transdermal and rectal delivery. *Adv. Drug. Deliv. Rev.*, *36*, 81Y99.

31. Ansari, M.T., Iqbal, I., Sunderland, V.B. (2009). Dihydroartemisinin–cyclodextrin complexation: solubility and stability. *Arch. Pharm. Res.*, *32*(1), 155–165.

32. Hirlekar, R., Kadam, V. (2009). Preformulation study of the inclusion complex irbesartan–β-cyclodextrin. *AAPSPharmSciTech.*, *10*(1), 276–281.

33. Asbahr A.C., Franco, L., Barison, A., Silva, C.W., Ferraz, H. G., Rodrigues LN. (2009). Binary and ternary inclusion complexes of finasteride in HP-β-CD and polymers: preparation and characterization. *Bioorg. Med. Chem.*, Feb 26.

34. Kalogeropoulos, N., Konteles, S., Mourtzinos, I., Troullidou, E., Chiou, A., Karathanos, V.T. (2009) Encapsulation of complex extracts in β-cyclodextrin: an application to propolis ethanolic extract. *J. Microencapsul.*, Mar. 6, pp. 1–11.

35. Mejri, M., Bensouissi, A., Aroulmoji, V., Rogé, B. (2009). Hydration and self-association of caffeine molecules in aqueous solution: comparative effects of sucrose and β-cyclodextrin. *Spectrochim. Acta A*, Jan. 24.

36. Daoud-Mahammed, S., Couvreur, P., Bouchemal, K., Chéron, M., Lebas, G., Amiel, C., Gref, R. (2009). Cyclodextrin and polysaccharide-based nanogels: entrapment of two hydrophobic molecules, benzophenone and tamoxifen. *Biomacromolecules*, Jan. 27.

37. Klein, S., Wempe, M.F., Zoeller, T., Buchanan, N.L., Lambert, J.L., Ramsey, M.G., Edgar, K.J., Buchanan, C.M. (2009). Improving glyburide solubility and dissolution by complexation with hydroxybutenyl-β-cyclodextrin. *J. Pharm. Pharmacol.*, *61*(1), 23–30.

38. Cirri, M., Righi, M.F., Maestrelli, F., Mura, P., Valleri, M. (2009). Development of glyburide fast-dissolving tablets based on the combined use of cyclodextrins and polymers. *Drug Dev. Ind. Pharm.*, *35*(1), 73–82.

39. Al-Marzouqi, A.H., Elwy, H.M., Shehadi, I., Adem, A. (2009). Physicochemical properties of antifungal drug–cyclodextrin complexes prepared by supercritical carbon dioxide and by conventional techniques. *J. Pharm. Biomed. Anal.*, *49*(2), 227–233.

40. Shah, M., Karekar, P., Sancheti, P., Vyas, V., Pore, Y. (2009). Effect of PVP K30 and/or L-arginine on stability constant of etoricoxib-HP-β-CD inclusion complex: preparation and characterization of etoricoxib-HP-β-CD binary system. *Drug Dev. Ind. Pharm.*, *35*(1), 118–129.

41. Franco, C., Schwingel, L., Lula, I., Sinisterra, R.D., Koester, L.S., Bassani, V.L. (2009). Studies on coumestrol/β-cyclodextrin association: inclusion complex characterization. *Int. J. Pharm.*, *369*(1–2), 5–11.

42. Joudieh, S., Bon, P., Martel, B., Skiba, M., Lahiani-Skiba, M. (2009). Cyclodextrin polymers as efficient solubilizers of albendazole: complexation and physico-chemical characterization. *J. Nanosci. Nanotechnol.*, *9*(1), 132–140.

43. Tao, T., Zhao, Y., Wu, J., Zhou, B. (2009). Preparation and evaluation of itraconazole dihydrochloride for the solubility and dissolution rate enhancement. *Int. J. Pharm.*, *367*(1–2), 109–114.

44. Lopedota, A., Trapani, A., Cutrignelli, A., Chiarantini, L., Magnani, F., Curci, R., Manuali, E., Trapani, G. (2009). The use of Eudragit((R)) RS 100/cyclodextrin nanoparticles for the transmucosal administration of glutathione. *Eur. J. Pharm. Biopharm.* Mar. 9.

45. Figueiras, A., Hombach, J., Veiga, F., Bernkop-Schnürch, A. (2009). In vitro evaluation of natural and methylated cyclodextrins as buccal permeation enhancing system for omeprazole delivery. *Eur. J. Pharm. Biopharm.*, *71*(2), 339–345.

46. Rezende, B.A., Cortes, S.F., De Sousa, F.B., Lula, I.S., Schmitt, M., Sinisterra, R.D., Lemos, V.S. (2009). Complexation with β-cyclodextrin confers oral activity on the flavonoid dioclein. *Int. J. Pharm.*, *367*(1–2), 133–139.

47. Zhang, M., Li, J., Zhang, L., Chao, J. (2009) Preparation and spectral investigation of inclusion complex of caffeic acid with hydroxypropyl-β-cyclodextrin. *Spectrochim. Acta A, 71* (5), 1891–1895.

48. Borghetti G.S., Lula, I.S., Sinisterra, R.D., Bassani, V.L. (2009). Quercetin/β-cyclodextrin solid complexes prepared in aqueous solution followed by spray-drying or by physical mixture. *AAPS PharmSciTech*, *10*(1), 235–242.

49. Sathigari, S., Chadha, G., Lee, Y.H., Wright, N., Parsons, D.L., Rangari, V.K., Fasina, O., Babu, R.J. (2009). Physicochemical characterization of efavirenz-cyclodextrin inclusion complexes, *AAPS PharmSciTech*, *10*(1), 81–87.

50. De Araujo, M.V., Macedo, O.F., Nascimento Cda, C., Conegero, L.S., Barreto, L.S., Almeida, L.E., da Costa, NB, Jr., Gimenez, I.F. (2009). Characterization, phase solubility and

molecular modeling of α-cyclodextrin/pyrimethamine inclusion complex. *Spectrochim. Acta A*, *72*(1), 165–170.

51. Wang, L.J., Zhu, Z.J., Che, K.K., Ju F.G. (2008). Characterization of microstructure of ibuprofen-hydroxypropyl-beta-cyclodextrin and ibuprofen-β-cyclodextrin by atomic force microscopy. *Yao Xue Xue Bao*, *43*(9), 969–973.

52. Pourgholami, M.H., Wangoo, K.T., Morris, D.L. (2008). Albendazole–cyclodextrin complex: enhanced cytotoxicity in ovarian cancer cells. *Anticancer Res.*, *28*(5A), 2775–2779.

53. Skold, M.E., Thyne, G.D., Drexler, J.W., Macalady, D.L., McCray, J.E. (2008). Enhanced solubilization of a metal–organic contaminant mixture (Pb, Sr, Zn, and perchloroethylene) by cyclodextrin. *Environ. Sci. Technol.*, *42*(23), 8930–8934.

54. Wan, J., Yuan, S., Chen, J., Li, T., Lin, L., Lu, X. (2008). Solubility-enhanced electrokinetic movement of hexachlorobenzene in sediments: a comparison of cosolvent and cyclodextrin. *J. Hazard. Mater.*, Nov. 18.

55. Thi, T.H., Azaroual, N., Flament, M.P. (2008). Characterization and in vitro evaluation of the formoterol/cyclodextrin complex for pulmonary administration by nebulization. *Eur. J. Pharm. Biopharm.*, Nov. 5.

56. Alexanian, C., Papademou, H., Vertzoni, M., Archontaki, H., Valsami, G. (2008). Effect of pH and water-soluble polymers on the aqueous solubility of nimesulide in the absence and presence of β-cyclodextrin derivatives. *J. Pharm. Pharmacol.*, *60*(11), 1433–1439.

57. Kojima, C., Toi, Y., Harada, A., Kono, K. (2008). Aqueous solubilization of fullerenes using poly(amidoamine) dendrimers bearing cyclodextrin and poly(ethylene glycol). *Bioconjugate Chem.*, *19*(11), 2280–2284.

58. Majumdar, S., Srirangam, R. (2008). Solubility, stability, physicochemical characteristics and in vitro ocular tissue permeability of hesperidin: a natural bioflavonoid. *Pharm. Res.*, Sept. 23.

59. Kusmin, A., Lechner, R.E., Kammel, M., Saenger, W. (2008). Native and methylated cyclodextrins with positive and negative solubility coefficients in water studied by SAXS and SANS. *J. Phys. Chem. B*, *112*(41), 12888–12898.

60. Shinde, V.R., Shelake, M.R., Shetty, S.S., Chavan-Patil, A.B., Pore, Y.V., Late, S.G. (2008). Enhanced solubility and dissolution rate of lamotrigine by inclusion complexation and solid dispersion technique. *J. Pharm. Pharmacol.*, *60*(9), 1121–1129.

61. Alvarez, C., Calero, J., Menéndez, J.C., Torrado, S., Torrado, J.J. (2008). Effects of hydroxypropyl-β-cyclodextrin on the chemical stability and the aqueous solubility of thalidomide enantiomers. *Pharmazie*, *63*(7), 511–513.

62. Echezarreta-López, M.M., Otero-Mazoy, I., Ramírez, H.L., Villalonga, R., Torres-Labandeira, J.J. (2008). Solubilization and stabilization of sodium dicloxacillin by cyclodextrin inclusion. *Curr. Drug Discov. Technol.*, *5*(2), 140–145.

63. Yang, Z.X., Chen, Y., Liu, Y. (2008). Inclusion complexes of bisphenol A with cyclomaltoheptaose (β-cyclodextrin): solubilization and structure. *Carbohydr. Res.*, *343*(14), 2439–2442.

64. Das, S., Lin, H.S., Ho, P.C., Ng, K.Y. (2008). The impact of aqueous solubility and dose on the pharmacokinetic profiles of

resveratrol. *Pharm. Res.*, *25*(11), 2593–2600 (Epub July 16, 2008). Erratum in: *Pharm. Res.*, *25*(12), 2983.

65. Miyawaki, A., Kuad, P., Takashima, Y., Yamaguchi, H., Harada, A. (2008). Molecular puzzle ring: pseudo[1]rotaxane from a flexible cyclodextrin derivative. *J. Am. Chem. Soc.*, *130* (50), 17062–17069.

66. Ribeiro, A., Figueiras, A., Santos, D., Veiga, F. (2008). Preparation and solid-state characterization of inclusion complexes formed between miconazole and methyl-β-cyclodextrin. *AAPS PharmSciTech*, *9*(4), 1102–1109.

67. Zhang, L., Hsieh, Y.L. (2008). Ultrafine cellulose acetate fibers with nanoscale structural features. *Nanosci. Nanotechnol.*, *8*(9), 4461–4469.

68. Pourgholami, M.H., Wangoo, K.T., Morris, D.L. (2008). Albendazole–cyclodextrin complex: enhanced cytotoxicity in ovarian cancer cells. *Anticancer Res.*, *28*(5A), 2775–2779.

69. Manosroi, J., Saowakhon, S., Manosroi, A. (2008). Enhancement of 17α-hydroxyprogesterone production from progesterone by biotransformation using hydroxypropyl-β-cyclodextrin complexation technique. *J. Steroid Biochem. Mol. Biol.*, *112*(4–5), 201–204.

70. Oishi, K., Toyao, K., Kawano, Y. (2008). Suppression of estrogenic activity of 17-β-estradiol by β-cyclodextrin. *Chemosphere*, *73*(11), 1788–1792.

71. Selvam, A.P., Geetha, D. (2008). Ultrasonic studies on lamivudine: β-cyclodextrin and polymer inclusion complexes. *Pak. J. Biol. Sci.*, *11*(4), 656–659.

72. He, J., Deng, L., Yang, S. (2008). Synthesis and characterization of β-cyclodextrin inclusion complex containing di(8-hydroxyquinoline)magnesium. *Spectrochim. Acta A*, *70*(4), 878–883.

73. Hegge, A.B., Schüller, R.B., Kristensen, S., Tønnesen, H.H. (2008). Studies of curcumin and curcuminoides: XXXIII. In vitro release on curcumin from vehicles containing alginate and cyclodextrin. *Pharmazie*, *63*(8), 585–592.

74. Daruházi, A.E., Szente, L., Balogh, B., Mátyus, P., Béni, S., Takács, M., Gergely, A., Horváth, P., Szoke, E., Lemberkovics, E. (2008). Utility of cyclodextrins in the formulation of genistein: 1. Preparation and physicochemical properties of genistein complexes with native cyclodextrins. *J. Pharm. Biomed. Anal.*, *48*(3), 636–640.

75. Cui, Q.H., Cui, J.H., Zhang, JJ. (2008). Preparation of coated tablets of glycyrrhetic acid–HP-β-cyclodextrin tablets for colon-specific release. *Zhongguo Zhong Yao Za Zhi*, *33* (20), 2339–2343.

76. Scalia, S., Tursilli, R., Iannuccelli, V. (2007). Complexation of the sunscreen agent, 4-methylbenzylidene camphor with cyclodextrins: effect on photostability and human stratum corneum penetration. *J. Pharm. Biomed. Anal.*, *44*(1), 29–34 (Epub Jan. 16, 2007).

77. Bouquet, W., Ceelen, W., Fritzinger, B., Pattyn, P., Peeters, M., Remon, J.P., Vervaet, C. (2007). Paclitaxel/β-cyclodextrin complexes for hyperthermic peritoneal perfusion: formulation and stability. *Eur. J. Pharm. Biopharm.*, *66*(3), 391–397 (Epub Dec. 13, 2006).

78. Tomren, M.A., Másson, M., Loftsson, T., Tønnesen, H.H. (2007). Studies on curcumin and curcuminoids: XXXI.

Symmetric and asymmetric curcuminoids: stability, activity and complexation with cyclodextrin. *Int. J. Pharm.*, *338*(1–2), 27–34.

79. Hamada, H., Ishihara, K., Masuoka, N., Mikuni, K., Nakajima, N. (2006). Enhancement of water-solubility and bioactivity of paclitaxel using modified cyclodextrins. *J. Biosci. Bioeng.*, *102*(4), 369–371.

80. Lahiani-Skiba, M., Barbot, C., Bounoure, F., Joudieh, S., Skiba, M. (2006). Solubility and dissolution rate of progesterone–cyclodextrin–polymer systems. *Drug Dev. Ind. Pharm.*, *32*(9), 1043–1058.

81. Cai, X., Zhang, A., Liu, W. (2006). Environmental significance of the diclofop-methyl and cyclodextrin inclusion complexes. *J. Environ. Sci. Health B*, *41*(7), 1115–1129.

82. Ammar, H.O., Salama, H.A., Ghorab, M., Mahmoud, A.A. (2006). Implication of inclusion complexation of glimepiride in cyclodextrin-polymer systems on its dissolution, stability and therapeutic efficacy. *Int. J. Pharm.*, *31;* 320 (1–2), 53–57.

83. Vertzoni, M., Kartezini, T., Reppas, C., Archontaki, H., Valsami, G. (2006). Solubilization and quantification of lycopene in aqueous media in the form of cyclodextrin binary systems. *Int. J. Pharm.*, *309*(1–2), 115–122.

84. Liu, L., Zhu, S. (2006). Preparation and characterization of inclusion complexes of prazosin hydrochloride with β-cyclodextrin and hydroxypropyl-β-cyclodextrin. *J. Pharm. Biomed. Anal.*, *40*(1), 122–127.

85. Gibáud, S., Zirar, S.B., Mutzenhardt, P., Fries, I., Astier, A. (2005). Melarsoprol–cyclodextrins inclusion complexes. *Int. J. Pharm.*, *306*(1–2), 107–121.

86. Holvoet, C., Heyden, Y.V., Plaizier-Vercammen, J. (2005). Inclusion complexation of diazepam with different cyclodextrins in formulations for parenteral use. *Pharmazie*, *60*(8), 598–603.

87. Pinto, L.M., Fraceto, L.F., Santana, M.H., Pertinhez, T.A., Júnior, S.O., de Paula, E. (2005). Physicochemical characterization of benzocaine–β-cyclodextrin inclusion complexes. *J. Pharm. Biomed. Anal.*, *39*(5), 956–963.

88. Zhang, X.N., Yan, X.Y., Tang, L.H., Gong, J.H., Zhang, Q. (2005). Preparation, identification and inclusion actions of irisquinone hydroxypropyl-β-cyclodextrin inclusion complex. *Yao Xue Xue Bao*, *40*(4), 369–372.

89. Liu, Y., Chen, G.S., Chen, Y., Lin, J. (2005). Inclusion complexes of azadirachtin with native and methylated cyclodextrins: solubilization and binding ability. *Bioorg. Med. Chem.*, *13*(12), 4037–4042.

90. Zheng, Y., Haworth, I.S., Zuo, Z., Chow, M.S., Chow, A.H. (2005). Physicochemical and structural characterization of quercetin–β-cyclodextrin complexes. *J. Pharm. Sci.*, *94*(5), 1079–1089.

91. Yap, K.L., Liu, X., Thenmozhiyal, J.C., Ho, P.C. (2005). Characterization of the 13-*cis*-retinoic acid/cyclodextrin inclusion complexes by phase solubility, photostability, physicochemical and computational analysis. *Eur. J. Pharm. Sci.*, *25*(1), 49–56.

92. Babu, M.K., Godiwala, T.N. (2004). Toward the development of an injectable dosage form of propofol: preparation and evaluation of propofol-sulfobutyl ether 7-β-cyclodextrin complex. *Pharm. Dev. Technol.*, *9*(3), 265–275.

93. Chandra Sekhara Rao, G., Satish Kumar, M., Mathivanan, N., Bhanoji Rao, M.E. (2004). Improvement of physical stability and dissolution rate of celecoxib suspensions by complexation with β-cyclodextrins. *Pharmazie*, *59*(8), 627–630.

94. Hartell, M.G., Hicks, R., Bhattacharjee, A.K., Koser, B.W., Carvalho, K., Van Hamont, J.E. (2004). Nuclear magnetic resonance and molecular modeling analysis of the interaction of the antimalarial drugs artelinic acid and artesunic acid with β-cyclodextrin. *J. Pharm. Sci.*, *93*(8), 2076–2089.

95. Dilova, V., Zlatarova, V., Spirova, N., Filcheva, K., Pavlova, A., Grigorova, P. (2004). Study of insolubility problems of dexamethasone and digoxin: cyclodextrin complexation. *Boll. Chim. Farm.*, *143*(1), 20–23.

96. Hanna, K., de Brauer, Ch., Germain, P. (2004). Cyclodextrin-enhanced solubilization of pentachlorophenol in water. *J. Environ. Manage.*, *71*(1), 1–8.

97. Rawat, S., Jain, S.K. (2004). Solubility enhancement of celecoxib using β-cyclodextrin inclusion complexes. *Eur. J. Pharm. Biopharm.*, *57*(2), 263–267.

98. Mukne, A.P., Nagarsenker, M.S. (2004). Triamterene-β-cyclodextrin systems: preparation, characterization and in vivo evaluation. *AAPS PharmSciTech*, *5*(1), E19.

99. Sridevi, S., Chauhan, A.S., Chalasani, K.B., Jain, A.K., Diwan, P.V. (2003). Enhancement of dissolution and oral bioavailability of gliquidone with hydroxy propyl-β-cyclodextrin. *Pharmazie*, *58*(11), 807–810.

100. Tirucherai, G.S., Mitra, A.K. (2003). Effect of hydroxypropyl-β-cyclodextrin complexation on aqueous solubility, stability, and corneal permeation of acyl ester prodrugs of ganciclovir. *AAPS PharmSciTech*, *4*(3), E45.

101. Koontz, J.L., Marcy, J.E. (2003). Formation of natamycin: cyclodextrin inclusion complexes and their characterization. *J. Agric. Food Chem.*, *51*(24), 7106–7110.

102. Chowdary, K.P., Kamalakara, R.G. (2003). Controlled release of nifedipine from mucoadhesive tablets of its inclusion complexes with β-cyclodextrin. *Pharmazie*, *58*(10), 721–724.

103. Liu, X., Lin, H.S., Thenmozhiyal, J.C., Chan, S.Y., Ho, P.C. (2003). Inclusion of acitretin into cyclodextrins: phase solubility, photostability, and physicochemical characterization. *J. Pharm. Sci.*, *92*(12), 2449–2457.

104. Ribeiro, L., Loftsson, T., Ferreira, D., Veiga, F. (2003). Investigation and physicochemical characterization of vinpocetine-sulfobutyl ether β-cyclodextrin binary and ternary complexes. *Chem. Pharm. Bull. (Tokyo)*, *51*(8), 914–922.

105. Rawat, S., Jain, S.K. (2003). Rofecoxib-β-cyclodextrin inclusion complex for solubility enhancement. *Pharmazie*, *58*(9), 639–641.

106. Hanna, K., de Brauer, C., Germain, P. (2003). Solubilization of the neutral and charged forms of 2,4,6-trichlorophenol by β-cyclodextrin, methyl-β-cyclodextrin and hydroxypropyl-β-cyclodextrin in water. *J. Hazard. Mater.*, *100*(1–3), 109–116.

107. Holvoet, C., Plaizier-Vercammen, J., Vander Heyden, Y., Gabriëls, M., Camu, F. (2003). Preparation and in-vitro release

rate of fentanyl–cyclodextrin complexes for prolonged action in epidural analgesia. *Int. J. Pharm.*, *265*(1–2), 13–26.

108. Ribeiro, L.S., Ferreira, D.C., Veiga, F.J. (2003). Physicochemical investigation of the effects of water-soluble polymers on vinpocetine complexation with β-cyclodextrin and its sulfobutyl ether derivative in solution and solid state. *Eur. J. Pharm. Sci.*, *20*(3), 253–266.

109. Ghorab, M.K., Adeyeye, M.C. (2003). Enhanced bioavailability of process-induced fast-dissolving ibuprofen cogranulated with β-cyclodextrin. *J. Pharm. Sci.*, *92*(8), 1690–1697.

110. Rodrigues Júnior, J.M., de Melo Lima, K., de Matos Jensen, C.E., de Aguiar, M.M., da Silva Cunha Júnior, A. (2003). The effect of cyclodextrins on the in vitro and in vivo properties of insulin-loaded poly(D,L-lactic-*co*-glycolic acid) microspheres. *Artif. Organs*, *27*(5), 492–497.

111. Francois, M., Snoeckx, E., Putteman, P., Wouters, F., De Proost, E., Delaet, U., Peeters, J., Brewster, M.E. (2003). A mucoadhesive, cyclodextrin-based vaginal cream formulation of itraconazole. *AAPS PharmSci*, *5*(1), E5.

112. Salem, I.I., Düzgünes, N. (2003). Efficacies of cyclodextrin-complexed and liposomen encapsulated clarithromycin against *Mycobacterium avium* complex infection in human macrophages. *Int. J. Pharm.*, *250*(2), 403–414.

113. Martín, L., León, A., Olives, A.I., Del Castillo, B., Martín, M.A. (2003). Spectrofluorimetric determination of stoichiometry and association constants of the complexes of harmane and harmine with β-cyclodextrin and chemically modified β-cyclodextrins. *Talanta*, *60*(2–3), 493–503.

114. Sridevi, S., Diwan, P.V. (2002). Optimized transdermal delivery of ketoprofen using pH and hydroxypropyl-β-cyclodextrin as co-enhancers. *Eur. J. Pharm. Biopharm.*, *54*(2), 151–154.

115. Tomono, K., Goto, H., Suzuki, T., Ueda, H., Nagai, T., Watanabe, J. (2002). Interaction of iodine with 2-hydroxypropyl-α-cyclodextrin and its bactericidal activity. *Drug Dev. Ind. Pharm.*, *28*(10), 1303–1309.

116. Bayomi, M.A., Abanumay, K.A., Al-Angary, A.A. (2002). Effect of inclusion complexation with cyclodextrins on photostability of nifedipine in solid state. *Int. J. Pharm.*, *243*(1–2), 107–117.

117. Granero, G., de Bertorello, M.M., Longhi, M. (2002). Solubilization of a naphthoquinone derivative by hydroxypropyl-β-cyclodextrin (HP-β-CD) and polyvinylpyrrolidone (PVP-K30): the influence of PVP-K30 and pH on solubilizing effect of HP-β-CD. *Boll. Chim. Farm.*, *141*(1), 63–66.

118. Larrucea, E., Arellano, A., Santoyo, S., Ygartua, P. (2002). Study of the complexation behavior of tenoxicam with cyclodextrins in solution: improved solubility and percutaneous permeability. *Drug Dev. Ind. Pharm.*, *28*(3), 245–252.

119. Kang, J., Kumar, V., Yang, D., Chowdhury, P.R., Hohl, R.J. (2002). Cyclodextrin complexation: influence on the solubility, stability, and cytotoxicity of camptothecin, an antineoplastic agent. *Eur. J. Pharm. Sci.*, *15*(2), 163–170.

120. Fernandes, C.M., Teresa Vieira, M., Veiga, F.J. (2002). Physicochemical characterization and in vitro dissolution behavior of nicardipine–cyclodextrins inclusion compounds. *Eur. J. Pharm. Sci.*, *15*(1), 79–88.

121. Faucci, M.T., Mura, P. (2001). Effect of water-soluble polymers on naproxen complexation with natural and chemically modified β-cyclodextrins. *Drug Dev. Ind. Pharm.*, *27*(9), 909–917.

122. Mura, P., Faucci, M.T., Bettinetti, G.P. (2001) The influence of polyvinylpyrrolidone on naproxen complexation with hydroxypropyl-β-cyclodextrin. *Eur. J. Pharm. Sci.*, *13*(2), 187–194.

123. Veiga, F., Fernandes, C., Maincent, P. (2001). Influence of the preparation method on the physicochemical properties of tolbutamide/cyclodextrin binary systems. *Drug Dev. Ind. Pharm.*, *27*(6), 523–532.

124. Pose-Vilarnovo, B., Perdomo-López, I., Echezarreta-López, M., Schroth-Pardo, P., Estrada, E., Torres-Labandeira, J.J. (2001). Improvement of water solubility of sulfamethizole through its complexation with β- and hydroxypropyl-β-cyclodextrin: characterization of the interaction in solution and in solid state. *Eur. J. Pharm. Sci.*, *13*(3), 325–331.

125. Xie, W., Xu, W., Feng, G. (2001). Study on the inclusion interaction of β-cyclodextrin with phosphatidylcholine by UV spectra. *Guang Pu Xue Yu Guang Pu Fen Xi*, *21*(5), 707–709.

126. Latrofa, A., Trapani, G., Franco, M., Serra, M., Muggironi, M., Fanizzi, F.P., Cutrignelli, A., Liso, G. (2001). Complexation of phenytoin with some hydrophilic cyclodextrins: effect on aqueous solubility, dissolution rate, and anticonvulsant activity in mice. *Eur. J. Pharm. Biopharm.*, *52*(1), 65–73.

127. Ran, Y., Zhao, L., Xu, Q., Yalkowsky, S.H. (2001). Solubilization of cyclosporin A. *AAPS PharmSciTech*, *2*(1), E2.

128. Filipović-Grcić, J., Voinovich, D., Moneghini, M., Becirevic-Lacan, M., Magarotto, L., Jalsenjak, I. (2000). Chitosan microspheres with hydrocortisone and hydrocortisone–hydroxypropyl-β-cyclodextrin inclusion complex. *Eur. J. Pharm. Sci.*, *9*(4), 373–379.

129. Ajisaka, N., Hara, K., Mikuni, K., Hara, K., Hashimoto, H. (2000). Effects of branched cyclodextrins on the solubility and stability of terpenes. *Biosci. Biotechnol. Biochem.*, *64*(4), 731–734.

130. Bodor, N., Drustrup, J., Wu, W. (2000). Effect of cyclodextrins on the solubility and stability of a novel soft corticosteroid, loteprednol etabonate. *Pharmazie*, *55*(3), 206–209.

131. Lin, H.S., Chean, C.S., Ng, Y.Y., Chan, S.Y., Ho, P.C. (2000). 2-Hydroxypropyl-β-cyclodextrin increases aqueous solubility and photostability of *all-trans*-retinoic acid. *J. Clin. Pharm. Ther.*, *25*(4), 265–269.

132. Bećirević-Laćan, M., Filipović-Grcić, J. (2000). Effect of hydroxypropyl-β-cyclodextrin on hydrocortisone dissolution from films intended for ocular drug delivery. *Pharmazie*, *55* (7), 518–520.

133. Ozkan, Y., Atay, T., Dikmen, N., Işimer, A., Aboul-Enein, H.Y. (2000). Improvement of water solubility and in vitro dissolution rate of gliclazide by complexation with β-cyclodextrin. *Pharm. Acta Helv.*, *74*(4), 365–370.

134. Lutka, A. (2000). Effect of cyclodextrin complexation on aqueous solubility and photostability of phenothiazine. *Pharmazie*, *55*(2), 120–123.

135. Usuda, M., Endo, T., Nagase, H., Tomono, K., Ueda, H. (2000). Interaction of antimalarial agent artemisinin with cyclodextrins. *Drug Dev. Ind. Pharm.*, *26*(6), 613–619.

136. Castillo, J.A., Palomo-Canales, J., Garcia, J.J., Lastres, J.L., Bolas, F., Torrado, J.J. (1999). Preparation and characterization of albendazole β-cyclodextrin complexes. *Drug Dev. Ind. Pharm.*, *25*(12), 1241–1248.

137. Orienti, I., Cerchiara, T., Zecchi, V., Arias Blanco, M.J., Gines, J.M., Moyano, J.R., Rabasco Alvarez, A.M. (1999). Complexation of ursodeoxycholic acid with β-cyclodextrin–choline dichloride coprecipitate. *Int. J. Pharm.*, *15;* 190(2) 139–153.

138. Miyake, K., Hirayama, F., Uekama, K. (1999). Solubility and mass and nuclear magnetic resonance spectroscopic studies on interaction of cyclosporin A with dimethyl-α- and β-cyclodextrins in aqueous solution. *J. Pharm. Sci.*, *88*(1), 39–45.

139. Shao, W., Wang, D., Mi, G., Wang, C. (1998). Study of the inclusion compound of rutin with β-cyclodextrin. *Zhong Yao Cai*, *21*(1), 31–33.

140. McCandless, R., Yalkowsky, S.H. (1998). Effect of hydroxypropyl-β-cyclodextrin and pH on the solubility of levemopamil HCl. *J. Pharm. Sci.*, *87*(12), 1639–1642.

141. Kim, Y., Oksanen, D.A., Massefski, W., Jr., Blake, J.F., Duffy, E.M., Chrunyk, B. (1998). Inclusion complexation of ziprasidone mesylate with β-cyclodextrin sulfobutyl ether. *J. Pharm. Sci.*, *87*(12), 1560–1567.

142. El-Arini, S.K., Giron, D., Leuenberger, H. (1998). Solubility properties of racemic praziquantel and its enantiomers. *Pharm. Dev. Technol.*, *3*(4), 557–564.

143. Trapani, G., Latrofa, A., Franco, M., Lopedota, A., Sanna, E., Liso, G. (1998). Inclusion complexation of propofol with 2-hydroxypropyl-β-cyclodextrin: physicochemical, nuclear magnetic resonance spectroscopic studies, and anesthetic properties in rat. *J. Pharm. Sci.*, *87*(4), 514–518.

144. Lee, B.J., Choi, H.G., Kim, C.K., Parrott, K.A., Ayres, J.W., Sack, R.L. (1997). Solubility and stability of melatonin in propylene glycol and 2-hydroxypropyl-β-cyclodextrin vehicles. *Arch. Pharm. Res.*, *20*(6), 560–565.

145. Loukas, Y.L., Vraka, V., Gregoriadis, G. (1997). Novel non-acidic formulations of haloperidol complexed with β-cyclodextrin derivatives. *J. Pharm. Biomed. Anal.*, *16*(2), 263–268.

146. Ammar, H.O., el-Nahhas, S.A., Ghorab, M.M. (1996). Improvement of some pharmaceutical properties of drugs by cyclodextrin complexation: 6. Ampicillin. *Pharmazie*, *51*(8), 568–570.

147. Jarho, P., Urtti, A., Järvinen, K., Pate, D.W., Järvinen, T. (1996). Hydroxypropyl-β-cyclodextrin increases aqueous solubility and stability of anandamide. *Life Sci.*, *58*(10), PL181–PL185.

148. Tanaka, M., Iwata, Y., Kouzuki, Y., Taniguchi, K., Matsuda, H., Arima, H., Tsuchiya, S. (1995). Effect of 2-hydroxypropyl-β-cyclodextrin on percutaneous absorption of methyl paraben. *J. Pharm. Pharmacol.*, *47*(11), 897–900.

149. Sharma, U.S., Balasubramanian, S.V., Straubinger RM. (1995). Pharmaceutical and physical properties of paclitaxel (Taxol) complexes with cyclodextrins. *Pharm. Sci.*, *84*(10), 1223–1230.

150. Jarho, P., Urtti, A., Järvinen, T. (1995). Hydroxypropyl-β-cyclodextrin increases the aqueous solubility and stability of pilocarpine prodrugs. *Pharm. Res.*, *12*(9), 1371–1375.

151. Vincieri, F.F., Mazzi, G., Mulinacci, N., Bambagiotti-Alberti, M., Dall' Acqua, F., Vedaldi, D. (1995). Improvement of dissolution characteristics of psoralen by cyclodextrins complexation. *Farmaco*, *50*(7–8), 543–547.

152. Krenn, M., Gamcsik, M.P., Vogelsang, G.B., Colvin, O.M., Leong, K.W. (1992). Improvements in solubility and stability of thalidomide upon complexation with hydroxypropyl-β-cyclodextrin. *J. Pharm. Sci.*, *81*(7), 685–689.

153. Brewster, M.E., Anderson, W.R., Estes, K.S., Bodor, N. (1991). Development of aqueous parenteral formulations for carbamazepine through the use of modified cyclodextrins. *J. Pharm. Sci.*, *80*(4), 380–383.

154. Abdel-Rahman, S.I., el-Sayed, A.M. (1990). Interaction of some benzothiadiazine diuretics with β-cyclodextrin. *Acta Pharm. Hung.*, *60*(2–3), 69–75.

155. Yoshida, A., Yamamoto, M., Itoh, T., Irie, T., Hirayama, F., Uekama, K. (1990). Utility of 2-hydroxypropyl-β-cyclodextrin in an intramuscular injectable preparation of nimodipine. *Chem. Pharm. Bull. (Tokyo)*, *38*(1), 176–179.

156. Green, A.R., Guillory, J.K. (1989). Heptakis(2, 6-di-O-methyl)-β-cyclodextrin (DIMEB) complexation with the antitumor agent chlorambucil. *J. Pharm. Sci.*, *78*(5), 427–431.

157. Mielcarek, J. (1997). Photochemical stability of the inclusion complexes formed by modified 1, 4-dihydropyridine derivatives with betacyclodextrin. *J. Pharm. Biomed. Anal.*, *15*, 681–686.

158. Scalia, S., Villani, S., Casolari, A. (1999). Inclusion complexation of the sunscreen agent 2-ethylhexyl-p-dimethylaminobenzoate with hydroxypropyl-β-cyclodextrin: effect on photostability. *J. Pharm. Pharmacol.*, *51*(12), 1367–1374.

159. Lutka, A., Koziara, J. (2000). Interaction of trimeprazine with cyclodextrins in aqueous solution. *Chem. Pharm. Bull. (Tokyo)*, *57*, 369–374.

160. Lutka, A. (2000). Effect of cyclodextrin complexation on aqueous solubility and photostability of phenothiazine. *Pharmazie*, *55*(2), 120–123.

161. Babu, R., Pandit, J.K. (1999). Effect of aging on the dissolution stability of glibenclamide/β-cyclodextrin complex. *Drug Dev. Ind. Pharm.*, *25*, 1215–1219.

162. Cwiertnia, B., Hladon, T., Stobiecki, M. (1999). Stability of diclofenac sodium in the inclusion complex in the β-cyclodextrin in the solid state. *J. Pharm. Pharmacol.*, *51*, 1213–1218.

163. Li, J., Guo, Y., Zografi, G. (2002). The solid-state stability of amorphous quinapril in the presence of β-cyclodextrins. *J. Pharm. Sci.*, *91*, 229–243.

164. Croyle, M.A., Cheng, X., Wilson, J.M. (2001). Development of formulations that enhance physical stability of viral vectors for gene therapy. *Gene Ther.*, *8*, 1281–1290.

165. Nagase, Y., Hirata, M., Wada, K. (2001). Improvement of some pharmaceutical properties of DY-9760e by sulfobutyl ether β-cyclodextrin. *Int. J. Pharm.*, *229*, 163–172.

166. Ma, D.Q., Rajewski, R.A., Vander Velde, D., Stella, V.J. (2000). Comparative effects of (SBE)7m-β-CD and HP-β-CD on the stability of two anti-neoplastic agents, melphalan and carmustine. *J. Pharm. Sci.*, *89*(2), 275–287.

167. Dwivedi, A.K., Kulkarni, D., Khanna, M., Singh, S. (1999). Effect of cyclodextrins on the stability of new antimalarial compound N1-31–acetyl/-41,51–dihydro–21 furanyl-N^4-(6-methoxy, 8-quinolinyl)–1, 4-pentane diamine. *Ind. J. Pharm. Sci.*, *61*, 175–177.

168. Singla, A.K., Garg, A., Aggarwal, D. (2002). Paclitaxel and its formulations. *Int. J. Pharm.*, *235*, 179–192.

169. Jarho, P., Vander Velde, D., Stella, VJ. (2000). Cyclodextrin-catalyzed deacetylation of spironolactone is pH and cyclodextrin dependent. *J. Pharm. Sci.*, *89*, 241–249.

170. Sortino, S., Giuffrida, S., De Guldi, G. (2001). The photochemistry of flutamide and its inclusion complex with β-cyclodextrin: dramatic effect of the microenvironment on the nature and on the efficiency of the photodegradation pathways. *Photochem. Photobiol.*, *73*, 6–13.

171. Koester, L.S., Guterres, S.S., Le Roch, M., Lima, V.L.E., Zuanazzi, J.A., Bassani, V.l. (2001). Ofloxacin/β-cyclodextrin complexation. *Drug Dev. Ind. Pharm.*, *27*, 533–540.

172. Connors, K.A. (1997). The stability of cyclodextrin complexes in solution, *Chem. Rev.*, *97*, 1325.

173. Díaz, D., Escobar Llanos, C.M., Bernad, M.J., Mora, J.G. (1998). Binding, molecular mechanics, and thermodynamics of cyclodextrin inclusion complexes with ketoprofen in aqueous medium. *Pharm. Dev. Tech.*, *3*(3), 307–313.

174. Wang, T., Bradshaw, J.S., Izatt, R.M. (1994). Applications of NMR spectral techniques for the study of macrocycle host-organic guest interactions. A short review. *J. Heterocyclic Chem.*, *31*, 1097.

175. Schneider, H.-J. (1993). NMR spectroscopy and molecular-mechanics calculations in supramolecular chemistry. *Recl. Trav. Chim. Pay-Bas*, *112*, 412.

176. Rajeswari, C., Ahuja, A., Ali, J. Khar, R.K. (2005). Cyclodextrins in drug delivery: an updated review. *AAPS PharmSciTech*, *6*(2), Art. 43.

177. French, D. (1957). The Schardinger dextrins. *Adv. Carbohydr. Chem.*, *12*, 189–260.

178. Koontz, J.L., Marcy, J.E., O'Keefe, S.F., Duncan, S.E. (2009). Cyclodextrin inclusion complex formation and solid-state characterization of the natural antioxidants α-tocopherol and quercetin. *J. Agric. Food Chem.*, *57*(4), 1162–1171.

179. Shanker, G., Kumar, C.K., Gonugunta, C.S., Kumar, B.V., Veerareddy, P.R. (2009). Formulation and evaluation of bioadhesive buccal drug delivery of tizanidine hydrochloride tablets. *AAPS PharmSciTech*, May 8.

180. Yujuan, C., Runhua, L.H. (2009). NMR titration and quantum calculation for the inclusion complexes of *cis*-cyclooctene, *cis,cis*-1,3-cyclooctadiene and *cis,cis*-1, 5-cyclooctadiene with β-cyclodextrin. *Spectrochim Acta A*, Mar. 28.

181. Matilainen, L., Maunu, S.L., Pajander, J., Auriola, S., Jääskeläinen, I., Larsen, K.L., Järvinen, T., Jarho, P. (2009). The stability and dissolution properties of solid glucagon/γ-cyclodextrin powder. *Eur. J. Pharm. Sci.*, *36*(4–5), 412–420 (Epub Nov. 19, 2008).

182. Nishi, Y., Tanimoto, T. (2009). Preparation and characterization of branched β-cyclodextrins having α-L-fucopyranose and a study of their functions. *Biosci. Biotechnol. Biochem.*, *73*(3), 562–569.

183. Rodriguez, J., Elola, M.D. (2009). Encapsulation of small ionic molecules within α-cyclodextrins. *J. Phys. Chem. B*, *113* (5), 1423–1428.

184. Stancanelli, R., Crupi, V., De Luca, L., Ficarra, P., Ficarra, R., Gitto, R., Guardo, M., Iraci, N., Majolino, D., Tommasini, S., Venuti, V. (2008). Improvement of water solubility of noncompetitive AMPA receptor antagonists by complexation with β-cyclodextrin. *Bioorg. Med. Chem.*, *16*(18), 8706–8712.

185. Neoh, T.L., Yamauchi, K., Yoshii, H., Furuta, T. (2008). Kinetic study of thermally stimulated dissociation of inclusion complex of 1-methylcyclopropene with α-cyclodextrin by thermal analysis. *J. Phys. Chem. B*, *112*(49), 15914–15920.

186. Mourtzinos, I., Makris, D.P., Yannakopoulou, K., Kalogeropoulos, N., Michali, I., Karathanos, V.T. (2008). Thermal stability of anthocyanin extract of *Hibiscus sabdariffa* L. in the presence of β-cyclodextrin. *J. Agric. Food Chem.*, *56*(21), 10303–10310.

187. Bernad-Bernad, M.J., Gracia-Mora, J., Díaz, D., Castillo-Blum, S.E. (2008). Thermodynamic study of cyclodextrins complexation with benzimidazolic antihelmintics in different reaction media. *Curr. Drug Discov. Technol.*, *5*(2), 146–153.

188. Van der Schoot, S.C., Vainchtein, L.D., Nuijen, B., Gore, A., Mirejovsky, D., Lenaz, L., Beijnen, J.H. (2008). Purity profile of the indoloquinone anticancer agent EO-9 and chemical stability of EO-9 freeze dried with 2-hydroxypropyl-β-cyclodextrin. *Pharmazie*, *63*(11), 796–805.

189. Cirri, M., Maestrelli, F., Mennini, N., Mura, P. (2008). Influence of the preparation method on the physical–chemical properties of ketoprofen–cyclodextrin–phosphatidylcholine ternary systems. *J. Pharm. Biomed. Anal.*, Nov. 13.

190. El-Maradny, H.A., Mortada, S.A., Kamel, O.A., Hikal, A.H. (2008). Characterization of ternary complexes of meloxicam–HP-β-CD and PVP or L-arginine prepared by the spray-drying technique. *Acta Pharm.*, *58*(4), 455–466.

191. Jullian, C., Moyano, L., Yañez, C., Olea-Azar, C. (2007). Complexation of quercetin with three kinds of cyclodextrins: an antioxidant study. *Spectrochim Acta A*, *67*(1), 230–234.

192. Denora, N., Potts, B.C., Stella, V.J. (2007). A mechanistic and kinetic study of the β-lactone hydrolysis of Salinosporamide A (NPI-0052), a novel proteasome inhibitor. *J. Pharm. Sci.*, *96* (8), 2037–2047.

193. Pralhad, T., Rajendrakumar, K. (2004). Study of freeze-dried quercetin–cyclodextrin binary systems by DSC, FT-IR, x-ray diffraction and SEM analysis. *J. Pharm. Biomed. Anal.*, *34*(2), 333–339.

194. Zhou, C.X., Zou, J.K., Zhao, Y., Chen, Y.Z. (2002). Application of β-cyclodextrin inclusion technique in new dosage form of *Angelica sinensis* essential oil. *Zhongguo Zhong Yao Za Zhi*, *27*(11), 832–834, 845.

195. Bongiorno, D., Ceraulo, L., Mele, A., Panzeri, W., Selva, A., Turco Liveri, V. (2002). Structural and physicochemical characterization of the inclusion complexes of cyclomaltooligosaccharides (cyclodextrins) with melatonin. *Carbohydr. Res.*, *337*(8), 743–754.

196. Suihko, E., Korhonen, O., Järvinen, T., Ketolainen, J., Jarho, P., Laine, E., Paronen, P. (2001). Complexation with tolbutamide modifies the physicochemical and tableting properties of hydroxypropyl-β-cyclodextrin. *Int. J. Pharm.*, *215*(1–2), 137–145.

197. Pomponio, R., Gotti, R., Bertucci, C., Cavrini, V. (2001). Evidences of cyclodextrin-mediated enantioselective photodegradation of rac-nicardipine by capillary electrophoresis. *Electrophoresis*, *22*(15), 3243–3250.

198. Emara, S., Morita, I., Tamura, K., Razee, S., Masujima, T., Mohamed, H.A., El-Gizawy, S.M., El-Rabbat, N.A. (2000). Effect of cyclodextrins on the stability of adriamycin, adriamycinol, adriamycinone and daunomycin. *Talanta*, *51*(2), 359–364.

199. Chowdary, K.P., Nalluri, B.N. (2000). Nimesulide and β-cyclodextrin inclusion complexes: physicochemical characterization and dissolution rate studies. *Drug Dev. Ind. Pharm.*, *26*(11), 1217–1220.

200. Ma, D.Q., Rajewski, R.A., Stella, V.J. (1999). New injectable melphalan formulations utilizing (SBE)(7m)-β-CD or HP-β-CD. *Int. J. Pharm.*, *189*(2), 227–234.

201. Mielcarek, J., Daczkowska, E. (1999). Photodegradation of inclusion complexes of isradipine with methyl-β-cyclodextrin. *J. Pharm. Biomed. Anal.*, *21*(2), 393–398.

202. Narisawa, S., Stella, V.J. (1998). Increased shelf-life of fosphenytoin: solubilization of a degradant, phenytoin, through complexation with (SBE)7m-β-CD. *J. Pharm. Sci.*, *87*(8), 926–930.

203. Mielcarek, J. (1995). Inclusion complexes of nifedipine and other 1.4-dihydropyridine derivatives with cyclodextrins: IV. The UV study on photochemical stability of the inclusion complexes of nisoldipine, nimodipine, nitrendipine and nicardipine with β-cyclodextrin in the solution. *Acta Pol. Pharm.*, *52*(6), 459–463.

204. Saleh, S.I., Rahman, A.A., Aboutaleb, A.E., Nakai, Y., Ahmed, M.O. (1993). Effect of dimethyl-β-cyclodextrin on nitrazepam stability. *J. Pharm. Belg.*, *48*(5), 383–388.

205. Yamamoto, M., Hirayama, F., Uekama, K. (1992) Improvement of stability and dissolution of prostaglandin E1 by maltosyl-β-cyclodextrin in lyophilized formulation. *Chem. Pharm. Bull. (Tokyo)*, *40*(3), 747–751.

3

CYCLODEXTRINS AS BIOAVAILABILITY ENHANCERS

FÜSUN ACARTÜRK AND NEVIN ÇELEBI

Department of Pharmaceutical Technology, Gazi University, Etiler-Ankara, Turkey

1. INTRODUCTION

A total of 30 to 50% of the existing or newly discovered active ingredients used in the pharmaceutical industry have quite low water solubility [1–3].

Depending on the level of water solubility, the following concerns are addressed [4]: (1) low oral bioavailability, (2) varying bioavailability, (3) effect of fast–fed status on bioavailability, and (4) variability of dose–response ratio. Increased bioavailability reduces variability in the systemic drug effect and levels. Therefore, in drug development, it is of great importance to increase solubility, dissolution rate, and bioavailability.

Among the many methods used to increase the solubility, the method of forming inclusion complexes with cyclodextrins (CDs) plays an important role. CDs form complexes by encapsulating the hydrophobic drug molecules into hydrophilic cavities. Thus, they increase the solubility [5–12], dissolution rate [13–15], and bioavailability [16–21] of poorly water-soluble drugs and also improve the stability [22–26] of the drugs. Parent CDs such as α, β, and γ have been widely used in pharmaceutical research and development studies. However, in general terms it is observed that the cavity size of α-CD is not sufficient for encapsulating many drugs and γ-CD is expensive; β-CD is the most widely used type of CD, due to its availability and sufficient cavity size for many drugs [17,27–29].

Chemically modified CDs are synthesized to increase inclusion capacity and enhance the physicochemical properties of parent CDs. With this aim, various hydrophilic, hydrophobic, and ionic CD derivatives have been developed [30–32]. The most widely used CD derivatives used in drug development studies include methyl (M), dimethyl (DM), diethyl (DE), hydroxyethyl (HE), hydroxypropyl (HP), carboxymethyl (CM), carboxymethyl ethyl (CME), sulfobutyl ether (SBE), glucosyl (G1), and maltosyl (G2) derivative groups (Table 1).

Hydrophilic CDs undertake a role in drug delivery in immediate-release formulations, modify the release rate of the active ingredient, and increase absorption of the drugs through the biological barriers. On the other hand, hydrophobic CDs undertake a role in sustained-release delivery for water-soluble drugs. With the combined use of two different types of CDs, prolonged-release products can be prepared which enhance oral bioavailability. Table 2 shows the use of CDs in solid-dosage forms.

Various methods are used in the formation of CD complexes:

- Co-precipitation
- Slurry complexation
- Paste complexation
- Damp mixing
- Heating method
- Extrusion method
- Dry mixing
- Freeze-drying

Water amount, mixing duration, speed, and heating duration used for all methods should be optimized for each drug. CDs increase the bioavailability of water-insoluble

Cyclodextrins in Pharmaceutics, Cosmetics, and Biomedicine: Current and Future Industrial Applications, First Edition. Edited by Erem Bilensoy.
© 2011 John Wiley & Sons, Inc. Published 2011 by John Wiley & Sons, Inc.

Table 1. Derivatives of CDs

Derivative[a]	Characteristic
Hydrophilic Derivatives	
Methylated β-CD	
M-β-CD	Soluble in cold water and in organic solvents
DM-β-CD	Surface active, hemolytic
TM-β-CD	
DMA-β-CD	Soluble in water, low hemolytic
Hydroxyalkylated β-CD	
2-HE-β-CD	
2-HP-β-CD	Amorphous mixture with different derivatives (Encapsin)
3-HP-β-CD	Highly water-soluble (>50%), low toxicity
Branched β-CD	
G1-β-CD	Highly water-soluble (>50%)
G2-β-CD	Low toxicity
Hydrophobic Derivatives	
Alkylated β-CD	
DE-β-CD	Water-insoluble, soluble in organic solvents, surface active
TE-β-CD	
Acylated β-CD	
TA-β-CD	Water-insoluble, soluble in organic solvents
TB-β-CD	Mucoadhesive
TV-β-CD	Film formation
Ionizable Derivates	
Anionic β-CD	
CME-β-CD	$pK_a = 3$–4, soluble at pH > 4
β-CD sulfate	$pK_a > 1$, water-soluble
SBE4-β-CD	Water-soluble
SBE7-β-CD	Water-soluble (Captisol)
Al-β-CD sulfate	Water-insoluble

Source: Adapted from [31].

[a] M,: randomlymethylated; DM, 2,6-di-*O*-methyl; TM, 2,3,6-tri-*O*-methyl; DMA, acetylated DM-β-CD; 2-HE, 2-hydroxyethyl; 2-HP, 2-hydroxypropyl; 3 HP, 3-hydroxypropyl; G1:glycosyl; G2:maltosyl; DE, 2,6-di-*O*-ethyl; TB, 2,3,6-tri-*O*-ethyl; TA, 2,3,6-tri-*O*-acyl(C$_2$-C$_{18}$); TB, 2,3,6-tri-*O*-butanoyl; TV, 2,3,6-tri-*O*-valery; SBE4, derivative 4 of sulfobutyl ether group; SBE7, derivative 7 of sulfobutyl ether group; CME, *O*-carboxymethyl-*O*-ethyl.

Table 2. Use of CDs in Solid-Dosage Forms with Different Release Profiles

Use of CD	Aim	Release Pattern
HP-β-CD SBE-β-CD Methylated ß-CD Branched β-CD Ethylated β-CD	Enhanced dissolution and absorption of poorly water-soluble drugs	Immediate release
Per-*O*-acylated-β-CD	Sustained release of water-soluble drugs	Prolonged release
CME-β-CD	pH-dependent (enteric) release of unstable drug or stomach-irritating drug	Delayed release
Simultaneous use of CDs and pharmaceutical excipients	More balanced bioavailability with prolonged therapeutic effect	Modified release
Drug–CD conjugate	Colonic delivery	Site-specific release
Dendrimer–CD conjugate	Gene delivery	

Source: Adapted from [31].

integrity of the lipid layer of the biological membranes, CDs increase the availability of the drug at the surface of the biological membranes [23,24]. In addition, it is important to use a CD that is capable of solving the drug in aqueous media. An excessive amount reduces the availability of the drug.

Parent CDs and as well as CD derivatives have been used to increase oral [36–39], sublingual [40], nasal [41–44], ocular [45,46], buccal [47], rectal [48], vaginal [48,49], dermal–transdermal [50–54], and subcutaneous [55] bioavailability in drug development studies. However, this section is focused on the effects of CDs on oral bioavailability.There are many approved and commercially available products containing CD (Table 3) [16,56,57]. These products are used through the oral, parenteral, ocular, or nasal routes. CD preparations in the market have been used widely, particularly thanks to their low oral and local toxicities and low eye and mucosa irritations, as well as their availability.

2. INCREASING ORAL BIOAVAILABILITY OF DRUGS WITH CDs

As mentioned previously, the most important property of CDs is that they increase the bioavailability of drugs. CDs increase the oral bioavailability of drugs in the following ways [21,23,24,35,36,58]: They (1) enhance the solubility of

drugs by increasing their wettability, dissolution rate, and/or permeability [33–35].

CDs serve as a carrier for hydrophobic drug molecules. By delivering the molecule to the surface of the biological membranes, they make the drug available. Partition of CDs on the surface of these membranes takes place without disrupting the integrity of the lipid barrier. While conventional penetration enhancers are effective by disrupting the

Table 3. Approved and Commercially Available CD-Containing Products

Drug	CD Type	Product Name	Indication	Formulation	Company/Country
Alprostadil	α - CD	Rigidur	Erectile dysfunction	Intravenous (IV) solution	Ferring/Denmark
Aripiprazole	SBE-β-CD	Abilify	Antipsychotic and antidepressant	Intramuscular (IM) solution	Bristol-Myers Squibb/U.S. Otsuka Pharm Co. Japan
Benexate	β-CD	Ulgut Lonmiel	Antiulcerant	Capsule	Teikoku/Japan Shionogi/Japan
Cefotiam-hexetil	α-CD	Pansporin T	Antibiotic	Tablet	Takeda/Japan
Cephalosporin (ME 1207)	β-CD	Meiact	Antibiotic	Tablet	Meiji Seika/Japan
Cetirizine	β-CD	Cetirizin	Antiallergic	Chewing tablet	Losan Pharma/ Germany
Chloramphenicol	M-β-CD	Clorocil	Antibiotic	Eye drop solution	Oftalder/Portugal
Chlordiazepoxide	β-CD	Transillium	Tranquilizer	Tablet	Godor/Argentina
Cisapride	HP-β-CD	Coordinax Prepulsid	Gastrointestinal mobility stimulant	Suppository	Janssen/Belgium
Dexamethasone	β-CD	Glymesason	Analgesic, anti-inflammatory	Ointment	Fujinaga/Japan
Dextromethorphan	β–CD	Rynathisol	Antitussive	Syrup	Synthelabo/Italy
Diclofenac Na	HP-γ-CD	Voltaren ophtha	Nonsteorid anti-inflammatory	Eyedrop	Novartis/Switzerland
Diphenhydramin HCl, chlortheophyllin	β-CD	Stada-Travel	Travel sickness	Chewing tablet	Stada/Germany
Garlic oil	β-CD	Xund, Tegra, Allidex, Garlessence	Antiartherosclerotic	Dragees	Bipharm, Hermes/ Germany Pharmafontana/U.S.
Hydrocortisone	HP-β-CD	Dexocort	Mouthwash against aphta, gingivitis	Solution	Actavis/Iceland
Indomethacin	HP-β-CD	Indocid	Anti-inflammatory	Eyedrop solution	Chauvin/France
Iodine	β-CD	Mena-Gargle	Throat disinfectant	Solution	Kyushin/Japan
Itraconazole	HP-β-CD	Sporanox	Esophageal candidiosis	Oral and IV solution	Janssen/Belgium, U.S.
Meloxicam		Mobitil	Anti-inflammatory	Tablet and suppository	Medical Union Pharm/Egypt
Mitomycin	HP-β-CD	MitoExtra Mitozytrex	Anticancer	IV infusion	Novartis/Switzerland
Nicotine	β-CD	Nicorette	Treatment of tobacco dependence	Sublingual tablet	Pharmacia/Sweden
		Nicogum		Chewing gum	Pierre Fabre/France
Nimesulide	β-CD	Nimedex Mesulid Fast	NSAID	Tablet Oral sachet	Novartis/Italy
Nitroglycerin	β-CD	Nitropen	Coronary dilator	Sublingual tablet	Nippon Kayaku/ Japan
Omeprazol	β-CD	Omebeta	Proton pump inhibitor	Tablet	Betafarm/Germany
OP-1206	γ-CD	Opalman	Buerger's disease	Tablet	Ono/Japan
PGE_1	α-CD	Prostavasin Edex	Chronic arterial Occlusive disease	Intraenterial infusion Intracavernous injection	Ono/Japan Schwarz/Germany, U.S.
PGE_1	α-CD	Prostandin 500	Controlled hypotension during surgery	Infusion	Ono/Japan

(*continued*)

Table 3. (*Continued*)

Drug	CD Type	Product Name	Indication	Formulation	Company/Country
PGE$_2$	β-CD	Prostarmon E	Induction of labor	Sublingual tablet	Ono/Japan
Piroxicam	β-CD	Brexin	Anti-inflammatory analgesic	Tablet	Chiesi/Italy
		Flogene		Suppository	Ono/Japan
		Cicladol		Liquid	Ache/Brazil, Belgium, France, Germany, The Netherlands, Scandinavia, Switzerland
Tc-^{99} Teoboroxime	HP-γ-CD	Cardiotec	Radioactive imaging agent	IV solution	Bracco/U.S.
Tiaprofenic acid	β-CD	Surgamyl	Analgesic	Tablet	Roussel-Maestrelli/ Italy
Voriconazole	SBE-β-CD	Vfend	Antimycotic	IV solution	Pfizer/U.S.
Ziprasidone mesylate	SBE-β-CD	Zeldox Geodon	Antischizophenic	IM solution	Pfizer/U.S. and Europe

Source: Adapted from [16,56,57].

a drug, (2) increase the dissolution rate and amount, (3) enhance stability in the absorption region, (4) reduce drug-related irritation, and (5) mask bitter tastes. However, several studies have reported that CDs do not always enhance the bioavailability of drugs. It is important that CDs be selected correctly to enhance the bioavailability of drugs.

In general terms, the solubility of active ingredients, dissolution rate, and intestinal absorption rate affect the bioavailability of oral drugs [17]. Figure 1 shows a schematic kinetic model indicating dissolution–absorption procedures of an orally administered drug. If $k_d > k_a$, the drug dissolves quickly and absorption is a rate-limiting step. In this case, complexation with CD does not increase absorption, and can even reduce it. In poorly water-soluble drugs, $k_d < k_a$, and solubility is therefore a rate-limiting step. However, k_d increases after decreasing the particle size of an active ingredient, and following preparation of solid dispersions and solutions and their complexation with CD.

With the inclusion of hydrophobic drugs into the CD cavity, the solubility and dissolution rates in gastrointestinal (GI) fluids increase; thus, blood level increases. With the complexation of an active ingredient with CD, the time required for dissolving of the drug from solid form to GI fluids and then diffusion to blood circulation decreases [17]. Improvement in the bioavailability of a poorly soluble drug by CD complexation is depicted in Fig. 2.

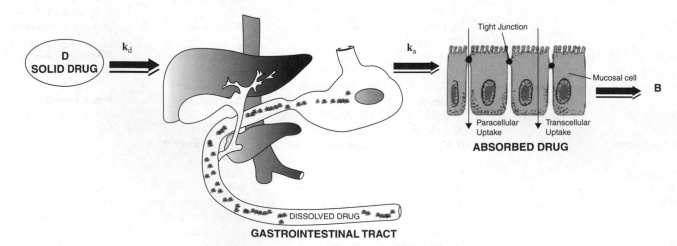

Figure 1. Dissolution–absorption process of an orally administered drug. D (solid drug), drug in the orally administered dosage form; B, concentration of drug in blood; k_a, rate constant of absorption; k_d, rate constant of dissolution.

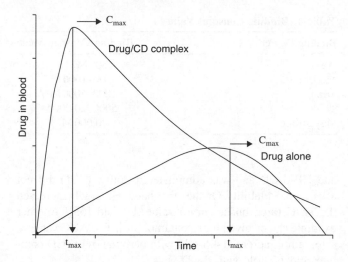

Figure 2. Improvement in the bioavailability of a poorly soluble drug by CD complexation. (Adapted from [17,60].)

2.1. Bioavailability-Enhancing Mechanisms of CDs in Drugs

The mechanisms that possibly enhance the bioavailability of drugs with complexation of CDs are as follows [31]:

- Hydrophilic CDs increase solubility, dissolution rate, and wettability of poorly water-soluble drugs.
- CDs prevent degradation on disposition of chemically unstable drugs in the GI tract as well as during storage.
- CDs increase the permeation of peptides and proteins through the nasal and rectal mucosa by changing the membrane fluidity.
- With tertiary compounds such as bile acid, cholesterol, and lipids, they provide competitive inclusion

complexation for drug release and thus increase drug release.

- In recent years, a new mechanism has been introduced to enhance the oral bioavailability of hydrophilic drugs [59]. DM-β-CD has an effect on the efflux pump activity of P-glucoprotein (P-Gp) and multidrug resistant-associated protein 2 (MRP2) and enhances the bioavailability of drugs (e.g., tacrolimus and vinblastine).

A CD complex of an active ingredient wets more easily than does the free drug itself. This means that complex formation provides rapid dissolution of the active substance under physiological conditions. The dissolution and dissociation equilibrium of CD complex in aqueous media is important in the absorption of drugs. Figure 3 indicates a schema of dissolution–dissociation–absorption steps of a CD-active substance complex [17,60]. The degree of dissociation of a complex depends on the binding constant (stability constant) and the concentration in the solution [17,58,60–62].

When a CD complex of an active ingredient with low water solubility is administered orally, two conditions arise. At low binding constant levels, a saturated concentration of the drug easily reaches GI fluids, and a higher drug concentration is obtained than in the initial free drug. In other cases, where the dissociation equilibrium is shifted toward complex formation, the concentration of the absorbable free drug is lower. The absorption of CDs is insignificant, as they cannot get into blood circulation either in free or complex form [17]. The absorption rate depends on free drug concentration, and its absorption rate increases or decreases depending on the dissociation equilibrium and/or solubility of the complex. After oral administration of CD complexes, different blood levels are related to the free active ingredient concentration in the GI tract [17].

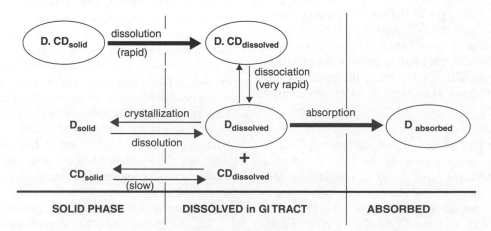

Figure 3. Dissolution–dissociation–absorption process of a drug–CD complex.

2.2. Effect on Bioavailability of Factors Related to CD Complexation

Various factors related to CD complexation affect the bioavailability of drugs: (1) the properties of the active ingredient, (2) the type of CDs, (3) the binding constant, and (4) the complexation methods.

In addition, other inactive ingredients used in a formulation have an effect. These factors are detailed in the next sections.

2.2.1. Properties of the Active Ingredient
The properties of an active ingredient required to form a complex with CD are as follows [56,58]:

- More than five atoms (C, P, S, and N) form the skeleton of the drug molecule.
- Solubility in water is less than 10 mg/mL.
- The melting-point temperature of the substance is below 250°C (otherwise, the cohesive forces between its molecules are too strong).
- The guest molecule consists of fewer than five condensed rings.
- A molecular weight between 100 and 400 (with smaller molecules the drug content of the complex is too low; large molecules do not fit the CD cavity of one CD unit).
- Hydrophobicity of the active ingredient should be log P > 2.5.

Apart from these, crystal structure, solubility, hydrophobicity, dose, ionized/nonionized form, and the pK_a of drugs have an effect. Drug–CD complexation generally includes hydrophobic interactions. The ionized or nonionized structure and the charge of active ingredients affect the binding constant of a drug–CD complex. It was observed that the binding constant was low in ionized drugs, although other studies have reported contrary findings. This indicates that the charge of active substances is effective in binding to CDs. For example, SBE-β-CD derivatives interact with a drug of opposite charge [63,64]. Active ingredients in salt form become ionized in the GI tract. The salt form of an active ingredient has a higher binding constant with CD than that of the free base. Thus, CDs affect the solubility and dissolution rate of active ingredients in different regions of the GI tract.

2.2.2. Types of CDs
Natural CDs and their derivatives have different effects on enhancing the bioavailability of drugs. HP-β-CD, SBE-β-CD, M-β-CD, and β-CD are used to enhance the water solubility and absorption of poorly water-soluble drugs. CD derivatives have a higher solubility than branched natural CDs and produce different binding constants. For example, when rutin formed a complex with β-CD

Table 4. Binding Constant Values

Binding Degree	Binding Constant Value
Weak	$<500 \, M^{-1}$
Moderate	$500–1000 \, M^{-1}$
Strong	$1000–5000 \, M^{-1}$
	$5000–20{,}000 \, M^{-1}$
Very strong	$>20{,}000 \, M^{-1}$

and HP-β-CD and was administered orally, β-CD did not affect bioavailability. On the other hand, HP-β-CD increased the AUC (area under curve) value 2.9-fold [65]. Another example can be given for salbutamol [66]. The AUC value of salbutamol increased 4.6-fold with perbutanoyl-β-CD complex and 1.7-fold with β-CD.

2.2.3. Effect of Binding Constant
Binding constant is an important parameter in the formation of complexes of active ingredients with CD; it affects dissolution rate and the bioavailability of drugs. The value of the binding constant varies in the range 0 to $100{,}000 \, M^{-1}$. In cases where the binding constant is zero, it is understood that active ingredients do not bind with CD and that no complex is formed. Binding constant values can be termed as weak, moderate, strong, or very strong (Table 4) [17,58,60–62]. As the binding constant increases, a decrease can be observed in the AUC increase. There have been a number of studies to investigate the effect of the binding constant on bioavailability associated with AUC increase. Szejtli indicated that there was no correlation between binding constant and AUC increase [17]. The majority of the previous studies made an evaluation by using low binding constant values.

Szejtli developed a theoretical model with computer simulation to indicate the relationship between binding constant and blood-level curves [17,67]. According to this model, when complexes prepared in a molar ratio of 1 : 1 (CD–drug) are administered orally, plasma blood levels are increased. If the binding constants of the complexes are high and if they contain a high molar ratio of CD, it is possible theoretically to estimate that low bioavailability will result. If the complex has a very high binding constant, the degree of dissociation and free drug concentration in GI fluids decreases. In GI fluids, high free drug concentration is obtained with a low binding constant and a high degree of dissociation. Change in the CD–drug molar ratio in GI fluids has an effect on peak height and the shape of plasma-level curves. If the binding constant is low, dissolved free drug concentration increases. Thus, higher blood levels can be reached in a shorter time. As for the high binding constant, the plasma-level peak (C_{max}) decreases and the time required to reach maximum blood concentration (t_{max}) increases.

The absorption rate of the complex-forming drug through the membranes decreases due to a high binding constant;

Table 5. Effect of CD Type and Binding Constant on the Bioavailability of Some Drugs

Drug	CD Type	CD–Drug Molar (ratio)	Binding Constant (M^{-1})	AUC Increase ($\times$AUC Control)	c_{max} Change (mg/mL)	t_{max} Change (h)	Species
Albendazole	HP-β-CD	1:1	18,106	1.4(0–inf)	1.2–2.8	9.7–2.3	Sheep
Artemisin	β-CD	1:1	63	1.7(0–inf)	0.27–0.65	2–1.6	Human
Carbamazepine	HP-β-CD	1:1	655	1.2(0–12)	10.7–12.2	—	Rat
Cinnarizine	SBE4-CD	1:1	4,276	7.8(0–inf)	0.018–0.22	1.3–0.75	Dog
Diazepam	γ-CD	3:2	120[a]	1.4(0–8)	0.59–1.05	0.50–0.40	Rabbit
Digoxin	γ-CD	4:1	12,200[a]	5.4(0–24)	3×10^{-4}–4×10^{-4}	No change	Dog
Flufenamic acid	β-CD	1:1	1,380	1.6 (% recovery in urine)	—	—	Rabbit
Flurbiprofen	HP-β-CD	1:1	5,300	Increase	7.2–13.6	2–0.5	Human
Glibenclamide	β-CD	1:1	827 (pH 7.4)	4.9 (Abs BA)	0.084–0.50	4.5–2.0	Dog
	SBE-CD	1:1,2:1	585 (1:1) 29 (1:2)	5.4 (Abs BA)	0.084–0.57	4.5–1.8	Dog
Nifedipine	β-CD	1:1	121.9	2.6(0–10)	0.39–0.90	3.3–5.5	Rabbit
Piroxicam	β-CD	2:1	28×10^4(pH 3.5)	1.1(0–24)	17.0–23.0	2–1	Rat
	DM-β-CD	2:1	240	1.2(0–inf)	4.0–7.0	2.5–1.3	Rabbit
Prednisolone	β-CD	2:1	3,600	1.2(0–8)	0.36–0.54	2.2–0.8	Human
Spirinolactone	HP-β-CD	2:1	15,700[a]	3.6(0–24)	1.8–6.2	No change	Dog
Tacrolimus	DM-β-CD	1:1,2:1	6060[a]	4.5(0–12)	0.0016–0.033	5.0–0.3	Rat

Source: Adapted from [58].

[a] Apparent 1:1 binding constant.

only the free drug is absorbed. Therefore, when the binding constant of the complex is high, since the release of the drug and permeation through the membrane will be difficult, bioavailability may not be increased; it can even be decreased. In this case, by adding a competitive agent with high interaction with CD, the active ingredient can be released from the drug–CD complex. Thus, the problem of absorption can be eliminated. The effect of CD type and binding constant on the bioavailability of some drugs is shown in Table 5.

Tokumura et al. [68] investigated the effect of competitive agent (DL-phenylalanine) on the bioavailability of drug CD complex. They reported that after the dissolution of complex and competitive agent, the drug became free and absorption can be increased. Tablets containing 25 mg of cinnarizine (CN) alone and CN/β-CD tablets equivalent to 25 mg of CN with DL-phenylalanine in a gelatin capsule were administered to male dogs and plasma levels were monitored. The C_{max} value of the CN/β-CD complex was 166.9 ± 22.4 ng/mL for 30 min. This value was six to eight times higher than that of CN alone. However, there was no statistically significant difference between the AUC values of CN and CN/β-CD complex for 8 h. When the reason behind this was investigated, it was concluded that the binding constant was high ($6.2 \times 10^3 M^{-1}$), and in parallel to this, the release of free active ingredient was difficult [68]. When CN was administered with DL-phenylalanine, although there was no significant difference on plasma level

and AUC values, it was observed that the plasma level and AUC values increased significantly in the case of the administration of DL-phenylalanine with CN/β-CD complex. These results indicate that particularly for complexes with high binding constants, competitive agents can enhance absorption of drug complexes and thus might increase bioavailability.

The same researchers also used l-leucine and l-isoleucine competitive agents to enhance the bioavailability of CN. Whereas l-isoleucine increased the bioavailability of CN/β-CD, l-leucine did not have a significant effect on the bioavailability of CN [69].

2.2.4. Complexation Methods

As indicated at the beginning of this section, various methods are used to produce inclusion complexes of drugs with CDs. These methods are influential on the effect of CD on increasing the dissolution rate and bioavailability of drugs. The methods used enable the particle size to change and an amorphic structure to form. Small particle size and amorphic structure cause faster dissolution of drugs. Each method might not yield successful results for each active ingredient [70,71].

2.2.5. Effect of Polymers and Other Inactive Ingredients Used in Formulation

Water-soluble polymers used in the formulation of solid-drug forms enhance the solubilizing effect of CDs. For example, polymers such as hydroxypropyl methylcellulose (HPMC) and poly(vinylpyrrolidone) (PVP) have a synergistic effect on CDs, enhancing the bioavailability of drugs [72–76]. In addition, water-soluble cellulose derivative polymers form complexes with CD that show various physicochemical properties. Thus, the solubility of CDs and the binding constant of complex between the drug and CD might also change. These polymers increase the bioavailability of active ingredients by interacting with them.

The literature indicates that when CDs form complexes with acidic substances, solubility, dissolution rate, and bioavailability increase. Compounds such as salts, surfactants, preservatives, and organic solvents in formulations reduce the binding efficiency between CD and the drugs [75,77]. For example, since nonionic surfactants bond to CD competitively, they reduce the binding capacity of diazepam with CD [58].

3. BIOPHARMACEUTICS CLASSIFICATION SYSTEM AND CDs

The Biopharmaceutics Classification System (BCS) was introduced by Amidon et al. [78,79] based on the aqueous solubility and permeability of active ingredients. This system is included in the guideline issued by the U.S. Food and Drug Administration [80] and currently approved by the European Medicines Agency [81] and the World Health Organization [82]. In this system, active ingredients are classified into four groups according to their solubility and permeability. Table 6 shows the BCS and the effect of drug–CD complexation on oral bioavailability of drugs.

3.1. Class I Drugs

Class I drugs include active ingredients with high aqueous solubility and high permeability [78]. The permeation of these drugs through the membrane is difficult, and their absorption decreases when they form complexes with CDs [36]. In other words, forming complexes with CDs does

not enhance the bioavailability. It is known that CDs might be effective in enhancing the solubility of active ingredients with low aqueous solubility. For example, piroxicam is not practically water soluble; however, according to the dose–solubility (D : S) ratio (low), it is classified as a class I drug [83]. CD complexes have no significant effect on the bioavailability of class I drugs. However, faster drug absorption is observed and tolerability in the gastrointestinal tract increases. In general terms, oral bioavailability of nonsteroidal anti-inflammatory drugs (NSAIDs) in humans is >90%. Despite their good bioavailability, many NSAIDs ($pK_a \sim 4.5$), such as ketoprofen, naproxen, and tiaprofenic acid, are classified as class II based on their solubility at pH1.0. However, based on the solubility at pH > 5.0 (e.g., pH of duodenum), they can be classified as class I. When water-soluble complexes of these drugs are formed, although there is no significant increase in their bioavailability, faster absorption, slight variability in bioavailability, and reduced irritation in the GI tract were observed [83]. It was found that oral bioavailability of ketoprofen (pK_a 4.5; D : S $\sim$ 4000 mL), fenbufen (pK_a 4.5; log $K_{oct/water}$ 3.2), and 4-biphenylacetic acid, which is the active metabolite of fenbufen, increased with CD complexation [36]. Rofecoxib, which is a nonionizable NSAID drug, has an oral bioavailability of 93%. However, there is a high variability in absorption rate. Although there is no dramatic increase in bioavailability with CD complexation, a decrease was observed in the variability in absorption [36].

3.2. Class II Drugs

The drugs in class II have low aqueous solubility and high permeability [78]. Carbamazepine is an example for this group. Carbamazepine (log $K_{oct/water}$ 2.5; D : S approximately 1000 mL) has various polymorphs. Furthermore, its bioavailability shows a high variation [84,85]. A significant increase in oral bioavailability was observed after forming complexes with CD. Another example is that after the CD complexation of digoxin [86,87], which has low solubility (70 μg/mL) and whose absorption is limited to dissolution

Table 6. Biopharmaceutics Classification System and the Effect of Drug–CD Complexation on the Oral Bioavailability of Drugs

BCS Class	Aqueous Solubility[a]	Permeability[b]	In Vitro–In Vivo Correlation	Absorption Rate Control	Effect of CDs on Drug on Bioavailability
I	High	High	Can be good	Gastric emptying	Can decrease
II	Poor	High	Good	Dissolution	Can enhance
III	High	Poor	Poor	Permeability	Can decrease
IV	Poor	Poor	Poor	Dissolution and permeability	Can enhance

Source: Adapted from [36].

[a] Solubility of the drug dose in aqueous solution (high: D : S ≤ 250 mL; poor: D : S > 250 mL).

[b] Permeability of a drug through a lipophilic biomembrane.

rate, bioavailability increases. Glibenclamide [72] is a typical class II drug. It has a solubility of $6\,\mu g/mL$ at pH 7.4; $\log K_{oct/water}$ 4.8; D : S ratio $\sim 2500\,mL$. Due to its low solubility, its absorption is limited to dissolution rate. It has an absolute bioavailability of 14.7% in dogs. It was found that after complexation with CD, bioavailability increased to 90.5% and it was stated that glibenclamide could be classified as a class I drug. The water solubility of spironolactone is $28\,\mu g/mL$, $\log K_{oct/water}$ 2.8; the D : S ratio is approximately $10,000\,mL$ and the oral bioavailability is 25%. After complexation with CD, oral bioavailability increased 2.4-fold. It was found that after forming complexes with CD, the bioavailability of gliclazide [88], gliquidone [89], miconazole [90], phenytoin [91], tolbutamide [92], and itraconazole [93], which are classified as class II, increased. Water-soluble CD complexes increase the apparent C_s value (saturation solubility of the drug in the aqueous fluid) of class II drugs and provide mucosal surface diffusion and thus enhance bioavailability.

3.3. Class III Drugs

Class III drugs have a D : S ratio of $<250\,mL$. However, their permeation through the biological membranes is difficult; thus, they have low bioavailability. Therefore, forming complexes with CD does not enhance their bioavailability. For example, diphenhydramine hydrochloride [36] is highly water soluble, its D : S ratio is approximately $50\,mL$, and its oral bioavailability is approximately 75%. Due to high solubility and low permeability, they are classified as class III and have low bioavailability. Due to these properties, bioavailability is not enhanced with CD complexation. Another example is acyclovir [36]. The rate-limiting factor in absorption is the permeability of the drug through the GI membranes. After complexation with CDs, permeation of the drug through the diffusion barrier increases; however, there is no significant increase in oral bioavailability.

3.4. Class IV Drugs

The drugs in this group are water insoluble and do not permeate the biological membranes. Cyclosporine A has a high molecular weight and has a water-insoluble peptide structure. It has a very low oral bioavailability. It has a high intra- and inter-subject variation. The drug is subjected to a first-pass effect and shows a low level of permeability. Therefore, it is classified as class IV. Studies on rats indicated that as a result of forming complexes with DM-β-CD, oral bioavailability of cyclosporine A increased fivefold [36]. In conclusion, CDs increase the dissolution rate of drugs that have limited solubility in water, and as a result of this, they enable fast drug absorption and fast drug effect.

4. EVALUATION OF BIOEQUIVALENCE OF GENERIC DRUGS CONTAINING CD COMPLEXES

Although forming complexes with CD has many advantages for the administration of drugs, there are a limited number of generic drugs prepared with CD complexes. There are problems in proving the bioequivalency of products developed using CD complexes with the original products. A drug that is formulated with CD might not be equivalent to the original reference products approved previously. However, since these products have higher solubility and faster absorption, their pharmacological activities might increase [56].

In the literature, it was reported that a product formulated with CD is considered a *supergeneric drug* instead of a simple generic drug. In this case, clinical studies would be required by the relevant regulatory authorities [56]. In supergeneric CD formulations, the t_{max} value is lower and the C_{max} and AUC values are greater. Pharmacodynamic studies revealed that a greater and faster therapeutic effect was achieved. Even if their doses are decreased, supergeneric formulations are not bioequivalent with the original product.

A product developed with a CD complex is considered original. All preclinical studies involving stability, toxicology, and pharmacology studies and all clinical studies must be performed.

4.1. Examples from Reported Studies

There are many published studies on enhancing the bioavailability of poorly water-soluble drugs using parent CDs and as well as CD derivatives. The examples selected from previous studies are explained below according to the pharmacological groups of the active ingredients.

4.1.1. Anti-inflammatory Drugs Nonsteroidal anti-inflammatory drugs are most widely investigated to form complexes with CDs. *Fenbufen* is an effective, nonsteroidal, anti-inflammatory drug used in rheumatoid arthritis treatment. Fenbufen was designed as the prodrug of 4-biphenylacetic acid. To enhance solubility and oral bioavailability of fenbufen, its solid complexes were prepared with α-, β-, and γ-CDs [20]. Fenbufen alone and α- and γ-CD complexes equivalent to $30\,\mu g/kg$ fenbufen were administered to rabbits. Pharmacokinetic parameters were calculated by measuring fenbufen and two major metabolites: 3-(4-biphenylhydroxymethy)propionic acid and 4-biphenylacetic acid. Following oral administration of CD complexes, the serum concentrations of fenbufen and its metabolites were found to be higher than those of fenbufen alone. AUC values of CD complexes were found to be approximately 2.5 to 4 times higher. The complexes prepared with α- and γ-CD enhanced drug absorption. Furthermore, complexation had no effect on drug metabolism and increased the concentration of active meta-bolites. In conclusion, with the complexes of fenbufen prepared with α- and γ-CD, the bioavailability of fenbufen and its two metabolites was enhanced and the bitter taste of the drug was masked.

Meloxicam, which is another NSAID, has low solubility and dissolution rate and thus low bioavailability. Meloxicam complex was prepared with β-CD and tablets were com-pressed by direct compression method using 6 parts active ingredients to 94 parts β-CD/spray-dried lactose [94]. Bio-availability studies were carried out in eight male volunteers. The tablets prepared with β-CD complex were compared with commercially available products. When the pharmaco-kinetic parameters of meloxicam were evaluated, it was found that the C_{max} and $AUC_{0-\infty}$ values of the complex prepared with β-CD was significantly higher than those of the commercially available tablets, and the t_{max} value de-creased; in other words, the maximum serum level was rapidly attained.

Flurbiprofen is an effective NSAID used in rheumatoid arthritis treatment. It has low watersolubility and low absorption. Using β-CD, a flurbiprofen-β-CD (1 : 1 molar ratio) complex containing 17% flurbiprofen was prepared by the co-precipitation method [95]. The rats were administered 1-, 3-, 10-, and 30-mg/kg doses of flurbiprofen alone and complex. Pharmacokinetic parameters were compared. At 1-mg/kg dose, the C_{max} value of the complex increased significantly; however, there was no significant increase in AUC. At 3- and 10-mg/kg doses, there was no significant difference found for C_{max} and AUC values of drug alone and its complex. When the 30-mg/kg dose was administered, C_{max} and AUC values of the complex were two times higher than those of drug alone. At a dose of 10 mg/kg of flurbi-profen, it was saturated due to its poor solubility. However, when the complex form of drug was administered, it was not saturated and, in parallel with an increased dissolution rate at low pH, the absorption window is widened. A linear dose–response relationship was observed for the four dose levels (1, 3, 10, 30 mg/kg) in the case of drug alone and CD complexes.

In another study conducted by the same researcher on flurbiprofen [96], cinnarizine (CN) was used as a competing agent. Rats were administered flurbiprofen (flur) alone,

flur/β-CD complex, flur/β-CD + CN complex, and flur + CN mixture. It was observed that the C_{max} value of flur/β-CD + CN complex prepared using cinnarizine was sig-nificantly higher than that of the other complexes; however, there was no difference in terms of AUC.

Indometacin has low water solubility; to enhance bio-availability, β-CD, HE-β-CD and HP-β-CD complexes were prepared [97]. The complexes and physical mixture were filled into gelatin capsules, and their bioavailability was compared to that of commercially available product, 50-mg Indocin capsules, in rabbits. Table 7 shows pharma-cokinetic parameters. When the pharmacokinetic para-meters were compared, it was found that there was no significant difference for C_{max} and t_{max} values of the complex and commercially available product. However, the AUC values of β-CD complex were found to be signif-icantly higher than those of the other capsule formulations. Thus, the bioavailability of indomethacine was enhanced with β-CD complex. However, there was no correlation between bioavailability, solubility, and dissolution results since the solubility of HE-β-CD and HP-β-CD complexes was found to be higher than those of β-CD.

Nimesulide is a weakly acidic, nonsteroidal, anti-inflam-matory drug. Depending on its low watersolubility and wet-tability, there are problems in preparing oral and parenteral formulations. β-, γ-, HP-β- and permethylated-β-CD com-plexes of nimesulide were prepared and ternary solid disper-sions with PEG 600 were formed [98]. The bioavailability of all formulations and pure nimesulide was investigated in rats. When pharmacokinetic parameters were evaluated, it was observed that the C_{max} value of the complex prepared with *HP*-β-CD was higher than those of the other complexes; however, the solid dispersions prepared by a second hydro-philization process with PEG did not yield good results.

4.1.2. Antidiabetics *Glipizide* is a second-generation sulfonylurea antidiabetic. Depending on its poor water sol-ubility, it has a low absorption in the GI system. To enhance the bioavailability of glipizide, complexes were prepared with β-CD by the kneading method in a 1 : 2 (drug–CD) molar ratio [99]. The effects on the blood glucose levels of mice were investigated for three different tablet formula-tions: containing glipizide alone, glipizidel/β-CD complex,

Table 7. Pharmacokinetic Parameters of Indomethacin Released from β-CD, HE-β-CD, HP-β-CD, and Indocin Capsules

Parameter	β-CD Capsules	HE-β-CD Capsules	HP-β-CD Capsules	Indocin Capsules
Dose	40	40	40	50
K_a (h)	0.19 ± 0.11	4.07 ± 3.85	6.79 ± 4.10	0.10 ± 0.03
AUC (mg.h/L)	2090 ± 1063	1655 ± 1066	921 ± 204	1617 ± 512
C_{max} (μg/mL)	171 ± 211	60 ± 25	100 ± 92	61 ± 19
T_{max} (h)	6.7 ± 3.0	3.0 ± 3.0	1.7 ± 2.2	10.3 ± 3.9

Source: Adapted from [97].

Table 8. Pharmacokinetic Parameters on Oral Administration of Binary Systems of Gliquidone/HP-β-CD and Pure Gliquidone to Rats

System	C_{max} (μg/mL)	T_{max} (h)	$AUC_{0-\infty}$ (μg·h/mL)
Gliquidone	3.34 ± 0.215	4	23.96 ± 1.53
Physical mixture	3.25 ± 0.379	4	27.84 ± 2.24
Kneaded mixture	3.47 ± 0.198	4	32.86 ± 2.9
Co-evaporated mixture	4.55 ± 0.365	4	38.82 ± 2.1
Co-lyophilized mixture	5.27 ± 0.339	4	48.10 ± 2.84

Source: Adapted from [89].

and sodium carboxymethylcellulose (NaCMC). When relative bioavailability was compared, it was found that relative bioavailability values of glipizide alone, glipizide/β-CD, and a glipizide/β-CD + NaCMC mixture were 1%, 1.63%, and 1.92%, respectively. Particularly, the addition of NaCMC to the complex resulted in high solubility and bioavailability.

Gliquidone [89] is another of the sulfonylurea group of drugs. Depending on its very low wettability, it has a low water solubility and its bioavailability shows variations. To enhance solubility and dissolution rate and to improve the bioavailability of gliquidone, solid complexes were prepared with HP-β-CD by physical mixing, kneading, co-evaporation, and co-lyophilization methods. The bioavailability of the prepared complexes in comparison with the pure drug was investigated in rats. Table 8 shows the pharmacokinetic parameters obtained.

FPFS-410 ((2-(*N*-cyanoimino)-5-[(E)-4-styrylbenzylidene]-4-oxothiazolidine) is a new antidiabetic and lipid-lowering agent. It has a very low water solubility (0.0054 μg/mL). Solid complexes of this drug were prepared with 2-HP-β-CD (1:2 ratio) by spray-drying and co-grinding methods [100]. It was observed that 2-HP-β-CD increased the solubility of FPFS-410 200,000-fold. The oral bioavailability of the complexes prepared was compared to the active ingredient alone and it was also administered via the IV route. After evaluation of the pharmacokinetic parameters, it was

found that C_{max} increased from 0.25 to 0.68; AUC increased from 0.74 to 2.05; t_{max} decreased from 1.67 h to 1 h. Absolute bioavailability increased approximately threefold, from 5% to 14%. When blood glucose levels were investigated, no difference was found between the solid complexes in terms of pharmacodynamic effectiveness. However, in terms of bioavailability, it was observed that the preparation method was an important factor and that the best result was obtained with co-evaporation and co-lyophilization methods.

To enhance the oral bioavailability of *tolbutamide*, inclusion complexes were prepared with β-CD and HP-β-CD by the freeze-drying method [92]. Inclusion complexes equivalent to a 20-mg/kg dose of tolbutamide and commercially available tolbutamide product were administered orally to New Zealand rabbits, and their bioavailability was compared. In addition to the assaying of tolbutamide levels in blood, glucose levels were also measured. The pharmacokinetic parameters obtained are shown in Table 9. Oral absorption of tolbutamide is faster and more effective with inclusion complexes. There was a relationship between plasma level and the hypoglycemic effect of tolbutamide. In this case, depending on the increased bioavailability and hypoglycemic effect, administration of reduced dose in the treatment may be possible with the complexes prepared with cyclodextrin.

4.1.3. Antihypertensives *Rutin* is a natural phenolic flavonoid glycoside. It is used clinically in capillary hemorrhages and as a hypertensive. This molecule has very low water solubility. To enhance bioavailability in oral formulations, β-CD and HP-β-CD complexes at a ratio of 1 : 1 were prepared by the kneading method [65]. The formulations prepared were compressed as tablets in such way to contain a formulation equivalent to 200 mg of active ingredient, and in vivo tests were carried out in male beagle dogs. To calculate pharmacokinetic parameters, homovanillic acid, which is the major metabolite of rutin, was used. Table 10 shows the pharmacokinetic parameters obtained.

Furthermore, after intravenous administration of rutin, absolute bioavailability was calculated. Absolute bioavailability values were found to be 25.7, 23.2 and 73.5 for rutin, β-CD complex, and HP-β-CD complex, respectively. Solubility, dissolution rate, and stabilizing effects of HP-β-CD were close to those of β-CD; under in vivo conditions and an

Table 9. Pharmacokinetic Parameters and Relative Bioavailability of Tolbutamide After Oral Administration of Tolbutamide Alone, and Its β-CD and HP-β-CD Complexes

System	C_{max} (μg/mL)	$AUC_{0-\infty}$ (μg·h/mL)	t_{max} (h)	Relative Bioavailability (%)
Tolbutamid	18.58 ± 3.27	443.96 ± 63.33	8.50 ± 1.15	
β-CD complex	36.26 ± 2.17	621.62 ± 52.73	3.83 ± 0.65	165.00
HP-β-CD complex	34.99 ± 1.36	654.63 ± 62.42	4.00 ± 0.45	163.21

Source: Adapted from [92].

Table 10. Pharmacokinetic Parameters of Homovanillic Acid After Oral Administration of Rutin Alone and Its β-CD and HP-β-CD Complexes to Beagle Dogs

System	C_{max} (ng/mL)	AUC (ng· h/mL)	t_{max} (h)
Rutin alone	72.43 ± 37.73	224.70 ± 98.34	4.00 ± 0.76
β-CD complex	61.13 ± 32.94	202.58 ± 100.97	3.33 ± 1.45
HP-β-CD complex	322.00 ± 68.95^a	642.62 ± 100.02^a	2.88 ± 0.52

Source: Adapted from [65].

$^a p < 0.05$ compared to rutin alone.

approximately threefold increase was obtained in bioavailability compared to β-CD. It was concluded that this result was due to the differences between in vitro and in vivo conditions. In conclusion, it was observed that after oral administration of rutin, its solubility, dissolution rate and bioavailability increased with HP-β-CD.

Digoxin is one of the most problematic drugs in terms of formulation and bioavailability, due to its very narrow therapeutic index. The water solubility of digoxin increased 2000 times by forming a complex with HP-β-CD [101]. The bioavailability of oral solution prepared with solid complexes formed with HP-β-CD was compared with commercial tablets in rabbits and humans. When the pharmacokinetic parameters obtained after the tests on humans were investigated, it was found that there was no difference for the C_{max} and AUC values of the two formulations. The t_{max} value of the solution decreased by twofold when compared to the tablet. The tests on rabbits were similar to those on humans in terms of pharmacokinetic parameters, and there was no difference between the bioavailability of the two formulations.

Nitrendipine is a calcium channel blocker. To enhance the solubility, dissolution rate, and bioavailability of nitrendipine, inclusion compounds were prepared with HP-β-CD by the solvent evaporation method [102]. To investigate bioavailability, a 10-mg/kg dose of inclusion compound, physical mixture, and nitrendipine powder was orally administered to rats. Table 11 shows the pharmacokinetic parameters calculated. As indicated in the table, oral bioavailability of nitrendipine was enhanced, depending on the increase in dissolution rate, by forming complexes with HP-β-CD.

Table 11. Bioavailability Parameters After Oral Administration of Various Forms of Nitrendipine in Rats

System	C_{max} (µg/mL)	AUC (µg· min/mL)	t_{max} (min)
Inclusion compound	0.581 ± 0.085	75.37 ± 12.24	63.75 ± 10.61
Physical mixture	0.407 ± 0.036	49.76 ± 7.94	67.5 ± 13.89
Nitrendipine powder	0.272 ± 0.051	39.88 ± 5.72	78.25 ± 22.32

Source: Adapted from [103].

4.1.4. Antitumor Drugs   *Tacrolimus* is an immunosuppressive agent, a hydrophobic macrolide lactone, produced by *Streptomyces tsukubaensis*. Tacrolimus has a very low oral bioavailability, and its absorption shows high variation (4 to 89%). To enhance the water solubility and bioavailability of tacrolimus, complexes were prepared using α-, β-, γ-, DM-β-, RM-β-, HP-β-, and and SBE-β-CD, and the bioavailability of these complexes were compared to those of the commercially available product, Prograf [103].

Tacrolimus–CD complexes were prepared by the kneading method in the molar ratio 1 : 50. For bioavailability studies, 5-mg/kg doses of aqueous suspensions with and without CD and Prograf capsules were administered to rats. Furthermore, 1-mg/kg-dose IV drug was administered. When pharmacokinetic parameters were investigated, it was found that C_{max} and AUC values of drug increased by approximately 4.5 and 20.3 times with DM-β-CD complexation. However, there was no significant difference compared with the commercially available product. Absolute bioavailability increased from 0.85 to 3.8 with DM-β-CD. Tacrolimus has a very narrow therapeutic window (5 to 20 ng/mL). A more constant blood level was obtained with DM-β-CD complex. In addition, 133 mg/kg DM-β-CD was administered to rats for 7 days, and no damage was observed in GI mucosa. It was concluded that DM-β-CD enhanced the oral bioavailability of tacrolimus and reduced the variability in absorption.

Ro 28-2653 (5-biphenyl-4-yl-5-[4-(4-nitrophenyl)piperazin-1-yl]pyrimidine-2,4,6-trione) is a synthetic matrix metalloproteinase inhibitor. Depending on its poor water solubility (0.56 µg/mL), it has a very low and variable oral bioavailability [104]. To enhance the bioavailability of the drug, the effect of HP-β-CD was investigated. To make comparisons, three different formulations were prepared and were administered to sheep. The content of the administered formulations was as follows:

Oral solution
  Ro 28-2653 (15 mg/mL)
  HP-β-CD (200 mM)
  l-Lysine (50 mM)
  Water
Oral suspension
  Ro 28-2653 (15 mg/mL)
  Polysorbate 80 (0.1 mg/mL)
  Simaldrate (Veegum HV, 1% m/v)
  Methylcellulose (Methocel A 400, 0.4% m/v)
  Water
IV solution
  Ro 28-2653 (10 mg/mL)
  HP-β-CD (200 mM)
  l-Lysine (20 mM)
  Water for injection

Table 12. Pharmacokinetic Parameters of Ro 28-2653 Obtained Following Oral Administration of Its Suspension and Oral Solution to Sheep

Pharmacokinetic Parameter	Suspension	Oral Solution
C_{max} (µg/mL)	4.84 ± 1.95	51.84 ± 23.73
$AUC_{0-\infty}$ (µg·h/mL)	214.65 ± 103.04	2070.13 ± 943.79
t_{max} (h)	12.34 ± 5.99	3.59 ± 1.52
F_{abs}	0.08	0.80

Source: Adapted from [104].

The pharmacokinetic parameters obtained are shown in Table 12. A synergistic effect was observed between l-lysine and HP-β-CD. With HP-β-CD, the C_{max} and absolute bioavailability values increased 10-fold.

4.1.5. Antihelmentics

Albendazole is a benzimidazole derivative drug with broad-spectrum activity against human and animal helminth parasites. It has a water solubility of approximately 0.2 µg/mL. In studies performed with albendazole, inclusion complexes were prepared with HP-β-CD and citric acid [105]. The bioavailability of inclusion complexes was compared with the suspension form of commercially available product (Valbazen) in sheep. Albendazole is absorbed through the GI system with passive diffusion. Combined use of citric acid and CD provided a synergistic effect, and the solubility increased 10000-fold. Table 13 shows pharmacokinetic parameters. The C_{max} value of the solution prepared with β-CD increased approximately twofold when compared to the commercial suspension. It was observed that the active ingredient from the solution was rapidly absorbed and that the t_{max} value decreased approximately threefold.

In a study by Garcia et al. [106], albendazole was dissolved in 20% HP-β-CD solution and the bioavailability of the prepared solution was compared to that of commercial suspension (Panreac) in mice. Depending on the fast degradation of albendazole, it has a low plasma concentration. Therefore, the concentration of albendazole sulfoxide, which is its active metabolite, was also measured in this study. The blood concentration of both albendazole and albendazole sulfoxide increased with CD complexation. In addition,

Table 13. Pharmacokinetic Parameters of Albendazole Obtained Following Oral Administration of Its Suspension and Oral Solution Containing HP-β-CD to Sheep

Pharmacokinetic Parameter	Solution	Suspension
C_{max} (µg/mL)	2.8 ± 0.6	1.2 ± 0.2
$AUC_{0-\infty}$ (µg·h/mL)	36.7 ± 3.4	26.7 ± 7.9
t_{max} (h)	2.3 ± 0.5	9.7 ± 2.0

Source: Adapted from [105].

antihelmentic activity against enteral and parenteral stages of *Trichinella spiralis* was studied in mice. CD containing albendazole solution was found to be more effective in both cases. However, in cases of migrating larvae, there was no significant difference between the two formulations.

4.1.6. Antiepileptics

Phenytoin is one of the most problematic drugs in terms of bioavailability, due to its low water solubility and dissolution rate. To enhance the bioavailability of phenytoin, charged and neutral CDs were used. Captisol [sulfobutyl ether-β-CD (SBE_{7m}-β-CD)] was selected as charged, and Encapsin (HP-β-CD) was selected as a neutral CD [107]. Solid complexes were prepared by dissolving phenytoin in 0.05 M NaOH (pH 11.0) solution containing 72.3 mM CD and were then freeze-dried. The product obtained was filled into gelatin capsules.

In addition, the freeze-dried form of phenytoin was also prepared. Phenytoin alone, a physical mixture, and CD complexes equivalent to 300 mg of phenytoin were administered to four male beagle dogs. It was observed that the solid complex and physical mixture prepared with CDs increased the C_{max} and AUC values approximately twofold when compared to phenytoin alone and its freeze-dried form. There was no difference between the solid complexes and the physical mixture prepared with CD in terms of the oral pharmacokinetic of phenytoin (Table 14). CDs increased the C_{max} and AUC values, but they did not affect the t_{max} value. In other words, the absorption rate did not change. It was found that anionic and neutral CD had similar bonding properties.

Carbamazepine, which is another antiepileptic agent, shows varying bioavailability depending on its low dissolution rate. To enhance the dissolution rate of carbamazepine,

Table 14. Pharmacokinetic Parameters of Phenytoin in Plasma After Oral Administraion in Various Formulations to Beagle Dogs

System	C_{max} (µg/mL)	AUC_{0-24h} (ng·h/mL)	t_{max} (h)
Crystal phenytoin	4.22 ± 0.56[a]	34.90 ± 4.06[a]	3.0 ± 0.6
Lyophilized phenytoin	4.29 ± 0.49[a]	35.72 ± 3.65[a]	3.5 ± 0.5
Physical mixture	6.92 ± 1.30	68.88 ± 12.31	3.5 ± 0.5
Phenytoin/ HP-β-CD	7.06 ± 0.85	70.33 ± 11.01	3.0 ± 0.6
Phenytoin/ SBE_{7m}-β-CD	6.87 ± 0.78	74.22 ± 8.60	4.0 ± 0.0

Source: Adapted from [107].

[a] Significantly different from the values for the capsules containing phenytoin/HP-β-CD, phenytoin/SBE_{7m}-β-CD, and the physical mixture ($p < 0.05$).

various agents such as poly(ethylene glycol), phospholipids, and HP-β-CD, were used [108]. CD complexes were prepared in a molar ratio of 1 : 1 by the solvent method. In vivo tests were conducted on rabbits and the bioavailability values of the systems prepared were compared with a suspension of the commercially available product Tegretol. It was observed that CD complexes prepared with HP-β-CD gave the highest AUC and C_{max} values and that they had faster absorption than that of the commercially available product.

In another study carried out with carbamazepine, the bioavailability of complexes prepared with β-CD was investigated in beagle dogs [109]. Carbamazepine/β-CD complexes in a ratio of 1 : 1, prepared by the spray-drying technique, were compressed to tablet form and their bioavailability was compared with the reference product, Tegretol CR 200. Although a high intersubject variability was observed, it was found that β-CD enhanced the bioavailability of carbamazepine sixfold compared to the commercial product.

4.1.7. Antimycotic Drugs

Griseofulvin is a drug that is known to present a bioavailability problem. To enhance its bioavailability, griseofulvin β-CD complexes were prepared in a molar ratio of 1 : 1 and these complexes were administered to rabbits and humans [110]. When t_{max} values were compared, it was found that in both rabbits and humans, the complexes prepared with β-CD had shorter t_{max} values compared to that of pure griseofulvin and, as a result of this, had a faster absorption.

Clotrimazole is another antimycotic agent. To enhance the oral bioavailability of clotrimazole, inclusion compounds with β-CD were prepared by the spray-drying method [111]. Inclusion compounds that were suspended in 1% povidone solution and a 80-mg/kg dose of clotrimazole powder were administered orally to rats. Total plasma concentration of inclusion compounds was found to be higher than those of the powder form. Table 15 shows the pharmacokinetic parameters. It was found that the AUC and C_{max} values of the inclusion compounds were approximately three times higher than those of clotrimazole alone. In addition, the t_{max} value was found to decrease. It was concluded that inclusion compounds could be more easily absorbed orally than could the drug alone.

Table 15. Pharmacokinetic Parameters of Clotrimazole After Oral Administration of the Suspension Form of Clotrimazole Powder and Inclusion Complex to Rats

Parameter	Clotrimazole Powder	Inclusion Compound
AUC (μg·h/mL)	18.90 ± 6.24	56.15 ± 35.70^a
C_{max} (μg/mL)	1.93 ± 0.46	6.29 ± 3.39^a
t_{max} (h)	4.16 ± 0.98	1.00 ± 0.00^a

Source: Adapted from [111].

$^a p < 0.05$ compared with clotrimazole powder.

Table 16. Pharmacokinetic Parameters of Itraconazole After Oral Administration of Itraconazole/β-CD Complex

Preparation Method	AUC_{0-8h} (μg·min/mL)	C_{max} (μg/mL)	t_{max} (min)
Physical mixture	5.75 ± 0.09^a	1.30 ± 0.05^a	120
Co-precipitation	8.51 ± 0.07^a	2.93 ± 0.01^a	120
SC CO_2	12.07 ± 0.17	4.54 ± 0.12	120

Source: Adapted from [113].

$^a p < 0.001$ vs. SC CO_2 group.

Itraconazole is an antimycotic agent. The bioavailability and bioequivalence of a commercial itraconazole solution (Sporanox) containing HP-β-CD and two capsule formulations were evaluated on 30 male volunteers by a crossover study [112]. When pharmacokinetic parameters were investigated, it was found that the bioavailability of itraconazole and hydroxyitraconazole solutions were 30 to 33% and 35 to 37% higher than those of capsule formulations, respectively. When the two capsule formulations were compared, they were found to be equivalent. It was found that when administered in solution form, the bioavailability of itraconazole was enhanced.

In another study carried out with itraconazole, the dissolution rate and bioavailability of complexes formed with β-CD were investigated [113]. Itraconazole/β-CD inclusion complexes in a molar ratio of 1 : 2 were prepared by the physical mixing, conventional co-precipitation, and supercritical carbon dioxide (SC CO_2) methods. Complexes containing drug equivalent to 10 mg/kg itraconazole were administered to rats. Table 16 shows the pharmacokinetic data calculated. The AUC and C_{max} values of the complex prepared by the supercritical CO_2 method were found to be significantly higher than those of the other two methods, and bioavailability was found to increase.

4.1.8. Steroid Hormones

To enhance the solubility and bioavailability of *dehydroepiandrosteron* (DHEA), solid complexes were prepared with α-CD by the high-energy co-grinding method [114]. To enhance the effectiveness of the system, ternary products were prepared using some inactive ingredients, such as glycine, biomaltodextrin, poly(vinylpyrrolidone), and poly(ethylene glycol 400). It was found that α-CD-glycine (1 : 2 : 3 molar ratio) was the most effective in terms of solubility, dissolution rate, and bioavailability. When pharmacokinetic parameters were evaluated, it was found that C_{max} and AUC values increased approximately twofold compared to the pure active ingredient. It was found that a DHEA/α-CD/glycine system can be used successfully in putative hormone replacement treatment.

Thanks to their estrogen-like structure, *isoflavones* have an estrogenic effect and are referred to as phytoestrogens.

Table 17. Pharmacokinetic Parameters for Total Isoflavones After the Oral Administration of Isoflavone and Isoflavone/β-CD Complex to Rats

Isoflavone	C_{max} (μg/mL)		$AUC_{0-360\ min}$ (μg·min/mL)	
	Isoextract	β-CD/Isoextract	Isoextract	Isoextract/β-CD
Daidzein	1379 ± 502	2320 ± 114	340 ± 111	430 ± 71
Glycitein	90.5 ± 76.2	485.7 ± 297.6	28 ± 4	48 ± 9
Genistein	33.6 ± 14.3	123.7 ± 68.0	11 ± 1	20 ± 3

Source: Adapted from [115].

They have very low water solubility. To enhance the solubility and bioavailability of soy isoflavone extracts, inclusion complexes were prepared with β-CD and their bioavailability was investigated in rats [115]. Isoflavones constitute a large group; in this study daidzin, genistein and glycitein forms were investigated. After complexation with β-CD, the solubility of isoflavones increased 26-fold compared to the pure drug. For a comparison of bioavailability, the concentrations of daidzin, genistein, and glycitein in blood were monitored. When the plasma concentrations of pure drug and β-CD complex were analyzed, it was found that C_{max} values of β-CD complexes were approximately five times higher than that of the pure drug. The pharmacokinetic parameters are shown in Table 17. When AUC values were compared, a 126% increase was found for daidzin, but this was not statistically significant. For glycitein and genistein, there was a significant increase of 170% and 180%, respectively.

4.1.9. Others

Artemisinin is a plant-origin antimalarial drug obtained from *Artemisia annua*. Due to its low water solubility, it has low absorption after oral administration. To enhance the bioavailability of artemisinin, complexes were prepared with β- and γ-CD in a molar ratio of 1:1 (CD:artemisinin) by the slurry method [116]. Prepared complexes were compared with capsules of the commercially available product Artemisinin 250, and bioavailability studies were conducted.

Twelve health volunteers participated in the in vivo studies. Table 18 shows pharmacokinetic parameters obtained after the administration of commercially available product and CD complexes. As indicated in the table, the

Table 18. Pharmacokinetic Parameters of Artemisinin After Oral Administration of its CD Complexes and Commercially Available Product

Formulation	C_{max} (ng/mL)	$AUC_{0-\infty}$ (ng·h/mL)	t_{max} (h)
Artemisinin 250	271.7 ± 174.6	782.3 ± 392.0	2.0 ± 0.6
β-CD complex	651.5 ± 604.9	1329.4 ± 562.4	1.6 ± 0.8
γ-CD complex	458.1 ± 182.2	1131.3 ± 456.6	1.4 ± 0.6

Source: Adapted from [115].

C_{max} and AUC values of both complexes increased compared to the reference product. The absorption rate and ratio of both complexes were found to be higher than those of the commercially available product.

Raloxifene hydrochloride is a selective estrogen receptor modulator used in osteoporosis treatment. It is a water-insoluble, highly metabolized drug. To enhance the oral bioavailability of raloxifen, a derivative of hydroxybutenyl-β-CD (HB-β-CD), a new cyclodextrin, was used [117]. Solid powder complexes were prepared by the freeze-drying method. Furthermore, by adding PEG 400 to the system, liquid systems were prepared and in vivo tests were carried out in rats. It was found that compared to the pure drug, C_{max} increased twofold with solid complexes prepared with HB-β-CD, AUC increased threefolds, and relative bioavailability increased twofold. The addition of PEG 400 to the system increased relative bioavailability compared to the pure active ingredient, but it was found to be lower than that of the solid complex. When the pharmacokinetic parameters of raloxifen were investigated for gluronide metabolite, it was observed that C_{max} increased 12-fold and AUC increased 6.5-fold.

Another example is baicealin a flavonoid with antibacterial and anti-HIV properties. Although its pharmacological properties have a broad spectrum due to its low water solubility, it has limited pharmaceutical use. To enhance the solubility and bioavailability of baicalein, complexes with HP-β-CD were prepared by the freeze-drying method. With HP-β-CD, the solubility of baicalein increased 5.5-fold [118]. The bioavailability of free baicalein and HP-β-CD complexes of drug were investigated in rats after the bolus IV injection and oral administration. Free baicalein and its complex were quickly metabolized in in vivo and baicalein converted into glucuronide metabolite. Absolute bioavailability was calculated according to oral pharmacokinetic parameters and was found to be 52.3% for free drug and 86.4% for an HP-β-CD complex.

Coenzyme-Q10 (Co-Q10), which has an important role in energy metabolism, is a fat-soluble substance and has oral bioavailability problems. Pat complexes were prepared with β-CD [119]. The bioavailability of the complexes in dogs was compared with that of the soft gelatin capsule form commercially available containing 30 mg Co-Q10. When pharmacokinetic parameters were investigated, it was found that in the complexes prepared with β-CD, the C_{max} value increased twofold, the AUC_{0-48h} value increased threefold, and the t_{max} value decreased and thus absorption increased.

5. CONCLUSIONS

It was concluded that in most studies to improve the bioavailability of drugs with CD complexation, pharmacokinetic parameters such as C_{max} and AUC increased significantly while t_{max} decreased compared with plain drugs. CDs are particularly effective in increasing the bioavailability of class II drugs.

Products developed with CD complexation of drugs are considered super generics, and it has been suggested that clinical studies should be carried out on them. Therefore, the number of commercially available generic products developed by CD complexation has not increased. Most previous studies have focused on improving the oral bioavailability of small molecules using CDs. In the near future, investigation of the appropriateness of macromolecules such as peptide–protein drugs for CD complexation to improve oral bioavailability will enhance the importance of CDs in the pharmaceutical industry.

Acknowledgment

We thank Dr. Fatmanur Tugcu Demiroz for her kind help.

REFERENCES

1. Junghans, J.A.H., Müller, R.H. (2008). Nanocrystal technology, drug delivery and clinical applications. *Int. J. Nanomed.* 3, 295–309.

2. Chen, H., Khemtong, C., Yang, X., Chang, X., Gao, J. (2010). Nanonization strategies for poorly water-soluble drugs. *Drug Discov. Today*, doi: 10.1016/j.drudis.2010.02.009.

3. Katteboinaa, S., Chandrasekhar, P., Balaji, S. (2009). Drug nanocrystals: a novel formulation approach for poorly soluble drugs. *Int. J. PharmTech Res.*, 1, 682–694.

4. http://www.elandrugtechnologies.com./nanocrystal_technology.

5. Çelebi, N., Shirakura, O., Machida, Y., Nagai, T. (1987). The inclusion complex of piromidic acid with dimethyl-β-cyclo-dextrin in aqueous solution and in the solid state. *J. Inclusion Phenom.*, 5, 407–413.

6. Acartürk, F., Imai, T., Saito, H., Ishikawa, M., Otagiri, M. (1993). Comparative study on inclusion complexation of maltosyl-β-cyclodextrin, heptakis(2,6-di-O-methyl)-β-cyclo-dextrin and β-cyclodextrin with fucosterol in aqueous and solid state. *J. Pharm. Pharmacol.*, 45, 1028–1032.

7. Çelebi, N. (1987). Cyclodextrins: I. Properties, methods of preparation and its clathrate compounds. *FABAD J. Pharm. Sci.*, 12, 5–15.

8. Çelebi, N. (1987). Cyclodextrins: II. Application in pharmaceutical sciences and interactions of cyclodextrin with drugs. *FABAD J. Pharm. Sci.*, 12, 16–25.

9. Erden, N., Çelebi, N. (1988). A study of the inclusion complex of naproxen with β-cyclodextrin. *Int. J. Pharm.*, 48, 83–89.

10. Çelebi, N., Erden, N. (1992). Interaction of naproxen with β-cyclodextrin in ground mixture. *Int. J. Pharm.*, 78, 183–187.

11. Cappello, B., Maio, C.D., Iervolino, M., Miro, A. (2006). Improvement of solubility and stability of valsartan by hydroxypropyl-β-cyclodextrin. *J. Inclusion Phenom. Macrocyclic Chem.*, 54, 289–294.

12. Jadhav, G.S., Vavia, P.R. (2008). Physicochemical, in silico and in vivo evaluation of a danazol–β-cyclodextrin complex. *Int. J. Pharm.*, 352, 5–16.

13. Çelebi, N., Nagai, T. (1987). Enhancement of dissolution properties of nalidixic acid from ground mixtures with γ-cyclodextrin. *STP Pharma Sci.*, 3, 868–871.

14. Çelebi, N., Nagai, T. (1988). Improvement of dissolution characteristics of piromidic acid by dimethyl-β-cyclodextrin complexation. *Drug Dev. Ind. Pharm.*, 14, 63–75.

15. Ozdemir N., Ordu, S. (1998). Improvement of dissolution properties of furosemide by complexation with β-cyclodextrin. *Drug Dev. Ind. Pharm.*, 24, 19–25.

16. Mosher, G., Thompson, D.O. (2002). Compexation and cylodextrins. In: Swarbrick, J.S., Boylan, J.C. Eds. *Encyclopedia of Pharmaceutical Technology*, 2nd ed., Marcel Dekker, New York, pp. 531–568.

17. Szejtli, J. (1988) *Cyclodextrin Technology.* Kluwer, Boston, pp. 186–306.

18. Uekama, K., Otagiri, M. (1987). Cyclodextrins in drug carrier systems. *CRC Crit. Rev. Ther. Drug Carrier Syst.*, 3, 1–40.

19. Uekama, K., Hirayama, F., Irie, T. (1998). Cyclodextrin drug carrier systems. *Chem. Rev.*, 98, 2045–2076.

20. Miyaji, T., Inoune, Y., Acartürk, F., Imai, T., Otagiri, M., Uekama, K. (1992). Improvement of oral bioavailability of fenbufen by cyclodextrin complexations. *Acta Pharm. Nord.*, 4, 17–22.

21. Erden, N., Çelebi, N. (1988). Cyclodextrins: III. Enhancement of bioavailability of drugs by cyclodextrins. *FABAD J. Pharm. Sci.*, 13, 381–392.

22. Davis, M.E., Brewster, M.E. (2004). Cyclodextrin-based pharmaceutics: past, present and future. *Nat. Rev. Drug Discov.*, 3, 1023–1035.

23. Del Valle, M.E.M. (2004). Cyclodextrins and their uses: a review. *Process Biochem.*, 39, 1033–1046.

24. Challa, R., Ahuja, A., Ali, J., Khar, R.K. (2005). Cyclodextrins in drug delivery: an updated review. *AAPS PharmSciTech*, 6, E329–E357.

25. Acartürk, F. (1993). The effect of cyclodextrins on the physical and chemical stability of drugs: I. Effect on solid state stability. *FABAD J. Pharm. Sci.*, 18, 77–85.

26. Acartürk, F. (1993). The effect of cyclodextrins on the physical and chemical stability of drugs: II. Effect on aqueous solution stability. *FABAD J. Pharm. Sci.*, *18*, 87–94.

27. Szejtli, J. (1998). Introduction and general overview of cyclodextrin chemistry. *Chem. Rev.*, *98*, 1743–1753.

28. Duchêne, D., Vaution, C., Glomot, F. (1986). Cyclodextrins, their value in pharmaceutical technology. *Drug Dev. Ind. Pharm.*, *12*, 2193–2215.

29. Duchêne, D., Wouessidjewe, D., Ponchel, G. (1999). Cyclodextrins and carrier systems. *J. Control. Release*, *62*, 263–268.

30. Uekama, K. (2002). Recent aspects of pharmaceutical application of cyclodextrins. *J. Inclusion Phenom. Macrocyclic Chem.*, *44*, 3–7.

31. Uekama, K., Hirayama, F., Arima, H. (2006). Recent aspect of cyclodextrin-based drug delivery system. *J. Inclusion Phenom. Macrocyclic Chem.*, *56*, 3–8.

32. Fernandes, C.M., Ramos, P., Falcão, A.C., Veiga, F.J. (2003). Hydrophilic and hydrophobic cyclodextrins in a new sustained release oral formulation of nicardipine: in vitro evaluation and bioavailability studies in rabbits. *J. Control. Release*, *88*, 127–134.

33. Stella, V.J., Rao, V.M., Zannou, E.A., Zai, V. (1999). Mechanisms of drug release from cyclodextrin complexes. *Adv. Drug Deliv. Rev.*, *368*(1), 3–16.

34. Shimpi, S., Chauhan, B., Shimpi, P. (2005). Cyclodextrins: application in different routes of drug administration. *Acta Pharm.*, *55*, 139–156.

35. Loftsson, T., Duchêne, D. (2007). Cyclodextrins and their pharmaceutical applications. *Int. J. Pharm.*, *329*, 1–11.

36. Loftsson, T., Brewster, M.E., Másson, M. (2004). Role of cyclodextrins in improving oral drug delivery. *Am. J. Drug Deliv.*, *2*, 261–275.

37. Wang, L., Jiang, X., Xu, W., Li, C. (2007). Complexation of tanshinonc IIA with 2-hydroxypropyl-β-cyclodextrin: effect on aqueous solubility, dissolution rate, and intestinal absorption behavior in rats. *Int. J. Pharm.*, *341*, 58–67.

38. Corti G., Cirri M., Maestrelli F., Mennini N., Mura P. (2008). Sustained-release matrix tablets of metformin hydrochloride in combination with triacetyl-β-cyclodextrin. *Eur. J. Pharm. Biopharm.*, *68*, 303–309.

39. Barillaro, V., Evrard, B., Delattre, L., Piel, G. (2005). Oral bioavailability in pigs of a miconazole/hydroxypropyl-γ-cyclodextrin/L-tartaric acid inclusion complex produced by supercritical carbon dioxide processing. *AAPS J.*, *7*, E149–E155.

40. Salehian, B., Wang, C., Alexander, G., Davidson, T., McDonald, V., Berman, N., Dudley, R.E., Ziel, F., Swerdloff, R.S. (1995). Pharmacokinetics, bioefficacy, and safety of sublingual testosterone cyclodextrin in hypogonadal men: comparison to testosterone enanthate. A clinical research center study. *J. Clin. Endocrinol. Metab.*, *80*, 3567–3575.

41. Sinswat, P., Tengamnuay, P. (2003). Enhancing effect of chitosan on nasal absorption of salmon calcitonin in rats: comparison with hydroxypropyl and dimethyl-β-cyclodextrins. *Int. J. Pharm.*, *257*, 15–22.

42. Yetkin, G., Çelebi, N., Ağabeyoğlu, İ., Gökçora, N. (1999). The effect of dimetil-β-cyclodextrin and sodium taurocholate on nasal bioavailability of salmon calcitonin in rabbits. *STP Pharma Sci.* *9*, 249–252.

43. Yetkin, G., Çelebi, N., Özoğul, C., Demiryürek, A.T. (2001). Enhancement of nasal absorption of salmon calcitonin in rabbits using absorption enhancer. *STP Pharma Sci.*, *11*, 187–191.

44. Tas, C., Ozkan, C.K., Savaser, A., Ozkan, Y., Tasdemir U., Altunay, H. (2009). Nasal administration of metoclopramide from different dosage forms: in vitro, ex vivo and in vivo evaluation. *Drug Deliv.*, *16*, 167–175.

45. Cappello, B., Carmignani, C., Iervolino, M., Immacolata La Rotonda, M., Fabrizio Saettone, M. (2001). Solubilization of tropicamide by hydroxypropyl-β-cyclodextrin and water-soluble polymers: in vitro/in vivo studies. *Int. J. Pharm.*, *213*, 75–81.

46. Aktaş, Y., Ünlu, N., Orhan, M., Irkeç, M., Hıncal, A.A. (2003). Influence of hydroxypropyl-β-cylodextrin on the corneal permeation of pilocarpine. *Drug Dev. Ind. Pharm.*, *29*, 223–230.

47. Jug, M., Bećirević-Laćan, M. (2004). Influence of hydroxypropyl-β-cyclodextrin complexation on piroxicam release from buccoadhesive tablets. *Eur. J. Pharm. Sci.*, *21*, 251–260.

48. Değim, Z., Değim, T., Acartürk, F., Erdoğan, D., Ozoğul, C., Köksal, M. (2005). Rectal and vaginal administration of insulin–chitosan formulations: an experimental study in rabbits. *J. Drug Target.*, *13*, 563–572.

49. Cevher, E., Sensoy, D., Zloh, M., Mülazimoglu, L. (2008). Preparation and characterization of natamycin:γ-cyclodextrin inclusion complex and its evaluation in vaginal mucoadhesive formulations. *J. Pharm. Sci.*, *97*, 4319–4335.

50. Duchéne, D., Wouessidjewe, D. (1991) Dermal use of cyclodextrins and derivatives. In: Duchine, D.,(Ed.), *New Trends in Cyclodextrins and Derivatives*. Editions de Sante, Paris, pp. 449–481.

51. Anadolu, R.Y., Sen, T., Tarimci, N., Birol, A., Erdem, C. (2004). Improved efficacy and tolerability of retinoic acid in acne vulgaris: a new topical formulation with cyclodextrin complex. *J. Eur. Acad. Dermatol. Venereol.*, *18*, 416–421.

52. Çelebi, N., Kışilal, Ö., Tarimci, N. (1993). The effects of β-cyclodextrin and penetration additives on release of naproxen from ointment bases. *Pharmazie*, *48*, 914–917.

53. Çelebi, N., Gül, Z.İ., Ocak, F., Yildiz, S., Acartürk, F. (1996). Investigation of the effect of β-CD on in vitro release of ketocanozole from different gel bases. In: Szejtli, J., Szente, L. (Eds.), *Proceedings of the Eighth International Symposium on Cyclodextrins*. Kluwer, Boston, pp. 461–464.

54. Kear, C.L., Yang, J., Godwin, D.A., Felton, L.A. (2008). Investigation into the mechanism by which cyclodextrins influence transdermal drug delivery. *Drug Dev. Ind. Pharm.*, *34*, 692–697.

55. Tokihiro, K., Arima, H., Tajiri, S., Irie, T., Hirayama, F., Uekama, K. (2000). Improvement of subcutaneous bioavailability of insulin by sulphobutyl ether β-cyclodextrin in rats. *J. Pharm. Pharmacol.*, *52*, 911–917.

56. Szejtli, J. (2005). Cyclodextrin complexed generic drugs are generally not bio-equivalent with the reference products: therefore the increase in number of marketed drug/cyclodextrin formulations is so slow. *J. Inclusion Phenom. Macrocyclic Chem.*, *52*, 1–11.

57. Cylodextrin containing, products. http://www3.hi.is/~thorstlo/cyclodextrin.pdf.

58. Carrier, RL., Miller, LA., Ahmed, I. (2007). The utility of cyclodextrins for enhancing oral bioavailability. *J. Controll. Release*, *123*(2), 78–99.

59. Arima, H., Yunomae, K., Hirayama, F., Uekama, K. (2001) Contribution of P-glycoprotein to the enhancing effects of dimethyl-β-cyclodextrin on oral bioavailability of tacrolimus. *J. Pharmacol. Exp. Ther.*, *297*, 544–555.

60. Frömming, K., Szejtli, J. (1994) *Cyclodextrins in Pharmacy*, Kluwer, Dordrecht, The Netherlands, pp. 105–126.

61. Rajewski, R.A., Stella, V.J. (1996). Pharmaceutical applications of cyclodextrins: 2. In vivo drug delivery. *J. Pharm. Sci.*, *85*, 1142–1169.

62. Connors, K.A. (1997). The stability of cyclodextrin complexes in solution. *Chem. Rev.*, *97*, 1325–1358.

63. Okimoto, K., Rajewski, R.A., Uekama, K., Jona, J.A., Stella, V.J. (1996). The interaction of charged and uncharged drugs with neutral (HP-β-CD) and anionically charged (SBE7-β-CD) beta-cyclodextrins. *Pharm. Res.*, *13*, 256–264.

64. Zia, V., Rajewski, R.A., Stella, V.J. (2001). Effect of cyclodextrin charge on complexation of neutral and charged substrates: comparison of (SBE)7m-β-CD to HP-β-CD. *Pharm. Res.*, *18*, 667–673.

65. Miyake, K., Arima, H., Hirayama, F., Yamamoto, M., Horikawa, T., Sumiyoshi, H., Noda, S., Uekama, K. (2000). Improvement of solubility and oral bioavailability of rutin by complexation with 2-hydroxypropyl-β-cyclodextrin. *Pharm. Dev. Technol.*, *5*, 399–407.

66. Hirayama, F., Uekama, K. (1999). Cyclodextrin-based controlled drug release system. *Adv. Drug Deliv. Rev.*, *36*, 125–141.

67. Szejtli, J. (1994). Medical applications of cyclodextrins. *Med. Res. Rev.*, *14*, 353–386.

68. Tokumura, T., Nanbu, M., Tsushima, Y., Tatsuishi, K., Kayano, M., Machida, Y., Nagai, T. (1986). Enhancement of bioavailability of cinnarizine from its β-cyclodextrin complex on oral administration with *dl*-phenylalanine as a competing agent. *J. Pharm. Sci.*, *75*, 391–394.

69. Tokumura, T., Tsushima, Y., Tatsuishi, K., Kayano, M., Machida, Y., Nagai, T. (1986). Enhancement of the bioavailability of cinnarizine from its β-cyclodextrin complex for oral administration with L-isoleucine as a competing agent. *Chem. Pharm. Bull.*, *34*, 1275–1279.

70. Emara, L.H., Badr, R.M., Elbary, A.A. (2002). Improving the dissolution and bioavailability of nifedipine using solid dispersions and solubilizers. *Drug Dev. Ind. Pharm.*, *28*(7), 795–807.

71. Soliman, O.A.E., Kimura, K., Hirayama, F., Uekama, K., El-Sabbagh, H.M., El-Gawad, A.E.H.A., Hashim, F.M. (1997). Amorphous spironolactone–hydroxypropylated cyclodextrin complexes with superior dissolution and oral bioavailability. *Int. J. Pharm.*, *149*(1), 73–83.

72. Savolainen, J., Järvinen, K., Taipale, H., Jarho, P., Loftsson, T., Järvinen, T. (1998). Coadministration of a water-soluble polymer increases the usefulness of cyclodextrins in solid oral dosage forms. *Pharm. Res.*, *15*, 1696–1701.

73. Loftsson, T., Fridriksdóttir, H. (1998). The effect of water soluble polymers on the aqueous solubility and complexing abilities of β-cyclodextrin. *Int. J. Pharm.*, *163*, 115–121.

74. Koester, L.S., Bertuola, J.B., Grochb, K.R., Xaviera, C.R., Moellerkeb, R., Mayorgaa, P., Costaa, T.D., Bassani, V.L. (2003). Bioavailability of carbamazepine:β-cyclodextrin complex in beagle dogs from hydroxypropylmethylcellulose matrix tablets. *Eur. J. Pharm. Sci.*, *55*, 85–91.

75. Rao, V.M., Haslam, J.L., Stella, V.J. (2001). Controlled and complete release of a model poorly water-soluble drug, prednisolone, from hydroxypropyl methylcellulose matrix tablets using (SBE)7m-cyclodextrin as a solubilizing agent. *J. Pharm. Sci.*, *90*, 807–816.

76. Loftsson, T., Brewster, M.E. (1996). Pharmaceutical applications of cyclodextrins: 1. Drug solubilization and stabilization. *J. Pharm. Sci.*, *85*, 1017–1025.

77. Rodenti, E. Szente, L., Szejtli, J. (2000). Drug/cyclodextrin/hydroxy acid multicomponent systems: Properties and pharmaceutical applications. *J. Pharm. Sci.*, *89*, 1–8.

78. Amidon, G.L., Lennernas, H., Shah, V.P., Crison, J.R. (1995). A theoretical basis for a biopharmaceutic drug classification: the correlation of in vitro drug product dissolution and in vivo bioavailability. *Pharm. Res.*, *12*, 413–420.

79. Yu, L.X., Amidon, G.L., Polli, J.E., Zhao, H., Mehta, M.U., Conner, D.P., Lesko, L.J., Lee, M.L., Hussain, A.S. (2002) Biopharmaceutics classification system: the scientific basis for biowaiver extensions. *Pharm. Res.*, *19*, 921–925, 2002.

80. U.S. FDA (2000). *Guidance for Industry: Waiver of In Vivo Bioavailability and Bioequivalence Studies for immediate Release Solid Oral Dosage Forms Based on a Biopharmaceutics Classification System*, CDER/FDA, Aug.

81. EMA (2010). *Guideline on the Investigation of Bioequivalence.* CPMP/EWP/QWP/EMA, Jan.

82. WHO (World Health Organization) (2005) Proposal to waive in vivo bioequivalence requirements for the WHO model list of essential medicines.

83. Yazdanian, M., Briggs, K., Jankovsky, C., Hawi, A. (2004). The "high solubility" definition of the current FDA Guidance on Biopharmaceutical Classification System may be too strict for acidic drugs. *Pharm. Res.*, *21*(2), 293–299.

84. Betlach, C.J., Gonzalez, M.A., McKiernan, B.C., Neff-Davis, C., Bodor, N. (1993). Oral pharmacokinetics of carbamazepine in dogs from commercial tablets and a cyclodextrin complex. *J. Pharm. Sci.*, *82*, 1058–1060.

85. El-Gindy, G.A., Mohammed, F.A., Salem, S.Y. (2002). Preparation, pharmacokinetic and pharmacodynamic evaluation of carbamazepine inclusion complexes with cyclodextrins. *STP Pharma Sci.*, *12*, 369–378.

86. Löbenberg, R., Amidon, G.L. (2004). Modern bioavailability, bioequivalence and biopharmaceutics classification system:

new scientific approaches to international regulatory standards. *Eur. J. Pharm. Biopharm.*, *50*, 3–12.

87. Uekama, K., Fujinaga, T., Hirayama, F., Otagiri, M., Yamasaki, M., Seo, H., Hashimoto, T., Tsuruoka, M. (1983). Improvement of the oral bioavailability of digitalis glycosides by cyclodextrin complexation. *J. Pharm. Sci.*, *72*, 1338–1341.

88. Aggarwal, S., Singh, P.N., Mishra, B. (2002). Studies on solubility and hypoglycemic activity of gliclazide β-cyclodextrin-hydroxypropylmethylcellulose complexes. *Pharmazie*, *57*, 191–193.

89. Sridevi, S., Chauhan, A.S., Chalasani, K.B., Jain, A.K., Diwan, P.V. (2003). Enhancement of dissolution and oral bioavailability of gliquidone with hydroxypropyl-β-cyclodextrin. *Pharmazie*, *58*, 807–810.

90. Tenjarla, S., Puranajoti, P., Kasina, R., Mandal, M. (1998). Preparation, characterization, and evaluation of miconazole-cyclodextrin complexes for improved oral and topical delivery. *J. Pharm. Sci.*, *87*(4), 425–429.

91. Tanino, T., Ogiso, T., Iwaki, M. (1999). Effect of sugar-modified β-cyclodextrins on dissolution and absorption characteristics of phenytoin. *Biol. Pharm. Bull.*, *22*, 298–304.

92. Veiga, F., Fernandes, C., Teixeira, F. (2000). Oral bioavailability and hypoglycaemic activity of tolbutamide/cyclodextrin inclusion complexes. *Int. J. Pharm.*, *202*, 165–171.

93. Peeters, J., Neeskens, P., Tollenaere, J.P., Van Remoortere, P., Brewster, M.E. (2002). Characterization of the interaction of 2-hydroxypropyl-β-cyclodextrin with itraconazole at pH 2, 4, and 7. *J. Pharm. Sci.*, *91*, 1414–1422.

94. Ghorab, M.M., Abdel-Salam, H.M., El-Sayad, M.A., Mekhel, M.M. (2004). Tablet formulation containing meloxicam and β-cyclodextrin: mechanical characterization and bioavailability evaluation. *AAPS PharmSciTech.*, *5*(4), E59.

95. Muraoka, A., Tokumura, T., Machida, Y. (2004). Evaluation of the bioavailability of flurbiprofen and its β-cyclodextrin inclusion complex in four different doses upon oral administration to rats. *Eur. J. Pharm. Biopharm.*, *58*, 667–671.

96. Tokumura, T., Muraoka, A., Machida, Y. (2009). Improvement of oral bioavailability of flurbiprofen from flurbiprofen/β-cyclodextrin inclusion complex by action of cinnarizine. *Eur. J. Pharm. Biopharm.*, *73*, 202–204.

97. Jambhekar, S., Casella, R., Maher, T. (2004). The physicochemical characteristics and bioavailability of indomethacin from β-cyclodextrin, hydroxyethyl-β-cyclodextrin, and hydroxypropyl-β-cyclodextrin complexes. *Int. J. Pharm.*, *270*, 149–166.

98. Dutet, J., Lahiani-Skiba, M., Didier, L., Jezequel, S., Bounoure, F., Barbot, C., Arnaud, P., Skiba, M. (2007). Nimesulide/cyclodextrin/PEG 6000 ternary complexes: physico-chemical characterization, dissolution studies and bioavailability in rats. *J. Inclusion Phenom. Macrocyclic Chem.*, *57*, 203–209.

99. Aly, A.A., Qato, M.K., Ahmad, M.O. (2003). Enhancement of the dissolution rate and bioavailability of glipizide through cyclodextrin inclusion complex. *Pharm. Technol.*, *2*, 54–66.

100. Hara, T., Hirayama, F., Arima, H., Yamaguchi, Y., Uekama, K. (2006). Improvement of solubility and oral bioavailability of 2-(N-cyanoimino)-5-[(E)-4-styrylbenzylidene]-4-oxothiazolidine (FPFS-410) with antidiabetic and lipid-lowering activities in dogs by 2-hydroxypropyl-β-cyclodextrin. *Chem. Pharm. Bull. Tokyo*, *54*, 344–349.

101. He, Z.G., Li, Y.S., Zhang, T.H., Tang, X., Zhao, C., Zhang, R. H. (2004). Effects of 2-hydroxypropyl-β-cyclodextrin on pharmacokinetics of digoxin in rabbits and humans. *Pharmazie*, *59*, 200–202.

102. Choi, H.G., Kim, D.D., Jun, H.W., Yoo, B.K., Yong, C.S. (2003). Improvement of dissolution and bioavailability of nitrendipine by inclusion in hydroxypropyl-β-cyclodextrin. *Drug Dev. Ind. Pharm.*, *29*, 1085–1094.

103. Arima, H., Yunomae, K., Miyake, K., Irie, T., Hirayama, F., Uekama, K. (2001). Comparative studies of the enhancing effects of cyclodextrins on the solubility and oral bioavailability of tacrolimus in rats. *J. Pharm. Sci.*, *90*, 690–701.

104. Piette, M., Evrard, B., Frankenne, F., Chiap, P., Bertholet, P., Castagne, D., Foidart, J.M., Delattre, L., Piel, G. (2006). Pharmacokinetic study of a new synthetic MMP inhibitor (Ro 28-2653) after IV and oral administration of cyclodextrin solutions. *Eur. J. Pharm. Sci.*, *28*, 189–95.

105. Evrard, B., Chiap, P., DeTullio, P., Ghalmi, F., Piel, G., Van Hees, T., Crommen, J., Losson, B., Delattre, L. (2002). Oral bioavailability in sheep of albendazole from a suspension and from a solution containing hydroxypropyl-β-cyclodextrin. *J. Controll. Release*, *85*, 45–50.

106. García, J.J., Bolás, F., Torrado, J.J. (2003). Bioavailability and efficacy characteristics of two different oral liquid formulations of albendazole. *Int. J. Pharm.*, *250*, 351–358.

107. Savolainen, J., Järvinen, K., Matilainen, L., Järvinen, T. (1998). Improved dissolution and bioavailability of phenytoin by sulfobutylether-β-cyclodextrin ((SBE)$_{7m}$-β-CD) and hydroxypropyl-β-cyclodextrin (HP-β-CD) complexation. *Int. J. Pharm.*, *165*, 69–78.

108. El-Zein, H., Riad, L., El-Bary, A.A. (1998). Enhancement of carbamazepine dissolution: in vitro and in vivo evaluation. *Int. J. Pharm.*, *168*, 209–220.

109. Koester, L.S., Bertuol, J.B., Groch, K.R., Xavier, C.R., Moellerke, R., Mayorga, P., Costa, T.D., Bassani, V.L. (2004). Bioavailability of carbamazepine: β-cyclodextrin complex in beagle dogs from hydroxypropylmethylcellulose matrix tablets. *Eur. J. Pharm. Sci.*, *22*, 201–207.

110. Dhanaraju, M.D., Kumaran, K.S., Baskaran, T., Moorthy, M. S. (1998). Enhancement of bioavailability of griseofulvin by its complexation with β-cyclodextrin. *Drug Dev. Ind. Phar.*, *24*, 583–587.

111. Prabagar, B., Yoo, B.K., Woo, J.S., Kim, J.A., Rhee, J.D., Piao, M.G., Choi, H.G., Yong, C.S. (2007). Enhanced bioavailability of poorly water-soluble clotrimazole by inclusion with β-cyclodextrin. *Arch. Pharm. Res.*, *30*, 249–254.

112. Barone, J.A., Moskovitz, B.L., Guarnieri, J., Hassell, A.E., Colaizzi, J.L., Bierman, R.H., Jessen, L. (1998). Enhanced bioavailability of itraconazole in hydroxypropyl-β-cyclodex-

trin solution versus capsules in healthy volunteers. *Antimicrob. Agents Chemother.*, *42*, 1862–1865.

113. Hassan, H.A., Al-Marzouqi, A.H., Jobe, B., Hamza, A.A., Ramadan, G.A. (2007) Enhancement of dissolution amount and in vivo bioavailability of itraconazole by complexation with β-cyclodextrin using supercritical carbon dioxide. *J. Pharm. Biomed. Anal.*, *45*, 243–250.

114. Mora, P.C., Cirri, M., Guenther, S., Allolio, B., Carli, F., Mura, P. (2004). Enhancement of dehydroepiandrosterone solubility and bioavailability by ternary complexation with α-cyclodextrin and glycine. *J. Pharm. Sci.*, *92*, 2177–2184.

115. Lee, S.H., Kim, Y.H., Yu, H.J., Cho, N.S., Kim, T.H., Kim, D. C., Chung, C.B., Hwang, Y.I., Kim, K.H. (2007). Enhanced bioavailability of soy isoflavones by complexation with β-cyclodextrin in rats. *Biosci. Biotechnol. Biochem.*, *71*, 2927–2933.

116. Wong, J.W., Yuen, K.H. (2001). Improved oral bioavailability of artemisinin through inclusion complexation with β- and γ-cyclodextrins. *Int. J. Pharm.*, *227*, 177–185.

117. Wempe, M.F., Wacher, V.J., Ruble, K.M., Ramsey, M.G., Edgar, K.J., Buchanan, N.L., Buchanan, C.M. (2008). Pharmacokinetics of raloxifene in male Wistar–Hannover rats: influence of complexation with hydroxybutenyl-β-cyclodextrin. *Int. J. Pharm.*, *346*, 25–37.

118. Liu, J., Qiu, L., Gao, J., Jin, Y. (2006). Preparation, characterization and in vivo evaluation of formulation of baicalein with hydroxypropyl-β-cyclodextrin. *Int. J. Pharm.*, *312*, 137–143.

119. Prosek, M., Butinar, J., Lukanc, B., Fir, M.M., Milivojevic, L., Krizman, M., Smidovnik, A. (2008). Bioavailability of water-soluble CoQ10 in beagle dogs. *J. Pharm. Biomed. Anal.*, *47*, 918–922.

4

CYCLODEXTRINS AS SMART EXCIPIENTS IN POLYMERIC DRUG DELIVERY SYSTEMS

Agnese Miro, Francesca Ungaro, and Fabiana Quaglia

Department of Pharmaceutical and Toxicological Chemistry, University of Naples, Naples, Italy

1. INTRODUCTION

An ideal dosage form should be able to release the drug at a rate dictated by the needs of the body throughout the therapy regimen and deliver the drug at its target site. A time-controlled drug delivery system (DDS) is especially designed to exert control of the drug dose administered over time with the ultimate goal to have release from the dosage form be the rate-limiting step for drug availability: in other words, the kinetics of drug release rather than absorption control drug bioavailability, although some few exceptions to this general consideration exist. As depicted in Fig. 1, control over amounts of drug released can be different, depending on specific therapeutic requirements. In fact, drugs could need to be delivered at constant rates when a relationship exits between drug steady-state plasma levels and the resulting pharmaceutical response (extended or prolonged or sustained release), or at a variable rate for drugs needing peak or valley plasma levels or acting on rhythmic functions of the body (pulsatile systems). Nevertheless, a pharmacological response could be needed after the dosage form is administered (delayed systems).

To control the rate at which a drug is delivered from a dosage form, a polymeric material that hinders drug dissolution in body fluids is employed. In a DDS the drug and the polymer are arranged in a reservoir or matrix system, although other architectures can be especially suited for the purpose of attaining specific delivery requirements. In a reservoir device, a polymeric membrane wraps a drug reservoir (solid or dispersed in a suitable liquid), whereas in a matrix system a drug is distributed uniformly in the polymer (Fig. 2). In general, drug release from a DDS evolves in time according to different kinetics, which can be mainly zero order (constant with time) or first order (which means that the amount delivered depends on the time elapsed from initial activation).

Polymer selection (hydrophilic, lipophilic, degradable, not degradable or eliminable, natural, synthetic) is generally related to the administration route, need for a specific delivery site, rate, and duration (hours, days, weeks) (Fig. 2). Thus, to achieve a delayed delivery, for example, reservoir architectures and soluble coatings can be selected. On the other hand, if sustained delivery is required, an erodible matrix offering a near-zero order release will work perfectly. At that point, a polymer platform will be selected depending on how long the release should be. For the oral route, drug absorption is related to transit time of the dosage form, which sets an approximately 10-h limit for the delivery of drugs absorbed from the small intestine region. A polymer eroding in this time frame will be eligible as an oral platform, whereas in the case of parenteral administration, materials with slow erosion rates will be preferred to ensure a persistent therapeutical response. Although DDS behavior in a biological environment is dictated primarily by polymer properties, physico-chemical features of the incorporated drug, such as solubility, lipophilicity, and diffusion coefficient, would, at least in part, play a role. Central to the development of a successful DDS, therefore, is an understanding of the variables that affect transport processes and how these parameters can be controlled selectively. This means that it is important to know the mechanisms underlining release behavior, and in turn to single out the major physico-chemical

Cyclodextrins in Pharmaceutics, Cosmetics, and Biomedicine: Current and Future Industrial Applications, First Edition. Edited by Erem Bilensoy.
© 2011 John Wiley & Sons, Inc. Published 2011 by John Wiley & Sons, Inc.

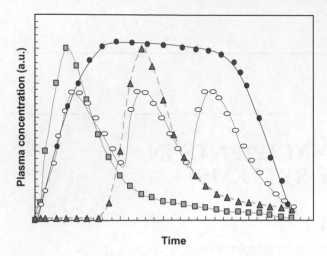

Figure 1. Plasma concentration obtained after administration of a conventional drug dosage form (squares), an extended DDS (filled circles), a pulsatile DDS (open circles), and a delayed DDS (triangles).

processes involved in control of the release rate as affected by DDS architecture and drug–polymer composition. Depending on polymer characteristics, one or more than one mechanism can concur in regulating drug delivery rate. Diffusion, degradation, osmosis, and dissolution can be considered the main mechanisms driving drug release from a polymeric DDS. Special emphasis should be given to diffusion, a process encountered very often in elucidating a delivery process.

It is worth to emphasize that once a DDS is designed, which means that the rate-controlling polymer is selected and the drug level fixed, only device architecture (matrix or reservoir), microstructure, and geometry can be manipulated to attain specific delivery requirements, and in many cases this is not enough. Since cyclodextrins (CDs) may form inclusion complexes with different chemical entities and in so doing alter their physico-chemical properties, it is expected that CDs alter the release properties of a given drug–polymer system. This approach is of utmost interest since CDs can act as an additional tool to modulate (i.e.,

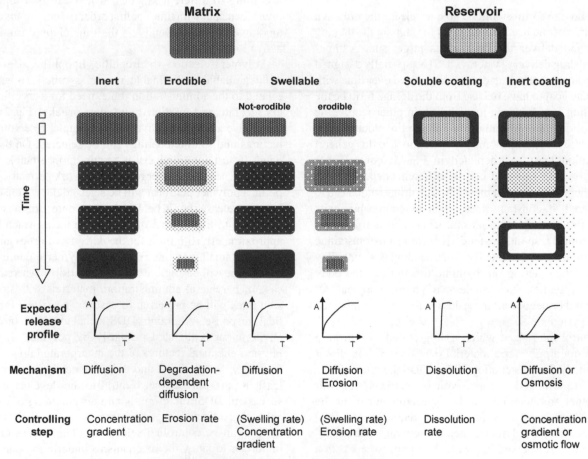

Figure 2. Possible architecture of a polymeric DDS and its expected time evolution and release profile when in contact with a biological fluid. Main mechanisms underlining release as well as controlling steps in drug release are reported. This schema does not report all the possible designs and mechanisms for DDSs.

speed up or slow down) drug release rate without changing polymeric platform composition. For example, if a blend of hydrophilic polymers displays optimal mucoadhesive properties for a specific application but is not capable of sustaining the delivery rate as desired, one could use a specific CD to modify the delivery rate, preserving the overall beneficial properties of a DDS.

Nevertheless, the effect of CD addition in a polymeric DDS depends strongly on the way a CD is introduced during DDS manufacturing and its effect on both drug and polymer properties. For example, it is well documented that a CD can form complexes with several drugs and increase their dissolution rate as well as stabilize a drug toward chemical or physical degradation, whereas much less is known on the capability of CDs to complex polymers. In this chapter we provide a critical review of the state of the art on DDSs comprising CDs, with the special intention of modulating drug delivery rate and attempting to suggest their rational use.

2. BASIC CONCEPTS OF DRUG DIFFUSION IN POLYMERIC SYSTEMS

Diffusion is probably the main mechanism involved in drug delivery for several systems based on polymers with different physico-chemical properties and thus behaving differently once in contact with an aqueous fluid. For this reason, it is treated separately here, and some basic concepts useful for understanding the release behavior of different DDSs and how CDs can eventually affect them, are noted.

Diffusion is a process by which molecules are transported from inside a membrane to outside as a result of random movements in the absence of mixing. A diffusive flux (J) develops under a concentration gradient ($\partial C / \partial x$) as the driving force according to Fick's first law:

$$J = \mathcal{D} \frac{\partial C}{\partial X} \qquad (1)$$

The proportionality constant between the diffusional flux and the diffusional potential is termed the diffusion coefficient ($\mathcal{D}$) or diffusivity.

Solute diffusion in polymers is strictly dependent on the polymer microstructure, which can be classified according to Swan and Peppas [1] as macroporous (pore size in the range 0.1 to 1.0 μm), microporous (pore size in the range 10 to 50 nm), and nonporous. The diffusion coefficient for porous systems is solely dependent on the pore structure, and for nonporous systems, on the polymer network.

Nonporous polymers are constituted by a homogeneous polymer phase or network. The space between macromolecular chains (mesh) is the only area available for the diffusion of solutes (Fig. 3). In this case, solute transport is

presumed to occur by a process involving dissolution of the solute within the polymer, followed by its diffusion in the polymer. These mechanisms include a network of hydrophobic polymers with macromolecular meshes between approximately 2 and 10 nm and most types of hydrogels (i.e., water-swollen polymer networks). Any interaction between the diffusant and the polymer macromolecule is expected to affect diffusion coefficients since macromolecular meshes create a screening effect on solute diffusion through the polymer. For uncross-linked polymers, this screening is provided by the meshes formed by entangled chains, whereas in semicrystalline polymers the crystallites act as physical cross-links. Thus, the degree of crystallinity, glassy–rubbery state, degree of swelling, and mesh size must all be considered.

Taking into account that the drug is in equilibrium with the respective surface layer of the membrane at both sides of the membrane (Fig. 3A), a partition coefficient (K) can be introduced in the equation as well as the membrane thickness (l). At steady state, the number of molecules leaving the membrane is constant with time, and equation (1) can be integrated to give

$$J = \frac{\mathcal{D} K \Delta C}{l} \qquad (2)$$

where ΔC is the difference in concentration between the interior and exterior of the system. K can be expressed as $Cm_{(i)}/C_{(i)}$ and $Cm_{(e)}/C_{(e)}$ at the upstream and downstream surfaces and is assumed to be constant with time. $\mathcal{D}$ is a function of the permeant mobility in the polymer and depends on the size, nature, and membrane properties; the term $K \Delta C$ indicates the number of permeant molecules diffusing in the membrane; and l represents the distance that each molecule must walk to leave the membrane. In biological environments, passive drug absorption is also accounted for by Fick's first law and the biological membrane considered as a nonporous membrane.

For microporous membranes, Fick's law cannot be directly applied since diffusion occurs in the water-filled pores (Fig. 3B) [2]. In this case, the diffusion coefficient estimated in a liquid lowers down, depending on the porosity (void volume fraction, ε), pore tortuosity (τ), and partition coefficient (the ratio of permeant concentration inside the pore to that in the polymer, K_p). Thus, equation (2) can be rewritten

$$J = \frac{\mathcal{D}_p \varepsilon K_p}{\tau l} \Delta C \qquad (3)$$

where $\mathcal{D}_p$ is the diffusion coefficient of the solute through the solvent-filled pores. For microporous polymers, a term accounting for additional steric hindrance and frictional resistance of the pores (k_r), which occurs when solute and pore

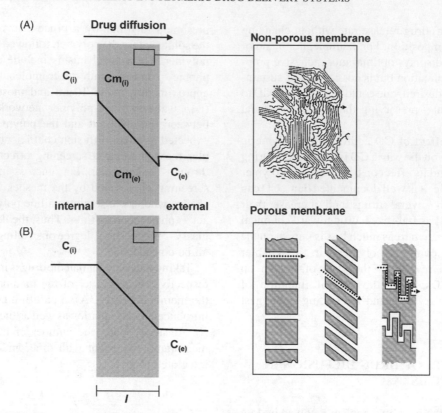

Figure 3. Drug diffusion through a nonporous membrane (A) or microporous membranes (B) with increasing tortuosity. $Cm_{(i)}/C_{(i)}$ and $Cm_{(e)}/C_{(e)}$ represent a partition coefficient [(K in equation (2)] at the upstream and downstream surfaces, which are assumed to be constant with time.

sizes are of comparable magnitude, also needs to be considered.

It is expected that when diffusion is the main mechanism controlling delivery rate, each factor altering diffusion coefficient can potentially contribute to affect the drug release kinetics and overall DDS behavior in a biological environment.

3. CD ADDITION IN POLYMERIC DRUG DELIVERY SYSTEMS: THE STATE OF THE ART

3.1. Gels

The word *gel* refers to a liquid that sets to a solidlike material that does not flow but is elastic and retains some liquid characteristics [3]. Gels can thereby be defined as semisolid systems consisting of either suspensions made up of small inorganic particles (two-phase gels, such as in the case of bentonite or silica) or large organic molecules interpenetrated by a liquid (single-phase gels). Single-phase gels consist of organic hydrophilic macromolecules distributed uniformly throughout a liquid so that no boundary exists between the liquid and the dispersed macromolecules. Single-phase gels comprise macromolecules capable of extensive solvatation (protein, polysaccharide, synthetic

hydrophilic materials) to a liquid, which is generally water rich (water and hydroalcoholic solutions are widely employed). Depending on polymer chemistry, sol–gel transition (*sol* refers to a fluid phase containing dispersed or dissolved macromolecules) and three-dimensional network formation can occur upon an increase in polymer concentration, change of pH or temperature, ion addition, and so on. Gels behave as elastic solids at low stresses even though they consist primarily of a liquid.

Physically bonded gels are reversible systems formed primarily by natural polymers (proteins and polysaccharide), semisynthetic derivatives of cellulose, and synthetic hydrophilic materials (Table 1) [3]. A gel network is formed by cooperative association of several polymer segments in highly ordered regions called *junction zones*. A number of junction zones dispersed throughout a network confer mechanical strength to the gel. Thus, the microstructure of such a system is very complex and depends on the type of junction zone formed. This is related to the chemical nature of the polymer repeated unit, which in turn affects some physical properties of the gel, such as rigidity. The junction zone can be (1) microcrystalline, due to chain bundles present in a polymer chain (carboxymethylcellulose) or in stereoregular polymers (polyvinyl alcohol); (2) micelle-like, due to supramolecular organization of hydrophobic polymer

Table 1. Polymers Used to Produce Gel and Their Gelling Concentration

Hydrophilic Polymers	Gel-Forming Concentrations (wt%)	Required Additive
Cellulose derivatives		
Hypromellose (hydroxypropyl methylcellulose)	2–10	
Hydroxypropylcellulose	8–10	
Methylcellulose	2–4	
Sodium carboxymethyl cellulose	4–6	Sodium ions
	10–25	Sodium ions
Noncellulosic		
Gums/polysaccharides		
K-carrageenan	1–2	Potassium ions
Guar gum	2.5–10	
	0.25	Borate iones
Gellum gum (low acetyl)	0.5–1	Calcium ions
Pectin (low methyloxy)	0.8–2	Calcium ions
Sodium alginate	0.5–1	Calcium ions
	5–10	Sodium ions
Others		
Carbomer	0.5–2	
Poloxamer	15–50	
Poly(vinyl alcohol)	10–20	
Inorganic substances		
Aluminum hydroxide	3–5	
Bentonite	5	

Source: Adapted from [98].

segments (methylcellulose or poloxamers); or (3) due to ion bridging (egg-crate structure of alginate). Finally, some physically bonded gel networks are held together by simple entanglements occurring between polymer chains. As in the case of hyaluronic acid or carbomers, above a critical concentration, long chains are forced through the domain of other chains because of their large molecular volume, giving a highly intertwined network. In this last case, molecular association between polymer segments takes place through the cooperation of several intermolecular forces, such as hydrogen bonding, van der Waals forces, and electrostatic attractive/repulsive forces. Gelation of a polymer solution occurs at a specific concentration or under specific conditions. For example, sodium alginate forms a gel by either reducing pH or by electrostatic interaction with divalent cations such as calcium, whereas poly(acrylic acid) undergoes gelation at pH 7 following neutralization of the pendant carboxylic groups with alkali such as thryethanolamine or sodium hydroxide. Furthermore, some gels are thermoreversible, as is true for xunthan gum at concentrations above 0.5% or for hydroxypropyl methylcellulose (HPMC) and methylcellulose at temperatures above 50°C.

Every perturbation of junction zones is expected to affect the macroscopic behavior of the gel. For this reason, drugs can interfere with molecular bonds at junction zones and there are several cases of drug-to-polymer interactions that affect gelation properties [3]. On the other hand, CD can promote gelation of polymer solutions, giving physically bonded gel networks. In this regard, it has been reported that poly(ethylene oxide) (PEO)/α-CD solutions undergo sol–gel transition due to interaction of PEO chains with the hydrophobic CD cavity, forming necklace-like supramolecular structures [4–7]. On the other hand, poloxamers, which are block copolymers of PEO and poly(propylene oxide) (PPO) with a PEO–PPO–PEO structure, are capable of interacting with different cyclodextrins. More β-CD would selectively thread the middle PPO block to form a polyrotaxane [8,9], whereas smaller α-CD selectively includes the flanking PEO block [10–13]. An extensive review of the formation of polymer–CD supramolecular aggregates is provided by Li and Li [14].

The clear expectation is that CD affect the macroscopic–microscopic properties of a polymeric gel network due to the perturbation of junction zones and in so doing alter some of the features of the gel, such as release behavior. In this sense, changes in solution turbidity, which are diagnostic of changes in the hydration state of a polymer in solution, and in turn of polymer–CD interactions, can be evaluated by cloud point [15]. Actually, cloud point variations can be related to a competition of additives, including CDs, with the polymer for water binding and to changes in the conformation of macromolecular chains [16]. In this respect, the presence of β-CD or HP-β-CD did not significantly modify the cloud point and cinematic viscosity of HPMC K4M dispersions, suggesting that the hydrophilicity and water "structure" around the cellulose ether did not induce significant changes in the system analyzed [17]. On the other hand, the apparent viscosity of similar HPMC gels was decreased by the presence of β-CD and DM-β-CD as the shear stress increased [18]. Analogously, Boulmedarat et al. [19] observed that the addition of DM-β-CD in Carbopol 974P NF gels determined a pronounced decrease in the zero-shear rate viscosity as a function of DM-β-CD concentration. This finding was attributed to the hydrophobic character of DM-β-CD, which interacted with polymer chains, resulting in a reduction in the polymer chain unfolding. This effect modified polymer affinity for the hydration medium, hence decreasing its swelling. In the same study it was observed also that DM-β-CD had no effect on the same gels made in HEPES/NaCl solution, suggesting that cationic electrolytes, which are well known to reduce gel thickening efficiency, did not induce additional effects on viscosity decrease. Finally, HP-β-CD decreased the viscosity, and also the bioadhesive force, of a thermosensitive poloxamer gel containing a rhEGF/HP-β-CD complex as a function of the protein/CD ratio concentration in the gel only at temperatures above critical gelation temperature [20]. The authors hypothesized that the hydrogen bonding in a cross-linked reticular gel is weaker in the presence of CD, due to perturbation of micelle interactions.

Thus, when examining the effect of CD in polymeric gels, at least two different aspects should be taken into account: the effect on the physico-chemical properties of the drug (especially the apparent solubility), and perturbation of the polymer network. It is worth noting that release results can vary widely depending on the experimental setup employed to perform experiments, which, of course, should reproduce conditions close to those occurring in vivo. In this sense it has been demonstrated that when drug release into the receptor medium is consecutive to either drug diffusion in the gel or gel erosion, drug diffusion is predominant in membrane models, whereas polymer dissolution is the major mechanism observed in membraneless models [21]. As a consequence, the role of a CD as a release modulator can change greatly, depending on experimental conditions and can be suitably adapted if a specific effect is highlighted.

The first example of using CDs in polymeric gels to modulate delivery rate dates back to 1994, when Samy and Safwat introduced β-CD in various cellulosic and aerosol gels [22]. The release of different nonsteroidal anti-inflammatory drugs speeded up or slowed down, depending on the type of gel and physical state of the drug in the gel (solubilized or dispersed). Similar results were obtained when studying the release of diclofenac sodium and sulfamethizole from HPMC gels [17]. In a recent study [23], it was observed that in gels made of Carbopol 974 containing progesterone and β-CD, the effect of CD addition depended on the drug solubility inside the gel. Since progesterone is not completely soluble in the gel, the effect of CD was mainly on drug dissolution inside the gel and increased drug apparent solubility, which in turn increases the number of diffusible species. Analogously, Bilensoy et al. [24] found that the incorporation of a

clotrimazole/β-CD complex in a mucoadhesive thermosensitive Pluronic F127/HPMC gel allowed to avoid drug precipitation and obtain a clear gel. Furthermore, besides the fact that the addition of β-CD allowed the formation of a platform gelling at 33°C, which is very close to body temperature, a remarkable decrease in the release rate, independent of the composition of the platform was observed. A decreased delivery rate was observed again on poloxamer gels containing a soluble protein, rhEGF, and HP-β-CD as a preformed complex [20]. The retarding effect of HP-β-CD was linearly related to its amount in the complex. Nevertheless, the introduction of a physical mixture did not have any significant effect on the release profile. This effect was explained assuming that the effective diffusivity of the drug in the gel was decreased upon CD addition.

From these data it is clear that CDs do not act in an univocal fashion in polymeric gel since an accelerated or reduced release rate is experienced depending on polymeric platform composition, drug physico-chemical properties, especially diffusivity and solubility, as well as drug–polymer interactions.

3.2. Swellable and Swelling-Controlled Systems

According to the classification of Peppas [25], the category of solvent-activated DDSs comprises two subcategories: swellable and swelling-controlled devices, both showing a matrix architecture. These systems are prepared by incorporation of a drug into a hydrophilic, glassy polymer forming a nonporous solid of variable shape (film, tablets, spheres) and can be swollen by contact with an aqueous fluid. The swelling may or may not be the controlling mechanism for diffusional release, depending on the magnitude of macromolecular relaxation of the polymer. A schematic presentation of processes occurring upon contact of a hydrophilic matrix with an aqueous fluid is represented in Fig. 4. When brought

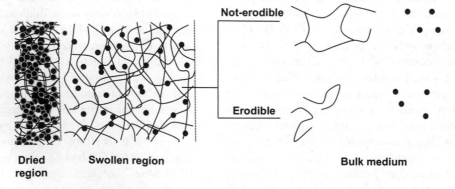

Figure 4. Processes occurring upon contact of a hydrophilic matrix with an aqueous fluid. In the case of a nonerodible matrix (hydrogel), the system swells up to equilibrium, the drug is solubilized in the uptaken aqueous fluid and diffuses out in the bulk medium. At the end of release, an empty hydrogel remains. In the case of erodible matrices, the matrix swells, the drug is solubilized in the swollen layer, and the drug is released due to diffusion and polymer erosion. At the end of release, drug and polymer are solubilized in the medium.

in contact with an aqueous medium, a distinct water-penetration front (interface) is formed which separates the glassy from the rubbery (gel-like) state of materials. Under these conditions, macromolecular relaxation takes place and controls the rate of diffusion of the dissolved drug. Swelling proceeds up to equilibrium, depending on the water activity and nature of the polymer. If the polymer is highly hydrophilic and can solubilize in water, as in the case of hydrophilic erodible matrices, the equilibrium state is represented by a polymer aqueous solution. If the polymer is cross-linked so that chain entanglement can be maintained in the hydrated state, as in the case of hydrogels, the equilibrium state is a water-swollen polymer. The drug is immobile in the glassy matrix but begins to solubilize and diffuse out as the polymer swells in water.

Drug release thus depends on the rate of two simultaneously occurring processes: water migration in the matrix and drug diffusion in the bulk medium through the gel layer. Gel layer thickness evolves with time and generally offers a significant resistance to transport, and the overall drug release rate depends on both the rate of water uptake and drug diffusion in the gel layer. To affect the right parameter and modulate the delivery rate, it is important to know which mechanism plays a prominent role. For example, if release is controlled by water penetration, it is unworthy to affect drug diffusion in the gel.

3.2.1. Hydrogels

Hydrogels can be defined as irreversible gels constituting a covalently cross-linked three-dimensional network made up of hydrophilic polymers [25]. Hydrogels are generally prepared by covalent cross-linking of linear or branched polymer chains in the presence of a cross-linking agent. Cross-links can be biodegradable or not, thus establishing biodegradable or nonbiodegradable networks. Hydrogels will be characterized by a number of parameters—the equilibrium swelling ratio, the molecular weight between cross-links (mesh size), and the cross-linking density—all of them having a strong impact on release behavior.

Drug release from a nonbiodegradable hydrogel occurs according to the mechanism already described for hydrophilic matrices, which involves water uptake and swelling of the glassy polymer, drug solubilization in the gel layer, and its diffusion in the external medium by walking through the meshes of the swollen gel. The critical parameters affecting the release process are in this case the rate of water uptake and drug counterdiffusion. In general, the process with the highest "characteristic time" is also the rate-limiting step for drug release.

In the first contribution of our research group toward an understanding of the applicability of CD as a modulator of drug release, we investigated the influence of β-CD on the release of nicardipine hydrochloride (NIC) from cross-linked PEG monolithic devices [25]. β-CD formed an inclusion

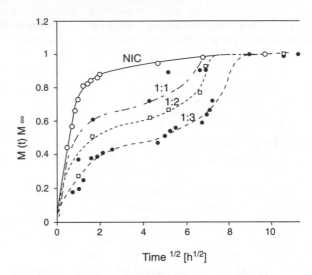

Figure 5. Nicardipine release from hydrogels made of cross-linked PEO 2000 Da and containing pure NIC or NIC/CD at different molar ratios (1:1, 1:2, 1:3). (From [26]. Copyright © Elsevier.)

complex with NIC, showing a low stability constant. Diffusivity measurements performed on swollen membranes evidenced a much higher value of NIC diffusivity than that of β-CD, probably due to the differences in molecular weight of the two molecules. As shown in Fig. 5, release kinetics of NIC from slabs loaded with pure NIC (below drug saturation limit) or NIC/β-CD at different molar ratios (1:1, 1:2, 1:3) were significantly dependent on the amount of loaded β-CD: namely, the higher the amount of loaded β-CD the longer the time to complete release. All the release curves referring to NIC/β-CD loaded membranes first showed a faster release regimen, followed by a slower stage. Drug release from the slabs loaded with the drug alone were determined substantially by the swelling kinetics, and the release kinetics strictly followed the swelling profile. In the case of matrices loaded with both NIC and β-CD, it was hypothesized that drug complexation reduced the concentration of "free" drug molecules, resulting in a decrease in the average mobility of the drug in the swollen matrix. Thus, the resulting overall drug mobility depends on both the diffusivity of free species and the diffusivity of the complex, which is lower than that of free guest molecules.

A mathematical treatment made it possible to introduce an effective drug diffusivity of NIC in the hydrogel when CD is present, which is related in a simple way to the stability constant of the complex (K) and the total CD concentration in the gel (CD^*):

$$\mathcal{D}_{\text{eff}} = \frac{\mathcal{D}_{\text{dr}}}{1 + K \cdot CD^*} \qquad (4)$$

The effective diffusivity was obviously equivalent to $\mathcal{D}_{\text{dr}}$ when CD^* or K are equal to zero. $\mathcal{D}_{\text{eff}}$ decreased as K and/or CD^* increased. Thus, the characteristic time for diffusion was lower the higher the CD concentration. This relationship held true only when:

1. $\mathcal{D}_{CD} \ll \mathcal{D}_{dr}$ ($\mathcal{D}_{com} \ll \mathcal{D}_{dr}$), which means that CD^* is constant.
2. $K \cdot \mathcal{D}_{dr} \ll 1$.
3. $K \cdot CD^* \leq 1$.
4. There is instantaneous equilibrium between the complexed and free forms of drug–CD.

On this basis, it was hypothesized that if CD is present, due to the effect of complexant on drug effective diffusivity that we have discussed, NIC release was not complete when swelling was over, and two release stages could be evidenced: an initial fast release and a slower second stage. It was speculated that the fractional amount of drug released during the first stage was related to the fraction of free NIC. In fact, the relative amount of NIC released during the swelling increased as the NIC/β-CD ratio decreased. Nevertheless, the following release phase was dominated by an essentially diffusive mechanism as affected by the amount of CD present; namely, the higher the NIC/β-CD ratio, the lower the release rate during this second stage.

However, it is worth noting that this mechanistic view of drug release from hydrogel should also take into account the interaction of both drug and CD with the polymer network, which can strongly affect the effective drug diffusivity in the gel and in turn suppress the role of CD as release modulator [27].

3.2.2. Hydrophilic Erodible Matrices

A matrix tablet is the simplest and the most cost-effective way to fabricate a sustained-release dosage form. In its simplest form, a typical extended-release matrix formulation consists of a drug, release-retardant polymer (hydrophilic or hydrophobic or both), one or more excipients (as fillers or binders), flow aid (glidants), and a lubricant. Other functional ingredients, such as buffering agents, stabilizers, solubilizers, and surfactants may also be included to improve or optimize the release and/ or stability performance of the formulation. Components are mixed together, eventually granulated and then tabletted. Table 2 shows a list of materials commonly used for fabrication of matrices [28,29]. Various water-soluble or water-swellable polymers with high molecular weight can be used in hydrophilic matrices, such as HPMC, hydroxypropylcellulose, and PEO. HPMC is identified as the most popular polymer in matrix applications because of a number of key features and advantages [30]. Hydrophobic materials can be also used either alone (hydrophobic matrix systems) or in combination with hydrophilic matrix systems (hydrophilic–hydrophobic matrix systems).

The mechanism of drug release from hydrophilic matrix tablets is based on diffusion of the drug through, and erosion

Table 2. Polymers Commonly Studied for the Fabrication of Extended-Release Hydrophilic Matrices

Hydrophilic Polymers	Water-Insoluble and Hydrophobic Polymers
Cellulosic	Ethylcellulose
Methylcellulose	Hypromellose acetate succinate
Hypromellose (hydroxypropylmethylcellulose)	Cellulose acetate
Hydroxypropylcellulose	Cellulose acetate propionate
Hydroxyethylcellulose	Methycrylic acid copolymers
Sodium carboxymethylcellulose	Poly(vinyl acetate)
Noncellulosic	
Gums/polysaccharides	
Sodium alginate	
Xanthan gum	
Carrageenan	
Chitosan	
Guar gum	
Pectin	
Others	
Polyethylene oxide	
Homopolymers and copolymers of acrylic acid	

Source: Adapted from [30].

of, the outer hydrated polymer layer. Typically, when the matrix is exposed to an aqueous solution, the surface of the tablet is wetted and the polymer hydrates to form a gelly-like structure around the matrix (the *gel layer*). This process leads to relaxation and swelling of the matrix periphery, whereas the core remains essentially dry at this stage. In the case of a highly soluble drug, this phenomenon may lead to an initial burst release due to the presence of the drug on the surface of the matrix. The thickness of the gel layer increases with time as more water permeates the core of the matrix, thereby providing an evolving diffusional barrier to drug release. Simultaneously, as the outer layer becomes fully hydrated, the polymer chains become completely relaxed and can no longer maintain the integrity of the gel layer, thereby leading to disentanglement and erosion of the surface of the matrix. Water continues to penetrate toward the core of the tablet, through the gel layer, until it has been eroded completely.

The key parameters affecting the drug release rate from hydrophilic matrices, with special emphasis on those based on HPMC, were recently reviewed by Tiwari and Rajabi-Siahboomi [30]. Drug solubility and dose are the most important factors to consider in the design of matrices. In general, design of extended-release formulations for drugs with extreme solubilities coupled with a high dose is challenging. For highly water-soluble drugs, dissolution within a gel layer (even with small amounts of free water) is fast and sustained release too brief. On the other hand, drugs with very low solubility may dissolve slowly and have slow diffusion through the gel layer of a hydrophilic matrix. When drug concentration in the gel layer is above the saturation level, equilibrium is attained between solid and dissolved drug in the hydrated matrix. As a consequence, drug release would be dictated mainly by erosion of the surface. In this case, incomplete drug release is experienced due to the fact that undissolved drug particles are released in the medium upon matrix erosion.

Due to their ability to enhance drug solubility, it is expected that CDs can be of help in formulating hydrophilic matrices for poorly soluble drugs by increasing their dissolution rate in the gel layer, and give extensive release. In this respect, many examples are available in the literature. The use of CDs in a hydrophilic matrix with this strategy in mind was first reported by the group of Conte [31,32], who proposed β-CD and HP-β-CD as accelerators of the release rate of naftazone or diazepam from tablets made of HPMC (Methocel K4M and K15M). It was also found that due to CD addition, a zero-order release rate was attained. In later studies, the addition of CDs in hydrophilic erodible platforms containing poorly water-soluble drugs continued to be considered as an additional tool to modulate, and specifically increase, the drug release rate.

A rationalization of the effect of CDs in hydrophilic matrices for poorly soluble drugs was attempted by Sangalli et al. [33] and Rao et al. [34], who clarified the mechanism by which CDs act in the matrix, analyzing the release behavior of HPMC tablets in a variety of conditions (replacing CDs with hydrophilic fillers or osmotic agents). The main conclusion drawn by Rao et al. was that increase in the drug delivery rate in the presence of a sulfobutyl ether/β-CD derivative (SBE-β-CD) was ascribed to an enhanced dissolution rate of the drug in the gel layer of the matrix due to in situ complex formation, with only a minor effect of enhanced water uptake. As a consequence of drug complexation inside the hydrated polymeric matrix and depending on the polymer network properties, CDs provided extensive and faster releases of poorly soluble drugs by enhancing the concentration of diffusable species. Sangalli et al. [33] drew more or less similar conclusions, although they found that the efficiency of β-CD as a release modulator was much poorer than that of SBE-β-CD, probably due to its lower solubility. On the other hand, HPMC tablets containing β-CD showed a higher hardness than that of tablets containing other excipients, such as lactose, but also different β-CD derivatives, which resulted in a slower water uptake and, in turn, a slower release rate.

On this basis, later studies have focused on the use of a more hydrophilic CD such as HP-β-CD as a release accelerator in hydrophilic matrices. Nevertheless, it was found that the effect of HP-β-CD depended on the polymer platform employed to produce the matrix. For example, HP-β-CD was found to be effective in increasing the delivery rate of piroxicam from HPMC tablets, whereas no effect was observed for HPMC/Carbopol 940 tablets [35]. Authors hypothesized that in this case, a restricted diffusivity of piroxicam–CD complex occurred in the swollen external layer. Thus, if the gel layer is strong or a less soluble CD is employed, the overall release rate is unaffected by the presence of CD.

The observations reported so far provide for the addition of CD in the tablet as powder; that is, no drug–CD precomplexation is realized. However, another strategy to further accelerate the delivery rate from hydrophilic matrices is to prepare a drug–CD complex which is then incorporated in the matrix along with retarding polymer and other excipients. It has been observed that the higher the dissolution rate of the complex, the higher the drug release rate from the tablet [17,34–41].

Thus, it is quite clear that the mechanism of release from monolithic devices made of swellable and erodible polymers loaded with poorly water-soluble drugs and CDs as a release modulator is rather complex in view of several physical phenomena involved. In fact, reliable predictive models of drug delivery from these systems entail the solution of moving boundary problems, which should take into account (1) water penetration inside the tablet with concurrent swelling, solubilization and erosion of the matrix, (2) dissolution of both the drug and the CD in the swollen layer, (3) CD–drug complex formation, and (4) counterdiffusion of drug, CD, and complex in the swollen layer.

In our previous work we tried to explain the release behavior of PEO tablets containing the poorly soluble drug carvedilol (CAR) and HP-β-CD as a release modulator [38]. We found that when CAR was incorporated in PEO tablets in the presence of HP-β-CD as a CAR/HP-β-CD physical mixture (PM) or freeze-dried complex (FD), the CAR release rate increased significantly (Fig. 6A). The release profile of tablets without HP-β-CD was linear with time, whereas in the case of HP-β-CD-containing tablets, linear behavior was attained after a transient. Tablets loaded with both PM and FD displayed similar HP-β-CD release profiles (Fig. 6A), indicating that the same boundary conditions were established for both systems in terms of external HP-β-CD concentration. As a consequence, the observed differences in CAR release profiles for HP-β-CD-containing tablets could be attributed to a different CAR solubility in the release medium. The erosion profiles of the tablets based on PEO or PEO/HP-β-CD (Fig. 6B) also highlighted the fact that the presence of HP-β-CD speeded up the solubilization rate of the matrix as whole, probably due to an increase in polymer hydrophilicity [34]. On the basis of the behavior observed experimentally, a qualitative view of a delivery process that could supply a simplified approach for the interpretation of results was provided. It was assumed that three relevant

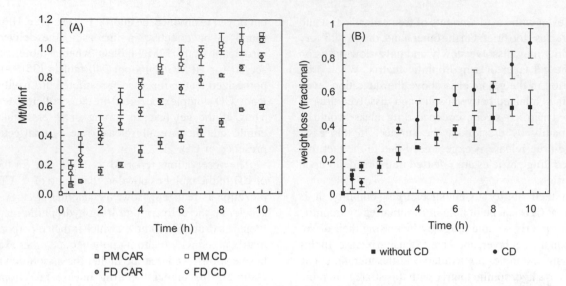

Figure 6. Effect of HP-β-CD addition on the properties of CAR-loaded PEO matrices: (A) release profiles of carvedilol (CAR) and HP-β-CD from PEO tablets incorporating a CAR/HP-β-CD physical mixture (PM) or a CAR/HP-β-CD freeze-dried product (FD) in phosphate buffer at pH 6.8 and 37°C: (B) weight loss of PEO tablets without and with HP-β-CD in phosphate buffer at pH 6.8 and 37°C. (From [38]. Copyright © Elsevier.)

moving fronts are established in a hydrophilic tablet in contact with a water solution (Fig. 7): (1) the swelling front, S, which separates the unpenetrated core from the swollen and dissolving shell; (2) the erosion or solubilization front, E, which separates the swollen polymer layer from the external

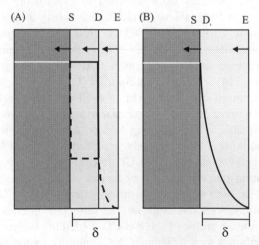

Figure 7. Moving fronts established in a hydrophilic tablet in contact with a water solution: (A) drug profiles in CD-loaded tablets at pseudo-steady state [total drug profile (dissolved and undissolved), solid line; dissolved drug profile, dashed line]; (B) CD profile in the tablet (CD has been assumed as completely soluble in the water-swollen layer). δ represents the thickness of the swollen layer. Profiles of drug, CD, and in situ–formed complex in the swollen outer shell are assumed to be linear. (From [38]. Copyright © Elsevier.)

medium; and (3) the eventual drug dissolution front, D, which separates an inner oversaturated region, where both dispersed and solubilized drugs are present, from an outer water-swollen polymer layer, where only solubilized drug is present (this layer is superimposed on the S front in the case of water-soluble species, as is the case with CD) [42]. The moving rates of these fronts are substantially dependent on the nature of the components and their concentration in the tablet. After a transient, if the thickness of the tablet is large enough, a pseudo-steady state is established when a synchronization of S and E fronts occurs and a constant drug profile is established in the swollen layer. The drug concentration profile that develops in the swollen layer depends on drug diffusivity and solubility and its dissolution rate in the swollen polymer as well as on the thickness of the developing swollen layer.

On this basis, CAR release was contributed by both the movement of E front and the diffusive flux related to the drug concentration profile at the E interface. The rate of the E front as well as the diffusive flux changed with time during the initial transient to attain values constant with time at the pseudo-steady state. The presence of HP-β-CD increased the erosion rate as well as drug concentration at the D interface at pseudo-steady state (C*) and in turn speeded up CAR release rate, assuming that HP-β-CD does not affect the mobility of the drug in the swollen layer (HP-β-CD, CAR, and their complex have similar diffusivities in the gel layer). Hence, the nonlinearity observed in the transient was due to the fact that before the synchronization of E and D fronts, both contributions to the release are time dependent. In the case of HP-β-CD-containing tablets, the movement of the E front was faster than for tablets without HP-β-CD (see Fig. 7). Furthermore, a higher diffusive flux developed as a consequence of the increase in dissolution rate operated by HP-β-CD, which increased the amount of mobile species. Both of these effects resulted in an enhanced release rate of drug compared to the case of tablets without HP-β-CD. The further increase in delivery rate experienced when the tablet was loaded with a preformed complex well fitted with this view, highlighting the fact that an increase in the drug–CD complex dissolution rate also takes place in the water-swollen polymer.

It is worthy of note that acceleration of the drug release rate can also occur when CD does not affect the drug dissolution rate in a bulk medium. In this respect, we observed that PEO/HP-β-CD hydrophilic matrices loaded with the sparingly soluble drug diclofenac acid (DicH) showed an increased drug release rate compared to tablets formulated in the absence of HP-β-CD, although a lack of increase in the dissolution rate of DicH/HP-β-CD physical mixtures was observed in water. Since HP-β-CD operates in hydrophilic matrices as rate-controlling agent in a swollen gel layer and can be assumed to be instantaneously solubilized in a confined hydrated gel, it is likely that its dissolution enhancer properties toward the drug could be amplified inside the tablet.

In summary, the introduction of CD in erodible sustained-release tablets supplies an additive tool to tailor the release rate of poorly soluble drugs by modulating its dissolution rate in the swollen outer layer as well as matrix erosion rate.

CD addition in films based on hydrophilic polymers was also demonstrated to be useful to speed up the release rate of poorly soluble drugs. Since release was performed through a semipermeable membrane allowing only drug transport, the effect of CD addition was in this case enlarged [43]. In fact, films of poly(vinyl alcohol) and HPMC added with a RAMEB–atenolol complex accelerated the release rate, giving first-order kinetics, whereas only minor effects were observed in HPMC films. An accelerating effect was again obtained for chitosan films containing β-CD and glimepiride [44].

CD addition also had an impact on the formulation of hydrophilic matrices for highly water-soluble drugs (i.e., drugs that are completely solubilized inside the gel layer), although much less is known in this respect. It has been observed that the release rate of a hydrophilic drug from erodible and swellable matrices of PEO or HPMC slightly decreased or remained unaltered in the presence of CD [17,45]. This behavior is analogous to that observed for cross-linked PEO hydrogel described above and attributed to the influence of CDs on the effective drug diffusivity in the gel layer as a function of complex stability constant and the loaded-drug/complexant ratio [26] as well as to the CD effect on the properties of the gel layer.

3.2.3. Impact of CD-Containing Hydrophilic Matrices on Drug Absorption Through a Buccal Membrane

Buccal administration of drugs is attracting considerable attention since it has the advantage of giving high blood levels, bypass first-pass metabolism and avoiding degradation in the gastrointestinal tract by enzymes and bacteria [46]. Buccal systems are generally based on bioadhesive polymers which, once hydrated, adhere to the buccal mucosa and withstand salivation, tongue movements, and swallowing for a significant period of time. A successful design of a buccal delivery system should guarantee, beside an intimate contact with the mucosa for an adequate time interval, proper release rates. Actually, before a drug passes through the mucosal barrier and reaches blood circulation, it should dissolve in the medium penetrating the buccal tablet and then partition to mucosa. A contribution to drug transport through buccal membrane is certainly given by the presence of an unstirred water layer (UWL) lining the buccal epithelium, formed mainly by mucin and water (∼95%) and with a thickness of approximately 70–100 μm. Considering that the resistance to transport of UWL is higher than that of the buccal membrane, transport through UWL substantially controls drug absorption. The dissolution step is generally critical for the buccal absorption of lipophilic drugs, which although being well absorbed through oral epithelia, exhibit too low fluxes due to a low chemical potential gradient, which is the driving force for transport. Since the addition of CDs in DDSs based on hydrophilic polymers can allow a modulation of drug release rate, their use as smart excipients in specific buccal systems has been proposed to enhance the level of dissolved molecules available for permeation. Therefore, the addition of a CD in a buccal system can be expected to change the concentration gradient between UWL and the membrane as well as to alter the transport properties of a given drug in UWL [47].

It was found that upon HP-β-CD addition in PEO buccal tablets containing the poorly water-soluble drug CAR, drug permeation through porcine buccal mucosa occurred at a rate progressively higher as the CAR release rate from the tablet increased [38]. This enhancing effect of HP-β-CD on permeation was much more evident than that observed on the drug release rate in an aqueous fluid. Similar results were observed on PEO/HP-β-CD tablets containing the sparingly soluble drug DicH [45]. Permeation experiments performed on PEO tablets stuck to porcine buccal mucosa clearly showed that PEO–CD–DicH tablets allowed a drug flux higher than that observed for PEO–DicH tablets (Fig. 8A upper graph). Very interestingly, when the drug was incorporated in the tablet as freely soluble sodium salt (DicNa), an opposite behavior was experienced (Fig. 8A, lower graph). Although the presence of HP-β-CD in tablets containing soluble DicNa did not significantly alter the drug release rate, the drug flux through porcine mucosa from

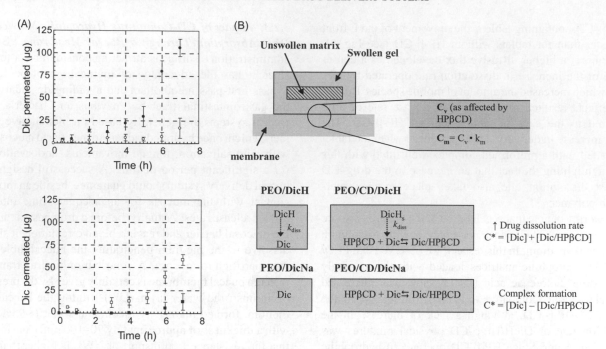

Figure 8. Effect of HP-β-CD addition in adhesive tablets on permeation through buccal mucosa: (A) permeation profiles through porcine buccal mucosa of diclofenac from PEO–DicH (□) and PEO–CD–DicH (■) tablets (upper graph) and from PEO–DicNa () and PEO–CD–DicNa (○) (lower graph); (B) drug partioning into a lipophilic membrane at the tablet–membrane interface. C_v is the drug concentration at the swollen tablet–membrane interface, k_m is the drug partition coefficient between the vehicle and the membrane, $k_m C_v$ represents the drug-amount partitioning in the membrane at the donor side (C_m), and C^* is the effective drug concentration in the swollen gel. (Adapted from [45]. Copyright © Elsevier.)

PEO–DicNa tablets was found to be lower when the tablet contained HP-β-CD.

On the basis of a simplified mathematical treatment, the different behavior experienced for the tablets containing differently water-soluble forms of diclofenac was explained as schematized in Fig. 8B. Assuming that (1) effective drug concentration in the swollen gel (C^*) is contributed by free and complexed drug molecules, (2) HP-β-CD is freely soluble in the swollen gel layer, (3) a drug/HP-β-CD complex is formed, and (4) HP-β-CD concentration in the swollen gel at the membrane interface can be considered constant since it is unable to partition into the membrane, the only species that can partition in the membrane is free drug. In the case of PEO–CD–DicH tablets, the effective drug concentration in the swollen gel (C^*) was increased compared to PEO–DicH tablets due to the formation of a soluble drug/HP-β-CD complex in the swollen layer. If at every position and time, the equilibrium between complexed and free drug molecules exists and is attained instantaneously in the gel, the availability of drug molecules from the complex is faster than the availability of drug molecules from the solid. Although complexation does not alter the chemical potential (i.e., the amount of free drug molecule that can partition into the membrane), the presence of a drug/HP-β-CD soluble

"reservoir" allows faster transport of solubilized drug molecules to the membrane surface and results in an increased drug flux. On the other hand, in PEO–CD–DicNa tablets, the drug dissolution rate is high and the drug dissolves rapidly in the gel layer. In this situation, the formation of a Dic/HP-β-CD complex inside swollen PEO decreases C^*. Since the complex is unable to partition into the membrane, an overall decrease in the drug cumulative flux is attained due to the progressive increase in CD molecules at the gel layer/mucosal membrane interface. It was concluded that the role of CD is very relevant when the tablet is employed as a transmucosal system since, different from solution conditions, a very limited contribution to delivery derives from matrix erosion. Thus, the usefulness of CDs in erodible hydrophilic matrices depends strongly on the environment where release occurs. Transferring these results to an in vivo situation, in the case of gastrointestinal delivery, where release occurs in a bulk fluid, HP-β-CD addition can be suggested to accelerate the release rate of poorly soluble drugs. On the other hand, when considering applications where matrix is applied to a mucosal membrane, HP-β-CD diffusion out of the system is very limited and drug transport is accelerated for poorly soluble drugs and slowed down for highly soluble drugs.

It is worthy of note that an important prerequisite of a CD-containing buccal DDS is that its overall mucoadhesive properties, interdiffusion of polymer chains, and mucus components at the interface are not impaired, which seems to depend on polymer and CD type [38,44].

3.2.4. *A Mechanistic Simplified Interpretation of the CD Effect in Drug Delivery Systems Based on Hydrophilic Polymers*

On the basis of previous findings and the model proposed by Bibby et al. [48], the situation occurring in a gel system can be depicted as reported in Fig. 9, assuming that the effect of CD addition on drug-release features depends mainly on drug loading of the matrix and both drug and CD diffusivities in the hydrated polymer. A prediction of CD effect could be done on the basis of known network properties (porosity, mesh size, polymer solubility, etc.).

When drug concentration in the polymeric matrix is below the saturation level (case I in Fig. 9), two situations can occur: (1) when all the species have comparable diffusivities, there is no change in drug diffusion rate, whereas (2) when drug diffusivity is much higher than CD diffusivity, the formation of drug–CD complex in the hydrated matrix results in a decrease of concentration gradient and, in turn, of the drug release rate. On the other hand, when drug concentration in the polymeric matrix is above the saturation level (case II of Fig. 9), an equilibrium is attained between solid and dissolved drug in the hydrated matrix. Upon the addition of CD in the hydrated matrix, the number of drug molecules solubilized in the hydrated polymer increases and the final effect on drug release will depend on drug and CD diffusiv-

ities inside the hydrated matrix. If drug diffusivity is similar to CD diffusivity ($\mathcal{D}_{dr} \approx \mathcal{D}_{CD}$), so that both free and complexed drug can diffuse out of the matrix, the release rate is increased. If the drug diffusivity is much higher than the CD diffusivity ($\mathcal{D}_{dr} \gg \mathcal{D}_{CD}$), complexed drug cannot diffuse out of the polymeric matrix. Thus, the final effect depends on how much CD improves the amount of free drug molecules for systems with a drug level well above saturation and decreases the amount of diffusable species due to complexation in the hydrated polymer for systems with a drug level below saturation.

3.3. Osmotic Systems

Osmotic systems utilize the principle of osmotic pressure for the delivery of drugs and are generally called as osmotic pumps. Osmosis refers to the process of movement of solvent molecules from lower concentration to higher concentration across a semipermeable membrane, that is, a membrane permeable to water but completely impermeable to osmotic agent. Thus, the basic architecture of an osmotic pump, in particular an osmotic tablet, comprises a drug reservoir, eventually containing an osmogen (electrolyte, sugar, etc.) coated with a semipermeable membrane drilled with one or more orifices (Fig. 10). When the system comes in contact with a biological fluid, an osmotic pressure generates due to imbibitions of fluid from external environment into the dosage form, which regulates the delivery of the drug through the orifice. The hydrostatic pressure can be created by osmotic agents, the drug itself, or a tablet component, after

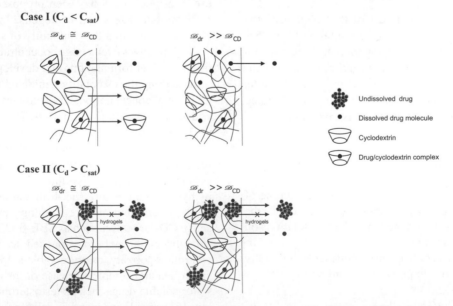

Figure 9. Equilibria occurring in a hydrated polymeric system with different diffusivity upon CD addition. In case I, the drug level is below saturation, whereas in case II, the drug level is above saturation. CD diffusivity ($\mathcal{D}_{CD}$) as compared to drug diffusivity ($\mathcal{D}_{dr}$) differs depending on the gel–hydrogel mesh size (i.e., CD mobility in the polymer network).

OSMOTIC PUMP TABLET

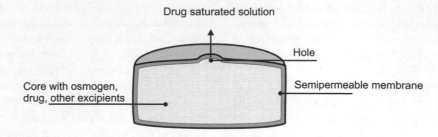

CONTROLLED POROSITY-OSMOTIC PUMP TABLET

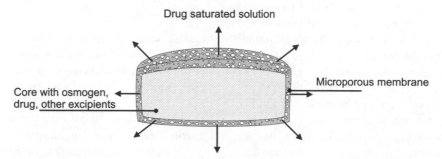

Figure 10. Osmotic tablets. In controlled-porosity osmotic tablets, drug flow occurs through macropores generated in the semipermeable membrane upon contact with an aqueous medium.

water is imbibed across the semipermeable membrane. As a colligative property, the osmotic pressure (Π) of a solution is dependent on the number of drug molecules present in the solution:

$$\Pi = \Phi CRT$$

where Φ is the osmotic coefficient of the solution, C the molar concentration of osmogen in the solution, R the gas constant, and T the absolute temperature. Experimental measurements of osmolality (ξ_m in Osm/kg) gives an accurate description of solvent activity in the presence of solute.

As a consequence of Π, the osmotic water flow generated through the semipermeable membrane is given by

$$\frac{dv}{dt} = \frac{A\,Q\,\Delta\Pi}{L}$$

where dv/dt is the water flow rate across the membrane of area A, thickness L, and permeability Q, and $\Delta\Pi$ is the difference in osmotic pressure between the two solutions on either side of the membrane.

The rate of drug delivery from an osmotic tablet is directly proportional to the osmotic pressure according to

$$\frac{dM_t}{dt} = \frac{dV}{dt}\,C_s$$

where C_s represents drug solubility.

Drug release from these systems is strongly dependent on drug aqueous solubility and to a large extent independent of pH and other physiological parameters. Thus, it is possible to modulate the release characteristics by optimizing the properties of the drug and the device.

Since some CDs generate high osmotic pressures, their use in osmotic DDS has been proposed in the double role of both solubilizers and osmogens. In particular, Zannou et al. [49] measured the osmolality of solutions with β- and γ-CD derivatives over a large concentration range relevant to osmotic tablet formulations and developed a model capable of describing the osmolality generated in CD solutions as a function of concentration. The study was conducted in water with eight CDs: HP-β-CD, HP-γ-CD, SBE-β-CD, and SBE-γ-CD at various degrees of substitution.

In the evolution of osmotic tablets, the semipermeable membrane contains leachable pore-forming materials. In this system, the drug, after dissolution inside the core, is released by hydrostatic pressure and diffusion through pores created by the dissolution of pore-forming materials. Osmotically active CD derivatives such as SBE-β-CD, also acting as drug solubilizers, have been proposed as smart excipients in controlled-porosity osmotic tablets [50] able to allow a controlled and complete release of the drug dose for poorly water-soluble drugs such as testosterone [51] and prednisolone [52] and pH-independent release for cationic drugs such as chlorpromazine [53,54]. Appropriate composition ratios of the osmotic systems allow excess drug/CD molar ratios to work properly.

SBE-β-CD was also used to formulate controlled-porosity osmotic pellets for the delivery of prednisolone to achieve both osmotic and solubilizing properties [54–58]. Pellet performance could be designed to attain a desired delivery rate by modifying the drug–CD ratio as well as the coating thickness and coating composition.

3.4. Biodegradable Drug Delivery Systems Based on Lipophilic Polymers

Biodegradable DDSs releasing the active ingredient at controlled rates have been developed especially for parenteral administration to overcome the inconvenient surgical insertion and removal of nondegradable systems. This category of devices is formed by enzymatically or chemically degradable and erodible materials [59,60]. For clarity, it should be emphasized that the term *degradation* indicates simply that a chain scission of polymer in shorter chains occurs, whereas the term *erosion* refers to the occurrence of material loss from the system. Depending on the rate of water uptake in the system, erosion can involve either the surface of the delivery system (*surface erosion*) if the rate of polymer cleavage is higher than the water intake rate, or occurs more or less homogeneously throughout the system (*bulk erosion*) if the water penetration rate is higher than the polymer degradation rate (Fig. 11) [61]. For surface-eroding DDSs, the device shrinks with time and the drug is released in correspondence to the disappearance of the surrounding polymer matrix. Meanwhile, the inner structure (e.g., internal porosity, relative drug content, and distribution) remains

essentially unaltered. For bulk-eroding DDS, the entire device is rapidly wetted upon contact with aqueous medium, the drug becomes mobile and diffuses out of the device. Device dimensions remain nearly unaltered, whereas the inner structure changes significantly since the porosity increases and the drug concentration decreases.

Degradable and erodible DDSs may be developed by making use of different polymers [59,60]. The polymer nature and composition would directly dictate the DDS physico-chemical properties, including bulk hydrophilicity, morphology, structure, and thermal (i.e., transition and/or melting temperature) and chemical (e.g., the presence of charged groups) attributes. These properties will all influence the performance of the delivery system by changing the mass transport and degradation rate of both the polymer and the device. In addition, depending on polymer properties such as glass-transition temperature, crystallinity, hydrophilicity, and device shape and dimension, a surface erosion rather than a bulk erosion can take place. A great deal of work has been devoted to the development of safe and effective DDSs based on synthetic polymers undergoing hydrolytic degradation, characterized primarily by a hydrophobic nature. Actually, the main advantage of these synthetic materials is that they can be easily tailored to offer properties for specific applications. Furthermore, as compared to degradable polymers via enzymatic activity (e.g., chitosan, alginate), their release features are not dependent on physiological factors; that is, the therapeutic response to the drugs released would not depend on the level of enzyme expression at the site of action, probably resulting in less inter- and intraindividual variability.

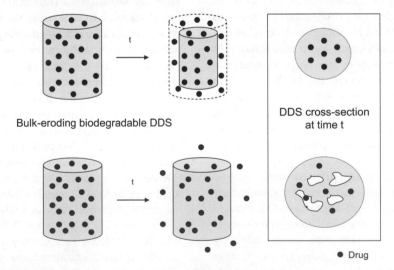

Figure 11. Typical mechanisms governing material loss during polymer erosion. A surface-eroding biodegradable DDS shrinks with time, but the inner structure (see the cross section) is unaltered. In contrast, the shape of a bulk-eroding DDS remains approximately unaltered during erosion, whereas the inner structure, in term of internal porosity, changes significantly with time.

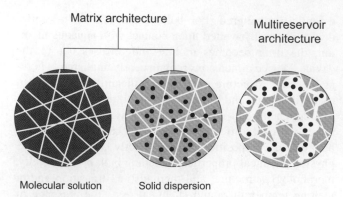

Figure 12. Typical architectures of biodegradable DDSs. For a matrix DDS, hydrophobic drugs (black dots) can be dissolved (molecular solution) in the polymer matrix, whereas hydrophilic and hydrophobic drugs can be dispersed in solid form within the polymer matrix (solid dispersion). The CD is expected to locate preferentially within polymer micropores (white solid lines). For a multireservoir DDS, hydrophilic drug and CD are located mainly within the DDS macroporosity (white solid circles).

Drug release from biodegradable DDS is controlled by (1) water uptake in the device, (2) drug solubilization inside the hydrated system, (3) diffusion of dissolved drug through internal pores and (4) degradation rate. Beyond the physicochemical properties of the polymer, additional factors, such as shape/geometry, physical form, and size of the delivery system will inevitably affect its release features. In particular, the ratio "rate of polymer chain cleavage/rate of water penetration" will inevitably influence the polymer degradation rate, whereas the rate at which water enters the system depends strongly on device shape and dimensions [61]. Furthermore, the preparation technique will play a crucial role in determining the internal microstructure of the systems, essential in controlling the drug-release features. In general, two different DDS architectures can be achieved: biodegradable matrix DDSs, where a molecular solution or dispersion of the drug throughout the polymer matrix is realized, or multireservoir biodegradable DDS, where the drug is confined within macropores formed throughout the matrix during specific processing. Due to the low solubility of drugs in the polymer, solid dispersions are often encountered for matrix DDSs. A schematization of the possible device designs is reported in Fig. 12.

Among synthetic biodegradable polymers, thermoplastic aliphatic polyesters such as poly(lactic acid) and their derivatives poly(lactic-*co*-glycolic acid) (PLGA) have generated tremendous interest due to their excellent biocompatibility as well as the possibility of tailoring their biodegradability by varying the composition (e.g., lactide/glycolide ratio), molecular weight, and chemical structure (e.g., capped or uncapped end groups) [62]. Drug release from PLGA-based DDSs is governed by a combined diffusion-erosion mechanism. Upon immersion in an aqueous medium, polymer

hydration occurs and activates the diffusion of dissolved drug through the innate micropores of PLGA and the eventual macroporous structure of the matrix. Since water penetration into PLGA-based DDSs is much more rapid than the subsequent polymer chain cleavage, these systems typically undergo bulk erosion. As a consequence, matrix porosity evolves over time, contributing strongly to the overall drug release rate (Fig. 13). Since CDs are not soluble in the polymer, it is expected that they are located in DDS internal pores. After the hydration phase, free or complexed CDs dissolve, diffuse, and are released through DDS interconnected pores. Thus, their effect upon DDS release features may be very different for systems characterized by different microstructures and thus internal porosity.

3.4.1. Implants Biodegradable polymer depots, providing local and sustained delivery of the active ingredient, may be achieved by drug entrapment within an implant, which can be formed in situ or inserted within the body by surgical intervention. Implant performance depends on many interrelated factors, including device geometry, which is generally designed to fit the site of implantation.

Great research efforts have been devoted to the design and development of PLGA-based cylindrical implants intended for site-specific controlled release of cytotoxic agents [63]. Nonetheless, for hydrophobic drugs, such as most anticancer drugs, direct incorporation within PLGA matrices is not successful, since solid dispersions are generated, which exhibit an extremely slow release of small drug quantities due to drug low water solubility. These delivery features are generally unsuitable since limited drug availability is achieved at the tumor site throughout therapeutic treatment. In analogy to the most intuitive application of CDs in DDSs, drug complexation with different CDs has been shown to be a useful strategy to accelerate the release kinetics of hydrophobic drugs from PLGA depots [64–66]. In particular, the addition of CDs to PLGA millirods for the controlled release of poorly soluble cytotoxic agents (β-lapachone) resulted in a convenient increase in the drug release rate [65,66]. Similarly, the incorporation of HP-β-CD in injectable cylindrical PLGA–lauryl ester implants for 2-methoxyestradiol reduced the initial lag phase and moderately increased the drug release rate [67]. CD complexation efficiency and CD water solubility played a crucial role in determining the drug release rate. In fact, a faster release was observed for β-lapachone/HP-β-CD than for β-lapachone/αCD and β-lapachone/γ-CD complexes, due to the lower binding affinity of CD for drug (HP-β-CD > γ-CD > α-CD). At the same time, the higher release rate observed for HP-β-CD-containing than for β-CD-containing formulations was ascribable primarily to their considerably different water solubilities [65].

3.4.2. Microspheres Injectable biodegradable and biocompatible microspheres less than 250 μm in size releasing

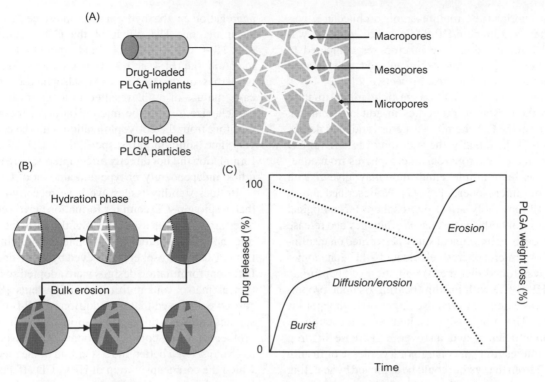

Figure 13. Mechanism of drug release from a PLGA-based DDS. (A) Drug is released by diffusion through a porous microenvironment formed by intrinsic copolymer micropores, evolving in mesopores upon polymer degradation, and eventual macropores derived from the preparation process. (B) After a hydration phase, drug diffusion out of water-filled pores and progressive PLGA bulk erosion (i.e., progressive increase of the internal pore diameter) are activated. This mechanism will result in a typical triphasic drug release profile (C). After an initial *burst*, due to the rapid dissolution of drug molecules localized on a pore close to the surface, a second slow diffusive/erosive release phase is expected, during which a continuous loss of PLGA weight occurs. When massive PLGA degradation occurs, a third rapid drug release phase should be observed (i.e., a pure erosion phase). (Adapted from [99]. Copyright © John Wiley and Sons, Inc.)

small drugs or macromolecules at controlled rates have been especially developed to overcome the inconvenient surgical insertion of large implants. The first studies on the role of CDs in microparticle preparation were carried out by Loftsson et al. on nonbiodegradable polymers (e.g., ethylcellulose) (see citations in [68]). Since they aimed to obtain slow-release oral dosage forms, most of the drugs tested displayed a lipophilic character (e.g., hydrocortisone, 17β-estradiol, carboplatin). This resulted in an easy micro-encapsulation process, but also in a release depending mainly on water solubility of the drug, which obviously could be accelerated by selecting the proper CD type.

The co-encapsulation of CDs has been recently attempted in biodegradable microspheres, with particular regard to PLGA microspheres. Actually, drug encapsulation within PLGA copolymers is regarded as a powerful means of achieving its sustained release for long time frames and, in the case of labile drugs, effectively protecting the molecule from in vivo degradation occurring at the administration site [69–72].

CD incorporation in PLGA microspheres was first attempted to control the encapsulation efficiency and release rate of low-molecular-weight drugs [73–75], where the CD seems to play a role in controlling the release of both hydrophilic and hydrophobic species. The encapsulation of rhodium citrate/HP-β-CD complexes in a multireservoir device was useful to improve encapsulation efficiency and prolong the release duration of the hydrophilic antitumor agent from the particles as compared to PLGA microspheres containing rhodium citrate alone [73]. Using a CD of different lipophilicity (e.g., HP-β-CD and M-β-CD), the same authors successfully controlled encapsulation efficiency and release properties of PLGA microspheres containing different chlorexidine derivatives (e.g., poorly water-soluble chlorhexidine free base and its water-soluble digluconate derivative) [74]. Not only CD physico-chemical properties, but also formulation conditions would determine the effect of CD addition on microsphere release features [75]. In particular, a more rapid release of zolpidem from HP-β-CD-containing PLGA microspheres with a macroporous interior (i.e., multireservoir architecture) was achieved when a zolpidem/HP-β-CD solid complex was encapsulated directly within the macropores compared to microspheres bearing drug and excipient separately, located in PLGA micro- and macropores, respectively [75].

A particular effect of CDs on particle microstructure derives from their osmotic properties in aqueous solution, which depend on the CD chemical structure and total degree of substitution [49]. Actually, the addition of free HP-β-CD in the preparation of macroporous microspheres from aqueous dispersions—a multiple emulsion—may give rise to typically porous microspheres [75–77]. As described above, this effect will inevitably affect microsphere technological features in terms of both encapsulation efficiency and release properties. A deep technological study performed on insulin-loaded PLGA microspheres elucidated this phenomenon [76]. It was found that the combination of appropriate amounts of HP-β-CD with insulin (bearing osmotic properties as well) was essential to achieve large and highly porous microspheres. This obviously resulted in a reduction of encapsulation efficiency and an overall increase in drug release rate. Interestingly, a contemporary release of insulin and HP-β-CD from the system could be achieved by selecting appropriate formulation conditions, highlighting the fact that the microparticles prepared have the potential of acting in vivo as a sustained delivery systems of insulin/HP-β-CD complexes. Taking advantage of this effect, large porous particles with flow properties and size suitable for aerosolization and deposition in deep regions of the lung have been developed [77]. Notably, PLGA/HP-β-CD/insulin large porous particles (LPPs) reach alveoli and release insulin, which is absorbed in its bioactive form, thus representing a new frontier in the development of sustained-release dry powders for pulmonary delivery.

CDs have also been used into PLGA microparticles for the controlled release of proteins to stabilize and/or solubilize the macromolecule, improving its therapeutic efficacy [78–82]. Once CDs form complexes with proteins by including hydrophobic side chains inside their cone-shaped cavity, protein three-dimensional structure and chemical–biological properties, comprising stability, can be affected [83]. Proper selection of both the CD derivative to be used and the microencapsulation technique to be adopted is essential to improve the properties of protein-loaded PLGA microspheres [84]. For protein microencapsulation in the solid form, which allows formation of a PLGA solid dispersion, the main determinant of protein stabilization was found to be formation of the solid complex by freeze-drying prior to encapsulation. As a matter of fact, when solid α-chymotrypsin was simply suspended in a M-β-CD-containing organic phase, protein integrity was not preserved upon microencapsulation. More than to a M-β-CD-mediated lyoprotectant mechanism, the authors ascribe the results to the fact that the α-chymotrypsin/M-β-CD complex might decrease protein aggregation at the hydrophobic interfaces formed during encapsulation. The effect of the CD chemical structure (i.e., HP-β-CD or M-β-CD) and amount added to the formulation (i.e., 1:4 and 1:20 mass ratios) on the overall properties of the particles was also investigated. In each case, the use of a CD resulted in a significant decrease in particle size, due to the micronization of the solid protein resulting from the co-lyophilization with the excipient (i.e., very fine solid in oil suspensions). CD addition to the formulation did not directly affect microsphere release properties, independently of type and amount of CD used.

Besides stability issues, we have recently demonstrated that the protein–CD complex formation plays a crucial role in determining the release features of the particles and thus can offer an additional tool to modulate the protein release rate from PLGA microspheres achieved by spray-drying [85,86]. Different formulation designs were adopted so as to achieve typical matrix and multireservoir systems [86]. In both cases we observed that the addition of HP-β-CD to microspheres allowed full optimization of the insulin release profile, affording constant amounts of released peptide up to 45 days. For a better understanding of the mechanisms by which the co-encapsulation of HP-β-CD affects the protein release rate, protein–CD interactions directly within PLGA-based particles prepared by spray-drying after microencapsulation and during release were assessed by Fourier transform infrared (FTIR) spectroscopy [85]. Independent of the formulation design, FTIR spectra acquired on nonhydrated particles clearly showed a rearrangement of insulin β-structures upon addition of HP-β-CD within the initial formulation, probably due to complex formation. Nonetheless, release studies clearly showed that the modulator effect of HP-β-CD depended on the initial formulation conditions. In the case of microspheres with a macroporous interior, HP-β-CD was able to reduce insulin burst, whereas when a homogeneous dispersion of the complex was achieved within the PLGA matrix (i.e., matrix architecture), the overall insulin release rate slowed down (Fig. 14A). This effect corresponded to a difference in the relative concentration of β-sheet components (i.e., insulin/HP-β-CD interactions during the release stage), as suggested by FTIR spectra of hydrated microspheres (Fig. 14B). Thus, the control of the CD loading and release rate by an appropriate formulation choice, modifying final polymer micro- and macroporosity, was crucial for an effective modulation of protein release rate.

From these data it is clear that the effect of CD on drug-loaded PLGA microspheres is not univocal and always foreseeable but results from the complex interplay of many factors, such as drug physico-chemical properties (e.g., solubility, molecular weight), drug–CD affinity, and formulation design (e.g., encapsulation technique, formulation conditions), which strongly affect the architecture and microstructure of the particles.

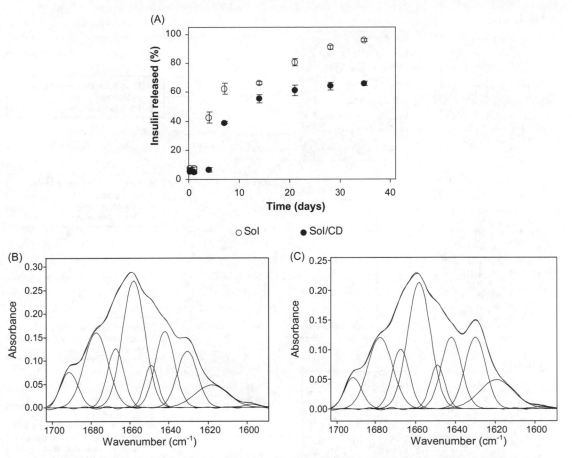

Figure 14. Effect of HP-β-CD on the release properties of spray-dried insulin-loaded PLGA microspheres with a porous interior. The addition of HP-β-CD within the particles (i.e., sol/CD formulations) allows a modulation of insulin release rate (A) as compared to PLGA microspheres prepared in the absence of HP-β-CD (B). FTIR spectra of insulin within HP-β-CD-containing hydrated particles show differences in the relative concentration of β-sheet components (band at 1642 nm) indicating insulin/HP-β-CD interactions during the release stage (C). (From [85] with permission. Copyright © Elsevier.)

3.4.3. Nanoparticles Nanoparticles are submicrometer-sized polymeric colloidal particles bearing a therapeutic agent, which can be entrapped within the polymer matrix or adsorbed or conjugated onto a particle surface. Due to their submicrometer dimensions, nanoparticles offer a number of advantages over microparticulate systems, the most important being their ability to provide targeted drug delivery at the tissue and even cellular levels. Biodegradable nanoparticles based on different polymers, such as poly(alkyl cyanoacrylates) (PACA) or PLGA copolymers, have been investigated extensively for sustained and targeted/localized delivery of various agents, including plasmid DNA, peptides, proteins, and low-molecular-weight compounds [87–90]. In particular, numerous studies have shown that submicrometer colloidal systems represent the choice to control both tissue and cell distribution of anticancer drugs, thus strongly increasing their therapeutic efficacy [68].

In light of their complexing properties, CD co-encapsulation in biodegradable nanoparticles has recently been investigated to improve the properties of the particles in many different ways. Some work has been done on PACA nanoparticles by Duchêne and colleagues [68]. PACAs have gained increasing research interest not only because of their biodegradability, but also for their simple polymerization process, occurring in an aqueous medium without any initiator. Despite these advantages, the major drawback of PACA-based nanoparticles is related to the encapsulation of hydrophobic drugs, such as the cytotoxic agents commonly used in cancer treatment, which can be dissolved only difficulty in a polymerization aqueous medium. Choosing progesterone as a model hydrophobic molecule, da Silveira et al. [91] succeeded in achieving a 50-fold increase in encapsulation efficiency when HP-β-CD was added to the polymerization medium. Along this direction, poly(ethyl cyanoacrylate) (PECA) nanoparticles were prepared in the

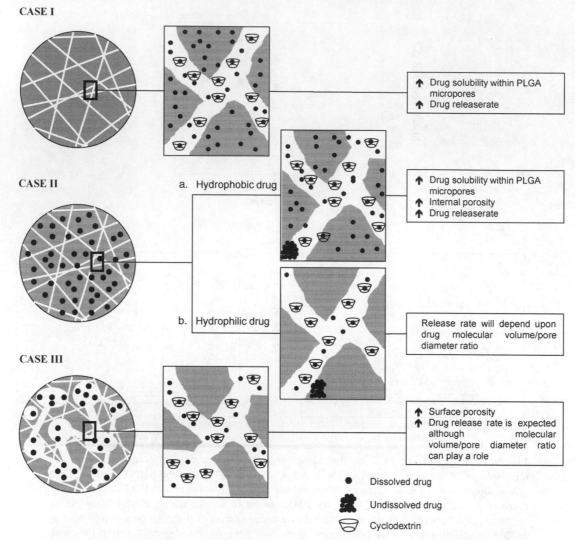

CASE I

↑ Drug solubility within PLGA micropores
↑ Drug releaserate

CASE II

a. Hydrophobic drug

↑ Drug solubility within PLGA micropores
↑ Internal porosity
↑ Drug releaserate

b. Hydrophilic drug

Release rate will depend upon drug molecular volume/pore diameter ratio

CASE III

↑ Surface porosity
↑ Drug release rate is expected although molecular volume/pore diameter ratio can play a role

● Dissolved drug

✿ Undissolved drug

▽ Cyclodextrin

Figure 15. Expected effect of CD addition on the release properties of biodegradable DDS based on hydrophobic polymers such as PLGA. The different cases of molecular solution (case I), solid dispersion (case II), and multireservoir (case III) DDS architectures are analyzed. In a molecular solution, CD is mainly located primarily inside micropores and increases the apparent solubility of the entrapped hydrophobic drug—dispersed in the amorphous region of the polymer—increasing the amount of diffusible species. In the case of a solid dispersion for a hydrophobic drug, the effect of CD addition is similar to that reported in case I. In the case of hydrophilic drugs, both drug and CD are located within the porosity and the overall effect on drug release will depend on the effect of complex formation on drug diffusivity. For a multireservoir system, both drug and CD are located inside the porosity. Surface porosity can increase for osmotic CD, and an increase in the drug release rate is expected.

presence of HPßCD to improve acyclovir solubility in the polymerization medium [92]. Notably, the amount of drug entrapped was not influenced by the preparation conditions, highlighting the fact that the drug-loading capacity of PECA nanoparticles depended also on drug physicochemical characteristics and drug–polymeric matrix affinity and interaction.

To increase the encapsulation efficiency of nanoparticles, CDs have recently been tested also in nanoparticles based on

preformed polymers, such as PLGA, developed to control release of the tripeptide dalargin [93]. Since incorporation of an ionic drug (dalargin was cationically charged under the experimental conditions adopted) can be increased appropriately tailoring polymer opposite charges, anionic sulfobutyl ether-β-cyclodextrin (SB-β-CD) was added in the nanoparticle formulation. A twofold increase in entrapment efficiency was observed, even if limited compared to that achieved by highly charged anionic polymers such as dextran sulfate. Interestingly, no significant difference on dalargin release profiles from nanoparticles was observed when a CD was added to the formulation. Making use of polymer–CD interactions, Alonso's group has developed a novel technique to produce chitosan nanoparticles cross-linked with carboxymethyl-β-cyclodextrin for the fast or slow delivery of macromolecules [94–96].

In the light of these few results, it can be concluded that the CD potential in the formulation of biodegradable nanoparticles has not yet been fully elucidated. CDs remain choice excipients to increase nanoparticle encapsulation efficiency, but whether CDs can play a role in nanoparticle technological and consequent release features is still unknown.

3.4.4. Mechanistic Simplified Interpretation of the CD Effect in DDSs Based on Hydrophobic Biodegradable Polymers

In the light of the literature findings described above and current knowledge on parameters controlling drug release from DDS based on biodegradable polymers, especially PLGA, a simplified mechanistic interpretation of physical and chemical phenomena governing the CD effect on release properties of this class of DDS has been attempted (Fig. 15). The CD effect could be predicted on the basis of fixed parameters (i.e., DDS architecture, drug physicochemical properties). To simplify the interpretation, the effect of CDs in the early release stage is highlighted, whereas the contribute of progressive polymer erosion to DDS interior porosity, and thus in the later release stage, is neglected.

When molecular solutions (case I) or solid dispersions (case IIa) of a hydrophobic drug within the polymer matrix are produced, drug dissolved in the hydrated matrix (i.e., after the hydration step) can be released by (1) direct partitioning from polymer to release medium and (2) dissolution in water-filled micropores and subsequent diffusion out of the system. When CD is added, instant in situ formation of drug–CD inclusion complexes within pores increases drug water solubility. If the drug diffusivity is similar to that of the drug–CD complex ($\mathcal{D}_{dr} \cong \mathcal{D}_{CD}$), the release rate is increased. For solid dispersions, a concomitant factor may contribute to accelerate drug release from a DDS: that is, the enhancement of pore formation within the polymer matrix operated by the CD. Actually, free hydrophilic CD molecules, which dissolve rapidly in the hydration medium, may be released rapidly from the DDS, thus leaving open pores in the matrix which facilitate drug release.

When a solid dispersion of a hydrophilic drug within a polymer matrix is attained (case IIb), drug dissolves rapidly within PLGA micropores and, depending on its molecular volume, may diffuse out of the DDS. A CD can play a role in the drug release rate only if the drug diffusivity is much higher than the CD diffusivity ($\mathcal{D}_{dr} \gg \mathcal{D}_{CD}$); that is, the drug release rate will depend on the complex molecular volume/pore diameter ratio. This phenomenon is expected to decrease the drug release rate, with particular regard to the initial diffusion phase.

The same phenomenon may also occur in multireservoir biodegradable DDS (case III), where hydrophilic species are located within the polymer macroporosity. Nonetheless, the release properties of CD-containing DDSs would also depend on the formulation design and, in particular, on the nature and concentration of the CD used. Actually, a CD displaying osmotic properties in water may act as a pore-forming agent during DDS preparation, thus increasing the surface porosity. This phenomenon is expected to increase the drug release rate. On the other hand, for CDs which are not osmogens, the effect on the release rate will again depend on the complex molecular volume/pore diameter ratio.

Nevertheless, the presence of CD–polymer interactions should also be taken into account since it is well established that some biodegradable polymers can complex CD and in so doing vary some physico-chemical properties [97].

4. CONCLUSIONS

A few years ago it was realized that CDs can be used as aid-excipients in polymeric DDSs to increase drug loading and modify release properties. This effect has been ascribed primarily to CDs complexation ability toward poorly soluble drugs and consequent improvement of their dissolution properties inside a DDS. From a look at the literature produced so far, it is immediately clear that this concept has severely limited their application in drug delivery technologies, in some cases prompting toward "evidence-based" rather than rationalized use. In this chapter, we have clarified the effect of CD introduction in different DDSs on the basis of their possible interactions with both drug and polymer, thus providing a strong rationale to CD selection and proper use. Thus, the achievement of a speeding-up or slowing-down effect on the drug release rate upon CD addition could be less fortuitous and designed *ab initio*. This approach, in our opinion, could contribute to a more extensive and systematic use of CDs in DDS engineering and possibly highlight their great potential in this field.

REFERENCES

1. Swan, E.A., Peppas, N.A. (1981). Drug release kinetics from hydrophobic porous monolithic devices. *Proc. Symp. Control. Release Bioact. Mater.*, 8, 18.

2. Siegel, R.A. (1998). Modeling of drug release from porous systems. In *Controlled Release of Drugs*. VCH, Weinheim, Germany, p. 51.

3. Ofner, C.M., III, Klech-Gelotte, C.M. (2002). Gels and jellies. In: *Encyclopedia of Pharmaceutical Technology*, 2nd ed.

Swarbrick, J., Boylan, J. C., Eds., Marcel Dekker, New York, pp. 1327–1343.

4. Harada, A., Kamachi, M. (1990). Complex-formation between poly(ethylene glycol) and α-cyclodextrin. *Macromolecules, 23,* 2821–2823.

5. Li, J., Harada, A., Kamachi, M. (1994). Formation of inclusion complexes of oligoethylene and its derivatives with α-cyclodextrin. *Bull. Chem. Soc. Jpn., 67,* 2808–2818.

6. Li, J., Harada, A., Kamachi, M. (1994). Sol–gel transition during inclusion complex-formation between α-cyclodextrin and high-molecular-weight poly(ethylene glycol)s in aqueous-solution. *Polym. J., 26,* 1019–1026.

7. Li, J., Ni, X.P., Leong, K.W. (2003). Injectable drug-delivery systems based on supramolecular hydrogels formed by poly(ethylene oxide) and α-cyclodextrin. *J. Biomed. Mater. Res. A, 65,* 196–202.

8. Fujita, H., Ooya, T., Yui, N. (1999). Synthesis and characterization of a polyrotaxane consisting of β-cyclodextrins and a poly(ethylene glycol)–poly(propylene glycol) triblock copolymer. *Macromol. Chem. Phys., 200,* 706–713.

9. Fujita, H., Ooya, T., Yui, N. (1999). Thermally induced localization of cyclodextrins in a polyrotaxane consisting of β-cyclodextrins and poly(ethylene glycol)–poly(propylene glycol) triblock copolymer. *Macromolecules, 32,* 2534–2541.

10. Agnely, F., Djedour, A., Bochot, A., Grossiord, J.L. (2006). Properties of various thermoassociating polymers: pharmaceutical and cosmetic applications. *J. Drug Deliv. Sci. Technol., 16,* 3–10.

11. Mayer, B., Klein, C.T., Topchieva, I.N., Kohler, G. (1999). Selective assembly of cyclodextrins on poly(ethylene oxide)–poly(propylene oxide) block copolymers. *J. Comput.-Aided Mol. Des., 13,* 373–383.

12. Li, J., Loh, X.J. (2008). Cyclodextrin-based supramolecular architectures: syntheses, structures, and applications for drug and gene delivery. *Adv. Drug Deliv. Rev., 60,* 1000–1017.

13. Olson, K., Chen, Y.Y., Baker, G.L. (2001). Inclusion complexes of α-cyclodextrin and (AB)(n) block copolymers. *J. Polym. Sci. A Polym. Chem., 39,* 2731–2739.

14. Li, X., Li, J. (2008). Supramolecular hydrogels based on inclusion complexation between poly(ethylene oxide)–β-poly(ε-caprolactone) diblock copolymer and α-cyclodextrin and their controlled release property. *J. Biomed. Mater. Res. A, 86,* 1055–1061.

15. Sarkar, N. (1979). Thermal gelation properties of methyl and hydroxypropyl methylcellulose. *J. Appl. Polym. Sci., 24,* 1073–1087.

16. Mitchell, K., Ford, J.L., Armstrong, D.J., Elliott, P.N.C., Hogan, C. (1990). The influence of additives on the cloud point, disintegration and dissolution of hydroxypropylmethyl cellulose gels and matrix tablets. *Int. J. Pharm., 66,* 233–242.

17. Pose-Vilarnovo, B., Rodriguez-Tenreiro, C., Rosa dos Santos, J.F., Vazquez-Doval, J., Concheiro, A., Alvarez-Lorenzo, C., Torres-Labandeira, J.J. (2004). Modulating drug release with cyclodextrins in hydroxypropyl methylcellulose gels and tablets. *J. Control. Release, 94,* 351–363.

18. Jug, M., Bećirević-Laćan, M., Kwokal, A., Cetina-Cizmek, B. (2005). Influence of cyclodextrin complexation on piroxicam gel formulations. *Acta Pharm., 55,* 223–236.

19. Boulmedarat, L., Grossiord, J.L., Fattal, E., Bochot, A. (2003). Influence of methyl-β-cyclodextrin and liposomes on rheological properties of Carbopol 974P NF gels. *Int. J. Pharm., 254,* 59–64.

20. Kim, E.Y., Gao, Z.G., Park, J.S., Li, H., Han, K. (2002). rhEGF/HP-β-CD complex in poloxamer gel for ophthalmic delivery. *Int. J. Pharm.s, 233,* 159–167.

21. Dumortier, G., Grossiord, J.L., Agnely, F., Chaumeil, J.C. (2006). A review of Poloxamer 407 pharmaceutical and pharmacological characteristics. *Pharm. Res., 23,* 2709–2728.

22. Samy, E.M., Safwat, S.M. (1994). In vitro release of anti-inflammatory drugs with β-cyclodextrin from hydrophilic gel bases. *STP Pharma. Sci., 4,* 458–465.

23. Rathnam, G., Narayanan, N., Ilavarasan, R. (2008). Carbopol-based gels for nasal delivery of progesterone. *AAPS PharmSciTech, 9,* 1078–1082.

24. Bilensoy, E., Rouf, M.A., Vural, I., Şen, M., Hincal, A.A. (2006). Mucoadhesive, thermosensitive, prolonged-release vaginal gel for clotrimazole:β-cyclodextrin complex. *AAPS PharmSciTech., 7,* E38.

25. Peppas, N.A., Ed. (1987) *Hydrogels in Medicine and Pharmacy.* CRC Press, Boca Raton, FL.

26. Quaglia, F., Varricchio, G., Miro, A., La Rotonda, M.I., Larobina, D., Mensitieri, G. (2001). Modulation of drug release from hydrogels by using cyclodextrins: the case of nicardipine/beta-cyclodextrin system in crosslinked polyethylenglycol. *J. Control. Release, 71,* 329–337.

27. Tomic, K., Veeman, W.S., Boerakker, M., Litvinov, V.M., Dias, A.A. (2008). Lateral and rotational mobility of some drug molecules in a poly(ethylene glycol) diacrylate hydrogel and the effect of drug–cyclodextrin complexation. *J. Pharma. Sci., 97,* 3245–3256.

28. Inactive ingredient search for approved drug products. http://www.accessdata.fda.gov/scripts/cder/iig/index.cfm, 2009.

29. Rowe, R.C., Sheskey, P.J., Qinn, M.E., Eds. (2006). *Handbook of Pharmaceutical Excipients,* 5th ed. Pharmaceutical Press, Chicago.

30. Tiwari, S.B., Rajabi-Siahboomi, A.R. (2008). Extended-release oral drug delivery technologies: monolithic matrix systems. *Methods Mol. Biol., 437,* 217–243.

31. Conte, U., Giunchedi, P., Maggi, L., La Manna, A. (1993). Erodible matrixes containing hydroxypropyl β-cyclodextrin for linear release of a water-insoluble drug (diazepam). *STP Pharma Sci., 3,* 242–249.

32. Giunchedi, P., Maggi, L., La Manna, A., Conte, U. (1994). Modification of the dissolution behaviour of a water-insoluble drug, naftazone, for zero-order release matrix preparation. *J. Pharm. Pharmacol., 46,* 476–480.

33. Sangalli, M.E., Zema, L., Maroni, A., Foppoli, A., Giordano, F., Gazzaniga, A. (2001). Influence of β-cyclodextrin on the release of poorly soluble drugs from inert and hydrophilic heterogeneous polymeric matrices. *Biomaterials, 22,* 2647–2651.

34. Rao, V.M., Haslam, J.L., Stella, V.J. (2001). Controlled and complete release of a model poorly water-soluble drug, prednisolone, from hydroxypropyl methylcellulose matrix tablets using (SBE)(7m)-β-cyclodextrin as a solubilizing agent. *J. Pharm. Sci.*, *90*, 807–816.

35. Jug, M., Bećirević-Laćan, M. (2004). Influence of hydroxypropyl-β-cyclodextrin complexation on piroxicam release from buccoadhesive tablets. *Eur. J. Pharm. Sci.*, *21*, 251–260.

36. Singh, S., Jain, S., Muthu, M.S., Tilak, R. (2008). Preparation and evaluation of buccal bioadhesive tablets containing clotrimazole. *Curr. Drug Deliv.*, *5*, 133–141.

37. Shanker, G., Kumar, C.K., Gonugunta, C.S., Kumar, B.V., Veerareddy, P.R. (2009). Formulation and evaluation of bioadhesive buccal drug delivery of tizanidine hydrochloride tablets. *AAPS PharmSciTech*, *10*, 530–539.

38. Cappello, B., De Rosa, G., Giannini, L., La Rotonda, M.I., Mensitieri, G., Miro, A., Quaglia, F., Russo, R. (2006). Cyclodextrin-containing poly(ethyleneoxide) tablets for the delivery of poorly soluble drugs: potential as buccal delivery system. *Int. J. Pharm.*, *319*, 63–70.

39. Koester, L.S., Xavier, C.R., Mayorga, P., Bassani, V.L. (2003). Influence of β-cyclodextrin complexation on carbamazepine release from hydroxypropyl methylcellulose matrix tablets. *Eur. J. Pharm. Biopharm.*, *55*, 85–91.

40. Miro, A., Quaglia, F., Giannini, L., Cappello, B., La Rotonda, M.I. (2006). Drug/cyclodextrin solid systems in the design of hydrophilic matrices: a strategy to modulate drug delivery rate. *Curr. Drug Deliv.*, *3*, 373–378.

41. Shivakumar, H.N., Desai, B.G., Pandya, S., Karki, S.S. (2007). Influence of β-cyclodextrin complexation on glipizide release from hydroxypropyl methylcellulose matrix tablets. *PDA J. Pharm. Sci. Technol.*, *61*, 472–491.

42. Bettini, R., Catellani, P.L., Santi, P., Massimo, G., Peppas, N.A., Colombo, P. (2001). Translocation of drug particles in HPMC matrix gel layer: effect of drug solubility and influence on release rate. *J. Control. Rel.*, *70*, 383–391.

43. Jug, M., Bećirević-Laćan, M., Bengez, S. (2009). Novel cyclodextrin-based film formulation intended for buccal delivery of atenolol. *Drug Dev. Ind. Pharm.*, *35* (7), 786–807.

44. Ammar, H.O., Salama, H.A., El Nahhas, S.A., Elmotasem, H. (2008). Design and evaluation of chitosan films for transdermal delivery of glimepiride. *Curr. Drug Deliv.*, *5*, 290–298.

45. Miro, A., Rondinone, A., Nappi, A., Ungaro, F., Quaglia, F., La Rotonda, M.I. (2009). Modulation of release rate and barrier transport of diclofenac incorporated in hydrophilic matrices: role of cyclodextrins and implications in oral drug delivery. *Eur. J. Pharma. Biopharm.*, *72*, 76–82.

46. Rathbone, M.J., Ponchel, G. Chazali, F.A. (1996). Systemic oral mucosal drug delivery and delivery systems. In: Rathbone, M. J., Ed., *Oral Mucosal Drug Delivery*. Marcel Dekker, New York.

47. Loftsson, T., Vogensen, S.B., Brewster, M.E., Konradsdóttir, F. (2007). Effects of cyclodextrins on drug delivery through biological membranes. *J. Pharm. Sci.*, *96*, 2532–2546.

48. Bibby, D.C., Davies N.M., Tucker, I.G. (2000). Mechanisms by which cyclodextrins modify drug release from polymeric drug delivery systems. *Int. J. Pharm.*, *197*, 1–11.

49. Zannou, E.A., Streng, W.H., Stella, V.J. (2001). Osmotic properties of sulfobutylether and hydroxypropyl cyclodextrins. *Pharm. Res.*, *18*, 1226–1231.

50. Okimoto, K., Tokunaga, Y., Ibuki, R., Irie, T., Uekama, K., Rajewski, R.A., Stella, V.J. (2004). Applicability of (SBE)7m-β-CD in controlled-porosity osmotic pump tablets (OPTs). *International Journal of Pharmaceutics*, *286*, 81–88.

51. Okimoto, K., Rajewski, R.A., Stella, V.J. (1999). Release of testosterone from an osmotic pump tablet utilizing (SBE)7m-β-cyclodextrin as both a solubilizing and an osmotic pump agent. *J. Control. Rel.*, *58*, 29–38.

52. Okimoto, K., Miyake, M., Ohnishi, N., Rajewski, R.A., Stella, V.J., Irie, T., Uekama, K. (1998). Design and evaluation of an osmotic pump tablet (OPT) for prednisolone, a poorly water soluble drug, using (SBE)7m-β-CD. *Pharm. Res.*, *15*, 1562–1568.

53. Okimoto, K., Ohike, A., Ibuki, R., Aoki, O., Ohnishi, N., Irie, T., Uekama, K., Rajewski, R.A., Stella, V.J. (1999). Design and evaluation of an osmotic pump tablet (OPT) for chlorpromazine using (SBE)7m-β-CD. *Pharm. Res.*, *16*, 549–554.

54. Okimoto, K., Ohike, A., Ibuki, R., Aoki, O., Ohnishi, N., Rajewski, R.A., Stella, V.J., Irie, T., Uekama, K. (1999). Factors affecting membrane-controlled drug release for an osmotic pump tablet (OPT) utilizing (SBE)(7m)-β-CD as both a solubilizer and osmotic agent. *J. Control. Rel.*, *60*, 311–319.

55. Sotthivirat, S., Haslam, J.L., Stella, V.J. (2007). Controlled porosity-osmotic pump pellets of a poorly water-soluble drug using sulfobutylether-β-cyclodextrin, (SBE) 7m-β-CD, as a solubilizing and osmotic agent. *J Pharm. Sci.*, *96*, 2364–2374.

56. Sotthivirat, S., Lubach, J.W., Haslam, J.L., Munson, E.J., Stella, V.J. (2007). Characterization of prednisolone in controlled porosity osmotic pump pellets using solid-state NMR spectroscopy. *J. Pharm. Sci.*, *96*, 1008–1017.

57. Sotthivirat, S., Haslam, J.L., Stella, V.J. (2007). Evaluation of various properties of alternative salt forms of sulfobutylether-β-cyclodextrin, (SBE)7m-β-CD. *Int. J. Pharm.*, *330*, 73–81.

58. Sotthivirat, S., Haslam, J.L., Lee, P.I., Rao, V.M., Stella, V.J. (2009). Release mechanisms of a sparingly water-soluble drug from controlled porosity-osmotic pump pellets using sulfobutylether-β-cyclodextrin as both a solubilizing and osmotic agent. *J. Pharma. Sci.*, *98*, 1992–2000.

59. Uhrich, K.E., Cannizzaro, S.M., Langer, R.S., Shakesheff, K. M. (1999). Polymeric systems for controlled drug release. *Chem. Rev.*, *99*, 3181–3198.

60. Markland, P., Yang, V.C. (2002). Biodegradable polymers as drug carriers. In: Swarbrick, J., Boylan, J.C.Eds., *Encyclopedia of Pharmaceutical Technology*, 2nd Ed. Marcel Dekker, New York, pp. 136–155.

61. Siepmann, J., Siepmann, F. (2009). Time-controlled drug delivery systems. In: Florence, A.T., Siepmann, J.Eds., *Modern Pharmaceutics*, Vol. 2, *Application and Advances*. Informa Healthcare, New York, pp. 1–22.

62. Brannon-Peppas, L., Vert, M. (2000). Polylactic and polyglycolic acids as drug delivery carriers. In: Wise, D.L.Ed., *Handbook of Pharmaceutical Controlled Release Technology*. Marcel Dekker, New York, pp 99–130.

63. Weinberg, B.D., Blanc, E., Ga, J.M. (2008). Polymer implants for intratumoral drug delivery and cancer therapy. *Journal of Pharmaceutical Sciences*, *97*, 1681–1702.

64. Blanco, E., Weinberg, B.D., Stowe, N.T., Anderson, J.M., Gao, J.M. (2006). Local release of dexamethasone from polymer millirods effectively prevents fibrosis after radiofrequency ablation. *J. Biomed. Materials Research Part A*, *76*, 174–182.

65. Wang, F., Blanco, E., Ai, H., Boothman, D.A., Gao, J. (2006). Modulating β-lapachone release from polymer millirods through cyclodextrin complexation. *J. Pharm. Sci.*, *95*, 2309–2319.

66. Wang, F.J., Saidel, G.M., Gao, J.M. (2007). A mechanistic model of controlled drug release from polymer millirods: effects of excipients and complex binding. *J. Control. Release*, *119*, 111–120.

67. Desai, K.G.H., Mallery, S.R., Schwendeman, S.P. (2008). Formulation and characterization of injectable poly(DL-lactide-*co*-glycolide) implants loaded with *N*-acetylcysteine, a MMP inhibitor. *Pharm. Res.*, *25*, 586–597.

68. Duchêne, D., Ponchel, G., Wouessidjewe, D. (1999). Cyclodextrins in targeting: application to nanoparticles. *Adv. Drug Deliv. Rev.*, *36*, 29–40.

69. Jiang, W., Gupta, R.K., Deshpande, M.C., Schwendeman, S.P. (2005). Biodegradable poly(lactic-*co*-glycolic acid) microparticles for injectable delivery of vaccine antigens. *Adv. Drug Deliv. Rev.*, *57*, 391–410.

70. O'Hagan, D.T., Singh, M., Ulmer, J.B. (2004). Microparticles for the delivery of DNA vaccines. *Immunol. Rev.*, *199*, 191–200.

71. Pawar, R., Ben Ari, A., Domb, A.J. (2004). Protein and peptide parenteral controlled delivery. *Expert Opin. Biol. Ther.*, *4*, 1203–1212.

72. Tamber, H., Johansen, P., Merkle, H.P., Gander, B. (2005). Formulation aspects of biodegradable polymeric microspheres for antigen delivery. *Adv. Drug Deliv. Rev.*, *57*, 357–376.

73. Sinisterra, R.D., Shastri, V.P., Najjar, R., Langer, R. (1999). Encapsulation and release of rhodium(II) citrate and its association complex with hydroxypropyl-β-cyclodextrin from biodegradable polymer microspheres. *J. Pharm. Sci.*, *88*, 574–576.

74. Yue, I.C., Poff, J., Cortes, M.E., Sinisterra, R.D., Faris, C.B., Hildgen, P., Langer, R., Shastri, V.P. (2004). A novel polymeric chlorhexidine delivery device for the treatment of periodontal disease. *Biomaterials*, *25*, 3743–3750.

75. Trapani, G., Lopedota, A., Boghetich, G., Latrofa, A., Franco, M., Sanna, E., Liso, G. (2003). Encapsulation and release of the hypnotic agent zolpidem from biodegradable polymer microparticles containing hydroxypropyl-β-cyclodextrin. *Int. J. Pharm.*, *268*, 47–57.

76. Ungaro, F., De Rosa, G., Miro, A., Quaglia, F., La Rotonda, M.I. (2006). Cyclodextrins in the production of large porous particles: development of dry powders for the sustained release of insulin to the lungs. *Eur. J. Pharm. Sci.*, *28*, 423–432.

77. Ungaro, F., d'Emmanuele, V, Giovino, C., Miro, A., Sorrentino, R., Quaglia, F., La Rotonda, M.I. (2009). Insulin-loaded PLGA/cyclodextrin large porous particles with improved aerosolization properties: in vivo deposition and hypoglycaemic activity after delivery to rat lungs. *J. Control. Release*, *135*, 25–34.

78. Morlock, M., Kissel, T., Li, Y.X., Koll, H., Winter, G. (1998). Erythropoietin loaded microspheres prepared from biodegradable LPLG–PEO–LPLG triblock copolymers: protein stabilization and in-vitro release properties. *J. Control. Release*, *56*, 105–115.

79. Meinel, L., Illi, O.E., Zapf, J., Malfanti, M., Peter, M.H., Gander, B. (2001). Stabilizing insulin-like growth factor-I in poly(D,L-lactide-*co*-glycolide) microspheres. *J. Control. Release*, *70*, 193–202.

80. Murillo, M., Goni, M.M., Irache, J.M., Arangoa, M.A., Blasco, J.M., Gamazo, C. (2002). Modulation of the cellular immune response after oral or subcutaneous immunization with microparticles containing *Brucella ovis* antigens. *J. Control. Release*, *85*, 237–246.

81. Rodrigues Junior, J.M., de Melo, L.K., Matos Jensen, C.E., de Aguiar, M.M., Silva Cunha, J.A. (2003). The effect of cyclodextrins on the in vitro and in vivo properties of insulin-loaded poly (D,L-lactic-*co*-glycolic acid) microspheres. *Artif. Organs*, *27*, 492–497.

82. Morlock, M., Koll, H., Winter, G., Kissel, T. (1997). Microencapsulation of rh-erythropoietin, using biodegradable poly (d,l-lactide-*co*-glycolide): protein stability and the effects of stabilizing excipients. *Eur. J. Pharm. Biopharm.*, *43*, 29–36.

83. Irie, T., Uekama, K. (1999). Cyclodextrins in peptide and protein delivery. *Adv. Drug Deliv. Rev.*, *36*, 101–123.

84. Castellanos, I.J., Flores, G., Griebenow, K. (2006). Effect of cyclodextrins on α-chymotrypsin stability and loading in PLGA microspheres upon s/o/w encapsulation. *J. Pharm. Sci.*, *95*, 849–858.

85. De Rosa, G., Larobina, D., Immacolata, L.R., Musto, P., Quaglia, F., Ungaro, F. (2005). How cyclodextrin incorporation affects the properties of protein-loaded PLGA-based microspheres: the case of insulin/hydroxypropyl-β-cyclodextrin system. *J. Control. Release*, *102*, 71–83.

86. Quaglia, F., De Rosa, G., Granata, E., Ungaro, F., Fattal, E., Immacolata, L.R. (2003). Feeding liquid, non-ionic surfactant and cyclodextrin affect the properties of insulin-loaded poly (lactide-*co*-glycolide) microspheres prepared by spray-drying. *J. Control. Release*, *86*, 267–278.

87. Bala, I., Hariharan, S., Kumar, M.N. (2004). PLGA nanoparticles in drug delivery: the state of the art. *Crit. Rev. Ther. Drug Carrier Syst.*, *21*, 387–422.

88. Barratt, G. (2003). Colloidal drug carriers: achievements and perspectives. *Cellu. Mol. Life Sci.*, *60*, 21–37.

89. Kayser, O., Lemke, A., Hernandez-Trejo, N. (2005). The impact of nanobiotechnology on the development of new drug delivery systems. *Curr. Pharm. Biotechnol.*, *6*, 3–5.

90. Couvreur, P., Vauthier, C. (2006). Nanotechnology: intelligent design to treat complex disease. *Pharm. Res.*, *23*, 1417–1450.

91. da Silveira, A.M., Ponchel, G., Puisieux, F., Duchêne, D. (1998). Combined poly(isobutylcyanoacrylate) and cyclodextrins nanoparticles for enhancing the encapsulation of lipophilic drugs. *Pharm. Res.*, *15*, 1051–1055.

92. Fresta, M., Fontana, G., Bucolo, C., Cavallaro, G., Giammona, G., Puglisi, G. (2001). Ocular tolerability and in vivo bioavailability of poly(ethylene glycol) (PEG)-coated polyethyl-2-cy-

anoacrylate nanosphere-encapsulated acyclovir. *J. Pharm. Sci,.*, *90*, 288–297.

93. Chen, Y., Wang, F., Benson, H.A. (2008). Effect of formulation factors on incorporation of the hydrophilic peptide dalargin into PLGA and mPEG-PLGA nanoparticles. *Biopolymers, 90*, 644–650.

94. Krauland, A.H., Alonso, M.J. (2007). Chitosan/cyclodextrin nanoparticles as macromolecular drug delivery system. *Int. J. Pharm.*, *340*, 134–142.

95. Teijeiro-Osorio, D., Remunan-Lopez, C., Alonso, M.J. (2009). Chitosan/cyclodextrin nanoparticles can efficiently transfect the airway epithelium in vitro. *Eur. J. Pharm. Biopharm.*, *71*, 257–263.

96. Teijeiro-Osorio, D., Remunan-Lopez, C., Alonso, M.J. (2009). New generation of hybrid poly/oligosaccharide nanoparticles as carriers for the nasal delivery of macromolecules. *Biomacromolecules*, *10*, 243–249.

97. Shin, K.M., Dong, T., He, Y., Taguchi, Y., Oishi, A., Nishida, H., Inoue, Y. (2004). Inclusion complex formation between α-cyclodextrin and biodegradable aliphatic polyesters. *Macromol. Biosci.*, *4*, 1075–1083.

98. Swarbrick, J., Boylan, J.C.Eds. (2002). *Encyclopedia of Pharmaceutical Technology*, 2nd ed., Marcel Dekker, New York.

99. Batycky, R.P., Hanes, J., Langer, R., Edwards, D.A. (1997). A theoretical model of erosion and macromolecular drug release from biodegrading microspheres. *J. Pharm. Sci.*, *86*, 1464–1477.

5

RECENT FINDINGS ON SAFETY PROFILES OF CYCLODEXTRINS, CYCLODEXTRIN CONJUGATES, AND POLYPSEUDOROTAXANES

Hidetoshi Arima, Keiichi Motoyama, and Tetsumi Irie

Graduate School of Pharmaceutical Sciences, Kumamoto University, Kumamoto, Japan

1. INTRODUCTION

Cyclodextrins (CDs, Fig. 1), cyclic oligosaccharides consisting of several glucopyranose units, are host molecules that form inclusion complexes. The benefits of the use of CDs and their complexes can be employed in the formulation of food, cosmetics, and pharmaceuticals [1–3]. So far, the usefulness of three parent CDs (α-CD, β-CD, and γ-CD, Fig. 1) in drug delivery has been reported with respect to drug stabilization, improvement, and modulation in drug release and enhancement of drug bioavailability and alleviation of local irritation [4–6]. In addition, the structural and physicochemical properties of macrocyclic CDs such as δ-CD and ε-CD have been characterized [7].

Various types of chemically modified CD derivatives have been prepared to extend the physicochemical properties and inclusion capacity of parent CDs [8,9]. Table 1 summarizes the pharmaceutically useful β-CD derivatives, classified into hydrophilic, hydrophobic, and ionizable derivatives [10]. Among these compounds, methylated CDs such as 2,6,-di-O-methy-β-CD (DM-β-CD) and 2,3,6-tri-O-methyl-β-CD (TM-β-CD) are surface-active and have superior inclusion ability to β-CD. In addition, methyl-β-CD (M-β-CD) has the high solubilizing ability of cholesterol and is used widely as a lipid raft–disrupting agent, as described below. Hydroxypropyl-β-CD (HP-β-CD) and branched β-CD have received special attention because their toxicity is extremely low and their aqueous solubility is very high, promising parenteral use [11–13]. Among branched CDs, a glucuronyl-glucosyl-β-CD (GUG-β-CD) is a new entry in the branched CDs which contains a carboxyl group in the glucuronyl group and is known to have a molecular chaperonlike activity for protein aggregation [11]. Notably, CD derivatives having a targeting ability for cell-specific drug delivery have been developed. For example, heterobranched CDs having a galactose, fucose, or mannose moiety to target CDs toward lectin on cell surfaces have been reported [14–16], as well as folate-appended CD derivatives to deliver drugs to tumor cells that overexpress folate receptors on cell surfaces [17,18]. The potential use of various new CD derivatives as drug carriers has also been reported [19]. The hydrophobic CDs include both alkylated and acylated CDs. Of the alkylated CDs, ethylated CDs such as 2,6-di-O-ethyl-β-CD (DE-β-CD), which can retard the dissolution rate of water-soluble drugs, may serve as sustained release carriers for water-soluble drugs, including peptides and proteins. Of the acylated CDs, tri-O-valeryl-β-CD (TV-β-CD) is known to form an adhesive film, and films that include water-soluble drugs such as molsidomine and isosorbide dinitrate have the potential to be prolonged-release systems [20]. Furthermore, the ionizable CDs include sulfobutyl ether β-CD (SBE7-β-CD) having a degree of substitution of a sulfobutyl ether group of approximately 7, especially Captisol, which can improve the inclusion capacity, a modification of dissolution rate, and the alleviation of local irritation by drugs. Thus, these CDs are known to alter various properties of drugs, pharmaceutical formulations, and biomembranes, resulting in enhancement and/or modulation of bioavailability in various routes

Cyclodextrins in Pharmaceutics, Cosmetics, and Biomedicine: Current and Future Industrial Applications, First Edition. Edited by Erem Bilensoy.
© 2011 John Wiley & Sons, Inc. Published 2011 by John Wiley & Sons, Inc.

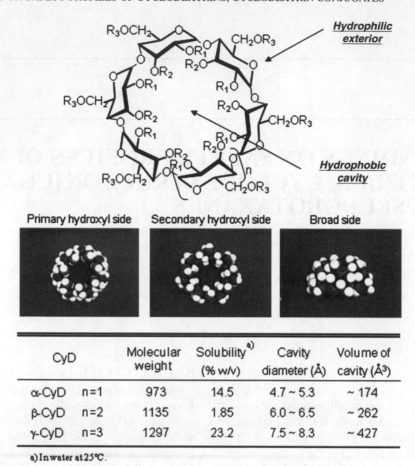

CyD		Molecular weight	Solubility[a] (% w/v)	Cavity diameter (Å)	Volume of cavity (Å³)
α-CyD	n=1	973	14.5	4.7 ~ 5.3	~ 174
β-CyD	n=2	1135	1.85	6.0 ~ 6.5	~ 262
γ-CyD	n=3	1297	23.2	7.5 ~ 8.3	~ 427

a) In water at 25°C.

Figure 1. Structures and properties of natural CDs.

Table 1. Pharmaceutically Useful β-CD Derivatives

Derivative[a]	Characteristic	
Hydrophilic derivatives		
Methylated β-CD	Soluble in cold water and in organic solvents, surface-active, hemolytic	Oral, dermal, mucosal[b]
DM-β-CD		
TM-β-CD		
DMA-β-CD	Soluble in water, poorly hemolytic	Parenteral
Hydroxyalkylated β-CD		
2-HE-β-CD	Amorphous mixture with different degrees of substitution, highly water-soluble (>50%), low toxicity	Oral, dermal, mucosal, parenteral (intravenous)
2-HP-β-CD		
3-HP-β-CD		
2,3-DHP-β-CD		
2-HB-β-CD		
Branched β-CD		
G$_1$-β-CD	Highly water-soluble (>50%), low toxicity	Oral, mucosal, parenteral (intravenous)
G$_2$-β-CD		
GUG-β-CD		
Hydrophobic derivatives		
Alkylated β-CD		
DE-β-CD	Poorly water-soluble, soluble in organic solvents, surface-active	Oral, parenteral (subcutaneous) (slow release)

Table 1. (*Continued*)

Derivative[a]	Characteristic	
TE-β-CD		
Acylated β-CD(C$_2$–C$_{18}$)		
TA-β-CD	Poorly water-soluble, soluble in organic solvents	Oral, dermal (slow release)
TV-β-CD	Film formation	
Ionizable derivatives		
Anionic β-CD		
CME-β-CD	pK_a = 3 to 4, soluble at pH > 4	
S-β-CD	pK_a > 1, water soluble	
SBE-β-CD		
β-CD-phosphate		
Aluminium-β-CD·sulfate	Water-insoluble	

[a] DM, 2,6-di-*O*-methyl; TM, per-*O*-methyl; DMA, acetylated DM-β-CD;2-HE, 2-hydroxyethyl; 2-HP,2-hydroxypropyl; 3-HP, 3-hydroxypropyl; 2,3-DHP,2,3-dihydroxypropyl; 2-HB,2-hydroxybutyl; G$_1$, glucosyl; G$_2$, maltosyl; GUG, glucuronyl–glucosyl; DE, 2, 6-di-*O*-ethyl; TE, per-*O*-ethyl; TA, per-*O*-acetyl, TV, per-*O*-valeryl; CME, *O*-carboxymethyl–*O*-ethyl; S, sulfate; SBE, sulfobutyl ether.
[b] Mucosal; nasal, sublingual, ophthalmic, pulmonary, rectal, vaginal, etc.

[21–24]. Actually, it is estimated that there are over 30 CD-containing commercially available pharmaceutical products (Table 2). It should be noted that drug complexes with HP-β-CD and SBE7-β-CD can be widely prescribed, and those containing randomly methyl-β-CD (RM-β-CD) and hydro-xypropyl-γ-CD (HP-γ-CD) have been used in hospitals as well.

The conjugates of CDs with drugs or polymers have been designed and evaluated. CD complex is in equilibrium with guest and host molecules in aqueous solution, with the degree

Table 2. Marketed Pharmaceutical Products Containing CDs

Drugs	Trade Name	Formulation	Company/Country
α-CD			
Alprostadil (PGE)	Prostavastin, Caverject, Edex	Intravenous	Ono/Japan, Pfizer/U.S. Schwarts/EU
Cefotiam hexetil HCI	Pansporin T	Tablet	Takeda/Japan
Limaprost	Opalmon	Tablet	Ono/Japan
β-CD			
Benexate HCl	Ulgut, Lonmiel	Capsule	Teikoku/Japan, Shionogi/Japan
Betahistidine	Betahist	Tablet	Geno Pharmaceuticals/India
Cephalosporin	Meiact	Tablet	Meiji Seikai/Japan
Cetirizine	Cetirizine	Chewing tablet	Losan Pharma/Germany
Chlordiazepoxide	Transillium	Tablet	Gador/Argentina
Dexamethasone	Glymesason	Ointment, tablet	Fujinaga/Japan
Dextromethorphan	Rynathisol	Tablet	Synthelabo/EU
Dinoprostone (PGE$_2$)	Prostarmon E	Sublingual tablet	Ono/Japan
Diphenhydramin, chlortheophyllin	Stada-Travel	Chewing tablet	Stada/EU
Flunarizine	Fluner	Tablet	Geno Pharmaceuticals/India
Garlic oil	Xund, Tegra	Dragée	Bipharm, Hermes/EU
Iodine	Mena-Gargle	Solution	Kyushin/Japan
Meloxicam	Mobitil	Tablet, suppository	Medical Union Phamaceuticals/Egypt
Nicotine	Nicorette	Sublingual tablet	Pfizer/EU
Nimesulide	Nimedex, Mesulid	Tablet	Novartis/EU
Nitroglycerin	Nitropen	Sublingual tablet	Nihon Kayakui/Japan
Omeprazole	Omebeta	Tablet	Betafarm/EU
Piroxicam	Brexin, Cicladol, Flogene	Tablet, suppository	Chiesi/EU, Ranbaxy/India, Aché/Brazil
Rofecoxib	Rofizgel	Tablet	Wockhardt/India
Tiaprofenic acid	Surgamyl	Tablet	Roussel-Maestrell/EU

Table 2. (*Continued*)

Drugs	Trade Name	Formulation	Company/Country
γ-CD			
Minoxidil	Alopexy	Solution	Pierre Fabre/IJK
Morphine	Moraxen	Suppository	Schwarz/UK
Hydroxypropyl-β-CD			
Cisapride	Propulsid	Suppository	Janssen/EU
Hydrocortisone	Dexocort	Solution	Actavis/EU
Indomethacin	Indocid	Eyedrops	Chauvin/EU
Itraconazole	Sporanox	Oral, intravenous	Janssen/EU, U.S., Japan
Mitomycin	MitoExtra, Mitozytrex	Intravenous	Novartis/EU, SuperGen/U.S.
Hydraxypropyl-γ-CD			
Dicrofenac sodium	Voltaren optha	Eyedrops	Novartis/EU
Tc-99m teoboroxime	CardioTec	Intravenous solution	Bracco/U.S.
Aripiprazole	Abilify	intramuscular solution	Bristol-Myers Squibb/U.S., Otsuka Pharm/Japan, U.S.
Sutfobutyl ether-β-CD			
Maropitant	Cerenia	Parenteral solution	Pfizer Animal Health/U.S.
Voliconazole	Vfend	Intravenous solution	Pfizer/U.S., EU, Japan
Ziprasidone mesylate	Geodon, Zeldox	intramuscular solution	Pfizer/Europe, U.S.
Methyl β-CD			
Chloramphenicol	Clorocil	Eyedrops	Oftalder/EU
17β-Estradiol	Aerodiol	Nasal spray	Servier/EU

of dissociation being dependent on the magnitude of the stability constants of the complex. This property is desirable because the complex dissociates to give free CD and drug at the absorption site, and thus only drug in a free form enters the systemic circulation. However, the inclusion equilibrium is sometimes disadvantageous when drug targeting is to be attempted, because the complex dissociates before it reaches the organ or tissues to which it is to be delivered. Meanwhile, CDs interact only very slightly with nucleic acid drugs such as plasmid DNA, antisense DNA, decoy DNA, small interfering RNA (siRNA), and micro-RNA (miRNA). Thus, CDs deliver nucleic acid drugs into cells inefficiently. To overcome these problems, a method used to prevent dissociation is to bind a drug or a carrier covalently to CD. Therefore, the CD conjugate approach can provide a versatile means of drug delivery. That is, CD–drug conjugates have the potential to act as a colon-targeting system of drugs, including a steroid drug [25–27], nonsteroidal anti-inflammatory drugs (NSAIDs) [28,29], and butyric acid [30]. Meanwhile, CD–cationic polymer conjugates may be novel candidates for nonviral vectors to enhance gene and siRNA transfer [31–36] and bioconjugates of folic acid (FA) to β-CDs through a poly (ethylene glycol) spacer (CD–PEG–FA) and to delivery drugs to tumor cells [17,18]. Furthermore, we have demonstrated that polypseudorotaxanes of CDs with pegylated proteins can be used as a sustained system of proteins [37,38]. Also, polypseudorotaxanes of CDs with PEG have been reported to work as sustained-release carriers for proteins, DNA, and RNA [39]. In this review, we introduce primarily the safety profiles of CD;

CD derivatives, including Sugammadex; CD conjugates with drugs and polymers; and polypseudorotaxanes containing CDs.

2. SAFETY PROFILES OF CDs IN VITRO

2.1. Hemolysis

One of the most substantial requirements for drug carriers is that they have either no or acceptably low levels of intrinsic cytotoxicity. The hemolytic data are known to prove a simple and reliable measure for the estimation of CD-induced membrane damage or cytotoxicity, because the interaction of CDs with plasma membranes must be the initial step in cell damage. In contrast, CDs are well known to suppress hemolysis induced by amphiphilic drugs such as phenothiazine neuroleptics, flufenamic acid, imipramine, bile acids, and DY-9760e through the inclusion complexation of drugs with CDs, depending on the magnitude of the stability constants of the complexes [40–45]. Actually, at first step CDs induce morphological changes in erythrocytes followed by hemolysis. Parent CDs at higher concentrations cause hemolysis of human and rabbit erythrocytes in the order γ-CD < α-CD < β-CD in isotonic solution, and macrocyclic CDs over γ-CD have lower hemolytic activity than γ-CD. The potencies of CDs for solubilizing various components of erythrocytes were α-CD greater than β-CD much greater than γ-CD for phospholipids, β-CD much greater than γ-CD greater than α-CD for cholesterol, and β-CD much greater than γ-CD greater than α-CD for proteins. It is of importance that the

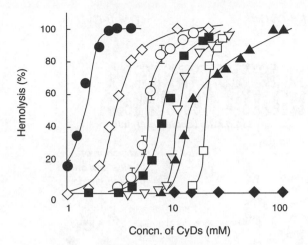

Figure 2. Hemolytic curves of hydrophilic β-CDs on rabbit erythrocytes in 0.1 M phosphate buffer solution (pH 7.4) at 37°C.○, β-CD; ●, DM-β-CD; ■, G₂-β-CD; ▲, HP-β-CD (DS 4.8); ◇, M-β-CD; □, SBE-β-CD (DS 6.3); ◆, DMA-β-CD (DS 7.0); ▽, GUG-β-CD. Each point represents the mean ± S.E. of three to seven experiments.

foregoing processes occurred without entry of the solubilizer into the membrane, since (1) [^{14}C]β-CD did not bind to erythrocytes and (2) CDs did not enter the cholesterol monolayer [46]. A study of [^{3}H]cholesterol transport between erythrocytes indicated that β-CD extracted this lipid from membrane into a new compartment located in the aqueous phase which could equilibrate rapidly with additional erythrocytes. Therefore, the effects of CDs differ from those of detergents, which first incorporate themselves into membranes, then extract membrane components into supramolecular micelles [46]. Interestingly, per(6-deoxy)β-CD, which is removed from an internal crown consisting of six primary alcohol groups, formed no complex with membrane components but interacted with the surface of the membrane, suggesting the different hemolytic mechanism of per(6-deoxy)β-CD from that of β-CD [47]. Figure 2 shows the hemolysis curves of hydrophilic β-CD derivatives on rabbit erythrocytes. Of various β-CD derivatives, methylated CD derivatives have strong hemolytic activity. Hemolytic activity of hydrophobic CDs cannot be shown here because of low solubility in aqueous solution and buffer. In the series of α-CDs, the hemolytic activity increased in the order hydroxypropyl-α-CD (HP-α-CD) < α-CD < 2,6-di-O-methyl-α-CD (DM-α-CD). Similarly, in the β-CD series, the activity increased in the order polysulfated CD ≃ β-CD sulfate (S-β-CD) ≃ 2,6-di-O-methyl-3-O-acetyl-β-CD, having a degree of substitution of an acetyl group of 7 (DMA7-β-CD) < SBE7-β-CD < dihydroxypropyl-β-CD (DHP-β-CD) ≃ hydroxyethyl-β-CD (HE-β-CD) ≤ 6-O-α-D-di-maltosyl-CD [(G₂)₂-β-CD] ≤ 6-O-α-D-maltosyl-CD (G₂-β-CD) ≤ 6-O-α-D-glucosyl-CD (G₁-β-CD) ≤ 3-hydroxypropyl-β-CD (3-HP-β-CD) ≤ HP-β-CD < β-CD < methyl-β-CD (M-β-CD) ≤ 2,3,6-tri-O-methyl-β-CD (TM-β-CD) < 2,6-di-O-methyl-

β-CD (DM-β-CD) [48–53]. Additionally, the hemolytic activities of heterogeneous and homogeneous branched CDs were lower than those of each parent nonbranched CD, and the hemolytic activity became weaker in the order nonbranched CD > 6-O-α-D-mannosyl-CD > G₁-β-CD > 6-O-α-D-galactosyl-CD in each series of α-, β-, and γ-CDs [54]. Furthermore, the hemolytic activity of 6(1),6(3),6(5)-tri-O-α-maltosyl-CD and 6(I),6(n)-di-O-(β-L-fucopyranosyl)-CD was lower than β-CD [16,55]. Also, Attioui et al. [56] reported that the substitutions of the β-CD ring by sugar antennas reduced the undesirable physicochemical properties of the β-CD (i.e., their hemolytic activities). In the series of γ-CD derivatives, the activity increased in the order HP-γ-CD < γ-CD < TM-γ-CD, but not enough data from the γ-CD series have yet been reported [49]. Taken together, any methylated CDs have the strongest hemolytic activity. It is evident that these differences are ascribed to the differential solubilization effects of membrane components by each CD rather than their intrinsic solubility or surface activity [49,57].

The pattern of morphological changes in erythrocytes induced by CDs may provide crucial information regarding the cytotoxicity of CDs. Two types of morphological changes are well known: stomatocyte and echinocyte formations. Under physiological conditions, erythrocytes show discocyte, but erythrocytes change to either stomatocyte or echinocyte through various stimulations. Figure 3 shows the proposed mechanism of morphological changes in erythrocytes induced by methylated CDs. DM-α-CD-induced morphological changes in rabbit erythrocytes from discocyte to stomatocyte [50], while M-β-CD and DM-β-CD induced changes from discocyte to echinocyte, indicating the totally different morphological types, depending on the cavity sizes of CDs [51]. It is likely that these differences can be ascribed to the distinct interaction mode between CDs and membrane components.

To gain insight into the mechanism by which CDs induced the differential morphological changes, we envisaged that CDs interact with different lipid rafts, depending on the cavity size of the CDs. Lipid rafts are characterized by their insolubility in nonionic detergents such as Triton X-100 at 4°C, and are microdomains in the cell membranes, which contain cholesterol, glycolipids, and sphingomyelin. Caveolae are relatively stable because caveolin, an integral protein, supports the structure. On the other hand, lipid rafts are considered to be unstable and dynamically produced and degraded. Recent studies have reported that lipid rafts contain many signaling molecules, such as GPI-anchored proteins, acylated proteins, G-protein-coupled receptors, and trimeric and small G-proteins and their effectors, suggesting that the lipid rafts have an important role in receptor-mediated signal transduction [58]. Interestingly, lipid rafts on cell membranes have heterogeneity such as cholesterol-rich microdomains and sphingolipid-rich microdomains [51,59]. Considering the findings, we envisaged the following

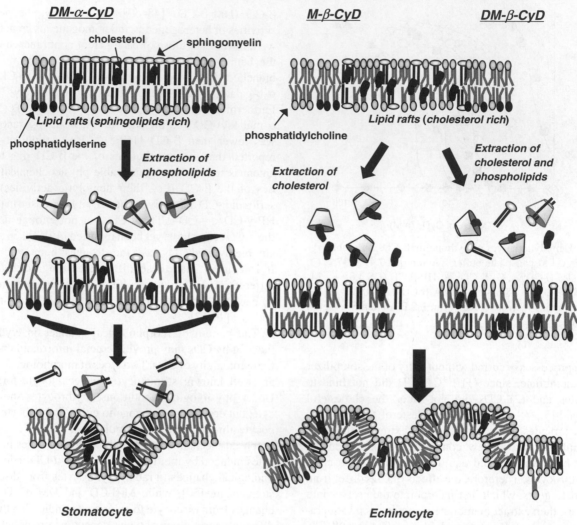

Figure 3. Mechanism proposed for morphological changes in red blood cells induced by methylated CDs. (*See insert for color representation of the figure.*)

mechanism for the different morphological changes induced by methylated CDs. The morphological changes induced by DM-α-CD may be due to extraction of sphingolipids from sphingolipid-rich lipid rafts of erythrocyte membranes, while those induced by M-β-CD and DM-β-CD may be due to the extraction of both cholesterol and proteins from cholesterol-rich lipid rafts of erythrocyte membranes. This hypothesis may be supported by the following findings. Fauvelle et al. [60] reported that the interaction of CDs with the lipid components of erythrocyte membranes is the determining factor in the hemolysis induced by these cyclic oligosaccharides, and α-CD has the strongest affinity with phosphatidylinositol among membrane lipids. In addition, we reported that there is a positive correlation between the hemolytic

activity of several CDs and their capacity to solubilize cholesterol, which acts as the main rigidifier in a lipid bilayer [49]. Furthermore, M-β-CD is well known to be not only a cholesterol-depleting agent but also a lipid raft–disrupting agent. Collectively, CDs, especially methylated CDs, act on lipid rafts of plasma membrane in a different mode, depending on the cavity sizes of CDs.

2.2. Cytotoxicity

A cell culture system is very useful in evaluating the safety profile of drugs and excipients. Recently, the effects of CDs not only on eukaryotic cells but also on prokaryotic cells have also been studied using a cell culture system of eukaryotic

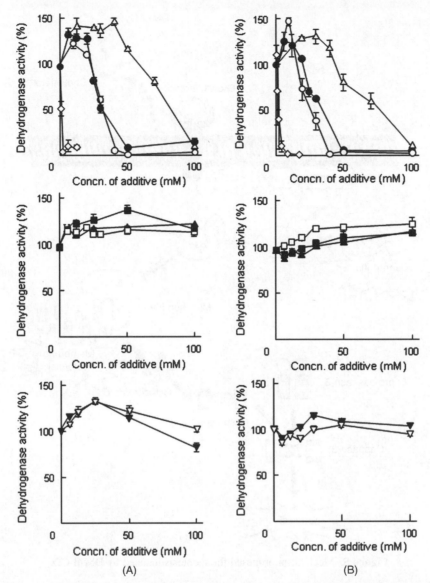

Figure 4. Cytotoxicity of CDs and Tween 20 in Caco-2 (A) and Caco-2R (B) cells in HBSS at 37°C. These Caco-2 cells were washed three times with HBSS (pH 7.4) and then incubated for 60 min with 100 μL of HBSS containing CDs (○, DM-α-CD; ●, DM-β-CD; △, M-β-CD; ▲, HP-α-CD; □, HP-β-CD; ■, HP-γ-CD; ▽, SBE4-β-CD; ▼, SBE7-β-CD) and Tween 20 (◇) at various concentrations at 37°C. After three washes with HBSS to remove CDs and Tween 20 again, cell viability was assayed using the WST-1 method. Each point represents the mean ± S.E. of six experiments.

cells. Evidently, CDs are known to induce not only hemolysis but also cytotoxicity at high concentrations. Generally, cytotoxic activity of CDs increased in the order γ-CD < β-CD < α-CD. For example, in an in vitro model of the blood–brain barrier (BBB), the cytotoxicity of native CDs increased in the order γ-CD < β-CD < α-CD. Here α-CD removed phospholipids and β-CD extracted phospholipids and cholesterol, and γ-CD was less lipid selective than the other CDs [61]. Thus, the magnitude of the cytotoxicity of parent CDs is inconsistent with the order of magnitude of hemolytic activity. This difference could be ascribed to the differential

cholesterol content because the content of cholesterol in erythrocytes is markedly higher than in other cells. Methylated CDs induced cell death in various cells, but HP-CDs or SBE7-β-CD did not, being proportional to the magnitude of the hemolytic activity of each CD. Figure 4 shows the cytotoxicity of CDs and Tween 20 in Caco-2 cells and vinblastin-resistant Caco-2 (Caco-2R) cells by the WST-1 method. As the tentative measure of cytotoxicity, a typical nonionic surfactant Tween 20 was used as a positive control, which induced almost complete cell death even at 2 mM. Methylated CDs at concentrations higher than 25 mM

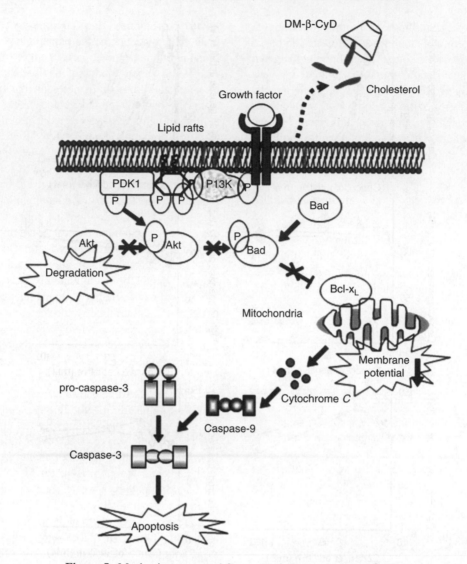

Figure 5. Mechanism proposed for apoptosis induced by DM-β-CD.

provide cytotoxicity, the cytotoxic effects of which were weaker than those of Tween 20. In contrast to methylated CDs, other CDs were fairly bioadaptable even at higher concentrations [62]. Similarly, higher cytotoxic activity in methylated CDs than in parent CDs and other CD derivatives can be demonstrated in other cells, such as TR146 and PC-12 cells [63–65]. Most recently, we reported a mechanism for cell death induced by methylated CDs: DM-β-CD and TM-β-CD caused marked apoptosis in NR8383, A549, and Jurkat cells, through cholesterol depletion in cell membranes [66]. Interestingly, cell death induced by DM-β-CD and TM-β-CD was found to be apoptosis, resulting from inhibition of the activation of the PI3K-Akt-Bad pathway (Fig. 5). In sharp contrast, DM-α-CD induced cell death in a nonapoptotic mechanism, probably a necrosis [66]. A similar finding regarding cell death induced by methylated CDs was reported by Ulloth et al. [67] and Hipler et al. [68]. That is, nerve growth factor–differentiated PC12 cells exposed to

0.25% M-β-CD died via apoptosis [67]. Also, M-β-CD triggers the activity of the effectors caspase-3 and caspase-7 in HaCaT keratinocytes, indicating an apoptotic cell death [68]. From the findings described above, we presumed that the apoptosis and necrosis induced by CDs may be attributed to a different interaction of methylated CDs with cholesterol-rich lipid rafts than with sphingolipid-rich lipid rafts, respectively. Collectively, these findings suggest that similar to erythrocytes, lipid rafts on the plasma membrane of cells would be involved in cell death and cellular function. In fact, the widespread use of M-β-CD has been made to disrupt the function of caveolae and lipid rafts (e.g., aelolysin oligomerization) [69,70]; Ca^{2+} mobilization leading to airway smooth muscle contraction induced by submaximal concentrations of acetylcholine [71]; oxidative stress–induced signal transduction in human renal proximal tubule cells [72]; the relationship between CD36 function, nitric oxide synthase activation, caveolae integrity, and blood pressure regulation [73]; neurological symptoms and liver cholesterol storage [74]; invasion and metastatic potentials [75]; and atherogenesis [76].

However, it is clear that the magnitude of cytotoxicity of CDs depends on the experimental conditions (i.e., the cell type and CD concentrations and incubation time with CDs). For example, the toxicity of RM-β-CD on buccal mucosa using a reconstituted human oral epithelium model composed of TR146 cells is concentration dependent. In fact, 10% RM-β-CD shows cytotoxic and inflammatory effects depending on time exposure, whereas 2% and 5% RM-β-CD do not induce tissue damage even after 5 days of repeated exposure. Consequently, the highly water-soluble RM-β-CD is thought to be a safe candidate as an excipient for buccal mucosal drug delivery [64]. Meanwhile, branched CDs and anionic CDs show low hemolytic activity and cytotoxicity. Ono et al. [63] reported that the cytotoxicity of CDs at the same concentration increased in the order γ-CD < G_2-β-CD < G_2-α-CD << α-CD in Caco-2 cells. In addition, SBE7-β-CD and DMA7-β-CD show no cytotoxicity up to a concentration of 10 mM of the CD in RAW264.7 cells [77]. Also, Kiss et al. [65] demonstrated that in HeLa cells the cytotoxicity decreased in the order DM-β-CD > TM-β-CD > or = RM-β-CD > (2-hydroxy-3-N,N,N-trimethylamino)propyl-β-CD > carboxymethyl-β-CD (CM-β-CD) [degree of substitution (DS) = 3.5], and that the cationic quaternary amino-β-CD was less toxic than the methylated CDs. Most of the second-generation CD derivatives, which a contain ionic substituent in addition to the methyl groups, show less cytotoxicity than do the parent compounds.

Importantly, CDs change the membrane permeability of drugs, nutrition, and ions, mediated by transporters or not without cytotoxicity. Totterman et al. [78] reported that exposure to HP-β-CD and SBE-β-CD solutions had only minor effects on the integrity of Caco-2 cell monolayers. In contrast, DM-β-CD clearly increased the epithelial permeability for the hydrophilic marker [^{14}C]mannitol across Caco-2 monolayers, decreased transepithelial electrical resistance (TEER), and showed a dose-dependent cytotoxicity. Histological observations revealed that DM-β-CD increased the permeability of the apical cell membrane without discernible effects on cytoskeletal actin. Thus, HP-β-CD and SBE-β-CD appear to be safer additives for use in enteral spironolactone preparations with respect to their acute local effects on epithelial integrity [78]. Interestingly, hydrophilic CDs may increase the permeability of nucleic acid drugs. Zhao et al. [79] reported that HP-β-CD, HE-β-CD, and a mixture of various HP-β-CDs (e.g., Encapsin) increased in phosphorothioate oligodeoxynucleotide uptake up to two- to threefold in 48 h. Under the experimental conditions, CD itself was not toxic at the concentration used. These studies suggest that CDs and their analogs might be used successfully as carriers for hydrophilic drugs and oligonucleotides.

In cancer chemotherapy the side effects of antineoplastic agents and drug resistance are major problems, so new ideas and methodologies have been developed. Two major ideas utilizing CDs for improvement of cancer therapy will be introduced: (1) the enhancing effects of CDs on anticancer drugs through an increase in membrane permeability, and (2) the inhibitory effects of CDs on efflux transporters such as P-glycoprotein (P-Gp). Regarding the first idea, when M-β-CD was used at 500 and 1000 μM in combination with doxorubicin (DOX), M-β-CD significantly potentiated the activity of DOX on both sensitive and multidrug-resistant cell lines (HL-60 S and HL-60 R); 50% growth-inhibitory (IC_{50}) ratios (IC_{50} M-β-CD-DOX/IC_{50} DOX) were about 3:4 and 1.6:4 for HL-60 S and HL-60 R, respectively. Similar results were obtained using other paired MCF 7 sensitive and resistant cell lines. These data provide a basis for the potential therapeutic application of M-β-CD in cancer therapy [80]. Additionally, the increase in cytotoxicity of camptothecin (CPT) in the presence of CDs has been demonstrated through an increase in its stability [81]. Regarding the second idea, CDs increase the membrane permeability of drugs through the inhibition of efflux transporters. Figure 6 shows a scheme for a possible inhibitory mechanism of DM-β-CD on P-Gp and multidrug resistance–associated protein 2 (MRP2) in Caco-2 cell monolayers. In both Caco-2 as well as vinblastine-resistant Caco-2 (Caco-2R) cell monolayers, pretreatment of the apical membranes of the monolayers with DM-β-CD decreased the efflux of tacrolimus and rhodamine 123 without any associated cytotoxicity. DM-β-CD decreased the P-Gp level in the apical membranes of both Caco-2 and Caco-2R cell monolayers, probably by allowing the release of P-Gp from the apical membrane into the medium buffer. Under the experimental conditions, DM-β-CD, however, did not decrease *MDR1* gene expression in Caco-2 or Caco-2R cells [62]. Additionally, DM-β-CD was found to impair significantly the efflux activity not only of P-Gp but also of multidrug resistance–associated protein 2 (MRP2). Interestingly, DM-β-CD released P-Gp and MRP2 from the monolayers into the apical side medium buffer and decreased the contents of cholesterol as well as P-Gp and MRP2 in caveolae of Caco-2 cell monolayers, but not caveolin and flotillin-1, without cytotoxicity. Therefore, these results suggest that the inhibitory effect of DM-β-CD on P-Gp and MRP2 function, at least in part, could be attributed to the release of these transporters from the apical membranes into the medium as secondary effects through cholesterol depletion in caveolae after treatment of Caco-2 cell monolayers with DM-β-CD [82]. The similar results were reported by Fenyvesi et al. [83]; P-Gp inhibition by CD treatments arises through modulation of its membrane microenvironment rather than as a result of concomitant cytotoxicity.

As described above, CDs, especially methylated CDs, inhibit the function of efflux transporters. These lines of evidence make it tempting to speculate that these CDs change

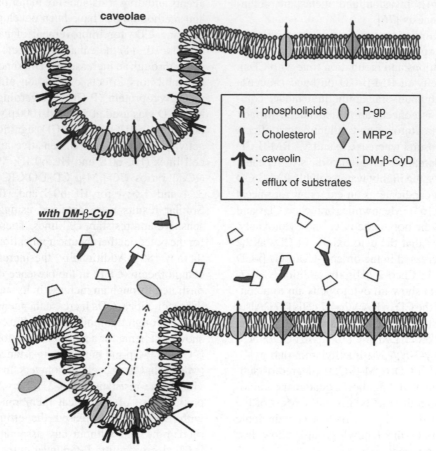

Figure 6. Scheme for possible inhibitory mechanism of DM-β-CD on P-Gp and MRP2 in Caco-2 cell monolayers.

the other transports and/or channels. For example, localization of cardiac L-type Ca^{2+} channels to a caveolar macromolecular signaling complex is required for β_2-adrenergic regulation [84]. Also, the inhibitory effects of CDs on *N*-methyl-D-aspartic acid (NMDA) receptor resulted in neuroprotection of cortical neuronal cultures against ischemic and exitotoxic insults. Since cholesterol-rich membrane domains exist in neuronal postsynaptic densities, these results imply that synaptic NMDA receptor subpopulations underlie excitotoxicity, which can be targeted by CDs without affecting overall neuronal Ca^{2+} levels [85]. Additionally, M-β-CD added apically to mouse cortical collecting duct [(MPK) CCD(14)] cells resulted in a slow decline in amiloride-sensitive sodium transport with short-circuit current reductions of $38.1 \pm 9.6\%$ after 60 min, suggesting that M-β-CD inhibits the epithelial sodium channel in lipid rafts [86]. Moreover, DM-β-CD, TM-β-CD, and M-β-CD inhibited the function of peptide transporter 1 (PepT-1), L-type amino acid transporter 1 (LAT-1), and sodium-coupled glucose co-transporter (SGLT-1) in Caco-2 and Coco-2R cells. On the contrary, a few reports demonstrated that CDs can no longer

increase the permeability of drugs. Ahsan et al. [87] reported that addition of $[^{125}I]$insulin to the apical side of 16HBE14o$^-$ cells in the absence or presence of 1% DM-β-CD resulted in little or no $[^{125}I]$insulin movement to the basolateral chamber or degradation in the apical chamber, suggesting that DM-β-CD is not sufficient to stimulate transepithelial insulin movement. Monnaert et al. [88] demonstrated that in an in vitro model of the BBB, endothelial permeability of DOX was relatively low, and addition of γ-CD or HP-γ-CD, up to 15 and 35 mM, respectively, decreased the DOX delivery, probably due to the low complex penetration across the BBB and the decrease in free DOX concentration. Additionally, higher CD concentrations increased the DOX delivery to the brain, but this effect is due to a loss of BBB integrity. Hence, CDs are not able to increase the delivery of DOX across our in vitro model of BBB. Taken together, the effects of CDs on the membrane permeability of drugs may depend on cell types and CD concentrations, so further investigation is required.

Considering the safety profile of CDs, the effects of CDs on immune response and microorganisms should be clarified. There are several reports regarding the effects of CDs on prokaryotic cells. For example, Bar et al. [89] demonstrated that CDs provided the bacterial toxicity and Zhang et al. [90] reported that growth of alkaliphilic *Bacillus halodurans* C-125 both on agar plates and in liquid culture was inhibited by M-β-CD and α-CD. In sharp contrast, the inhibition was not observed with gram-negative and gram-positive bacteria except for *Bacillus* strains through disruption of cell membranes by M-β-CD. On the other hand, the presence of β-CD derivatives resulted in a fourfold increment of the transformation rate for in-house cells. Since CDs have little or no effect on DNA uptake by noncompetent cells or no interactions between CDs and DNA-like molecules, the effects of CDs should be related to interaction with the bacterial cell wall [91]. Also, M-β-CDs have been shown to enhance sterol conversion to 4-androstene-3,17-dione (AD) and 1,4-androstadiene-3,17-dione (ADD) by growing *Mycobacterium* supernatants [92].

It is well known that various toxins exhibit toxic activity in lipid rafts on plasma membrane of eukaryotic cells. CDs interact with both bacterial toxins and bacteria and suppress their toxin activity in most cases. M-β-CD reduced the cytotoxicity of *Clostridium septicum* α-toxin through inhibition of its oligomerization [93], and M-β-CD suppressed the binding and internalization of *Clostridium perfringens* ι-toxin in lipid rafts [94]. Several amino acid derivatives of β-CD inhibit the activity of *S. aureus* α-hemolysin and *B. anthraces* lethal toxin in cell-based assays at low micromolar concentrations [95]. Moreover, many toxins and bacteria are associated with lipid rafts: *Clostridium difficile* toxin A [96], *Clostridium perfringens* ε-toxin [97], Cry1A toxin [98], *Actinobacillus actinomycetemcomitans* cytolethal-distending toxin [99], *Shigella* [100], BK virus [101], *P. aeruginosa* [102], and uropathogenic *Escherichia coli* [103]. In fact, *Helicobacter pylori* vacuolating toxin, which is associated with lipid rafts, should be inhibited by the addition of M-β-CD [104,105]. Therefore, M-β-CD should affect the

activity of the toxin through disrupting lipid raft structure. However, it should be noted that M-β-CD enhances the activity of *Botulinum neurotoxin* serotype A (BoNT/A), one of seven serotypes of botulinum neurotoxin [106]. To develop potent antianthrax drugs, CD-based inhibitors of anthrax toxins were searched. As a result, per-6-(3-aminopropylthio)-β-CD interacts strongly with the protective antigen (PA) pore lumen, blocking PA-induced transport at subnanomolar concentrations and completely protecting the highly susceptible Fischer F344 rats from lethal toxin [107,108]. Also, per-6-(3-aminopropylthio)-β-CD was shown to inhibit the toxicity of lethal factor (LF) in vitro and in vivo. Taken together, these data suggest the potential usefulness of per-6-(3-aminopropylthio)-β-CD in combination with a catalytically inactive fragment of LF-decorated liposomes loaded with the fluorescent dye 8-hydroxypyrene-1,3,6-trisulfonic acid and antianthrax drugs against intracellular targets [109].

CDs affect Toll-like receptor (TLR) activity. Innate immune receptors are promising targets to regulate the complex cascade that will lead to cytokine production [110]. These receptors recognize not only pathogenic but also endogenous ligands through their molecular pattern and then use several signaling pathways to alter gene expression. TLRs are a family of innate immune-recognition receptors that recognize molecular patterns associated with microbial pathogens and induce antimicrobial immune responses [111]. At least 11 TLRs have been identified in the mammalian genome and are classified by the ligands that initially activate TLR-dependent signaling, including highly conserved pathogen proteins, cell wall components, and nucleic acids [112]. CDs may inhibit TLR signaling in a different mechanism. We demonstrated that DM-α-CD inhibited excess activation of macrophages stimulated with LPS (TLR4 ligand) and poly I : C (TLR3 ligand), but not CpG DNA (TLR9 signaling), by which DM-α-CD acted to lipid rafts, possibly sphingolipid-rich lipid rafts (Fig. 7) [113]. In contrast, DMA7-β-CD had

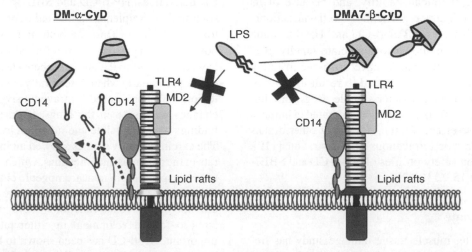

Figure 7. Inhibitory mechanism proposed for LPS binding to LPS receptors by DM-α-CD and DMA7-β-CD.

greater inhibitory activity than other CDs against the production of nitric oxide (NO) and various proinflammatory cytokines, including TNF-α in murine macrophages stimulated with two serotypes of LPS and lipid A through complexation with LPS, resulting in suppression of the binding of LPS to an LPS receptor on macrophages (Fig. 7). Thus, DMA7-β-CD may have promise as a new therapeutic agent for endotoxin shock induced by LPS [114].

CDs can be useful as a medium component for cell culture and remediation of pollution. For example, DM-β-CD enhanced pertussis toxin production 100 times more in synthetic media, such as Stainer–Scholte medium, than in DM-β-CD-free medium in 2-day shake cultures [115]. In addition, growth medium containing CDs and a low concentration of horse serum for cultivation of *H. pylori* [116], and *Brucella* broth supplemented with CDs is an improved medium for bacterial culture and industrial production of *H. pylori* antigens [117]. Also, the use of HP-CDs in the remediation of environmental pollution has been attempted [118].

3. IN VIVO SAFETY PROFILE OF CDs

β-CD and other CDs have utility for solubilizing and stabilizing drugs, however, some are nephrotoxic when administered parenterally. A number of workers have attempted to identify, prepare, and evaluate various CD derivatives with superior inclusion complexation and maximal in vivo safety for various biomedical uses. A systematic study led to SBE7-β-CD and HP-β-CD. SBE7-β-CD and HP-β-CD have undergone extensive safety studies and are currently used in six products approved by the U.S. Food and Drug Administration (four for SBE7-β-CD and two for HP-β-CD). They are also in use in numerous clinical and preclinical studies. SBE7-β-CD interacts very well with neutral drugs to facilitate solubility and chemical stability, and because of its polyanionic nature, it interacts particularly well with cationic drugs. Complexes between SBE7-β-CD and HP-β-CD and various drugs have been shown to dissociate rapidly after parenteral drug administration, to have no tissue-irritating effects after intramuscular dosing, and to result in superior oral bioavailability of poorly water-soluble drugs. In addition, SBE7-β-CD and HP-β-CD are well tolerated in humans and have no adverse effects on the kidneys or other organs following either oral or intravenous administration [119]. Excellent reviews on safety profiles of HP-β-CD and SBE7-β-CD are available [5,9,119,120].

3.1. Parenteral Route

In a parenteral route, muscle tissue damage study has frequently been used to evaluate local irritation of drugs and excipients. When the muscle tissue damage after the intramuscular injection (100 mg/mL) of hydrophilic CDs into the M. vastus lateralis of rabbits was compared with that of mannitol and nonionic surfactants, α-CD and DM-β-CD showed a relatively high irritation reaction, the degree of which corresponded to that of Tween 80. On the other hand, G₂-β-CD, HP-β-CD, SBE-β-CD, DMA-β-CD, and S-β-CD showed no or only slight irritation reaction, the degree of which was comparable to those of γ-CD, mannitol, and HCO-60 [121]. Meanwhile, the most critical issues of the parent CDs for parenteral use are their toxic effect on the kidneys, which is the main organ for the excretion of CDs from systemic circulation and for concentrating CDs in the proximal convoluted tubule after glomerular filtration. The nephrotoxicity of α- and β-CDs is manifested as a series of alterations in the vacuolar organelles of the proximal tubule [122]. For acute intravenous administration, γ-CD is safer than α- and β-CDs; that is, the intravenous doses that are lethal to 50% of population (LD₅₀ values) are 1000, 788, and >3750 mg/kg for α-, β- and γ-CDs, respectively, in rats. Meanwhile, subcutaneous injection of β-CD increased the number and size of renal tubular cell tumors in inbred Wistar rats treated with 1000 ppm of N-ethyl-N-hydroxyethylnitrosamine (EHEN). The incidence of renal tumors at the end of the 32-week experiment was 50% in rats treated with 1000 ppm EHEN for 2 weeks and 100% in rats treated with 1000 ppm EHEN for 2 weeks and then given daily subcutaneous injections of β-CD for 1 week. The incidence of renal tumors more than 3 mm in diameter was 70% in rats treated with 1000 ppm EHEN before β-CD, but 0% in rats treated with EHEN alone. In addition, β-CD promoted the development of renal tumors in rats treated with 500 ppm EHEN, which is a subthreshold dose for renal tubular cell tumorigenesis. These results show that β-CD promotes EHEN-induced renal tubular cell tumorigenesis [123].

In an attempt to alleviate these properties of parent CDs, various CD derivatives have been developed. Among various CD derivatives, HP-β-CD and SBE7-β-CD can be used for solubilizing excipients. Gould and Scott [120] reviewed the toxicity of HP-β-CD, using both literature information and novel data, and presented new information. That is, HP-β-CD is well tolerated in the animal species tested (rats, mice, and dogs), particularly when dosed orally, and shows only limited toxicity. When dosed intravenously, histopathological changes were seen in the lungs, liver, and kidney, but all findings were reversible and no effect levels were achieved. The carcinogenicity studies showed an increase in tumors in rats in the pancreas and intestines, which are both considered to be rat specific. No noncarcinogenic changes were noted in the urinary tract, but these changes were also reversible and did not impair renal function. There were no effects on embryo–fetal development in either rats or rabbits. Most important, HP-β-CD has been shown to be well tolerated in humans, the main adverse event being diarrhea, and no adverse events on kidney function have been documented to date [120]. Meanwhile, the sulfoalkyl ether-CD (SAE-β-CD) derivatives did not produce mortality in mice following intraperitoneal injection at doses exceeding 5.45 mmol/kg. No significant histological lesions were observed in the kidney tissue of mice receiving the CD derivatives. The SAE-β-CD derivatives were excreted faster and to a greater extent than were β-CD derivatives, and at rates comparable to those for HP-β-CD. The hemolytic potential of these derivatives was less than that of β-CD and comparable to or better than that of HP-β-CD. The SAE-β-CD derivatives did not increase APTT clotting times, indicating that these derivatives have no significant anticoagulant activity. The toxicological profile of these derivatives suggests that these molecules may have applications as biologically safe β-CD derivatives [124]. The intravenous application of SBE7-β-CD complex with drugs is known well: voriconazole [125], etomidate [126], maropitant, and ziprasidone mesylate [127]. Among methylated CDs, DMA7-β-CD can be used parenterally. Simultaneous administration of DMA7-β-CD at a dose of approximately 4 g/kg not only intraperitoneally but also intravenously, and intraperitoneal injection of aqueous solution containing LPS (50 ng) and D-galactosamine (1.25 g/kg) in murine endotoxin shock model suppressed fatality. Also, DMA7-β-CD decreased the blood level of TNFα as well as serum levels of aspartate transaminase (AST) and alanine transaminase (ALT) in mice. Thus, DMA7-β-CD may have promise as a new therapeutic agent for endotoxin shock induced by LPS [114]. Other routes, except intravenous application of CD derivatives, are well known. HP-β-CD increases the solubility of molecules by inclusion of the agent in the lipophilic interior of the ring. This property is of particular use for the administration of molecules by the intracerebral or intrathecal routes, indicating the utility of complexation with HP-β-CD for intracerebral drug delivery and compatibility with brain and spinal tissue [128].

Recently, various hydrophilic and ionizable CD derivatives have been prepared and evaluated for practical use in various fields. Per(3,6-anhydro-2-O-carboxymethyl)-α-CD is a polydentate analog of EDTA, a well-known cation chelating reagent, is not hemolytic, and exhibits no lethal properties in mice (LD$_{50}$ = 42 mM). In vivo injection at supralethal amounts of uranyl complex of this CD prevents immediate death in mice but is unable to protect against later death [129].

Sugammadex (designation Org 25969, Bridion, Fig. 8) is a novel agent for reversal of neuromuscular blockade by the agent rocuronium in general anesthesia and is the first selective relaxant binding agent (SRBA) [130]. Interestingly, the intravenous administration of sugammadex creates a concentration gradient favoring the movement of rocuronium molecules from the neuromuscular junction back into the plasma, which results in a fast recovery of neuromuscular function. Sugammadex is biologically inactive, does not bind

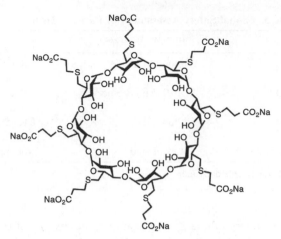

Figure 8. Structure of sugammadex.

to plasma proteins, and appears to be safe and well tolerated. Additionally, it has no effect on acetylcholinesterase or any receptor system in the body. The efficacy of the compound as an antagonist does not appear to rely on renal excretion of the CD–relaxant complex. Human and animal studies have demonstrated that sugammadex can reverse very deep neuromuscular blockade induced by rocuronium without muscle weakness [130]. Actually, when sugammadex 2.0 mg/kg was administered by intravenous bolus injection in patients, mean recovery time was 1.8 min after both propofol and sevoflurane anesthesia, and the 95% confidence interval for the difference in recovery time between the two groups was well within the predefined equivalence interval, indicating that recovery from the neuromuscular blockade was unaffected by maintenance anesthesia. Importantly, there were no treatment-related serious adverse events and no discontinuations or deaths and no residual paralysis occurred. Also, sugammadex provides a rapid and dose-dependent reversal of profound neuromuscular blockade induced by high-dose rocuronium (1.0 or 1.2 mg/kg) in adult surgical patients [131–134].

3.2. Oral Route

Since the data available suggest that CDs given orally are not absorbed intact and the products of whatever limited digestion that may take place are probably single and multiple units of glucose, all CDs are likely to be safe when taken orally. Table 3 summarizes the results of a global safety assessment of parent CDs for food applications. As shown in the table, α- and γ-CD are not specified, but β-CD is specified. α-CD does not pose a safety concern at the proposed use levels and resulting predicted consumption as a food ingredient and food additive. The acceptable daily intake (ADI) established in 2001 specified α-CD for use as a carrier and stabilizer for flavors, colors, and sweeteners; as a water solubilizer for fatty acids and certain vitamins; as a flavor modifier in soya milk; and as an absorbent in

Table 3. Global Safety Assessment of Parent CDs for Food Use

CD	JECFA (WHO/FAO)	United States	Japan	EU
α-CD	Without ADI[a]	GRAS[b] approved in 2004 (wide application)	Available	Novel food
β-CD	ADI (0–5 mg/kg per day)	GRAS approved in 2001 (carrier for food and cosmetics)	Available	Available (1 g/1 kg food)
γ-CD	Without ADI	GRAS approved in 2000 (wide application)	Available	Novel food
Branched-CD	Not evaluated	Not approved	Available	Not approved

[a] ADI; Acceptable daily intake.
[b] GRAS; Generally recognized as safe.

confectionery (Table 3). For example, the toxicity of α-CD was examined in a 4-week range finding study and a 13-week oral toxicity study in rats. As a result, the ingestion of α-CD for 13 weeks at dietary levels of up to 20% (corresponding to intakes of 12.6 and 13.9 g/kg body weight per day in male and female rats, respectively) did not produce any signs of toxicity or adverse effects [135]. In addition, when α-CD was fed at dietary concentrations of 0, 1.5, 5, 10, or 20% to groups of 25 pregnant female rats from day 0 to 21 of gestation in Wistar Crl : (WI)WU BR rats, no adverse effects were observed at α-CD intakes of up to 20% of the diet, the highest dose level tested at which the rats consumed about 13 g/kg body weight per day [136]. Similar results of α-CD were observed in New Zealand White rabbits [137]. However, dietary α-CD lowers low-density lipoprotein cholesterol and alters plasma fatty acid profile in low-density lipoprotein receptor knockout mice on a high-fat diet.

It should be noted that the story of β-CD is totally different from that of α-CD. The ADI of β-CD is now set at 0 to 5 mg/kg body weight (i.e., 0 to 0.3 g/60 kg). The oral LD_{50} values for β-CD are reported to be >12.5, 18.8, and 5 kg/kg for mice, rats, and dogs, respectively [138]. No significant toxic effects were observed in rats fed a 10% β-CD diet for 90 days [49]. A 52-week toxicity study by dietary administration in Sprague–Dawley rats and in pure-bred beagle dogs with β-CD showed no pathological evidence of systemic toxicity, although there were minor changes in urinalysis and biochemical parameters and a slightly higher incidence of liquid faces, but these changes were considered to be of no toxicological importance. As a result, the nontoxic effect level was 12,500 ppm in the rat (equivalent to 654 or 864 mg/kg per day for males or females, respectively) and 50,000 ppm in the dog (equivalent to 1831 or 1967 mg/kg per day for males or females, respectively) [139]. When groups of 50 males and 50 females of in Fischer 344 (F344) rats were given β-CD in their diet at concentrations of 0 (control), 2.5, or 5% for 104 weeks, dose-dependent inhibitory effects of β-CD on growth were observed in both sexes of the groups treated. The survival rates, mean survival times, and range, however, demonstrated no significant differences between the control and treated groups. A variety of tumors developed in all

groups, including the control group, but all the neoplastic lesions were histologically similar to those known to occur spontaneously in this strain of rat, and no statistically significant increase in the incidence of any tumor was found for either sex of the treated groups. Thus, under the present experimental conditions, the high dose, about 340 to 400 times higher than the current daily human intake from ingestion as a food additive and from pharmaceutical use, does not have any carcinogenic potential in F344 rats [140]. Also, the results of oncogenicity studies of β-CD in inbred Fischer 344 rats and CD-1 outbreed mice are presented. Chronic feeding of β-CD to Fischer 344 rats and CD-1 mice did not cause any treatment-related carcinogenic effects. The only toxic effect was seen in mice as macroscopic distension of the large intestine with soft or fluid contents, histologically associated with the mucosa covered by mucous secretion containing exfoliated cells, and mucosal flattening and intestinal gland atrophy. Despite these observations, no differences between control and treated groups were observed concerning mortality, clinical observations, or body weight and food consumption [141]. However, when a 13-week oral toxicity study of β-CD was carried out in F344 rats at the dose levels of 0, 0.6, 1.25, 2.5, 5, and 10% in powdered diet, all animals survived at the end of the experiment, but a slight decrease in body-weight gain was observed in males of the 10% and 5% groups. Dose-dependent increases in serum levels of AST, ALT, and alkaline phosphatase were observed as well as increases in serum levels of urea nitrogen and relative liver weights in treated males. Histopathologically, a dose-dependent increase in the severity of inflammatory cell infiltration was seen in the liver of treated animals, focal hepatocellular necrosis being detected in both sexes of the 10% group and females of the 5% group. These findings indicate that β-CD causes hepatocellular injury to rats when it is administered orally [142]. Furthermore, a three-generation study with two mating phases per generation and a teratology phase was performed in the rat to assess the reproductive and developmental toxicity of β-CD in the diet. Transient neonatal growth retardation occurred with 5% β-CD: a similar but equivocal effect was also observed with 2.5%. No permanent defects or other indications of developmental toxicity were found. There was no significant maternal toxicity. However, the dietary level of 1.25% was found to be a nonobserved adverse effect level (NOAEL) for developmental toxicity. Further investigations showed the growth retardation to be specific to dietary administration during lactation: It was not produced by intraperitoneal administration and was not influenced by treatment of the dams or litters during gestation. Slight maternal nutritional deficiency, caused by physicochemical interactions of β-CD with nutrients in the gut is proposed as the mechanism of action [143]. Additionally, the effects of β-CD on cholesterol and bile acid metabolism in hypercholesterolaemic rats were examined. Male Wistar rats were divided into four groups that received during 7 weeks: control diet, 2% cholesterol diet (A), A + 2.5% β-CD (B), and A + 5% β-CD (C). The cholesterol-rich diet induced hepatomegaly and fatty liver and significantly reduced cholesterol, bile acid, and phospholipid secretion. Addition of β-CD normalized biliary lipid secretion. Moreover, when compared to A, β-CD significantly lowered plasma phospholipid concentration (B: −21%; C: −29%) and the liver free/total cholesterol molar ratio (B: −40%; C: −38%), increased bile acid fecal output (B: +17%; C: +62%) and enhanced cholesterol 7α-hydroxylase activity (B: +50%; C: +100%) and mRNA levels (B: +14%; C: +29%). 5% β-CD also reduced plasma triglycerides concentration (−38%). However, ALT and AST activities were significantly increased (B: +140% and +280%; C: +72% and +135%) and there was a high incidence of cell necrosis with portal inflammatory cell infiltration. Addition of β-CD to a cholesterol-rich diet results in a triglyceride-lowering action, enhancement of bile acid synthesis and excretion, and normalization of biliary lipid secretion, but produces a marked hepatotoxic effect [144].

Among α-, β- and γ-CDs, drugs including α-CD and β-CD have been commercialized until the 1980s in Japan, but γ-CD has not. A number of safety studies of γ-CD have been performed. In foods, γ-CD may be used as a carrier for flavors, vitamins, polyunsaturated fatty acids, and other ingredients. It also has useful properties as a stabilizer in different food systems. The daily intake from all its intended uses in food at the highest feasible concentrations has been estimated at 4.1 g/person per day for consumers of γ-CD-containing foods. The toxicity studies consist of standard genotoxicity tests, subchronic rat studies with oral and intravenous administration of γ-CD for up to three months, a subchronic (three-month) toxicity study in dogs, a (one-year) oral toxicity study in rats, and embryotoxicity/teratogenicity studies in rats and rabbits. In the studies with oral administration, γ-CD was given at dietary concentrations of up to 20%. All these studies demonstrated that γ-CD is well tolerated and elicits no toxicological effects. Metabolic studies in rats showed that γ-CD is rapidly and essentially completely digested by salivary and pancreatic amylase.

Therefore, the metabolism of γ-CD closely resembles that of starch and linear dextrins. A human study with ingestion of single doses of 8 g of γ-CD or 8 g of maltodextrin did not reveal a difference in gastrointestinal tolerance of these two products. An interaction of ingested γ-CD with the absorption of fat-soluble vitamins or other lipophilic nutrients is not to be expected. On the basis of these studies it is concluded that γ-CD is generally recognized as safe (GRAS) for its intended uses in food [145].

The embryotoxicity/teratogenicity of γ-CD (γ-CD) was examined in Wistar Crl : (WI)WU BR rats and artificially inseminated New Zealand White rabbits. In rats, γ-CD was fed at dietary concentrations of 0, 1.5, 5, 10, and 20% to groups of 25 pregnant female rats from day 0 to 21 of gestation. Generally, γ-CD was well tolerated and no deaths occurred in any group. Weight gain and food consumption were similar in all groups during gestation, except for a slightly reduced food intake in the 20% γ-CD group from day 0 to 16. Water intake was similar in all γ-CD groups; in the lactose group, it was significantly higher than in the control group. Reproductive performance was not affected by the γ-CD treatment. Examination of the fetuses for external, visceral, and skeletal alterations did not reveal any fetotoxic, embryotoxic, or teratogenic effects of γ-CD. In conclusion, no adverse effects were observed at γ-CD intakes of up to about 20% of the diet (approximately 11 g/kg body weight per day) [146]. When γ-CD was administered to groups of 16 rabbits at dietary concentrations of 0, 5, 10, or 20%, the dietary γ-CD is well tolerated by pregnant rabbits and had no adverse effect on reproductive performance, and is not embryotoxic, fetotoxic, or teratogenic at dietary concentrations of up to 20% [147]. From these findings, ADI of γ-CD is not specified (Table 3).

The new opinion about the oral safety of the natural CDs and the early work showing the safety of HP-β-CD in animals and humans has created a favorable environment for expeditious approval of HP-β-CD-containing formulations. Additional clinical studies using heretofore intractable but potent drugs have been completed using HP-β-CD as the drug delivery vehicle in phase I studies, investigational new drug (IND) applications, and in completed new drug applications (NDAs). From the pharmaceutical standpoint in the United States and Europe, no CDs have GRAS status for any route of dosing, and no lifetime carcinogenicity studies conducted in a GLP fashion for any CD have been reported. Even so, drugs containing CDs have reached the marketplace.

The relative effectiveness of two β-CD derivatives, DM-β-CD and HP-β-CD, in enhancing enteral absorption of insulin was evaluated in the lower jejunal/upper ileal segments of the rat by means of an in situ closed loop method. The incorporation of 10% w/v DM-β-CD to a 0.5-mg/mL porcine–zinc insulin solution dramatically increased insulin bioavailability from a negligible value (approximately

0.06%) to 5.63% when administered enterally at a dose of 20 U/kg. However, addition of 10% w/v HP-β-CD did not improve enteral insulin uptake significantly with a bioavailability of only 0.07%. Similarly, the pharmacodynamic relative efficacy values obtained after the enteral administration of 20 U/kg insulin, 20 U/kg insulin with 10% HP-β-CD, and 20 U/kg insulin with 10% DM-β-CD were 0.24%, 0.26%, and 1.75%, respectively. Biodegradation studies of 0.5 mg/mL insulin hexamers by 0.5 μM α-chymotrypsin revealed no inhibitory effect on the enzymatic activity by the two CDs. On the contrary, the apparent first-order rate constant increased significantly in the presence of 10% DM-β-CD, suggesting insulin oligomer dissociation by DM-β-CD. Histopathological examination of the rat intestine was performed to detect tissue damage following enteral administration of the β-CD derivatives. Light microscopic inspection indicated no observable tissue damage, thereby arguing for direct membrane fluidization as the primary mechanism for enhanced insulin uptake. This study indicates the feasibility of using CDs as mucosal absorption promoters of proteins and peptide drugs [148]. In addition, after the administration of 10 mg of cholesterol to mice through a gastric tube, the cholesterol level increased about 125–130%, and only 15–20%, if the cholesterol was administered together with 20 mg of DM-β-CD. That means that the DM-β-CD formed complexes with approximately 80 to 85% of the cholesterol administered in the mice gastrointestinal tract [149]. HP-β-CD is well tolerated in the animal species tested (i.e., rats, mice, and dogs), particularly when dosed orally, and shows only limited toxicity [120]. Furthermore, SBE7-β-CD has a sufficient safety profile after oral administration [5,150]. Thus, HP-β-CD and SBE7-β-CD are well tolerated when administered orally.

It is believed that when CDs are administered orally, they cannot be absorbed in an intact form [4]. Smooth muscle cell migration and proliferation are important regulatory processes in the development of intimal thickening after vascular injury. Beyond expectation, β-CD tetradecasulfate, an orally active synthetic heparin mimic, is effective in inhibiting rabbit aortic smooth muscle cell proliferation in vitro and in limiting restenosis in an experimental angioplasty restenosis model in rabbits [151]. However, its effects on migration are unknown, as are its effects on human vascular smooth muscle cell biology in general. Thus, β-CD tetradecasulfate may be an effective agent in inhibiting intimal thickening after vascular injury by limiting both smooth muscle cell migration and proliferation via the oral route [151].

3.3. Nasal Route

In recent years, extensive research into novel forms of drug delivery has suggested that mucosal approaches offer a promising therapeutic alternative, especially for systemically

acting drugs. Transmucosal drug delivery offers many benefits, including noninvasive administration, convenience, and rapid onset, as well as elimination of hepatic first-pass metabolism. The investigated absorptive surfaces consist of the nasal, buccal, ocular, vaginal, and rectal mucosa. Among these, the nasal and buccal routes have proved the most promising to date. The bioavailability achieved depends primarily on the pathophysiological state of the mucosa and the properties of both the drug and delivery systems. Various agents can increase the efficacy of transmucosal drug delivery. These include CDs, bile salts, surfactants, fusidic acid derivatives, microspheres, liposomes, and bioadhesive agents [152].

The nasal mucosa offers numerous benefits as a target issue for drug delivery, such as a large surface area for delivery, rapid drug onset, potential for central nervous system delivery, and no first-pass metabolism. A wide variety of therapeutic compounds can be delivered in the nasal route, including relatively large molecules such as peptides and proteins, particularly in the presence of permeation enhancers [153]. The potential use of CDs in a nasal route has been reviewed [154]. Regarding a safety profile, the cilio-inhibitory effect of a series of CDs was examined using a human cell suspension culture system exhibiting in vitro ciliogenesis. Among the CDs investigated (γ-CD, HP-β-CD, anionic-β-CD polymer, DM-β-CD, and α-CD), γ-CD (10% w/v), HP-β-CD (10% w/v), and anionic-β-CD polymer (8% w/v) showed no significant cilio-inhibitory effects after 30 min of exposure. Similarly, ciliary beat frequency remained stable upon cell exposure to α-CD (2% w/v) and DM-β-CD (1% w/v). However, higher concentrations of α-CD and DM-β-CD resulted in mild to severe cilio inhibition after 45 min of exposure. The effect of α-CD (5% w/v) was partially reversible, while DM-β-CD (10% w/v) was irreversible [155]. By utilizing a 5-min exposure of each CD solution to the nasal mucosa, no tissue damage was visible for 1.5% w/v β-CD and 5 and 20% w/v HP-β-CD, and the effects were quite similar to those of controls. However, using 20% w/v RM-β-CD showed severe damage to the integrity of nasal mucosa. The severity was similar to 1% w/v polyoxyethylene-9-lauryl ether or 1% w/v sodium deoxycholate. Meanwhile, 30 or 60 min of exposure to 10% w/v HP-β-CD or RM-β-CD resulted in no obvious mucosal damage. In addition, in vivo repeated dosing of RM-β-CD did not show any toxicity up to 20% w/v. These results suggest that at least less than 10% w/v CD solutions do not induce gross tissue damage and can keep the histological integrity of the nasal mucosa [156].

The effects of chemically modified CDs on the nasal absorption of buserelin, an agonist of luteinizing hormone–releasing hormone, were investigated in anesthetized rats. Of the CDs tested, DM-β-CD was most effective in improving the rate and extent of the nasal bioavailability of buserelin. The CDs increased the permeability of the nasal mucosa, which was the primary determinant based on the multiple regression analysis of the nasal absorption enhancement of buserelin. Scanning electron microscopic observations revealed that DM-β-CD induced no remarkable changes in the surface morphology of the nasal mucosa at the minimal concentration necessary to achieve substantial absorption enhancement. The present results suggest that DM-β-CD could improve the nasal bioavailability of buserelin and is well tolerated by the nasal mucosa of the rat [157]. In addition, Abe et al. [158] reported that the rate and extent of nasal bioavailability of buserelin were increased remarkably by coadministration of oleic acid and HP-β-CD, compared with the sole use of the enhancer, through the lowering of both the enzymatic and physical barriers of the nasal epithelium to the peptide, probably resulting from the facilitated transmucosal penetration of oleic acid solubilized in HP-β-CD. Similarly, Arsan et al. [159] reported that mixing DM-β-CD and dodecylmaltoside resulted in mutual inhibition of their ability to enhance systemic absorption of insulin following nasal delivery, consistent with the formation of an inclusion complex between dodecylmaltoside and DM-β-CD, which lacks the ability to enhance nasal insulin absorption. Importantly, Zhao et al. [160] compared the effect on the lasting time of the ciliary movement of various surfactants and CDs. The lasting time of ciliary movement shortened in the order 1% sodium dodecyl sulfate (SDS) > 1% sodium deoxycholate (SDC) > 1% Brij35 > 5% Tween 80 > 0.1% EDTA > 5% HP-β-CD > 1% lecithin, and the effect on the velocity of ciliary movement decreased in the order 1% Brij35 > 1% SDC > 1% SDS > 0.1% EDTA > 1% lecithin > 5% Tween 80 > 5% HP-β-CD. In addition, the effect on ciliary structural and specific cellular changes of nasal mucosa decreased in the order 1% SDS approximately 1% SDC approximately 1% Brij35 > 5% Tween 80 > 0.1% EDTA approximately 5% HP-β-CD approximately 1% lecithin. Furthermore, SBE7-β-CD has the potential as a novel excipient for nasal drug delivery: A nasal spray of midazolam formulated in aqueous SBE7-β-CD buffer solution was tested in 12 healthy volunteers and compared to intravenous midazolam in an open crossover trial. As a result, clinical sedative effects were observed within 5 to 10 min and lasted for about 40 min, and no serious side effects were observed, although mild to moderate, transient irritation of nasal and pharyngeal mucosa was reported [161].

3.4. Pulmonary Route

Targeting drug delivery into the lungs has become one of the most important aspects of systemic or local drug delivery systems. Consequently, in the last few years, techniques and new drug delivery devices intended to deliver drugs into the lungs have been widely developed. Currently, the main drug targeting regimens include direct application of a drug into the lungs, mostly by inhalation therapy using either pressurized metered dose inhalers or dry powder inhalers. Intratracheal administration is commonly used as a first approach

in lung drug delivery in vivo. To convey a sufficient dose of drug to the lungs, suitable drug carriers are required [162]. In a cell culture system RM-β-CD evoked cell death and membrane damage in Calu-3 cells at lower concentrations compared to the other CDs tested. Based on the cumulative penetrated amount at 4 h, the apparent permeability coefficients for α-, β-, and γ-CD were $6.77 \pm 2.23 \times 10^{-8}$, $6.68 \pm 0.84 \times 10^{-8}$ and $6.71 \pm 0.74 \times 10^{-8}$ cm/s, respectively. As a result, this study indicates that (1) in terms of their local safety, hydroxypropylated CDs and natural γ-CD seem to be the safest of the tested CDs in pulmonary drug delivery, and (2) CDs may be absorbed into the systemic circulation from the lungs [163]. Actually, CDs may be used in inhalation powders to improve pharmaceutical and biopharmaceutical properties of drugs without lowering their pulmonary deposition [164]. In addition, the toxicity of CD complexes in dry powder formulations was investigated in the rat by monitoring blood urea nitrogen (BUN) and urinary creatinine, as well as determining the hemolysis of human red blood cells. As a result, γ-CD and DM-β-CD were found to be able to promote salbutamol delivery in dry powder inhaler, compared to a formulation containing lactose. Moreover, γ-CD is relatively safe in the rat if the amount of γ-CD in the formulation is similar to that in this experiment [165]. Hence, CDs were known to markedly improve the poor solubility of cyclosporine A (CsA). The ciliostatic and hemolytic activities of G2-α-CD were the weakest of all the CDs tested. Inhalation of the complex of CsA with G2-α-CD, where the dose of CsA was approximately ninefold less than that of CsA inhaled alone, also inhibited eosinophil accumulation significantly, with a longer duration of action than with the response to CsA alone. Thus, the effective dose of CsA could be reduced by formation of a complex with G2-α-CD, and a wider therapeutic safety margin by inhalation of CsA as a complex with G2-α-CD could be expected [166].

3.5. Rectal Safety

The potential use of CDs in a rectal route has been reviewed [3,150]. The rectal route can be an extremely useful route for delivery of drugs to infants, young children, and patients, where difficulties can arise from oral administration because of swallowing, nausea, and vomiting. Also, this route offers several potential opportunities for drug delivery, including the avoidance of hepatic first-pass elimination, absorption enhancement, and the possibility of rate-controlled drug delivery. However, the rectal route also has some potential disadvantages: (1) poor or erratic absorption across the rectal mucosa of many drugs, (2) a limiting absorbable surface area, and (3) a dissociation problem due to the small fluid content in the rectum. To overcome these problems, many attempts to enhance rectal absorption of various drugs have been made [167]. To improve the rectal delivery of an anti-inflammatory drug, biphenylylacetic acid (BPAA), the use of HP-β-CD and DM-β-CD was investigated. Both in vivo and in situ studies demonstrated that a rather high amount of HP-β-CD (about 20% of dose) was absorbable from the rat's rectum, compared with DM-β-CD (less than 5% of dose), suggesting the possibility of the permeation of BPAA through the rectal membrane in the form of HP-β-CD complex. Furthermore, DM-β-CD and HP-β-CD significantly reduced the irritation of the rectal mucosa caused by BPAA after administration of the suppositories to rats [168]. An attempt was made to optimize the rectal delivery of morphine, using CDs as an absorption enhancer and polysaccharides as a swelling hydrogel in Witepsol H-15 hollow-type suppositories, and this was tested in rabbits. α- and β-CDs enhanced the rate and extent of bioavailability, the former being more effective; γ-CD decreased the absorption of morphine. Importantly, gross and microscopic observations suggested that this preparation was less irritating to the rectal mucosa [169].

3.6. Ocular Safety

The anatomy and physiology of the eye make it a highly protected organ. Designing an effective therapy for ocular diseases, especially for the posterior segment, has been considered to be a formidable task. Limitations of the topical and intravitreal route of administration have challenged scientists to find an alternative mode of administration, such as periocular routes. Therefore, novel ocular drug delivery systems must be developed to improve therapies [170]. However, topical and systemic administration of drugs to the eye is highly inefficient, and there is a need for controlled, sustained release, particularly for conditions that affect the posterior segment. Possible advantages in the ophthalmic use of CDs are the increase in solubility and/or stability and avoidance of imcompatibilities of drugs, such as irritation and discomfort. In an ocular route, α-CD, HP-β-CD, and SBE7-β-CD have been used to solubilize lipophilic drugs. An early study performed by Kanai et al. [171] demonstrated that 0.025% of cyclosporine (Cs) with α-CD (40 mg/mL) resulted in the least corneal toxicity and penetrated the cornea five- to tenfold more than did a lipophilic vehicle with Cs. In addition, a pilocarpine prodrug, *O,O'*-dipropionyl-(1,4-xylylene) bispilocarpic acid diester, showed decreased peak and prolonged duration of miosis compared to pilocarpine, but it caused ocular irritation. HP-β-CD decreased both the ocular delivery of pilocarpine and the irritation by the prodrug, but the net effect was positive. Thus, administering 1% of pilocarpine as a prodrug with 15% (w/v) HP-β-CD, the irritation was at the same level as that of the commercial pilocarpine eyedrop, but the ocular delivery was improved substantially. Eventually, ocular delivery of the pilocarpine prodrug may be enhanced in relation to its local irritation by a proper combination of the buffer, viscosity, and HP-β-CD [172]. Also, the HP-β-CD-based drug delivery system

enhanced both the solubility of disulfiram in aqueous eyedrops and the permeability of the drug into the rabbit eye. Similarly, SBE7-β-CD coadministered with *O,O'*-dipropionyl-(1,4-xylylene) bispilocarpate eliminated the eye irritation due to the pilocarpine prodrug, but also decreased the miotic response [173]; namely, ocular absorption of the prodrug was improved by increasing the viscosity of prodrug/SBE7-β-CD solution with poly(vinyl alcohol) without inducing eye irritation. As a result, administration of pilocarpine prodrug in viscous SBE7-β-CD solution decreased eye irritation substantially, whereas ocular absorption is not affected [174]. Similarly, it was reported that eye irritation of the pilocarpine prodrug was prevented by levels of SBE4-β-CD that do not affect the apparent ocular absorption of the prodrug [175]. Meanwhile, there are a few reports regarding methylated CDs in an ocular route. Jansen et al. [176] reported that DM-β-CD at concentrations of 5 and 12.5% was not a suitable vehicle for ophthalmic formulations, since it was toxic to the corneal epithelium, and that HP-β-CD at a concentration of 12.5% is well tolerated by the rabbit eye and was not toxic to the corneal epithelium.

3.7. Dermal Safety

Transdermal drug delivery systems have been gaining increasing popularity. Many drugs have been delivered successfully by this route for both local and systemic action. However, human skin shows a protective function by imposing physicochemical limitations to the types of pentrants that can traverse the barrier. Recently, the delivery of drugs of differing lipophilicity and molecular weight, including proteins, peptides, and oligonucletides, has been shown to be improved by active methods such as iontophoresis, electroporation, mechanical perturbation, and other energy-related techniques, such as ultrasound and needleless injection [177]. Few studies have been performed to assess the risk of skin damage by CDs, and they have yielded contradictory results. Piel et al. [178] reported the use of corneoxenometry bioassay on human stratum corneum to compare the skin compatibility of CDs currently used in pharmaceutical preparations (β-CD, γ-CD, RM-β-CD, DM-β-CD, TM-β-CD, HP-β-CD, and HP-γ-CD) and that of new amphiphilic CD derivatives, namely, amphiphilic CDs carrying a phospholipidyl chain [dimyristoylphosphatidylethanolamine (DMPE), DMPE-DM-β-CD, and DMPE-TM-β-CD]. All the CDs tested were well tolerated by the stratum corneum at a concentration of 5%. However, interindividual reactivity was larger for DMPE-DM-β-CD, suggesting a more aggressive trend for this compound. Cutaneous index of mildness values obtained confirm that DM-β-CD is able to extract some skin components and shows that DMPE-DM-β-CD performs similarly. Meanwhile, CDs release some components, such as cholesterol, phospholipids, and proteins from the stratum corneum of skin, and thus may change the barrier function of skin and the permeability of drugs or the other xenobiotics. Thus, particular attention should be directed to the possible irritation effects to skin. In fact, Uekama et al. [179] demonstrated that parent CDs at sufficiently higher concentrations caused skin irritation in guinea pigs in the order γ-CD < α-CD < β-CD, a result that depends largely on their ability to extract lipids from the stratum corneum. Additionally, it is evident that DM-β-CD, not HP-β-CD, causes the removal of fatty acids and cholesterol from stratum corneum, especially when higher CDs are applied to the skin [180–182], but there are reports that CDs have a significant safety margin in dermal application [9,183]. Thus, the effects of CDs on mutinous irritation seem to be too ambiguous, due to differences in the experimental conditions. Further elaborate studies should be required.

3.8. Vaginal Safety

The human vagina represents a potential accessible space that offers a valuable route for drug delivery through the use of specifically designed carrier systems for both local and systemic applications. Intravaginal drug delivery is particularly appropriate for drugs associated with women's health issues, but it may also have applications in general drug delivery within the female population [184]. In addition, the development of vaginal medications, especially antifungal medications, requires that the drug be solubilized as well as retained at or near the mucosa for sufficient periods of time to ensure adequate bioavailability. Primary irritation studies and subchronic toxicity studies using a rabbit vaginal model indicated that a formulation containing itraconazole, a broad-spectrum antifungal agent, and HP-β-CD was safe, well tolerated, and retained in the vaginal space. Clinical investigations indicated that application of 5 g of a 2% itraconazole cream was very well tolerated and that itraconazole was not systemically absorbed. Additional studies in women found that the cream was highly effective in reducing or eliminating fungal cultures with few adverse effects. These studies suggested that a mucoadhesive, HP-β-CD-based vaginal cream formulation of itraconazole was a useful and effective dosage form for treating vaginal candidiasis [185]. In addition, another application of CDs to vaginal drug delivery was reported by Cevher et al.: the formation of a complex between natamycin and γ-CD and effective combination with polymers to attain a bioadhesive and sustained-release formulation of natamycin suitable for vaginal delivery and the effective treatment of *Candida* infections [186].

4. SAFETY PROFILE OF CD CONJUGATES

Common CD complexes with lipophilic drugs have stability constants of $10^4 M^{-1}$ at the most, while CDs interact with nucleic acid drugs such as plasmid DNA (pDNA) and siRNA

only very slightly [187,188]. Especially in a systemic circulation, the complexes are easily dissociated, due to the dilution and competitive inclusions with endogenous lipids and albumin, which has association constants greater than $10^5 M^{-1}$. Thus, CDs are poor at delivering lipophilic drugs or nucleic acid drugs to specific target cells after administration.

4.1. CD Conjugates with Drugs

In a gastrointestinal tract, CDs are known to be barely capable of being hydrolyzed and are only slightly absorbed in passage through the stomach and small intestine, however, they are fermented to small saccharides by colonic microflora and thus absorbed as maltose or glucose in the large intestine. This biological property of CDs is useful as a source of site-specific delivery of drugs to colon and as a promoiety for reducing an adverse effect. Taking these factors into account, we have designed CD conjugates with various drugs, such as NSAIDs, biphenylylacetic acid [28], and ketoprofen [29], a short-chain fatty acid, *n*-butylic acid [30], and a steroidal drug, prednisolone [25–27,189], anticipating new candidates for colon-specific delivery drugs. All of the conjugate having ester linkage between CD and drugs and can deliver drugs efficiently to the large intestine. It is certain that the CD–drug conjugates should be degraded in the colon; thus, the local side effect of the drug in the gastrointestinal tract is reduced by the conjugation with CD. However, detailed data regarding the safety profiles of various CD conjugates remain unclear. The potential use of CD conjugates with drugs for colon delivery has been reviewed [4,190].

4.2. CD Conjugates with Polymers

Gene therapy is emerging as a potential strategy for the treatment of genetic diseases, cancers, cardiovascular diseases, and infectious diseases. Clinical trials employing over 1500 gene therapy protocols have been carried out for various diseases [191]. Recently, gene silencing induced by siRNA, RNA interference (RNAi), became a powerful tool of gene analysis and gene therapy [192]. Similarly, vector-based short-hairpin RNA (shRNA) expression systems have been developed to prolong the RNAi effect [193]. However, standard therapeutic use of DNA (gene) and siRNA in clinical settings in humans has been hampered by the lack of effective methods to deliver these nucleic acid drugs into diseased organs and cells [194]. For these reasons, improvement in the transfer activity of a nonviral vector (carrier) is of utmost importance [195]. To improve the gene transfer activity of a nonviral vector, we synthesized the starburst polyamidoamine (PAMAM) dendrimer [from the second generation (G2) to the fourth generation (G4)] conjugates with α-, β- and γ-CDs (CDE), expressing the synergistic effect of dendrimer and CDs. From the in vitro and in vivo

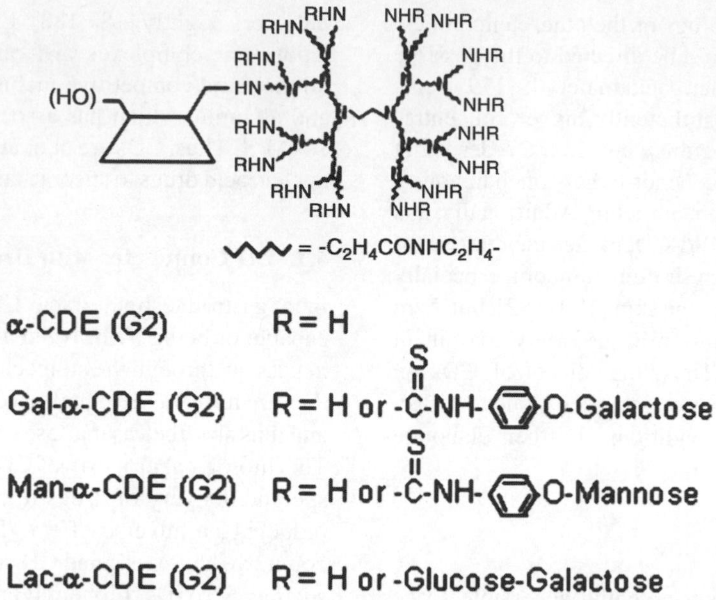

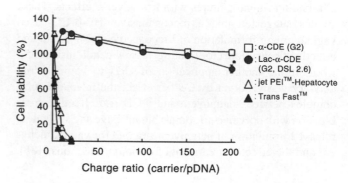

$\bigvee\bigvee\bigvee$ = $-C_2H_4CONHC_2H_4-$

α-CDE (G2)   R = H

Gal-α-CDE (G2)   R = H or $-\overset{\overset{\text{S}}{\|}}{\text{C}}$-NH-⬡-O-Galactose

Man-α-CDE (G2)   R = H or $-\overset{\overset{\text{S}}{\|}}{\text{C}}$-NH-⬡-O-Mannose

Lac-α-CDE (G2)   R = H or -Glucose-Galactose

Figure 9. Chemical structures of α-CDE (G2) and sugar-appended α-CDEs (G2).

evaluation, it could be concluded that conjugate having G3 dendrimer and an average degree of substitution (DS) of 2.4 of α-CD is the best compound among the nine conjugates tested [196–198]. It should be noted that the optimal α-CDE (G3, DS2.4) showed no cytotoxicity up to a charge ratio of carrier to plasmid DNA (an N/P ratio) of 100 in vitro and no side effects after intravenous administration of the pDNA complex at an N/P ratio of 20 [198].

α-CDE (G3, DS2) possesses the potential to be a novel carrier for nucleic acid drugs, but a lack of cell-specific gene transfer activity of α-CDEs has been shown. A carrier system needs to fulfill the following requirements to be a promising candidate for in vivo gene delivery. The carrier should be able to accumulate efficiently in specific target tissues with a lack of toxicity and immunogenicity, and deliver the intact gene into the nucleus of the target cell to acheive high levels of gene expression. Then we prepared mannosylated α-CDE (G2), galactosylated α-CDE (G2), and lactosylated α-CD (G2) (Fig. 9) [199]. Of these sugar-appended α-CDEs, α-CDEs bearing α-lactose, a disaccharide formed from α-glucose and α-galactose (Lac-α-CDE) without using the spacer, especially Lac-α-CDE (G2, DSL3), was found to have higher gene transfer activity than dendrimer and α-CDE (G3) to HepG2 cells, asialoglycoprotein receptor (AgpR)-positive cells, but not to A549 cells, AgpR-negative cells. Importantly, the AgpR-dependent gene delivery of Lac-α-CDE (G2, DSL3) was observed in vivo: Lac-α-CDE (G2, DSL3) provided gene transfer activity much higher than that of α-CDE (G2) in parenchymal cells and much lower than that in spleen 12 h after intravenous injection in mice. In addition, it should be noted that pDNA complexes with α-CDE (G2) and Lac-α-CDE (G2, DSL3) provided no

cytotoxicity up to an N/P ratio of 150, but a complex with commercially available transfection reagents such as JetPEI-Hepatocyte and TransFast, elicited severe cytotoxicity in HepG2 cells (Fig. 10). Hence, these results hold promise for the potential use of Lac-α-CDE (G2, DSL3) as a hepatocyte-selective nonviral vector with negligible cytotoxicity (Fig. 11). Moreover, targeting of the folate receptor (FR) received much attention in recent years, since the FR has been shown to be overexpressed in human cancer cells [84]. In an attempt to develop FR-overexpressing cancer cell–specific gene transfer carriers, we prepared folate-appended α-CDEs [Fol-α-CDE (G3)] and folate-PEG-appended α-CDEs [Fol-PαC (G3)] (Fig. 12) and evaluated the potential as a novel cell-specific gene transfer carrier. As a result,

Figure 10. Cytotoxicity of pDNA complexes with various carriers in HepG2 cells. The cells were incubated with carrier–pDNA complexes for 24 h. Cell viability was assayed using the WST-1 method. The amount of pDNA was 2.0 μg. The culture medium was supplemented with 7.5% FCS. Each point represents the mean ± S.E. of four experiments. $^{*}p < 0.05$, compared with α-CDE (G2).

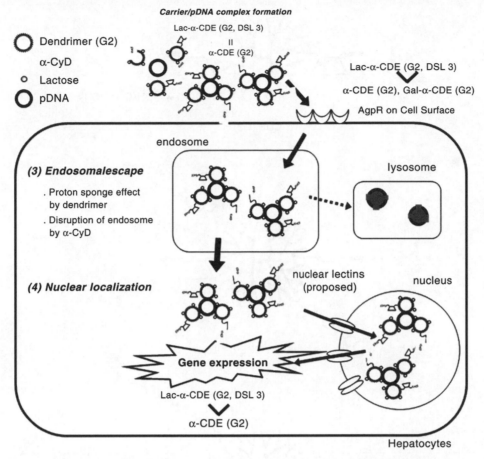

Figure 11. Scheme proposed for improved effects of gene transfer activity by Lac-α-CDE (G2, DSL 3).

potentially, Fol-PαC (G3, DSF5) could be used as a FR-overexpressing cancer cell–selective gene transfer carrier because of FR-mediated gene delivery and extremely low cytotoxicity [199]. Recently, a number of nonviral vectors containing CDs (but not α-CDEs) for DNA, siRNA, and anticancer drug delivery have been developed. CD conjugates, CD polymers, and CD nanoparticles used to delivery DNA are as follows: polyethyleneimine (PEI) conjugates with β-CD [200]; cationic supramolecular polyrotaxanes with multiple cationic α-CD rings threaded and blocked on a poly[(ethylene oxide)-*ran*-(propylene oxide)] random copolymer chain [201]; conjugates having multiple oligoethylenimine (OEI) arms onto an α-CD core [202]; bis(guanidinium)-tetrakis-β-CD tetrapod [203]; supramolecular hydrogels based on self-assembly of inclusion complexes between CDs with biodegradable block copolymers [204]; polypseudorotaxane consisting of a linear PEI (molecular weight 22,000) and γ-CDs [39]; low-molecular-weight PEIs with β-CD [205], polymers of low-molecular-weight PEI cross-linked by HP-β-CD or HP-γ-CD [206]; linear CD-containing polymers [207]; supramolecular inclusion complexes by threading α-CD molecules over PEG and poly(ε-caprolactone) chains of ternary block copolymers of PEG, poly

(ε-caprolactone) (PCL), and PEI [208]; linear PEI and branched PEI grafted with β-CD [209]; amidine-based polycations that contain the carbohydrates *d*-trehalose and β-CD within the polycation backbone [210]; 3(A),3(B)-dideoxy-3(A),3(B)-diamino-β- and γ-CD monomers [211]; and nanoparticles consisting of a linear CD-containing polycation including transferrin protein [212]. Recently, CD conjugates, CD polymers, and CD nanoparticles used to delivery siRNA have been reported: α-CDE (G3) [34], bis(guanidinium)-tetrakis-(β-CD) dendrimeric tetrapod [203], linear CD-containing polymers [207], and CD-containing polycations and their nanoparticles of adamantane–PEG conjugates [213]. Furthermore, CD conjugates, CD polymers, and CD nanoparticles used to delivery antitumor drugs include nanoparticles made of the amphiphilic CD heptakis (2-*O*-oligo(ethyleneoxide)-6-hexadecylthio-)-β-CD–entrapping docetaxel [214], amphiphilic β-CD derivatives such as 6-*N*-CAPRO-β-CD modified on the primary face or secondary face with a 6C aliphatic amide [215–217], and γ-CD-containing liposomes [218].

Almost all the studies on CD conjugates, CD polymers, and CD nanoparticles for nucleic acid drug carriers deal with in vitro cytotoxicity, but only a few showed the in vivo

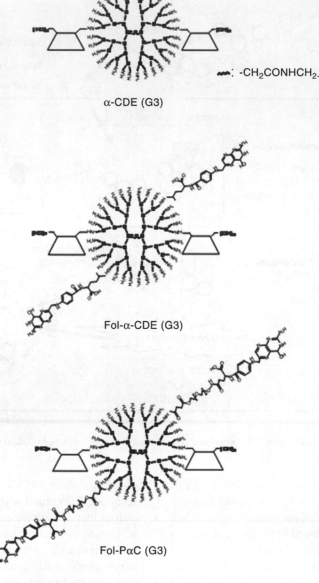

Figure 12. Chemical structures of α-CDE (G3), Fol-α-CDE (G3), and Fol-PαC (G3).

safety data on complexes with CD-containing nonviral vectors. We evaluated safety profiles after intravenous bolus administration (not hydrodynamic injection) of pDNA complexes with α-CDE (G3, DS2.4), Man-α-CDE (G2), and Lac-α-CDE (G2) in mice. Fortunately, various blood chemistry values (i.e., creatinine, BUN, AST, ALT, lactate dehydrogenase) were not changed by the intravenous injection of pDNA complexes [198,219]. Additionally, linear cationic β-CD-based polymers (βCDPs) are capable of forming polyplexes with nucleic acids and transfectioning cultured cells. The βCDPs are synthesized by the condensation of a diamino-CD monomer with a diimidate comonomer. In vitro toxicity varied by one order of magnitude, and the lowest toxicity was observed with βCDP8. The

LD_{40} of the βCDP6 to mice is 200 mg/kg, making this polymer a promising agent for in vivo gene delivery applications [220]. Also, amphiphilic β-CD nanospheres (mean diameter 90 to 110 nm) prepared by the solvent displacement method were developed as a colloidal drug delivery system. After a single intravenous injection of labeled nanoparticles in mice, rapid clearance of ^{125}I-labeled amphiphilic β-CD nanospheres from the blood circulation to the mononuclear phagocyte system was visualized using a noninvasive planar imaging study. Radioactivity measurements in organs showed that the nanospheres concentrated primarily in the liver and spleen, where 28% and 24% of the radioactivity per gram of organ, respectively, was found 10 min after injection. In contrast, the blood activity was low at that time and become negligible thereafter. Finally, no particular sign of toxicity was observed after intravenous administration of β-CD nanoparticles [221]. Furthermore, whether multiple systemic doses of targeted β-CDP/adamantane/PEG/transferrin nanoparticles containing nonchemically modified siRNA can safely be administered to nonhuman primates was evaluated in cynomolgus monkeys. When administered to monkeys at doses of siRNA (3 and 9 mg/kg), the β-CDP/adamantane/PEG/transferrin nanoparticles were well tolerated, although elevation of BUN, creatinine, ALT, and AST was observed at 27 mg/kg siRNA. Overall, no clinical signs of toxicity clearly attributable to treatment were observed. The multiple administrations, spanning a period of 17 to 18 days, enabled assessment of antibody formation against the human transferrin component of the formulation. Everything considered, multiple systemic doses of targeted nanoparticles containing

nonchemically modified siRNA can safely be administered to nonhuman primates [212].

5. SAFETY PROFILE OF POLYPSEUDOROTAXANES

Recently, supramolecular chemistry has been expanding to supramolecular polymer chemistry. The combination of cyclic molecules and linear polymers has provided many types of intriguing supramolecular architectures, such as rotaxanes and catenanes [222]. The supramolecular structures formed between CDs and polymers, especially, have inspired interesting developments of novel supramolecular biomaterials [204].

CD polypseudorotaxanes have potential as a sustained-release carrier of pegylated proteins. As shown in Fig. 13,

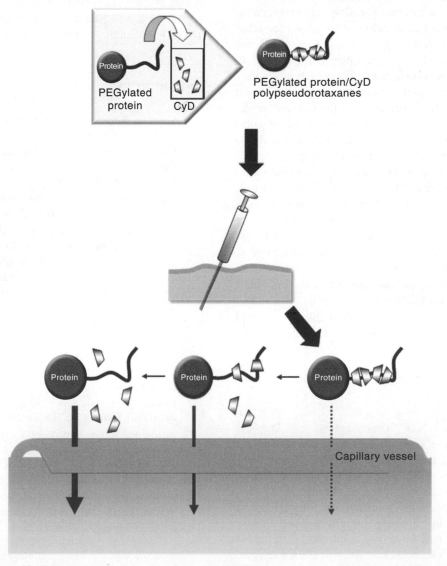

Figure 13. Scheme proposed for the release of PEGylated protein from CD polypseudorotaxane.

when a suspension containing CD polypseudorotaxane with pegylated proteins was injected subcutaneously, CDs were gradually released from the supramolecules, and then pegylated proteins were dissolved in subcutaneous fluids, followed by sustained translocation into the bloodstream. In fact, by inserting two PEG chains in the γ-CD cavity Higashi et al. [38] demonstrated that the pegylated insulin formed polypseudorotaxane with γ-CD. The pegylated insulin/γ-CD polypseudorotaxanes were less soluble in water, and the release rate of the drug from the polypeudorotaxanes was slower than that from the drug alone [38]. The plasma levels of the pegylated insulin after subcutaneous administration of γ-CD polypseudorotaxane to rats were significantly prolonged, accompanying an increase in the area under the plasma drug concentration–time curve, which was clearly reflected in the prolonged hypoglycemic effect [38]. The results indicated that the pegylated insulin/CD polypseudorotaxanes act as a sustained drug release system [38]. In addition, the potential use of polypseudorotaxane formation with CDs as a sustained drug delivery technique for other pegylated proteins, such as pegylated lysozyme [223] and randomly pegylated insulin [224], has been demonstrated. In the in vivo study, no side effects were observed, but the safety profiles of the CD polypseudorotaxanes remain unknown because of the low aqueous solubility of the supramolecule. Meanwhile, a series of novel cationic supramolecular polyrotaxanes with multiple cationic α-CD rings threaded and blocked on a poly[(ethylene oxide)-ran-(propylene oxide)] random copolymer chain were synthesized, and their usefulness as gene delivery carriers has been investigated. In particular, cytotoxicity studies showed that cationic polyrotaxanes, all with linear multiple oligoethylenimine chains of molecular weights up to 423 Da, exhibited much less cytotoxicity than did high-molecular-weight branched polyethylenimine (25 kDa) in both HEK293 and COS7 cell lines [201]. In the future, various supramolecular biomaterials that have undergone CD inclusion complexation, shown to have many advantages over conventional polymers, should be developed for use in designing novel drug and gene delivery systems.

6. PERSPECTIVE

A number of CD derivatives, CD conjugates, and CD (polypseudo)rotaxanes have been designed and evaluated for practical use in various fields. As described above, some hydrophilic CDs, such as HP-β-CD, HP-γ-CD, SBE7-β-CD, and M-β-CD, have been employed in pharmaceutical formulations. In addition, CD nanoparticles for pDNA and siRNA delivery are also approaching clinical use. In July 2008, Sugammadex had been approved in the European Union, but the U.S. Food and Drug Administration rejected the application in August 2008. Again, this topic made us

recognize the importance of safety issues regarding CD-based pharmaceutical products in clinical applications. No matter how good in vitro and in vivo pharmacological data are, in vivo safety data on CD derivatives, CD conjugates, CD polymers, and CD polypseudorotaxanes and their complexes are absolutely imperative for clinical and practical use. We hope that a number of CD-based products with excellent efficacy and safety histories will soon appear on the market.

Acknowledgments

We would like to express sincere thanks to Dr. K. Uekama and Dr. F. Hirayama, Faculty of Pharmaceutical Sciences, Sojo University, for their valuable advice, warm support, and kind help. We thank colleagues in our laboratory in the Graduate School of Pharmaceutical Sciences, Kumamoto University, for their excellent contribution to this study. We appreciate the assistance of Dr. Hattori and Dr. Takeuchi, Tokyo Polytechnic University, with SPR measurement.

REFERENCES

1. Slain, D., Rogers, P.D., Cleary, J.D., Chapman, S.W. (2001). Intravenous itraconazole. *Ann. Pharmacother.*, *35*, 720–729.
2. Szejtli, J., Szente, L. (2005). Elimination of bitter, disgusting tastes of drugs and foods by cyclodextrins. *Eur. J. Pharm. Biopharm.*, *61*, 115–125.
3. Matsuda, H., Arima, H. (1999). Cyclodextrins in transdermal and rectal delivery. *Adv. Drug Deliv. Rev.*, *36*, 81–99.
4. Uekama, K., Otagiri, M. (1987). Cyclodextrins in drug carrier systems. *Crit. Rev. Ther. Drug Carrier Syst.*, *3*, 1–40.
5. Thompson, D.O. (1997). Cyclodextrins—enabling excipients: their present and future use in pharmaceuticals. *Crit. Rev. Ther. Drug Carrier Syst.*, *14*, 1–104.
6. Brewster, M.E., Loftsson, T. (2007). Cyclodextrins as pharmaceutical solubilizers. *Adv. Drug Deliv. Rev.*, *59*, 645–666.
7. Akasaka, H., Endo, T., Nagase, H., Ueda, H., Kobayashi, S. (2000). Complex formation of cyclomaltononaose δ-cyclodextrin (δ-CD) with macrocyclic compounds. *Chem. Pharm. Bull.*, *48*, 1986–1989.
8. Uekama, K. (2004). Design and evaluation of cyclodextrin-based drug formulation. *Chem. Pharm. Bull.*, *52*, 900–915.
9. Albers, E., Muller, B.W. (1995). Cyclodextrin derivatives in pharmaceutics. *Crit. Rev. Ther. Drug Carrier Syst.*, *12*, 311–337.
10. Uekama, K., Hirayama, F., Irie, T. (1998). Cyclodextrin drug carrier systems. *Chem. Rev.*, *98*, 2045–2076.
11. Tavornvipas, S., Hirayama, F., Arima, H., Uekama, K., Ishiguro, T., Oka, M., Hamayasu, K., Hashimoto, H. (2002). 6-O-α-(4-O-α-D-glucuronyl)-D-glucosyl-β-cyclodextrin: solubilizing ability and some cellular effects. *Int. J. Pharm.*, *249*, 199–209.

12. Okada, Y., Kubota, Y., Koizumi, K., Hizukuri, S., Ohfuji, T., Ogata, K. (1988). Some properties and the inclusion behavior of branched cyclodextrins. *Chem. Pharm. Bull.*, *36*, 2176–2185.

13. Koizumi, K., Okada, Y., Kubota, Y., Utamura, T. (1987). Inclusion complexes of poorly water-soluble drugs with glucosyl-cyclodextrins. *Chem. Pharm. Bull.*, *35*, 3413–3418.

14. Oda, Y., Yanagisawa, H., Maruyama, M., Hattori, K., Yamanoi, T. (2008). Design, synthesis and evaluation of D-galactose-β-cyclodextrin conjugates as drug-carrying molecules. *Bioorg. Med. Chem.*, *16*, 8830–8840.

15. Nishi, Y., Tanimoto, T. (2009). Preparation and characterization of branched β-cyclodextrins having α-L-fucopyranose and a study of their functions. *Biosci. Biotechnol. Biochem.*, *73*, 562–569.

16. Nishi, Y., Yamane, N., Tanimoto, T. (2007). Preparation and characterization of 6*I*,6*n*-di-*O*-(L-fucopyranosyl)-β-cyclodextrin (*n* = II–IV) and investigation of their functions. *Carbohydr. Res.*, *342*, 2173–2181.

17. Salmaso, S., Bersani, S., Semenzato, A., Caliceti, P. (2007). New cyclodextrin bioconjugates for active tumour targeting. *J. Drug Target.*, *15*, 379–390.

18. Salmaso, S., Semenzato, A., Caliceti, P., Hoebeke, J., Sonvico, F., Dubernet, C., Couvreur, P. (2004). Specific antitumor targetable β-cyclodextrin-poly(ethylene glycol)–folic acid drug delivery bioconjugate. *Bioconjugate Chem.*, *15*, 997–1004.

19. Liu, Y., Chen, G.S., Chen, Y., Cao, D.X., Ge, Z.Q., Yuan, Y.J. (2004). Inclusion complexes of paclitaxel and oligo(ethylenediamino) bridged bis(β-cyclodextrin)s: solubilization and antitumor activity. *Bioorg. Med. Chem.*, *12*, 5767–5775.

20. Hirayama, F. (1993). Development and pharmaceutical evaluation of hydrophobic cyclodextrin derivatives as modified-release drug carriers. *Yakugaku Zasshi*, *113*, 425–437.

21. Szejtli, J. (1994). Medicinal applications of cyclodextrins. *Med. Res. Rev.*, *14*, 353–386.

22. Loftsson, T., Jarho, P., Másson, M., Järvinen, T. (2005). Cyclodextrins in drug delivery. *Expert Opin. Drug Deliv.*, *2*, 335–351.

23. Shimpi, S., Chauhan, B., Shimpi, P. (2005). Cyclodextrins: application in different routes of drug administration. *Acta Pharm.*, *55*, 139–156.

24. Challa, R., Ahuja, A., Ali, J., Khar, R.K. (2005). Cyclodextrins in drug delivery: an updated review. *AAPS PharmSciTech*, *6*, E329–E357.

25. Yano, H., Hirayama, F., Kamada, M., Arima, H., Uekama, K. (2002). Colon-specific delivery of prednisolone-appended α-cyclodextrin conjugate: alleviation of systemic side effect after oral administration. *J. Control. Release*, *79*, 103–112.

26. Yano, H., Hirayama, F., Arima, H., Uekama, K. (2001). Prednisolone-appended α-cyclodextrin: alleviation of systemic adverse effect of prednisolone after intracolonic administration in 2,4,6-trinitrobenzenesulfonic acid-induced colitis rats. *J. Pharm. Sci.*, *90*, 2103–2112.

27. Yano, H., Hirayama, F., Arima, H., Uekama, K. (2001). Preparation of prednisolone-appended α-, β-, and γ-cyclodextrins: substitution at secondary hydroxyl groups and in vitro hydrolysis behavior. *J. Pharm. Sci.*, *90*, 493–503.

28. Minami, K., Hirayama, F., Uekama, K. (1998). Colon-specific drug delivery based on a cyclodextrin prodrug: release behavior of biphenylylacetic acid from its cyclodextrin conjugates in rat intestinal tracts after oral administration. *J. Pharm. Sci.*, *87*, 715–720.

29. Kamada, M., Hirayama, F., Udo, K., Yano, H., Arima, H., Uekama, K. (2002). Cyclodextrin conjugate-based controlled release system: repeated- and prolonged-releases of ketoprofen after oral administration in rats. *J. Control. Release*, *82*, 407–416.

30. Hirayama, F., Ogata, T., Yano, H., Arima, H., Udo, K., Takano, M., Uekama, K. (2000). Release characteristics of a short-chain fatty acid, *n*-butyric acid, from its β-cyclodextrin ester conjugate in rat biological media. *J. Pharm. Sci.*, *89*, 1486–1495.

31. Yang, C., Li, H., Wang, X., Li, J. (2009). Cationic supramolecules consisting of oligoethylenimine-grafted α-cyclodextrins threaded on poly(ethylene oxide) for gene delivery. *J. Biomed. Mater. Res. A*, *89*, 13–23.

32. Burckbuchler, V., Wintgens, V., Leborgne, C., Lecomte, S., Leygue, N., Scherman, D., Kichler, A., Amiel, C. (2008). Development and characterization of new cyclodextrin polymer–based DNA delivery systems. *Bioconjugate Chem.*, *19*, 2311–2320.

33. Tsutsumi, T., Hirayama, F., Uekama, K., Arima, H. (2008). Potential use of polyamidoamine dendrimer/α-cyclodextrin conjugate (generation 3, G3) as a novel carrier for short hairpin RNA-expressing plasmid DNA. *J. Pharm. Sci.*, *97*, 3022–3034.

34. Tsutsumi, T., Hirayama, F., Uekama, K., Arima, H. (2007). Evaluation of polyamidoamine dendrimer/α-cyclodextrin conjugate (generation 3, G3) as a novel carrier for small interfering RNA (siRNA). *J. Control. Release*, *119*, 349–359.

35. Arima, H. (2004). Polyfection as nonviral gene transfer method: design of novel nonviral vector using α-cyclodextrin. *Yakugaku Zasshi*, *124*, 451–464.

36. Davis, M.E., Brewster, M.E. (2004). Cyclodextrin-based pharmaceutics: past, present and future. *Nat. Rev. Drug Discov.*, *3*, 1023–1035.

37. Higashi, T., Hirayama, F., Arima, H., Uekama, K. (2007). Polypseudorotaxanes of pegylated insulin with cyclodextrins: application to sustained release system. *Bioorg. Med. Chem. Lett.*, *17*, 1871–1874.

38. Higashi, T., Hirayama, F., Misumi, S., Arima, H., Uekama, K. (2008). Design and evaluation of polypseudorotaxanes of pegylated insulin with cyclodextrins as sustained release system. *Biomaterials*, *29*, 3866–3871.

39. Yamashita, A., Choi, H.S., Ooya, T., Yui, N., Akita, H., Kogure, K., Harashima, H. (2006). Improved cell viability of linear polyethylenimine through γ-cyclodextrin inclusion for effective gene delivery. *ChemBioChem*, *7*, 297–302.

40. Uekama, K., Irie, T., Sunada, M., Otagiri, M., Iwasaki, K., Okano, Y., Miyata, T., Kase, Y. (1981). Effects of

cyclodextrins on chlorpromazine-induced haemolysis and central nervous system responses. *J. Pharm. Pharmacol.*, *33*, 707–710.

41. Irie, T., Sunada, M., Otagiri, M., Uekama, K. (1983). Protective mechanism of β-cyclodextrin for the hemolysis induced with phenothiazine neuroleptics in vitro. *J. Pharmacobio-Dyn.*, *6*, 408–414.

42. Uekama, K., Irie, T., Sunada, M., Otagiri, M., Tsubaki, K. (1981). Protective effects of cyclodextrins on drug-induced hemolysis in vitro. *J. Pharmacobio-Dyn.*, *4*, 142–144.

43. Panini, R., Vandelli, M.A., Leo, E., Salvioli, G., Cameroni, R. (1996). The influence of 2-hydroxypropyl-β-cyclodextrin on the haemolysis induced by bile acids. *J. Pharm. Pharmacol.*, *48*, 641–644.

44. Funasaki, N., Okuda, T., Neya, S. (2001). Mechanisms and surface chemical prediction of imipramine-induced hemolysis suppressed by modified cyclodextrins. *J. Pharm. Sci.*, *90*, 1056–1065.

45. Nagase, Y., Hirata, M., Arima, H., Tajiri, S., Nishimoto, Y., Hirayama, F., Irie, T., Uekama, K. (2002). Protective effect of sulfobutyl ether β-cyclodextrin on DY-9760e-induced hemolysis in vitro. *J. Pharm. Sci.*, *91*, 2382–2389.

46. Ohtani, Y., Irie, T., Uekama, K., Fukunaga, K., Pitha, J. (1989). Differential effects of α-, β- and γ-cyclodextrins on human erythrocytes. *Eur. J. Biochem.*, *186*, 17–22.

47. Debouzy, J.C., Crouzier, D., Dabouis, V., Gadelle, A. (2008). [Per(6-deoxy) derivative of β-cyclodextrin (B6): physico-chemical characterization, haemolytic activity and membrane properties]. *Ann. Pharm. Fr.*, *66*, 19–27.

48. Weisz, P.B., Kumor, K., Macarak, E.J. (1993). Protection of erythrocytes against hemolytic agents by cyclodextrin polysulfate. *Biochem. Pharmacol.*, *45*, 1011–1016.

49. Irie, T., Uekama, K. (1997). Pharmaceutical applications of cyclodextrins: III. Toxicological issues and safety evaluation. *J. Pharm. Sci.*, *86*, 147–162.

50. Motoyama, K., Arima, H., Toyodome, H., Irie, T., Hirayama, F., Uekama, K. (2006). Effect of 2,6-di-O-methyl-α-cyclodextrin on hemolysis and morphological change in rabbit's red blood cells. *Eur. J. Pharm. Sci.*, *29*, 111–119.

51. Motoyama, K., Toyodome, H., Onodera, R., Irie, T., Hirayama, F., Uekama, K., Arima, H. (2009). Involvement of lipid rafts of rabbit red blood cells in morphological changes induced by methylated β-cyclodextrins. *Biol. Pharm. Bull.*, *32*, 700–705.

52. Arima, H., Hirayama, F., Okamoto, C.T., Uekama, K. (2002). Recent aspects of cyclodextrin-based pharmaceutical formulations. *Recent Res. Dev. Chem. Pharm. Sci.*, *2*, 155–193.

53. Arima, H., Uekama, K. (2007). In: Touitou, E., Barry, B.W. Eds., *Enhancement in Drug Delivery*. CRC Press, London, pp. 147–172.

54. Okada, Y., Matsuda, K., Hara, K., Hamayasu, K., Hashimoto, H., Koizumi, K. (1999). Properties and the inclusion behavior of 6-O-α-D-galactosyl- and 6-O-α-D-mannosyl-cyclodextrins. *Chem. Pharm. Bull.*, *47*, 1564–1568.

55. Okada, Y., Semma, M., Ichikawa, A. (2007). Physicochemical and biological properties of 6(1),6(3),6(5)-tri-O-α-maltosyl-

cyclomaltoheptaose (6(1),6(3),6(5)-tri-O-α-maltosyl-β-cyclodextrin). *Carbohydr. Res.*, *342*, 1315–1322.

56. Attioui, F., al-Omar, A., Leray, E., Parrot-Lopez, H., Finance, C., Bonaly, R. (1994). Recognition ability and cytotoxicity of some oligosaccharidyl substituted β-cyclodextrins. *Biol. Cell*, *82*, 161–167.

57. Irie, T., Uekama, K. (1999). Cyclodextrins in peptide and protein delivery. *Adv. Drug Deliv. Rev.*, *36*, 101–123.

58. Ohkubo, S., Nakahata, N. (2007). Role of lipid rafts in trimeric G protein–mediated signal transduction. *Yakugaku Zasshi*, *127*, 27–40.

59. Delmas, O., Breton, M., Sapin, C., Le Bivic, A., Colard, O., Trugnan, G. (2007). Heterogeneity of raft-type membrane microdomains associated with VP4, the rotavirus spike protein, in Caco-2 and MA 104 cells. *J. Virol.*, *81*, 1610–1618.

60. Fauvelle, F., Debouzy, J.C., Crouzy, S., Goschl, M., Chapron, Y. (1997). Mechanism of α-cyclodextrin–induced hemolysis: 1. The two-step extraction of phosphatidylinositol from the membrane. *J. Pharm. Sci.*, *86*, 935–943.

61. Monnaert, V., Tilloy, S., Bricout, H., Fenart, L., Cecchelli, R., Monflier, E. (2004). Behavior of α-, β-, and γ-cyclodextrins and their derivatives on an in vitro model of blood–brain barrier. *J. Pharmacol. Exp. Ther.*, *310*, 745–751.

62. Arima, H., Yunomae, K., Hirayama, F., Uekama, K. (2001). Contribution of P-glycoprotein to the enhancing effects of dimethyl-β-cyclodextrin on oral bioavailability of tacrolimus. *J. Pharmacol. Exp. Ther.*, *297*, 547–555.

63. Ono, N., Arima, H., Hirayama, F., Uekama, K. (2001). A moderate interaction of maltosyl-α-cyclodextrin with Caco-2 cells in comparison with the parent cyclodextrin. *Biol. Pharm. Bull.*, *24*, 395–402.

64. Boulmedarat, L., Bochot, A., Lesieur, S., Fattal, E. (2005). Evaluation of buccal methyl-β-cyclodextrin toxicity on human oral epithelial cell culture model. *J. Pharm. Sci.*, *94*, 1300–1309.

65. Kiss, T., Fenyvesi, F., Pasztor, N., Feher, P., Varadi, J., Kocsan, R., Szente, L., Fenyvesi, E., Szabo, G., Vecsernyes, M., Bacskay, I. (2007). Cytotoxicity of different types of methylated β-cyclodextrins and ionic derivatives. *Pharmazie*, *62*, 557–558.

66. Motoyama, K., Kameyama, K., Onodera, R., Araki, N., Hirayama, F., Uekama, K., Arima, H. (2009). Involvement of PI3K–Akt–Bad pathway in apoptosis induced by 2,6-di-O-methyl-β-cyclodextrin, not 2,6-di-O-methyl-α-cyclodextrin, through cholesterol depletion from lipid rafts on plasma membranes in cells. *Eur. J. Pharm. Sci.*, *38*, 249–261.

67. Ulloth, J.E., Almaguel, F.G., Padilla, A., Bu, L., Liu, J.W., De Leon, M. (2007). Characterization of methyl-β-cyclodextrin toxicity in NGF-differentiated PC12 cell death. *Neurotoxicology*, *28*, 613–621.

68. Hipler, U.C., Schonfelder, U., Hipler, C., Elsner, P. (2007). Influence of cyclodextrins on the proliferation of HaCaT keratinocytes in vitro. *J. Biomed. Mater. Res. A*, *83*, 70–79.

69. Abrami, L., van Der Goot, F. G. (1999). Plasma membrane microdomains act as concentration platforms to facilitate intoxication by aerolysin. *J. Cell Biol.*, *147*, 175–184.

70. Nelson, K.L., Buckley, J.T. (2000). Channel formation by the glycosylphosphatidylinositol-anchored protein binding toxin aerolysin is not promoted by lipid rafts. *J. Biol. Chem.*, *275*, 19839–19843.

71. Gosens, R., Stelmack, G.L., Dueck, G., Mutawe, M.M., Hinton, M., McNeill, K.D., Paulson, A., Dakshinamurti, S., Gerthoffer, W.T., Thliveris, J.A., Unruh, H., Zaagsma, J., Halayko, A.J. (2007). Caveolae facilitate muscarinic receptor-mediated intracellular Ca^{2+} mobilization and contraction in airway smooth muscle. *Am. J. Physiol. Lung Cell Mol. Physiol.*, *293*, L1406–L1418.

72. Han, W., Li, H., Villar, V.A., Pascua, A.M., Dajani, M.I., Wang, X., Natarajan, A., Quinn, M.T., Felder, R.A., Jose, P.A., Yu, P. (2008). Lipid rafts keep NADPH oxidase in the inactive state in human renal proximal tubule cells. *Hypertension*, *51*, 481–487.

73. Kincer, J.F., Uittenbogaard, A., Dressman, J., Guerin, T.M., Febbraio, M., Guo, L., Smart, E.J. (2002). Hypercholesterolemia promotes a CD36-dependent and endothelial nitric-oxide synthase–mediated vascular dysfunction. *J. Biol. Chem.*, *277*, 23525–23533.

74. Camargo, F., Erickson, R.P., Garver, W.S., Hossain, G.S., Carbone, P.N., Heidenreich, R.A., Blanchard, J. (2001). Cyclodextrins in the treatment of a mouse model of Niemann–Pick C disease. *Life Sci.*, *70*, 131–142.

75. Zhang, Q., Furukawa, K., Chen, H.H., Sakakibara, T., Urano, T., Furukawa, K. (2006). Metastatic potential of mouse Lewis lung cancer cells is regulated via ganglioside GM_1 by modulating the matrix metalloprotease-9 localization in lipid rafts. *J. Biol. Chem.*, *281*, 18145–18155.

76. Kritharides, L., Kus, M., Brown, A.J., Jessup, W., Dean, R.T. (1996). Hydroxypropyl-β-cyclodextrin-mediated efflux of 7-ketocholesterol from macrophage foam cells. *J. Biol. Chem.*, *271*, 27450–27455.

77. Arima, H., Nishimoto, Y., Motoyama, K., Hirayama, F., Uekama, K. (2001). Inhibitory effects of novel hydrophilic cyclodextrin derivatives on nitric oxide production in macrophages stimulated with lipopolysaccharide. *Pharm. Res.*, *18*, 1167–1173.

78. Totterman, A.M., Schipper, N.G., Thompson, D.O., Mannermaa, J.P. (1997). Intestinal safety of water-soluble β-cyclodextrins in paediatric oral solutions of spironolactone: effects on human intestinal epithelial Caco-2 cells. *J. Pharm. Pharmacol.*, *49*, 43–48.

79. Zhao, Q., Temsamani, J., Agrawal, S. (1995). Use of cyclodextrin and its derivatives as carriers for oligonucleotide delivery. *Antisense Res. Dev.*, *5*, 185–192.

80. Grosse, P.Y., Bressolle, F., Pinguet, F. (1997). Methyl-β-cyclodextrin in HL-60 parental and multidrug-resistant cancer cell lines: effect on the cytotoxic activity and intracellular accumulation of doxorubicin. *Cancer Chemother. Pharmacol.*, *40*, 489–494.

81. Kang, J., Kumar, V., Yang, D., Chowdhury, P.R., Hohl, R.J. (2002). Cyclodextrin complexation: influence on the solubility, stability, and cytotoxicity of camptothecin, an antineoplastic agent. *Eur. J. Pharm. Sci.*, *15*, 163–170.

82. Yunomae, K., Arima, H., Hirayama, F., Uekama, K. (2003). Involvement of cholesterol in the inhibitory effect of dimethyl-β-cyclodextrin on P-glycoprotein and MRP2 function in Caco-2 cells. *FEBS Lett.*, *536*, 225–231.

83. Fenyvesi, F., Fenyvesi, E., Szente, L., Goda, K., Bacso, Z., Bacskay, I., Varadi, J., Kiss, T., Molnar, E., Janaky, T., Szabo, G., Jr., Vecsernyes, M. (2008). P-glycoprotein inhibition by membrane cholesterol modulation. *Eur. J. Pharm. Sci.*, *34*, 236–242.

84. Balijepalli, R.C., Foell, J.D., Hall, D.D., Hell, J.W., Kamp, T.J. (2006). Localization of cardiac L-type Ca^{2+} channels to a caveolar macromolecular signaling complex is required for $β_2$-adrenergic regulation. *Proc. Natl. Acad. Sci. USA*, *103*, 7500–7505.

85. Abulrob, A., Tauskela, J.S., Mealing, G., Brunette, E., Faid, K., Stanimirovic, D. (2005). Protection by cholesterol-extracting cyclodextrins: a role for *N*-methyl-D-aspartate receptor redistribution. *J. Neurochem.*, *92*, 1477–1486.

86. Hill, W.G., Butterworth, M.B., Wang, H., Edinger, R.S., Lebowitz, J., Peters, K.W., Frizzell, R.A., Johnson, J.P. (2007). The epithelial sodium channel (ENaC) traffics to apical membrane in lipid rafts in mouse cortical collecting duct cells. *J. Biol. Chem.*, *282*, 37402–37411.

87. Ahsan, F., Arnold, J.J., Yang, T., Meezan, E., Schwiebert, E.M., Pillion, D.J. (2003). Effects of the permeability enhancers, tetradecylmaltoside and dimethyl-β-cyclodextrin, on insulin movement across human bronchial epithelial cells (16HBE14o-). *Eur. J. Pharm. Sci.*, *20*, 27–34.

88. Monnaert, V., Betbeder, D., Fenart, L., Bricout, H., Lenfant, A.M., Landry, C., Cecchelli, R., Monflier, E., Tilloy, S. (2004). Effects of γ- and hydroxypropyl-γ-cyclodextrins on the transport of doxorubicin across an in vitro model of blood–brain barrier. *J. Pharmacol. Exp. Ther.*, *311*, 1115–1120.

89. Bar, R., Ulitzur, S. (1994). Bacterial toxicity of cyclodextrins: luminuous *Escherichia coli* as a model. *Appl. Microbiol. Biotechnol.*, *41*, 574–577.

90. Zhang, H.M., Li, Z., Uematsu, K., Kobayashi, T., Horikoshi, K. (2008). Antibacterial activity of cyclodextrins against *Bacillus* strains. *Arch. Microbiol.*, *190*, 605–609.

91. Aachmann, F.L., Aune, T.E. (2009). Use of cyclodextrin and its derivatives for increased transformation efficiency of competent bacterial cells. *Appl. Microbiol. Biotechnol.*, *83*, 589–596.

92. Donova, M.V., Nikolayeva, V.M., Dovbnya, D.V., Gulevskaya, S.A., Suzina, N.E. (2007). Methyl-β-cyclodextrin alters growth, activity and cell envelope features of sterol-transforming mycobacteria. *Microbiology*, *153*, 1981–1992.

93. Hang'ombe, M.B., Mukamoto, M., Kohda, T., Sugimoto, N., Kozaki, S. (2004). Cytotoxicity of Clostridium septicum α-toxin: its oligomerization in detergent resistant membranes of mammalian cells. *Microb. Pathog.*, *37*, 279–286.

94. Nagahama, M., Yamaguchi, A., Hagiyama, T., Ohkubo, N., Kobayashi, K., Sakurai, J. (2004). Binding and internalization of *Clostridium perfringens* ι-toxin in lipid rafts. *Infect. Immun.*, *72*, 3267–3275.

95. Karginov, V.A., Nestorovich, E.M., Schmidtmann, F., Robinson, T.M., Yohannes, A., Fahmi, N.E., Bezrukov,

S.M., Hecht, S.M. (2007). Inhibition of *S. aureus* α-hemolysin and *B. anthracis* lethal toxin by β-cyclodextrin derivatives. *Bioorg. Med. Chem.*, *15*, 5424–5431.

96. Giesemann, T., Jank, T., Gerhard, R., Maier, E., Just, I., Benz, R., Aktories, K. (2006). Cholesterol-dependent pore formation of *Clostridium difficile* toxin A. *J. Biol. Chem.*, *281*, 10808–10815.

97. Miyata, S., Minami, J., Tamai, E., Matsushita, O., Shimamoto, S., Okabe, A. (2002). *Clostridium perfringens* ε-toxin forms a heptameric pore within the detergent-insoluble microdomains of Madin–Darby canine kidney cells and rat synaptosomes. *J. Biol. Chem.*, *277*, 39463–39468.

98. Zhuang, M., Oltean, D.I., Gomez, I., Pullikuth, A.K., Soberon, M., Bravo, A., Gill, S.S. (2002). *Heliothis virescens* and *Manduca sexta* lipid rafts are involved in Cry1A toxin binding to the midgut epithelium and subsequent pore formation. *J. Biol. Chem.*, *277*, 13863–13872.

99. Boesze-Battaglia, K., Besack, D., McKay, T., Zekavat, A., Otis, L., Jordan-Sciutto, K., Shenker, B.J. (2006). Cholesterol-rich membrane microdomains mediate cell cycle arrest induced by *Actinobacillus actinomycetemcomitans* cytolethal-distending toxin. *Cell Microbiol.*, *8*, 823–836.

100. Schroeder, G.N., Hilbi, H. (2007). Cholesterol is required to trigger caspase-1 activation and macrophage apoptosis after phagosomal escape of *Shigella. Cell Microbiol.*, *9*, 265–278.

101. Moriyama, T., Marquez, J.P., Wakatsuki, T., Sorokin, A. (2007). Caveolar endocytosis is critical for BK virus infection of human renal proximal tubular epithelial cells. *J. Virol.*, *81*, 8552–8562.

102. Zaidi, T., Bajmoczi, M., Zaidi, T., Golan, D.E., Pier, G.B. (2008). Disruption of CFTR-dependent lipid rafts reduces bacterial levels and corneal disease in a murine model of *Pseudomonas aeruginosa* keratitis. *Invest. Ophthalmol. Vis. Sci.*, *49*, 1000–1009.

103. Chassin, C., Vimont, S., Cluzeaud, F., Bens, M., Goujon, J.M., Fernandez, B., Hertig, A., Rondeau, E., Arlet, G., Hornef, M. W., Vandewalle, A. (2008). TLR4 facilitates translocation of bacteria across renal collecting duct cells. *J. Am. Soc. Nephrol.*, *19*, 2364–2374.

104. Schraw, W., Li, Y., McClain, M.S., van der Goot, F.G., Cover, T.L. (2002). Association of *Helicobacter pylori* vacuolating toxin (VacA) with lipid rafts. *J. Biol. Chem.*, *277*, 34642–34650.

105. Patel, H.K., Willhite, D.C., Patel, R.M., Ye, D., Williams, C.L., Torres, E.M., Marty, K.B., MacDonald, R.A., Blanke, S.R. (2002). Plasma membrane cholesterol modulates cellular vacuolation induced by the *Helicobacter pylori* vacuolating cytotoxin. *Infect. Immun.*, *70*, 4112–4123.

106. Petro, K.A., Dyer, M.A., Yowler, B.C., Schengrund, C.L. (2006). Disruption of lipid rafts enhances activity of botulinum neurotoxin serotype A. *Toxicon*, *48*, 1035–1045.

107. Karginov, V.A., Nestorovich, E.M., Moayeri, M., Leppla, S.H., Bezrukov, S.M. (2005). Blocking anthrax lethal toxin at the protective antigen channel by using structure-inspired drug design. *Proc. Natl. Acad. Sci. USA*, *102*, 15075–15080.

108. Karginov, V.A., Nestorovich, E.M., Yohannes, A., Robinson, T.M., Fahmi, N.E., Schmidtmann, F., Hecht, S.M., Bezrukov, S.M. (2006). Search for cyclodextrin-based inhibitors of anthrax toxins: synthesis, structural features, and relative activities. *Antimicrob. Agents Chemother.*, *50*, 3740–3753.

109. Backer, M.V., Patel, V., Jehning, B.T., Claffey, K.P., Karginov, V.A., Backer, J.M. (2007). Inhibition of anthrax protective antigen outside and inside the cell. *Antimicrob. Agents Chemother.*, *51*, 245–251.

110. Takeda, K., Akira, S. (2007). Toll-like receptors. In: *Current Protocols in Immunology.* Unit 14.12. Coligan JE, Ed. Hoboken, NJ: Wiley.

111. Parkinson, T. (2008). The future of Toll-like receptor therapeutics. *Curr. Opin. Mol. Ther.*, *10*, 21–31.

112. Lancaster, G.I., Khan, Q., Drysdale, P., Wallace, F., Jeukendrup, A.E., Drayson, M.T., Gleeson, M. (2005). The physiological regulation of Toll-like receptor expression and function in humans. *J. Physiol.*, *563*, 945–955.

113. Motoyama, K., Hashimoto, Y., Hirayama, F., Uekama, K., Arima, H. (2009). Inhibitory effects of 2,6-di-*O*-methyl-α-cyclodextrin on poly I : C signaling in macrophages. *Eur. J. Pharm. Sci.*, *36*, 285–291.

114. Arima, H., Motoyama, K., Matsukawa, A., Nishimoto, Y., Hirayama, F., Uekama, K. (2005). Inhibitory effects of dimethylacetyl-β-cyclodextrin on lipopolysaccharide-induced macrophage activation and endotoxin shock in mice. *Biochem. Pharmacol.*, *70*, 1506–1517.

115. Imaizumi, A., Suzuki, Y., Ono, S., Sato, H., Sato, Y. (1983). Effect of heptakis (2,6-*O*-dimethyl) β-cyclodextrin on the production of pertussis toxin by *Bordetella pertussis. Infect. Immun.*, *41*, 1138–1143.

116. Morshed, M.G., Karita, M., Konishi, H., Okita, K., Nakazawa, T. (1994). Growth medium containing cyclodextrin and low concentration of horse serum for cultivation of *Helicobacter pylori. Microbiol. Immunol.*, *38*, 897–900.

117. Marchini, A., d'Apolito, M., Massari, P., Atzeni, M., Copass, M., Olivieri, R. (1995). Cyclodextrins for growth of *Helicobacter pylori* and production of vacuolating cytotoxin. *Arch. Microbiol.*, *164*, 290–293.

118. Lin, Z., Kong, D., Zhong, P., Yin, K., Dong, L. (2005). Influence of hydroxypropylcyclodextrins on the toxicity of mixtures. *Chemosphere*, *58*, 1301–1306.

119. Stella, V.J., He, Q. (2008). *Cyclodextrins. Toxicol. Pathol.*, *36*, 30–42.

120. Gould, S., Scott, R.C. (2005). 2-Hydroxypropyl-β-cyclodextrin (HP-β-CD): a toxicology review. *Food Chem. Toxicol.*, *43*, 1451–1459.

121. Uekama, K., Hirayama, F. (2003). *Improvement of Drug Properties by Cyclodextrins.* Elsevier, Amsterdam, pp. 671–696.

122. Frank, D.W., Gray, J.E., Weaver, R.N. (1976). Cyclodextrin nephrosis in the rat. *Am. J. Pathol.*, *83*, 367–382.

123. Hiasa, Y., Ohshima, M., Kitahori, Y., Konishi, N., Fujita, T., Yuasa, T. (1982). β-Cyclodextrin: promoting effect on the development of renal tubular cell tumors in rats treated with *N*-ethyl-*N*-hydroxyethylnitrosamine. *J. Natl. Cancer Inst.*, *69*, 963–967.

124. Rajewski, R.A., Traiger, G., Bresnahan, J., Jaberaboansari, P., Stella, V.J., Thompson, D.O. (1995). Preliminary safety evaluation of parenterally administered sulfoalkyl ether β-cyclodextrin derivatives. *J. Pharm. Sci.*, *84*, 927–932.

125. von Mach, M.A., Burhenne, J., Weilemann, L.S. (2006). Accumulation of the solvent vehicle sulphobutylether β cyclodextrin sodium in critically ill patients treated with intravenous voriconazole under renal replacement therapy. *BMC Clin. Pharmacol.*, *6*, 6.

126. McIntosh, M.P., Schwarting, N., Rajewski, R.A. (2004). *In vitro* and *in vivo* evaluation of a sulfobutyl ether β-cyclodextrin enabled etomidate formulation. *J. Pharm. Sci.*, *93*, 2585–2594.

127. Kim, Y., Oksanen, D.A., Massefski, W., Jr., Blake, J.F., Duffy, E.M., Chrunyk, B. (1998). Inclusion complexation of ziprasidone mesylate with β-cyclodextrin sulfobutyl ether. *J. Pharm. Sci.*, *87*, 1560–1567.

128. Yaksh, T.L., Jang, J.D., Nishiuchi, Y., Braun, K.P., Ro, S.G., Goodman, M. (1991). The utility of 2-hydroxypropyl-β-cyclodextrin as a vehicle for the intracerebral and intrathecal administration of drugs. *Life Sci.*, *48*, 623–633.

129. Debouzy, J.C., Gadelle, A., Tymen, H., Le Gall, B., Millot, X., Moretto, P., Fauvelle, F., Le Peoc, H.M., Dabouis, V., Martel, B. (2003). In vitro uranyle affinity of per(3,6-anhydro-2-*O*-carboxymethyle)-α-cyclodextrin and conditions required for in vivo application. *Ann. Pharm. Fr.*, *61*, 62–69.

130. Naguib, M. (2007). Sugammadex: another milestone in clinical neuromuscular pharmacology. *Anesth. Analg.*, *104*, 575–581.

131. Vanacker, B.F., Vermeyen, K.M., Struys, M.M., Rietbergen, H., Vandermeersch, E., Saldien, V., Kalmar, A.F., Prins, M.E. (2007). Reversal of rocuronium-induced neuromuscular block with the novel drug sugammadex is equally effective under maintenance anesthesia with propofol or sevoflurane. *Anesth. Analg.*, *104*, 563–568.

132. Puhringer, F.K., Rex, C., Sielenkamper, A.W., Claudius, C., Larsen, P.B., Prins, M.E., Eikermann, M., Khuenl-Brady, K.S. (2008). Reversal of profound, high-dose rocuronium-induced neuromuscular blockade by sugammadex at two different time points: an international, multicenter, randomized, dose-finding, safety assessor-blinded, phase II trial. *Anesthesiology*, *109*, 188–197.

133. Cammu, G., De Kam, P.J., Demeyer, I., Decoopman, M., Peeters, P.A., Smeets, J.M., Foubert, L. (2008). Safety and tolerability of single intravenous doses of sugammadex administered simultaneously with rocuronium or vecuronium in healthy volunteers. *Br. J. Anaesth.*, *100*, 373–379.

134. Mirakhur, R.K. (2009). Sugammadex in clinical practice. *Anaesthesia*, *64* (Suppl. 1), 45–54.

135. Lina, B.A., Bar, A. (2004). Subchronic oral toxicity studies with α-cyclodextrin in rats. *Regul. Toxicol. Pharmacol.*, *39* (Suppl. 1), S14–S26.

136. Waalkens-Berendsen, D.H., Bar, A. (2004). Embryotoxicity and teratogenicity study with α-cyclodextrin in rats. *Regul. Toxicol. Pharmacol.*, *39* (Suppl. 1), S34–S39.

137. Waalkens-Berendsen, D.H., Smits-Van Prooije, A.E., Bar, A. (2004). Embryotoxicity and teratogenicity study with α-cyclodextrin in rabbits. *Regul. Toxicol. Pharmacol.*, *39* (Suppl. 1), S40–S46.

138. Szejtli, J. (1988). In: Davies, J.E.D. Ed., *Cyclodextrin Technology*. Kluwer, Dordrecht, The Netherlands.

139. Bellringer, M.E., Smith, T.G., Read, R., Gopinath, C., Olivier, P. (1995). β-Cyclodextrin:52-week toxicity studies in the rat and dog. *Food Chem. Toxicol.*, *33*, 367–376.

140. Toyoda, K., Shoda, T., Uneyama, C., Takada, K., Takahashi, M. (1997). Carcinogenicity study of β-cyclodextrin in F344 rats. *Food Chem. Toxicol.*, *35*, 331–336.

141. Waner, T., Borelli, G., Cadel, S., Privman, I., Nyska, A. (1995). Investigation of potential oncogenetic effects of β-cyclodextrin in the rat and mouse. *Arch. Toxicol.*, *69*, 631–639.

142. Toyoda, K., Hayashi, S., Uneyama, C., Kawanishi, T., Takada, K., Takahashi, M. (1995). Hepatic lesions in F344 rats treated orally with β-cyclodextrin for 13 weeks. *Eisei Shikenjo Hokoku*, 36–43.

143. Barrow, P.C., Olivier, P., Marzin, D. (1995). The reproductive and developmental toxicity profile of β-cyclodextrin in rodents. *Reprod. Toxicol.*, *9*, 389–398.

144. Garcia-Mediavilla, V., Villares, C., Culebras, J.M., Bayon, J. E., Gonzalez-Gallego, J. (2003). Effects of dietary β-cyclodextrin in hypercholesterolaemic rats. *Pharmacol. Toxicol.*, *92*, 94–99.

145. Munro, I.C., Newberne, P.M., Young, V.R., Bar, A. (2004). Safety assessment of γ-cyclodextrin. *Regul. Toxicol. Pharmacol.*, *39* (Suppl. 1), S3–S13.

146. Waalkens-Berendsen, D.H., Verhagen, F.J., Bar, A. (1998). Embryotoxicity and teratogenicity study with γ-cyclodextrin in rats. *Regul. Toxicol. Pharmacol.*, *27*, 166–171.

147. Waalkens-Berendsen, D.H., Smits van Prooije, A.E., Bar, A. (1998). Embryotoxicity and teratogenicity study with γ-cyclodextrin in rabbits. *Regul. Toxicol. Pharmacol.*, *27*, 172–177.

148. Shao, Z., Li, Y., Chermak, T., Mitra, A.K. (1994). Cyclodextrins as mucosal absorption promoters of insulin: II. Effects of β-cyclodextrin derivatives on α-chymotryptic degradation and enteral absorption of insulin in rats. *Pharm. Res.*, *11*, 1174–1179.

149. Somogyi, G., Posta, J., Buris, L., Varga, M. (2006). Cyclodextrin (CD) complexes of cholesterol: their potential use in reducing dietary cholesterol intake. *Pharmazie*, *61*, 154–156.

150. Rajewski, R.A., Stella, V.J. (1996). Pharmaceutical applications of cyclodextrins: 2. In vivo drug delivery. *J. Pharm. Sci.*, *85*, 1142–1169.

151. Okada, S.S., Kuo, A., Muttreja, M.R., Hozakowska, E., Weisz, P.B., Barnathan, E.S. (1995). Inhibition of human vascular smooth muscle cell migration and proliferation by β-cyclodextrin tetradecasulfate. *J. Pharmacol. Exp. Ther.*, *273*, 948–954.

152. Song, Y., Wang, Y., Thakur, R., Meidan, V.M., Michniak, B. (2004). Mucosal drug delivery: membranes, methodologies,

and applications. *Crit. Rev. Ther. Drug Carrier Syst.*, *21*, 195–256.

153. Costantino, H.R., Illum, L., Brandt, G., Johnson, P.H., Quay, S.C. (2007). Intranasal delivery: physicochemical and therapeutic aspects. *Int. J. Pharm.*, *337*, 1–24.

154. Merkus, F.W., Verhoef, J.C., Marttin, E., Romeijn, S.G., van der Kuy, P.H., Hermens, W.A., Schipper, N.G. (1999). Cyclodextrins in nasal drug delivery. *Adv. Drug Deliv. Rev.*, *36*, 41–57.

155. Uchenna Agu, R., Jorissen, M., Willems, T., Van den Mooter, G., Kinget, R., Verbeke, N., Augustijns, P. (2000). Safety assessment of selected cyclodextrins: effect on ciliary activity using a human cell suspension culture model exhibiting in vitro ciliogenesis. *Int. J. Pharm.*, *193*, 219–226.

156. Asai, K., Morishita, M., Katsuta, H., Hosoda, S., Shinomiya, K., Noro, M., Nagai, T., Takayama, K. (2002). The effects of water-soluble cyclodextrins on the histological integrity of the rat nasal mucosa. *Int. J. Pharm.*, *246*, 25–35.

157. Matsubara, K., Abe, K., Irie, T., Uekama, K. (1995). Improvement of nasal bioavailability of luteinizing hormone-releasing hormone agonist, buserelin, by cyclodextrin derivatives in rats. *J. Pharm. Sci.*, *84*, 1295–1300.

158. Abe, K., Irie, T., Uekama, K. (1995). Enhanced nasal delivery of luteinizing hormone releasing hormone agonist buserelin by oleic acid solubilized and stabilized in hydroxypropyl-β-cyclodextrin. *Chem. Pharm. Bull.*, *43*, 2232–2237.

159. Ahsan, F., Arnold, J.J., Meezan, E., Pillion, D.J. (2001). Mutual inhibition of the insulin absorption-enhancing properties of dodecylmaltoside and dimethyl-β-cyclodextrin following nasal administration. *Pharm. Res.*, *18*, 608–614.

160. Zhao, Y., Zhang, D.W., Zheng, A.P., Yu, S.Y., Wu, F.L. (2004). Influence of seven absorption enhancers on nasal mucosa: assessment of toxicity. *Beijing Da Xue Xue Bao*, *36*, 417–420.

161. Gudmundsdóttir, H., Sigurjónsdóttir, J.F., Másson, M., Fjalldal, O., Stefansson, E., Loftsson, T. (2001). Intranasal administration of midazolam in a cyclodextrin based formulation: bioavailability and clinical evaluation in humans. *Pharmazie*, *56*, 963–966.

162. Courrier, H.M., Butz, N., Vandamme, T.F. (2002). Pulmonary drug delivery systems: recent developments and prospects. *Crit. Rev. Ther. Drug Carrier Syst.*, *19*, 425–498.

163. Matilainen, L., Toropainen, T., Vihola, H., Hirvonen, J., Jarvinen, T., Jarho, P., Järvinen, K. (2008). In vitro toxicity and permeation of cyclodextrins in Calu-3 cells. *J. Control. Release 126*, 10–16.

164. Kinnarinen, T., Jarho, P., Järvinen, K., Järvinen, T. (2003). Pulmonary deposition of a budesonide/γ-cyclodextrin complex in vitro. *J. Control. Release*, *90*, 197–205.

165. Srichana, T., Suedee, R., Reanmongkol, W. (2001). Cyclodextrin as a potential drug carrier in salbutamol dry powder aerosols: the in-vitro deposition and toxicity studies of the complexes. *Respir. Med.*, *95*, 513–519.

166. Fukaya, H., Iimura, A., Hoshiko, K., Fuyumuro, T., Noji, S., Nabeshima, T. (2003). A cyclosporin A/maltosyl-α-cyclodextrin complex for inhalation therapy of asthma. *Eur. Respir. J.*, *22*, 213–219.

167. Arima, H., Uekama, K. (2007). *Cyclodextrins and Other Enhancers in Rectal Delivery*. CRC Press, Boca Raton, FL, pp. 147–172.

168. Arima, H., Kondo, T., Irie, T., Hirayama, F., Uekama, K., Miyaji, T., Inoue, Y. (1992). Use of water-soluble β-cyclodextrin derivatives as carriers of anti-inflammatory drug biphenylylacetic acid in rectal delivery. *Yakugaku Zasshi*, *112*, 65–72.

169. Uekama, K., Kondo, T., Nakamura, K., Irie, T., Arakawa, K., Shibuya, M., Tanaka, J. (1995). Modification of rectal absorption of morphine from hollow-type suppositories with a combination of α-cyclodextrin and viscosity-enhancing polysaccharide. *J. Pharm. Sci.*, *84*, 15–20.

170. Gaudana, R., Jwala, J., Boddu, S.H., Mitra, A.K. (2009). Recent perspectives in ocular drug delivery. *Pharm. Res.*, *26*, 1197–1216.

171. Kanai, A., Alba, R.M., Takano, T., Kobayashi, C., Nakajima, A., Kurihara, K., Yokoyama, T., Fukami, M. (1989). The effect on the cornea of α cyclodextrin vehicle for cyclosporin eye drops. *Transplant. Proc.*, *21*, 3150–3152.

172. Suhonen, P., Järvinen, T., Lehmussaari, K., Reunamaki, T., Urtti, A. (1995). Ocular absorption and irritation of pilocarpine prodrug is modified with buffer, polymer, and cyclodextrin in the eyedrop. *Pharm. Res.*, *12*, 529–533.

173. Wang, S., Li, D., Ito, Y., Nabekura, T., Wang, S., Zhang, J., Wu, C. (2004). Bioavailability and anticataract effects of a topical ocular drug delivery system containing disulfiram and hydroxypropyl-β-cyclodextrin on selenite-treated rats. *Curr. Eye Res.*, *29*, 51–58.

174. Jarho, P., Järvinen, K., Urtti, A., Stella, V.J., Järvinen, T. (1996). Modified β-cyclodextrin (SBE7-β-CyD) with viscous vehicle improves the ocular delivery and tolerability of pilocarpine prodrug in rabbits. *J. Pharm. Pharmacol.*, *48*, 263–269.

175. Järvinen, T., Järvinen, K., Urtti, A., Thompson, D., Stella, V.J. (1995). Sulfobutyl ether β-cyclodextrin (SBE-β-CD) in eyedrops improves the tolerability of a topically applied pilocarpine prodrug in rabbits. *J. Ocul. Pharmacol. Ther.*, *11*, 95–106.

176. Jansen, T., Xhonneux, B., Mesens, J., Borgers, M. (1990). β-cyclodextrins as vehicles in eye-drop formulations: an evaluation of their effects on rabbit corneal epithelium. *Lens Eye Toxic. Res.*, *7*, 459–468.

177. Brown, M.B., Traynor, M.J., Martin, G.P., Akomeah, F.K. (2008). Transdermal drug delivery systems: skin perturbation devices. *Methods Mol. Biol.*, *437*, 119–139.

178. Piel, G., Moutard, S., Uhoda, E., Pilard, F., Pierard, G.E., Perly, B., Delattre, L., Evrard, B. (2004). Skin compatibility of cyclodextrins and their derivatives: a comparative assessment using a corneoxenometry bioassay. *Eur. J. Pharm. Biopharm.*, *57*, 479–482.

179. Uekama, K., Irie, T., Sunada, M., Otagiri, M., Arimatsu, Y., Nomura, S. (1982). Alleviation of prochlorperazine-induced primary irritation of skin by cyclodextrin complexation. *Chem. Pharm. Bull.*, *30*, 3860–3862.

180. Vollmer, U., Muller, B.W., Wilffert, B., Peters, T. (1993). An improved model for studies on transdermal drug absorption in-vivo in rats. *J. Pharm. Pharmacol.*, *45*, 242–245.

181. Vitoria, M., Bentley, L.B., Vianna, R.F., Wilson, S., Collett, J. (1997). Characterization of the influence of some cyclodextrins on the stratum corneum from the hairless mouse. *J. Pharm. Pharmacol.*, *49*, 39–402.

182. Bentley, M.V., Vianna, R.F., Wilson, S., Collett, J.H. (1997). Characterization of the influence of some cyclodextrins on the stratum corneum from the hairless mouse. *J. Pharm. Pharmacol.*, *49*, 397–402.

183. Duchêne, D., Wouessidjewe, D., Poelman, M.C. (1991). In: Duchene, D. Ed., *New Trends in Cyclodextrins and Derivatives*. Editions de Sante, Paris, pp. 447–481.

184. Woolfson, A.D., Malcolm, R.K., Gallagher, R. (2000). Drug delivery by the intravaginal route. *Crit. Rev. Ther. Drug Carrier Syst.*, *17*, 509–555.

185. Francois, M., Snoeckx, E., Putteman, P., Wouters, F., De Proost, E., Delaet, U., Peeters, J., Brewster, M.E. (2003). A mucoadhesive, cyclodextrin-based vaginal cream formulation of itraconazole. *AAPS PharmSci*, *5*, E5.

186. Cevher, E., Sensoy, D., Zloh, M., Mulazimoglu, L. (2008). Preparation and characterisation of natamycin: γ-cyclodextrin inclusion complex and its evaluation in vaginal mucoadhesive formulations. *J. Pharm. Sci.*, *97*, 4319–4335.

187. Miyake, K., Irie, T., Arima, H., Hirayama, F., Uekama, K., Hirano, M., Okamaoto, Y. (1999). Characterization of itraconazole/2-hydroxypropyl-β-cyclodextrin inclusion complex in aqueous propylene glycol solution. *Int. J. Pharm.*, *179*, 237–245.

188. Uekama, K. (2004). Pharmaceutical application of cyclodextrins as multi-functional drug carriers. *Yakugaku Zasshi*, *124*, 909–935.

189. Yano, H., Hirayama, F., Arima, H., Uekama, K. (2000). Hydrolysis behavior of prednisolone 21-hemisuccinate/β-cyclodextrin amide conjugate: involvement of intramolecular catalysis of amide group in drug release. *Chem. Pharm. Bull.*, *48*, 1125–1128.

190. Hirayama, F., Uekama, K. (1999). Cyclodextrin-based controlled drug release system. *Adv. Drug Deliv. Rev.*, *36*, 125–141.

191. Lowenstein, P.R. (2008). Clinical trials in gene therapy: ethics of informed consent and the future of experimental medicine. *Curr. Opin. Mol. Ther.*, *10*, 428–430.

192. Castanotto, D., Rossi, J.J. (2009). The promises and pitfalls of RNA-interference-based therapeutics. *Nature*, *457*, 426–433.

193. Rao, D.D., Vorhies, J.S., Senzer, N., Nemunaitis, J. (2009). siRNA vs. shRNA: similarities and differences. *Adv. Drug Deliv. Rev.*, *61*, 746–759.

194. Grimm, D. (2009). Small silencing RNAs: state-of-the-art. *Adv. Drug Deliv. Rev.*, *61*, 672–703.

195. Akhtar, S. (2006). Non-viral cancer gene therapy: beyond delivery. *Gene Ther.*, *13*, 739–740.

196. Arima, H., Kihara, F., Hirayama, F., Uekama, K. (2001). Enhancement of gene expression by polyamidoamine dendrimer conjugates with α-, β-, and γ-cyclodextrins. *Bioconjugate Chem.*, *12*, 476–484.

197. Kihara, F., Arima, H., Tsutsumi, T., Hirayama, F., Uekama, K. (2002). Effects of structure of polyamidoamine dendrimer on gene transfer efficiency of the dendrimer conjugate with α-cyclodextrin. *Bioconjugate Chem.*, *13*, 1211–1219.

198. Kihara, F., Arima, H., Tsutsumi, T., Hirayama, F., Uekama, K. (2003). In vitro and in vivo gene transfer by an optimized α-cyclodextrin conjugate with polyamidoamine dendrimer. *Bioconjugate Chem.*, *14*, 342–350.

199. Arima, H., Motoyama, K. (2009). Recent findings concerning PAMAM dendrimer conjugates with cyclodextrins as carriers of DNA and RNA. *Sensors*, *9*, 6346–6361.

200. Forrest, M.L., Gabrielson, N., Pack, D.W. (2005). Cyclodextrin–polyethylenimine conjugates for targeted in vitro gene delivery. *Biotechnol. Bioeng.*, *89*, 416–423.

201. Yang, C., Wang, X., Li, H., Goh, S.H., Li, J. (2007). Synthesis and characterization of polyrotaxanes consisting of cationic α-cyclodextrins threaded on poly[(ethylene oxide)-*ran*-(propylene oxide)] as gene carriers. *Biomacromolecules*, *8*, 3365–3374.

202. Yang, C., Li, H., Goh, S.H., Li, J. (2007). Cationic star polymers consisting of α-cyclodextrin core and oligoethylenimine arms as nonviral gene delivery vectors. *Biomaterials*, *28*, 3245–3254.

203. Menuel, S., Fontanay, S., Clarot, I., Duval, R.E., Diez, L., Marsura, A. (2008). Synthesis and complexation ability of a novel bis-(guanidinium)-tetrakis-(β-cyclodextrin) dendrimeric tetrapod as a potential gene delivery (DNA and siRNA) system. Study of cellular siRNA transfection. *Bioconjugate Chem.*, *19*, 2357–2362.

204. Li, J., Loh, X.J. (2008). Cyclodextrin-based supramolecular architectures: syntheses, structures, and applications for drug and gene delivery. *Adv. Drug Deliv. Rev.*, *60*, 1000–1017.

205. Tang, G.P., Guo, H.Y., Alexis, F., Wang, X., Zeng, S., Lim, T.M., Ding, J., Yang, Y.Y., Wang, S. (2006). Low molecular weight polyethylenimines linked by β-cyclodextrin for gene transfer into the nervous system. *J. Gene Med.*, *8*, 736–744.

206. Huang, H., Tang, G., Wang, Q., Li, D., Shen, F., Zhou, J., Yu, H. (2006). Two novel non-viral gene delivery vectors: low molecular weight polyethylenimine cross-linked by (2-hydroxypropyl)-β-cyclodextrin or (2-hydroxypropyl)-γ-cyclodextrin. *Chem. Commun.*, Issue 22, 2382–2384.

207. Heidel, J.D. (2006). Linear cyclodextrin-containing polymers and their use as delivery agents. *Expert Opin. Drug Deliv.*, *3*, 641–646.

208. Shuai, X., Merdan, T., Unger, F., Kissel, T. (2005). Supramolecular gene delivery vectors showing enhanced transgene expression and good biocompatibility. *Bioconjugate Chem.*, *16*, 322–329.

209. Pun, S.H., Bellocq, N.C., Liu, A., Jensen, G., Machemer, T., Quijano, E., Schluep, T., Wen, S., Engler, H., Heidel, J., Davis, M.E. (2004). Cyclodextrin-modified polyethylenimine polymers for gene delivery. *Bioconjugate Chem.*, *15*, 831–840.

210. Reineke, T.M., Davis, M.E. (2003). Structural effects of carbohydrate-containing polycations on gene delivery. 1. Carbohydrate size and its distance from charge centers. *Bioconjugate Chem.*, *14*, 247–254.

211. Popielarski, S.R., Mishra, S., Davis, M.E. (2003). Structural effects of carbohydrate-containing polycations on gene delivery: 3. Cyclodextrin type and functionalization. *Bioconjugate Chem.*, *14*, 672–678.

212. Heidel, J.D., Yu, Z., Liu, J.Y., Rele, S.M., Liang, Y., Zeidan, R. K., Kornbrust, D.J., Davis, M.E. (2007). Administration in non-human primates of escalating intravenous doses of targeted nanoparticles containing ribonucleotide reductase subunit M2 siRNA. *Proc. Natl. Acad. Sci. USA*, *104*, 5715–5721.

213. Bartlett, D.W., Davis, M.E. (2007). Physicochemical and biological characterization of targeted, nucleic acid–containing nanoparticles. *Bioconjugate Chem.*, *18*, 456–468.

214. Quaglia, F., Ostacolo, L., Mazzaglia, A., Villari, V., Zaccaria, D., Sciortino, M.T. (2009). The intracellular effects of nonionic amphiphilic cyclodextrin nanoparticles in the delivery of anticancer drugs. *Biomaterials*, *30*, 374–382.

215. Memisoglu-Bilensoy, E., Vural, I., Bochot, A., Renoir, J.M., Duchêne, D., Hincal, A.A. (2005). Tamoxifen citrate loaded amphiphilic β-cyclodextrin nanoparticles: in vitro characterization and cytotoxicity. *J. Control. Release*, *104*, 489–496.

216. Memisoglu-Bilensoy, E., Dogan, A.L., Hincal, A.A. (2006). Cytotoxic evaluation of injectable cyclodextrin nanoparticles. *J. Pharm. Pharmacol.*, *58*, 585–589.

217. Bilensoy, E., Gurkaynak, O., Dogan, A.L., Hincal, A.A. (2008). Safety and efficacy of amphiphilic β-cyclodextrin nanoparticles for paclitaxel delivery. *Int. J. Pharm.*, *347*, 163–170.

218. Arima, H., Hagiwara, Y., Hirayama, F., Uekama, K. (2006). Enhancement of antitumor effect of doxorubicin by its complexation with γ-cyclodextrin in pegylated liposomes. *J. Drug Target.*, *14*, 225–232.

219. Wada, K., Arima, H., Tsutsumi, T., Chihara, Y., Hattori, K., Hirayama, F., Uekama, K. (2005). Improvement of gene delivery mediated by mannosylated dendrimer/α-cyclodextrin conjugates. *J. Control. Release*, *104*, 397–413.

220. Hwang, S.J., Bellocq, N.C., Davis, M.E. (2001). Effects of structure of β-cyclodextrin-containing polymers on gene delivery. *Bioconjugate Chem.*, *12*, 280–290.

221. Geze, A., Chau, L.T., Choisnard, L., Mathieu, J.P., Marti-Batlle, D., Riou, L., Putaux, J.L., Wouessidjewe, D. (2007). Biodistribution of intravenously administered amphiphilic β-cyclodextrin nanospheres. *Int. J. Pharm.*, *344*, 135–142.

222. Harada, A., Takashima, Y., Yamaguchi, H. (2009). Cyclodextrin-based supramolecular polymers. *Chem. Soc. Rev.*, *38*, 875–882.

223. Higashi, T., Hirayama, F., Yamashita, S., Misumi, S., Arima, H., Uekama, K. (2009). Slow-release system of pegylated lysozyme utilizing formation of polypseudorotaxanes with cyclodextrins. *Int. J. Pharm.*, *374*, 26–32.

224. Higashi, T., Hirayama, F., Misumi, S., Motoyama, K., Arima, H., Uekama, K. (2009). Polypseudorotaxane formation of randomly-pegylated insulin with cyclodextrins: slow release and resistance to enzymatic degradation. *Chem. Pharm. Bull.*, *57*, 541–544.

6

REGULATORY STATUS OF CYCLODEXTRINS IN PHARMACEUTICAL PRODUCTS

A. Atilla Hincal

IDE Drug Regulatory Affairs, Education, Consultancy Ltd. Co., Ankara, Turkey, Department of Pharmaceutical Technology, Hacettepe University, Ankara, Turkey

Hakan Eroğlu

Department of Basic Pharmaceutical Sciences, Hacettepe University, Ankara, Turkey

Erem Bilensoy

Department of Pharmaceutical Technology, Hacettepe University, Ankara, Turkey

1. INTRODUCTION

Cyclodextrins (CDs) are cyclic oligosaccharides composed of six α-cyclodextrin, seven β-cyclodextrin, and eight or more γ-cyclodextrin glucopyranoside units, linked by α-(1,4) bonds. They are produced through exposure of starch to intermolecular transglycosylation with the enzyme CD glucanotransferase [1]. The first record of CDs goes back nearly 119 years to a study conducted by Villiers, in which he described isolation of 3 g of crystallized substance after bacterial ingestion of 1000 g of starch [2]. Studies to date by various researchers on CDs have accelerated, revealing increasingly wider areas of use. There are mainly three groups of CDs—α-, β-, and γ-CD—which provide the subject matter of more than 1000 articles published in international journals every year, a large majority of which relate to pharmaceuticals or products related to pharmaceutical products. These three types of CDs are also called the first-generation or principal CDs [3].

Despite their limited industrial use, owing to high production costs and impurity problems, CDs are now used more widely, thanks to advancements in biotechnology which have led to improved efficiency in production and enabled production of highly purified CD. In the pharmaceutical field, CDs are used mainly as complexing agents, the intent being to increase the solubility of poorly water-soluble drugs in order to improve their bioavailability and stability. The first industrial use of CDs was in the early 1970s, primarily in food and cosmetic applications. They have been used in the food industry to improve the stability of sweetening agents, thereby helping to mask their poor taste and odor. The first attempts to use them in cosmetics have been to enhance the stability of chemically susceptible compounds and thus to prolong their action time [4].

CDs saw industrial-scale use in Japan in the early 1980s, with annual production amounts increasing every year and making Japan the largest consumer of CDs by the early 2000s. The food industry consumed 78.8% of the total volume, with agricultural chemicals and pharmaceuticals accounting for a mere 4.6% of the market [5]. Also in Europe and the United States, industrial-scale use of CDs advanced, albeit more slowly, beginning with the launch of a CD-based fabric softener by Procter & Gamble in the early 1990s [6].

While a large number of modified derivatives of CDs have been developed through the years, those most commonly used in the market as a pharmaceutical excipient are derivatives with the substituents 2-hydroxypropyl, sulfobutyl

Cyclodextrins in Pharmaceutics, Cosmetics, and Biomedicine: Current and Future Industrial Applications, First Edition. Edited by Erem Bilensoy.
© 2011 John Wiley & Sons, Inc. Published 2011 by John Wiley & Sons, Inc.

ether, and methyl. Although they differ in electronic characteristics and size, these substituents bind to the main structure by reacting with one or more 3-hydroxy groups of the glucopyranoside structure [7].

A drug substance must dissolve to a certain level in water in order to exert its pharmaceutical effect. In drugs where solubility is an issue, the ways of overcoming this problem include combining different solvents, buffering agents, or surfactants during formulation studies. However, these may cause irritation or other adverse reactions during use. In general, for drug-filing procedures, the excipients included in the formulation bear much relevance. In formulations using CD, however, thanks to the complex formed with the drug substance, the solubility of the substance is improved and even its stability controlled in the formulation or until delivery to the site of action. Maintaining its complex form with CD until it reaches the biological surface where absorption will take place, the active substance, at that point, partitions and permeates the biological membrane. Owing to its low lipophilic character, CD does not permeate the biological membrane and remains within the aqueous medium. The traditional absorption enhancers used in drug formulations act by altering or distorting the membrane structure, in contrast to CDs, which serve their function solely by enhancing the presence of the drug substance at the site of absorption [8,9]. In formulation studies, excipients with such favorable properties obviously provide specific advantages for both formulators and pharmaceutical product application dossiers compiled for registration purposes.

CDs have numerous applications in the pharmaceutical field: to enhance the bioavailability of poorly water-soluble drug substances and to improve the stability of substances against such factors as hydrolysis, oxidation, heat, or light through the inclusion compounds that they form among other uses. An important issue of interest that formulators focus on during formulation studies is formulation with CD derivatives to mask poor taste or odor characteristics of a variety of substances [10]. Reports of their low toxicity profile, attributed to the high resistance of CDs used in formulations against starch-digestive enzymes, have recently promoted their use in drug formulations. They are digested by enzyme α-amylase, which is present in the colonic flora [11]. Animal studies conducted by different groups have found that absorption of CDs ranged from 0.02 to 3% of the oral dose [12–15]. The principal CDs, α- and β-CDs, are known to be resistant against gastric acid, saliva, and pancreatic enzymes, compared to γ-CDs, which are partially digested by gastrointestinal enzyme amylase. Absorbed CDs are, however, excreted in urine without undergoing any obvious metabolization. Taken together, these findings elucidate the low-toxicity profile of CDs after oral administration and support their use in drug formulations as a new excipient for various purposes [9,16].

2. REGULATORY STATUS OF EXCIPIENTS IN FORMULATIONS

Hesitancy on the part of industrial centers undertaking drug research and development to include a new excipient in formulations is understandable. It should be remembered that this is the underlying reason that CDs and similar other excipients do not have a wider industrial application today. An important reason is that health authorities, in their capacity as the regulatory organs of approval, subject these new excipients to evaluation during new drug applications (NDAs). The extremely cautious and meticulous manner in which the regulatory authorities deal with these excipients against the possibility of potential problems or issues often leads to delays in the evaluation process or even results in rejection of an application.

International industrial excipient companies involved in the research, development, control, and approval of medicinal products established the International Pharmaceutical Excipients Council in 1991 to address issues surrounding excipients used in the formulation of pharmaceutical formulations. This organization is comprised of four associations: IPEC–America (United States of America), IPEC–Europe, IPEC–Japan, and the recently established IPEC–China. The core objective of this institution is to ensure harmonization of international standards for excipients, to provide new excipients suitable for industrial applications, and to compile existing practical and theoretical scientific knowledge of excipients. In this premise, IPEC has led the way for development of an examination and evaluation instrument known as the *new excipient evaluation procedure*, to ensure independent inspection of new excipients prior to application for official approval and for the establishment of a forum that would perform preevaluation of excipients according to the aforesaid procedure. This organ is composed of experts from academic, industry, and regulatory circles who undertake evaluation of new excipient candidates in terms of their safety and chemical characterization. Although the protocols resulting from these studies emerge through processes also involving U.S. Food and Drug Administration (FDA) officials, these protocols and officials do not provide regulatory approval. They are aimed at assisting pharmaceutical manufacturers and researchers who undertake formulation development in submitting an application with a more complete dossier for formulations that contain new excipients, helping them in obtaining regulatory approval, and encouraging them in using new excipients during their new formulation studies.

2.1. Regulatory Status in ICH Countries

2.1.1. United States of America   When we look at the situation regarding excipients used in pharmaceutical formulations, we see that the International Conference on

Harmonization (ICH) has not undertaken any studies toward evaluating the safety of new excipients that may be included in drug formulations, and has not published nor is it working on any specific "guideline" in this respect. However, in the *Guidance for Industry: Nonclinical Studies for the Safety Evaluation of Pharmaceutical Excipients*, issued in May 2005, the FDA refers to the overall safety testing guidelines for drugs of the ICH (e.g., ICH S1A, S2B, S3A, S5A, S7A, and M3) [17]. On the other hand, in some instances, supportive data for a new excipient are included in the drug application dossier or in the fully referenced drug master file (DMF), where the data provided are arranged in a manner similar to that for drug substances and contain chemical, manufacture, safety, and quality control (CMC) data. In recent years, the approval of new excipients by U.S. regulatory authorities takes an indirect approach: by granting approval for their inclusion in the formulation of a drug product, meaning that the approval is not for the excipient itself, but solely for its use in the formulation of that drug product. Consequently, using a new excipient, found to be compliant with the scientific criteria required by the regulatory authority, in the formulation of other drug formulations is allowed, provided that the level and duration of exposure specified in the approved drug formulation are not exceeded. If the new drug formulation containing the new excipient will have a different route of administration than that of the previously approved drug product, the FDA requires submission of toxicology data for the new route of administration proposed [18–20].

In the guidance document for excipients cited earlier [17], the FDA gives the following definition of a new excipient: "any inactive ingredients that are intentionally added to therapeutic and diagnostic products, but that: (1) we believe are not intended to exert therapeutic effects at the intended dosage, although they may act to improve product delivery (e.g., enhance absorption or control release of the drug substance); and (2) are not fully qualified by existing safety data with respect to the currently proposed level of exposure, duration of exposure, or route of administration." The strategies recommended for excipients proposed to be used for the first time in the formulation of a drug potentially to be approved by the national health authority (which may include CDs used as an excipients) are grouped under six headings in the guidance document:

1. *Safety pharmacology:* it is required that these studies be performed according to the standard tests specified in ICH S7A. These evaluations can be performed during the course of toxicology studies or as independent studies.
2. *Potential excipients intended for short-term use:* includes tests for excipients that are intended for use in products that are limited by labeling to clinical use for 14 or fewer consecutive days.
3. *Potential excipients intended for intermediate use:* includes tests for excipients that are intended for use in drug products that are labeled for clinical use of more than two weeks but less than three months.
4. *Potential excipients intended for long-term use:* includes tests for excipients that are intended for use in drug products labeled for clinical use of more than three months.
5. *Potential excipients for use in pulmonary, injectable, or topical product:* includes safety tests for excipients that are intended for use in injectable, topical (dermal, intranasal, intraoral, ophthalmic, rectal or vaginal), or pulmonary drug products.
6. *Photosafety data:* it is recommended that either the excipient or the complete drug product be evaluated as described in the CDER guidance *Photosafety Testing*.

2.1.2. European Union The *Guideline on Excipients in the Dossier for Application for Marketing Authorisation of a Medicinal Product* [21], issued by the European Medicines Agency, gives the following definition of an excipient: "Excipients are the constituents of a pharmaceutical form apart from the active substance." This guideline aims to provide guidance on the applications for drug product registration/licensing and marketing authorization or variations relating to an excipient in registered medicinal products. As mentioned previously, the approval sought here is not for the excipient but for the drug product that contains it. Therefore, the approval granted is solely for the level and duration of exposure as specified for that particular drug product. The information that should be included with respect to excipients in the appropriate modules of drug product registration dossiers in the Common Technical Document (CTD) format that is required in EU countries and candidates are briefly as follows:

1. Description and Composition of the Drug Product (3.2. P.1)
2. Pharmaceutical Development (3.2.P.2)
3. Specifications (3.2.P.4.1)

   In this module where the specifications are examined of excipients intended for use in drug products, if, example, it is included in the *European Pharmacopoeia* or in a national pharmacopeia of a member state of the European Union, references made in the application to such monographs are deemed acceptable. If the monograph for an excipient is not registered in the *European Pharmacopoeia* or in that of any member state, but is in the pharmacopoeia of any of the ICH non-EU tripartite countries (e.g., *United States Pharmacopeia/National Formulary* and *Japanese Pharmacopoeia*), conformance to such monograph is also deemed acceptable.

However, the applicant is required to document compliance of the proposed excipient not only with the aforesaid pharmacopeia, but also with the specifications prescribed in the section "Substances for Pharmaceutical Use" of the *European Pharmacopoeia*. If the excipient proposed for use in the drug product is not registered in any pharmacopeia, a full specification listing should be applied based on the following data:

Physical characteristics

Identification tests

Purity tests (totally or individually per each impurity)

Assay or limit tests

Other relevant test

4. Justification of Specifications (3.2.P.4.4)
5. Excipients of Human or Animal Origin (3.2.P.4.5)
6. Novel Excipients (3.2.P.4.6)

The stance toward new excipients is that detailed descriptions of the manufacture, characterization, and control processes, a detailed description of the excipient, along with an explanation of its conditions and purpose of use, must be submitted to support the safety data for the new excipient. If the excipient is a complex or a combination of multiple ingredients, each of its constituents should be specified quantitatively and qualitatively. A chemical documentation of the new excipients used should be performed based on "CPMP Guideline on the Chemistry of New Active Substances" (CPMP/QWP/130/96). Furthermore, bibliographic data on the excipient's chemistry, toxicological characteristics, and current areas of use should also be submitted.

7. Control of Drug Product (3.2.P.5)
8. Stability (3.2.P.8)
9. Labeling

Moreover, the control requirements for excipients are provided in Module 3, Section 3.2.2.4, as prescribed in Directive 2003/63/EC on the technical and scientific particulars of dossiers required during new drug registration applications [22]. It is provided that: "For excipient(s) used for the first time in a medicinal product or by a new route of administration, full details of manufacture, characterization, and controls, with cross references to supporting safety data, both non-clinical and clinical, shall be provided according to the active substance format previously described" and "Additional information on toxicity studies with the novel excipient shall be provided in Module 4. Clinical studies shall be provided in Module 5." Further requirements are that "any novel excipient shall be the subject of a specific safety assessment" [Module 2 (2.4. Non-clinical Overview)] and that "the toxicology and pharmacokinetics of an excipient used for the first time in the pharmaceutical field shall be investigated" [Module 4, Section 4.2.(3)].

2.1.3. Japan   According to the evaluation system implemented by the Japanese Ministry of Health and Welfare, drug products containing excipients that have previously been in use are evaluated by the Pharmaceuticals and Medical Devices Evaluation Center (PMDE), and products containing new excipients that are being used for the first time are evaluated by the Subcommittee on Pharmaceutical Excipients of the Central Pharmaceutical Affairs Council (CPAC). Any excipient, even if previously included in the formulation of an existing drug product in another country, is qualified as "new" if it will be used for the first time in Japan, its route of administration is different than any previous route of administration, or the level of exposure is above the previous level used [23]. It may be concluded that these standards, governing excipients in general, are also applicable for CDs.

2.2. Regulatory Status of CDs on GRAS Substances

According to the U.S. Federal Food, Drug and Cosmetic Act (FDCA), Sections 201(s) and 409, a food additive is defined as any substance that is added to any food to serve a specific function. These substances are subject to FDA evaluation prior to marketing and are approved if so deemed by experts after a series of evaluations to be safe under the conditions envisaged for use. These evaluations take place independent of the FDA, and their results are reported to the FDA for listing. After completion of the evaluations in accordance with the FDCA, the additives used are included in the category *generally recognized as safe* GRAS in one of the following two ways:

1. After evaluation through scientific procedures: Scientific evidence, in the same quantity and quality as that required for the approval of a food additive, is required (supported by published or unpublished studies).
2. In the case of an additive that has been in use since before 1958, there must be a history of considerable consumption by a sufficient number of consumers, resulting from experience with its use during that time frame.

It is suggested by some that a food additive included in the GRAS determination may be included in drug formulations intended for oral administration, up to the quantities specified therein [7]. Accordingly, after a review of the FDA's GRAS list for CDs, it has been noted that the three CDs were included in the table. γ-CD was listed first, in May 2000 (GRAS No. 46). The intended use of γ-CDs in food products is as a stabilizer, emulsifier, carrier, or formulation aid. β-CD (GRAS No. 74) (as an aroma carrier and preservative) and α-CD (GRAS No. 155) (as an aroma adjuvant, colorant carrier, and stabilizer for vitamins) later followed the GRAS listing of γ-CD. These GRAS determinations with respect to CDs include information on the quantities and areas of use, physicochemical and characteristic properties, impurity, results from biological studies (absorption, disposition, metabolism, and excretion), results from toxicology studies (acute, subchronic, and chronic studies), embryotoxicity/teratogenicity, genotoxicity, irritation/sensitization studies, and tolerability in humans [24].

2.3. Regulatory Status of CDs in Different Pharmacopoeias

From a regulatory perspective, α- and β-CD monographs are included in the *U.S. Pharmacopeia/National Formulary* (USP/NF25) [25] and the *European Pharmacopoeia* (EP 6.0) [26]. A monograph for CDs is also provided in the *Handbook of Pharmaceutical Excipients*, a compendial source, where CDs are classified as solubilizing and stabi-

lizing agents [27]. As mentioned earlier, validation of CD-containing pharmaceutical formulations for commercialization has begun with the approval of pharmaceutical products by U.S., European, and Japanese authorities. Cyclodextrins may be present in formulations in the form of complexed or uncomplexed structures. After administration, the pseudo-complex structure that a CD forms with an active substance disintegrates in the biological environment, releasing the drug substance to exert its pharmaceutical action. Thus, the regulatory approval for these formulations is granted not for the complex structure of active substance and CD, but solely for the active substance. CD-containing products whose regulatory approval procedures have been completed by health authorities in a number of countries in addition to the ICH tripartite countries (the United States, the EU, and Japan) are listed in Table 1.

Table 1. Drug Products Containing CDs as Excipients

Active Pharmaceutical Ingredient/Drug Molecule	Type of CD[a]	Trade Name	Formulation	Country of Registration
17β-Estradiol	M-β-CD	Aerodiol	Nasal spray	Europe
Alprostadil	α-CD	Prostavastin	Intravenous (IV) solution	Europe
		Rigidur		Japan
		Edex		U.S.
		Caverject		U.S.
Amiodarone	SBE-β-CD	Nexterone	IV	U.S.
Aripiprazole	SBE-β-CD	Abilify	Intramuscular (IM) solution	Japan, U.S.
Benexate	β-CD	Ulgut	Capsule	Japan
		Lonmiel		
Betahistidine	β-CD	Betahist	Tablet	India
Cefotiam-hexetil	α-CD	Pansporin T	Tablet	Japan
Cephalosporin	β-CD	Meiact	Tablet	Japan
Cetirizine	β-CD	Cetirizine	Chewable tablet	Switzerland, U.S.
Chloramphenicol	M-β-CD	Clorocil	Eyedrop	Europe
Chlordiazepoxide	β-CD	Transillium	Tablet	Argentina
Cisapride	HP-β-CD	Propulsid	Suppository	Europe
Dexamethasone	β-CD	Glymesason	Ointment, tablet	Japan
Dextromethorphan	β-CD	Rynathisol		Europe
Diclofenac sodium	HP-γ-CD	Voltaren ophtha	Eyedrop solution	Europe
Diphenhydramin HCl, chlortheophyllin	β-CD	Stada-Travel	Chewing tablet	Europe
Flunarizine	β-CD	Fluner	Tablet	India
Hydrocortisone	HP-β-CD	Dexocort	Solution	Europe
Indomethacine	HP-β-CD	Indocid	Eyedrop	Europe
Iodine	β-CD	Mena-Gargle	Solution	Japan
Itraconazole	HP-β-CD	Sporanox	Oral and IV solutions	Europe, U.S.
Maropitant[b]	SBE-β-CD	Cerenia	Parental solution	USA
Meloxicam	β-CD	Mobitil	Tablet, suppository	Egypt
Minoxidil	γ-CD	Alopexy	Solution	Europe
Mitomycin	HP-β-CD	Mitozytrex Mitoextra	IV infusion	Europe, U.S.
Modified porcine[b] (circovirus inactivated)	SL-CD	Suvaxyn PCV	Suspension for injection	Europe
Nicotine	β-CD	Nicorette, Nicogum	Sublingual tablet, chewing gum	U.S., Europe, Japan
Nimesulide	β-CD	Nimedex	Tablet	Europe
		Mesulid Fast		

(*continued*)

Table 1. (*Continued*)

Active Pharmaceutical Ingredient/Drug Molecule	Type of CD[a]	Trade Name	Formulation	Country of Registration
Nitroglycerin	β-CD	Nitropen	Sublingual tablet	Japan
Omeprazol	β-CD	Omebeta	Tablet	Europe
OP-1206	α-CD	Opalmon	Tablet	Japan
PGE_2	β-CD	Prostarmon E	Sublingual tablet	Japan
Piroxicam	β-CD	Brexin Cicladon Flogene	Tablet, suppository, liquid	Europe, India, Brazil
Rofecoxib	β-CD	Rofizgel	Tablet	India
Sugammadex (as sodium salt)	(It is a modified γ-CD)	Bridion	IV solution	Europe
^{99}Tc teoboroxime	HP-γ-CD	Cardiotec	IV solution	U.S.
Telavancin	HP-β-CD	Vibativ	IV infusion	U.S.
Thiomersal	β-CD	Vitaseptol	Eyedrop	Monaco
Tiaprofenic acid	β-CD	Surgamyl	Tablet	Europe
Voriconazole	SBE-β-CD	Vfend	IV solution	Europe, U.S., Japan
Ziprasidone mesylate	SBE-β-CD	Geodon Zeldox	IM solution	Europe, U.S.

Source: The information listed in this table is based partly on Szejtli [30] and Loftsson [31].

[a] α-CD, α-cyclodextrin; β-CD, β-cyclodextrin; γ-CD, γ-cyclodextrin; HP-β-CD, 2-hydroxypropyl-β-cyclodextrin; SBE-β-CD, sulfobutyl ether β-cyclodextrin; HP-γ-CD, 2-hydroxypropyl-γ-cyclodextrin; SL-CD, sulfolipo-cyclodextrin.
[b] For veterinary use.

2.4. Current Regulatory Perspectives on CDs

The Inactive Ingredients Database, which is updated continuously by the FDA, includes a list of inactive pharmaceutical ingredients contained in products that have been granted regulatory approval. When used in new drug formulations, the inactive pharmaceutical ingredients contained in the formulation of approved drug products are not qualified as new excipients, resulting in a less comprehensive examination during evaluation of the second drug product. The use of substances that have been approved for a specific quantity in a specific dosage form is considered safe when used in a similar product in a similar quantity. The list contains information on the inactive ingredient's full name; approved product's route of administration, and dosage form; CAS (Chemical Abstracts Service) number; UNII, unique ingredient identifier; and the quantity of inactive substance (excipient) used in the specified route of administration and dosage form. The results from a review of the Inactive Ingredients Database in January 2010 for CD are given in Table 2.

The EMEA Committee for Human Medicinal Products has issued the guideline, *Opinion on the Potential Risks of Carcinogens, Mutagens and Substances Toxic to Reproduc-*

Table 2. Inactive Ingredient Search Results for CDs Used in the Pharmaceutical Industry

Inactive Ingredient	Route; Dosage Form	CAS Number[a]	UNII[b]	Maximum Potency
γ-Cyclodextrin	Intravenous; injection	17465860	KZJ0BYZ5VA	5.00%
Sulfobutyl ether-β-cyclodextrin	Intramuscular; powder, for injection solution, lyophilized, with additives		2PP9364507	44.14%
Sulfobutyl ether-β-cyclodextrin	Intravenous (infusion); powder, for injection solution, lyophilized		2PP9364507	

Source: Adapted from [28].

[a] Chemical Abstracts Service (CAS), a division of the American Chemical Society, provides comprehensive electronic chemical information services.
[b] The UNII (unique ingredient identifier) is a part of the joint USP/FDA Substance Registration System (SRS), which has been designed to support health information technology initiatives by providing unique identifiers for substances in drugs, biologics, foods, and devices based on molecular structure and/or descriptive information.

tion When These Substances Are Used as Excipients of Medicinal Products for Human Use [29], a scientific article on excipients destined for use in pharmaceutical products. One provision of this guideline is that each new excipient undergo a full evaluation, like that for a new active substance. With respect to an excipient used for the first time, the intention is to ensure submission of information on the manufacture, quality control, reproducibility of impurities, relevant functional characteristics, and animal toxicity. An example has been given of an issue experienced with sulfobutyl ether-β-cyclodextrin sodium (SBE-β-CD) used in an antifungal product formulated as a powder for infusion. A potentially genotoxic carcinogen impurity was detected in the SBE-β-CD, used as an excipient in the formulation. At the conclusion of this investigation, the CHMP required the manufacturer to minimize the level of this impurity since no other CD existed that could be substituted for this specific excipient, making its removal from the formulation impossible. In the final assessment of the overall risk–benefit balance of the product, it was concluded that the purpose had been served and the use of the excipient, a CD derivate, in the product was approved [29]. Indeed, the conclusion reached by the EMEA has been a striking one with respect to the use of CDs in drug formulations.

3. FUTURE PERSPECTIVES ON THE REGULATORY STATUS OF CDs

With regard to the facts stated above, pharmaceutical manufacturers and regulatory authorities agree on the necessity of developing a common regulation to globally govern excipients that are used in the manufacture of drug products. Regretfully, the inability thus far to develop such a regulation for excipients remains the greatest obstacle in the way of development of new excipients and a barrier to development and regulatory approval of formulations with new excipients that will put these excipients in the service of human health. In fact, new excipients are indirectly granted approval merely through their inclusion in new drug application dossiers for that particular formulation, route of administration, and quantity. From the perspective of manufacturers, formulating with a new and not-yet-compendial excipient entails substantial risks, since scientific studies of excipients are not as widely available and comprehensive as those for active substances, and results from some of these studies are not released. Any concern that may arise in connection with an excipient included in the formulation may delay the approval procedures or even lead to rejection.

In a large majority of the evaluations performed in both ICH tripartite countries and elsewhere of formulations containing CD derivatives as an inactive pharmaceutical ingredient, the safety evaluations used have been those prescribed by ICH generally for active ingredients, which is the correct approach not only scientifically but also for human health and life.

4. CONCLUSIONS

It is an established fact that excipients, while pharmacologically inert, bear the utmost relevance for the efficacy of a drug product in obtaining optimal results from a pharaceutical formulation developed to serve the well-being of humans. The point that we have reached today in the course of research and development of pharmaceutical formulations makes it evident that use of more effective excipients in newer pharmaceutical forms is inevitable. The situation facing pharmaceutical researchers today is a challenge, as obviously, formulators have indeed shown their merit in surmounting the scientific stages that are required to identify and utilize an excipient that will help to develop a pharmaceutical formulation that is both more presentable, and more convenient, stable, and effective. However, a common point has yet to be reached by all stakeholders as to the standards of examination and evaluation which are indicative of quality, safety, and efficacy from a regulatory perspective. In other words, the unavailability of comprehensive studies in this field, for whatever reason, is among the underlying causes of the present situation. And for this reason it remains impossible with respect to new excipients to state confidently and based on scientific evidence, in particular from a safety perspective, that a particular excipient can be used in pharmaceutical formulations. This leads health authorities who carry out regulatory approval procedures to be much more cautious and meticulous when dealing with these substances, and prolongs the application procedure for registration. In particular, the issue of making new excipients confidently usable in pharmaceutical formulations for human health can be resolved by a common commitment on the part of academic, industrial, and regulatory circles as to their respective responsibilities. Modalities jointly developed by stakeholders will be instrumental in developing international standards and overcoming the difficulties encountered during the development of pharmaceutical formulations that employ new excipients, in light of comprehensive and specific scientific studies to be conducted on excipients utilizing the financial resources to be so generated.

REFERENCES

1. Szejtli, J. (1998). Introduction and general overview of cyclodextrin chemistry. *Chem. Rev.*, *98*(5), 1743–1754.

2. Villiers, A. (1891). Sur la fermentation de la fécule par l'action du ferment butyrique. *C. Rend. Acad. Sci.*, *112*, 536–538.

3. Dass, C.R., Jessup, W. (2000). Apolipoprotein A-I, cyclodextrins and liposomes as potential drugs for the reversal of

atherosclerosis. a review. *J. Pharm. Pharmacol.*, *52*(7), 731–761.

4. Vaution, C., Hutin, M., Glomot, F., Duchêne, D. (1987). The use of cyclodextrins in various industries. In: Duchêne, D., Ed., *Cyclodextrins and Their Industrial Uses*. Editions de Santé, Paris, pp. 299–350.

5. Hashimoto, H. (2002). Present status of industrial application of cyclodextrins in Japan. *J. Inclusion Phenom. Macrocyclic Chem.*, *44*, 57–62.

6. Loftsson, T., Duchêne, D. (2007). Cyclodextrins and their pharmaceutical applications. *Int. J. Pharm.*, *329*(1–2), 1–11.

7. Thompson, D.O. (1997). Cyclodextrins—enabling excipients: their present and future use in pharmaceuticals. *Crit. Rev. Ther. Drug Carrier Syst.*, *14*(1), 1–104.

8. Rajewski, R.A., Stella, V.J. (1996). Pharmaceutical applications of cyclodextrins. 2. In vivo drug delivery. *J. Pharm. Sci.*, *85*(11), 1142–1169.

9. Del Valle, E.M.M. (2004). Cyclodextrins and their uses: a review. *Process Biochem.*, *39*, 1033–1046.

10. Singh, M., Sharma, R., Banerjee, U.C. (2002). Biotechnological applications of cyclodextrins. *Biotechnol. Adv.*, *20*(5–6), 341–359.

11. Szejtli, J. (1989). Downstream processing using cyclodextrins. *Trends Biotechnol.*, *7*(7), 170–174.

12. Gerloczy, A., Antal, S., Szathmari, I., Muller-Horvath, R., Szejtli, J. (1990). Absorption, distribution and excretion of ^{14}C-labelled hydroxypropyl β-cyclodextrin in rats following oral administration, Minutes, 5th International Symposium on Cyclodextrins. Editions de Santé, Paris, pp. 507–513.

13. Kubota, Y., Fukuda, M., Muroguchi, M., Koizumi, K. (1996). Absorption, distribution and excretion of β-cyclodextrin and glucosyl-β-cyclodextrin in rats. *Biol. Pharm. Bull.*, *19*(8), 1068–1072.

14. De Bie, A.T., Van Ommen, B., Bar, A. (1998). Disposition of [^{14}C]γ-cyclodextrin in germ-free and conventional rats. *Regul. Toxicol. Pharmacol.*, *27*(2), 150–158.

15. Van Ommen, B., De Bie, A.T., Bar, A. (2004). Disposition of ^{14}C-α-cyclodextrin in germ-free and conventional rats. *Regul. Toxicol. Pharmacol.*, *39*(Suppl. 1), 57–66.

16. Stella, V.J., Rajewski, R.A. (1997). Cyclodextrins: their future in drug formulation and delivery. *Pharm. Res.*, *14*(5), 556–567.

17. U.S., FDA (2005). *Guidance for Industry: Nonclinical Studies for the Safety Evaluation of Pharmaceutical Excipients*. U.S. Department of Health and Human Services, Food and Drug Administration Center for Drug Evaluation and Research (CDER), Center for Biologics Evaluation and Research (CBER).

18. Osterberg, R.E., See, N.A. (2003). Toxicity of excipients: a Food and Drug Administration perspective. *Int. J. Toxicol.*, *22*(5), 377–380.

19. Pifferi, G., Restani, P. (2003). The safety of pharmaceutical excipients. *Farmaco*, *58*(8), 541–550.

20. Steinberg, M., Silverstein, I. (2003). The use of unallowed excipients. *Int. J. Toxicol.*, *22*(5), 373–375.

21. EMA (2007). *Guideline on Excipients in the Dossier for Application for Marketing Authorisation of a Medicinal Product*. EMEA/CHMP/QWP/396951/2006, European Medicines Agency Committee for Medicinal Products for Human Use.

22. EC (2003). Commission Directive 2003/63/EC Amending Directive 2001/83/EC of the European Parliament.

23. Baldrick, P. (2000). Pharmaceutical excipient development: the need for preclinical guidance. *Regul. Toxicol. Pharmacol.*, *32*(2), 210–218.

24. Wacker (2000). *GRAS Notification Claim for Gamma Cyclodextrin*. Wacker Biochem Corporation, Adrian, MI.

25. USP (2007). *The United States Pharmacopeia 30*. Port City Press, Baltimore, pp. 1056, 1071.

26. COE/EDQM (2008). The European Pharmacopoeia 6.0. Council of Europe European (COE)–European Directorate for the Quality of Medicines (EDQM).

27. Rowe, R.J., Sheskey, P.J., Owen, S.C. (2006). *Handbook of Pharmaceutical Excipients*, 5th ed., Pharmaceutical Press, London, and American Pharmacists Association, Washinton, DC, p. 217.

28. U.S. FDA (2010). *Inactive Ingredients Search for Approved Drug Products*. U.S. Department of Health and Human Services, Food and Drug Administration. http://www.accessdata.fda.gov/scripts/cder/iig/index.cfm.

29. EMA (2007). Opinion on the potential risks of carcinogens mutagens and substances toxic to reproduction when these substances are used as excipients of medicinal products for human use. CHMP Scientific Article 5(3). Committee for Human Medicinal Products, European Medicines Agency.

30. Szejtli, J. (2005). Cyclodextrin complexed generic drugs are generally not bio-equivalent with the reference products: therefore the increase in number of marketed drug/cyclodextrin formulations is so slow. *J. Inclusion Phenom. Macrocyclic Chem.*, *52*, 1–11.

31. Loftsson, T. (2004). Role of cyclodextrins in improving oral drug delivery. *Am. J. Drug Deliv.*, *2*, 261–275.

7

CYCLODEXTRINS IN THE COSMETIC FIELD

NILÜFER TARIMCI

Department of Pharmaceutical Technology, Ankara University, Ankara, Turkey

1. INTRODUCTION

Cyclodextrins (CDs) are cyclic oligosaccharides that consist of α-D-glucopyranose units that are bound with α-(1,4) bonds and have hydrophobic inner cavities and hydrophilic outer surfaces. Cyclodextrins are derived as a result of the intramolecular transglucosylation reaction that occurs in an anhydrous environment during degradation of the starch in a cyclodextrin glycocyltransferase (CGT) enzyme [1–4].

There are three principal types of natural CDs, also known as first-generation CDs: α-CD, β-CD, and γ-CD. These CDs consist of six, seven, and eight glucopyranose units, respectively. β-CD is the most commonly used, based on its cheapness, availability, and complex-forming capacity. Various derivatives of these natural CDs have been prepared in order to enhance their inclusion complex–forming capacities and physicochemical properties. These derivatives are usually obtained with the amination, esterification, or etherification of primary and secondary hydroxyl groups of CDs.

The most important characteristic of CDs is their ability to form inclusion complexes with hydrophobic molecules by retaining them within their inner cavities with their nonpolar cavities and hydrophilic outer surfaces. With the formation of inclusion complexes, the physicochemical properties of the guest molecules bound temporarily or entrapped in the inner cavity can be changed [5,6]. As a result of the preparation of the inclusion compound, positive modifications are provided in the properties of guest molecules, such as enhancement of the dissolution of insoluble substances, stabilization of nonstable substances against the degrading effects of light and heat, sublimation and control of the volatility, physical

isolation of the incompatible compounds, chromatographic separations, masking of undesired odor and taste, and controlled release of drugs and fragrance–taste substances [7]. The CDs and their inclusion complexes are therefore used as delivery systems in the food, drug, cosmetics, packaging, and textile industries [8].

2. CDs AND THE SKIN

There are many studies concerning dermal and transdermal application of CDs [9–12]. However, CD–skin interactions are well known. The clinical efficacy of a drug applied topically, depends on the drug availability at the target site in addition to its pharmacological properties. The target site can extend to a very wide region, depending on the lesions on the skin. It should be noted, however, that the stratum corneum layer provides a significant protective barrier in the penetration of drugs and other chemical substances. Compounds that are very hydrophobic, especially, cannot pass easily through the stratum corneum and viable epidermis. The stratum corneum layer, however, can provide easier delivery, depending on the hydration level, due to the use of a diffusion mechanism for skin penetration [13]. As is known, hydrophilic compounds with a maximum molecular weight of 300 Da have the easiest skin penetration abilities. But the CD molecules are too large (molecular weight range ~1000 to 2000 Da), and delivery of the CD molecules and complexes through the biological membranes under normal conditions is quite difficult [14]. For example, the effect of HP-β-CD was reviewed upon the penetration of a preservative, methylparaben, through the excised hairless mouse skin.

Cyclodextrins in Pharmaceutics, Cosmetics, and Biomedicine: Current and Future Industrial Applications, First Edition. Edited by Erem Bilensoy.
© 2011 John Wiley & Sons, Inc. Published 2011 by John Wiley & Sons, Inc.

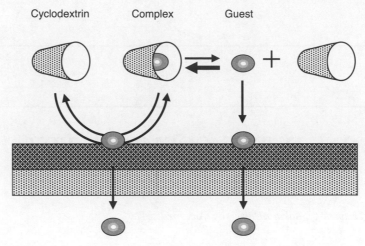

Figure 1. Apparent main mechanism of penetration enhancement.

It was found that the methylparaben/HP-β-CD complex reduced the cutaneous permeability of methylparaben [15].

Furthermore, the CDs can cause irritation on the skin in certain cases. For example, parent CDs and chemically modified CDs in high concentration cause skin irritation in guinea pigs. It was concluded that this is caused by extraction of the skin lipids by the CDs [16].

But CDs can reduce the side effects of certain drugs by decreasing the cutaneous penetration and therefore increasing the bioavailability of the entrapped molecules [17]. Various methods can be adopted to increase the cutaneous penetration of the drugs. Substances defined as *penetration enhancers* can be added to formulations for this purpose. CDs are also used as penetration enhancers. Whereas conventional penetration enhancers such as alcohols and fatty acids affect the skin barrier by damaging it, hydrophilic CDs positively influence the penetration by enhancing the availability of the drug on the skin surface, as shown in Fig. 1 [18].

Penetration enhancers added to semisolid vehicles increase the percutaneous absorption by damaging the stratum corneum barrier of the skin temporarily, thus providing a more efficient topical treatment [19]. Various unsaturated fatty acids, fatty alcohols, and glycerol monoethers are used as penetration enhancers. Several studies suggest that these substances enhance the dermal and transdermal delivery of many drugs [20,21].

3. CDs AS A DELIVERY SYSTEM

CDs and their derivatives are used widely in the drug, cosmetics, food, and textile industries. Their most important advantages include enhancing drug dissolution and thus having a positive effect on bioavailability, increasing the drug's stability and providing controlled release in the pharmaceutical industry [22]. Many studies are conducted with CDs in oral, parenteral, ocular, nasal, dermal, and rectal drug administrations. Table 1 presents examples of CD application through various routes. CDs also play important roles in colon-specific drug delivery, peptide and protein delivery, and gene and oligonucleotide delivery [23–26].

Another area of CD use is as excipients in pharmaceutical dosage forms. There are several studies concerning their use as filler and disintegrant in tablet manufacturing, to mask of the bitter taste of a drug in liquid dosage forms, or as a pelletization agent in an extrusion/spheronization process [27–29]. In a study where cross-linked carboxymethylcellulose and CD polymer are compared as the disintegrant in spironolactone tablet formulations, it was concluded that the CD polymer provides a capillary effect and is as good a disintegrant as are cellulose derivatives, and that it increases the dissolution rate of the drug [30].

3.1. CDs and Dermal/Transdermal Delivery

CDs and their inclusion complexes have varying effects on the behavior of the drugs in ointments and other semisolid formulations. CD complexes are used to reduce the irritant effect of a drug on the skin or the mucosa, preventing the metabolism of the skin site of application, increasing absorption, and ensuring extended release from the excipient [44–47].

Encapsulation of active agents with CDs does not cause a change in the intrinsic pharmacological properties of a drug while changing its physicochemical properties. One of the most important properties of CDs is the lack of effect on the barrier feature of the stratum corneum layer in dermal and transdermal drug delivery. As is known, the CDs are molecular delivery systems. The lipophilic drug molecules carried by CDs are released on the skin surface. The free drug fraction here depends on the dissolution rate of the drug.

A proper vehicle is essential for the CDs to exhibit their functions. For example, the release of corticosteroids from the hydrophilic vehicles increases significantly with hydro-

Table 1. Applications of CDs in Various Dosage Forms

Drug	CD	Application Route	Observations	Ref.
Griseofulvin	β-CD	Oral	Improvement in drug bioavailability due	[31]
Ketoprofen	β-CD	Oral	to increased rates of both dissolution	[32]
Digoxin	γ-CD	Oral	and permeation	[33]
Diltiazem	DE-β-CD, TE-β-CD	Oral	Obtained slow-release characteristic of drug	[34]
Dexamethasone	Sugar-branched β-CD	Parenteral	Stable complexation of drug in blood	[35]
Paclitaxel	HP-β-CD	Parenteral	Enhanced solubility and stability of paclitaxel in solution	[36]
Hydrocortisone	HP-β-CD	Ocular	Increased aqueous solubility	[37]
Pilocarpine nitrate	HP-β-CD	Ocular	Enhanced permeability and miotic response of drug	[38]
Morphine	α-CD	Rectal	Enhanced rectal absorption of drug	[39]
Midazolam	SBE-β-CD	Nasal	Improved nasal absorption of drug	[40]
Insuline	β-CD/DM-β-CD	Nasal	Reduced nasal toxicity of sodium desoxycholate added as an enhancer in a formulation	[41]
Naproxene	2-HP-β-CD	Dermal	Improved drug release	[42]
Dexamethasone	HP-β-CD	Dermal	Increased skin permeability of drug	[43]

philic CDs, while the drug release from fatty vehicles is delayed [48]. The dermocorticoids, which are used extensively in dermatological treatment, nonsteroid anti-inflammatory drugs, and retinoids, are among the substances whose inclusion complexes are studied most. Currently, there is a commercially available ointment formulation on the market, called Glymesason (Fujinaga, Japan), which is a dexamethasone/β-CD complex and has analgesic anti-inflammatory effects [28].

Retinoic acid is used successfully in the treatment of acne vulgaris, due to its keratolytic activity. But the side effects, such as the sensitivity that it causes on the skin, erythema, and xerosis, limit its use. Anadolu et al. [49] investigated the efficacy and tolerability of the retinoic acid/β-CD inclusion complex on 66 acne vulgaris patients. Retinoic acid/β-CD complex was applied to the patients in hydrogel and moisturizer cream formulations, and the treatment efficacy was determined to be higher than that of the commercial product, which contains twice as much retinoic acid in the concentration. Furthermore, the topical side effects (erythema, desquamation, xerosis, and irritation) have a proportion of 92% in the commercial product group and 27% in patients that use the formulation with β-CD complex, as shown in Fig. 2. On the other hand, the low water solubility and stability of retinoic acid were enhanced with the CD complexes in other studies [50,51].

Increasing irritant activity of drug molecules in tissues can be prevented using CDs. It was demonstrated using

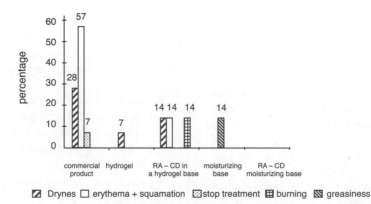

Figure 2. Side effects of treatment.

histological studies that the invasive effect of celecoxib on the stratum corneum layer is reduced using complexes prepared with DM-β-CD and HP-β-CD [52]. Other research has shown that when the nitroglycerin release increases with an inclusion complex prepared with DE-β-CD from the ointment formulation, better percutaneous absorption was obtained [53].

Corticosteroids are among the most topically used drugs. Depending on the purpose of the treatment, they are applied in ointment or cream dosage forms. While the water solubility of corticosteroids is low, the release rate of the drug from preparations depends on the vehicles selected. Therefore, selection of the right vehicle is essential. There are studies that have investigated the effect of CDs on the permeability and release of dermal corticosteroids such as betamethasone and beclamethasone. When complexes of these drugs prepared with β-CD and γ-CD are formulated in the hydrophilic ointment base, the release rates increase significantly compared to those of free drugs [54,55].

CDs do not affect drug release from oily bases such as vaseline. This is attributed to the lower solubility of the drug in these bases. While the release of the hydrocortisone from the oil/water cream and hydrogel bases increases with β-CD and HP-β-CD complexes, it decreases in vaseline or water/oil cream [56]. CDs also enhance the dermal delivery of non-steroid anti-inflammatory drugs (NSAIDs). A study reports that the complexes of 4-biphenylacetic acid, which are prepared with β-CD, DM-β-CD, and HP-β-CD, enhanced the in vivo absorption of the drug in mice [57]. Another study with healthy volunteers reported that indomethacin complexes, which are prepared with β-CD, DM-β-CD, and HP-β-CD, enhanced the anti-inflammatory effect, depending on the increase in release from hydroxyethylcellulose hydrogels [58]. The reason for this may be the increase of permeability through improving the thermodynamic activity of drugs of a lipophilic nature in bases with water contents.

3.2. CDs as a Cosmetic Delivery System

Many cosmetic products are manufactured in conventional dispersion systems such as emulsion and suspension. But the change in the cosmetic perception from one person to another and the differentiation of their expectations with the development of cultural and economic conditions, has led cosmetic formulators to new active agents and new techniques. These new active agents, called *cosmeceuticals* or *dermacosmetics*, lie between drugs and cosmetics. The main purpose of cosmetic delivery systems is control of cutaneous penetration of the cosmeceutical substances, description of their duration of action, and their delivery to the target skin layer [59]. Cosmetic delivery systems also improve the appearance and stability of products and ensure sustained release of the active agent. The most important aspect of the delivery of active agents through controlled release is the

Table 2. Novel Cosmetic Delivery Systems

I. Emulsion Systems	II. Vesicular Systems	III. Particulate Systems	IV. Molecular Systems
Micro-emulsions	Liposomes	Microcapsules	Cyclodextrins
Multiple emulsions	Niosomes	Nanocapsules Nanoparticles Matrix particles	

localization of the substance in the epidermis or dermis and its failure to proceed with the systemic circulation. For example, sunscreen products are required to retain their sun filters on the skin for the longest period possible and to have the least possible transdermal delivery of the substances. The main target of antiperspirants and deodorants are the bacteria and fungi on the skin surface, and their availability at these sites must be high. On the other hand, antioxidants, skin lighteners, and antiaging compounds should provide their activity on viable epidermis and dermis. To fulfill the intended performance, products containing cosmeceuticals should be prepared in a proper delivery system.

As indicated in Table 2, cosmetic delivery systems can be classified into four groups. CDs are molecular delivery systems. They retain guest molecules of varying shapes, sizes, or polarities in their hydrophobic inner cavities; thus, the properties of the bound substance change. Since the substance entrapped in the inner cavity of a CD is protected against oxidative, photolytic, and thermal degradations, it becomes more durable against environmental conditions and its stability increases. CD encapsulation also decreases the cutaneous penetration of the guest molecule and thereby decreases the undesired side effects [60].

4. APPLICATION OF CDs IN COSMETIC FORMULATIONS

CDs have been widely used recently for the molecular encapsulation of active cosmetic agents. CDs appear today in many cosmetic product formulations, including creams, lotions, shampoos, toothpastes, and perfumes [61,62]. The main purposes and advantages of the use of CDs in cosmetic formulations can be summarized as follows [63–65].

- Enhancement of the physical and chemical stability of guest molecules (protection of guest molecules against oxidation, hydrolysis, heat- and light-induced decomposition reactions, and other organic compounds and reactions)
- Increase or decrease of cutaneous absorption of several compounds
- Increase or decrease in skin irritation

- Reduction or elimination of unwanted body odor
- Sustained release of fragrance materials and cosmeceuticals
- Establishment or increase in stabilization for emulsions and suspensions
- Inhibition of foam formation with surface-active agents
- Masking or reduction of unwanted odor and taste
- Prevention of interactions between the compounds in a formulation
- Conversion of fats and fatty substances and liquid substances to microcrystal or amorphous powders and hence the handling for these substances for use in cosmetic formulations
- Reduction in hygroscopicity
- Reduction in volatility of the volatile substances
- Decreased use of preservatives in formulations (unlike starch, due to the failure of CDs and their derivatives to create a nourishing environment for microorganisms).

CDs are employed in cosmetics as both inclusion complexes and empty molecules.

4.1. Cosmetic Usage of Empty CDs

An uncomplexed CD can itself be used as an active agent in cosmetic formulations for several purposes.

4.1.1. Entrapment of Unpleasant Odors
Empty CDs are used in various cosmetic products, such as deodorants, due to their ability to retain volatile molecules in their inner cavities. They are used with fragrance substances, antiseptics, and antimicrobial preservatives in various deodorant formulations in order to eliminate mouth, body, and hair odor. There are many patents for products that have been prepared for this purpose [66–69]. When an odor absorber formulation that contains empty CD in a aqueous or alcoholic solution is applied to the skin, CD forms nonvolatile complexes with substances with bad odor and thereby prevents the volatilization of these substances from the skin surface [18].

Another formulation that was designed for the prevention of underarm and foot odor contains HP-β-CD in addition to other substances [70]. Powder CDs, in which the particle size is smaller than 12 mm, are provided in diapers, menstrual products, and paper towels in order to control the smell. Hydroxypropyl-β-CD also enhances the antimicrobial activity of the product on its own or with other substances [71].

4.1.2. Absorption of Fatty Compounds [51]
CDs are used in cleaner skin care products and acne-healing preparations with their property of absorbing fatty compounds and sebum

release. It was reported that a preparation prepared by the spray-drying method, containing calcium phosphate, β-CD, and squalene as moisturizers have high levels of sebum-absorbing capabilities [72]. Shiseido has prepared cosmetic packs that contain poly(vinyl alcohol), carboxymethylcellulose, methyl β-CD, ethanol, and perfume in order to clean the skin without causing irritation [73].

4.1.3. Other Applications of Empty CDs

Stabilization of Emulsions with CDs Emulsions can be stabilized with CDs without using any surface-active agents [74]. Cyclodextrins form a robust film layer on the disperse phase surface of the emulsion. It was reported that the stabile emulsions can be formed in a simple three-component system consisting of liquid paraffin, water, and β-CD, and that these emulsions can remain stabile for the long-term under varying storage conditions [75].

Scrubbing Particles [65] Water-insoluble CD polymer beads, which are prepared as the treatment of β-CD with the epichlorohydrin, can be used as a facial cleanser, due to their scrubbing effects [76].

4.2. Cosmetic Use of Inclusion Complexes

Molecular encapsulation of CDs with the cosmeceuticals introduces many advantages for the guest molecule.

4.2.1. Enhancement of the Solubility of Guest Molecules
Many cosmetic active agents are much too hydrophobic and not very soluble in the water. Therefore, it is a common practice to use surfactants as dissolution enhancers. Economically speaking, surfactants are very attractive compounds for the aqueous cosmetic formulations, since they are cheap and effective in low concentrations. But the use of surfactants also bring along several disadvantages, including skin irritation or light sensitivity, clouding problems in the preparation of transparent formulations, and foam formation. CDs and their derivatives form water-soluble inclusion complexes with several lipophile substances. Complexation significantly enhances the water solubility of the active agent. CDs are not toxic or allergenic. It is considered that they do not cause skin irritation, and even in some cases, complex formation decreases the possible irritation effect of the guest molecule [60].

Water solubility of triclosan, which is used widely as an antibacterial agent against plaque formation in toothpastes and other mouth care products, is very low. This also affects the biological activity of the substance in the oral cavity. In inclusion complexes prepared with β-CD and HP-β-CD, the water solubility of triclosan is increased 2000- to 4000-fold [77]. Addition of triclosan–CD complexes to the toothpaste formulations also enhances the availability of triclosan

Table 3. Effect of HP-β-CD, PVP, and Lysine on the Water Solubility of Triclosan[a]

HP-β-CD (%a/h)	PVP (%a/h)	Lysine (mM)	Solubility of Triclosan (mg/mL)
0	0.00	0	0.00
0	0.25	0	0.00
0	0.00	50	0.00
10	0.00	0	5.93
10	0.25	0	6.51
10	0.00	50	3.75
10	0.25	50	3.67

Compound of Aqueous Complexation Environment

[a] At 22–23°C and pH 5.60 ± 0.44.

as a result of the increase in solubility of the substance. The preference of triclosan/β-CD/CMC complex over free triclosan in toothpastes has increased the initial concentration of triclosan in the saliva after brushing approximately threefold and extended the duration of action from 0.7 h to 1.5 h [78].

Duan et al. [79] studied the effect of 2-HP-β-CD and RM-β-CD on the water solubility of triclosan and triclocarban and concluded that the formation of inclusion complex with the CDs improved the solubility of these substances. While PVP, which is added to the aqueous HP-β-CD solutions, causes an improvement in the solubility of triclosan, which is induced with HP-β-CD, the addition of lysine to the aqueous complexation environment affected the solubility of the substance negatively, as shown in Table 3. The same additives also increased the CD solubility of triclocarban. Addition of PVP and Mg^{2+} to the aqueous complexation environment increases the solubility of triclocarban, which is provided with RM-β-CD to some extent, whereas addition of ascorbic acid or lysine resulted in a significant solubility enhancement [79]. Minoxidil, an agent that stimulates keratinocyte growth and hair growth, increased water solubility by preparing an inclusion complex with α-CD [80].

p-Hydroxybenzoic acid esters (parabens) are preservatives that are commonly used in cosmetics. Extension of the alkyl chain of the parabens increases their antiseptic effect and clinical reliabilities, but their low water solubility limits the use of long-alkyl-chain parabens. There are several studies concerning the enhancement of paraben solubility with the help of CDs [81,82]. The usability of 2-HP-β-CD was investigated to increase the solubility of methylparaben and suppress the percutaneous absorption, and compared to HCO-60 [poly(oxyethylene hydrogen) and castor oil 60 EO], which is a nonionic surfactant, HP-β-CD significantly increased the water solubility of methylparaben. In vitro cutaneous delivery of excised hairless mouse skin is also suppressed by HP-β-CD, and therefore conversion to the

p-hydroxybenzoic acid, a less toxic metabolite in the epidermis, increases [83].

An investigation of the interactions between HP-β-CD and p-hydroxybenzoic acid esters in an aqueous solution leads to an increase of solubility for p-hydroxybenzoic acid esters in pH 7.4 phosphate buffer with the addition of HP-β-CD of varying amounts. In solutions containing 12% HP-β-CD, the solubility of methylparaben increases six fold (from 2.25 mg/mL to 12.7 mg/mL), while the solubility of benzylparaben increases 308-fold (from 28.2 mg/mL to 8.7 mg/mL) [81]. Cyclodextrins are used for the formulation of oil-soluble vitamins such as vitamin A, E, D_3, and K_3 in aqueous systems. Vitamin A (retinol) and vitamin A acid (retinoic acid) are cosmeceutical agents used against the aging of the skin. Due to low water solubility, however, it is a problem to prepare them in aqueous bases. This problem has been resolved by preparing CD complexes of the retinoic acid. The water solubility of retinoic acid increases 11.7- and 17.37-fold by preparing an inclusion complex with β-CD and HP-β-CD in 1:1 molar ratios, respectively, as shown in Table 4 and Fig. 3 [84]. Stable gel formulations of retinoic acid can thus be prepared [85].

To provide perfumes dissolved in an aqueous medium without using surface-active agents, inclusion complexes are prepared with CDs. Synthetic perfumes are used widely in various commercial products. Insufficient solubility of the perfume in this medium would lead to an inadequate fragrance. The solubilities of the synthetic perfumes have increased significantly when their inclusion complexes were prepared with sulfoalkyl ether-β-CD [86].

4.2.2. Increased Stability of Guest Molecules CDs protect guest molecules against the attacks of several reactive molecules through complexation. Therefore, CDs can prevent or reduce the hydrolysis, oxidation, racemization,

Table 4. Results of Solubility Experiments of Retinoic Acid[a]

	Water Solubility (μg/mL)	Enhancement	Complex Efficacy
RA	2.77 ± 0.37		
RA/β-CD physical mixture	20.20 ± 1.02	7.29	
RA/β-CD complex	30.67 ± 9.00	11.07	0.846
RA/HP β-CD physical mixture	9.98 ± 0.31	4.40	
RA/HP β-CD complex	48.12 ± 5.33	17.37	0.657

[a] $n = 6$.

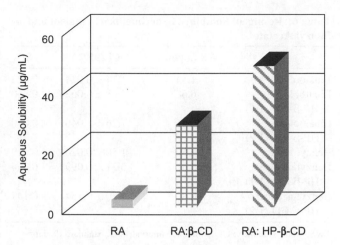

Figure 3. Aqueous solubility of RA with β-CD and HP-β-CD complexes.

isomerization, polymerization, and enzymatic decomposition of the encapsulated molecule. Thus, the shelf life of skin care products can be extended. Complexed compound is also isolated from the outer environment, and this leads to a significant reduction of the interaction between the incompatible compounds of formulations.

Resveratrol (*trans*-3,5,4'-trihydroxystilben) is a triphenolic phytoalexin found in several plants, including grapevines, mulberries, and peanuts. It provides antioxidant properties due to this phenolic structure. Resveratrol is a hydrophobic substance that is sensitive against heat and oxidative enzymes. The stability and bioavailability of the resveratrol increase with the molecular encapsulation within the CD [87].

Hydroquinone is used as a skin whitening agent in cosmetic creams. It is stable within a certain pH interval in aqueous solutions. Its stability was improved significantly with CD complexes prepared to protect the hydroquinone from oxidation. The efficacy of the complex prepared is greater than that of free hydroquinone [88]. Kojic acid is another substance used as a skin whitening agent. In cases of exposure to heat and light, it decomposes and turns yellowish brown. Preparation of an inclusion complex for the kojic acid with the CDs increases the water solubility and stability of the substance, prevents its gradual coloring, and results in an increase in the skin-whitening effect [89].

Linoleic acid is an essential fatty acid that is important for skin care. It is used in many skin care creams prepared for dry and sensitive skins. But due to its lack of resistance against oxidization, it is easily degraded. An oxidization-resistant compound of linoleic acid was formed, due to an inclusion compound prepared with α-CD [90].

Retinol, a substance used in topical antiaging formulations, reduces wrinkles and supports the rejuvenation of damaged tissues along with ultraviolet (UV) light. However,

UV light and oxygen lead to the initiation of chemical reactions in the retinol, and several peroxydic toxic intermediate products are formed during oxidization. With the inclusion complex of retinol, which was prepared with 2-HP-β-CD, both its photostability and efficacy were enhanced [91].

The most important aspect of the prevention of photoaging is adequate protection of the skin against the sun. Skin filters efficient against UVA and UVB light of the sun are used in many cosmetic product formulations. Sun filters are used to protect the skin against the harmful effects of the sun. However, UV radiation induces partial degradation or change in these filters and leads to a reduction or loss of their capability to protect the skin, a phenomena called *photodegradation*. Photodegradation of the sun filter may also lead to the creation of molecules which are toxic to the skin and can result in skin irritation and sensitivity. Three different sun filters (octyl methoxycinnamate, butyl methoxydibenzoylmethane, and bezofenon-3) used in cosmetics were encapsulated with phospholipids and β-CDs and researchers attempted to establish the photostability of these substances. The sun filters that were complexed with the CDs presented much higher photostability than that of filters that were in free form or were encapsulated with phospholipids [92].

2-Ethylhexyl-*p*-dimethylaminobenzoate (EH-DMAB) is another UVB filter used widely in sunscreen products. The inclusion compound of EH-DMAB with HP-β-CD, increases its water solubility and photostability significantly, as shown in Table 5. In these formulations the type of vehicle is also very important [93].

The BM-DBM9/HP-β-CD complex of butylmethoxydibenzoylmethane (BM-DBM9), which is a UVA filter, leads to a significant increase in the water solubility of BM-DBM and reduces the photodegradation of the UV filter both in the solution and in the emulsion bases [94]. In another study, the inclusion complex of the *trans*-ethylhexyl-*p*-methoxycinnamate (*trans*-EHMC) formed with β-CD significantly reduces decomposition of the UV filter in alkali solutions [95].

Phenyl benzimidazole sulfonic acid is another alternative as a sunscreen agent. But it is dissolved to a small extent in an acidic medium, and when it is exposed to radiation it produces active oxygen and free radicals. The chemical

Table 5. Comparative Photodegradation Data Derived for EH-DMAB in Free and Complex Form in Various Excipients

Product Type	% Loss of the Sunscreen Agent[a]	
	EH-DMAB	Complex
Solution	54.6 ± 5.5	$25.5 \pm 8.8^*$
Lotion	32.8 ± 6.8	28.5 ± 3.2
Lotion + 5% HP-β-CD	33.4 ± 5.3	$25.1 \pm 1.8^*$

[a]Each value is the average $\pm$ S.D. of the results taken from six analyses.
*, $p < 0.05$.

stability of phenyl benzimidazole sulfonic acid can be increased via complexation with RM-β-CD and HP-β-CD [96].

Evaluation of encapsulations of several commercial sun filters prepared with various CDs (α-, β-, and γ-CD, and dimethyl-β-CD) resulted in compounds with better sun-filtering properties. Systemic absorptions of these new compounds that commercial sun filters form with CDs are much lower, but their water resistance is higher [97].

α-Lipoic acid, which is used as an antiaging agent, is a very good antioxidant and free-radical scavenger. It has no solubility in water, but it is soluble in fat. A complex of heat-sensitive α-lipoic acid with CDs (especially α-CD) is resistant against heat and light, and the unpleasant odor of the substance also decreases [98].

4.2.3. Conversion of Liquid Substances to Powder Form

Whatever the physical state of a guest molecule may be—gas, liquid, or solid—the resulting inclusion complex is always a solid powder. Usage of a stable powder is easier than, for example, the use of highly volatile and nonstable aromatic oils. Thus the production processes can become repeatable. To sum up, complexation of cosmetic ingredients with CDs causes a general technological advantage.

The addition of perfumes to cosmetic preparations is a preference factor for consumers. However, the low stability of perfumes, the irritation and bad odor of the degradation products, and the lack of solubility in aqueous solutions are major problems. With the complexes of perfumes prepared with CDs, these undesirable features can be overcome. The encapsulation products of various fragrance materials formed with 2-HP-β-CD results in much better solubility than that of the pure fragrance substance [99].

The majority of compact powders contain fragrance substances. But the volatility and rapid degradation of fragrances limit the shelf life of powders. Szejtli et al. [100] prepared inclusion complexes of fragrances and volatile oils with β-CD. The study suggested that the fragrances, which were volatile and sensitive to oxidization, were stabilized with

Table 6. Results of Solubility Experiments of Linalool and Benzyl Acetate[a]

	WS (mg/mL)	CI (95%)	SD
Linalool	1.14	1.14 ± 0.024	0.023
Linalool/2-HP-β-CD (1:1)	6.68	6.68 ± 0.063	0.060
Linalool/2-HP-β-CD (1:2)	7.26	7.26 ± 0.040	0.040
Benzyl acetate	1.50	1.50 ± 0.020	0.019
Benzyl acetate/2-HP-β-CD (1:1)	6.31	6.31 ± 0.099	0.094
Benzyl acetate/2-HP-β-CD (1:2)	6.80	6.80 ± 0.154	0.147

[a] WS, water solubility; CI, confidence interval; S.D., standard deviation ($n = 6$).

complexation, and the resulting inclusion complexes in powder form could be mixed homogeneously with the other powders in the formulation.

Both the water solubility and the stability of linalol and benzyl acetate are increased in inclusion compounds prepared with β-CD and 2-hydroxypropyl-β-CD as shown in Table 6 and Fig. 4 [101]. Peroxyacetic acid is a liquid used in aqueous solutions. It forms solid complexes with α-cyclodextrin and β-cyclodextrin. These stable powders provide ease of use and can be used as skin whiteners in cosmetic formulations [102].

4.2.4. Correction of Undesired Properties

CDs can decrease the cutaneous penetration by encapsulating a guest molecule, and they protect the skin against the undesired side effects of the guest molecule [49]. Elimination of undesired and bad odors can also be achieved through complexation with CDs. Bad odor has two sources in cosmetic products. The first is the formulation of substances with a bad odor, and the second is the formation of substances with a bad odor during use of the cosmetic product. When dihydroxyaceton, a

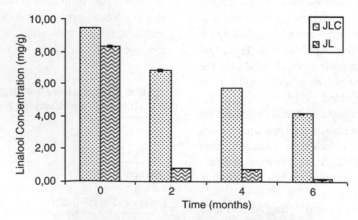

Figure 4. Results of stability studies on JL and JLC. JL, gel formulation containing uncomplexed linalool; JLC, gel formulation containing linalool/2-HP-β-CD(1:1) complex.

tanning agent, reacts with the skin, it creates an undesired odor, which is difficult to mask with perfumes. Preparation of its inclusion complex with CDs can prevent this odor. Slow release of the dihydroxyacetone from the complex also provides a homogeneous tanning of the skin [62].

N,N-Diethyl-3-methylbenzamide (DEET) is a substance used as an insect repellent. It leads to undesired side effects in its topical application. Ethyl alcohol, which is used as a solvent in conventional DEET formulation, increases the absorption of the substance and lead to the occurrence of toxic effects due to its penetration-enhancing nature. As a novel approach, inclusion complexes of DEET were prepared with CDs. These complexes increase the stability of DEET and side effects on the skin are reduced when CD is used as an alternative to ethanol [103].

Mercapto compounds are used widely in permanent preparations. These compounds are the cause of the bad odor that arises during application of the preparation. This odor is sometimes suppressed with perfume. Another alternative is the elimination of undesired odor through preparation of inclusion complexes of the mercapto compounds with CDs [62].

Menthol is used as a cooling agent in various cosmetic preparations. Due to the low water solubility of menthol, alcohol is usually added to the formulation as the solvent. A CD complex of the menthol, on the other hand, can be dissolved easily in water and forms a clear solution without alcohol [48]. On the other hand, the bitter taste and characteristic smell of menthol were resolved with the preparation of complexes with CD and HP-β-CD [104,105].

4.2.5. Establishment of Controlled Release

Complexation of the fragrances used intensively in cosmetics with CDs makes possible the establishment of controlled release since only the free fragrance fraction, which is in balance with the complex-forming fraction, is released from the skin substance. A fragrance–CD inclusion complex which is formulated in a body lotion is in balance with the surrounding environment. Since this lotion, which is applied to the body, is in contact with air, it loses its aromatic oil. Depending on the vaporization, reduction of the fragrance concentration, which is in a dissolved state within the lotion, leads to release of the fragrance from the CD cavity in accordance with the existing balance. Thus, the sustained release of the fragrances is achieved [60]. There are several patents concerning this practice [106,107].

Three types of HPCD use (HP-α-CD, HP-β-CD, and HP-γ-CD) were investigated to increase the solubility and delay the release of fragrances, and then compared to HCO-60. HP-β-CD significantly increases the water solubility of fragrances. The release rates of d-camphor and 3, 1-metoxypropane 1,2-diol, on the other hand, were decreased significantly with the addition of HPCDs. It was concluded that HP-β-CD is more efficient than other HPCDs and HCO-60 in establishing a constant release of fragrances [108].

Glycolic acid is released slower than β-CD complex, and this increases its moisturizer and exfoliant effect in cosmetic use. Since it is also released more slowly than the glycolic acid complex, irritations caused by the free glycolic acid are also reduced [109]. In addition to its cosmetic use, CDs have been also used in the textile industry in recent years. Permanent fixation of CDs on the surface of textile materials provides several advantages over unmodified textile products. The organic compounds of sweat forms a complex with CDs during contact of the textile product with skin. It therefore becomes possible to prevent or reduce the microbiological degradation of these substances and resulting formation of body odors. Several cosmetic active agents can also be complexed with CDs before the use of textile products. Complex formation allows the establishment of stability for oxygen and light-sensitive molecules. Furthermore, the substances that are encapsulated within the CDs are released from the CD cavities in the presence of water, which is normally found in the skin, during textile–product contact. The substance released can be absorbed directly by the skin. Textile products such as underwear, T-shirts, and jackets, which have fixed CDs, are available in the German market. Bed sheets with fixed CDs have also been launched commercially. The use of curtains with fixed CDs is recommended to filter odorous substances from the air [110].

4.3. Use of CDs in Several Cosmetic Products

CDs have been used in the cosmetics industry since the mid-1990s. The first studies were conducted in Hungary, Japan, and France. Many product are available in the market that utilize CDs for various purposes.

Salicylic acid is used for to clean skin due to its antibacterial and keratolytic effects. The solubility of this substance and its derivatives in aqueous solutions is low. Its complexes are prepared with CDs and its water solubility is increased. Complexation with CDs leads to an enhancement of the disinfectant, bacteriostatistic, and keratolytic properties of the substance. An inclusion complex of salicylic acid and hydroxypropyl-β-CD is available commercially as Lipo CD-SA (Lipo Chemicals, Inc.). Another product from the same company, referred to commercially as Lipo CD-E, is a delivery system for tocopherol. This product, which encapsulates tocopherol within CDs, contains 30% active agent and is in powder form. Delivery systems including cyclodextrins for triclosan and ethylhexylmethoxycinnamite, called Lipo CD-TC and Lipo CD-OMC, are produced by the same company.

Another product is Bioclin Sebo Care Impure Skin Cream® (Ganassini). Cyclodextrin, which is found in the formula, facilitates removal of follicular obstruction.

Table 7. Examples of Marketed Cosmetic–CD Formulations

Trade Name	Active Substances	CD	Indication	Company	Country
Lipo CD-SA	Salicylic acid	HP-β-CD	Delivery system for salicylic acid (keratolytic)	Lipo Chemicals, Inc.	United States
Lipo CD-E	Tocopherol	CDs	Delivery system for tocopherol (antiaging)	Lipo Chemicals, Inc.	United States
Lipo CD-OMC	Ethylhexyl methoxycinnamite	CDs	Delivery system for the sun screen	Lipo Chemicals, Inc.	United States
Lipo CD-TC	Triclosan	HP-β-CD	Delivery system for triclosan (antimicrobial)	Lipo Chemicals, Inc.	United States
Biolin Sebo Care Impure Skin Cream	—	CDs	Acne-prone skin	Ganassini	Italy
Cellutex	L-Carnitine	β-CD	Anticellulite Cream	Regina Neu Cosmetic	Germany
Lyminys Cream	α-Tocopherol	β-CD	Makeup camouflage cream	Roan S.p.a	Italy
Mirakelle	Vitamin-A	β-CD	Antiaging	Distributor for Vor Laboraties, Inc.	Sweden
Novo Flex	Vitamins A and E	HP-β-CD	Antiaging	Revlon	South America
Gesichts tonic	—	β-CD	—	AnnaMaria Borlind	Germany
Eucerin Vital Retinol	Vitamin-A	CDs	Antiaging	Beiersdorf	Germany
Self-Action Super Tan for the Face	Dihydroxyacetone	CDs	Self-tanning	Estée Lauder	United States
Klorane Extra Gentle Dry Shampoo	—	CDs	Dry shampoo	Klorane, Pierre Fabre Dermo Cosmetique	France

Another CD-containing product of vitamin E (α-tocopherol), sold under the commercial name Luminys Cream (Roan S.p.a., Italy) contains vitamin E/β-CD complex. The cream is launched as an antiaging properties.

CDs also increase the solubility of the secretion agents of the skin and so are used in skin cleaning products. β-Cyclodextrin, which is used in preparations along with menthol and salicylic acid, forms a fatty complex with released skin fats and can easily get away from the skin. Such a product is available commercially as Gesichtstonic (Annamaria Börlind).

A cellulite cream known as cyclosome that contains β-CD complex with L-carnitine, is sold in the German market by Regana Ney Cosmetics under the brand name Cellutex.

Used in topical antiaging formulations, retinol decreases wrinkles and supports the renovation of tissue damaged with UV. Retinol–CD complexes are used in several commercial product compounds, including Eucerin Vital Retinol (Beiersdorf), Nutrients & Anti-aging Agents (Efal), and Dexol A (Collaborative Laboratories).

A dihydroxyaceton–CD complex used in self-tanning products is available in the commercial preparation Ultrasun Selftan (Ultrasun). The Cyclosystem Complex of the product provides better stability and controlled release of dihydroxyacetone. Another product with this agent is available commercially as Self-Action Super Tan for the Face (Estee Lauder).

An other interesting CD-containing product is Klorane's Extra Gentle Dry Shampoo with Oat Extract. CD within this product, which assists in the anhydrous cleaning of hair, contributes to the absorption of dirt and grease in the hair. Table 7 summarizes CD-containing commercial product samples.

5. CONCLUSIONS AND FUTURE OUTLOOK

CDs have been known for more than 100 years. In the last 10–15 years CD-containing drug formulations and cosmetic products have been approved and launched, first in Japan and Europe, then in the United States. Although their initial intended use was to increase the water solubility and chemical stability of drugs with the help of drug–CD complexes, they can now be used for various purposes. Although the safety evaluations have not been completed for many CDs, they play a major role as safe drug delivery agents in drug delivery systems. They also play a major role in stability improvement, prevention of side effects, elimination of undesired properties, and prevention of volatile compounds.

In conclusion, CDs show great promise for researchers in the field of biotechnology and in the development of drug and cosmetic drug formulations with their unique architecture and chelating properties. CD technologies are under constant development.

REFERENCES

1. Szejtli, J. (1998). Introduction and general overview of cyclodextrin chemistry. *Chem. Rev.*, *98*, 1743–1753.

2. Rowe, R.C., Sheskey, P.J., Owen, S.C. (2006). *Handbook of Pharmaceutical Excipient*, 5th ed. Pharmaceutical Press, London, pp. 217–221.

3. Loftsson, T., Brewster, M.E. (1996). Pharmaceutical applications of cyclodextrins: 1. Drug solubilization and stabilization. *J. Pharma. Sci.*, *85*(10), 1017–1025.

4. Hedges, A. (2009). *Starch: Chemistry and Technology*, 3rd ed. Elsevier, New York, pp. 844–851.

5. Stella, V.J., Rajewski, R.A. (1997). Cyclodextrins: their future in drug formulation and delivery. *Pharm. Res.*, *14*(5), 556–567.

6. Çelebi, N. (1988). Cyclodextrins: I. Properties, preparation methods, and inclusion compounds. *FABAD J. Pharm. Sci.*, *12*, 5–15.

7. Singh, M., Sharma, R., Banerjee, U.C. (2002). Biotechnological applications of cyclodextrins. *Biotechnol. Adv.*, *20*, 341–359.

8. Rogers, K. (1999). Controlled release technology and delivery systems. *Cosmet. Toiletries*, *114*(5), 53–60.

9. Uekama, K., Adachi, H., Irie, T., Yano, T., Saita, M., Noda, K. (1992). Improved transdermal delivery of prostaglandin E1 through hairless mouse skin: combined use of carboxymethylethyl-β-cyclodextrin and penetration enhancers. *J. Pharm. Pharmacol.*, *44*(2), 119–121.

10. Felton, A.L., Wiley, C.J., Godwin, D.A. (2002). Influence of hydroxypropyl-β-cyclodextrin on the transdermal permeation and skin accumulation of oxybenzone. *Drug Dev. Ind. Pharm.*, *28*(9), 1117–1124.

11. Preis, A., Mehnert, W., Frömming, K.H. (1995). Penetration of hydrocortisone into excied human skin under the influence of cyclodextrins. *Pharmazie*, *50*, 121–126.

12. Loftsson, T., Sigurdardóttir, A.M., Ólafsson, J.H. (1995). Improved acitretin delivery through hairless mouse skin by cyclodextrin complexation. *Int. J. Pharm.*, *115*, 255–258.

13. Rougier, A., Lotte, C., Maibach, H.I. (1991). *Percutaneous Absorbtion: Drug–Cosmetics–Mechanism–Methodology*, Marcel Dekker, New York, pp. 117–132.

14. Loftsson, T., Másson, M. (2001). Cyclodextrins in topical drug formulations theory and practice. *Int. J. Pharm.*, *225*, 15–30.

15. Tanaka, M., Iwata, Y., Kauzuki, Y., Taniguchi, K., Matsuda, H., Arima, H., Tsuchwja, S. (1995). Effect of 2-hydroxypropyl-β-cyclodextrin on percutaneous absorbtion of methyl paraben. *J. Pharm. Pharmacol*, *47*, 897–900.

16. Uekama, K., Irie, T., Sunada, M., Otagiri, M., Arimatsu, Y., Nomura, S. (1982). Alleviation of prochlorperozine-induced primary irritation of skin by cyclodextrin complexation. *Chem. Pharm. Bull.*, *30*, 3860–3862.

17. Morganti, P., Ruocco, E., Wolf, R., Ruocco, V. (2001). Percutaneous absorbtion and delivery systems. *Clin. Dermatol.*, *19*, 489–501.

18. Loftsson, T. (2000). Cyclodextrins in skin delivery. *Cosmet Toiletries*, *115*(10), 59–66.

19. Loftsson, T., Ólafsson, J.H. (1998). Cyclodextrins: new drug delivery systems in dermatology. *Int. J. Dermatol.*, *37*, 241–246.

20. Cooper, E.R. (1984). Increased skin permeability for lipophilia molecules. *Journal of Pharm. Sci.*, *73*, 1153–1156.

21. Loftsson, T., Petersen, D.S., Le Goffie, F., Ólafsson, J.H. (1997). Unsaturated glycerol monoethers as novel skin penetration enhancers. *Pharmazie*, *52*, 563–565.

22. Challa, R., Ahuja, A., Ali, J., Khar, R. (2005). Cyclodextrins in drug delivery: an updated review. *AAPS PharmSciTech*, *6*(2), 329–357.

23. Minami, K., Hirayama, F., Uekama, K. (1998). Colon-specific drug delivery based on a cyclodextrin prodrug: release behavior of biphenylylacetic acid from its cyclodextrin conjugates in rat intestinal tracts after oral administration. *J. Pharm. Sci.*, *87*, 715–720.

24. Lopez, M.E.V., Reyes, L.N., Igea, S.A., Espinar, F.J.O., Mendez, J.B. (1999). Formulation of triamcinolone acetonide pellets suitable for coating and colon targeting. *Int. J. Pharm.*, *79*, 229–235.

25. Irie, T., Uekama, K. (1999). Cyclodextrins in peptide and protein delivery. *Adv. Drug Deliv. Rev.*, *36*, 101–123.

26. Redenti, E., Pietra, C., Gerlozy, A., Szente, L. (2001). Cyclodextrins in oligonucleotide delivery. *Adv. Drug Deliv. Reviews*, *53*, 235–244.

27. Tarimci, N., Çelebi N. (1988). Studies on cyclodextrin polymer: I. The effect of CDP on indomethacin tablet formulation. *Pharmazie*, *43*(5), 323–325.

28. Szejtli, J. (2004). Past, present and future of cyclodextrin research. *Pure Appl. Chem.*, *76*(10), 1825–1845.

29. Gazzaniga, A., Sangalli, M.E., Bruni, G., Zema, L., Vecchio, C., Giordano, F. (1998). The use of β-cyclodextrin as a pelletization agent in the extrusion/spheronization process. *Drug Dev. Ind. Pharm.*, *24*, 869–873.

30. Çelebi, N., Tarimci, N., Iscanoglu, M., Doganay, T. (1994). Studies on direct compression of spironolactone tablets using cyclodextrin polymer. *Pharmazie*, *49*, 748–750.

31. Dhanaraju, M.D., Senthil Kumaran, K., Baskaran, T., Moorthy, M.S.R. (1998). Enhancement of bioavailability of griseofulvin by its complexation with β-cyclodextrin. *Drug Dev. Ind. Pharm.*, *24*, 583–587.

32. Mura, P., Faucci, M.T., Parrini, P.L., Furlanetto, S., Pinzauti, S. (1999). Influence of the preparation method on the physico-

chemical properties of ketoprofen cyclodextrin binary systems. *Int. J. Pharm.*, *179*, 117–128.

33. Uekama, K., Fujinaga, T., Hirayama, F. (1983). Improvement of the oral bioavailability of digitalis glycosides by cyclodextrin complexation. *J. Pharm. Sci.*, *72*, 1338–1341.

34. Horiuchi, Y., Hirayama, F., Uekama, K. (1990). Slow-release characteristics of diltiazem from ethylated β-cyclodextrin complexes. *J. Pharm. Sci.*, *79*, 128–132.

35. Shinoda, T., Katagani, S., Maeda, A., Konno, Y., Hashimoto, H., Hara, K., Fujita, K., Sonobe, T. (1999). Sugar-branched-cyclodextrins as injectable drug carriers in mice. *Drug Dev. Ind. Pharm.*, *25*(11), 1185–1192.

36. Dordunoo, K., Burt, H.M. (1996). Solubility and stability of taxol: effects of buffers and cyclodextrins. *Int. J. Pharm.*, *133*, 191–201.

37. Davies, N.M., Wang, G., Tucker, I.G. (1997). Evaluation of a hydrocortisone/hydroxypropyl-β-cyclodextrin solution for ocular drug delivery. *Int. J. Pharm.*, *156*, 201–209.

38. Aktas, Y., Ünlu, N., Orhan, M., Irkeç, M., Hincal, A.A. (2003). Influence of hydroxypropyl β-cyclodextrin on the corneal permeation of pilocarpine. *Drug Dev. Ind. Pharm.*, *29*, 223–230.

39. Uekama, K., Kondo, T., Nakamura, K. (1995). Modification of rectal absorption of morphine from hollow-type suppositories with a combination of α-cyclodextrin and viscosity-enhancing polysaccharide. *J. Pharm. Sci.*, *84*, 15–20.

40. Gudmundsdóttir, H., Sigurjónsdóttir, J.F., Másson, M., Fjalldal, O., Stefansson, E., Loftsson, T. (2001). Intranasal administration of midazolam in a cyclodextrin based formulation: bioavailability and clinical evaluation in humans. *Pharmazie*, *56*, 963–966.

41. Zhang, Y., Jiang, X.G., Yao, J. (2001). Nasal absorption enhancement of insulin by sodium deoxycholate in combination with cyclodextrins. *Acta Pharm. Sin.*, *22*, 1051–1056.

42. Tarimci, N., Kilinç, T., Ermiş, D. (1996). The effect of 2-hydroxypropyl β-cyclodextrin on release of naproxen from pluronic F-127 gel. *Proceedings of the 8th International Cyclodextrin Symposium*, Budapest, Hungary, Mar. 30–Apr 2 1996, pp. 453–456.

43. Lopez, R.F., Collett, J.H., Bently, M.V. (2000). Influence of cyclodextrin complexation on the in vitro permeation and skin metabolism of dexamethasone. *Int. J. Pharm.*, *200*, 127–132.

44. Kal, K., Centkowska, K. (2008). Use of cyclodextrins in topical formulations: practical aspects. *Eur. J. Pharm. Biopharm.*, *68*, 467–478.

45. Matsuda, H., Arima, H. (1999). Cyclodextrins in transdermal and rectal delivery. *Adv. Drug Deliv. Rev.*, *36*, 81–99.

46. Frömming, K.H., Szejtli, J. (1994). Cyclodextrins in various drug formulations. In: *Cyclodextrins in Pharmacy*. Kluwer, Dordrecht, The Netherlands, pp. 151–193.

47. Motwani, M., Zatz, J.L. (1997). Applications of cyclodextrins in skin products. *Cosmet. Toiletries*, *112*(7), 39–47.

48. Uekama, K., Hirayama, F., Irie, T. (1998). Cyclodextrins drug carrier systems. *Chem. Rev.*, *98*(5), 2045–2076.

49. Anadolu, R.Y., Şen, T., Tarimci, N., Birol, A., Erdem, C. (2004). Improved efficacy and tolerability of retinoic acid in acne vulgaris: a new topical formulation with cyclodextrin complex. *J. Eur. Acad. Dermatol. Venereol.*, *18*(4), 416–421.

50. Yap, K.L., Liu, X., Thenmozhiyal, J.C., Ho, P.C. (2005). Characterization of the 13-*cis*-retinoic acid/cyclodextrin inclusion complexes by phase solubility, photostability, physicochemical and computational analysis. *Eur. J. Pharm. Sci.*, *25*, 49–56.

51. Montasier, P., Duchêne, D., Poelman, M.C. (1997). Inclusion complexes of tretinoin with cyclodextrin. *Int. J. Pharm.*, *153*, 199–209.

52. Ventura, C.A., Tommasini, S., Falcone, A., Giannone, I., Paolino, D., Sdrafkakis, V., Mondello, M.R., Puglisi G. (2006). Influence of modified cyclodextrins on solubility and percutaneous absorption of celecoxib through human skin. *Int. J. Pharm.*, *314*, 37–45.

53. Umemura, M. (1990). Effect of diethyl-β-cyclodextrin on the release of nitroglicerin from formulations. *Drug Des. Deliv.*, *6*(4), 297–310.

54. Otagiri, M., Fujinaga, T., Sakai, A., Uekama, K. (1984). Effects of β- and γ-cyclodextrins on release of betamethasone from ointment bases. *Chem. Pharm. Bull.*, *32*, 2401–2405.

55. Uekama, K., Otagiri, M., Sakai, A., Irie, T., Matsuo N., Matsuoka, Y. (1985). Improvement in the percutaneous absorption of beclomethasone dipropionate by γ-cyclodextrin complexation. *J. Pharm. Pharm.*, *37*, 532–535.

56. Loftsson, T., Fridriksdóttir, H., Ingvarsdóttir, G., Jonsdóttir, B., Sigurdardóttir, A.M. (1994). The influence of 2-hydroxypropyl-β-cyclodextrin on diffusion rates and transdermal delivery of hydrocortisone. *Drug Dev. Ind. Pharm.*, *20*, 1699–1708.

57. Arima, H., Adachi, H., Irie, T., Uekama, K. (1990). Improved drug delivery through the skin by hydrophilic β-cyclodextrins: enhancement of anti-inflammatory effect of 4-biphenylylacetic acid in rats. *Drug Invest.*, *2*, 155–161.

58. Lin, S.Z., Wouessidjewe, D., Poelman, M.C., Duchene, D. (1994). In-vivo evaluation of indomethacin/cyclodextrin complexes gastrointestinal tolerance and dermal anti-inflammatory activity. *Int. J. Pharm.*, *106*, 63–67.

59. Tarimci, N. (2004). Modern kozmetik taş iyici sistemler. In: Yazan, Y., Ed. *Kozmetik Bilimi*, Nobel Tip Kitabevleri, Istanbul, Turkey, pp. 253–274.

60. Amann, M., Dressnandt, G. (1993). Solving problems with cyclodextrins in cosmetics. *Cosmet. Toiletries*, *108*(11), 90–95.

61. Citernesi, U., Sciacchitano, M. (1995). Cyclodextrins in functional dermocosmetics. *Cosmet. Toiletries*, *110*(3), 53–61.

62. Buschmann, H.J., Schollmeyer, E. (2002). Applications of cyclodextrins in cosmetic products: a review. *Int. J. Cosmet. Sci.*, *53*, 185–191.

63. Duchêne, D., Wouessidjewe, D. (1991). Dermal use of cyclodextrins and derivatives. In: Duchêne, M., Ed., *New Trends in Cyclodextrins and Derivatives*. Editions de Santé, Paris, pp. 449–481.

64. Del Vale, E.M.M. (2004). Cyclodextrins and their uses: a review. *Process Biochem.*, *39*, 1033–1046.

65. Duchêne, D., Wouessidjewe, D., Poelman, M.C. (1999). Cyclodextrins in cosmetics. In: *Novel Cosmetic Delivery Systems*, Marcel Dekker, New York, pp. 275–293.

66. Cho, H., Torii, M., Kanamori, T.(Lion Corp.). (1987). Deodorant controlling mouth odor. Japanese patent 63,264,516 (88,264,516), Apr. 20.

67. Trinh, J., Dodd, T.M., Bartolo, R., Lucas, J.M. (1999). Cyclodextrin based compositions for reducing body odour. U.S. patent 5,897,855.

68. Maekawa, A.(Sunstar Inc.) (1989). Antiperspirant aerosol compositions containing cyclodextrin. Japanese patent 03,170,415 (91,170,415), Nov. 30.

69. Yamagata, Y., Yoshibumi, S.(Lion Corp.) (1986). Hair preparations containing cationic surfactants and cyclodextrins. Japanese patent 62,267,220 (87,267,220), May 15.

70. Matsuda, H., Ito, K.(Shiseido Co. Ltd.) (1991). Body deodorants containing hydroxyalkylated cyclodextrins, Japanese patent 03,284,616 (91,284,616), Dec. 16.

71. Woo, R.A.M., Trinh, T., Cobb, D.S., Schneiderman, E., Wolff, A.M., Rosenbalm, E.L., Ward, T.E., Chung, A.H., Reece, S. (1998). Uncomplexed cyclodextrin compositions for odour control. U.S. patent 5,942,217.

72. Saeki, T., Morifuji, T.(Sekisui Plastics, Tanaka Narikazu, Japan) (1996). Moisturizer-containing cyclodextrin composite particles for cosmetics. Japanese patent 08,151,317 (96,151,317), June 11.

73. Matsuda, H., Ito, K.(Shiseido Co. Ltd.) (1991). Cosmetics packs containing hydroxyalkylated cyclodextrin. Japanese patent 03,287,512 (91,287,512), March 31.

74. Shimada, K., Ohe, Y., Ohguni, T., Kawano, K., Ishii, J., Nakamura, T. (1991). Emulsifying properties of α-, β-, γ-cyclodextrins. *J. Jpn. Soc. Food Sci. Technol.*, *38*, 16.

75. Laurent, S., Serpelloni, M., Pioch, D. (1999). A study of β-cyclodextrin-stabilized paraffin oil/water emulsions. *Int. J. Cosmet. Sci.*, *50*, 15–22.

76. Imamura, K., Tsuchama, Y., Tsunakawa, H., Okamura, K., Okamoto, R., Harada, K.(Merushan Kk. Kogyo Gijutsuib) (1991). Cosmetics containing water-soluble cyclodextrin polymers as scrubbing particles. Japanese patent 05,105, 619 (93,105,619), Feb. 14.

77. Grove, C., Liebenberg, W., Du Perez, J.L., Yang, W., Devilliers, M.M. (2003). Improving the aqueous solubility of triclosan by solubilization, complexation and in situ salt formation. *Int. J. Cosmet. Sci.*, *54*, 537–550.

78. Loftsson, T., Leeves, N., Bjornsdóttir, B., Duffy, L., Mason, M. (1999). Effect of cyclodextrins and polymers on triclosan availability and substantivity in toothpastes in vivo. *J. Pharm. Sci.*, *88*(12), 1254–1258.

79. Duan, M.S., Zhao, N., Ossurardóttir, .I.B., Thorsteinsson, T., Loftsson, T. (2005). Cyclodextrin solubilization of the antibacterial agents triclosan and triclocarban: formation of aggregates and higher-order complexes. *Int. J. Pharm.*, *297*, 213–222.

80. Navarro, R., Delaunois, M. (1995). Minoxidil-based hair care composition. Fabre Pierre Dermo-Cosmétique, Pattent WO 95 25,500.

81. Lehner, S.J., Müller, B.W., Seydel, J.K. (1993). Interactions between *p*-hydroxybenzoic acid esters and hydroxypropyl-β-cyclodextrin and their antimicrobial effect against *Candida albicans. Int. J. Pharm.*, *93*, 201–208.

82. Loftsson, T., Stefansdóttir, O., Frioriksdóttir, H., Guomundsson, O. (1992). Interactions between preservatives and 2-hydroxypropyl-β-cyclodextrin. *Drug Dev. Ind. Pharm.*, *18*(13), 1477–1484.

83. Tanaka, M., Iwata, Y., Kouzuki, Y. (1995). Effect of 2-hydroxypropyl-β-cyclodextrin on percutaneous absorption of methyl paraben. *J. Pharm. Pharm.*, *47*, 897–900.

84. Şen, T., Tarimci, N. (2000). Improvement of retinoic acid solubility by means of cyclodextrins, *Proceedings of the 3rd World Meeting APV/APGI*, Berlin, Apr. 3–6, pp. 947–948.

85. Tarimci, N., Şen, T. (2000). Optimization of gel formulations of retinoic acid-β-cyclodextrin complexation. Presented at the 6th European Congress of Pharmaceutical Sciences, Budapest, Hungary, Sept. 16–19, 2000, PO-82. *Eur. J. Pharm. Sci.*, *11*(Suppl. 1), 55.

86. Qu, Q., Tucker, E., Christian, S.D. (2003). Solubilization of synthetic perfumes by nonionic surfactants and by sulfoalkyl ether β-CDs. *J. Inclusion Phenom. Macrocyclic Chem.*, *45*(1–2), 83–88.

87. Lucas-Abellan, C., Fortea, I., Lopez-Nicolas, M., Nunez-Delicado, E. (2007). Cyclodextrins as resveratrol carrier system. *Food Chem.*, *104*, 39–44.

88. Tsomi, V. (1991). Cosmetic skin-lightening composition based on a hydroquinone/2,6-dimethyl-[3-cyclodextrin complex. Patent WO 9l/18589.

89. Hatae, S., Nakashima, K.(Sansho Seiyaku KK.) (1987). Whitening cosmetic. Patent EP 0241572.

90. Marlies, R. (2007). Oxidation-stable linoleic acid by inclusion in α-cyclodextrin. *J. Inclusion Phenom. Macrocyclic Chem.*, *57*, 471–474.

91. Lin, H.S., Chean, C.S., Ng, Y.Y., Chan, S.Y., Ho, P.C. (2001). 2-Hydroxypropyl-β-cyclodextrin increases aqueous solubility and photostability of all-*trans*-retinoic acid. *J. Clin. Pharm. Ther.*, *25*(4), 265–269.

92. Citernesi, U. (2001). Photostability of sun filters complexed in phospholipids or β-cyclodextrin. *Cosmet. Toiletries*, *116*(9), 77–86.

93. Scalia, S., Villani, S., Casolari, A. (1999). Inclusion complexation of the sunscreen agent 2-ethylhexyl-*p*-dimethyl-aminobenzoate with hydroxypropyl-β-cyclodextrin: effect on photostability. *J. Pharm. Pharm.*, *51*, 1367–1374.

94. Scalia, S., Villani, S., Scatturin, A., Vandelli, M.A., Forni, F. (1998). Complexation of the sunscreen agent, butyl-methoxy dibenzoylmethane, with hydroxypropyl-β-cyclodextrin. *Int. J. Pharm.*, *175*, 205–213.

95. Scalia, S., Casolari, A., Iaconinoto, A., Simeoni, S. (2002). Comperative studies of the influence of cyclodextrins on the

stability of the sunscreen agent, 2-ethylhexyl-*p*-methoxycin-namate. *J. Pharm. Biomed. Anal.*, *30*, 1181–1189.

96. Scalia, S., Molinari, A., Casolari, A., Maldotti, A. (2004). Complexation of the sunscreen agent, phenylbenzimidazole sulphonic acid with cyclodextrins: effect on stability and photo-induced free radical formation. *Eur. J. Pharm. Sci.*, *22*, 241–249.

97. Coelho, G.L.N. (2008). Preparation and evaluation of inclusion complexes of commercial sunscreens in cyclodextrins and montmorillonites: performance and substantivity studies. *Drug Dev. Ind. Pharm.*, *34*(5), 536–546.

98. Reuscher, H., Bauer, M.A.(Wacker Chemie AG.) (2004). Process for preparing an α-lipoic acid/cyclodextrin complex and product prepared. Patent EP1707217 A1.

99. Matsuda, H., Ito, K., Fujiwara, Y., Tanaka, M., Taki, A., Oejima, O., Sumiyoshi, H. (1991). Complexation of various fragrence materials with 2-hydroxypropyl-β-cyclodextrin. *Chem. Pharm. Bull.*, *39*(4), 827–830.

100. Szejtli, J., Szente, L., Kulcsár, G. (1986). β-Cyclodextrin complexes in talc powder compositions. *Cosmet. Toiletries*, *101*(10), 74–79.

101. Numanoglu, U., Şen, T., Tarimci, N., Kartal, M., Koo, O.M., Onyuksel, H. (2007). Use of cyclodextrins as a cosmetic delivery system for fragrance materials: linalool and benzyl acetate. *AAPS PharmSciTech*, *8*(4), 34–42.

102. Kubo, S., Nakamura, F. (1985). Waving lotion for cold waving. U.S. Patent 45,48,811.

103. Proniuk, S., Liederer, B., Dixon, S.E., Rein, J., Kalen, A.M., Blanchard, J. (2002). Topical formulation studies with DEET (*N,N*-diethyl-3-metylbenzamide) and cyclodextrins. *J. Pharm. Sci.*, *91*, 101–110.

104. Meyers, A., Lutrario, C.A., Elliott, M., Gallagher, L.A. (2002). Breath freshening lipstick. U.S. Patent 6,383,475, May 7.

105. Ito, K., Nagai, I.(Shiseido Co. Ltd.) (1993). Cosmetic containing mentol derivatives, JP 06,329,528 [94,329,528].

106. Holland, L., Rizzi, G., Malton, P. (1999). Cosmetic compositions comprising cyclic oligosaccharides and fragrance. PCT International Application WO 67,716.

107. Prasad, N., Straus, D., Reichart, G. (1999). Cyclodextrin flavor delivery systems. U.S. Patent 6,287,603, Sept. 16.

108. Tanaka, M., Matsuda, H., Sumiyoshi, H. (1996). 2-Hydroxypropylated cyclodextrins as a sustained-release carrier for fragrance materials. *Chem. Pharm. Bull.*, *44*(2), 416–420.

109. Citernesi, U. (1996). Controlled release inclusion system of glycolic acid in β-cyclodextrin and process for the above system preparation. Patent EP 19960109132.

110. Buschmann, H.J., Schollmeyer, E. (2004). Cosmetic textiles: a new functionality of clothes. *Cosmet. Toiletries*, *11*(5), 105–112.

8

CYCLODEXTRIN-ENHANCED DRUG DELIVERY THROUGH MUCOUS MEMBRANES

PHATSAWEE JANSOOK

Faculty of Pharmaceutical Sciences, University of Iceland, Reykjavik, Iceland

MARCUS E. BREWSTER

Chemical and Pharmaceutical Development, Johnson & Johnson Pharmaceutical Research and Development, Janssen Pharmaceutica, Beerse, Belgium

THORSTEINN LOFTSSON

Faculty of Pharmaceutical Sciences, University of Iceland, Reykjavik, Iceland

1. INTRODUCTION

Over the past two decades, drug delivery to and through mucous membranes (mucosae) has gained significant attention, both in conventional oral drug delivery and in alternative delivery routes such as buccal, nasal, and pulmonary administration that allow for rapid uptake of drugs into the systemic circulation, avoiding first-pass liver metabolism and thus circumventing some of the body's natural defense mechanisms [1,2]. The main permeation mechanisms of drug uptake through the mucosa include passive diffusion, carrier-mediated diffusion, active transport, and pinocytosis or endocytosis. Recent evidence suggests that passive diffusion is the primary mechanism for drug transportation across the mucosa. There are two routes of passive transport: transcellular (intracellular) and paracellular (intercellular) translocation [3,4], wherein the transcellular route most often dominates the paracellular mechanism [5].

Mucus, the secreted viscous aqueous fluid associated with the external layers of mucous membranes (i.e., epithelial cell layers), coats various tissues, such as the eye surface, lung airway, intestinal tract, nasal cavity, and vaginal tissue. In the context of drug delivery through mucous membranes, the mucus layer is one of the primary barriers to penetration that the therapeutic agent must overcome [6]. The driving force associated with passive drug permeation through mucus is the drug concentration gradient across the mucus layer. The rapid turnover of mucus limits both drug permeation through this aqueous layer and drug partitioning from the aqueous mucus into the lipophilic epithelium. Nanoparticulate systems have been found to cross the mucosal barrier [7]. Furthermore, nanoscale systems can provide controlled drug delivery and efficient vaccine or gene delivery to mucosal tissues [8]. In some cases, nanoparticulate drug delivery systems are able to penetrate the aqueous mucosa at a faster rate than can the individual drug molecules [9].

The presence of the mucus layer may also provide an opportunity for sustained or prolonged drug delivery through the application of mucoadhesive dosage forms [10]. For drug delivery purposes, mucoadhesion (i.e., bioadhesion) may be defined as the attachment of a synthetic or natural macromolecule to mucus or mucous membranes. The presence of hydroxy, carboxyl, or amine groups on the macromolecule favors polymer–mucus interaction manifested mainly via hydrogen bonds and hydrophobic interactions, although disulfide covalent bonds are also known to participate in mucoadhesion [1,11].

Frequently, drug permeation through mucosa is too slow or incomplete, such that therapeutic drug plasma levels may not be acheived. One possible approach to overcoming the

Cyclodextrins in Pharmaceutics, Cosmetics, and Biomedicine: Current and Future Industrial Applications, First Edition. Edited by Erem Bilensoy.
© 2011 John Wiley & Sons, Inc. Published 2011 by John Wiley & Sons, Inc.

mucosal barrier is through the use of substances such as cyclodextrins (CDs) that enhance drug permeation through these ubiquitous barriers. The use of CDs as mucosal absorption enhancers is attracting growing attention among pharmaceutical scientists, owing to their unique characteristics, such as their ability to form drug complexes, to dissociate and deaggregate protein oligomers and polymers, their high tissue compatibility, and their wide commercial availability. CDs are oligosaccharides formed by α(1,4)-linked α-D-glucopyranose units, with a hydrophilic outer surface and a lipophilic central cavity [12,13]. It has been recognized that CDs act as true carriers by keeping the hydrophobic drug molecules in solution and deliver them through the aqueous mucus layer to the surface of lipophilic tissue barrier, where they can partition into the barrier [14,15]. Thus, CDs enhance drug permeation by increasing drug availability at the surface of the lipophilic epithelium. Furthermore, drug–CD complexes can form nanoscale aggregates in aqueous solutions. The diameter of the aggregate appears to increase gradually with increasing CD concentration. For example, the reported diameter of hydrocortisone–CD was found to be between 10 and 100 nm [16,17]. It is possible that the formation of such CD nanoparticulates is partly responsible for the ability of CD to enhance drug delivery through mucus.

In this chapter we review the effect of CDs on drug permeation through mucous membranes and discuss the possible mechanism of action as well as the role of CDs in various drug delivery routes. We intended to give a few examples of the use of CDs in this context, and the reader is referred to other chapters in the book for more detailed descriptions of CDs and their use as pharmaceutical excipients.

2. MUCOUS MEMBRANES

Mucous membranes consist of an inner connective tissue layer (the lamina propia) and an outer epithelial layer that is most often covered by an external mucus layer. The epithelia may consist of a single cell layer (e.g., stomach, small and large intestine, bronchi) or a multilayered/stratified cell layer (e.g., esophagus, vagina, cornea). Single-cell-layer systems contain globet cells, which secrete mucus directly onto the epithelial surfaces, while the multilayer systems contain, or are adjacent to tissues containing, specialized organs such as salivary glands that secrete mucus to protect their epithelia. Mucus is present as either an aqueous gel layer attached to the mucosal surface or as an aqueous luminal soluble or suspended form [1].

The main component of mucus is water (90 to 95 wt%). The remaining mass comprises glycoprotein fibers, oligosaccharides, lipids, migrating or sloughed cell and cell contents, enzymes, antibodies, DNA, and electrolytes. The thickness of the mucus layer depends on its location, varying from 50 to 450 μm in stomach to less than 1 μm in the oral cavity [1,6]. Mucins, highly glycosylated large proteins (10 to 40 MDa) secreted by epithelial cells, represent the principal component of the viscoelastic gel that protects the underlying epithelia from pathogens and toxins [18,19]. Mucus forms an aqueous diffusion barrier, while the cell-based epithelia form a lipophilic membrane barrier. An example of a biological membrane is the cornea, which consists of a lipophilic membrane (epithelium) with an exterior aqueous layer (tear film). The hydrophilic stroma is located below the epithelial layer (Fig. 1). The tear film

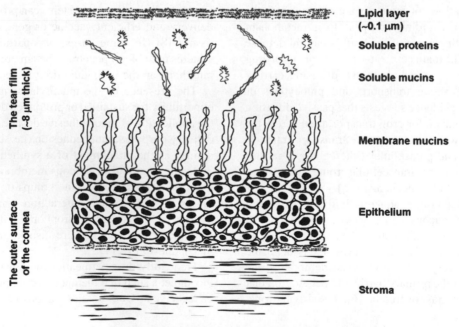

Figure 1. Tear film on the corneal surface of the eye. (Modified from [20].)

consists of three layers. The outermost layer is a lipid layer that retards water evaporation from the eye surface. The central aqueous layer contains mainly water and small amounts of other substances, such as proteins. The innermost layer is the mucous component, a gel-like fluid containing mainly water (~95%) and mucin [20].

Most biological barriers (or biomembranes) are lipophilic, and when they come in contact with an aqueous environment, a stagnant water layer forms at the membrane surface; this is frequently referred to as the *unstirred water layer* (UWL). Despite mechanical shearing actions of the blinking eye, swallowing, coughing, intestinal peristalsis, and copulation, the dynamic viscoelastic properties of mucus is maintained as an unstirred layer of the aqueous mucus adjacent to epithelial surfaces [8,21]. If drug permeation through the UWL is the rate-limiting step of drug permeation through the barrier, CDs can frequently enhance the permeation. However, hydrophilic CDs are in most cases unable to enhance drug permeation through a lipophilic membrane barrier, and excess CD (i.e., more than is needed to dissolve the drug) will hamper drug permeation through the membrane. Too high or too low amounts of CD in formulations can result in less than optimum drug bioavailability [22].

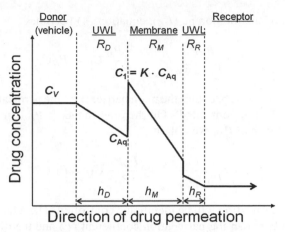

Figure 2. Drug permeation from a donor phase through UWL on the donor side, through a membrane, and finally through UWL on the receptor side. C_V, drug concentration in the donor (vehicle); C_{Aq}, drug concentration in the UWL immediate to the membrane surface; C_1, drug concentration within the membrane at the donor side; K, the drug partition coefficient between UWL and the membrane; h_D, thickness of the UWL on the donor side; h_M, thickness of the membrane; h_R, thickness of the UWL on the receptor side; R_D, R_M, and R_R, resistances in the UWL at the donor side, within membrane and in the UWL at the receptor side, respectively. (Modified from [23].)

3. MATHEMATICAL MODEL

Drug molecules must permeate the UWL to reach the lipophilic membrane [20,23]. Although many biological membranes contain specialized transport systems that assist certain compounds to pass the membranes, most drugs permeate these membranes through passive diffusion, either via transcellular or paracellular pathways [4,24]. The fundamental equation describing passive drug transport through membranes is based on Fick's first law:

$$J = P_T C_V \qquad (1)$$

where J is the flux of the compound through the membrane, P_T the overall permeability coefficient, and C_V the concentration of the compound in the vehicle (i.e., the donor phase).

The UWL adjacent to the lipophilic membrane can act as an aqueous diffusion barrier for rapidly permeating drug. The thickness of this diffusion barrier and its significance in the overall barrier function depends on the physicochemical properties of both the membrane and the permeating drug molecules [4]. For example, the aqueous diffusion barrier on the surface of the skin is usually quite thin, and thus its contribution to the overall permeation barrier of that organ is usually insignificant. On the other hand, the aqueous mucin layer on the surface of the eye and along the gastrointestinal tract can contribute significantly to the overall

barrier function. Assuming independent and additive total resistance, R_T of a simple membrane (Fig. 2) can be defined as

$$R_T = R_D + R_M + R_R \qquad (2)$$

where R_D, R_M, and R_R are the resistances of the UWL on the donor side, within the membrane and of the UWL on the receptor side, respectively. The flux (J) of a drug through the membrane can be described by the equation

$$J = P_T C_V = (R_D + R_M + R_R)^{-1} C_V$$
$$= \left(\frac{1}{P_D} + \frac{1}{P_M} + \frac{1}{P_R} \right)^{-1} C_V \qquad (3)$$

where P_D, P_M, and P_R are the corresponding permeability coefficients, respectively. If R_R is assumed to be negligible due to sink conditions (i.e., relatively rapid removal of drug molecules from the receptor side of the membrane), we obtain

$$J = \frac{P_D P_M}{P_D + P_M} C_V \qquad (4)$$

If the value of the permeability coefficient through the lipophilic membrane (P_M) is much greater than the value of the permeation coefficient through the UWL on the donor side of the membrane (P_D), equation (4) becomes

$$J = \frac{P_D P_M}{P_D + P_M} C_V \approx \frac{P_D P_M}{P_M} C_V = P_D C_V \tag{5}$$

and the UWL becomes the main barrier (i.e., the permeation is diffusion controlled). On the other hand, if P_D is much greater than P_M, we obtain

$$J = \frac{P_D P_M}{P_D + P_M} C_V \approx \frac{P_D P_M}{P_D} C_V = P_M C_V \tag{6}$$

and permeation will be membrane controlled. The relationship between the permeation coefficient (P) and the diffusion coefficient (D) is given by

$$P = \frac{DK}{h} \tag{7}$$

where h is the thickness of the layer (h_D or h_M in Fig. 2) and K is the partition coefficient between the aqueous phase and the membrane. The value of K is unity for the UWLs.

Finally, D can be estimated from the Strokes–Einstein equation:

$$D \approx \frac{RT}{6\pi\eta r N} \tag{8}$$

where R is the molar gas constant, T the absolute temperature, η the apparent viscosity within the UWL or lipophilic membrane, r the radius of the permeating drug molecule, and N is Avogadrós number. Thus, the diffusion constant within UWL will decrease with the increasing viscosity of the layer (i.e., increasing η value) as well as the increasing molecular weight of the drug (i.e., increasing r value). The presence of mucin in the mucus layer increases not only the thickness (h) of this UWL but also its viscosity (η), both of which can increase its resistance (R_D) and consequently decrease the permeability (P_D) [equations (7) and (8)]. The thickness of the UWL can be significant (Table 1), and mucus and other surface structures can enhance its barrier function by increasing its viscosity, leading to an overall decrease in the diffusion coefficient [D in equation (8)]. Studies have shown that drug diffusion through mucus is up to 100 times slower than diffusion through pure water [10].

Table 1. Membrane Properties and the Effect of CDs on Drug Flux Through Membranes[a]

Epithelium (in Human)	Structure	Surface Characteristics	UWL (h_D) (μm)	R_D/R_M Ratio[b]	Effect of Increasing CD Concentration[c]		Refs.
					Aqueous Drug Suspension	Unsaturated Drug Solution	
Eye cornea/ sclera	Collagen and elastic fibers (sclera) or flat epithelium cells with tight junctions in the intercellular space (cornea)	Mucus/tear fluid	~8 (in vivo)	>1	Increasing permeability	Decreasing permeability	[25–29]
Nasal mucosa	Partly ciliated epithelium covered with mucus	Mucus	~50	>1	Increasing permeability	Decreasing permeability	[30]
Intestinal mucosa	Surface gel layer (or mucus) over villi and microvilli	Mucus	30–100 (in vivo)	>1	Increasing permeability	Decreasing permeability	[10,28,31,32]
Buccal mucosa	Saliva over keratinized and nonkeratinized mucosa	Saliva	70–100 (in vivo)	>1	Increasing permeability	Decreasing permeability	[28,33]
Lung mucosa	Surface gel layer over a periciliary layer	Mucus	10–15 (in vivo)	<1	Increasing permeability	No or little effect	[10,28]

Source: Adapted from [23], with permission.

[a] Based on permeation of somewhat lipophilic drugs (MW < 700 Da) that are able to form hydrophilic CD complexes. Lipophilic CDs, such as the methylated β-CDs, are able to penetrate lipophilic membranes and change their barrier properties.

[b] The permeation resistance in the UWL on the donor side (R_D) and within the membrane (R_M) is a function of both the thickness (h_D or h_M) and the viscosity (η) of the barrier, as well as binding of the permeating molecules to, for example, mucin in the UWL.

[c] The effect will depend on the relative resistance of the membrane barrier R_D/R_M.

4. MUCOSAL DRUG DELIVERY

Several reviews of the literature describe the role of CDs in drug delivery when applied to various routes of administration, including oral, buccal, transdermal, rectal, ocular, nasal, and pulmonary administration [3,22,34–37]. Some examples of the use of CDs to enhance drug permeation through mucous membranes are provided in Table 2.

4.1. Ocular Drug Delivery

Aqueous eyedrop solutions (pH 7.4) containing dorzolamide (2.0 or 4.0% w/v) and 7.70 or 18.7% w/v randomly methylated β-cyclodextrin (RM-β-CD) were investigated in rabbits. The formulations were compared to the marketed product, Trusopt (aqueous 2% w/v dorzolamide solution, pH 5.6, Merck, United States). RM-β-CD-containing formulations were well tolerated by the rabbits and no

Table 2. Examples of CD-Containing Formulations for Drug Delivery Through Mucous Membranes

Drug	Cyclodextrin	Method	Brief Results	Refs.
Eye Cornea/Sclera				
Acetazolamide	β-CD, TM-β-CD, DM-β-CD	In vivo (rabbits)	Augments its intensity of action, bioavailability, and prolongs duration of action	[38]
	HP-β-CD	In vivo (rabbits) and human study	Improves corneal bioavailability, leading to effect in lowering the IOP	[39,40]
Dexamethasone	HP-β-CD	In vivo (rabbits)	Improves ocular bioavailability in conjunctiva, cornea, iris, and aqueous humor	[24,41,42]
	RM-β-CD	In vivo (rabbits)	Delivers a significant amount of drug to the rabbit retina	[43]
	HP-β-CD	Human study	Effective *trans*-ocular delivery of dexamethasone into the eye	[44,45]
Pilocarpine	α-CD, HP-β-CD	In vitro (bovine corneas)	Significantly increases permeation of drug by α-CD, but is slightly decreased by HP-β-CD	[46]
	HP-β-CD	In vivo (rabbits)	Significant increase in the miotic response	[47,48]
Nasal Mucosa				
Buserelin	α-CD, β-CD, γ-CD, HP-α-CD, DM-α-CD, CM-α-CD, CM-β-CD, HP-β-CD, DM-β-CD, G_2-β-CD, S-β-CD	In vivo (rats)	DM-β-CD could improve the nasal bioavailability of buserelin (60%) and is well tolerated by the nasal mucosa of the rat. Less effective were those from the others.	[49]
Insulin	α-CD, β-CD, γ-CD HP-β-CD, DM-β-CD DM-β-CD	In situ perfusion (rats)	The CDs are able to dissociate insulin hexamers into smaller aggregates, promoting their transport across the mucosa membranes.	[50]
		In vivo (rabbits)	Insulin/DM-β-CD powder formulation improves insulin absorption, leading to decrease blood glucose concentration.	[51]
Intestinal Mucosa				
Paclitaxel	HP-β-CD	In vitro (rat intestinal mucosa)	HP-β-CD enhances permeation of the drug in the intestinal mucosa with a minor change in membrane fluid.	[52]
Propanolol	β-CD, HP-β-CD, SBE-β-CD	In vitro (rat intestinal epithelium)	β-CDs serve as carriers rather than permeability enhancers.	[53]
Buccal Mucosa				
Carvediol	HP-β-CD	In vitro (porcine buccal mucosa)	Improves dissolution, resulting in an increase of the release rate and the amount of carvedilol permeated through the porcine buccal mucosa	[33]

(*continued*)

Table 2. (*Continued*)

Drug	Cyclodextrin	Method	Brief Results	Refs.
Danazol	HP-β-CD	In vivo (rats)	The absolute bioavailability of danazol from aqueous solution of the complexes is 26%.	[54]
	SBE-β-CD	In vivo (dogs)	Increases the bioavailability due to enhanced solubility by complexation and possible avoidance of first-pass metabolism	[55]
Lung Mucosa				
Insulin	DM-β-CD	In vivo (rats)	CD plays an important role in increasing insulin solubility and sustains the glycemia reduction when the complex is encapsulated in microspheres.	[56]
	DM-β-CD	In vivo (rats)	Increases the bioavailability of insulin by enhancing the mucosal absorption	[57,58]
	DM-β-CD	In vivo (rats)	Effective in enhancing pulmonary insulin absorption and causes increased insulin absorption by acting on the membrane rather than by interacting with insulin.	[59]
Recombinant human growth hormone	DM-β-CD	In vivo (rats)	Enhances absorption more than 2.5-fold higher than that of the dry power containing no absorption enhancer	[60]
Salmon calcitonin	DM-β-CD	In vivo (rats)	Significantly enhances the pulmonary absorption of calcitonin compared to that of calcitonin alone	[61]

α-CD, α-cyclodextrin; β-CD, β-cyclodextrin; γ-CD, γ-cyclodextrin; CM-α-CD, carboxymethyl-α-cyclodextrin; CM-β-CD, carboxymethyl-β-cyclodextrin; DM-α-CD, dimethyl-α-cyclodextrin; DM-β-CD, dimethyl-β-cyclodextrin; G$_2$-β-CD, maltosyl-β-cyclodextrin; HP-α-CD, 2-hydroxypropyl-α-cyclodextrin; HP-β-CD, 2-hydroxypropyl-β-cyclodextrin; RM-β-CD, randomly methylated β-cyclodextrin; SBE-β-CD, sulfobutyl ether-β-cyclodextrin sodium salt; S-β-CD, β-cyclodextrin sulfate; TM-β-CD, trimethyl-β-cyclodextrin.

macroscopic signs of irritation, redness, or other cage side signs and symptoms were observed [62]. The results indicated that after 1 and 2 h, the 4% w/v dorzolamide RM-β-CD solution was superior with regard to drug delivery to the back of the eye (i.e., retina and optic nerve), while Trusopt provided higher drug levels to the front of the eye (i.e., cornea, aqueous humor, iris, corpus ciliare).

In vivo studies of aqueous CD-containing eyedrop formulations of dexamethasone were compared and shown in Table 3. The reference eyedrop formulation was Maxidex (Alcon, United States), which contains 0.1% w/v dexamethasone in an alcoholic suspension [43,63]. The dexamethasone concentrations in aqueous humor were somewhat higher after topical administration of the drug in the RM-β-CD-based vehicle compared to formulations containing either γ-CD or 2-hydroxypropyl-β-cyclodextrin (HP-β-CD) and significantly higher than after administration of Maxidex. Since RM-β-CD is a lipophilic and surface-active excipient, it acts not only as a solubilizer enhancing drug delivery through the aqueous mucin layer on the eye surface

Table 3. Concentration of Dexamethasone in Aqueous Humor 2 h After Administration of Aqueous Dexamethasone Eyedrop Solutions or Suspensions Containing CD in Rabbits

Aqueous Eyedrop Solution[a]	Dexamethasone Concentration[b] (ng/g)	No. of Rabbits	Refs.
0.5% Dexamethasone/RM-β-CD solution	170 ± 76	6	[43]
1.5% Dexamethasone/RM-β-CD solution	576 ± 226	6	[63]
1.3% Dexamethasone/HP-β-CD solution	320 ± 230	4	[65,66]
1.5% Dexamethasone/γ-CD suspension	236 ± 67	8	[63]
Maxidex (0.1% dexamethasone suspension)	66 ± 20	4	[65,66]

Source: Adapted from [63].

[a] RM-β-CD, randomly methylated β-cyclodextrin; HP-β-CD, 2-hydroxypropyl-β-cyclodextrin; γ-CD, γ-cyclodextrin.

[b] Values are expressed as mean ± S.D.

but also as a conventional penetration enhancer, increasing drug permeability through biological membranes by decreasing their barrier function [34,64]. In addition, the results show that formulating the drug as a suspension, where the particles consist not only of the pure drug but also drug–CD inclusion complexes, can increase drug delivery to the posterior segment of the eye. In this instance, the delivery enhancement is over threefold when relevant retinal concentrations are compared.

4.2. Nasal Drug Delivery

Intranasal administration of drugs has received considerable attention because it is a noninvasive route that circumvents the gastrointestinal tract and liver first-pass metabolism. In addition, the route is advantageous due to its ease of administration and because it allows convenient self-medication for materials that may be difficult to dose orally. Furthermore, drugs are usually rapidly absorbed from the nasal cavity due to the relatively large surface area of the cavity and the highly vascularized mucosa [67]. In nasal drug delivery, methylated β-CDs have been found to be effective absorption enhancers with a good safety profile, and clinical studies have demonstrated that they are well tolerated in humans [49,68]. Methylated β-CDs have been reported to substantially increase nasal absorption of calcitonin (molecular weight 3300) in rats [68]. In addition, 2,6-di-*O*-methyl-β-cyclodextrin (DM-β-CD) has been reported to be a potent nasal absorption enhancer for some drugs, such as insulin, 17β-estradiol, and progesterone [51,69–71].

Permeation studies through bovine nasal mucosa of a solution of WIN 51711 (Disoxaril), an anti-picornaviral agent, in the presence of 1.5% w/v DM-β-CD, compared to a suspension of the drug alone, showed that DM-β-CD enhanced the delivery of the drug. It was shown that the lipophilic CD derivatives are able to increase drug permeability through biomembranes by the extraction and complexation of the lipid components of cells (phospholipids and cholesterol) [72]. One cause for increased permeability could be the opening of tight junctions. Alternative mechanisms may include reduction of the ciliary beat frequency, resulting in increased contact time between human growth hormone and mucosa [73,74]. In this way, the lipophilic CDs can act as a conventional penetration enhancer.

Midazolam intranasal (IN) delivery in a vehicle containing sulfobutyl ether β-cyclodextrin sodium salt (SBE-β-CD), was compared to intravenous (IV) administration of Dormicum (midazolam parenteral solution, Roche, Switzerland) [75]. Serum concentration–time profiles in healthy volunteers after IN and IV midazolam administrations are shown in Fig. 3. Midazolam was absorbed rapidly after IN administration reaching maximum plasma concentrations of 54.3 ± 5.0 ng/mL at 15 ± 2 min with a mean elimination

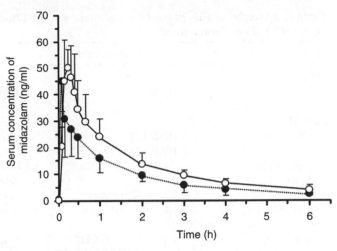

Figure 3. Serum concentration–time profiles in healthy volunteers after IN administration of a 0.06-mg/kg dose (○) or IV administration of a 2-mg fixed dose (●) of midazolam. Each point represents the mean value ($n = 6$) and the error bars represent S.E.M. (Modified from [75].)

half-life of 2.2 ± 0.3 h. The mean absolute bioavailability was $73 \pm 7\%$. It was concluded that midazolam nasal spray was rapidly absorbed from the nasal cavity into the systemic circulation without noticeable nasal irritation. Unlike methylated β-CD derivatives, SBE-β-CD is a very hydrophilic β-CD derivative that is unable to permeate nasal epithelia (i.e., the membrane barrier).

4.3. Oral Drug Delivery

The gastrointestinal mucosa with its thick mucus layer acts as a barrier to the diffusion and/or absorption of various drugs into the blood [76]. The effect of CDs on oral drug absorption can be explained in the context of the Biopharmaceutics Classification System of drugs according to their aqueous solubility characteristics and their ability to permeate the intestinal mucosa (Table 4) [22]. CDs enhance the bioavailability of class II drugs (i.e., compounds with poor aqueous solubility but high intestinal permeability). On the other hand, CDs do not enhance the bioavailability of class III drugs (high solubility, low permeability). Class III drugs are hydrophilic and do not, in general, form high-affinity inclusion complexes with CDs.

Furthermore, CD complexation of class III drugs can reduce their ability to partition from the bulk media or mucous layer into the lipophilic membrane [i.e., the CD complexation will lower their K value in Eq. (7)]. The negligible effect that CDs have on the bioavailability of BCS class III drugs (high solubility, low permeability), and the large effects they have on class II (low solubility, high permeability) and class IV (low solubility, low permeability) drugs, suggest that hydrophilic CDs do not enhance drug

Table 4. Examples of CDs in Oral Formulations and the Effect of CD Complexation on Absolute Bioavailability Compared with Identical CD-Free Formulation[a]

Drug	Cyclodextrin[b]	Formulation	Species	F_{rel}[c]	Refs.
Class I					
Piroxicam	β-CD	Tablet, capsule, and oral suspension	Human, rat, rabbit	≤1.4	[77–80]
Class II					
Albendazole	HP-β-CD	Oral powder	Mice	1.4	[81]
Carbamazepine	HP-β-CD, DM-β-CD	Oral powder and solution, tablet	Rabbit, dog, rat	≤5.6	[82–86]
Clotrimazole	β-CD	Oral suspension	Rat	3.0	[87]
Digoxin	γ-CD	Tablet	Dog	5.4	[88]
Glibenclamide	β-CD, SBE-β-CD	Capsule containing powder	Dog, rat	≤6.2	[89,90]
Miconazole	HP-β-CD, HP-γ-CD	Aqueous suspension, capsule containing powder	Rat, pig	≤3.9	[91,92]
Phenytoin	E-β-CD, Glu-β-CD, Mal-β-CD, SBE-β-CD, HP-β-CD	Suspension, capsule containing powder	Rat, dog	≤5	[96–98]
Spironolactone	β-CD, γ-CD, DM-β-CD, SBE-β-CD, HP-β-CD	Oral solution and powder	Rat, dog	≤3.6	[93–95]
α-Tocopheryl nicotinate	DM-β-CD	Capsule containing powder	Dog	~70	[101]
Tolbutamide	β-CD, HP-β-CD	Suspension, oral powder	Rabbit, dog	≤1.5	[99,100]
Vinpocetine	β-CD, SBE-β-CD	Tablet	Rabbit	≤2.7	[102]
Class III					
Acyclovir	β-CD	Oral suspension	Rat	1.1	[103]
Diphenhydramine HCl	DM-β-CD, HP-β-CD	Solution	Rat	≤0.9	[104]
Class IV					
Cyclosporin A	DM-β-CD	Oral suspension	Rat	4.7	[105]

Source: Adapted from [22].

[a] Tested in vivo in humans and/or animals.

[b] β-CD, β-cyclodextrin; γ-CD, γ-cyclodextrin; DM-β-CD, dimethyl-β-cyclodextrin; E-β-CD, β-cyclodextrin epichlorohydrin polymer; Glu-β-CD, glucosyl-β-cyclodextrin; HP-β-CD, 2-hydroxypropyl-β-cyclodextrin; Mal-β-CD, maltosyl-β-cyclodextrin; SBE-β-CD, sulfobutyl ether-β-cyclodextrin sodium salt.

[c] F_{rel} (relative bioavailability) is the AUC of the plasma concentration versus time profile when the CD-containing formulation was given divided by the AUC for the formulation containing no CD.

bioavailability by reducing the barrier properties of the lipophilic epithelium. Rather, the main effect appears to be an increase in drug solubility and enhanced drug permeation through the aqueous mucus upon formation of water-soluble drug–CD complexes. This is also supported by the observation that drug–CD complexes sometimes result in better drug bioavailability than physical mixtures of drugs and CDs. For example, spironolactone, phenytoin, and tolbutamide are all BCS class II drugs. In vivo studies of these drugs are presented in Fig. 4, wherein drug plasma concentration–time profiles after oral administration to dogs and rabbits (for spironolactone, its major metabolite, canrenone was measured) are illustrated. All three drugs showed a notable increase in the drug plasma concentration–time profiles upon administration of drug–CD complexes [93,96,99]. The bioavailability of the drugs was increased up to fivefold (Table 4). CD enhancement of drug bioavailability allows for a lower dose to be administered and for more consistent plasma levels profiles to be generated.

4.4. Buccal Drug Delivery

Drug administered through the buccal mucosa provides for direct absorption into the systemic circulation, thereby preventing drug degradation within the gastrointestinal tract as well as drug loss due to the first-pass hepatic metabolism [11,106–110]. Han et al. studied the effect of β-CD on nalbuphine enanthate release from Carbopol 934/hydroxypropylcellulose (CP/HPC)-based disks. It demonstrated that the drug release rate increased with increasing amount of β-CD in these disks [110]. A study of RM-β-CD toxicity and membrane damage of the buccal mucosa using 3-[4,4-dimethylthiazol-2-yl]-2,5-diphenyltetrazolium bromide (MTT) assay and lactic acid dehydrogenase (LDH) release showed that it can be a safe excipient for buccal mucosa drug delivery [111].

An in vivo study (dogs) wherein the mean plasma concentration–time profile after buccal dosing of the danazol/SBE-β-CD complex compared with that obtained after oral

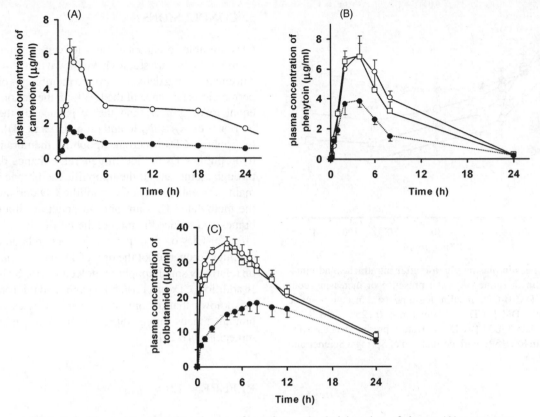

Figure 4. Plasma concentration–time profiles after oral administration of drugs: (A) canrenone following oral administration of spironolactone/HP-β-CD complex (○), spironolactone alone (●) (equivalent to 50 mg of spironolactone) to dogs; (B) phenytoin following oral administration of phenytoin/SBE$_{7m}$-β-CD complex (○), phenytoin/HP-β-CD complex (□), phenytoin alone (●) (equivalent to 300 mg of phenytoin) to dogs; (C) tolbutamide following the oral administration of tolbutamide/β-CD (○), tolbutamide/HP-β-CD (□), tolbutamide alone (●) (equivalent to tolbutamide 20 mg/kg) to rabbits. Each point represents mean ± S.D., n = 3 to 6. (From [93,96,99], with permission.)

administration of commercially available Danocrine capsules (200 mg) is shown in Fig. 5. The relative bioavailability of the buccal tablets was 13-fold higher than that of Danocrine; thus, buccal drug delivery with CD complexation increased the bioavailability of drug due to the increased solubility and potential avoidance of first-pass hepatic metabolism [55].

4.5. Pulmonary Drug Delivery

Coadministration of insulin with 5% DM-β-CD delivered intratracheally via instillation resulted in nearly complete insulin uptake from the pulmonary sacs. These data showed a significantly improved hypoglycemic response and tissue damage assessment, indicating relatively low acute mucotoxicity. The respiratory mucosa has, in general, much lower resistance toward drug permeation than do other mucosas, and thus pulmonary delivery of macromolecules usually results in much greater bioavailability than that of other

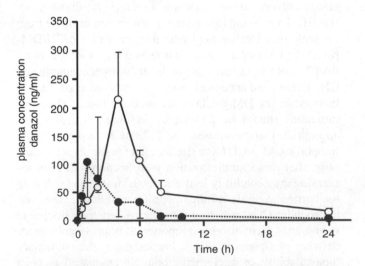

Figure 5. Mean concentration of danazol after drug administration via the per oral route for Danocrine capsules (200 mg of danazol) (●) and danazol/SBE7-β-CD complex buccal tablets (40 mg of danazol) ($n = 4 \pm$ S.D.) (○). (From [55], with permission.)

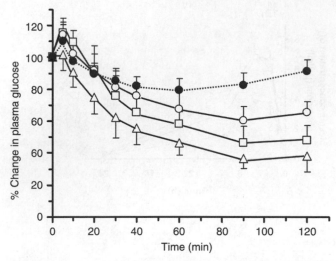

Figure 6. Changes in plasma glucose after intratracheal administration of insulin in saline or in the presence of increasing concentrations of DM-β-CD; insulin + saline (●), insulin + saline insulin + 0.06% DM-β-CD (○), insulin + 0.125% DM-β-CD (□), insulin + 0.25% DM-β-CD (△) Data represent mean ± S.D., $n = 3$ to 5. (From Ref. [59], with permission of Springer Science and Business Media.)

routes of administration [57]. Hussain et al. reported that the addition of DM-β-CD to insulin formulations for intratracheal administration exhibited a significant decrease in plasma glucose due to increases in plasma insulin compared to those obtained when insulin was administered in saline. Furthermore, increasing amounts of DM-β-CD resulted in a proportional decrease in plasma glucose (Fig. 6) [59]. The other authors showed that DM-β-CD enhanced the transmucosal delivery of recombinant human growth hormone (rhGH). They found that systemic absorption of rhGH from rat lung after intratracheal administration of a rhGH/DM-β-CD (1 : 100 molar ratio) mixture as dry powder was more than 2.5-fold higher than that of the dry powder containing no CD. Further enhancement was not observed upon further increase in the DM-β-CD concentration. Thus, CD in formulations should be considered in order to achieve an appropriate therapeutic outcome [22,60]. Most probably, the lipophilic DM-β-CD has a similar effect in the lungs as in the nose after nasal administration (see Section 4.2) both increasing drug solubility in the aqueous mucus and reducing the barrier properties of the lipophilic membrane. However, hydrophilic CDs, which are unable to penetrate lipophilic membranes, have also been reported to improve pulmonary delivery of lipophilic drugs. For example, the pulmonary bioavailability of beclomethasone dipropionate has been increased through complexation of the drug with HP-β-CD [112].

5. CONCLUSIONS

CDs are able to enhance drug delivery through biological membranes. In vitro studies have shown that hydrophilic CDs can enhance drug delivery through membranes only when the permeation resistance of the UWL on the donor side is about equal to or greater than the resistance of the membrane barrier [i.e., $R_D \geq R_M$ in equations (2) and (3)]. Little or no hydrophilic CDs penetrate lipophilic membranes, and thus hydrophilic CDs do not, in general, enhance drug delivery through membranes if the lipophilic membrane barrier is the main permeation barrier. Lipophilic CD derivatives such as the methylated CDs are able to penetrate lipophilic membranes and reduce the membrane barrier function [34]. The effect of CDs on drug permeation depends on the physicochemical properties of the drug. CDs have the greatest effect on relatively small lipophilic molecules (BCS class II drugs). In addition, CDs can form both nano- and microparticles, and it is known that in some cases nanoparticles can penetrate human mucus more rapidly than can individual drug molecules [18,23].

REFERENCES

1. Smart, J.D. (2005). The basics and underlying mechanisms of mucoadhesion. *Adv. Drug Deliv. Rev.*, *57*, 1556–1568.
2. Ahuja, A., Khar, R.K., Ali, J. (1997). Mucoadhesive drug delivery systems. *Drug Dev. Ind. Pharm.*, *23*, 489–515.
3. Senel, S., Hincal, A.A. (2001). Drug permeation enhancement via buccal route: possibilities and limitations. *J. Controll. Release*, *72*, 133–144.
4. Brewster, M.E., Noppe, M., Peeters, J., Loftsson, T. (2007). Effect of the unstirred water layer on permeability enhancement by hydrophilic cyclodextrins. *Int. J. Pharm.*, *342*, 250–253.
5. Shah, P., Jogani, V., Mishra, P., Mishra, A.K., Bagchi, T., Misra, A. (2007). Modulation of ganciclovir intestinal absorption in presence of absorption enhancers. *J. Pharm. Sci.*, *96*, 2710–2722.
6. Cu, Y., Saltzman, W.M. (2009). Mathematical modeling of molecular diffusion through mucus. *Adv. Drug Deliv. Rev.*, *61*, 101–114.
7. Prego, C., Garcia, M., Torres, D., Alonso, M.J. (2005). Transmucosal macromolecular drug delivery. *J. Controll. Release*, *101*, 151–162.
8. Cone, R.A. (2009). Barrier properties of mucus. *Adv. Drug Deliv. Rev.*, *61*, 75–85.
9. Lai, S.K., Wang, Y.Y., Hanes, J. (2009). Mucus-penetrating nanoparticles for drug and gene delivery to mucosal tissues. *Adv. Drug Deliv. Rev.*, *61*, 158–171.
10. Khanvilkar, K., Donovan, M.D., Flanagan, D.R. (2001). Drug transfer through mucus. *Adv. Drug Deliv. Rev.*, *48*, 173–193.

11. Salamat-Miller, N., Chittchang, M., Johnston, T.P. (2005). The use of mucoadhesive polymers in buccal drug delivery. *Adv. Drug Deliv. Rev.*, *57*, 1666–1691.

12. Brewster, M.E., Loftsson, T. (2007). Cyclodextrins as pharmaccutical solubilizers. *Adv. Drug Deliv. Rev.*, *59*, 645–666.

13. Loftsson, T., Duchêne, D. (2007). Cyclodextrins and their pharmaceutical applications. *Int. J. Pharm.*, *329*, 1–11.

14. Loftsson, T., Bodor, N. (1995). *Percutaneous Penetration Enhancers*. CRC Press, Boca Raton, FL, pp. 335–341.

15. Shimpi, S., Chauhan, B., Shimpi, P. (2005). Cyclodextrins: application in different routes of drug administration. *Acta Pharma.*, *55*, 139–156.

16. He, Y.F., Fu, P., Shen, X.H., Gao, H.C. (2008). Cyclodextrin-based aggregates and characterization by microscopy. *Micron*, *39*, 495–516.

17. Jansook, P., Kurkov, S.V., Loftsson, T. (2010). Cyclodextrin as solubilizers: formation of complex aggregates. *J. Pharm. Sci.*, *99*, 719–729.

18. Lai, S.K., O'Hanlon, D.E., Harrold, S., Man, S.T., Wang, Y.Y., Cone, R., Hanes, J. (2007). Rapid transport of large polymeric nanoparticles in fresh undiluted human mucus. *Proc. Natl. Acad. Sci. USA*, *104*, 1482–1487.

19. Maleki, A., Lafitte, G., Kjoniksen, A.L., Thuresson, K., Nystrom, B. (2008). Effect of pH on the association behavior in aqueous solutions of pig gastric mucin. *Carbohydr. Res.*, *343*, 328–340.

20. Loftsson, T., Sigurdsson, H.H., Konradsdóttir, F., Gisladóttir, S., Jansook, P., Stefánsson, E. (2008). Topical drug delivery to the posterior segment of the eye: anatomical and physiological considerations. *Pharmazie*, *63*, 171–179.

21. Lai, S.K., Wang, Y.Y., Wirtz, D., Hanes, J. (2009). Micro- and macrorheology of mucus. *Adv. Drug Deliv. Rev.*, *61*, 86–100.

22. Loftsson, T., Jarho, P., Másson, M., Järvinen, T. (2005). Cyclodextrins in drug delivery. *Expert Opin. Drug Deliv.*, *2*, 335–351.

23. Loftsson, T., Vogensen, S.B., Brewster, M.E., Konradsdóttir, F. (2007). Effects of cyclodextrins on drug delivery through biological membranes. *J. Pharm. Sci.*, *96*, 2532–2546.

24. Loftsson, T., Stefánsson, E. (2007). Cyclodextrins in ocular drug delivery: theoretical basis with dexamethasone as a sample drug. *J. Drug Deliv. Sci. Technol.*, *17*, 3–9.

25. Jarho, P., Urtti, A., Pate, D.W., Suhonen, P., Järvinen, T. (1996). Increase in aqueous solubility, stability and in vitro corneal permeability of anandamide by hydroxypropyl-β-cyclodextrin. *Int. J. Pharm.*, *137*, 209–216.

26. Bary, A.R., Tucker, I.G., Davies, N.M. (2000). Considerations in the use of hydroxypropyl-β-cyclodextrin in the formulation of aqueous ophthalmic solutions of hydrocortisone. *Eur. J. Pharm. Biopharm.*, *50*, 237–244.

27. Richter, T., Keipert, S. (2004). In vitro permeation studies comparing bovine nasal mucosa, porcine cornea and artificial membrane: androstenedione in microemulsions and their components. *Eur. J. Pharm. Biopharm.*, *58*, 137–143.

28. Washington, N., Washington, W.C., Wilson C.G. (2001). *Physiological Pharmaceutics: Barriers to Drug Absorption*. Taylor & Francis, London, pp. 37–58, 109–142, 221–270.

29. Mannermaa, E., Vellonen, K.S., Urtti, A. (2006). Drug transport in corneal epithelium and blood–retina barrier: emerging role of transporters in ocular pharmacokinetics. *Adv. Drug Deliv. Rev.*, *58*, 1136–1163.

30. Kublik, H., Bock, T.K., Schreier, H., Müller, B.W. (1996). Nasal absorption of 17-β-estradiol from different cyclodextrin inclusion formulations in sheep. *Eur. J. Pharm. Biopharm.*, *42*, 320–324.

31. Lennernas, H. (1998). Human intestinal permeability. *J. Pharm. Sci.*, *87*, 403–410.

32. Somogyi, G., Posta, J., Buris, L., Varga, M. (2006). Cyclodextrin (CD) complexes of cholesterol: their potential use in reducing dietary cholesterol intake. *Pharmazie*, *61*, 154–156.

33. Cappello, B., De Rosa, G., Giannini, L., La Rotonda, M.I., Mensitieri, G., Miro, A., Quaglia, F., Russo, R. (2006). Cyclodextrin-containing poly(ethyleneoxide) tablets for the delivery of poorly soluble drugs: potential as buccal delivery system. *Int. J. Pharm.*, *319*, 63–70.

34. Merkus, F.W.H.M., Verhoef, J.C., Marttin, E., Romeijn, S.G., van der Kuy, P.H.M., Hermens, W.A.J.J., Schipper, N.G.M. (1999). Cyclodextrins in nasal drug delivery. *Adv. Drug Deliv. Rev.*, *36*, 41–57.

35. Matsuda, H., Arima, H. (1999). Cyclodextrins in transdermal and rectal delivery. *Adv. Drug Deliv. Rev.*, *36*, 81–99.

36. Loftsson, T., Järvinen, T. (1999). Cyclodextrins in ophthalmic drug delivery. *Adv. Drug Deliv. Rev.*, *36*, 59–79.

37. Hussain, A., Arnold, J.J., Khan, M.A., Ahsan, F. (2004). Absorption enhancers in pulmonary protein delivery. *J. Controll. Release*, *94*, 15–24.

38. Ammar, H.O., El-Nahhas, S.A., Khalil, R.M. (1998). Cyclodextrins in acetazolamide eye drop formulations. *Pharmazie*, *53*, 559–562.

39. Loftsson, T., Fridriksdóttir, H., Thórisdóttir, S., Stefánsson, E., Sigurdardóttir, A.M., Gudmundsson, O., Sigthórsson, T. (1994). 2-Hydroxypropyl-β-cyclodextrin in topical carbonic–anhydrase inhibitor formulations. *Eur. J. Pharm. Sci.*, *1*, 175–180.

40. Loftsson, T., Stefánsson, E., Kristinsson, J.K., Fridriksdóttir, H., Sverrisson, T., Gudmundsdóttir, G., Thórisdóttir, S. (1996). Topically effective acetazolamide eye-drop solution in man. *Pharm. Sci.*, *2*, 277–279.

41. Usayapant, A., Karara, A.H., Narurkar, M.M. (1991). Effect of 2-hydroxypropyl-β-cyclodextrin on the ocular absorption of dexamethasone and dexamethasone acetate. *Pharm. Res.*, *8*, 1495–1499.

42. Loftsson, T., Stefánsson, E. (2002). Cyclodextrins in eye drop formulations: enhanced topical delivery of corticosteroids to the eye. *Acta Ophthalmol. Scand.*, *80*, 144–150.

43. Sigurdsson, H.H., Konradsdóttir, F., Loftsson, T., Stefánsson, E. (2007). Topical and systemic absorption in delivery of dexamethasone to the anterior and posterior segments of the eye. *Acta Ophthalmol. Scand.*, *85*, 598–602.

44. Kristinsson, J.K., Fridriksdóttir, H., Thorisdóttir, S., Sigurdardóttir, A.M., Stefánsson, E., Loftsson, T. (1996). Dexamethasone–cyclodextrin–polymer co-complexes in aqueous eye drops: aqueous humor pharmacokinetics in humans. *Invest. Ophthalmol. Vis. Sci.*, *37*, 1199–1203.

45. Saari, K.M., Nelimarkka, L., Ahola, V., Loftsson, T., Stefánsson, E. (2006). Comparison of topical 0.7% dexamethasone–cyclodextrin with 0.1% dexamethasone sodium phosphate for postcataract inflammation. *Graefes Arch. Clin. Exp. Ophthalmol.*, *244*, 620–626.

46. Siefert, B., Keipert, S. (1997). Influence of α-cyclodextrin and hydroxyalkylated β-cyclodextrin derivatives on the in vitro corneal uptake and permeation of aqueous pilocarpine-HCl solutions. *J. Pharm. Sci.*, *86*, 716–720.

47. Freedman, K.A., Klein, J.W., Crosson, C.E. (1993). β-Cyclodextrins enhance bioavailability of pilocarpine. *Curr. Eye Res.*, *12*, 641–647.

48. Aktas, Y., Ünlu, N., Orhan, M., Irkeç, M., Hincal, A.A. (2003). Influence of hydroxypropyl β-cyclodextrin on the corneal permeation of pilocarpine. *Drug Dev. Ind. Pharm.*, *29*, 223–230.

49. Matsubara, K., Abe, K., Irie, T., Uekama, K. (1995). Improvement of nasal bioavailability of luteinizing-hormone-releasing hormone agonist, buserelin, by cyclodextrin derivatives in rats. *J. Pharm. Sci.*, *84*, 1295–1300.

50. Shao, Z., Krishnamoorthy, R., Mitra, A.K. (1992). Cyclodextrins as nasal absorption promoters of insulin: mechanistic evaluations. *Pharm. Res.*, *9*, 1157–1163.

51. Schipper, N.G.M., Romeijn, S.G., Verhoef, J.C., Merkus, F.W.H.M. (1993). Nasal insulin delivery with dimethyl-β-cyclodextrin as an absorption enhancer in rabbits: powder more effective than liquid formulations. *Pharm. Res.*, *10*, 682–686.

52. Zhang, X.N., Xu, J., Tang, L.H., Gong, J., Yan, X.Y., Zhang, Q. (2007). Influence on intestinal mucous permeation of paclitaxel of absorption enhancers and dosage forms based on electron spin resonance spectroscopy. *Pharmazie*, *62*, 368–371.

53. Zheng, Y., Zuo, Z., Chow, A.H.L. (2006). Lack of effect of β-cyclodextrin and its water-soluble derivatives on in vitro drug transport across rat intestinal epithelium. *Int. J. Pharm.*, *309*, 123–128.

54. Badawy, S.I.F., Ghorab, M.M., Adeyeye, C.M. (1996). Bioavailability of danazol–hydroxypropyl-β-cyclodextrin complex by different routes of administration. *Int. J. Pharm.*, *145*, 137–143.

55. Jain, A.C., Aungst, B.J., Adeyeye, M.C. (2002). Development and in vivo evaluation of buccal tablets prepared using danazol-sulfobutylether 7 β-cyclodextrin (SBE 7) complexes. *J. Pharm. Sci.*, *91*, 1659–1668.

56. Rodrigues, J.M., Lima, K.D., Jensen, C.E.D., de Aguiar, M.M.G., Cunha, A.D. (2003). The effect of cyclodextrins on the in vitro and in vivo properties of insulin-loaded poly (D,L-lactic-co-glycolic acid) microspheres. *Artif. Organs*, *27*, 492–497.

57. Shao, Z.Z., Li, Y.P., Mitra, A.K. (1994). Cyclodextrins as mucosal absorption promoters of insulin: 3. Pulmonary route of delivery. *Eur. J. Pharm. Biopharm.*, *40*, 283–288.

58. Aguiar, M.M.G., Rodrigues, J.M., Cunha, A.S. (2004). Encapsulation of insulin–cyclodextrin complex in PLGA microspheres: a new approach for prolonged pulmonary insulin delivery. *J. Microencapsul.*, *21*, 553–564.

59. Hussain, A., Yang, T.Z., Zaghloul, A.A., Ahsan, F. (2003). Pulmonary absorption of insulin mediated by tetradecyl-β-maltoside and dimethyl-β-cyclodextrin. *Pharm. Res.*, *20*, 1551–1557.

60. Jalalipour, M., Najafabadi, A.R., Gilani, K., Esmaily, H., Tajerzadeh, H. (2008). Effect of dimethyl-β-cyclodextrin concentrations on the pulmonary delivery of recombinant human growth hormone dry powder in rats. *J. Pharm. Sci.*, *97*, 5176–5185.

61. Kobayashi, S., Kondo, S., Juni, K. (1996). Pulmonary delivery of salmon calcitonin dry powders containing absorption enhancers in rats. *Pharm. Res.*, *13*, 80–83.

62. Sigurdsson, H.H., Stefánsson, E., Gudmundsdóttir, E., Eysteinsson, T., Thorsteinsdóttir, M., Loftsson, T. (2005). Cyclodextrin formulation of dorzolamide and its distribution in the eye after topical administration. *J. Controll. Release*, *102*, 255–262.

63. Loftsson, T., Hreinsdóttir, D., Stefánsson, E. (2007). Cyclodextrin microparticles for drug delivery to the posterior segment of the eye: aqueous dexamethasone eye drops. *J. Pharm. Pharmacol.*, *59*, 629–635.

64. Yang, T., Hussain, A., Paulson, J., Abbruscato, T.J., Ahsan, F. (2004). Cyclodextrins in nasal delivery of low-molecular-weight heparins: in vivo and in vitro studies. *Pharm. Res.*, *21*, 1127–1136.

65. Loftsson, T., Frithriksdóttir, H., Thorisdóttir, S., Stefánsson, E. (1994). The effect of hydroxypropyl methylcellulose on the release of dexamethasone from aqueous 2–hydroxypropyl-β-cyclodextrin formulations. *Int. J. Pharm.*, *104*, 181–184.

66. Loftsson, T., Sigurdsson, H.H., Hreinsdóttir, D., Konradsdóttir, F., Stefánsson, E. (2007). Dexamethasone delivery to posterior segment of the eye. *J. Inclusion Phenom. Macrocyclic Chem.*, *57*, 585–589.

67. Behl, C.R., Pimplaskar, H.K., Sileno, A.P., deMeireles, J., Romeo, V.D. (1998). Effects of physicochemical properties and other factors on systemic nasal drug delivery. *Adv. Drug Deliv. Rev.*, *29*, 89–116.

68. Schipper, N.G.M., Verhoef, J.C., Romeijn, S.G., Merkus, F.W.H.M. (1995). Methylated β-cyclodextrins are able to improve the nasal absorption of salmon-calcitonin. *Calcif. Tissue Int.*, *56*, 280–282.

69. Merkus, F.W.H.M., Verhoef, J.C., Romeijn, S.G., Schipper, N.G.M. (1991). Absorption enhancing effect of cyclodextrins on intranasally administered insulin in rats. *Pharm. Res.*, *8*, 588–592.

70. Hermens, W.A.J.J., Deurloo, M.J.M., Romeyn, S.G., Verhoef, J.C., Merkus, F.W.H.M. (1990). Nasal absorption enhancement of 17-β-estradiol by dimethyl-β-cyclodextrin in rabbits and rats. *Pharm. Res.*, *7*, 500–503.

71. Schipper, N.G.M., Hermens, W.A.J.J., Romeyn, S.G., Verhoef, J., Merkus, F.W.H.M. (1990). Nasal absorption of

17-β-estradiol and progesterone from a dimethyl-cyclodextrin inclusion formulation in rats. *Int. J. Pharm.*, *64*, 61–66.

72. Ventura, C.A., Giannone, I., Musumeci, T., Pignatello, R., Ragni, L., Landolfi, C., Milanese, C., Paolino, D., Puglisi, G. (2006). Physico-chemical characterization of disoxaril–dimethyl-β-cyclodextrin inclusion complex and in vitro permeation studies. *Eur. J. Med. Chem.*, *41*, 233–240.

73. Merkus, F.W., Verhoef, J.C., Romeijn, S.G., Schipper, N.G. (1991). Absorption enhancing effect of cyclodextrins on intranasally administered insulin in rats. *Pharm. Res.*, *8*, 588–592.

74. Vermehren, C., Hansen, H.S., Thomsen, M.K. (1996). Time dependent effects of two absorption enhancers on the nasal absorption of growth hormone in rabbits. *Int. J. Pharm.*, *128*, 239–250.

75. Loftsson, T., Gudmundsdóttir, H., Sigurjónsdóttir, J.F., Sigurdsson, H.H., Sigfusson, S.D., Másson, M., Stefánsson, E. (2001). Cyclodextrin solubilization of benzodiazepines: formulation of midazolam nasal spray. *Int. J. Pharm.*, *212*, 29–40.

76. Larhed, A.W., Artursson, P., Bjork, E. (1998). The influence of intestinal mucus components on the diffusion of drugs. *Pharm. Res.*, *15*, 66–71.

77. Woodcock, B.G., Acerbi, D., Merz, P.G., Rietbrock, S., Rietbrock, N. (1993). Supermolecular inclusion of piroxicam with β-cyclodextrin: pharmacokinetic properties in man. *Eur. J. Rheumatol. Inflamm.*, *12*, 12–28.

78. Kimura, E., Bersani-Amado, C.A., Sudo, L.S., Santos, S.R., Oga, S. (1997). Pharmacokinetic profile of piroxicam β-cyclodextrin, in rat plasma and lymph. *Gen. Pharmacol.*, *28*, 695–698.

79. Deroubaix, X., Stockis, A., Allemon, A.M., Lebacq, E., Acerbi, D., Ventura, P. (1995). Oral bioavailability of CHF1194, an inclusion complex of piroxicam and β-cyclodextrin, in healthy subjects under single dose and steady-state conditions. *Eur. J. Clin. Pharmacol.*, *47*, 531–536.

80. McEwen, J. (2000). Clinical pharmacology of piroxicam-β-cyclodextrin: implications for innovative patient care. *Clin. Drug Invest.*, *19*, 27–31.

81. Castillo, J.A., Palomo-Canales, J., García, J.J., Lastres, J.L., Bolas, F., Torrado, J.J. (1999). Preparation and characterization of albendazole β-cyclodextrin complexes. *Drug Dev. Ind. Pharm.*, *25*, 1241–1248.

82. El-Gindy, G.A., Mohammed, F.A., Salem, S.Y. (2002). Preparation, pharmacokinetic and pharmacodynamic evaluation of carbamazepine inclusion complexes with cyclodextrins. *STP Pharma Sci.*, *12*, 369–378.

83. Brewster, M.E., Anderson, W.R., Meinsma, D., Moreno, D., Webb, A.I., Pablo, L., Estes, K.S., Derendorf, H., Bodor, N., Sawchuk, R., Cheung, B., Pop, E. (1997). Intravenous and oral pharmacokinetic evaluation of a 2-hydroxypropyl-β-cyclodextrin-based formulation of carbamazepine in the dog: comparison with commercially available tablets and suspensions. *J. Pharm. Sci.*, *86*, 335–339.

84. Betlach, C.J., Gonzalez, M.A., McKiernan, B.C., Neffdavis, C., Bodor, N. (1993). Oral pharmacokinetics of carbamazepine in dogs from commercial tablets and a cyclodextrin complex. *J. Pharm. Sci.*, *82*, 1058–1060.

85. Choudhury, S., Nelson, K.F. (1992). Improvement of oral bioavailability of carbamazepine by inclusion in 2-hydroxypropyl-β-cyclodextrin. *Int. J. Pharm.*, *85*, 175–180.

86. Koester, L.S., Ortega, G.G., Mayorga, P., Bassani, V.L. (2004). Mathematical evaluation of in vitro release profiles of hydroxypropylmethylcellulose matrix tablets containing carbamazepine associated to β-cyclodextrin. *Eur. J. Pharm. Biopharm.*, *58*, 177–179.

87. Prabagar, B., Yoo, B.K., Woo, J.S., Kim, J.A., Rhee, J.D., Piao, M.G., Choi, H.G., Yong, C.S. (2007). Enhanced bioavailability of poorly water-soluble clotrimazole by inclusion with β-cyclodextrin. *Arch. Pharm. Res.*, *30*, 249–254.

88. Uekama, K., Fujinaga, T., Hirayama, F., Otagiri, M., Yamasaki, M., Seo, H., Hashimoto, T., Tsuruoka, M. (1983). Improvement of the oral bioavailability of digitalis glycosides by cyclodextrin complexation. *J. Pharm. Sci.*, *72*, 1338–1341.

89. Savolainen, J., Järvinen, K., Taipale, H., Jarho, P., Loftsson, T., Järvinen, T. (1998). Co-administration of a water-soluble polymer increases the usefulness of cyclodextrins in solid oral dosage forms. *Pharm. Res.*, *15*, 1696–1701.

90. Babu, R.J., Pandit, J.K. (1995). Enhancement of dissolution rate and hypoglycemic activity of glibenclamide with β-cyclodextrin. *STP Pharma Sci.*, *5*, 196–201.

91. Tenjarla, S., Puranajoti, P., Kasina, R., Mandal, T. (1998). Preparation, characterization, and evaluation of miconazole–cyclodextrin complexes for improved oral and topical delivery. *J. Pharm. Sci.*, *87*, 425–429.

92. Barillaro, V., Evrard, B., Delattre, L., Piell, G. (2005). Oral bioavailability in pigs of a miconazole/hydroxypropyl-γ-cyclodextrin/L-tartaric acid inclusion complex produced by supercritical carbon dioxide processing. *AAPS J.*, *7*, E149–E155.

93. Soliman, O.A.E., Kimura, K., Hirayama, F., Uekama, K., El Sabbagh, H.M., Abd El Gawad, A.H., Hashim, F.M. (1997). Amorphous spironolactone–hydroxypropylated cyclodextrin complexes with superior dissolution and oral bioavailability. *Int. J. Pharm.*, *149*, 73–83.

94. Kaukonen, A.M., Lennernas, H., Mannermaa, J.P. (1998). Water-soluble β-cyclodextrins in paediatric oral solutions of spironolactone: preclinical evaluation of spironolactone bioavailability from solutions of β-cyclodextrin derivatives in rats. *J. Pharm. Pharmacol.*, *50*, 611–619.

95. Seo, H., Tsuruoka, M., Hashimoto, T., Fujinaga, T., Otagiri, M., Uekama, K. (1983). Enhancement of oral bioavailability of spironolactone by β-cyclodextrin and γ-cyclodextrin complexations. *Chem. Pharm. Bull.*, *31*, 286–291.

96. Savolainen, J., Järvinen, K., Matilainen, L., Järvinen, T. (1998). Improved dissolution and bioavailability of phenytoin by sulfobutylether-β-cyclodextrin ((SBE)(7m)-β-CD) and hydroxypropyl-β-cyclodextrin (HP-β-CD) complexation. *Int. J. Pharm.*, *165*, 69–78.

97. Uekama, K., Otagiri, M., Irie, T., Seo, H., Tsuruoka, M. (1985). Improvement of dissolution and absorption characteristics of phenytoin by a water-soluble β-cyclodextrin–epichlorohydrin polymer. *Int. J. Pharm.*, *23*, 35–42.

98. Tanino, T., Ogiso, T., Iwaki, M. (1999). Effect of sugar-modified β-cyclodextrins on dissolution and absorption characteristics of phenytoin. *Biol. Pharm. Bull.*, *22*, 298–304.

99. Veiga, F., Fernandes, C., Teixeira, F. (2000). Oral bioavailability and hypoglycaemic activity of tolbutamide/cyclodextrin inclusion complexes. *Int. J. Pharm.*, *202*, 165–171.

100. Kimura, K., Hirayama, F., Arima, H., Uekama, K. (2000). Effects of aging on crystallization, dissolution and absorption characteristics of amorphous tolbutamide–2-hydroxypropyl-β-cyclodextrin complex. *Chem. Pharm. Bull.*, *48*, 646–650.

101. Uekama, K., Horiuchi, Y., Kikuchi, M., Hirayama, F., Ijitsu, T., Ueno, M. (1988). Enhanced dissolution and oral bioavailability of α-tocopheryl esters by dimethyl-β-cyclodextrin complexation. *J. Inclusion Phenom.*, *6*, 167–174.

102. Ribeiro, L.S.S., Falcão, A.C., Patricio, J.A.B., Ferreira, D.C., Veiga, F.J.B. (2007). Cyclodextrin multicomponent complexation and controlled release delivery strategies to optimize the oral bioavailability of vinpocetine. *J. Pharm. Sci.*, *96*, 2018–2028.

103. Luengo, J., Aranguiz, T., Sepulveda, J., Hernandez, L., Von Plessing, C. (2002). Preliminary pharmacokinetic study of different preparations of acyclovir with β-cyclodextrin. *J. Pharm. Sci.*, *91*, 2593–2598.

104. Le Corre, P., Dollo, G., Chevanne, F., Le Verge, R. (1998). Influence of hydroxypropyl-β-cyclodextrin and dimethyl-β-cyclodextrin on diphenhydramine intestinal absorption in a rat in situ model. *Int. J. Pharm.*, *169*, 221–228.

105. Miyake, K., Arima, H., Irie, T., Hirayama, F., Uekama, K. (1999). Enhanced absorption of cyclosporin A by complexation with dimethyl-β-cyclodextrin in bile duct-cannulated and -noncannulated rats. *Biol. Pharm. Bull.*, *22*, 66–72.

106. Singh, B., Ahuja, N. (2002). Development of controlled-release buccoadhesive hydrophilic matrices of diltiazem hydrochloride: optimization of bioadhesion, dissolution, and diffusion parameters. *Drug Dev. Ind. Pharm.*, *28*, 431–442.

107. Rathbone, M.J. (1991). Human buccal absorption: 1. A method for estimating the transfer kinetics of drugs across the human buccal membrane. *Int. J. Pharm.*, *69*, 103–108.

108. Cassidy, J.P., Landzert, N.M., Quadros, E. (1993). Controlled buccal delivery of buprenorphine. *J. Controll. Release*, *25*, 21–29.

109. Guo, J.H. (1994). Bioadhesive polymer buccal patches for buprenorphine controlled delivery: formulation, in-vitro adhesion and release properties. *Drug Dev. Ind. Pharm.*, *20*, 2809–2821.

110. Han, R.Y., Fang, J.Y., Sung, K.C., Hu, O.Y.P. (1999). Mucoadhesive buccal disks for novel nalbuphine prodrug controlled delivery: effect of formulation variables on drug release and mucoadhesive performance. *Int. J. Pharm.*, *177*, 201–209.

111. Boulmedarat, L., Bochot, A., Lesieur, S., Fattal, E. (2005). Evaluation of buccal methyl-β-cyclodextrin toxicity on human oral epithelial cell culture model. *J. Pharm. Sci.*, *94*, 1300–1309.

112. Pinto, J.M.C.L., Marques, H.M.C. (1999). Beclomethasone/cyclodextrin inclusion complex for dry powder inhalation. *STP Pharma Sci.*, *9*, 253–256.

9

APPLICATIONS OF CYCLODEXTRINS FOR SKIN FORMULATION AND DELIVERY

AMÉLIE BOCHOT

School of Pharmacy, Université Paris–Sud, Paris, France

GÉRALDINE PIEL

Department of Pharmacy, University of Liège, Liège, Belgium

1. INTRODUCTION

Topically administered drugs act either dermally or transdermally. For that reason, they have to penetrate the deeper skin layers or permeate the skin. However, the efficacy of topically applied drugs is often limited by their poor penetration into the skin. Indeed, the outermost layer of the human skin, the stratum corneum, is responsible for its barrier function. In these cases, modulations of the skin penetration profiles of the drugs and skin barrier manipulations are necessary. In pharmaceutical products applied on skin, the use of cyclodextrins (CDs) is more generally investigated to improve drug properties, such as stability, innocuousness (no side effects), and bioavailability. The most frequently studied groups of drugs associated with CDs are retinoids, dermocorticosteroids, nonsteroidal anti-inflammatories, and hormones (see Table 1). In this chapter we provide an overview of the applications of CD for skin formulation and delivery. For further information on this topic, the reader is referred to several excellent reviews [1–11].

2. SAFETY OF CDs FOR THE SKIN

CDs were suspected to interact with some components of the skin, but contradictory results have been reported [4,12–18]. Generally, all types of CDs can be used in skin and mucosal formulations safely and without risk of irritation. Natural CDs and HP-β-CD are not able to induce disruption of the stratum corneum and are regarded as nonirritants to the skin [11]. Indeed, no irritating effect was observed for these CDs after application on $1\,cm^2$ of skin, at healthy volunteers, of equivalent quantities of 2 mg of CD dissolved in water or scattered in vaseline [19]. Methylated CD derivatives in low concentrations can also be applied safely. It seems that only methylated CD derivatives administered in high concentrations (10 to 20%) in aqueous solutions or suspensions can interact with stratum corneum components (cholesterol and triglycerides) and cause irritation [11,20,21]. Piel et al. [20] used the corneoxenometry bioassay on human stratum corneum to compare the skin compatibility of CDs currently used in pharmaceutical preparations (β-CD, γ-CD, RAMEB, Dimeb, Trimeb, HP-β-CD, and HP-γ-CD). All the CDs tested were well tolerated by the stratum corneum at 5% concentration. However, Dimeb has a lesser cutaneous index of mildness than the others.

It must be borne in mind that these studies concern empty CDs, and in normal use, CDs are employed in the form of an inclusion compound. Under normal conditions, such extraction would be suppressed by drug molecules and other lipophilic molecules usually present in dermatologic preparations. These lipophilic molecules will compete with the membrane constituents for a space in the CD cavity and in this way will reduce the abilities of CD to extract lipophilic compounds from the skin barrier [3].

Cyclodextrins in Pharmaceutics, Cosmetics, and Biomedicine: Current and Future Industrial Applications, First Edition. Edited by Erem Bilensoy.
© 2011 John Wiley & Sons, Inc. Published 2011 by John Wiley & Sons, Inc.

Table 1. Effect of CDs on Drug Absorption

Molecule	Refs.	CD[a]	Stoichiometry	K_c Stability Constant (M^{-1})	Form[b]	Method	Skin Type	Evaluation in:	Results[c]
Acitretin	[72]	RAMEB	—	—	Aq	Ex vivo	Hairless mice	Receptor medium	↑
Aspirin	[73]	β-CD	—	—	Anh	Ex vivo	Rat	Receptor medium	↓
		HP-β-CD	—	—	Anh	Ex vivo	Rat	Receptor medium	ns
		Dimeb	—	—	Anh	Ex vivo	Rat	Receptor medium	ns
Beclomethasone dipropionate	[74]	γ-CD	1:2	—	AqO	In vivo	Human	Biological effect	↑
Betamethasone	[75]	β-CD	1:2	5,420	AqO	In vitro	Artificial	Receptor medium	↑
		γ-CD	2:3	21,600	AqO	In vitro	Artificial	Receptor medium	↑
Biphenylacetic acid	[36]	β-CD	1:1	2,250	AqO	Ex vivo	Hairless mice	Receptor medium	↑
		HP-β-CD	1:1	1,780	AqO	Ex vivo	Hairless mice	Receptor medium	↑
		Dimeb	1:1	4,710	AqO	Ex vivo	Hairless mice	Receptor medium	↑
Bromhexine hydrochloride	[76]	β-CD	1:1	26	Aq	Ex vivo	Rat	Receptor medium	↑
Bupranolol	[21,77,78]	HP-β-CD	1:1	294	Aq	Ex vivo	Rat	Receptor medium	↑
		RAMEB	1:1	1,275	Aq	Ex vivo	Rat	Receptor medium	↑
		Pretreatment HP-β-CD 3 h	—	—	TTS	Bio	Albino rabbits	Biological effect	↑
		Pretreatment RAMEB 3 h	—	—	Aq	Ex vivo	Rat	Receptor medium	ns
Capsaicine	[79]	HP-β-CD	—	—	Aq	Ex vivo	Hairless mice, hamsters, rats, human	Receptor medium	ns
	[80]	HP-β-CD	1:1	1,822	Aq, AqG	Ex vivo	Rat	Receptor medium	↑
Celecoxib	[18]	HP-β-CD	1:1	190	Aq	Ex vivo	Human	Receptor medium	↑
		Dimeb	1:1	9,004	Aq	Ex vivo	Human	Receptor medium	↑

Chlorpromazine	[34]		1:2	141	Aq	Ex vivo	Porcine	Donor medium	↓
Corticosterone	[81]	β-CD	—	8,300	Aq	Ex vivo	Porcine	Donor medium	↓
		Dimeb	—	8,800	Aq	Ex vivo	Porcine	Donor medium	↓
		HP-β-CD	1:1	350	Aq	Ex vivo	Hairless mice	Receptor medium	ns
		Pretreatment HP-β-CD 3 h	—	—	—	Ex vivo	Hairless mice	Receptor medium	ns
Dehydraepi-androsterone	[82]	α-CD	1:2	—	Aq	Ex vivo	Porcine	Receptor medium	↑
	[83]	α-CD	—	—	ME	Ex vivo	Porcine	Receptor medium	↑
Dexamethasone acetate	[30]	β-CD	1:1	—	AqG	Ex vivo	Hairless mice	Receptor medium	↑
Diclofenac	[84]	HP-β-CD	1:1	—	AqG	In vitro	Artificial	Receptor medium	↑
Estradiol	[85]	HP-β-CD	1:1	2,000	Aq	Ex vivo	Hairless mice	Receptor medium	↑
	[86]	HP-β-CD	—	—	Aq	Ex vivo	Hairless mice	Receptor medium	↑
	[87]	HP-β-CD	—	—	AqO	Ex vivo	Human	Receptor medium	ns
Ethyl-4-biphenyl-acetate	[88]	HP-β-CD	1:1	7,790	o/w	Ex vivo	Hairless mice	Receptor medium, total skin	↑MR ↑TS
		HP-β-CD	—	—	PG	Ex vivo	Hairless mice	Receptor medium, total skin	↓MR ns TS
Flurbiprofen	[89]	HP-β-CD	1:1	7,452	Aq	Ex vivo	Human	Receptor medium	↑
Glipizide	[90]	Dimeb	1:2	—	AqG	Ex vivo	Rat	Receptor medium	↑
Gliquidone	[91]	HP-β-CD	1:1	1,200	Aq	In vivo	Rat	Biological effect	↑
					Aq	Ex vivo	Rat	Receptor medium	↑
Hydrocortisone	[92]	RAMEB + PVP	1:1	2,040	Aq	Ex vivo	Hairless mice	Receptor medium	↑
	[93]	HP-β-CD	1:1	890	Aq	Ex vivo	Hairless mice	Receptor medium	↑
	[94]	RAMEB	1:1	1,710	Aq	Ex vivo	Hairless mice	Receptor medium	↑

(continued)

Table 1. (*Continued*)

Molecule	Refs.	CD[a]	Stoichiometry	K_c Stability Constant (M^{-1})	Form[b]	Method	Skin Type	Evaluation in:	Results[c]
	[95]	β-CD	—	—	o/w	Ex vivo	Human	Stratum corneum, epidermis, dermis	↓SC ↓E, ↑D
					O	Ex vivo	Human	Stratum corneum, epidermis, dermis	nsSC ↓E, ↓D
					AqG	Ex vivo	Human	Stratum corneum, epidermis, dermis	↓SC ↓ E ns D
		HP-β-CD			o/w	Ex vivo	Human	Stratum corneum, epidermis, dermis	↓SC ↓E, ↑D
					O	Ex vivo	Human	Stratum corneum, epidermis, dermis	nsSC ↓E, nsD
					AqG	Ex vivo	Human	Stratum corneum, epidermis, dermis	↓SC ↓E, ↑D
	[96]	Pretreatment HP-β-CD	—	—	Aq	Ex vivo	Hairless mice	Receptor medium	ns
	[97]	HP-β-CD	—	—	Aq	Ex vivo	Human	Receptor medium, total skin	↑IP
Ibuprofen	[98]	HP-β-CD	1:1	—	Aq	Ex vivo	Human	Receptor medium	↑
	[99]	HP-β-CD	1:1	—	Aq	Ex vivo	Hairless mice	Receptor medium, total skin	↑
Indomethacin	[12]	β-CD	1:1	168	Aq	Ex vivo	Porcine	Receptor medium, total skin	→
		Dimeb	1:1	406	Aq	Ex vivo	Porcine	Receptor medium, total skin	→
Ketorolac	[100]	HP-β-CD	1:1	—	Aq	Ex vivo	Porcine	Receptor medium, total skin	↑
Ketoprofen	[101]	HP-β-CD	1:1	128–673	Aq	Ex vivo	Rat	Receptor medium	↑
Levosimendan	[102]	HP-β-CD			Aq	Ex vivo	Human	Receptor medium	→
Levothyroxine	[103]	Dimeb	1:1	0.085	Aq	Ex vivo	Rabbit ear	Receptor medium	↓

(continued)

Drug	Ref.	CD	Ratio	Value	Vehicle	Method	Species	Medium/parameter	Effect
Liarozole	[104]	HP-β-CD pH 4	—	—	Aq	In vivo	Rat	Blood	↑
		HP-β-CD pH 7	—	—	Aq	In vivo	Rat	Blood	→
		Dimeb pH 7	—	—	Aq	In vivo	Rat	Blood	→
	[13]	Pretreatment Dimeb 4 h	—	—	Aq	In vivo	Rat	Blood	↑
Loteprednol etabonate	[105]	Dimeb	—	—	Aq	Ex vivo	Hairless mice	Receptor medium	↑
Meglumine antimoniate	[106]	β-CD	1:1	104	Aq	Ex vivo	Hairless mice	Receptor medium	↑
Melanotonine	[107]	HP-β-CD	—	—	Aq	Ex vivo	Hairless mice	Receptor medium	↑
Metopimazine	[108]	HP-β-CD	1:1	658	Aq	Ex vivo	Porcine	Receptor medium	↑
			1:2	52					
		RAMEB	1:1	1512	Aq	Ex vivo	Porcine	Receptor medium	↑
			1:2	156					
Miconazole	[109]	α-CD	1:1	333	Aq	Ex vivo	Human	Receptor medium, epidermis	ns R
									↑E
		HP-β-CD	1:1	363	Aq	Ex vivo	Human	Receptor medium, total skin	ns R
									↑P
Minoxidil	[110]	TA-α-CD	—	—	Aq	In vivo	Mice	Biological effect	↑
	[111]	HP-β-CD	—	—	Aq	In vivo	Mice	Biological effect	ns
Nicotine	[112]	β-CD	1:1	—	TTS	Ex vivo	Rat	Receptor medium	→
			1:2	—					
			1:3	—					
Papaverine	[18]	HP-β-CD	—	78	Aq	Ex vivo	Rat	Receptor medium	↑
		Dimeb	—	154	Aq	Ex vivo	Rat	Receptor medium	↑
Piribedil	[15]	RAMEB	1:1	990	Aq	Ex vivo	Hairless rat	Receptor medium	↑
Piroxicam	[41]	HP-β-CD	1:1	166.7	AqG	Ex vivo	Hairless rat	Receptor medium, total skin	↑
		Pre-treatment, HP-β-CD 3 h	—	—	Aq	Ex vivo	Hairless rat	Receptor medium, total skin	ns
Prostaglandin E1	[113]	β-CD	1:1	—	Aq, ME	In vivo	Rat	Biological effect	ns
	[114]	CME-β-CD	1:5	—	O	In vivo	Hairless mice	Biological effect	↑
Retinoic acid	[24]	CME-β-CD	1:5	—	O	In vivo	Rabbit	Biological effect	↑
	[29]	CME-β-CD	1:5	—	O	In vivo	Hairless mice	Biological effect	↑
	[31]	β-CD	—	—	AqG, o/w	In vivo	Human	Biological effect	↑

Table 1. (*Continued*)

Molecule	Refs.	CD[a]	Stoichiometry	K_c Stability Constant (M^{-1})	Form[b]	Method	Skin Type	Evaluation in:	Results[c]
Sericoside	[22]	γ-CD	1:2	233	o/w	Ex vivo	Porcine	Receptor medium, skin strata	↑
Sulfanilic acid	[12]	β-CD	1:1	6.7	Aq	Ex vivo	Porcine	Receptor medium, total skin	ns
		Dimeb	1:1	1.2	Aq	Ex vivo	Porcine	Receptor medium, total skin	↑
		Pre-treatment Dimeb 24 h	—	—	Aq	Ex vivo	Porcine	Receptor medium, total skin	↑
Tenoxicam	[115]	RAMEB	1:1	57	AqG	Ex vivo	Rat	Receptor medium, total skin	↑
	[116]	RAMEB	—	—	AqG	Ex vivo	Rat	Receptor medium	ns
Testosterone	[2]	HP-β-CD	1:1	13,000	Aq	Ex vivo	Hairless mice	Receptor medium	↑
					o/w	Ex vivo	Hairless mice	Receptor medium	↑
Tolnaftate	[117]	β-CD polymer	1:2	17,000	Aq	In vivo	Mice	Total skin, blood	↑
Triamcinolone	[96]	Pretreatment HP-β-CD	—	—	Aq	Ex vivo	Hairless mice	Receptor medium	ns

[a] α-CD, α-cyclodextrin; β-CD, β-cyclodextrin; β-CD polymer, polymer of β-cyclodextrin; CME-β-CD, O-carboxymethyl-O-ethyl-β-cyclodextrin; Dimeb, 2,6-dimethyl-β-cyclodextrin; γ-CD, γ-cyclodextrin; HP-β-CD, hydroxypropyl-β-cyclodextrin; PVP, poly(vinylpyrrolidone); RAMEB, randomly methylated β-cyclodextrin; TA-α-CD, triamino-α-cyclodextrin.

[b] Anh, hydrocarbon gel base; Aq, aqueous solution or suspension; AqG, aqueous gel; AqO, hydrophilic ointment; ME, microemulsion; o/w, oil-in-water emulsion; O, ointment; PG, propylene glycol; TTS, transdermic therapeutic system.

[c] ↑, Increased absorption effect in the presence of cyclodextrin; ↓, decreased absorption effect in the presence of cyclodextrin; ns, nonsignificant effect in the presence of cyclodextrin.

3. POTENTIAL OF CDs IN DRUG STABILIZATION AND DRUG TOLERANCE

3.1. Drug Stabilization

3.1.1. Prevention of Drug Precipitation
CD molecules can first be employed to prevent the precipitation of poorly soluble drugs in the formulation during storage, as was shown with sericoside in extract form [22].

3.1.2. Stabilization of Labile Compounds
Stabilization of labile compounds into the formulation is essential to maintain the efficacy of drugs. It has been widely reported in the literature that CD molecules can improve the stability of several labile drugs against dehydration, hydrolysis (chemical or enzymatic), oxidation, and photodecomposition and thus increase the shelf life of drugs [9–11,23]. However, inclusion of drugs into CDs may also lead to faster degradation considering the characteristics of the inclusion complex [24,25]. For example, β-CD increases the profile degradation of vitamin A propionate, whereas γ-CD keeps a major part of this molecule unaffected for at least six months, even in the presence of light and oxygen [25]. In the γ-CD/ vitamin A propionate complex, the biggest part of the molecule is included in the CD, as the inner cavity is large enough and therefore protected from exterior environment. Only the terminal ester is affected by degradation. It may be supposed that this part of the molecule remains outside the cavity and is not protected or is attacked by the secondary hydroxyls of the γ-CD [25].

Human skin possesses metabolic capabilities that render some drugs unsuitable for skin delivery. Various enzymes, such as esterase, dehydrogenase, and glutathione-S-transferase, are known to be localized in the epidermis, particularly the basal cell layer [26–28]. CDs improve the protection of drugs such as prostaglandins [29] and corticoids [30] against skin metabolism [9,11]. As the ester group of dexamethasone acetate is at least partially enclosed in the cavity of HP-β-CD, the degradation of this compound by esterases located in the epidermis is reduced dramatically. In vitro, after a 2-h application on hairless mouse skin, about 30 and 65% of dexamethasone acetate was degraded by skin metabolism for dexamethasone acetate complexed and free, respectively [30]. However, the potential for CD molecules to affect degradation directly in the skin must be limited since most CDs are not absorbed by the skin, due to their hydrophilic character and high molecular weight [30].

3.2. Improvement of Drug Tolerance

Due to their inherent skin irritancy, some drugs applied to the skin provoke a local irritation, such as erythema, burning sensation, or desquamation. As a consequence of these unwanted side effects, patients usually discontinue therapy. During the development of a topical medication, it is necessary to consider this point since a well-tolerated formulation increases patient compliance and then the efficacy of the treatment. The incorporation of CD molecules into topical medications can be a satisfying alternative to limit side effects [3,9,11]. Thanks to their oligosaccharide composition, CDs are unlikely by themselves to cause any allergic reactions [17]. Treated with a retinoic acid commercial preparation, a drug employed successfully in the treatment of acne vulgaris and in some disorders of keratinization, all patients suffered from erythema and a burning sensation. In groups treated with hydrogel or moisturizing base containing retinoic acid/β-CD complexes, none of the patients reported marked local side effects [31]. Another study reported that irritation caused by retinoic acid/β-CD was 86% less than that caused by a free retinoic acid preparation [32]. The complex behaves as a controlled delivery system and limits the amount of free drug directly in contact with the skin. This effect was also observed with celecoxib, a nonsteroidal inflammatory drug employed in the treatment of inflammatory diseases [18]. This compound has an invasive action on stratum corneum, producing the destruction of desmosomes. In the presence of Dimeb or HP-β-CD/celecoxib complexes, the lesions to stratum corneum were less developed than those observed with the free drug. Because only free components can interact with stratum corneum, this trend can be the result of the complexation between the drug and the macrocycle of the drug. Moreover, as CD molecules permeate lipophilic membranes with difficulty [6] and drug–CD complexes are unable to permeate lipophilic biological membranes [6], only the solubilized free drug can cross the stratum corneum. Thus, a significant affinity between drug and CD limits the amount of free drug available for permeation.

CDs are also of interest in the formulation of drugs giving photosensitizing reactions in light-exposed tissues of patients treated in prolonged and high doses and in personnel in the health services. It is important to pay attention to the reduction of this type of dermatitis from the pharmaceutical and toxicological points of view. CD molecules were found efficient to reduce the photosensitizing potential of drugs such as chlorpromazine, an antipsychotic compound, and then skin irritation [33,34]. The photoallergic reactions to this drug may be induced by the photochemical reactions between the drug and proteins or other macromolecules present in the skin. Free chlorpromazine penetrates the skin easily, whereas the complexation with CD both alters the photochemical reactivity of the drug and limits its penetration into the skin. However, the efficacy of CD molecules depends on the free fraction of the drug, which is in equilibrium with the complexed fraction and then depends on the magnitude of the stability constant of the chlorpromazine/ CD complexes [33,34].

4. SKIN FORMULATIONS CONTAINING CDs

4.1. Conventional Formulations

Formulations applied on skin often consist of highly complex mixtures. The major vehicle components used in topical products are hydrophilic (water and cosolvents) and hydrophobic (petrolatum, soft and liquid paraffin, triglycerides, silicone oils, partial glyceride) base ingredients, emulsifiers, gelling agents, preservatives, and antioxidants [35]. One must be very careful with regard to the complexity of the formula, because CDs are able to modify the characteristics of the formulation but also to interact with formulation ingredients and cause physicochemical stability problems.

4.1.1. CDs and the Vehicle   The influence of the type of vehicle has been emphasized. For maximum delivery the drug–CD complex must be solubilized in the vehicle. Better results can be obtained from an aqueous vehicle than from a lipophilic vehicle. Indeed, the diffusion rate of ketoprofen from its β-CD and HP-β-CD inclusion complexes was in the order carbomer hydrogel > oil–water emulsion > fatty ointment. β- and HP-β-CD increase the in vitro release rate of hydrocortisone formulated in aqueous bases (oil/water cream and hydrogel) but slow down its release from nonaqueous bases (petrolatum vehicle or water–oil cream) [10]. Similarly, complexation with β-CD, Dimeb, and HP-β-CD increases the release of 4-biphenylacetic acid from hydrophilic ointment [36], and β- and HP-β-CD significantly enhanced the anti-inflammatory effects of indomethacin in hydroxyethylcellulose hydrogels in healthy volunteers. Conversely, the release of prednisolone from non-water-containing ointment bases was abated on complexation with Dimeb [9].

The excipient must not have too strong an affinity for the drug, so as not to displace the drug from the inside of the CD cavity to the bulk medium of the vehicle. As CDs are used most often to improve the bioavailability of poorly water-soluble ingredients (lipophilic), it seems preferable to use aqueous ointments (e.g., hydrogels). The use of oil/water emulsions, often selected as ointment bases for their pleasant touch, might be interesting but are probably also the origin of many dissociations. The drug can diffuse from the hydrophilic phase (where the inclusion complex is dissolved) to the lipophilic phase (depending on its stronger affinity).

4.1.2. CDs and the Polymer   The presence of β-CD associated with clotrimazole provokes for a few hours the precipitation of Carbomer 974, a mucoadhesive polymer used widely in the formulation of hydrogels. The mechanism of this incompatibility is not clear but is probably due to interaction of the polymer with the uncomplexed clotrimazole rather than with β-CD [37]. CDs may also act as inhibitors of polymer–drug interactions (especially with cationic compounds). Increasing concentrations of β-CD reduce interactions between the polymer and propanolol hydrochloride [38]. However, changes in rheological properties of hydrogels can also be observed in the presence of CD. The viscosity of Carbomer 974P gels prepared in water is decreased by RAMEB, whereas the viscosity is increased in the presence of γ-CD [39]. As RAMEB is more lipophilic than natural CD, hydrophobic interactions could occur between the polymer chains and the RAMEB, resulting in a reduction in the polymer chains unfolding. Consequently, it may modify the polymer affinity for the hydration medium, hence decreasing its swelling. Surprisingly, when Carbomer 974P NF gels were made in RAMEB/HEPES/NaCl solutions, no significant change in gel viscosity, even with 5% of RAMEB, was obtained compared to control gel prepared in the same buffer. Cationic electrolytes and RAMEB added in the same preparation do not induce additional effects on the viscosity decrease [39]. The incorporation of CD decreases the viscosity of poloxamer [40]. The bonding force of cross-linked poloxamer gel becomes weaker when HP-β-CD is incorporated in the formulation.

4.1.3. CDs and Cosolvents   Cosolvents such as propylene glycol and poly(ethylene glycol) are commonly used in skin formulations to increase the solubility of drugs. These molecules may interact with CD cavity [11]. For example, propylene glycol molecules displace piroxicam from inclusion with HP-β-CD cavity and then reduce complex stability [41] whereas interactions between α-CD and poly(ethylene glycol) result in the formation of crystalline inclusion complexes [42].

4.1.4. CDs and Preservatives   The formulations employed in cosmetic or dermatological practice require a low bacterial count. Given the hygiene risk associated with aqueous preparations, use of the product must be limited to a certain period of time after opening, or preservatives must be added to ensure microbiological quality, particularly if they are used in multidose applications. CD, due to their origin, favor the development of contamination with microorganisms. Among preservatives employed in skin formulations, we first mention the well-known family of p-hydroxybenzoic esters (methyl-, ethyl-, propyl-, and butylparabens). α- and β-CD [43,44], but also HP-β-CD [45,46], reduce the antimicrobial activity of parabens against *Candida albicans* (yeast). In the presence of β-CD, methylparaben showed a higher extent of interaction than ethyl- and propylparaben because the small size of the methylparaben molecule fits well into the cavity of β-CD. The loss of antimicrobial activity depends only on the fraction of preservative that is included in the CD molecule. Only the free part is used for antimicrobial attack [45]. Higher concentrations of parabens are then required for the same antimicrobial effect [45,46]. With HP-β-CD, the total amount of paraben must be increased two- to fourfold fold to preserve the antimicrobial effect. However, high concentrations of preservatives may cause some toxic effects for the user, due to their allergic potential. Finally, the interaction of methylparaben with HP-β-CD also results in a decrease of methylparaben permeability in vitro [47]. This promotes the bioconversion of methylparaben to the less toxic metabolite, *p*-hydroxybenzoic acid, in the epidermis.

The interactions between HP-β-CD and other preservatives having different chemical structures than that of parabens have also been investigated [46,48]. Thimerosal, phenylmercury acetate, and bronopol are well suited even if higher concentrations of HP-β-CD are used [46]. The antimicrobial activities of phenolic substances and aliphatic aryl–substituted alcohols are reduced too much and should not be applied in the presence of HP-β-CD. Interactions between benzalkonium chloride and HP-β-CD were also reported by Loftsson et al. [48]. In all cases, the loss in activity by complex formation correlated with the bound fraction.

Finally, the interaction of preservative molecules with CD may also displace the drug molecules from the CD cavity, thus reducing the solubilizing effect of the CD.

4.1.5. CDs and Emulsions

Emulsions are used widely in formulations applied on skin. However, they are thermodynamically unstable liquid–liquid dispersed systems that cause phase separation. The best way to stabilize the emulsion is through the use of an emulsifier which is an amphiphilic substance thermodynamically stable. CD molecules may interact with emulsifying agents inducing a coalescence of oil droplets [11]. Other works reported the ability of natural CDs to stabilize simple oil–water (o/w) [49–52] and multiple oil–water–oil (o/w/o) emulsions [53–55]. It is important to note that other emulsifying agents are not necessary to form these emulsions. The main interest in using CDs to stabilize emulsions is that their irritant potential is very weak in comparison with traditional surfactants, especially hydrophilic surface-active agents [56]. In works carried out with vegetable oils, increasing the concentration of natural CDs in the aqueous phase of the formulation results in an increase of emulsion stability [49]. Natural CDs themselves do not possess any surface-active properties [57]. A mechanism to explain the role played by CDs in emulsion stabilization has been proposed by Shimada and co-workers [57]. Vegetable oils are composed primarily of triglycerides. At the oil–water interface, a partial inclusion is formed between triglycerides and CDs. Indeed, only one fatty acid chain of the triglyceride can be entrapped with two or three CD molecules, depending on the length of the hydrocarbon chain. The two other fatty acid chains are not included in the CD. Due to its amphiphilic property (a hydrophilic head and a hydrophobic tail) the partial inclusion complex could play the role of a surface-active agent [57]. The best emulsifying effect is observed with α- and β-CD, whereas the

γ-CD is too wide to lead to an optimal interaction with the fatty acid chains [49,55]. Active ingredients may be added to emulsions formulated with CD and vegetable oils, but they must not interact with the CD cavity; otherwise, they displace the fatty acid chains and destabilize the emulsion. From this standpoint, high-molecular-weight active ingredients will probably not destabilize emulsions prepared with α-CD [50,54]. More recently, the formation of an emulsion consisting of *n*-alkane, water, and α-, β-, or γ-CD as an emulsifier was also reported [51,52,58,59]. The formation of a dense film at the oil–water interface and a three-dimensional structural network by precipitated complexes derived from CDs in the continuous phase is necessary for the formation of a stable emulsion. The type of emulsion is governed by the contact angle that the precipitated complex makes with the interface. The dissolved *n*-alkane/CD complex formed at low CD concentrations showed surface activity, but emulsions could not be prepared from these complexes. The precipitated complexes formed at a high CD level [59]. Finally, appropriate mixtures of paraffin or silicone oils with α- or β-CD can also result in emulsion formation [60].

As a conclusion, when skin formulations contained CD molecules, it seems preferable to use an aqueous phase in which the inclusion complex is soluble, in order to have free inclusion molecules capable of releasing the free active ingredient reversibly (according to the inclusion stability constant in the excipient). It is also important to consider a potential competition and/or interaction between excipient (polymer, cosolvent, emulsifying agent, preservative) and CD. In any case, it appears to be of the greatest importance to carry out serious aging tests to assess the stability of these systems.

4.2. New CD-Based Systems for Skin

Recent works report an interest in CDs in the design of novel drug carriers [61,62] such as beads, self-assembling CD-based systems, and liposomes, which can be promising for skin delivery.

4.2.1. Beads Made of Natural CDs and Oil

A new lipid carrier, beads made of natural CD and oil, has been reported in the literature [60,63–65]. A continuous external shaking for a few days of a mixture composed of α-CD, soybean oil, and water results in calibrated particles with a high fabrication yield (more than 80%) [63]. The resulting beads have a diameter of 1 to 2 mm and a semisolid consistency. They are made up of a partial crystalline matrix of CD molecules surrounding microdomains of oil. α-CD interacts with the triglycerides, which are the main components of vegetable oils, and participates in bead formation and organization [63]. Beads can be obtained from other ingredients using the method previously optimized for the soybean oil/α-CD/

water-based formulation. More particularly, γ-CD can also be used to formulate beads, and soybean oil can be replaced by mineral or synthetic oils [60]. However, soybean oil is the most useful option for bead formulation in terms of oil content and bead fabrication yield. The yield of encapsulation achieved for two retinoids employed in the treatment of acne (i.e., isotretinoin (a very poorly stable drug) [64] and adapalene [65]) reaches as high as 90 to 100%. Furthermore, the drug concentration in the beads can be controlled easily by varying the initial drug loading of the oily phase. After application to the skin of volunteers, no clinical reaction (erythema or desquamation) is observed after drug-free bead application for 72 h. The oil content of the beads is sufficient to provide an occlusive effect [65]. Finally, a tape-stripping procedure on pig skin shows that adapalene penetration into the stratum corneum is the same from beads as from gel and cream available on the market [65]. Beads can be proposed as a new, well-tolerated and efficient system for encapsulation and topical delivery of lipophilic drugs.

4.2.2. Self-Assembling CD-Based Systems
Daoud-Mahammed and co-workers showed the possibility of preparing in situ forming gel-like systems, based on the association of two hydrosoluble polymers [66], a β-CD polymer (*p*-β-CD) and a hydrophobically modified dextran (MD), by grafting alkyl side chains. These gels form spontaneously after mixing the two polymer aqueous solutions: Some alkyl moieties are included in some CD cavities, leaving CD available to include hydrophobic drugs such as tamoxifen and benzophenone with high loading efficiencies [67]. Both entrapped compounds are released in a sustained manner in vitro. The mild conditions of gel preparation (no organic solvents are required) as well as the presence of CD, known to modulate the release of drugs, are obviously advantages for their pharmaceutical applications. The soft consistency of MD-*p*-β-CD gels makes them compatible with injection through a syringe needle but also for skin application. MD-*p*-βCD gels present viscoelastic behavior under low shear [67]. Interestingly, colloidal nanoassemblies (nanogels) can also be formed employing *p*-β-CD and hydrophobically, MD, just by decreasing the concentration of the two polymers in aqueous solution. The ability of nanogels to entrap hydrophobic molecules opens new possibilities of applications of these systems, mainly in the cosmetic and dermatological fields [68].

New synthetic biocompatible materials able to deliver a protein to cultured cells via the use of an adenoviral delivery vector have been developed by Bellocq and co-workers [69]. The synthetic construct consists of linear CD-based poly (ethylene glycol) polymers and their inclusion complex formation with adamantane end-capped poly(ethylene glycol) polymers. When the two polymers are combined, they create an extended network by the formation of inclusion complexes between the CD and the adamantane. The

CD–adamantane constructs are highly biocompatible. Fibroblasts exposed to these synthetic constructs show proliferation rates and migration patterns similar to those obtained with collagen. Gene delivery to fibroblasts via the inclusion of adenoviral vectors in the synthetic construct is equivalent to levels observed with collagen. These in vitro results suggest that the synthetic constructs are very promising for local gene delivery to improve cutaneous wound healing [69].

4.2.3. CD-Based Liposomes
Liposomes are vesicles less than 1 μm in size composed of one or more phospholipid bilayers enclosing an aqueous phase. Liposomes can entrap hydrophilic molecules inside the vesicles, and hydrophobic molecules within the lipid bilayer. However, retention of lipophilic compounds in the lipid bilayer can be problematic because some molecules destabilize its structure, thus limiting the diversity and quantity of molecules that can be carried by liposomes [61]. CDs can improve the drug stability (against light or hydrolysis), enhance the loading efficiency of hydrophobic drugs, and modify their localization within liposomes. As a result, the drug included is predominantly incorporated within the aqueous core rather than in the lipid bilayer, changing its release profile [61]. However, it is also known that CDs extract lipid components from a bilayer of liposomes. This could undermine the potential benefits of liposomes as drug carriers. Phosphatidylcholine–cholesterol liposomes with various CDs may be stabilized by association with the amphiphilic polyelectrolyte poly(methacrylic acid-*co*-stearyl methacrylate) [70]. The polymer-associated liposome had the same vesicular form as liposome, and this structure is unaffected by the concentration and type of CD or the concentration of lipid components and drug. The improved stability was maintained when CDs were complexed with hydrophobic drugs. Skin permeability results showed that the stability of vesicles affects their function. The stability of vesicles could influence the skin permeability of CD–drug complexes [70].

In most cases, classical liposomes are of little or no value as carriers for transdermal drug delivery, as they do not penetrate skin deeply but, rather, remain confined to upper layers of the stratum corneum. Several studies have reported that deformable liposomes are able to improve in vitro skin delivery of various drugs. Recently, a new delivery system for cutaneous administration combining the advantages of CD inclusion complexes and those of deformable liposomes was developed by Gillet and co-workers [71], leading to a new concept: drug-in-CD-in-deformable liposomes. Betamethasone, chosen as the model drug, was encapsulated in the aqueous cavity of deformable liposomes (made of soybean phosphatidylcholine and sodium deoxycholate as edge activator) by the use of CD. CDs allow an increase in the aqueous solubility of β-methasone and thus the encapsulation efficiency in liposome vesicles. In comparison with nondeformable liposomes, these new vesicles showed improved encapsulation efficiency, higher deformability, a good stability, when stored at 4°C and higher in vitro diffusion percentages of encapsulated drug. These new vesicles are therefore promising for future use in ex vivo and in vivo experiments.

5. SKIN DELIVERY OF DRUGS WITH CDs [3,4,6,8–11]

5.1. Effects of CDs on Drug Absorption

Lack of unified valid pharmacopoeial in vitro release methods makes it difficult to compare a new semisolid formulation with CD addition to one without modification. The most popular and U.S. Food and Drug Administration–approved method for dissolution and permeation testing from topical preparations is with the use of a Franz diffusion cell. However, there is no unification of test parameters; the dimensions differ a lot between laboratories and manufacturers. This may lead to a discrepancy between results [11]. As shown in the nonexhaustive Table 1, there are numerous studies of the effects of CDs on topical drug availability in various CD-containing vehicles performed in different conditions. Comparison of the results is difficult, as some studies are performed in vivo, others ex vivo or in vitro, some with human skin, others with rat, mouse, or pig skin (Table 1). Moreover, evaluation methods are not the same; some evaluated penetration through the skin (in the receptor medium) while others evaluated the quantity in the whole skin or in strata.

Despite the complexity of the results, Loftsson et al. [6] were able to make some observations. Differences observed between skins used may be explained by the fact that under ex vivo conditions using hairless mouse skin, which is more permeable than human skin, and when the aqueous donor phase is unstirred, the permeation resistance in the unstirred water layer can be higher or equal to the resistance within the membrane. Under such conditions, CDs may enhance dermal and transdermal drug delivery. As a matter of fact, Table 1 shows that most of the studies are done with hairless mouse skin, and generally, this type of skin shows better results than those using human skin. However, it must be kept in mind that studies showing good results are more easily published.

Moreover, results show that generally hydrophilic CDs are [6]:

1. Unable to permeate biological membranes such as skin to any significant extent
2. Unable to enhance drug permeation from lipophilic environments
3. Unable to enhance permeation of hydrophilic drugs
4. Able to enhance permeation of lipophilic drugs

5. Able to reduce drug permeation through lipophilic membranes by decreasing drug partition from the exterior into the membrane
6. Able to increase the chemical stability of drugs at the aqueous membrane exterior

Thus, for CD molecules, the potential mechanism of action must be different from those of classical chemical penetration enhancers, which increase drug permeation through the membranes by penetrating the membranes and decreasing the barrier properties, making the membranes more permeable by, for example, increasing their hydration, modifying their intracellular lipid domains, or enhancing drug partition into the membranes by changing their solvent nature. Contrary to CDs, these chemical penetration enhancers improve the membrane permeation of both hydrophilic and lipophilic drugs, from both nonaqueous and aqueous donor phases.

5.2. Roles of CDs in Drug Absorption

Mechanisms are complexes but could be resumed as follows. In general, drugs permeate biomembranes such as skin by passive diffusion. Under such conditions there is a net flux of drug molecules from a donor phase, through the membrane, to the receptor phase. According to Fick's first law [equation (1)], where J is the drug flux across the membrane, $K_{m/d}$ the drug partition coefficient between donor phase and membrane, D_m the drug diffusion coefficient within the membrane, $\Delta[D]$ the difference in drug concentration between the donor phase and the receptor side of the membrane, and h_m is the thickness of the membrane [4], the driving force for the diffusion is the concentration gradient of drug molecules across the membrane:

$$J = \frac{K_{m/d}D_m}{h_m}\Delta[D]$$

As we will see, CDs may have an effect at different levels of Fick's first law, explaining the complexity of their action mechanism.

5.2.1. Effect of CDs on Drug Solubility Increased solubility by CDs has a direct effect on the $\Delta[D]$ parameter of Fick's equation. CDs increase the permeability of insoluble, hydrophobic drugs by making them available at the surface of the skin. The increase in bioavailability is due primarily to higher concentration in the site of administration, caused by higher aqueous solubility and thus improved availability onto the tissue surface than enhancement activity by the CD itself. By increasing solubility, CDs facilitate drug incorporation into formulation and thus increase the drug concentration in the formulation. Gels containing tenoxicam–rameb complexes enhanced the percutaneous penetration of the drug by improving its solubility. This increase in solubility makes it possible to double the amount of tenoxicam formulated, subsequently increasing the release and absorption parameters [115]. HP-β-CD increased the amount of piroxicam transported through skin, but pretreatment of skin with the CD showed no effect on drug retention in the skin. Hence, the CD's effect on the drug's skin permeability was reported to be due to increased drug concentration in gel [41].

CDs enhance the drug thermodynamic activity in vehicles and thus cause enhancement of drug release from vehicles. The thermodynamic activity is proportional to the solubility of drugs in a vehicle and becomes maximal in the saturated solution [9]. In such cases it is important to use just enough CD to solubilize the drug in the aqueous vehicle since excess may decrease the drug availability. Loftsson et al. [4,94] illustrated the effect of the CD concentration on the flux of hydrocortisone from aqueous vehicles containing hydrocortisone–CD complexes through hairless mouse skin. The hydrocortisone concentration was kept constant, but the CD concentration was increased from 1 to 20% w/v. When hydrocortisone was in suspension, an increase in the CD concentration resulted in an increased amount of dissolved drug, and this increase in solubility led to a larger flux through the skin. In contrast, when all the hydrocortisone was in solution, an increase in the CD concentration led to increased CD complexation of the drug molecules, and because the hydrated hydrocortisone–CD complex was unable to permeate the skin, this resulted in a decrease in the flux. The maximum flux through the skin was obtained when just enough CD was used to keep all hydrocortisone in solution. Comparable results were obtained from an oil-in-water cream [3,93].

In aqueous drug solutions, when the drug concentration is constant and below saturation, transdermal flux will decrease with increasing CD concentration. Thus, flux will decrease as the chemical potential of the drug decreases. In saturated solutions, the chemical potential is constant. The flux from an aqueous vehicle, which is saturated with the drug, will increase with increasing CD concentration if the drug is solubilized by CD.

5.2.2. Effect of the Affinity Constant

Because only the uncomplexed drug can be absorbed, the complex stability constant has a great influence on the drug bioavailability from inclusion complex. If the complex stability constant is too high, the complex may not release the free drug at the absorptive site and thus may decrease or inhibit drug absorption. In ointments, a drug in the CD complex may be displaced by ointment components, depending on the magnitude of the stability constant of the drug–CD complex. Hence, for optimum drug release, the vehicle or CD complex chosen should be such that the complex barely dissociates but still maintains a high drug thermodynamic activity in the

vehicle. For example, the order of the prednisolone release rate from a hydrophilic ointment was drug alone < γ-CD complex < β-CD complex < Dimeb complex, which was reflective of the order of the complex stability constants [9,10].

5.2.3. Effect of CDs on the Drug Partition Coefficient

Although the drug partition coefficient (e.g., a lipophilic drug) may be decreased on complexation with CD (e.g., with hydrophilic CD), the increased drug solubility and thermodynamic activity in vehicles can lead to increased drug permeability through skin (e.g., increased skin permeability of dexamethasone by HP-β-CD) [10].

5.2.4. Effect of CDs on the Drug Stabilization

As reported in Section 3.1.2, labile drug stabilization by CDs and their ability to ameliorate drug irritation, and thus to improve drug concentration and contact time at the absorption site in transdermal delivery, are some other important factors that contribute to the CD-improved bioavailability [2,8].

5.2.5. Effect of CDs as a Penetration Enhancer

Although only insignificant amounts of CD and drug–CD complexes can penetrate biological barriers because of their molecular weight and their hydrophilicity, CDs may interact with some of the skin components. Free CDs released on complex dissociation, due to their ability to remove some membrane surface components, can modify the membrane transport properties and thus can facilitate the absorption of drugs, especially water-soluble drugs [10]. Thus although CDs are able, under some specific conditions, to extract lipophilic components of biological membranes, it is highly unlikely that the principal mechanism of CD-enhanced transdermal drug delivery is disruption of the barrier.

Conventional penetration enhancers enhance drug permeability by causing some physicochemical changes within the barrier. Thus, combining CDs with conventional enhancers should result in an additive effect. This is exactly what has been observed. In one study, the effects of both HP-β-CD and a glycerol monoether extract on transdermal delivery of testosterone, from o/w cream through hairless mouse skin, was investigated [2]. About a 60% increase in the testosterone flux was observed when HP-β-CD was added to the cream, about a 40% increase was observed when the extract was added to the cream, but about an 80% increase in the flux was observed when both HP-β-CD and the extract were added to the cream.

5.3. Iontophoresis and Electroporation Techniques

Iontophoresis and electroporation are physical permeation-enhancing techniques which have been shown to be effective in delivering drug and macromolecules across the skin. Transdermal iontophoresis is the administration of therapeutic agents in ionic form through the skin by the application of a low-level electric current [118]. Doliwa et al. [119,120] and Chang and Banga [97] showed that iontophoresis might be used to increase piroxicam/HP-β-CD and hydrocortisone/HP-β-CD complex transdermal fluxes. In general, it was the free ionized piroxicam that was predominantly delivered iontophoretically through the skin. However, the presence of CDs allowed the incorporation of higher piroxicam into the gel, and this might result in increased delivery [119].

Electroporation is a process in which brief, intense electric charges create small pores in the phospholipid bilayer of cell membranes which can assist in the transdermal delivery of drugs and be effective in delivering drug and macromolecules across the skin. Electroporation appears to disrupt the lipid bilayers in the stratum corneum [118]. Electroporation also increased the permeation of CD-complexed drugs by several orders of magnitude relative to passive transport. The presence of β-CD enhanced the total transport of piroxicam from both permeant solutions and suspensions across the epidermis into the receiver compartment medium [121]. Based on an extensive review of the literature, it appears that CDs may enhance drug delivery through the skin by different routes. It is also possible that CDs may have other mechanisms of action that have not yet been elucidated.

6. BIOLOGICAL EFFECT OF CDs

The lipid components of biological membranes are important for normal cell function, and changes in the organization of lipids can have profound effects on cellular functions such as signal transduction and membrane trafficking. Cholesterol is one of the most important regulators of lipid organization. In the skin, lipids are involved in epidermal differentiation and pigmentation. Cholesterol has been shown to be involved in the late stages of epidermal differentiation by promoting cornified envelope formation [122].

Cholesterol depletion by β-CD or chemically modified β-CD from cellular membranes was used for studies on physiological roles of membrane cholesterol. Jin showed that cholesterol reduction induced by rameb inhibits melanogenesis in human melanocytes through activation of ERK (extracellular signal-regulated kinase) [122]. On the other hand, MMP-9 (matrix metalloproteinase-9) expression in human epidermal keratinocytes was increased by cholesterol depletion due to methyl-β-CD [123]. Gniadecki reports that disruption of lipid rafts by cholesterol-depleting compounds such as methyl-β-CD leads to a spontaneous clustering of Fas in the nonraft compartment of the plasma membrane, formati, on Fas–FADD complexes, activation of caspase-8, and apoptosis. In some cell types, exclusion of Fas from lipid rafts could lead to the spontaneous ligand-independent activation of this death receptor, a mechanism that can potentially be utilized in anticancer therapy [124]. Other studies report the effects of CDs on keratinocytes and other skin cells (see [125–132] and Chapter 16).

7. CONCLUSIONS

In this chapter we have seen that CDs present some advantages in the formulation of topical products (drug stabilization, reduction of local drug irritation, and in some cases, increased drug absorption). CDs are not able to penetrate biological membranes during topical application under normal conditions, although they can modify the bioavailability of lipophilic drugs mainly by increasing their concentration in the formulation. In topical formulations, CDs are generally used as solubilizers. However, the use of CDs for skin delivery is still limited. Currently, only one pharmaceutical product containing a CD is marketed in dermatology. It is an ointment containing dexamethasone and β-CD [133]. Their main limitations in skin formulations are due to their capacity to interact with some ingredients of the formulation but also to modify its characteristics and cause physicochemical stability problems. The development of these formulations is not simple and requires in-depth study.

REFERENCES

1. Loftsson, T., Brewster, M.E. (1996). Pharmaceutical applications of cyclodextrins: 1. Drug solubilization and stabilization. *J. Pharm. Sci.*, *85*, 1017–1025.

2. Loftsson, T., Másson, M., Sigurdsson, H.H., Magnusson, P., Le Goffic, F. (1998). Cyclodextrins as co-enhancers in dermal and transdermal drug delivery. *Pharmazie*, *53*, 137–139.

3. Loftsson, T., Ólafsson, J. H. (1998). Cyclodextrins: new drug delivery systems in dermatology. *Int. J. Dermatol.*, *37*, 241–246.

4. Loftsson, T., Másson, M. (2001). Cyclodextrins in topical drug formulations: theory and practice. *Int. J. Pharm.*, *225*, 15–30.

5. Loftsson, T., Máasson, M., Sigurdsson, H.H. (2002). Cyclodextrins and drug permeability through semi-permeable cellophane membranes. *Int. J. Pharm.*, *232*, 35–43.

6. Loftsson, T., Vogensen, S.B., Brewster, M.E., Konradsdóttir, F. (2007). Effects of cyclodextrins on drug delivery through biological membranes. *J. Pharm. Sci.*, *96*, 2532–2546.

7. Loftsson, T., Duchêne, D. (2007). Cyclodextrins and their pharmaceutical applications. *Int. J. Pharm.*, *329*, 1–11.

8. Másson, M., Loftsson, T., Másson, G., Stefánsson, E. (1999). Cyclodextrins as permeation enhancers: some theoretical evaluations and in vitro testing. *J. Control. Release*, *59*, 107–118.

9. Matsuda, H., Arima, H. (1999). Cyclodextrins in transdermal and rectal delivery. *Adv. Drug Deliv. Rev.*, *36*, 81–99.

10. Challa, R., Ahuja, A., Ali, J., Khar, R.K. (2005). Cyclodextrins in drug delivery: an updated review. *AAPS PharmSciTech*, 6, 329–357.

11. Cal, K., Centkowska, K. (2008). Use of cyclodextrins in topical formulations: practical aspects. *Eur. J. Pharm. Biopharm.*, 68, 467–478.

12. Okamoto, H., Komatsu, H., Hashida, M., Sezaki, H. (1986). Effects of β-cyclodextrin and di-O-methyl-β-cyclodextrin on the percutaneous-absorption of butylparaben, indomethacin and sulfanilic acid. *Int. J. Pharm.*, 30, 35–45.

13. Vollmer, U., Muller, B.W., Peeters, J., Mesens, J., Wilffert, B., Peters, T. (1994). A study of the percutaneous absorption-enhancing effects of cyclodextrin derivatives in rats. *J. Pharm. Pharmacol.*, 46, 19–22.

14. Hovgaard, L., Brøndsted, H. (1995). Drug delivery studies in Caco-2 monolayers: IV. Absorption enhancer effects of cyclodextrins. *Pharm. Res.*, 12, 1328–1332.

15. Legendre, J.Y., Rault, I., Petit, A., Luijten, W., Demuynck, I., Horvath, S., Ginot, Y.M., Cuine, A. (1995). Effects of β-cyclodextrins on skin: implications for the transdermal delivery of piribedil and a novel cognition enhancing-drug, S-9977. *Eur. J. Pharm. Sci.*, 3, 311–322.

16. Vitoria, M., Bentley, L.B., Vianna, R.F., Wilson, S., Collett, J.H. (1997). Characterization of the influence of some cyclodextrins on the stratum corneum from the hairless mouse. *J. Pharm. Pharmacol.*, 49, 397–402.

17. Irie, T., Uekama, K. (1997). Pharmaceutical applications of cyclodextrins: 3. Toxicological issues and safety evaluation. *J. Pharm. Sci.*, 86, 147–162.

18. Ventura, C.A., Tommasini, S., Falcone, A., Giannone, I., Paolino, D., Sdrafkakis, V., Mondello, M.R., Puglisi, G. (2006). Influence of modified cyclodextrins on solubility and percutaneous absorption of celecoxib through human skin. *Int. J. Pharm.*, 314, 37–45.

19. Bochot, A., Duchêne, D. (2006). Cyclodextrines. In: Doc, T., Ed., *Actifs et additifs en cosmétologie*, Lavoisier, Paris, pp. 993–1018.

20. Piel, G., Moutard, S., Uhoda, E., Pilard, F., Pierard, G.E., Perly, B., Delattre, L., Evrard, B. (2004). Skin compatibility of cyclodextrins and their derivatives: a comparative assessment using a corneoxenometry bioassay. *Eur. J. Pharm. Biopharm.*, 57, 479–482.

21. Babu, R.J., Pandit, J.K. (2004). Effect of cyclodextrins on the complexation and transdermal delivery of bupranolol through rat skin. *Int. J. Pharm.*, 271, 155–165.

22. Rode, T., Frauen, M., Muller, B.W., Dusing, H.J., Schonrock, U., Mundt, C., Wenck, H. (2003). Complex formation of sericoside with hydrophilic cyclodextrins: improvement of solubility and skin penetration in topical emulsion based formulations. *Eur. J. Pharm. Biopharm.*, 55, 191–198.

23. Loftsson, T., Petersen, D.S. (1997). Cyclodextrin solubilization of water-insoluble drugs: calcipotriol and EB-1089. *Pharmazie*, 52, 783–785.

24. Adachi, H., Irie, T., Hirayama, F., Uekama, K. (1992). Stabilization of prostaglandin-E1 in fatty alcohol propylene-glycol ointment by acidic cyclodextrin derivative, *ortho*-carboxymethyl-*ortho*-ethyl-β-cyclodextrin. *Chem. Pharm. Bull.*, 40, 1586–1591.

25. Weisse, S., Perly, B., Creminon, C., Ouvrard-Baraton, F., Djedaïni-Pilard, F. (2004). Enhancement of vitamin A skin absorption by cyclodextrins. *STP Pharm. Sci.*, 14, 77–86.

26. Potts, R.O., Mcneill, S.C., Desbonnet, C.R., Wakshull, E. (1989). Transdermal drug transport and metabolism: 2. The role of competing kinetic events. *Pharm. Res.*, 6, 119–124.

27. Chan, S.Y., Po, A.L.W. (1989). Prodrugs for dermal delivery. *Int. J. Pharm.*, 55, 1–16.

28. Liu, P.H., Higuchi, W.I., Ghanem, A.H., Kuriharabergstrom, T., Good, W.R. (1990). Quantitation of simultaneous diffusion and metabolism of β-estradiol in hairless mouse skin: enzyme distribution and intrinsic diffusion metabolism parameters. *Int. J. Pharm.*, 64, 7–25.

29. Uekama, K., Adachi, H., Irie, T., Yano, T., Saita, M., Noda, K. (1992). Improved transdermal delivery of prostaglandin-E1 through hairless mouse skin: combined use of carboxymethyl-ethyl-β-cyclodextrin and penetration enhancers. *J. Pharm. Pharmacol.*, 44, 119–121.

30. Lopez, R.F.V., Collett, J.H., Bentley, M.V.L.B. (2000). Influence of cyclodextrin complexation on the in vitro permeation and skin metabolism of dexamethasone. *Int. J. Pharm.*, 200, 127–132.

31. Anadolu, R.Y., Sen, T., Tarimci, N., Birol, A., Erdem, C. (2004). Improved efficacy and tolerability of retinoic acid in acne vulgaris: a new topical formulation with cyclodextrin complex psi. *J. Eur. Acad. Dermatol. Venereol.*, 18, 416–421.

32. Amdidouche, D., Darrouzet, H., Duchêne, D., Poelman, M.C. (1989). Inclusion of retinoic acid in β-cyclodextrin. *Int. J. Pharm.*, 54, 175–179.

33. Irie, T., Uekama, K. (1985). Protection against the photosensitized skin irritancy of chlorpromazine by cyclodextrin complexation. *J. Pharmacobio-Dyn.*, 8, 788–791.

34. Hoshino, T., Ishida, K., Irie, T., Uekama, K., Ono, T. (1989). An attempt to reduce the photosensitizing potential of chlorpromazine with the simultaneous use of β-cyclodextrins and dimethyl-β-cyclodextrins in guinea-pigs. *Arch. Dermatol. Res.*, 281, 60–65.

35. Daniels, R., Knie, U. (2007). Galenics of dermal products: vehicles, properties and drug release. *J. Dtsch. Dermatol. Gesellsch.*, 5, 367–383.

36. Arima, H., Miyaji, T., Irie, T., Hirayama, F., Uekama, K. (1996). Possible enhancing mechanism of the cutaneous permeation of 4-biphenylylacetic acid by β-cyclodextrin derivatives in hydrophilic ointment. *Chem. Pharm. Bull.*, 44, 582–586.

37. Bilensoy, E., Rouf, M.A., Vural, I., Şen, M., Hincal, A.A. (2006). Mucoadhesive, thermosensitive, prolonged-release vaginal gel for clotrimazole: β-cyclodextrin complex. *AAPS PharmSciTech*, 7, 1–7.

38. Blanco-Fuente, H., Esteban-Fernández, B., Blanco-Méndez, J., Otero-Espinar, F.J. (2002). Use of β-cyclodextrins to prevent modifications of the properties of carbopol hydrogels due to carbopol-drug interactions. *Chem. Pharm. Bull.*, 40, 40–46.

39. Boulmedarat, L., Grossiord, J.L., Fattal, E., Bochot, A. (2003). Influence of methyl-β-cyclodextrin and liposomes on rheological properties of Carbopol (R) 974P NF gels. *Int. J. Pharm.*, *254*, 59–64.

40. Kim, E., Gao, Z., Park, J., Li, H., Han, K. (2002). rhEGF/HP-β-CD complex in poloxamer gel for ophthalmic delivery. *Int. J. Pharm.*, *233*, 159–167.

41. Doliwa, A., Santoyo, S., Ygartua, P. (2001). Influence of piroxicam : hydroxypropyl-β-cyclodextrin complexation on the in vitro permeation and skin retention of piroxicam. *Skin Pharmacol. Appl. Skin Physiol.*, *14*, 97–107.

42. Harada, A. (1996). Preparation and structures of supramolecules between cyclodextrins and polymers. *Coord. Chem. Rev.*, *148*, 115–133.

43. Uekama, K., Ikeda, Y., Hirayama, F., Otagiri, M., Shibata, M. (1980). Inclusion complexation of *p*-hydroxybenzoic acid esters with α and β-cyclodextrins: dissolution behaviour and antimicrobial activities. *Yakugaku Zasshi*, *100*, 994–1003.

44. Chan, L.W., Kurup, T.R.R., Muthaiah, A., Thenmozhiyal, J.C. (2000). Interaction of *p*-hydroxybenzoic esters with β-cyclodextrin. *Int. J. Pharm.*, *195*, 71–79.

45. Lehner, S.J., Muller, B.W., Seydel, J.K. (1993). Interactions between *p*-hydroxybenzoic acid-esters and hydroxypropyl-β-cyclodextrin and their antimicrobial effect against *Candida albicans*. *Int. J. Pharm.*, *93*, 201–208.

46. Lehner, S.J., Muller, B.W., Seydel, J.K. (1994). Effect of hydroxypropyl-β-cyclodextrin on the antimicrobial action of preservatives. *J. Pharm. Pharmacol.*, *46*, 186–191.

47. Tanaka, M., Iwata, Y., Kouzuki, Y., Taniguchi, K., Matsuda, H., Arima, H., Tsuchiya, S. (1995). Effect of 2-hydroxypropyl-β-cyclodextrin on percutaneous absorption of methyl paraben. *J. Pharm. Pharmacol.*, *47*, 897–900.

48. Loftsson, T., Stefansdóttir, O., Frioriksdóttir, H., Guomundsson, O. (1992). Interactions between preservatives and 2-hydroxypropyl-β-cyclodextrin. *Drug Dev. Ind. Pharm.*, *18*, 1477–1484.

49. Shimada, K., Ohe, Y., Ohguni, T., Kawano, K., Ishii, J., Nakamura, T. (1991). Emulsifying properties of α-cyclodextrins, β-cyclodextrins and γ-cyclodextrins. *J. Jpn. Soc. Food Sci. Technol. (Nippon Shokuhin Kagaku Kogaku Kaishi)*, *38*, 16–20.

50. Yu, S.C., Bochot, A., Le Bas, G., Cheron, M., Grossiord, J.L., Seiller, M., Duchêne, D. (2001). Characteristics of o/w emulsions containing lipophilic molecules with cyclodextrins as emulsifiers. *STP Pharm. Sci.*, *11*, 385–391.

51. Hashizaki, K., Kageyama, T., Inoue, M., Taguchi, H., Ueda, H., Saito, Y. (2007). Study on preparation and formation mechanism of *n*-alkanol/water emulsion using α-cyclodextrin. *Chem. Pharm. Bull.*, *55*, 1620–1625.

52. Inoue, M., Hashizaki, K., Taguchi, H., Saito, Y. (2008). Emulsion preparation using β-cyclodextrin and its derivatives acting as an emulsifier. *Chem. Pharm. Bull.*, *56*, 1335–1337.

53. Yu, S.C., Bochot, A., Cheron, M., Seiller, M., Grossiord, J.L., Le Bas, G., Duchêne, D. (1999). Design and evaluation of an original o/w/o multiple emulsion containing natural cyclodextrins as the emulsifier. *STP Pharm. Sci.*, *9*, 273–277.

54. Yu, S.C., Bochot, A., Le Bas, G., Cheron, M., Mahuteau, J., Grossiord, J.L., Seiller, M., Duchêne, D. (2003). Effect of camphor/cyclodextrin complexation on the stability of o/w/o multiple emulsions. *Int. J. Pharm.*, *261*, 1–8.

55. Duchêne, D., Bochot, A., Yu, S.C., Pepin, C., Seiller, M. (2003). Cyclodextrins and emulsions. *Int. J. Pharm.*, *266*, 85–90.

56. Duchêne, D., Wouessijewe, D., Poelman, M.C. (1991). In: Duchêne, D., Ed., *New Trends in Cyclodextrins and Derivatives*. Editions de Santé, Paris, pp. 448–481.

57. Shimada, K., Kawano, K., Ishii, J., Nakamura, T. (1992). Structure of inclusion complexes of cyclodextrins with triglyceride at vegetable oil–water interface. *J. Food Sci.*, *57*, 655–656.

58. Inoue, M., Hashizaki, K., Taguchi, H., Saito, Y. (2008). Formation and characterization of emulsions using β-cyclodextrin as an emulsifier. *Chem. Pharm. Bull.*, *56*, 668–671.

59. Inoue, M., Hashizak, I.K., Taguchi, H., Saito, Y. (2009). Preparation and characterization of *n*-alkane/water emulsion stabilized by cyclodextrin. *J. Oleo Sci.*, *58*, 85–90.

60. Trichard, L., Fattal, E., Le Bas, G., Duchêne, D., Grossiord, J.L., Bochot, A. (2008). Formulation and characterisation of beads prepared from natural cyclodextrins and vegetable, mineral or synthetic oils. *Int. J. Pharm.*, *354*, 88–94.

61. Trichard, L., Bochot, A., Duchêne, D. (2006). Cyclodextrins in dispersed systems. In: Dodziuk, D. Ed., *Cyclodextrins and Their Complexes: Chemistry, Analytical Methods, Applications*. Wiley-VCH, Warsaw, Poland, pp. 423–446.

62. Vyas, A., Saraf, S., Saraf, S. (2008). Cyclodextrin based novel drug delivery systems. *J. Inclusion Phenom. Macrocyclic Chem.*, *62*, 23–42.

63. Bochot, A., Trichard, L., Le Bas, G., Alphandary, H., Grossiord, J.L., Duchêne, D., Fattal, E. (2007). α-cyclodextrin/oil beads: an innovative self-assembling system. *Int. J. Pharm.*, *339*, 121–129.

64. Trichard, L., Fattal, E., Besnard, M., Bochot, A. (2007). α-Cyclodextrin/oil beads as a new carrier for improving the oral bioavailability of lipophilic drugs. *J. Control. Release*, *122*, 47–53.

65. Trichard, L., Delgado-Charro, M.B., Guy, R.H., Fattal, E., Bochot, A. (2008). Novel beads made of α-cyclodextrin and oil for topical delivery of a lipophilic drug. *Pharm. Res.*, *25*, 435–440.

66. Daoud-Mahammed, S., Couvreur, P., Amiel, C., Besnard, M., Appel, M., Gref, R. (2004). Original tamoxifen-loaded gels containing cyclodextrins: in situ self-assembling systems for cancer treatment. *J. Drug Deliv. Sci. Technol.*, *14*, 51–55.

67. Daoud-Mahammed, S., Grossiord, J.L., Bergua, T., Amiel, C., Couvreur, P., Gref, R. (2007). Self-assembling cyclodextrin based hydrogels for the sustained delivery of hydrophobic drugs. *J. Biomed. Mater. Res. A*, *86*, 736–748.

68. Daoud-Mahammed, S., Couvreur, P., Bouchemal, K., Chéron, M., Le Bas, G., Amiel, C., Gref, R. (2009). Cyclodextrin and polysaccharide-based nanogels: entrapment of two hydrophobic molecules, benzophenone and tamoxifen. *Biomacromolecules*, *10*, 547–554.

69. Bellocq, N.C., Kang, D.W., Wang, X., Jensen, G.S., Pun, S.H., Schluep, T., Zepeda, M.L., Davis, M.E. (2004). Synthetic biocompatible cyclodextrin-based constructs for local gene delivery to improve cutaneous wound healing. *Bioconjugate Chem.*, *15*, 1201–1211.

70. Lim, H.J., Cho, E.C., Shim, J., Kim, D.H., An, E.J., Kim, J. (2008). Polymer-associated liposomes as a novel delivery system for cyclodextrin-bound drugs. *J. Colloid Interface Sci.*, *320*, 460–468.

71. Gillet, A., Grammenos, A., Compère, P., Evrard, B., Piel, G. (2009). Development of a new topical system: drug-in-cyclo-dextrin-in-deformable liposome. *Int. J. Pharm.*, *380*, 174–180.

72. Loftsson, T., Sigurdardóttir, A.M., Ólafsson, J. H. (1995). Improved acitretin delivery through hairless mouse skin by cyclodextrin complexation. *Int. J. Pharm.*, *115*, 255–258.

73. Ammar, H.O., Ghorab, M., El-Nahhas, S.A., Kamel, R. (2006). Design of a transdermal delivery system for aspirin as an antithrombotic drug. *Int. J. Pharm.*, *327*, 81–88.

74. Uekama, K., Otagiri, M., Sakai, A., Irie, T., Matsuo, N., Matsuoka, Y. (1985). Improvement in the percutaneous-absorption of beclomethasone dipropionate by γ-cyclodextrin complexation. *J. Pharm. Pharmacol.*, *37*, 532–535.

75. Otagiri, M., Fujinaga, T., Sakai, A., Uekama, K. (1984). Effects of β-cyclodextrins and γ-cyclodextrins on release of β-methasone from ointment bases. *Chem. Pharm. Bull.*, *32*, 2401–2405.

76. Ammar, H.O., Elnahhas, S.A. (1995). Improvement of some pharmaceutical properties of drugs by cyclodextrin complex-ation: 3. Bromhexine hydrochloride. *Pharmazie*, *50*, 408–410.

77. Babu, R.J., Dhanasekaran, M., Vaithiyalingam, S.R., Singh, P. N., Pandit, J.K. (2008). Cardiovascular effects of transder-mally delivered bupranolol in rabbits: effect of chemical penetration enhancers. *Life Sci.*, *82*, 273–278.

78. Babu, R.J., Pandit, J.K. (2005). Effect of penetration enhan-cers on the release and skin permeation of bupranolol from reservoir-type transdermal delivery systems. *Int. J. Pharm.*, *288*, 325–334.

79. Lee, B.J., Choi, H.G., Kim, C.K., Parrott, K.A., Ayres, J.W., Sack, R.L. (1997). Solubility and stability of melatonin in propylene glycol and 2-hydroxypropyl-β-cyclodextrin vehi-cles. *Arch. Pharm. Res.*, *20*, 560–565.

80. Zi, P., Yang, X.H., Kuang, H.F., Yang, Y.S., Yu, L.L. (2008). Effect of HP-β-CD on solubility and transdermal delivery of capsaicin through rat skin. *Int. J. Pharm.*, *358*, 151–158.

81. Shaker, D.S., Ghanem, A.H., Li, S.K., Warner, K.S., Hashem, F.M., Higuchi, W.I. (2003). Mechanistic studies of the effect of hydroxypropyl-β-cyclodextrin on in vitro transdermal perme-ation of corticosterone through hairless mouse skin. *Int. J. Pharm.*, *253*, 1–11.

82. Ceschel, G.C., Mora, P.C., Borgia, S.L., Maffei, P., Ronchi, C. (2002). Skin permeation study of dehydroepiandrosterone (DHEA) compared with its α-cyclodextrin complex form. *J. Pharm. Sci.*, *91*, 2399–2407.

83. Ceschel, G., Bergamante, V., Maffei, P., Borgia, S.L., Calabr-ese, V., Biserni, S., Ronchi, C. (2005). Solubility and

84. Dias, M.M.R., Raghavan, S.L., Pellett, M.A., Hadgraft, J. (2003). The effect of β-cyclodextrins on the permeation of diclofenac from supersaturated solutions. *Int. J. Pharm.*, *263*, 173–181.

85. Loftsson, T., Bodor, N. (1989). Effects of 2-hydroxypropyl-β-cyclodextrin on the aqueous solubility of drugs and trans-dermal delivery of 17-β-estradiol. *Acta Pharm. Nord.*, *1*, 185–194.

86. Loftsson, T., Fridriksdóttir, H., Olafsdóttir, B.J., Gudmundsson, O. (1991). Solubilization and stabilization of drugs through cyclodextrin complexation. *Acta Pharm. Nord.*, *3*, 215–217.

87. Williams, A., Shatri, S., Barry, B. (1998). Transdermal per-meation modulation by cyclodextrins: a mechanistic study. *Pharm. Dev. Technol.*, *3*, 283–296.

88. Arima, H., Miyaji, T., Irie, T., Hirayama, F., Uekama, K. (1998). Enhancing effect of hydroxypropyl-β-cyclodextrin on cutaneous penetration activation of ethyl 4-biphenylyl acetate in hairless mouse skin. *Eur. J. Pharm. Sci.*, *6*, 53–59.

89. Maitre, M.M., Longhi, M.R., Granero, G.G. (2007). Ternary complexes of flurbiprofen with HP-β-CD and ethanolamines characterization and transdermal delivery. *Drug Dev. Ind. Pharm.*, *33*, 311–326.

90. Ammar, H.O., Salama, H.A., Ghorab, M., El-Nahhas, S.A., Elmotasem, H. (2006). A transdermal delivery system for glipizide. *Curr. Drug Deliv.*, *3*, 333–341.

91. Sridevi, S., Diwan, P.V. (2002). Effect of pH and complexation on transdermal permeation of gliquidone. *Pharmazie*, *57*, 632–634.

92. Loftsson, T. (1998). Increasing the cyclodextrin complexation of drugs and drug bioavailability through addition of water-soluble polymers. *Pharmazie*, *53*, 733–740.

93. Loftsson, T., Sigurardóttir, A. (1994). The effect of polyvi-nylpyrrolidone and hydroxypropyl methylcellulose on HP-β-CD complexation of hydrocortisone and its permeability through hairless mouse skin. *Eur. J. Pharm. Sci. 2*, 297–301.

94. Siguroardóttir, A.M., Loftsson, T. (1995). The effect of poly-vinylpyrrolidone on cyclodextrin complexation of hydrocor-tisone and its diffusion through hairless mouse skin. *Int. J. Pharm.*, *126*, 73–78.

95. Preiss, A., Mehnert, W., Frömming, K.H. (1995). Penetration of hydrocortisone into excised human skin under the influence of cyclodextrins. *Pharmazie*, *50*, 121–126.

96. Kear, C.L., Yang, J., Godwin, D.A., Felton, L.A. (2008). Investigation into the mechanism by which cyclodextrins influence transdermal drug delivery. *Drug Dev. Ind. Pharm.*, *34*, 692–697.

97. Chang, S., Banga, A. (1998). Transdermal iontophoretic delivery of hydrocortisone from cyclodextrin solutions. *J. Pharm. Pharmacol.*, *50*, 635–640.

98. Iervolino, M., Cappello, B., Raghavan, S.L., Hadgraft, J. (2001). Penetration enhancement of ibuprofen from supersat-urated solutions through human skin. *Int. J. Pharm.*, *212*, 131–141.

transdermal permeation properties of a dehydroepiandroster-one cyclodextrin complex from hydrophilic and lipophilic vehicles. *Drug Deliv.*, *12*, 275–280.

99. Godwin, D.A., Wiley C.J., Felton, L.A (2006). Using cyclo-dextrin complexation to enhance secondary photoprotection of topically applied ibuprofen. *Eur. J. Pharm. Biopharm.*, *62*, 85–93.

100. Nagarsenker, M., Amin, L., Date, A. (2008). Potential of cyclodextrin complexation and liposomes in topical delivery of ketorolac: in vitro and in vivo evaluation. *AAPS PharmSciTech*, *9*, 1165–1170.

101. Sridevi, S., Diwan, P.V.R. (2002). Optimized transdermal delivery of ketoprofen using pH and hydroxypropyl-β-cyclo-dextrin as co-enhancers. *Eur. J. Pharm. Biopharm.*, *54*, 151–154.

102. Valjakka-Koskela, R., Hirvonen, J., Monkkonen, J., Kies-vaara, J., Antila, S., Lehtonen, L., Urtti, A. (2000). Transder-mal delivery of levosimendan. *Eur. J. Pharm. Sci.*, *11*, 343–350.

103. Padula, C., Pappani, A., Santi, P. (2008). In vitro permeation of levothyroxine across the skin. *Int. J. Pharm.*, *349*, 161–165.

104. Vollmer, U., Muller, B.W., Mesens, J., Wilffert, B., Peters, T. (1993). In-vivo skin pharmacokinetics of liarozole: percuta-neous-absorption studies with different formulations of cy-clodextrin derivatives in rats. *Int. J. Pharm.*, *99*, 51–58.

105. Bodor, N., Loftsson, T., Wu, W.M. (1992). Metabolism, distribution, and transdermal permeation of a soft corticoste-roid, loteprednol etabonate. *Pharm. Res.*, *9*, 1275–1278.

106. Martins, P.S., Ochoa, R., Pimenta, A.M.C., Ferreira, L.A.M., Melo, A.L., da Silva, J.B.B., Sinisterra, R.D., Demicheli, C., Frezard, F. (2006). Mode of action of β-cyclodextrin as an absorption enhancer of the water-soluble drug meglumine antimoniate. *Int. J. Pharm.*, *325*, 39–47.

107. Lee, B.J., Cui, J.H., Parrott, K.A., Ayres, J.W., Sack, R.L. (1998). Percutaneous absorption and model membrane varia-tions of melatonin in aqueous-based propylene glycol and 2-hydroxypropyl-β-cyclodextrin vehicles. *Arch. Pharm. Res.*, *21*, 503–507.

108. Bounoure, F., Skiba, M.L., Besnard, M., Arnaud, P., Mallet, E., Skiba, M. (2008). Effect of iontophoresis and penetration enhancers on transdermal absorption of metopimazine. *J. Dermatol. Sci.*, *52*, 170–177.

109. Tenjarla, S., Puranajoti, P., Kasina, R., Mandal, T. (1998). Preparation, characterization, and evaluation of miconazole–cyclodextrin complexes for improved oral and topical deliv-ery. *J. Pharm. Sci.*, *87*, 425–429.

110. Kim, J., Kim, M. (2007). After-rinsing hair growth promotion of minoxidil-containing amino α-cyclodextrins. *J. Microbiol. Biotechnol.*, *17*, 1965–1969.

111. Kim, J.C., Lee, M.H., Rang, M.J. (2003). Minoxidil-contain-ing dosage forms: skin retention and after-rinsing hair-growth promotion. *Drug Deliv.*, *10*, 119–123.

112. Davaran, S., Rashidi, M.R., Khandaghi, R., Hashemi, M. (2005). Development of a novel prolonged-release nicotine transdermal patch. *Pharmacol. Res.*, *51*, 233–237.

113. Dalmora, M.E., Dalmora, S.L., Oliveira, A.G. (2001). Inclu-sion complex of piroxicam with β-cyclodextrin and incorpo-ration in cationic microemulsion: in vitro drug release and

in vivo topical anti-inflammatory effect. *Int. J. Pharm.*, *222*, 45–55.

114. Adachi, H., Irie, T., Kaneto, U. (1993). Combination effects of *O*-carboxymethyl-*O*-ethyl-β-cyclodextrin and penetration en-hancer HPE-101 on transdermal delivery of prostaglandin E1 in hairless mice. *Eur. J. Pharm. Sci.*, *1*, 117–123.

115. Larrucea, E., Arellano, A., Santoyo, S., Ygartua, P. (2002). Study of the complexation behavior of tenoxicam with cyclo-dextrins in solution: improved solubility and percutaneous permeability. *Drug Dev. Ind. Pharm.*, *28*, 245–252.

116. Larrucea, E., Arellano, A., Santoyo, S., Ygartua, P. (2001). Combined effect of oleic acid and propylene glycol on the percutaneous penetration of tenoxicam and its retention in the skin. *Eur. J. Pharm. Biopharm.*, *52*, 113–119.

117. Szemán, J., Ueda, H., Szejtli, J., Fenyvesi, E., Watanabe, Y., Machida, Y., Nagai, T. (1987). Enhanced percutaneous ab-sorption of homogenized tolnaftate/β-cyclodextrin polymer ground mixture. *Drug Des. Deliv.*, *1*, 325–332.

118. Tanner, T., Marks, R. (2008). Delivering drugs by the trans-dermal route: review and comment. *Skin Res. Technol.*, *14*, 249–260.

119. Doliwa, A., Santoyo, S., Ygartua, P. (2001). Transdermal iontophoresis and skin retention of piroxicam from gels con-taining piroxicam: hydroxypropyl-β-cyclodextrin complexes. *Drug Dev. Ind. Pharm.*, *27*, 751–758.

120. Doliwa, A., Santoyo, S., Ygartua, P. (2001). Effect of passive and iontophoretic skin pretreatments with terpenes on the in vitro skin transport of piroxicam. *Int. J. Pharm.*, *229*, 37–44.

121. Murthy, S.N., Zhao, Y.L., Sen, A., Hui, S.W. (2004). Cyclo-dextrin enhanced transdermal delivery of piroxicam and carboxyfluorescein by electroporation. *J. Control. Release*, *99*, 393–402.

122. Jin, S., Lee, Y., Kang, H. (2008). Methyl-β-cyclodextrin, a specific cholesterol-binding agent, inhibits melanogenesis in human melanocytes through activation of ERK. *Arch. Der-matol. Res.*, *300*, 451–454.

123. Kim, S., Kim, Y., Lee, Y., Cho, K.H., Kim, K.H., Chung, J.H. (2007). Cholesterol inhibits MMP-9 expression in human epidermal keratinocytes and HaCaT cells. *FEBS Lett.*, *581*, 3869–3874.

124. Gniadecki, R. (2004). Depletion of membrane cholesterol causes ligand-independent activation of Fas and apoptosis. *Biochem. Biophys. Res. Commun.*, *320*, 165–169.

125. Leppimäki, P., Kronqvist, R., Slotte, J. (1998). The rate of sphingomyelin synthesis de novo is influenced by the level of cholesterol in cultured human skin fibroblasts. *Biochem. J.*, *335*, 285–291.

126. Sviridov, D., Fidge, N., Beaumier-Gallon, G., Fielding, C. (2001). Apolipoprotein A-I stimulates the transport of intra-cellular cholesterol to cell-surface cholesterol-rich domains (caveolae). *Biochem. J.*, *358*, 79–86.

127. Gniadecki, R., Christoffersen, N., Wulf, H.C. (2002). Cho-lesterol-rich plasma membrane domains (lipid rafts) in ker-atinocytes: importance in the baseline and UVA-induced generation of reactive oxygen species. *J. Invest. Dermatol.*, *118*, 582–588.

128. Jans, R., Atanasova, G., Jadot, M., Poumay, Y. (2004). Cholesterol depletion alters involucrin gene expression in epidermal keratinocytes through activation of p38. *J. Invest. Dermatol.*, *122*, 94–94.

129. Bang, B., Gniadecki, R., Gajkowska, B. (2005). Disruption of lipid rafts causes apoptotic cell death in HaCaT keratinocytes. *Exp. Dermatol.*, *14*, 266–272.

130. Zimina, E., Bruckner-Tuderman, L., Franzke, C. (2005). Shedding of collagen XVII ectodomain depends on plasma membrane microenvironment. *J. Biol. Chem.*, *280*, 34019–34024.

131. Lambert, S., Vind-Kezunovic, D., Karvinen, S., Gniadecki, R. (2006). Ligand-independent activation of the EGFR by lipid raft disruption. *J. Invest. Dermatol.*, *126*, 954–962.

132. Lisby, S., Faurschou, A., Gniadecki, R. (2007). The autocrine TNFα signalling loop in keratinocytes requires atypical PKC species and NF-κB activation but is independent of cholesterol-enriched membrane microdomains. *Biochem. Pharmacol.*, *73*, 526–533.

133. Duchêne, D., Bochot, A., Loftsson, T. (2009). Les cyclodextrines et leurs utilisations en pharmacie et cosmétologie. *STP Pharma Pratiques*, *19*, 15–27.

10

ORAL DRUG DELIVERY WITH CYCLODEXTRINS

FRANCISCO VEIGA, ANA RITA FIGUEIRAS, AND AMELIA VIEIRA
Faculty of Pharmacy, University of Coimbra, Coimbra, Portugal

1. INTRODUCTION

Oral drug administration is the route of choice for formulators and continues to dominate the area of drug delivery technologies. Although popular, this route is not free from limitations on absorption and bioavailability in the gastrointestinal (GI) tract [1]. These limitations have become even more prominent with the advent of protein and peptide drugs and the compounds emerging as a result of combinatorial chemistry and the technique of high-throughput screening [2]. Whenever a dosage form is administered orally, the drug in the formulation is released and dissolves in the surrounding GI fluid to form a solution. This process is solubility limited. Once the drug is in solution form, it passes across the membranes of the cells lining the GI tract. This process is permeability limited. Then, onward, the drug is absorbed into the systemic circulation. In summary, the oral absorption and hence bioavailability of drug is determined by the extent of drug solubility and permeability [3].

Cyclodextrins (CDs) can enhance the bioavailability of insoluble drugs by increasing drug solubility, dissolution, and/or drug permeability. CDs have played a very important role in the formulation of poorly water-soluble drugs by improving drug solubility and/or dissolution toward inclusion complexation or solid dispersion. They act as hydrophilic carriers for drugs with unfavorable molecular characteristics for complexation, or as tablet dissolution enhancers for formulation with a high drug-loading dose and are inappropriate to form inclusion complexes [4]. CDs have a wide range of applications in pharmaceutical technology. Hydrophilic CDs can increase the rate of drug release, which can be used for the enhancement of drug absorption across biological barriers, serving as a potent drug carrier in immediate-release formulations. On the other hand, hydrophobic and ionizable CDs are used in the design of controlled-release dosage formulations [5].

The objective of this contribution is to focus on the potential use of CDs as high-performance drug carriers in the development of advanced oral dosage forms. For this purpose, some topics are considered: in particular, the Biopharmaceuticals Classification System (BCS), in order to understand when CDs can increase drug solubility and permeability in oral administration in accord with drug physicochemical properties. Taking the latest topic, into account CD and drug permeability through biological membranes was also approached. CD complexation is the first step in increasing drug solubility, and for this reason strategies to increase CD complexation efficiency are described. Finally, the use of CDs in oral, buccal, and sublingual dosage forms is discussed and several examples are provided.

2. BIOPHARMACEUTICALS CLASSIFICATION SYSTEM

The in vivo performance of orally administered drug depends on its solubility and tissue permeability properties. Based on these characteristics, drug substances are divided into four classes in the BCS classification system [6]. This system acts as a guide to the development of various oral drug delivery technologies. The BCS is a useful tool not only in obtaining results for in vivo bioequivalence studies but also for decision making in the discovery and early development of new drugs [6–8].

Cyclodextrins in Pharmaceutics, Cosmetics, and Biomedicine: Current and Future Industrial Applications, First Edition. Edited by Erem Bilensoy.
© 2011 John Wiley & Sons, Inc. Published 2011 by John Wiley & Sons, Inc.

The BCS is a drug development tool that allows estimation of the contributions of three major factors—dissolution, solubility, and intestinal permeability—that affect oral drug absorption from immediate-release solid oral dosage forms [9]. To classify a drug according to the BCS, the solubility, dose, and permeability of the drug must be known. According to a U.S. Food and Drug Administration (FDA) guidance for the industry [9], in vivo bioavailability and bioequivalence studies for immediate-release solid oral dosage forms are based on the BCS. A biowaiver can currently be requested only for solid, orally administered immediate-release products (>85% release in 30 min), containing drugs with a high solubility over the pH range from 1 to 7.5 (dose/solubility (D:S) ratio <250 mL) and a high permeability (fraction absorbed >90%) [3]. In addition, only excipients that do not affect the rate or extent of absorption may be used. Further restrictions are that drugs with a narrow therapeutic range and drug products designed to be absorbed in the oral cavity may not be considered for biowaivers [10].

According to BCS, drug substances are classified as:

- Class I: high solubility and high permeability
- Class II: low solubility and high permeability
- Class III: high solubility and low permeability
- Class IV: low solubility and low permeability

This classification is associated with drug dissolution and the absorption model, which identifies the key parameters controlling drug absorption.

Class I drugs exhibit high absorption and a high dissolution number (e.g., metoprolol, diltiazem, verapamil, propranolol). The rate-limiting step is drug dissolution, and if dissolution is very fast, then the gastric-emptying rate becomes the rate-determining step. In general, hydrophilic CDs are not able to improve the bioavailability of class I drugs. However, they are used to reduce local drug irritation and to increase the rate of drug absorption [11]. In general, nonsteroidal anti-inflammatory drugs (NSAIDs) have an oral bioavailability greater than 90%, and in this type of system, CD complexation does not result in any drastic improvement of the absolute bioavailability. However, CD formulations can decrease drug absorption variability [12–16].

Class II drugs have high absorption but a low dissolution number. In vivo drug dissolution is the rate-limiting step for absorption except at a very high dose number. The absorption for class II drugs is usually slower than for class I and occurs over a longer period of time. In vitro–in vivo correlation is usually expected for class I and II drugs. Examples include phenytoin [17,18], itraconazole [19], and nifedipine [20,21]. Normally, drugs from this class are hydrophobic with very limited aqueous solubility. Complexation of these drugs with water-soluble CDs can increase drug solubility and enhance their diffusion to the mucosal surface, leading to enhanced oral bioavailability [22].

Class III drugs are water soluble but do not readily permeate biological membranes, due to their size and extent of hydration; consequently, permeability is the rate-limiting step for drug absorption (e.g., cimetidine [23], acyclovir [24], neomycin B [25], captopril [26]). These drugs exhibit a high variation in the rate and extent of drug absorption. Since the dissolution is fast, the variation is due to changes in physiology and membrane permeability rather than the dosage form factors. Thus, poor bioavailability of drugs in this class is due to the inability of the drugs to permeate biological membranes. Therefore, CDs do not enhance their bioavailability after oral administration [3], but they can increase the ability of the dissolved drug molecules to partition from the aqueous exterior into the GI mucosa.

Class IV drugs are water insoluble and do not readily permeate lipophilic biological membranes, consequently, they exhibit more problems for effective oral administration. Fortunately, class IV compounds are the exception rather than the rule and are rarely developed and reach the market. Some examples are water–insoluble zwitterions or relatively large lipophilic molecules. Hydrophilic water–insoluble compounds such as zwitterions do not readily form CD complexes, and thus hydrophilic CDs are not likely to improve their oral bioavailability. However, CDs are able to improve the aqueous solubility of some large lipophilic molecules, leading to increased drug availability at the mucosal surface. This will frequently leads to increased oral bioavailability. A good example is cyclosporine A, and CDs can turn this class IV drug into class I, increasing drug solubility [27], or into class II, increasing drug permeability [28].

In summary, high-throughput screening approaches for drug development have generated an increasing number of lipophilic water-insoluble drug candidates [29] or drugs whose clinical usefulness is hampered by their insolubility in water. It is possible to increase the apparent aqueous solubility of class II or IV drugs without decreasing their lipophilicity and consequently their absorption through biological membranes using complexation with CDs.

3. CDs AND DRUG PERMEABILITY THROUGH BIOLOGICAL MEMBRANES

It is not well known how CDs act as penetration enhancers and several possible mechanisms have been suggested. The potential use of CDs and methylated derivatives as a new class of transmucosal penetration enhancers for therapeutic drugs has been demonstrated [30,31].

CDs do not readily permeate biological membranes, due to its chemical structure (i.e., the large number of hydrogen donors and acceptors), their molecular weight (i.e., >970 Da), and their very low octanol/water partition coefficient (approximately log P between −3 and 0) [32]. Negligible amounts of hydrophilic CDs and drug–CD complexes

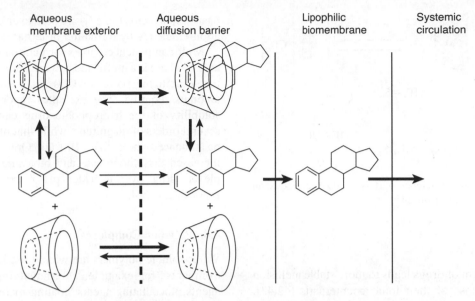

Figure 1. Drug permeation through a biomembrane from an aqueous vehicle containing CD–drug complex. (Adapted from [37].)

are able to permeate lipophilic membranes. Only the free form of the drug, which is in equilibrium with the drug–CD complex, is capable of penetrating lipophilic membranes [32–35].

In aqueous CD solutions saturated with drug (guest molecule), the drug flux through a biomembrane increases when the CD concentration increases. On the other hand, if the drug concentration is kept constant and below saturation, the flux will decrease with increasing CD concentration. These observations have been explained by the existence of an aqueous diffusion barrier at the membrane surface [30,36] (Fig. 1).

Biomembranes are lipophilic with an aqueous exterior that forms a structured water layer frequently referred as an *unstirred diffusion layer* at the membrane surface [37]. Hydrophilic CDs act as carriers by keeping hydrophobic drug molecules in solution in the diffusion layer and delivering them at the lipophilic membrane surface, where the drug molecules leave the CD cavity in the lipophilic membrane only if the drug permeation through the aqueous diffusion layer is the rate-limiting step.

CDs can, theoretically, enhance drug bioavailability by stabilizing drug molecules at the biomembrane surface [38,39]. However, methylated CDs act as absorption enhancers by different pathways. These CD derivatives probably act by transiently changing the membrane permeability, overcoming the aqueous diffusion barrier and opening tight junctions [30].

CDs do not, in general, enhance the permeability of hydrophilic water-soluble drugs through lipophilic biological membranes, and numerous studies have shown that excess CD will reduce drug permeability through biological membranes [32,36].

Additionally, the physicochemical properties of the drug (its solubility in water), the composition of the drug formulation (aqueous or nonaqueous), and the physiological composition of the membrane barrier (the presence of an aqueous diffusion layer) will determine whether the CDs will enhance or hinder drug absorption through a biological membrane [36].

4. STRATEGIES TO IMPROVE CD COMPLEXATION EFFICIENCY

For a variety of reasons, including cost, production capabilities, and toxicology, the amount of CDs that can be incorporated into drug formulations is limited. Even under ideal conditions, CD complexation will result in a 4- to 10-fold increase in the formulation bulk. This limits the use of CDs in, for example, solid oral dosage forms to potent drugs that possess good complexing properties [40].

The complexation efficiency of CDs is frequently low; hence large amounts of CD are needed to complex relatively small amounts of drug. It is therefore important to develop methods that can be applied to enhance the complexation efficiency of CDs [41]. Furthermore, inadequate optimization of CD containing drug formulations frequently leads to reduced drug availability.

If one drug molecule forms a complex with one CD molecule, the complexation efficiency will be equal to the intrinsic solubility of the drug (S_0) times the stability constant of the drug–CD complex (K_c). Thus, increased complexation efficiency can be obtained by increasing either S_0 or K_c, or by increasing both simultaneously (Fig. 2).

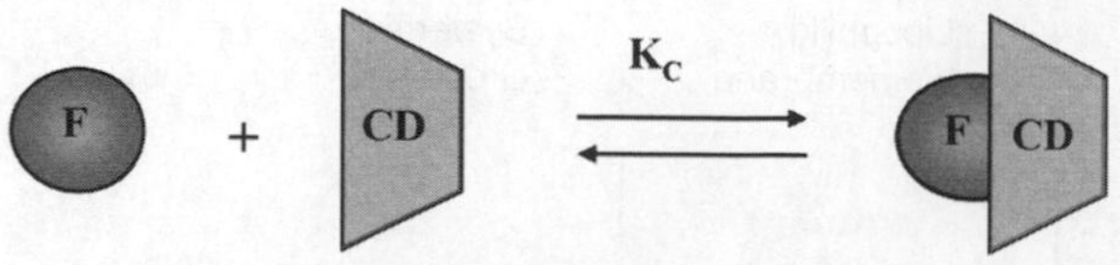

Figure 2. Complexation efficiency of 1:1 drug (F)/cyclodextrin (CD) complexes. (Adapted from [41].)

4.1. Drug Ionization

The nonionized form of drugs leads to more stable inclusion complexes than those of their ionic counterparts [42,43]. Ionized forms of drug molecules are more hydrophilic, thus having less propensity to form an inclusion complex by displacing water molecules from the CD cavity [44,45]. However, ionization is a common way to increase the aqueous solubility of ionizable drugs, and in aqueous solutions, CD complexation of ionized drug molecules can result in larger total drug solubilization (i.e., the solubilization of a drug due to both ionization and inclusion in the CD cavity) [46].

4.2. Salt Formation

Salt formation is an other tool used to increase drug solubility. It is possible to enhance the apparent intrinsic solubility of a drug through a combination of salt formation (i.e., forming a more water-soluble salt of the drug) with the formation of inclusion complexes without significantly reducing the ability of complexation [47–49]. Complexation with CDs and simultaneous salt formation can be used successfully to improve the properties of anionic drugs: in particular, to increase their water solubility and photo- or chemical stability. Also, this can result in fewer irritant effects, an improve rate and extent of absorption, and taste-masking enhancement [49].

4.3. Acid–Base Ternary Inclusion Complexes

Various organic acids, such as α-hydroxyl carboxylic acids (citric [50], tartaric [51,52], malic [53] acids), have been shown to increase complexation through noncovalent multicomponent (or ion pair) associations between CD, basic drug, and the acid. The addition of hydroxyl carboxylic acids yields freely water-soluble complexes that can be isolated by freeze- or spray-drying. The resulting amorphous compounds dissolve very quickly, producing supersaturated solutions that remain stable for several days [54]. In the same way, certain organic bases (hydroxylamines [47,55] and

basic amino acids [56,57]) are able to enhance the complexation efficiency by formation of ternary complexes.

Multicomponent complex formation in the presence of an acid or a base is a useful and powerful approach to improving the solubilizing power of CDs and therefore reducing the amounts required. The examples reported show that the solubility of the hydrophobic drug can be enhanced by several orders of magnitude, while that of CD solubility can be enhanced more than 10-fold. Oral formulations with increased dissolution rate and bioavailability of the complexed drug over a wider pH range can be prepared accordingly [54].

4.4. Polymer Complexes

Water-soluble polymers are widely used as pharmaceutical excipients: for example, as emulsifying and suspending agents, flocculating agents, coating materials, and binders. In addition, controlled-release dosage forms are frequently based on water-soluble polymers. Although the polymers are usually considered to be chemically inert, they are known to form complexes with small molecules in aqueous solutions [58,59].

A notable increase in drug solubility is frequently observed, due to the formation of water-soluble drug–polymer complexes. These polymers can form ternary systems with inclusion complexes, increasing the stability constant and consequently the complexation efficiency. Studies have shown that hydrophilic polymers can enhance the solubilizing effect of CDs and the aqueous solubility of β-cyclodextrin (β-CD) [60,61]. The polymers interact with drug molecules mainly via electrostatic bonds (i.e., ion-to-ion, ion-to-dipole, and dipole-to-dipole bonds); however other types of forces, such as van der Waals forces and hydrogen bridges, frequently are found in complex formation [62,63].

Addition of polymers increases the apparent stability constant of a drug–CD complex, and the entropy of the stability constant becomes more negative, indicating a more ordered complex structure. In aqueous solution, polymers can reduce the mobility of the CD molecules and enhance the solubility of formed complexes. When polymer and CD are mixed together, CD achieves a greater extent of solubilization than when the polymer and CD are used separately. This solubilization enhancement is more than a simply additive effect; it is more a synergistic effect. However, in the case of micelle-polymer complexes, the precise chemical structure of the drug–CD–polymer complexes is not known [64].

4.5. Solubilization of CD Aggregates

Studies by several research groups have shown that CDs form both inclusion and noninclusion complexes and that different types of complexes can coexist in aqueous solutions [41].

Furthermore, CDs and CD complexes can self–associate to form nanoscale aggregates [53,65]. These aggregates are able to solubilize drugs and other hydrophobic molecules through noninclusion complexation or micelle-like structures [66]. The formation of such structures is not easily detected. Their existence has been elucidated only recently.

In addition, common pharmaceutical excipients such as polymers and buffer salts can participate in the complex formation. Thus, stability constants obtained from phase-solubility diagrams are apparent stability constants describing the combined effect of the various complex structures on drug solubility [66]. The noninclusion complex formation of water-soluble microaggregates of drug and drug–CD complexes can contribute to the overall solubility. In such aqueous systems, water-soluble polymers such as hydroxypropyl methylcellulose (HPMC) appears to enhance the contribution of the noninclusion complexation to the overall drug solubility, as well as solubilize and stabilize drug/β-CD complexes [67]. Discovery of these aggregates, as well as the ability of CDs to form noninclusion complexes, is likely to have a profound influence on future CD research.

4.6. Combination of Two or More Methods

Frequently, the complexation efficiency can be enhanced by combining two or more of the aforementioned methods: for example, drug ionization and polymer addition or solubilization of the CD aggregates by adding both polymers and cations or anions to the aqueous complexation medium [41].

5. CDs IN ORAL DOSAGE FORMS

Controlled release is used as a collective term for any dosage forms of which the onset and/or rate of drug release is altered by pharmaceutical technology. Controlled-release dosage forms can be used to obtain more adequate plasma drug level–time profiles. Thus, plasma drug level–time profiles can be classified into two categories: rate controlled and time controlled.

Time-released formulations can be classified further as *delayed release*. These systems release the drug after a programmed lag time following oral administration [68]. Ionizable CDs derivatives such as carboxymethyl–β-cyclodextrin (CM-β-CD) have been used to obtain this release type. On the other hand, colon-specific drug delivery can be considered a delayed release and can be obtained using drug–CDs conjugates. In rate-controlled release, the system delivers the drug at a predetermined rate for a specific period of time. Rate-controlled release includes immediate release, prolonged release, and modified release.

Drug absorption from immediate-release systems in the GI tract consists of a series of processes, including tablet disintegration, drug dissolution in the aqueous GI fluids,

permeation of the drug molecules from the intestinal fluid through an aqueous diffusion layer immediately adjacent to mucosal surface, and permeation through the mucosa [3]. In the case of immediate-release dosage forms, the drug is released immediately after oral administration. This type of delivery is particularly useful in emergency situations, and it is applicable for analgesics, antipyretics, and coronary vasodilators. The hydrophilic CDs are useful for obtaining immediate-release profiles.

Prolonged-release provides slow drug release at a rate sufficient to cause a therapeutic response over a period of time after ingestion. Thus, prolonged release allows for a reduction in dosing frequency compared to a drug amount presented in a conventional dosage form [68]. Hydrophobic CDs, such as alkylated and acylated derivatives, with low aqueous solubility, are useful as slow-release carriers in prolonged-release formulations of water-soluble drugs.

Modified-release formulations can be obtained by combinations of the different release types. Various CDs derivates have been used in order to modify drug release in oral dosage forms. Moreover, modified-release formulations can be obtained by combinations of CDs with different release carriers.

5.1. Immediate Release

CDs have the ability to interact with poorly water-soluble drugs to form noncovalent dynamic inclusion complexes. The central cavity of CDs has a lipophilic microenvironment which has the ability to form inclusion complexes by taking the entire molecule or, rather, some nonpolar parts in its hydrophobic cavity. Thus, these inclusion complexes are more hydrophilic and consequently have higher solubility than that of the free drug itself. The increase in solubility can also increase the dissolution rate, so CD has been used as a potent drug carrier in an immediate-release formulations. Moreover, this increase in the dissolution performance can result in an improvement in the oral bioavailability of classes II and IV, increasing the pharmacological effect and allowing a reduction in the dose of the drug administered [22,69–71].

The use of CDs is often preferred to organic solvents (cosolvents and surfactants) from both a toxicological perspective and a mechanistic point of view. This is because although, the drug may be soluble in organic solvent, once the drug solvent mixture is diluted with an aqueous solvent, the poorly soluble drug often precipitates *en masse*. On the other hand, this nonlinearity is not a problem with the use of CDs, especially those that form 1 : 1 complexes [72]. CDs act as carrier materials and help to transport the drug through an aqueous medium to the lipophilic absorption surface in the GI tract. After dissolution of the inclusion complex, the guest molecule may be released through complex dissociation by replacement of the included guest by excipient and endogenous molecules (bile acids, cholesterol, lipids, and other

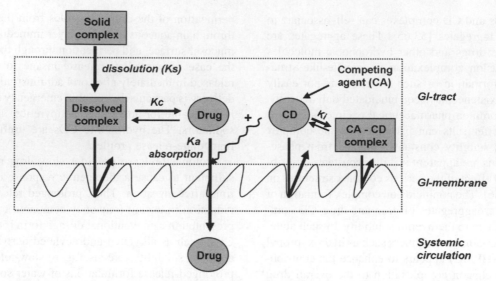

Figure 3. Overall process of drug absorption from an inclusion complex following dissolution and dissociation in the gastrointestinal tract. K_s, dissolution rate constant of drug–CD complex; K_c, stability constant of drug–CD complex; K_i, stability constant of competing agent–CD complex; K_a, absorption rate constant of drug. (Adapted from [5].)

compounds) which compete with the drug for the CD cavity at the absorption site. If the complex is located close to a biological membrane, the guest may be transferred to the matrix for which it has the highest affinity. These factors are responsible for the acceleration of drug absorption [5]. Additionally, CDs will operate locally on the hydrodynamic layer surrounding the particles of drug. This action results in an in situ inclusion process, producing a rapid increase in the amount of drug dissolved [73]. At the same time, this process removes some components from the membrane surface and perturbs the membrane fluidity, thereby modifying the transport properties of the membranes and facilitating drug absorption, especially for water-soluble drugs (Fig. 3). [5,71].

In fact, there are many studies related to the use of CDs to improve the dissolution kinetics of drugs with the use of different methods to form inclusion complexes. For example, an inclusion complex was developed between warfarin, a drug practically insoluble in acidic aqueous media, and β-CD to increase the dissolution rate of a commercial formulation. The dissolution rate of warfarin was increased in the order drug < physical mixture < kneaded product ≈ co-evaporated product < freeze-dried product [74] (Fig. 4).

Similarly, inclusion complexes between carvedilol, another drug with poor aqueous solubility, and methyl-β-cyclodextrin (M-β-CD) demonstrated a clear higher dissolution rate than dissolution of the drug alone. In this case, the highest drug dissolution rate was obtained for the kneading system (Fig. 5) [75]. Thus, dissolution profiles were influenced by the preparation method of binary systems, and this aspect has been demonstrated in numerous studies [73,75,76].

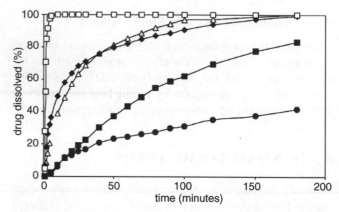

Figure 4. Dissolution profiles at pH 1.2 of pure warfarin (●), physical mixture (■), kneaded product (▲), co-evaporated product (◆), and freeze-dried product (□). (Adapted from [74].)

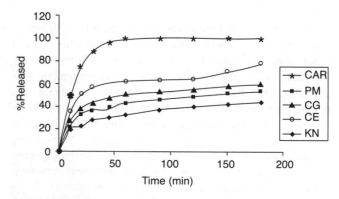

Figure 5. Dissolution diagram of CAR/-M-β-CD systems: CAR, carvedilol; PM, physical mixture; CG, co-ground; CE, co-evaporated; KN, kneaded. (Adapted from [75].)

The improvement in the drug dissolution rate was attributed to the surfactant-like properties of CDs, which can reduce the interfacial tension between water-insoluble drugs and the dissolution medium, leading to the formation of soluble inclusion complexes. The lower energetic state of the amorphous drug compared to the crystalline form results in increased solubility because water molecules break up amorphous material much easier than does the crystalline drug, leading to better wettability and a reduction in the drug particle size [74,75,77].

Among the CDs, β-CD has been widely used in the early stages of pharmaceutical applications because of its ready availability and cavity size suitable for the widest range of drugs, but its anomalous low aqueous solubility is a drawback to its wider use. To overcome this limitation, derivatives have been prepared to improve their solubility and their ability to dissolve hydrophobic compounds. Thus, hydroxypropyl-β-cyclodextrin (HP-β-CD), 6-O-maltosyl-β-cyclodextrin (G2-β-CD) and sulfobutyl ether- β-cyclodextrin (SBE-β-CD) were used in immediate-release formulations that dissolve readily in the GI tract and enhance the oral bioavailability of poorly soluble drugs. SBE-β-CD was shown to be an excellent solubilizer for several drugs and was more effective than β-CD but not as effective as DM-β-CD. Methylated CDs with a relatively low molar substitution appear to be the most powerful solubilizers [4].

Studies have been carried to compare the solubility potential of different CDs after formation of inclusion complexes with drugs. Figueiras et al. [73] verified through dissolution studies in artificial saliva media that after 6 min, the percentage of pure omeprazole dissolved was nearly 12% and above 90% when complexed with β-CD and M-β-CD. M-β-CD proved to have better solubilizing and complexing properties for omeprazol than did the parent β-CD. Badr-Eldin et al. [78], investigated inclusion complexes formation between tadalafil, a drug with limited water solubility, and three CDs: HP-β-CD, DM-β-CD, and β-CD. Results demonstrated that inclusion complexes prepared using HP-β-CD and DM-β-CD exhibited enhancement of tadalafil dissolution superior to that prepared using β-CD. Inclusion complexes of CD derivatives showed a burst effect of more than 75% during the first 5 min. Tadalafil-DM-β-CD systems also demonstrated higher dissolution than that of the corresponding systems with HP-β-CD. Authors suggested that it is, probably, due to the stronger interaction between tadalafil and DM-β-CD and/or better inclusion of the drug molecules into the CD hydrophobic cavity that was expanded by the two methyl groups.

Sathigari et al. [76], used β-CD, HP-β-CD, and randomly methylated-β-cyclodextrin (RM-β-CD) to improve the solubility and dissolution rate of efavirenz. They verified that the dissolution of this poorly water-soluble drug was substantially higher for HP-β-CD and RM-β-CD. However, the dissolution was faster for HP-β-CD than for RM-β-CD. The maximum drug release (~55%) was obtained within 30 and 180 min for HP-β-CD and RM-β-CD, respectively [76].

To improve the solubility of β-lapachone (β-LAP), four CDs have been selected: β-CD, HP-β-CD, SBE-β-CD, and RM-β-CD. The derivatives with high efficiency for β-LAP solubilization studied were SB-β-CD and especially RM-β-CD. Profiles with nearly 100% of the drug dissolved at 5 min were obtained for freeze-drying and kneading systems [79].

Moreover, different studies have showed that complexation with CDs is a motivating factor in increasing the oral bioavailability of drugs. For example, inclusion complexes between albendazole and HP-β-CD induce better drug bioavailability, characterized by double C_{max} (μg/mL) and lower T_{max} (h) values [80]. Similarly, inclusion complexes between tolbutamide-β-CD and HP-β-CD showed an increase in the oral bioavailability of the drug [81].

However, enhancement in dissolution kinetics is not always followed by improvement in drug bioavailability. An example was the case of complexation of piroxicam with CDs, which did not have any significant effect on the absolute bioavailability of the drug but resulted in faster drug absorption, faster onset of analgesia, and improved GI tolerability [3].

Regarding pharmaceutical dosage forms, CDs can enhance the rate of drug release from matrices systems by various mechanisms. According to Bibby et al. [82], when drug concentration in the hydrated polymeric matrix is above the saturation level, equilibrium is attained between solid and dissolved drugs in the hydrated matrix. The amount of drug molecules solubilized in the hydrated polymer increases upon the addition of CD. If diffusion of CDs and its complexes are comparable to drug diffusion, an increase in the concentration of mobile drug molecules occurs, resulting in increased drug release. The addition of CDs to polymeric systems may also enhance drug release by acting as a channeling or wicking agent or by promoting erosion of the matrix.

The influence of SBE-β-CD and β-CD on the prednisolone and carbamazepine release profiles from HPMC, a hydrophilic matrix, has been investigated [83,84]. It was found that CD increased drug dissolution rate, with the increase being greater when the drug and CD were formulated as a complex rather than a physical mixture. The authors concluded that the enhanced drug release in the presence of CDs appears to be due primarily to the improved solubility and dissolution rate as a result of in situ complex formation, with only a minor effect due to enhanced water uptake [84,85].

Pose-Vilarnovo et al. [85] reported that CDs can act as solubilizing agents, promoting drug release of sulfametizol from HPMC tablets, a hydrophobic drug. This effect was greater in the case of the most hydrophilic CD, HP-β-CD.

Sangalli et al. [86] found that β-CD increased the release rate of the poorly water-soluble drugs naproxen and ketoprofen from inert matrices (acrylic resins) and hydrophilic matrices. This increase was relatively much higher for the inert matrices (eightfold) than for swellable ones (2.5-fold). In inert matrices, the increase in release rate may be directly associated with the ability of CDs to form complexes rather than its channeling properties. On the other hand, in hydrophilic matrices, complexes showed a limited ability to diffuse, because they have higher molecular weight and larger size than the active ingredient alone, and transport could be hindered by the relatively small size of the macromolecular mesh due to the entanglement of the swollen hydrophilic polymeric chains.

The influence of HP-β-CD on carvedilol release from poly (ethylene oxide) (PEO) tablets showed that CDs are responsible for the improvement in the drug release rate because they promoted an increase in the erosion rate of the tablet and consequently improved the dissolution of the drug inside the polymeric matrix [87]. Recently, it was reported that incorporation of HP-β-CD in PEO tablets also resulted in an increased drug release for both forms of diclofenac (Dic): poorly soluble free acid (DicH) and freely water-soluble sodium salt (DicNa) [88].

5.2. Prolonged Release

Hydrophobic CDs such as ethylated and peracylated have been proposed as sustained-release carriers for highly soluble drugs with short biological half-lives, in virtue of poor water-soluble complexes. Among the alkylated CDs, diethyl–β-cyclodextrin (DE-β-CD) and triethyl-β-cyclodextrin (TE-β-CD) were the first to be used, and their hydrophobic complexes with diltiazem and isosorbide dinitrate provided slow drug release after administration in dogs [4].

To improve patient compliance with salbutamol, a broncodilator with a short biological half-life, Lemesle-Lamache et al. [89], prepared inclusion complexes with five types of ethyl-β-cyclodextrin (Et-β-CD) of different degrees of substitution and aqueous solubility (Fig. 6). They obtained different sustained-release profiles, and it was suggested that sustained release of salbutamol was due to its stronger complexation with the ethylated derivates than with β-CD. The ethylation of the hydroxyl groups may change the nature of host–guest interactions, because ethylation may expand the hydrophobic region of the CD cavity and enhance the binding of the substrate by means of hydrophobic effects. They concluded that the degree of substitution, the pattern of the CD derivatives, and consequently the dissociation of the complexes are responsible for the dissolution rates of the drug.

Similarly, studies have been developed with acylated CDs in order to obtain slowly dissolving complexes of drugs. Some of these studies are described below.

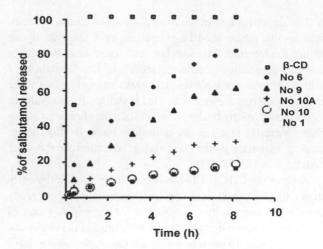

Figure 6. Release profiles of salbutamol with β-CD and different Et-β-CDs. (Adapted from [89].)

Nicardipine hydrochloride, a calcium channel blocking agent with very limited bioavailability (15 to 40%) and a short elimination half-life (about 1 h) is a weak basic drug with a pH-dependent solubility. Binary systems between nicardipine and triacetyl-β-cyclodextrin (TA-β-CD) retarded the dissolution rate of the drug in simulated gastric fluids (Fig. 7). The retarding effect was more evident for the spray-dried complex than for the kneading system. The drug release profile consisted of very slow release in the first stage, with zero-order kinetics and a faster release in the second phase. In simulated intestinal fluids (pH 6.8) the nicardipine/TA-β-CD system showed lower drug dissolution; consequently, the binary systems presented sustained drug release [90].

Metformin hydrochloride is an oral antihyperglycemic agent that is highly water soluble, with low bioavailability and a short and variable biological half-life (1.5 to 4.5 h); consequently, this drug needs frequent administrations to

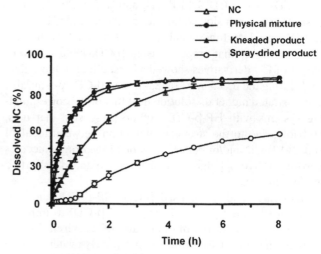

Figure 7. Dissolution profiles of NC and NC/TA-β-CD binary systems in simulated gastric fluid (pH 1.2). (Adapted from [90].)

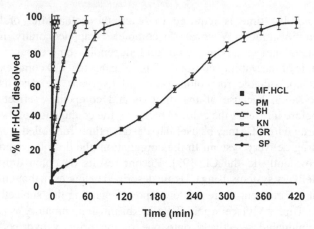

Figure 8. Dissolution curves of metformin·HCl (MF·HCl) alone or from its equimolar physical mixture (PM), sealed-heated (SH), kneaded (KN), co-ground (GR), and spray-dried (SP) products with TA-β-CD. (Adapted from [91].)

maintain effective plasma concentrations. To improve patient compliance Corti et al. [91] used TA-β-CD as a carrier to obtain a slow-dissolving complex of the drug. The spray-dried method showed the greatest retarding effect on the dissolution of metformin hydrochloride and made it possible to obtain an almost linear slow dissolution profile, reaching 100% of dissolved drug after only about 7 h (Fig. 8). Authors attributed the slow drug dissolution rate to the interactions established between the drug and the hydrophobic carrier during sample treatment, which were more or less intense, depending on the different conditions and methods used. Furthermore, CDs have been shown to decrease the rate of drug release from controlled-release devices under certain conditions.

Filípović-Grcić et al. [92], using cross-linked chitosan microspheres containing HP-β-CD, demonstrated a reduced release in the buffer of the poorly water-soluble drug nifedipine, compared to microspheres containing free drug (Fig. 9). This reduction occurred despite a nearly twofold increase in the aqueous solubility of nifedipine in the presence of HP-β-CD. Although diffusion of the CD molecule itself from the microspheres was considered unlikely, its release was not analyzed. The authors proposed that during the release studies, nifedipine dissociated rapidly from the complex, resulting in an increased concentration of free CD within the polymer matrix. A more hydrophilic matrix, chitosan–CD, was formed as a result, decreasing the drug–matrix permeability and slowing the release of drug. It is interesting that the microspheres were prepared using sunflower oil, whose lipid components have been shown to form strong complexes with CDs.

Bibby et al. [82] postulated that in hydrated polymeric matrices, when drug is present in concentrations below saturation, the addition of β-CD would decrease the free drug concentration. This reduction in the free concentration

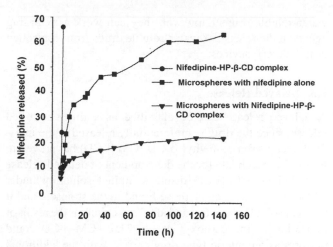

Figure 9. Release profile of drug complexed with HP-β-CD alone (●), and microspheres with incorporated nifedipine/HP-β-CD complexes (▼) in comparison to drug dissolution from microspheres with nifedipine alone (■). Points are experimental, and curves are the best fit for first-order kinetics of the biphasic type. (Adapted from [92].)

of drug would result in a decrease in the diffusion rate of free drug, due to the reduction in its concentration gradient. Consequently, any reduction in drug release observed would be attributed to the difference in the diffusion coefficient between the free and complexed drug.

Quaglia et al. [93] attributed the decrease in the release rate of nicardipine, after inclusion with β-CD from a poly (ethylene glycol) hydrogel, to the decrease in drug diffusion when it is complexed with CDs, due to an increase in effective molar size.

Similarly, Pose-Vilarnovo et al. [85] studied the influence of β-CD and HP-β-CD in the release rate of hydrophilic diclofenac from the HPMC gels and tablets. They verified a delay in the release profile of diclofenac. This effect was attributed to the hindering effect of the free CDs on drug diffusion. They considered that in contrast to the hydrophobic drug, the solubilizing effect of CDs is not the dominant effect in the hydrophilic drug release.

Ikeda et al. [94] verified that the release of metoprolol from metoprolol/HP-β-CD/ethylcellulose tablets was dependent on the amount of HP-β-CD in the tablets. It was observed that the release rate decreased when small amounts of HP-β-CD were added, whereas large amounts of HP-β-CD accelerated the release rate. Detailed studies on water penetration, scanning electron microscopic observations and physicochemical properties of HP-β-CD indicated that the retarding effect of HP-β-CD was attributable to a viscous gel formation in small pores on the surface of the tablets, HP-β-CD gels may work as a barrier to water penetration into the tablets and consequent drug release.

Therefore, hydrophilic CDs are used largely as solubilizers and fast-dissolving carriers for many poorly water-soluble drugs. However, they can work as retarding agents in the release of water-soluble drugs from controlled drug delivery devices.

5.3. Delayed Release

A delayed release can be classified as a time-controlled release, since the drug is preferentially released in the intestinal tract. Carboxy-methyl-β-CD (CM-β-CD) has proved to be an ideal candidate for the development of delayed-release formulations. CM-β-CD displays limited solubility under acidic conditions such as those found in the stomach, and it is freely soluble in neutral and alkaline regions as the result of ionization of the carboxyl group. Thus, CM-β-CD could perform as an enteric-type drug carrier, with the additional advantage of stabilizing the labile drugs because of its inclusion ability.

Diltiazem was complexed with CM-β-CD and then compressed into a tablet. The release rate of the drug was relatively low in low-pH solutions and increased when the pH rose. The release of water-soluble diltiazem·HCl from CM-β-CD was suppressed at low pH (Fig. 10). These studies indicated that the release rate of water-soluble drugs can be suppressed at low-pH regions of the stomach and increased at intestinal pH values, due to ionization of the carboxyl group of the drug molecule [95].

Horikawa et al. [96] studied the release of molsidomine, a water-soluble drug, from tablets, where the drug was complexed with CM-β-CD. The release rate of molsidomine was suppressed at a lower pH values and was increased following an increase in pH values, thereby showing a typical delayed-release pattern.

Colon-specific drug delivery systems can be classified as a delayed-release type of dosage form, although a fairly

long lag time is required to reach the colon after oral administration. When a CD complex is applied orally, it dissociates readily in the GI fluid, depending on the magnitude of the stability constant. This indicates that CD complex is not suitable for colon-specific delivery as the drug is released, because of the dilution and competitive effects before it reaches the colon. Therefore, the conjugation of the drug with CDs may be useful to design a time-controlled oral drug delivery system. In the conjugation, the drug is bound covalently to the CD [97]. Recent results on colon drug delivery systems using CDs are described below, even though, at the moment, there are few reports regarding this subject.

Biphenylylacetic acid (BPAA), an anti-inflammatory, was conjugated selectively onto one of the primary hydroxyl groups of α-, β-, and γ-CDs through an ester or amide linkage. In the rat cecum and colon, the ester prodrugs are subject to ring opening followed by hydrolysis to maltose and triose conjugates. The ester bond of the small saccharide conjugates is subsequently hydrolyzed to release the drug. In the case of amide prodrugs, the conjugate is hydrolyzed to small sacharide conjugates, but the amide bond resists hydrolysis and is retained as a maltose conjugate in the colon [98,99].

The serum levels of BPAA after oral administration of three ester conjugates with the drug alone and β-complex in rats were studied (Fig. 11). They showed higher serum levels than those of BPAA alone. The serum drug levels of α- and γ-CD conjugates increased after a lag time of about 3 h, and reached maximum levels at about 9 and 8 h, respectively, accompanied by a significant increase in the extent of bioavailability [98,99].

Hirayama et al. [100] prepared n-butyric acid/β-CD conjugate and studied the release behavior of the drug. They

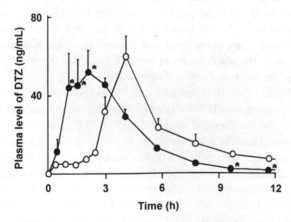

Figure 10. Plasma levels of diltiazem after the oral administration of tablets containing CME-β-CD complex (equivalent to diltiazem at 30 mg/body) in gastric acidity–controlled dogs; (○), high-gastric-acidity dogs; (●), low-gastric-acidity dogs. Each value represents the mean ± standard error of four dogs.*, $p < 0.05$ vs. high-gastric-acidity dogs. (Adapted from [5].)

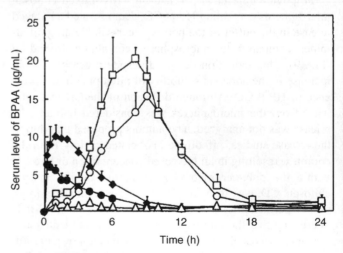

Figure 11. Serum levels of BPAA after oral administration of α-conjugate (○), β-conjugate (Δ), γ-conjugate (□); BPAA (●) or BPAA-β-CD complex (◆) (equivalent to 10 mg/kg BPAA) to rats. Each point represents the mean ± S.E. of three experiments. (Adapted from [98].)

suggested that two enzymes, sugar-degrading and ester-hydrolyzing, are necessary to release the drug from the conjugate in large intestinal contents. A recent study reached the same conclusion [101].

Prednisolone succinate/α-CD conjugate, ketoprofen/β-CD conjugate, and flurbiprofen/α- and β-CD conjugates may be degraded similar to those of the BPAA and *n*-butyric acid conjugates [102–104]. These conjugates may be useful as a delayed-release type of prodrug for colon specific delivery. Moreover, release of the drug from CD conjugate in the cecum and colon depends on the degree of substitution or the solubility of the conjugate. This was proved after preparation of conjugates between α, β, γ-CDs and 5-aminosalicylic acid with different degrees of substitution. The higher degree of substitution lowers the aqueous solubility and the hydrolysis rate [100,105].

Thus, the design of CD conjugates with nonsteroidal anti-inflammatory drugs such as biphenylylacetic acid [98,99], ketoprofen [103], acid mefenamic [106], flurbiprofen [104], a short-chain fatty acid, *n*-butyric acid [100], and a steroidal drug, prednisolone [102] may be presented as new candidates for colon-specific delivery prodrugs.

5.4. Modified Release

5.4.1. Combination of Drug with Hydrophilic and Hydrophobic CD Complexes and Drug–CD Conjugates
The critical combination of inclusion complexes obtained with different types of CDs derivatives could be a promising approach to developing new controlled drug release systems. To obtain the rapid appearance of nicardipine (NC) in the blood and to maintain a constant drug level for a long period of time, Fernandes et al. [107,108] combined a fast-releasing HP-β-CD/NC complex with a slow-releasing TA-β-CD/NC complex.

In a similar way, Ikeda et al. [109] concluded that a combination of HP-β-CD and perbutanoyl-β-CD (TB-β-CD) is useful to control the release of a water-soluble drug, captopril. HP-β-CD forms 1 : 1 solid complexes, while the hydrophobic form, TB-β-CD, forms a solid dispersion or solid solution with the drug. Both authors concluded that the drug release rate could be controlled by adjusting the molar ratio of its hydrophilic and hydrophobic complexes, depending on the sustained release frame time required [107–109].

In the ternary captopril/TB-β-CD/HP-β-CD system, the release rate was slowed by the addition of small amounts of HP-β-CD, whereas the rate become faster as the molar ratio of HP-β-CD increased further (>0.25 molar ratio). It happened because the pore size was dependent on the composition of HP-β-CD (i.e., small pores in lower HP-β-CD contents, larger pores in higher HP-β-CD contents). Thus, in the case of less HP-β-CD, after contact with water, HP-β-CD forms a gel on the surface or inside the tablets with tiny pores that prevent the penetration of water into tablets and consequent captopril release. In the case of high HP-β-CD

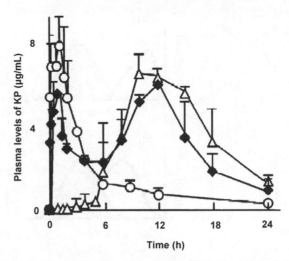

Figure 12. Plasma levels of ketoprofen (KP) after oral administrations of KP/HP-β-CD complex (O, equivalent to 2 mg/kg KP), KP/α-CD conjugate (△, equivalent to 5 mg/kg KP) and the combined preparation (◆, the complex + conjugate, containing the equivalent amounts of KP). Each point represents the mean ± S.E. of three to four experiments. (Adapted from [103].)

contents, captopril, HP-β-CD, or its complex dissolved or eroded from the surface, resulting in larger pores on a smooth surface [109].

These authors proposed that hydrophilic and hydrophobic CD systems seem to resemble a blend of water-soluble hydroxypropylcellulose (HPC) and water-insoluble ethylcellulose. Moreover, HP-β-CD has better stabilizing abilities than HPC [109].

Different drug-release profiles can be obtained based on a combination of CD complexes and CD conjugates. The immediate-release type of complex and the delayed-release type of conjugate are used simultaneously to obtain a repeated release profile (i.e., double peaks are observed in plasma drug levels) [97]. Kamada et al. [103] demonstrated this by a combination of the CD conjugate and the fast-dissolving ketoprofen/HP-β-CD complex [103] (Fig. 12).

On the other hand, a combined preparation of the conjugate with a slow-release fraction may provide a prolonged-release profile [97] (Fig. 13B). Furthermore, a combination of water-soluble complex, slow-release complex, and CD conjugates provide a gradient release profile [97] (Fig. 13C).

5.4.2. Combination of CDs with Others Controlled-Release Technologies
A pertinent combination of CD derivatives and pharmaceutical polymers is useful for advanced-controlled released systems [110]. Wang et al. [110] have developed a double-layer tablet in order to obtain a more balanced oral bioavailability with a prolonged therapeutic effect of nifedipine, a poorly water-soluble drug with a short elimination half-life. The double-layer tablet, consisting of a nifedipine/2-HP-β-CD/HCO-60 system as the fast-release portion and a nifedipine/HPC system as the slow-release

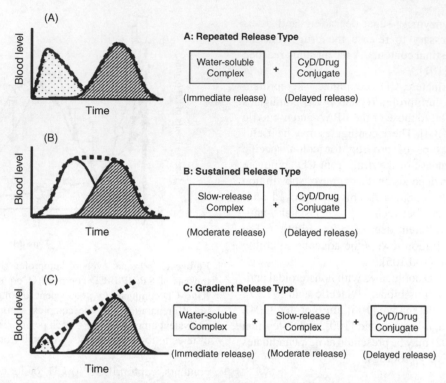

Figure 13. Drug release profiles based on a combination of CD complexes and CD conjugate. (Adapted from [97].)

portion, gave prolonged plasma drug levels with a more balanced bioavalibility. Thus a combination of HP-β-CD, HCO-60, and HPC serves as a modified-release carrier of nifedipine and can be applied to other poorly water-soluble drugs, with a short half-life [110,111].

Coadministration of the conjugate and the slow-releasing ketoprofen/ethylcellulose solid dispersion gave a typical sustained-release profile (i.e., a constant plasma level was maintained for at least 24 h) [103] (Fig. 14).

CDs have been used in osmotic systems. These systems use osmotic pressure as a driving force for controlled drug delivery. Drug release from these systems is, to a large extent, independent of the physiological variables factors of the GI tract and can be used for systemic as well as for local delivery of drugs. These systems are generally used only for water-soluble drugs. Poorly water-soluble drugs cannot dissolve adequately in the volume of water drawn into osmotic pump tablets (OPTs), making release from OPTs incomplete. Sulfobutyl ether-β-cyclodextrin (SBE$_{7m}$-β-CD) serves both as a solubility modulator and as an osmotic pumping agent for OPTs, from which the release rate of both water-soluble and poorly water-soluble drugs can be controlled. The osmotic pressure of SBE$_{7m}$-β-CD is due to the fact that it includes seven sulfonic acid sodium salt moieties per molecule. The advantages of controlled OPTs include controlled and complete release for poorly water-soluble drugs. For

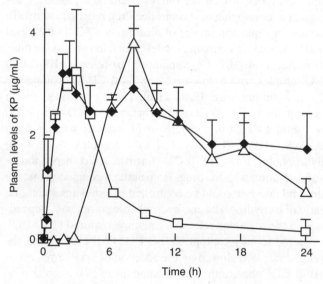

Figure 14. Plasma levels of ketoprofen after oral administrations of KP-EC solid dispersion (□, equivalent to 6 mg/kg KP), KP-α-CD conjugate (△, equivalent to 5 mg/kg KP) and the combined preparation (◆, EC solid dispersion + conjugate, containing the equivalent amounts of KP). The drugs were administered as powder filled in capsules. Each point represents the mean ± S.E. of three to four experiments. (Adapted from [103].)

soluble drugs, SBE_{7m}-β-CD acts primarily as an osmotic and an OPT control agent [112].

Okimoto et al. [113] compared the release of testosterone from the device in the presence of SBE_{7m}-β-CD and in the presence of HP-β-CD. The testosterone release was significantly faster with SBE_{7m}-β-CD than with HP-β-CD, which is consistent with greater osmotic pressure from SBE_{7m}-β-CD. Testosterone forms an inclusion complex with SBE_{7m}-β-CD. It appears that the major contribution from testosterone/SBE_{7m}-β-CD devices is osmotic, while from testosterone/HP-β-CD the major contribution may be diffusion [113,114].

OPTs using SBE_{7m}-β-CD were prepared for poorly soluble and two highly water-soluble drugs. The results confirmed that SBE_{7m}-β-CD serves as a solubility modulator and as an osmotic pumping agent for OPTs, from which the release rate of both water-soluble and poorly water-soluble drugs can be controlled [112].

Comparison of prednisolone and sodium chloride release, formulated as a complex with SBE_{7m}-β-CD from OPTs, demonstrated that the release rate was almost identical for the two drugs. The prednisolone was released from OPTs in the complexed form; however, NaCl, which has high polarity and a small molecular size, is not expected to form an inclusion complex. Thus, the release rates are controlled by the presence of SBE_{7m}-β-CD [112].

SBE_{7m}-β-CD and HP-β-CD increased the release rates of prednisolone (neutral drug) and chlorpromazine (basic drug) at pH 6.8 compared to a sugar mixture, which is consistent with their solubilizing effects, but decreased the release rate of the polar drugs clorpromazine at pH 1.2 and diltiazem. It is because the release rate is controlled by the osmotic, diffusion and viscosity properties of the CDs and not the drug themselves [112].

6. CDs IN BUCCAL AND SUBLINGUAL ADMINISTRATIONS

Buccal and sublingual routes for drug administration are two of the most efficient ways to bypass hepatic first-pass metabolism [115]. After contact with the buccal mucosa, the drug permeates the mucosal tissue until it reaches the systemic circulation. An important factor that preceeds drug permeation through the buccal mucosa is its solubility in the aqueous environment that surrounds the oral cavity (saliva). Due to the small volume of saliva in the mouth, the therapeutic dose has to be relatively small, and dissolution enhancers must usually be applied. In an attempt to improve the solubility of drugs with poor aqueous solubility, and considering that the majority of drugs present a hydrophobic structure, complexation with CDs appeared in the formulation of buccal drug delivery systems as a promising tool.

Several authors have directed their research to the application of CDs to enhance the aqueous solubility of drugs destined to be delivered through buccal mucosa. Jain et al. [116] conducted a study with the aim of improving the bioavailability of danazol. They prepared buccal tablets containing polycarbophil (PC) or HPMC as bioadhesive polymers, where they incorporated the complexed drug with SBE_{7m}-β-CD. In an initial phase, the authors verified that buccal tablets containing PC presented better mucoadhesive properties than the tablets containing HPMC. Afterward, they conducted an in vivo study in four female police dogs to compare the absolute bioavailability resulting from the oral administration of a marketed formulation, Danocrine (200 mg of danazol) and from buccal tablets containing the danazol/SBE_{7m}-β-CD complex. Table 1 illustrates the different pharmacokinetic parameters calculated after administration of the various formulations in four dogs.

Authors verified that the absolute bioavailability resulting from the oral administration of Danocrine to dogs was only 1.8% compared with the absolute bioavailability determined after the administration of buccal tablets (25%). In relation to the relative bioavailability of buccal tablets, it increased about 13- fold compared to the relative bioavailability of Danocrine, which is illustrated in Fig. 15.

They concluded that the increase in drug bioavailability observed when buccal tablets were administered in complexed form was due to the absence of the hepatic first effect when the drug was administered by the buccal route and to the increase in the drug aqueous solubility after complexation with SBE_{7m}-β-CD. This contributed to clarifying the cause of low drug bioavailability in oral administration and to noting

Table 1. Pharmacokinetic Parameters After Administration of Various Drugs

Formulation	C_{max} (ng/mL)	t_{max} (h)	AUC $_{0-24}$ (ng·h/mL)[a]	Absolute BD (F, %)	Relative BD
Intravenous	—	—	2154.64 (417.80)	100	
Danocrine, 200 mg	124.99 (82.06)	1.13 (0.63)	40.57 (30.44)	1.88	
Inclusion complex, 50 mg	205.19 (1046.95)[b]	0.75 (0.29)	137.00 (30.61)[c]	63.86	33.91
Buccal tablets, 40 mg	210.57 (79.18)	4.00 (0.00)	554.95 (133.13)[c]	25.76	13.67

[a] AUC values normalized to 2 mg/kg dog weight.
[b] Statistically significantly different from Danocrine and buccal formulation.
[c] Statistically significantly different from Danocrine.

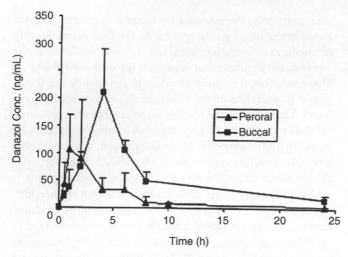

Figure 15. Average danazol concentrations after oral administration of the drug in capsules containing 200 mg of danazol (Danocrine) and in buccal tablets containing the complex between danazol and SBE7-β-CD, equivalent to 40 mg of danazol. (Adapted from [116].)

that the buccal administration of the danazol in the complexed form could be clinically useful, due to the reduction of secondary effects associated with the high clinical doses that are normally administered.

Jug and Bećirević-Laćan [117] developed a controlled release formulation for buccal administration of piroxicam. They prepared buccal tablets that contained piroxicam and hydrosoluble excipient (lactose or HP-β-CD) and as an alternative to the piroxicam/HP-β-CD complex, using as bioadhesive polymers HPMC or a mixture of HPMC and C940 (Carbomer 940). All of the formulations obtained contained an equal quantity of the drug sample. The authors verified that the dissolution of the piroxicam in lactose or in HP-β-CD without complexation, in buccal tablets, was comparatively slower than that of tablets containing the drug in a complexed form. The low aqueous solubility of the piroxicam permitted only a limited amount of noncomplexed drug to dissolve in the interior of the hydrated polymeric matrix. According to the results obtained, incorporation of the piroxicam in complexed form with HP-β-CD in the matrix increased the aqueous solubility of the drug, dissolving more easily in the hydrated polymeric environment, resulting in a high driving diffusion force and fast dissociation of the complexed drug. They also observed that by increasing the drug's solubility, the addition of HP-β-CD to the polymeric matrix caused an increase in the drug flow from the tablets or, rather, of the diffusible species of the drug in the interface membrane. Even though the complex does not penetrate the membrane, since the drug in the inclusion complex is in dynamic equilibrium with the molecules of the "free" drug, it promotes a continuous supply of molecules of the drug in the diffusible form to the membrane's surface. They also verified that all formulations presented an increase in weight due to

the water uptake. The rate of swelling decreased in the order lactose < HP-β-CD < drug complexed with CD, suggesting that the high water uptake by the complex could be a consequence of the faster hydration of the bioadhesive polymers in the presence of the amorphous complexes. Finally, the authors concluded that the CD did not interfere with the formation of mucoadhesive connection between the polymers and the mucin present in the saliva.

In the sublingual formulations the complexation of poorly water-soluble drugs with CD has been shown to increase the bioavailability of various lipophilic drugs. 2-HP-β-CD has been shown to increase the bioavailability of 17β-estradiol [118], androstenediol [119], clomipramine [120], and danazol [121]. In the case of lipophilic compounds, the aqueous solubility and dissolution rate of a drug is usually the rate-limiting step for drug absorption. The increased bioavailability achieved by CDs is due to the increased aqueous solubility and drug dissolution rate. One limitation in the use of CDs in sublingual administration is the effect of CDs on formulation bulk.

Mannila et al. [122] studied the effect of complexation with RM-β-CD in the bioavailability of the Δ^9-tetrahydrocannabinol (THC) and the canabidiol (CBD), two main cannabinoids present in marijuana and hashish, when administered by the sublingual route. They verified that the complexation with RM-β-CD increased significantly the dissolution rate in both (THC and CBD). Two equally important aspects were also observed; the absolute bioavailability of the complex when administered by the sublingual route was superior in relation to the bioavailability of the isolated cannabinoids when administered by the same route. The bioavailability of the complex when administered by the sublingual route was superior in relation to the bioavailability of the same complex when administered orally. The authors concluded that the complexation with RM-β-CD increased the aqueous solubility and the dissolution rate of the cannabinoids (THC and CBD) and the sublingual administration of the complex led to an increase in the absorption rate and the bioavailability of the active substances.

On the other hand, CDs can act as penetration enhancers in buccal mucosa. RM-β-CD, a more hydrophobic CD, can permeate the buccal mucosa and form inclusion complexes with hydrophobic molecules, lipids from the cellular membrane. This modified CD can change buccal mucosa permeability acting as a penetration enhancer for drugs administered by the buccal route [30]. Because the methylated CD derivatives have a highest affinity toward the most lipophylic components of the cell membranes (cholesterol and phospholipids), they have the highest hemolyzing capacity.

Boulmedarat et al. [123] carried out a study whose main objective was to evaluate the toxicological effects of RM-β-CD on the human oral epithelium, in order to verify the acceptance of these compounds as solubilizing agents and absorption promoters of formulations for buccal mucosa delivery. The study revealed that RM-β-CD does not present any toxic or inflammatory effects on the human buccal epithelium when exposed to concentrations between 2 and 5%. They verified that after 24 h of exposure using doses superior to 5%, the cellular layers presented alterations, although these were only of a superficial nature. They concluded that the methylated derivatives could be used safely as absorption promoters on buccal mucosa drug delivery in a concentration between 2 and 5% [124].

7. CONCLUSIONS

Solubility, stability, and membrane permeability issues continue to be major formulation barriers in the development of advanced dosage forms for the next generation of drugs. CDs are able to eliminate some of these undesirable properties of the drug candidates through inclusion complex formation, which consequently improves oral drug delivery. In the beginning, CDs were used to enhance the aqueous solubility and chemical stability of drugs, and their functionalities were related to their ability to form inclusion complexes with drugs. However, in recent years CDs have been shown to participate in various types of noninclusion complexes with organic salts or water-soluble polymers. They have also been shown to form aggregates, either alone or in combination with other excipients. These aggregates can form dispersed drug delivery systems such as micro- and nanoparticles.

We can conclude that CDs satisfied the requirements for a drug carrier in oral drug delivery systems very well. A desirable attribute for a carrier in oral drug delivery is the ability to control the rate and the time of drug release. For example, peracylated CDs act as novel hydrophobic carriers to control the release rate of water-soluble drugs. On the other hand, amphiphilic or ionizable CDs can modify the rate or time of drug release, and bind to the surface membrane of cells, which may be used for the enhancement of drug absorption across biological barriers. The final requirement of the drug carrier is its ability to deliver a drug to a target site. CD–drug conjugates may fulfill this requirement regarding the design of colon-specific delivery systems. A combined use of different CDs and pharmaceutical additives will provide more balanced oral bioavailability with prolonged therapeutic effects. The future promises a number of commercial products using various CD derivative-based advanced drug formulations.

8. FUTURE TRENDS

More than 30 different pharmaceutical products containing CDs are currently in the worldwide market. New CD-based technologies are constantly being developed, and thus 100 years after their discovery, CDs are still regarded as novel excipients of unexplored potential [125,126]. Thus, because of the multifunctional characteristics and bioadaptability, CDs are capable of overcome the undesirable properties of drug molecules in different areas of drug delivery: namely, in oral drug delivery. The final requirement of the drug carrier is its ability to delivery a drug to a target site, if desired. Owing to the increasingly globalized nature of the CD-related science and technology, development of CD-based drug formulation is also rapidly progressing [72].

CDs have applications in the design of some novel oral drug delivery systems, such as microspheres, osmotic pump, peptide and protein delivery, colon-specific drug delivery, and nanoparticles. The first studies on the role of CDs in microparticle preparation were carried out by Loftsson et al. [127].

Osmotically controlled oral drug delivery systems are available in various designs to control the drug release based on the principle of osmosis. Osmotic tablets offer many advantages, such as a zero-order delivery rate, improving patient compliance, a high degree of in vitro–in vivo correlation and they are simple of operation [11].

On the other hand, advances in biotechnology have increased the production of therapeutically active peptide- and protein-based drugs, making them more available for therapeutic uses [128]. However, there are considerable hurdles to be overcome before practical use can be made of therapeutic peptides and proteins because of chemical and enzymatic instability, poor absorption through biological membranes, rapid plasma clearance, peculiar dose–response curves, and immunogenicity. Many attempts have been made to address these problems by chemical modifications or by coadministration of adjuvants to eliminate undesirable properties. CD complexation seems to be an attractive alternative to these approaches [129].

Recently, intensive efforts have been made to design systems able to deliver drugs more efficiently to specific organs, tissues, or cells. CD complexes are in equilibrium with guest molecules in aqueous solution; however, the inclusion equilibrium is disadvantageous when drug targeting is to be attempted, because the complex dissociates before it reaches the specific site to which the drug is to be delivered. A new method to prevent this dissociation is to bind a drug covalently to CDs [130].

The major drawback of nanoparticles is associated with the drug-loading capacity of polymeric nanoparticles. CDs are used for this reason to improve water solubility and sometimes the hydrolytic or photolytic stability of drugs for better loading properties [131]. CD complexes act to solubilize or stabilize active excipients within the nanoparticles, resulting in increased drug concentration in the polymerization medium and increased hydrophobic sites in the nanosphere structure when large amounts of CDs are associated with the nanoparticles [11].

In summary, various controlled-release preparations can be designed by employing CD conjugated or complexed in combination with other carriers, displaying different releasing mechanisms [103]. The potential of CDs in oral drug delivery formulations has not been totally exploited, and the possibility of discovering new systems is present.

REFERENCES

1. Shojaei, A.H., Chang, R.K., Guo, X., Burnside, B.A., Couch, R.A. (2001). Systemic drug delivery via the buccal mucosa route. *Pharm. Technol.*, *1*, 70–81.

2. Senel, S., Kremer, M., Katalin, N., Squier, C.A. (2001). Delivery of bioactive peptides and proteins across oral (buccal) mucosa. *Curr. Pharm. Biotechnol.*, *2*, 175–186.

3. Loftsson, T., Brewster, M.E., Másson, M. (2004). Role of cyclodextrins in improving oral drug delivery. *Am. J. Drug Deliv.* *2*, 1–15.

4. Challa, R., Ahuja, A., Ali, J., Khar, R.K. (2005). Cyclodextrins in drug delivery: an updated review. *AAPS PharmSciTech*, *6*, E329–E357.

5. Hirayama, F., Uekama, K. (1999). Cyclodextrin-based controlled drug release system. *Adv. Drug Deliv. Rev.*, *36*, 125–141.

6. Bowen, W.E., Wang, Q., Wuelfing, W.P., Thomas, D.L., Nelson, E.D., Brian Hill, Y.M., Thompson, M., Gallagher, K., Reed, R.A. (2008). A Biopharmaceutical Classification System approach to dissolution: mechanisms and strategies. In: Krishna, R., Lawrence, Yu. Eds., *Biopharmaceutics Applications in Drug Development*, Springer-Verlag, New York.

7. Amidon, G.L., Lennernas, H., Shah, V.P., Crison, J.R.A. (1995). A theoretical basis for a biopharmaceutic drug classification: the correlation of in vitro drug product dissolution and in vivo bioavailability. *Pharm. Res.*, *12*, 413–420.

8. Ku, M.S. (2008). Use of the Biopharmaceutical Classification System in early drug development. *AAPS J.*, *10*, 208–212.

9. U.S., FDA (1995). *Guidance for Industry: Immediate Release Solid Oral Dosage Forms: Scale Up and Post Approval Changes.* CDER/FDA.

10. Lindenberg, M., Kopp, S., Dressman, J.B. (2004). Classification of orally administered drugs on the World Health Organization model list of essential medicines according to the biopharmaceutics classification system. *Eur. J. Pharm. Biopharm.*, *58*, 265–278.

11. Rasheed, A., Ashor Kumar, C.K., Sravanthi, V. (2008). Cyclodextrins as drug carrier molecule: a review. *Sci. Pharm.*, *76*, 567–598.

12. Puglisi, G., Santagati, N.A., Ventura, C.A., Pignatello, R., Panico, A.M., Spampinato, S. (1991). Enhancement of 4-biphenylacetic acid bioavailability in rats by its β-cyclodextrin complex after oral administration. *J. Pharm. Pharmacol.*, *43*, 430–432.

13. Puglisi, G., Ventura, C.A., Spadaro, A., Campana, G., Spampinato, S. (1995). Differential-effects of modified β-cyclodextrins om pharmacological activity and bioavailability of 4-biphenylacetic acid in rats after oral administration. *J. Pharm. Pharmacol.*, *47*, 120–123.

14. Ahn, H.J., Kim, K.M., Choi, J.S., Kim, C.K. (1997). Effects of cyclodextrin derivatives on bioavailability of ketoprofen. *Drug Dev. Ind. Pharm.*, *23*, 397–401.

15. Miyaji, T., Inoue, Y., Acartürk, F., Imai, T., Otagiri, M., Uekama, K. (1992). Improvement of oral bioavailability of fenbufen by cyclodextrin complexation. *Acta Pharm. Nord.*, *4*, 17–22.

16. Ahuja, N., Singh, A., Singh, B. (2003). Rofecoxib: an update on physicochemical, pharmaceutical, pharmacodynamic and pharmacokinetic aspects. *J. Pharm. Pharmacol.*, *55*, 859–894.

17. Savolainen, J., Jarvinen, K., Hatilain, L., Jarvinen, T. (1998). Improved dissolution and bioavailability of phenytoin by sulfobutylether-β-cyclodextrin and hydroxypropyl-β-cyclodextrin complexation. *Int. J. Pharm.*, *165*, 69–78.

18. Tanino, T., Ogiso, T., Iwaki, M. (1999). Effect of sugar-modified β-cyclodextrins on dissolution and absorption characteristics of phenytoin. *Biol. Pharm. Bull.*, *22*, 298–304.

19. Peeters, J., Neeskens, P., Tollenaere, J.P., Remoortere, P.V., Brewster, M.E. (2002). Characterization of the interaction of 2-hydroxypropyl-β-cyclodextrin with itraconazole at pH 2, 4 and 7. *J. Pharm. Sci.*, *91*, 1414–1422.

20. Chowdary, K.P.R., Kamalakara, R.G. (2003). Controlled release of nifedipine from mucoadhesive tablets of its inclusion complexes with β-cyclodextrin. *Pharmazie*, *58*, 721–724.

21. Emara, L.H., Badr, R.M., Abd Elbary, A. (2002). Improving the dissolution and bioavailability of nifedipine using solid dispersions and solubilizers. *Drug Dev. Ind. Pharm.*, *28*, 795–807.

22. Carrier, R.L., Miller, L.A., Ahmed, I. (2007). The utility of cyclodextrins for enhancing oral bioavailability. *J. Control. Release*, *123*, 78–99.

23. Frömming, K.-H., Wedelich, V., Mehnert, W. (1984). Influence of cyclodextrins on nitrosation of drugs. *J. Inclusion Phenom. Macrocyclic Chem.*, *2*, 605–611.

24. Chavanpatil, M., Vavia, P.R. (2002). Enhancement of nasal absorption of acyclovir via cyclodextrins. *J. Inclusion Phenom. Macrocyclic Chem.*, *44*, 137–144.

25. Huang, L., Taylor, H., Gerber, M., Orndorff, P.E., Horton, J.R., Tonelli, A. (1999). Formation of antibiotic, biodegradable/bioabsorbable polymers by processing with neomycin sulfate and its inclusion compound with β-cyclodextrin. *J. Appl. Polym. Sci.*, *74*, 937–947.

26. Ikeda, Y., Motoune, S., Matsuoka, T., Arima, H., Hirayama, F., Uekama, K. (2002). Inclusion complex formation of captopril with α- and β-cyclodextrins in aqueous solution: NMR spectroscopic and molecular dynamic studies. *J. Pharm. Sci.*, *91*, 2390–2398.

27. Ran, Y., Zhao, L., Xu, Q., Yalkowsky, S.H. (2001). Solubilization of cyclosporin A. *AAPS PharmSciTech*, *2*, 1–4.

28. Sharma, P., Varma, M.V.S., Chawla, H.P.S., Panchagnula, R. (2005). Relationship between lipophilicity of BCS class III and IV drugs and the functional activity of peroral absorption enhancers. *Il Farmaco*, *60*, 870–873.

29. Lipinski, C.A. (2000). Drug-like properties and the cause of poor solubility and poor permeability. *J. Pharmacol. Toxicol. Methods*, *44*, 235–249.

30. Másson, M., Loftsson, T., Másson, G., Stefánsson, E. (1999). Cyclodextrins as permeation enhancers: some theoretical evaluations and in vitro testing. *J. Control. Release*, *59*, 107–118.

31. Matilainen, L., Toropainen, T., Vihola, H., Hirvonen, J., Järvinen, T., Jarho, P., Järvinen, K. (2008). In vitro toxicity and permeation of cyclodextrins in Calu-3 cells. *J. Control. Release*, *126*, 10–16.

32. Loftsson, T., Sigfússon, S.D., Sigurosson, H.H., Másson, M. (2003). The effects of cyclodextrins on topical delivery of hydrocortisone: the aqueous diffusion layer. *STP Pharma Sci.*, *13*, 125–131.

33. Irie, T., Uekama, K. (1997). Pharmaceutical applications of cyclodextrins: III. Toxicological issues and safety evaluation. *J. Pharm. Sci.*, *86*, 147–161.

34. Uekama, K., Hirayama, F., Irie, T. (1998). Cyclodextrin drug carrier systems. *Chem. Rev.*, *98*, 2045–2076.

35. Matsuda, H., Arima, H. (1999). Cyclodextrins in transdermal and rectal delivery. *Adv. Drug Deliv. Rev. 36*, 81–99.

36. Loftsson, T., Másson, M. (2001). Cyclodextrins in topical drug formulations: theory and practice. *Int. J. Pharm.*, *225*, 15–30.

37. Loftsson, T., Jarho, P., Másson, M., Jarvinen, K. (2005). Cyclodextrins in drug delivery. *Expert Opin. Drug Deliv. 2*, 335–351.

38. Loftsson, T., Bodor, N. (1995) *Percutaneous Penetration Enhancers*, CRC Boca Raton, FL.

39. Irie, T., Wakamatsu, K., Arima, H., Aritomi, H., Uckama, K. (1992). Enhancing effects of cyclodextrins on nasal absorption of insulin in rats. *Int. J. Pharm.*, *84*, 129–139.

40. Loftsson, T. (1998). Cyclodextrins in pharmaceutical formulations: the effect of polymers on their complexation efficacy and drug availability. Report prepared for the Nordic Industrial Fund.

41. Loftsson, T., Hreinsdóttir, D., Másson, M. (2007). The complexation efficiency. *J. Inclusion Phenom. Macrocyclic Chem. 57*, 545–552.

42. Li, P., Tabibi, E., Yalkowsky, S.H. (1998). Combined effect of complexation pH on solubilization. *J. Pharm. Sci.*, *87*, 1535–1537.

43. Loftsson, T., Duchêne, D. (2007). Cyclodextrins and their pharmaceutical applications. *Int. J. Pharm.*, *329*, 1–11.

44. Hanna, K., Brauer, C.H., Germain, P. (2004). Cyclodextrin-enhanced solubilization of pentachlorophenol in water. *J. Environ. Manag.*, *71*, 1–8.

45. Zornoza, A., Martín, C., Sánchez, M., Piquer, A. (1998). Inclusion complexation of glisentide with α-, β- and γ-cyclodextrins. *Int. J. Pharm.*, *169*, 239–244.

46. Loftsson, T., Ólafsdóttir, B.J., Frioriksdóttir, H., Jónsdóttir, S. (1993). Cyclodextrin complexation of NSAIDs: physicochemical characteristics. *Eur. J. Pharm. Sci.*, *1*, 95–101.

47. Granero, G., Granero, C., Longhi, M. (2003). The effect of pH and triethanolamine on sulfisoxazole complexation with hydroxypropyl-β-cyclodextrin. *Eur. J. Pharm. Sci.*, *20*, 285–293.

48. Loftsson, T., Sigurosson, H.H., Másson, M., Schipper, N. (2004). Preparation of solid drug/cyclodextrin complexes of acidic and basic drugs. *Pharmazie*, *59*, 25–29.

49. Redenti, E., Szene, L., Szejtli, J. (2001). Cyclodextrin complexes of salts of acidic drugs: thermodynamic properties, structural features, and pharmaceutical applications. *J. Pharm. Sci.*, *90*, 979–986.

50. Esclusa-Díaz, M.T., Gayo-Otero, M., Pérez-Marcos, M.B., Vila-Jato, J.L., Torres-Labandeira, J.J. (1996). Preparations and evaluation of ketoconazole-β-cyclodextrin multicomponent complexes. *Int. J. Pharm.*, *142*, 183–187.

51. Ribeiro, L., Carvalho, R.A., Ferreira, D.C., Veiga, F.J.B. (2005). Multicomponent complex formation between vinpocetine, cyclodextrins, tartaric acid and water-soluble polymers monitored by NMR and solubility studies. *Eur. J. Pharm. Sci.*, *24*, 1–13.

52. Redenti, E., Ventura, P., Fronza, G., Selva, A., Rivara, S., Plazzi, P.V., Mor, M. (1999). Experimental and theoretical analysis on the interaction of $(+/-)$-cis-ketoconazole with β-cyclodextrin in the presence of $(+)$-L-tartaric acid. *J. Pharm. Sci.*, *88*, 599–607.

53. Faucci, M.T., Melani, F., Mura, P. (2000). H-NMR and molecular modeling techniques for the investigation of the inclusion complex of econazole with α-cyclodextrin in the presence of malic acid. *J. Pharm. Biomed. Anal.*, *23*, 25–31.

54. Redenti, E., Szente, L., Szejtli, J. (2000). Drug/cyclodextrin/hydroxy acid multicomponent systems: properties and pharmaceutical applications. *J. Pharm. Sci.*, *89*, 1–8.

55. Piel, G., Pirotte, B., Delneuville, I., Neven, P., Llabres, G., Delarge, J., Delattre, L. (1997). Study of the influence of both cyclodextrins and L-lysine on the aqueous solubility of nimesulide; isolation and characterization of nimesulide–L-lysine–cyclodextrin complexes. *J. Pharm. Sci.*, *86*, 475–480.

56. Mura, P., Maestrelli, F., Cirri, M. (2003). Ternary systems of naproxen with hydroxypropyl-β-cyclodextrin and aminoacids. *Int. J. Pharm.*, *260*, 293–302.

57. Figueiras, A., Sarraguça, J.M.G., Carvalho, R.A., Pais, A.A.C.C., Veiga, F.J.B. (2007). Interaction of omeprazole with a methylated derivative of β-cyclodextrin: phase solubility, NMR spectroscopy and molecular simulation. *Pharm. Res.*, *24*, 377–389.

58. Riley, C.M., Rytting, J.H., Kral, M.A. (1991) Takeru Higuhi, a memorial tribute: equilibria and thermodynamics. Kansas.

59. Loftsson, T., Fridriksdottir, H., Gudmundsdottir, T.K. (1996). The effect of water - soluble polymers on aqueous solubility of drugs. *Int. J. Pharm.*, *127*, 293–296.

60. Loftsson, T., Fridriksdóttir, H. (1998). The effect of water-soluble polymers on the aqueous solubility and complexing abilities of β-cyclodextrin. *Int. J. Pharm.*, *163*, 115–121.

61. Granero, G., Longhi, M. (2002). Thermal analysis and spectroscopy characterization of interactions between a naphthoquinone derivative with HP-beta-cyclodextrins and PVP. *Pharm. Dev. Technol.*, *7*, 381–390.

62. Rácz, I. (1989). *Drug Formulation*. Budapest, Hungary, pp. 212–242.

63. Mura, P., Faucci, M.T., Bettinetti, G.P. (2001). The influence of polyvinylpyrrolidone on naproxen complexation with hydroxypropyl-β-cyclodextrin. *Eur. J. Pharm. Sci.*, *13*, 187–194.

64. Loftsson, T. (1998). Increasing the cyclodextrin complexation of drugs and drug bioavailability through addition of water-soluble polymers. *Pharmazie*, *53*, 733–740.

65. Duan, M., Zhao, N., Ossurardóttir, I.B., Thorsteinsson, T., Loftsson, T. (2005). Cyclodextrin solubilization of the antibacterial agents triclosan and triclocarban: formation of aggregates and higher-order complexes. *Int. J. Pharm.*, *297*, 213–222.

66. Loftsson, T., Másson, M., Brewster, E.M. (2004). Self-association of cyclodextrins and cyclodextrin complexes. *J. Pharm. Sci.*, *93*, 1091–1099.

67. Loftsson, T., Matthíasson, K., Másson, M. (2003). The effects of organic salts on the cyclodextrin solubilization of drugs. *Int. J. Pharm.*, *262*, 101–107.

68. Qiu, Y., (2009). Rational design of oral modified-release drug delivery systems. In: Qiu, Y., Chen, Y., Zharg, G., Liu, L., Porter, W., Eds., *Developing Solid Oral Dosage Forms*. Academic Press, San Diego, CA, pp. 469–499.

69. Martin Del Valle, E.M. (2004). Cyclodextrins and their uses: a review. *Process Biochem.*, *39*, 1033–1046.

70. Brewster, M.E., Loftsson, T. (2007). Cyclodextrins as pharmaceutical solubilizers. *Adv. Drug Deliv. Rev.*, *59*, 645–666.

71. Vyas, A., Saraf, S., Saraf, S. (2008). Cyclodextrin based novel drug delivery systems. *J. Inclusion Phenom. Macrocyclic Chem. 62*, 23–42.

72. Stella, V.J., He, Q. (2008). *Cyclodextrins. Toxicol. Pathol. 36*, 30–42.

73. Figueiras, A., Carvalho, R.A., Ribeiro, L., Torres-Labandeira, J.J., Veiga, F.J.B. (2007). Solid-state characterization and dissolution profiles of the inclusion complexes of omeprazole with native and chemically modified β-cyclodextrin. *Eur. J. Pharm. Biopharm.*, *67*, 531–539.

74. Zingone, G., Rubessa, F. (2005). Preformulation study of the inclusion complex warfarin-β-cyclodextrin. *Int. J. Pharm.*, *291*, 3–10.

75. Hirlekar, R., Kadam, V. (2009). Preparation and characterization of inclusion complexes of carvedilol with methyl-β-cyclodextrin. *J. Inclusion Phenom. Macrocyclic Chem.*, *63*, 219–224.

76. Sathigari, S., Chadha, G., Lee, Y.-H.P., Wright, N., Parsons, D.L., Rangari, V.K., Fasina, O., Babu, R.J. (2009). Physico-chemical characterization of efavirenz–cyclodextrin inclusion complexes. *AAPS PharmSciTech*, *10*, 81–87.

77. Fernandes, C.M., Vieira, M.T., Veiga, F.J.B. (2002). Physicochemical characterization and in vitro dissolution behavior of nicardipine–cyclodextrins inclusion compounds. *Eur. J. Pharm. Sci.*, *15*, 79–88.

78. Badr-Eldin, S.M., Elkheshen, S.A., Ghorab, M.M. (2008). Inclusion complexes of tadalafil with natural and chemically modified β-cyclodextrins. I: Preparation and in-vitro evaluation. *Eur. J. Pharm. Biopharm.*, *70*, 819–827.

79. Cunha-Filho, M.S.S., Dacunha-Marinho, B., Torres-Labandeira, J.J., Martínez-Pacheco, R., Landín, M. (2009). Characterization of β-lapachone and methylated β-cyclodextrin solid-state systems. *AAPS PharmSciTech*, *8*, E68–E77.

80. Evrard, B., Chiap, P., DeTullioc, P., Ghalmid, F., Piela, G., Van Heesa, T., Crommen, J., Losson, B., Delattre, L. (2002). Oral bioavailability in sheep of albendazole from a suspension and from a solution containing hydroxypropyl-β-cyclodextrin. *J. Control. Release*, *85*, 45–50.

81. Veiga, F., Fernandes, C., Teixeira, F. (2000). Oral bioavailability and hypoglycaemic activity of tolbutamide/cyclodextrin inclusion complexes. *Int. J. Pharm.*, *202*, 165–171.

82. Bibby, D.C., Davies, N.M., Tucker, I.G. (2000). Mechanisms by which cyclodextrins modify drug release from polymeric drug delivery systems. *Int. J. Pharm.*, *197*, 1–11.

83. Rao, V.M., Haslam, J.L., Stella, V.J. (2001). Controlled and complete release of a model poorly water-soluble drug, prednisolone, from hydroxypropyl methylcellulose matrix tablets using (SBE)7β-cyclodextrin as a solubilizing agent. *J. Pharm. Sci.*, *90*, 807–816.

84. Koester, L.S., Xavier, C.R., Mayorga, P., Bassani, V.L. (2003). Influence of β-cyclodextrin complexation on carbamazepine release from hydroxypropyl methylcellulose matrix tablets. *Eur. J. Pharm. Biopharm.*, *55*, 85–91.

85. Pose-Vilarnovo, B., Rodríguez-Tenreiro, C., Santos, J.F.R., Vázquez-Doval, J., Concheiro, A., Alvarez-Lorenzo, C., Torres-Labandeira, J.J. (2004). Modulating drug release with cyclodextrins in hydroxypropyl methylcellulose gels and tablets. *J. Control. Release*, *94*, 351–363.

86. Sangalli, M.E., Zema, L., Maroni, A., Foppoli, A., Giordano, F., Gazzaniga, A. (2001). Influence of β-cyclodextrin on the release of poorly soluble drugs from inert and hydrophilic heterogeneous polymeric matrices. *Biomaterials*, *22*, 2647–2651.

87. Cappello, B., De Rosa, G., Giannini, L., La Rotonda, M.I., Mensitieri, G., Miro, A., Quaglia, F., Russo, R. (2006). Cyclodextrin-containing poly(ethyleneoxide) tablets for the delivery of poorly soluble drugs: potential as buccal delivery system. *Int. J. Pharm.*, *319*, 63–70.

88. Miro, A., Rondinone, A., Nappi, A., Ungaro, F., Quaglia, F., La Rotonda, M.I. (2009). Modulation of release rate and barrier transport of diclofenac incorporated in hydrophilic matrices: role of cyclodextrins and implications in oral drug delivery. *Eur. J. Pharm. Biopharm.*, *72*, 76–82.

89. Lemesle-Lamache, V., Wouessidjewe, D., Chéron, M., Duchêne, D. (1996). Study of β-cyclodextrin and ethylated β-cyclodextrin salbutamol complexes, in vitro evaluation of sustained-release behaviour of salbutamol. *Int. J. Pharm.*, *141*, 117–124.

90. Fernandes, C.M., Veiga, F.J.B. (2002). Effect of the hydrophobic nature of triacetyl-β-cyclodextrin on the complexation with nicardipine hydrochloride: physicochemical and dissolution properties of the kneaded and spray-dried complexes. *Chem. Pharm. Bull.*, *50*, 1597–1602.

91. Corti, G., Capasso, G., Maestrelli, F., Cirri, M., Mura, P. (2007). Physical–chemical characterization of binary systems of metformin hydrochloride with triacetyl-β-cyclodextrin. *J. Pharm. Biomed. Anal.*, *45*, 480–486.

92. Filípović-Grcić, J., Becirevic-Lacan, M., Skalko, N., Jalsenjak, I. (1996). Chitosan microspheres of nifedipine and nifedipine–cyclodextrin inclusion complexes. *Int. J. Pharm.*, *135*, 183–190.

93. Quaglia, F., Varricchio, G., Miro, A., La Rotonda, M.I., Larobina, D., Mensitieri, G. (2001). Modulation of drug release from hydrogels by using cyclodextrins: the case of nicardipine/β-cyclodextrin system in crosslinked polyethylenglycol. *J. Control. Release*, *71*, 329–337.

94. Ikeda, Y., Motoune, S., Marumoto, S., Sonoda, Y., Hirayaama, F., Arima, H., Uekama, K. (2004). Effect of 2-hydroxypropyl-β-cyclodextrin on release rate of metoprolol from ternary metoprolol/2-hydroxypropyl-β-cyclodextrin/ethylcellulose Tablets. *J. Inclusion Phenom. Macrocyclic Chem.*, *44*, 141–144.

95. Uekama, K., Horikawa, T., Horiuchi, Y., Hirayama, F. (1993). In vitro and in vivo evaluation of delayed-release behavior of diltiazem from its *O*-carboxymethyl-*O*-ethyl-β-cyclodextrin complex. *J. Control. Release*, *25*, 99–106.

96. Horikawa, T., Hirayama, F., Uekama, K. (1995). In-vivo and in-vitro correlation for delayed-release behaviour of a molsidomine/*O*-carboxymethyl-*O*-ethyl-β-cyclodextrin complex in gastric acidity–controlled dogs. *J. Pharm. Pharmacol.*, *47*, 124–127.

97. Uekama, K. (2004). Design and evaluation of cyclodextrin-based drug formulation. *Chem. Pharm. Bull.*, *52*, 900–915.

98. Uekama, K., Minami, K., Hirayama, F. (1997). 6A-*O*-[(4-Biphenylyl)acetyl]-α-, β-, and -γ-cyclodextrins and 6A-deoxy-6A-[[(4-biphenylyl)acetyl]amino]-α-, β-, and γ-cyclodextrins: potential prodrugs for colon-specific delivery. *J. Med. Chem. 40*, 2755–2761.

99. Minami, U., Hirayama, F., Uekama, K. (1998). Colon-specific drug delivery based on a cyclodextrin prodrug: release behavior of biphenylylacetic acid from its cyclodextrin conjugates in rat intestinal tracts after oral administration. *J. Pharm. Sci.*, *87*, 715–720.

100. Hirayama, F., Ogata, T., Yano, H., Arima, H., Udo, K., Takano, M., Uekama, K. (2000). Release characteristics of a short-chain fatty acid, *n*-butyric acid, from its β-cyclodextrin ester conjugate in rat biological media *J. Pharm. Sci.*, *89*, 1486–1495.

101. Feng Cao, Y.R., Weiyi, H. (2009). Cyclomaltoheptaose mixed esters of anti-inflammatory drugs and short-chain fatty acids and study of their enzymatic hydrolysis in vitro. *Carbohydr. Res.*, *344*, 526–530.

102. Yano, H., Hirayama, F., Kamada, M., Arima, H., Uekama, K. (2002). Colon-specific delivery of prednisolone-appended α-cyclodextrin conjugate: alleviation of systemic side effect after oral administration. *J. Control. Release*, *79*, 103–112.

103. Kamada, M., Hirayama, F., Udo, K., Yano, H., Arima, H., Uekama, K. (2002). Cyclodextrin conjugate–based controlled release system: repeated- and prolonged-releases of ketoprofen after oral administration in rats. *J. Control. Release*, *82*, 407–416.

104. El-Kamel, A.H., Abdel-Aziz, A.A.-M., Arnal, J., Fatani, A.J., El-Subbagh, H.I. (2008). Oral colon targeted delivery systems for treatment of inflammatory bowel diseases: synthesis, in vitro and in vivo assessment. *Int. J. Pharm.*, *358*, 248–255.

105. Zou, M., Okamoto, H., Cheng, G., Hao, X., Sun, J., Cui, F., Danjo, K. (2005). Synthesis and properties of polysaccharide prodrugs of 5-aminosalicylic acid as potential colon-specific delivery systems. *Eur. J. Pharm. Biopharm.*, *59*, 155–160.

106. Dev, S., Mhaske, D.V., Kadam, S.S., Dhaneshwar, S.R. (2007). Synthesis and pharmacological evaluation of cyclodextrin conjugate prodrug of mefenamic acid. *Indian J. Pharm. Sci.*, *69*, 69–72.

107. Fernandes, C.M., Ramos, P., Falcão, A.C., Veiga, F.J.B. (2003). Hydrophilic and hydrophobic cyclodextrins in a new sustained release oral formulation of nicardipine: in vitro evaluation and bioavailability studies in rabbits. *J. Control. Release*, *88*, 127–134.

108. Fernandes, C.M., Veiga, F.J.B. (2002). The cyclodextrins as modelling agents of drug controlled release. *J. Inclusion Phenom. Macrocyclic Chem. 44*, 79–85.

109. Ikeda, Y., Kimura, K., Hirayama F., Arima, H., Uekama, K. (2000). Controlled release of a water-soluble drug, captopril, by a combination of hydrophilic and hydrophobic cyclodextrin derivatives. *J. Control. Release*, *66*, 271–280.

110. Wang, Z., Hirayama, F., Ikegami, K., Uekama, K. (1993). Release characteristics of nifedipine from 2-hydroxypropyl-β-cyclodextrin complex during storage and its modification of hybridizing polyvinylpyrrolidone K-30. *Chem. Pharm. Bull.*, *45*, 1822–1826.

111. Zheng, W., Horikawa, T., Hirayama, T., Hirayama, F., Uekama, K. (1993). Design and in-vitro evaluation of a modified-release oral dosage form of nifedipine by hybridization of hydroxypropyl-β-cyclodextrin and hydroxypropylcellulose. *J. Pharm. Pharmacol.*, *45*, 942–946.

112. Okimoto, K., Tokunaga, Y., Ibuki, R., Irie, T., Uekama, K., Rajewski, R.A., Stella, V.J. (2004). Applicability of (SBE)7m-β-CD in controlled-porosity osmotic pump tablets (OPTs). *Int. J. Pharm. 286*, 81–88.

113. Okimoto, K., Rajewski, R.A., Stella, V.J. (1999). Release of testosterone from an osmotic pump tablet utilizing (SBE)7m-β-cyclodextrin as both a solubilizing and an osmotic pump agent. *J. Control. Release*, *58*, 29–38.

114. Zannou, E.A., Streng, W., Stella, V.J. (2001). Osmotic properties of sulfobutylether and hydroxypropyl cyclodextrins. *Pharm. Res.*, *18*, 1226–1231.

115. Harrs, D., Robinson, J.R. (1992). Drug delivery via the mucous membranes of the oral cavity. *J. Pharm. Sci.*, *82*, 1–10.

116. Jain, A.C., Aungst, B.J., Adeyeye, M.C. (2002). Development and in vivo evaluation of buccal tablets prepared using danazol–sulfobutylether 7 β-cyclodextrin (SBE 7) complexes. *J. Pharm. Sci.*, *91*, 1659–1668.

117. Jug, M., Bećirević-Laćan, M. (2004). Influence of hydroxypropyl-β-cyclodextrin complexation on piroxicam release from buccoadhesive tablets. *Eur. J. Pharm. Sci.*, *21*, 251–260.

118. Hoon, T.J., Dawood, M.Y., Khan-Dawood, F.S., Ramos, J., Batenhorst, R.L. (1993). Bioequivalence of 17 β-estradiol hydroxypropyl-β-cyclodextrincomplex in postmenopausal women. *J. Clin. Pharmacol.*, *33*, 1116–1121.

119. Brown, G.A., Martin, E.R., Roberts, B.S., Vukovich, M.D., King, D.S. (2002). Acute hormonal response to sublingual androstenediol in young men. *J. Appl. Physiol.*, *92*, 142–146.

120. Yoo, S.D., Yoon, B.M., Lee, H.S., Lee, K.C. (1999). Increased bioavailability of clomipramine after sublingual administration in rats. *J. Pharm. Sci.*, *88*, 1119–1121.

121. Badawy, S.I.F., Ghorab, M.M., Adeyeye, C.M. (1996). Bioavailability of danazolhydroxypropyl-β-cyclodextrin complex by different routes of administration. *Int. J. Pharm.*, *145*, 137–143.

122. Mannila, J., Järvinen, T., Järvinen, K., Tarvainen, M., Jarho, P. (2005). Effects of RM-β-CD on sublingual biovailability of Δ^9-tetrahydrocannabinol in rabbits. *Eur. J. Pharm. Sci.*, *26*, 71–77.

123. Boulmedarat, L., Bochot, A., Lesieur, S., Fattal, E. (2005). Evaluation of buccal methyl-β-cyclodextrin toxicity on human oral epithelial cell culture model. *J. Pharm. Sci.*, *94*, 1300–1309.

124. Merkus, F.W.H.M., Verhoef, J.C., Marttin, E., Romeijn, S.G., van der Kuy, P.H.M., Hermens, W.A.J.J., Schipper, N. G. M. (1999). Cyclodextrins in nasal drug delivery. *Adv. Drug Deliv. Rev. 36*, 41–57.

125. Szejtli, J. (2004). Past, present, and future of cyclodextrin research. *Pure Appl. Chem.*, *76*, 1825–1845.

126. Loftsson, T., Duchêne, D. (2007). Historical perspectives: cyclodextrins and their pharmaceutical applications. *Int. J. Pharm.*, *329*, 1–11.

127. Loftsson, T., Kristmundsdóttir, T., Ingvvarsdóttir, K., Ólafsdóttir, B.J., Baldvinsdóttir, J. (1992). Preparation and physical evaluation of micorcapsules of hydrophilic drug–cyclodextrin complexes. *J. Microencapsul. 9*, 375–382.

128. Soares, F.A., Carvalho, R.A., Veiga, F. (2007). Oral administration of peptides and porteins: nanoparticles and cyclodextrins as biocompatible delivery systems. *Nanomedicine*, *2*, 183–202.

129. Irie, T., Uekama, K. (1999). Cyclodextrins in peptide and protein delivery. *Adv. Drug Deliv. Rev.*, *36*, 101–123.

130. Hattori, K. (2007). Synthesis and evaluation of novel targeting cyclodextrins as drug carrier. Presented at the 4th Asian Cyclodextrin Conference.

131. Eguchi, M., Yong-Zhong, D., Ogawa, Y., Okada, T., Yumoto, N., Kodaka, M. (2006). Effects of conditions for preparing nanoparticles composed of aminoethylcarbamoyl-β-cyclodextrin and ethylene glycol diglycidyl ether on trap efficiency of a guest molecule. *Int. J. Pharm.*, *311*, 215–222.

PART II

NOVEL AND SPECIALIZED APPLICATIONS OF CYCLODEXTRINS

11

AMPHIPHILIC CYCLODEXTRINS: SYNTHESIS AND CHARACTERIZATION

FLORENT PERRET AND HÉLÈNE PARROT-LOPEZ

Institut de Chimie et Biochimie Moléculaire et Supramoléculaire, Équipe de Chimie Supramoléculaire Appliquée, Université de Lyon, Lyon, France

1. INTRODUCTION

The use of biologically active substances that are insoluble in water is a difficult problem to resolve in pharmaceutical formulation. Their solubilization at the molecular level as inclusion complexes inside cyclodextrins (CDs) is a good alternative and has been well described in the literature. Native CDs are therefore widely used as solubilizers and excipients, masking the physicochemical properties of the guest molecule (poor water solubility, stability problems, or undesired side effects). However, since dissociation takes place too readily upon dilution, inclusion complexes in simple water-soluble CD are not effective for drug delivery applications (binding constants of 102 to 104 M). Chemical modification of CDs has been the aim of many research groups, mainly to improve safety while maintaining the ability to form inclusion complexes with various substrates. Well-defined chemical modification allows some of them to self-organize as larger assemblies aimed at resolving this liability issue. Nanoparticles, for example, have been widely used with CDs for drug delivery systems, because of their active and passive targeting properties and because they do not require cosolvents, potentially toxics, for the delivery of nonsoluble drugs.

The most used CD derivatives in the pharmaceutical field are methyl cyclodextrins, which are obtained by methylation of cyclodextrins, on either all C_2 secondary and C_6 primary hydroxyl groups [dimethyl-cyclodextrins (DIMEBs)] or all the hydroxyl groups C_2, C_3, and C_6 [trimethyl-cyclodextrins (TRIMEBs)]. Hydroxypropylated cyclodextrins (HP-CDs) are also used in this field, but their structure is not well defined and they are more statistically substituted than the methylated cyclodextrins. They are commercially available as ocular collyrs, tablets, and as excipients under the trademarks Encapsin and Molecusol [1,2].

More than 20 years ago, amphiphilic CD synthesis began to interest research groups. Indeed, amphiphilic CDs can self-organize in water to form supramolecular assemblies such as vesicles [3], solid–lipid nanoparticles [4], nanospheres [5], liquid crystals [6], or micellar systems [7]. Emulsions and many complex structured particles can also be prepared with the help of CD-based surfactants. Such supramolecular assemblies often retain the complexation properties of the parent CD. They offer good opportunities for drug solubilization and possible sustained delivery and are expected to improve the vectorization of the entrapped drug in the organism.

2. SYNTHESIS OF AMPHIPHILIC CDs: GENERAL APPROACHES

Amphiphilic CDs have been synthesized for more than a decade, mainly to overcome problems of native CDs that limit their applications in the pharmaceutical field [8]: for example, to:

- Enhance the interaction of CDs with biological membranes

- Modify or enhance interaction of CDs with hydrophobic drugs arising from the large number of long aliphatic chains
- Allow self-assembly of CDs, allowing the formation of nanosize carriers and encapsulation of drugs

Indeed, amphiphilic CDs were believed to be promising carriers as nanoparticles because their structure gives them the ability to include active molecules in their hydrophobic cavity as well as within their long aliphatic chains [9,10]. Numerous reviews have recently been published on the synthesis and physical properties of amphiphilic CDs, particularly by Bilensoy [8], Sallas and Darcy [11], and Djedaïni-Pilard et al. [12–14]. In this chapter we focus on the synthesis of amphiphilic CDs.

Preparation of amphiphilic CDs involves modification of the primary face and/or the secondary face by lipophilic groups. Persubstituted amphiphilic CDs are obtained by persubstitution of the hydroxyl groups, on the secondary and/or primary face, but amphiphilic derivatives can also be obtained by grafting only one, two, three, or four hydrophobic anchors (Fig. 1).

In most cases, the hydrophilic CD architecture plays the role of the polar group of a surfactant-like molecule; nevertheless, increasing the hydrophilic character of the CD by grafting polar substituents is a supplementary possibility for controlling the physicochemical properties. The first amphiphilic CD derivatives were made by Kawabata et al. [15]. Hydrophobicity was introduced by substituting primary hydroxyl groups of β-CDs by alkylsulfinyl groups, of various lengths (Fig. 2). These molecules were not soluble in water and showed a high capacity to form monolayers at the air–water interface. The same amphiphiles could also form thermotropic liquid crystals, as demonstrated by Ling et al. [6].

Since then, numerous groups have synthesized amphiphilic CDs to form supramolecular assemblies able to complex and/or encapsulate drugs. Most of them studied β-CD substituted by alkyl chains; β-CD is slightly hydrophilic compared to α- and γ-CDs (the smallest water solubility: 18 g/L) and the grafted alkyl chains are in most cases short in length (from 6 to 12 carbon atoms). We will not describe the syntheses of all the known amphiphilic CDs, only examples of each type of amphiphilic CD. First we describe the synthesis and characterization of nonionic amphiphilic derivatives, among them the medusa-like [16,17], skirt-shaped [18–21], and bouquet-like [22,23] CD derivatives (Fig. 3), named directly from their shape. In these derivatives, all of the 6-hydroxyls and/or 2,3-hydroxyls are modified by lipophilic groups.

We then focus our attention on the synthesis of monosubstituted (lollipop [24] and cup-and-ball [25] cyclodextrins) and selectively functionalized derivatives, with special attention on di-, tri-, and tetra-substituted amphiphilic α-CDs, synthesis that has recently been developed, and on

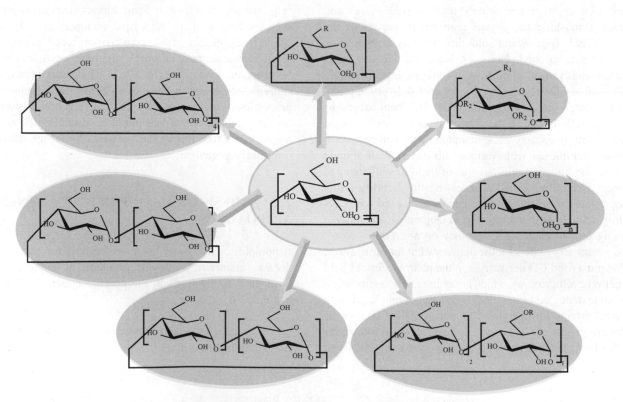

Figure 1. Common synthetic pathways to amphiphilic CDs.

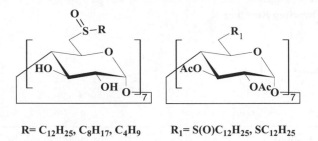

R= $C_{12}H_{25}$, C_8H_{17}, C_4H_9 R_1= $S(O)C_{12}H_{25}$, $SC_{12}H_{25}$

Figure 2. First amphiphilic CDs made by Kawabata et al. [15].

fluorinated derivatives. Finally, the synthesis of some cationic and anionic amphiphilic CDs is described.

3. NONIONIC AMPHIPHILIC CDs

3.1. Persubstituted Derivatives

Among the three types of hydroxyl groups present in CDs, those at the 6-position are the most basic and often most nucleophilic; those at the 2-position are the most acidic; and those at the 3-position are the most inaccessible. Thus, an electrophilic agent will attack the 6-position. This situation can be reversed by the use of a strong base. Because the hydroxyl groups at the 2-position are the most acidic, they will be deprotonated first and the resulting oxyanion is more nucleophilic than the nondeprotonated hydroxyl at the 6-position. Persubstituted amphiphilic CDs are thus obtained by substitution of all the

hydroxyl groups on the secondary and/or primary face. Among them we can distinguish three families.

3.1.1. Medusa-Like CDs   Due to their easy synthesis, medusa-like CDs (Fig. 3), persubstituted on the primary face, are abundant. They are generally obtained in two steps by nucleophilic substitution of a CD bearing a good leaving group. Another reason for the abundance of medusa-like amphiphilic CDs is that the primary face is more easily modifiable than the secondary one, thus avoiding the protection and deprotection steps of the primary face. Selective permodification of all the primary hydroxyl groups is relatively easier than mono-, di-, or tri-substitution because symmetrical substitution is achieved when the reaction is allowed to run longer with appropriate amounts of reagents. For the syntheses of these derivatives, activation of the hydroxyls groups is required.

Synthesis of Activated CDs   It has been necessary to wait more than a century (CDs were discovered in 1890) to see the emergence in the literature of selective and reproducible methods for the introduction of leaving groups in a one-step synthesis, leading only to persubstituted derivatives with good yields. These indispensable derivatives will be used for the synthesis of amphiphilic CDs. The main methods for the preparation of these activated CDs are outlined in Table 1.

Per-sulfonated derivatives, generally prepared directly from CDs and a large amount of sulfonyl chloride in pyridine,

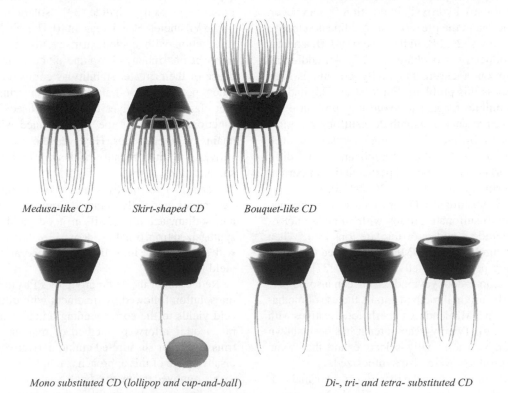

Medusa-like CD   *Skirt-shaped CD*   *Bouquet-like CD*

Mono substituted CD (lollipop and cup-and-ball)   *Di-, tri- and tetra- substituted CD*

Figure 3. Schematic structures of amphiphilic CDs.

Table 1. Methods for the Persubstitution of the Primary Face in a One-Step Reaction

$n = 6, 7, 8 ; \alpha, \beta, \gamma$

CDs	X	Conditions	Yield (%)	Ref.
β	OTs	TsCl, pyridine, H_2O, room temp., 5.5 h	24	[26]
γ	OTs	TsCl, pyridine, 10°C, 2 h	19	[27]
α	2-naphthalensulfonyl	2-Naphthalenesulfonyl chloride, $(Bu_3Sn)_2O$, toluene, 50°C, 24 h	78	[28]
α, β, γ	Br	CH_3SO_2Br, DMF, 65°C, 16 h	95–98	[29]
β	Br	CH_3SO_2Br, DMF, room temp., 24 h	80	[30]
β	Cl	CH_3SO_2Cl, imidazole, DMF, 70°C, 24 h	90	[31]
β	Br	PPh_3, Br_2, DMF, 80°C, 15 h	93	[32]
α, β	I	PPh_3, I_2, DMF, 80°C, 15 h	80–88	[32]
α, β, γ	Cl	Chloromethylenemorpholinium chloride, DMF, 60°C, 20 h	80–99	[33]
α, β, γ	Br	Bromomethylenemorpholinium bromide, DMF or DMAC, 45 to 55°C, 20 h	86–97	[33]
α, β, γ	Br, Cl, or I	N-Halosuccinimide, PPh_3, DMF, 3.5 h, 75°C (20 h, 90°C for iodo)	76–96	[34]

are usually obtained in lower yields than the halogeno derivatives (19 to 24% vs. 90%). The good leaving group behavior of the sulfonate ion poses a major limitation in this reaction because as the reaction proceeds, the tosylated groups at the 6-position tend to change to the 3,6-anhydro form, and thus the presence of by-products decreases the yield. Furthermore, in the presence of an excess of TsCl, the tosylation of secondary hydroxyls in position O_2 can occur, although the reaction in the presence of pyridine is selective for primary hydroxyls [26,27]. In the case of γ-CD, a hepta-tosylated by-product is also obtained [27]. An additional purification step on a chromataography column is then required. To bypass this problem, Fujita et al. [28] had the idea of using a bulkier reagent to avoid the substitution in position O_2. In the presence of naphthalenesulfonyl chloride, the derivative, per-substituted by leaving groups, is obtained with 78% yield. Despite all these complications and difficulties, many workers have reported pertosylation or permesylation of the primary side of CDs [26–28,35].

Per-6-deoxy-6-substituted CDs are then easily obtained by displacing the 6-sulfonate groups with suitable nucleophiles, such as halides, azides, or thiolate ions [36]. When treated with sodium halides in DMF, pertosylated CDs give the corresponding perhalogenated compounds [37]. Nevertheless, perhalogeno-6-deoxycyclodextrins can also be prepared directly: The original method entails the use of methanesulfonyl halide in DMF, leading to perhaloderivatives with excellent yields [29]. Twenty years later it has been shown that this reaction was not really selective and that some sulfonated side products were also synthesized.

The presence of imidazole in the reaction mixture makes it possible to avoid these side reactions [31]. The second method consists in activation of the primary hydroxyl groups by a Vilsmeier–Haack type of complex, generated in situ by the reaction of a dihalide (Br2 or I2) [32] or n-halosuccinimide [34] in the presence of triphenylphosphine and DMF. This is a very simple method that gives a very high yield (>90%) of pure product. This complex can also be prepared separately by the reaction of N-formylmorpholine in the presence of oxalyle halide [33]. Isolation and purification of the Vilsmeier–Haack reagent $[(CH_3)_2NCHBr]^+Br^-$ prior to reaction with cyclodextrin avoids severe problems of removal of triphenylphosphine oxide in this reaction. Because of their greater stability as compared to per-6-sulfonates, per-6-deoxy-6-halogenocyclodextrins are an important class of derivatives that can be used for the selective functionalizations of the primary face with hydrophobic chains or other groups. However, the use of per-halo derivatives is limited because of their insolubility in less polar solvents.

Per-azidocyclodextrins are also an important class of derivatives, generally obtained from per-6-mesitylated CDs and sodium azide in DMF in good yield. These are also synthesized from 6-halo-6-deoxy [38] derivatives by reacting with sodium azide in DMF at high temperatures in a very high yield (Fig. 4).

Reduction of the azido group by PPh_3 in aqueous ammonia solution followed by treatment with dilute hydrochloric acid yields to the corresponding amine salt [39]. This is a rather straightforward method of making aminocyclodextrins and other substituted amino derivatives, but the major disadvantage of this approach is that PPh_3/PPh_3O is difficult to be removed from the final crude product. Amino CD derivatives are very useful in the synthesis of amphiphilic

reagents and conditions : *i* PPh₃, Br₂, DMF, 80°C *ii* NaN₃, DMF, 60°C

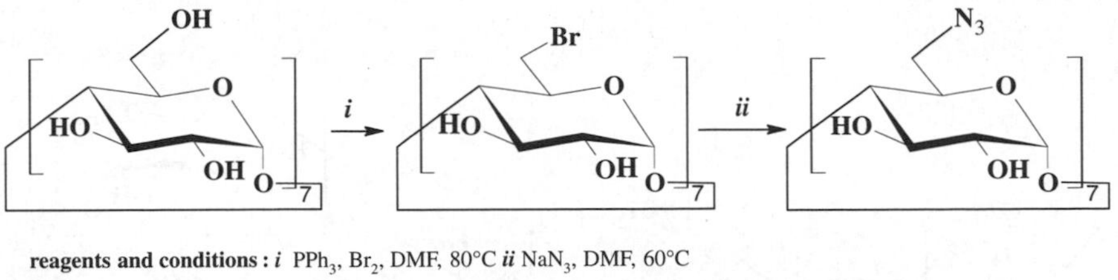

Figure 4. Synthesis of per azido β-CD.

CDs, where the chains are linked to the cavity by amide bonds.

Synthesis of Medusa-like Amphiphilic CDs Per-halogeno, per-amino, and per-azido CD derivatives are thus key intermediates for the synthesis of medusa-like amphiphilic derivatives by nucleophilic substitution: grafting amino, amido, sulfo, or thioalkyl aliphatic chains lead to the derivatives desired [16,17]. Per-iodocyclodextrins react, for example, with alkylamines at a high temperature to produce secondary amines in a very good yield (Fig. 5) [40]. Alkyl- [41] or aryl [42,43]-thiolate ions also displace the halogen groups from the 6-position to produce thio derivatives.

Among these medusa-like CDs, we can also cite the work published by Inamura et al. [44] on the synthesis of adipic-glucamine derivatives where peptide chemistry on peraminated β-CD derivative has been used (Fig. 6). D-Glucamine was connected with adipic acid monoethyl ester using DCC/ HOBT followed by the hydrolysis of the ester with aqueous sodium hydroxide solution to obtain a carboxylic acid. The latter was then reacted with 6-heptaamino-β-CD in the presence of EDC at pH 4.5 to 6.0, and the compound desired has been identified by elemental analysis, ^{1}H NMR spectra including two-dimensional NMR, and mass spectra. They have shown particularly that these adipic chains enhanced binding of aniline–naphthalene sulfonic acid and other guest molecules, as they act as an extension of the cavity.

Other CDs fully substituted with lipophilic chains only on the primary face have been described by Memisoglu et al. [45] and used as nanoparticles for encapsulation. They reported the synthesis and characterization of amphiphilic CD derivatives modified on the primary face with substitu-ents of different chain lengths (C$_6$ linear, C$_6$ branched, and

C$_{14}$ linear) and biodegradable bonds (hydrophilic amide or relatively hydrophobic esters bonds) (Fig. 7).

In these sets of experiments, the per-aminated β-CD has first been synthesized as described previously, then the amphiphilic derivatives with six, 14, or ramified carbon chains were obtained via an amide bond formation, by reacting the per-aminated β-CD with the corresponding anhydride or acid under peptide synthesis conditions. The effect of the nature of the linker has also been studied by linking the hydrophobic linear hexanoic chain via an ester bond, using periodo derivatives with cesium hexanoate (Fig. 7) [46]. All these compounds have been characterized by ^{1}H NMR, FAB mass spectroscopy, C, H, and N analysis, and DSC and used as nanoparticles for encapsulation [45]. In the case of caproyl derivatives, the authors demonstrated a partial inclusion of progesterone and bifonazoleand clotri-mazol in the CD cavity, leading to charged nanospheres.

Direct peralkylation of the primary face can also be achieved using an indirect method, implying many steps. The primary face of CDs is first protected by silyl groups, and the secondary side is then esterified. Desilylation of the primary face and reaction with an alkyl halide under strongly basic conditions gives the alkyl ether derivative, but with lower overall yields, due to these many steps [37,47].

Recently, an amphiphilic CD derivative has also been obtained in the group of Djedaïni-Pilard [48] by an original method, consisting of controlling the esterification of the primary face by lauric acid chloride. The original strategy of this synthesis was based on only one step, easy to apply, and

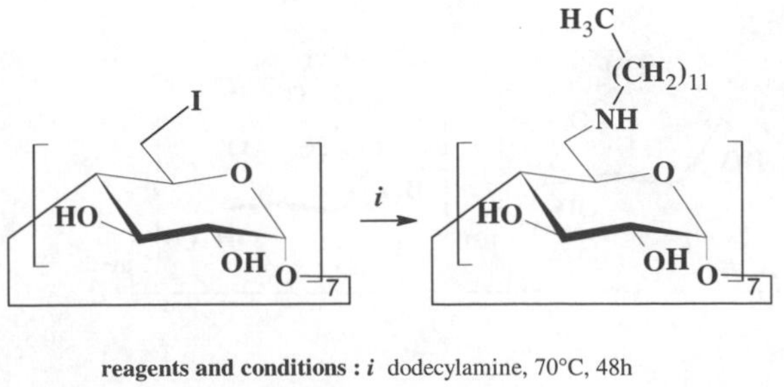

reagents and conditions : *i* dodecylamine, 70°C, 48h

Figure 5. Synthesis of heptakis(6-deoxy-6-dodecylamino)-β-CD.

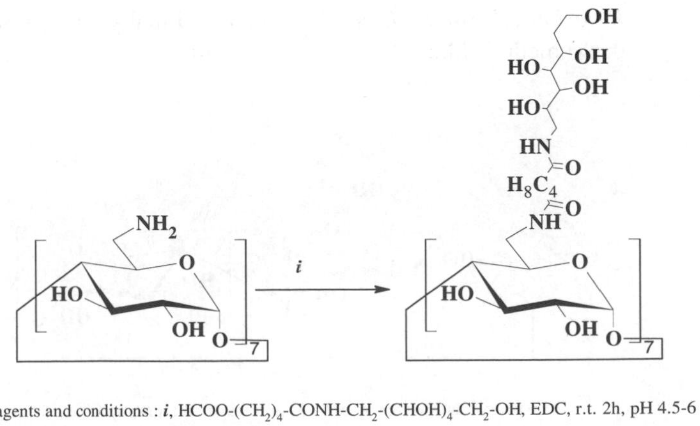

Reagents and conditions : *i*, HCOO-(CH₂)₄-CONH-CH₂-(CHOH)₄-CH₂-OH, EDC, r.t. 2h, pH 4.5-6

Figure 6. Synthesis of an adipic acid CD derivative.

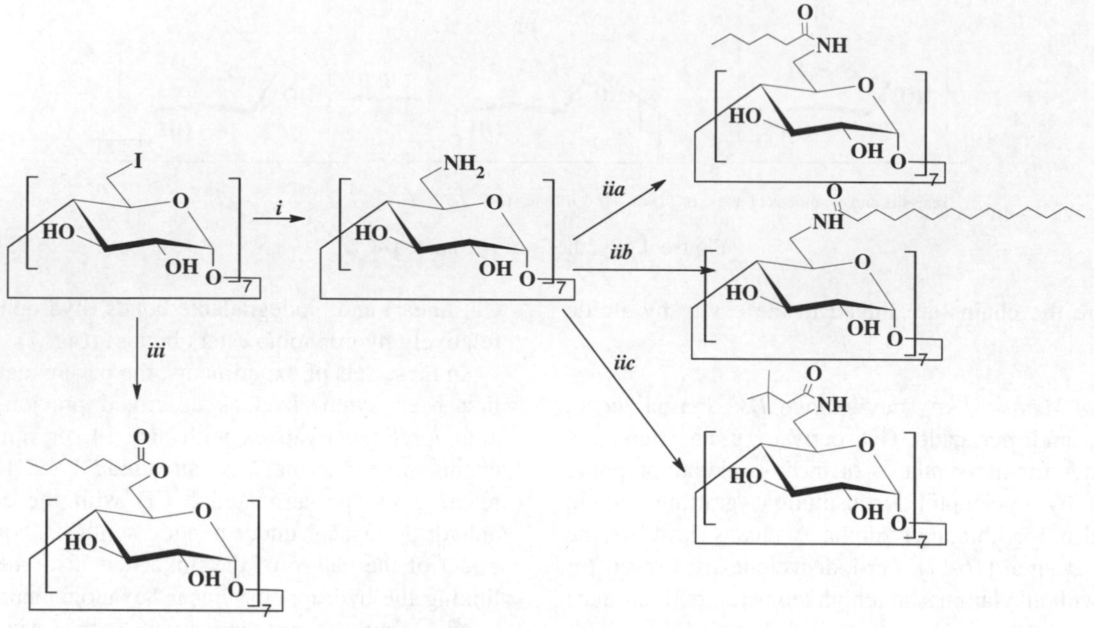

Reagents and conditions : *i*, NaN$_3$ then PPh$_3$, NH$_4$OH20%, DMF *ii*, a)hexanoic anhydride, DMF/MeOH; b) Myristic acid, DCC, HOBT, DMF, c) terbutyl acetic acid, DCC, HOBT, DMF, iii cesium hexanoate, DMF

Figure 7. Synthesis of myristil, caproyl, and ramified amphiphilic CD derivatives.

making it possible to obtain reproducible batches with a given average substitution rate permolecule that could be varied (Fig. 8).

The control of substitution of the primary face of the CD (from 0 to 7) was a function of three experimental parameters, such as an amount and ratio of reagents, temperature, and reaction time. The reaction was optimized with 10 equiv of lauric acid chloride for a reaction time of 65 h at room temperature. Purification by sublimation allows the removal of excess lauric acid and was followed by DSC. Product characterization were performed by mass spectroscopy that showed the presence of a mixture of grafted CDs with a degree of substitution ranging from one to seven chains per molecule and confirmed by ^{13}C NMR.

Fluorinated Amphiphilic Derivatives Finally, concerning these medusa-like CDs, fluorinated chains, known to have

an important lipophilicity (hydrophobic character of difluoromethylenic groups 1.75 higher than the hydrogenated analog [49]) have been anchored to CDs in order to stabilize the supramolecular architectures (e.g., nanospheres, nanocapsules) obtained with hydrocarbonated analogs. The first fluorophilic CD, per-subtituted on the primary face by seven fluorinated groups, has been synthesized by Diakur et al. [50] for the study of their complexation with biologically active molecules. The synthesis of the symmetrical heptakis(6-deoxy-6-fluoro)-β-CD was a five-step reaction. Protection of the primary face by 7.7 equiv of TBDMSCl in pyridine gave the persilylated derivatives, which was then acylated on the secondary face at high temperature and in high yields (84%). ^{13}C NMR spectrum of this intermediate indicated a highly symmetrical structure, meaning full substitution. Desilylation was accomplished with boron trifluoride

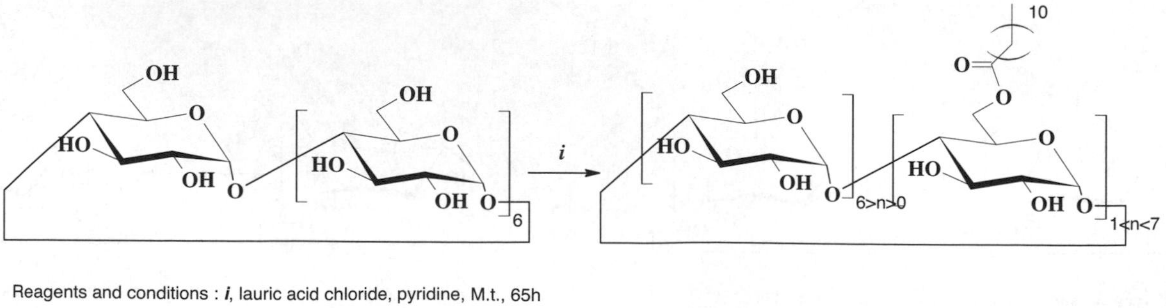

Reagents and conditions : *i*, lauric acid chloride, pyridine, M.t., 65h

Figure 8. Synthesis of lauric acid CD derivatives.

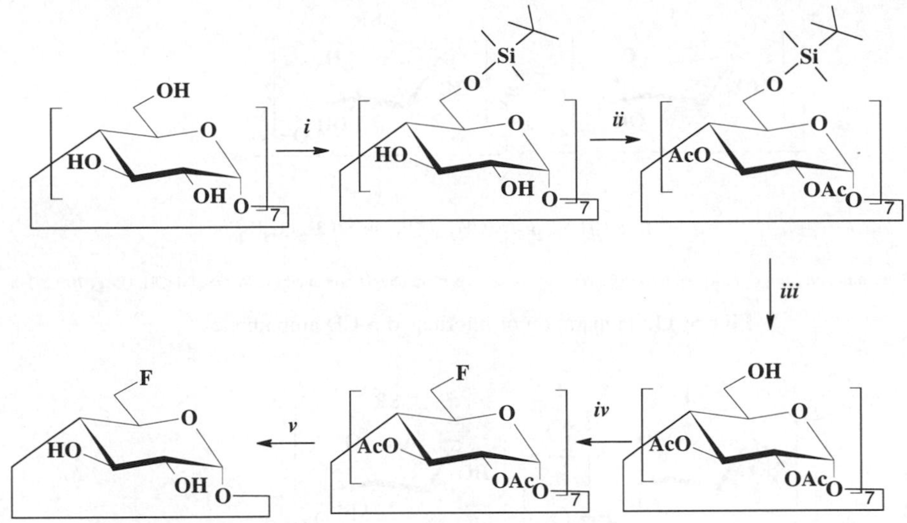

Reagents and conditions : *i* TBDMSiCl, Pyridine, *ii* 2:1 Pyridine Ac$_2$O, *iii* BF$_3$, Et$_2$O, CHCl$_3$ *iv* DAST, CHCl$_3$, *v* NaOMe/MeOH.

Figure 9. Synthesis of the first fluorinated amphiphilic CD on the primary face.

etherate in chloroform, and the heptafluoro-CD was obtained by reacting this last intermediate with excess diethylaminosulfur trifluoride (DAST) in chloroform. The yield for this transformation was low (35%). *O*-deacylation in methanolic sodium methoxide solution yielded the desired heptakis-(6-fluoro-6-deoxy)-β-CD (Fig. 9).

The ^{13}C NMR spectrum of this compound was highly symmetrical, meaning that this compound was persubsituted. Nevertheless, we cannot say that this derivative was amphiphilic! One year later, Granger et al. [51] prepared the 6-pertrifluoromethylthio-β-CD (Fig. 10), which, despite its short hydrophobic chain, formed monolayers at the air–water interface. To favor the hydrophobicity, the fluorinated chain length has also been increased.

These derivatives have been obtained by thio-alkylation of tosylated or iodinated CD derivatives. Thioether links have been chosen in this case, not only to avoid their hydrolysis in biological medium, but also because these thiolates have a high level of nucleophilic activity, thus compensating the high inductive effect of the fluorous chains, which decreases the reactivity [52–54].

These novel derivatives were first obtained by thioalkylation of β-CD derivatives having tosyl or iodo groups at the C$_6$ position (Fig. 11). Unfortunately, whatever the conditions used, these compounds do not allow nanoparticles formation as desired; it was argued that the origin of such failures was the noncompatibility of the sevenfold symmetry of the β-CD heptamers toward tight chain organization in the assemblies [55]. Attention has been then focused to α-CD derivatives so as to favor both the interactions with water and the organization of the molecular assemblies [56]. Amphiphilic α-CD derivatives substituted at the primary face by alkylthio and perfluoroalkylpropanethio groups, and their analogs substituted at the secondary face by methyl groups have thus been synthesized (Fig. 12) [53].

After activation of the primary face by an iodo or mesyl group, introduction of the alkyl or perfluoroalkyl chains has been done by adding perfluoroalkylpropyl- or

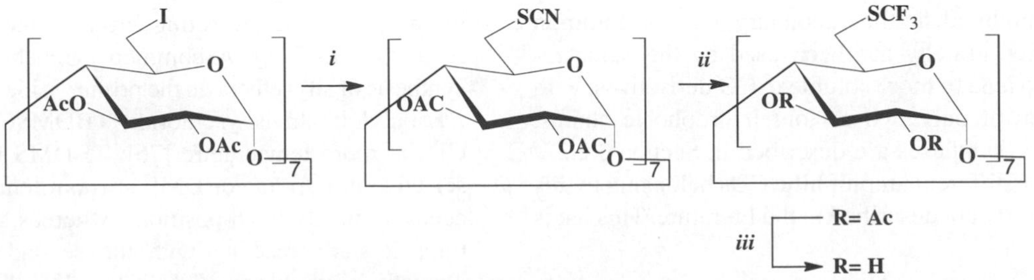

Reagents and conditions : *i*, KSCN, DMF, 100°C, 1h, *ii*, CF$_3$SiMe$_3$, TBAF, THF, r.t.., 3,5h; *iii*, NaOH, MeOH, r.t.., 12h

Figure 10. Synthesis of the 6-per-trifluoromethylthio-β-CD.

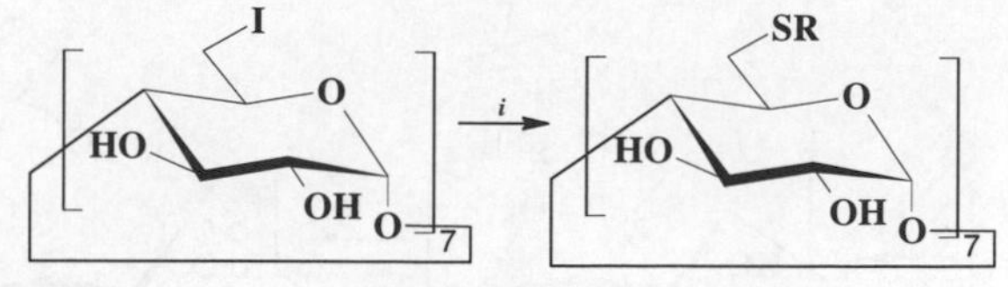

R=S(CH$_2$)$_3$C$_2$F$_5$, S(CH$_2$)$_3$C$_6$F$_{13}$ or S(CH$_2$)$_3$C$_8$F$_{17}$

Reagents and conditions : *i* 3-(perfluoroethyl)propanethiol or 3-(perfluorohexyl)propanethiol, MeONa/MeOH, DMF, 70°C, 24h

Figure 11. Preparation of fluorinated β-CD amphiphiles.

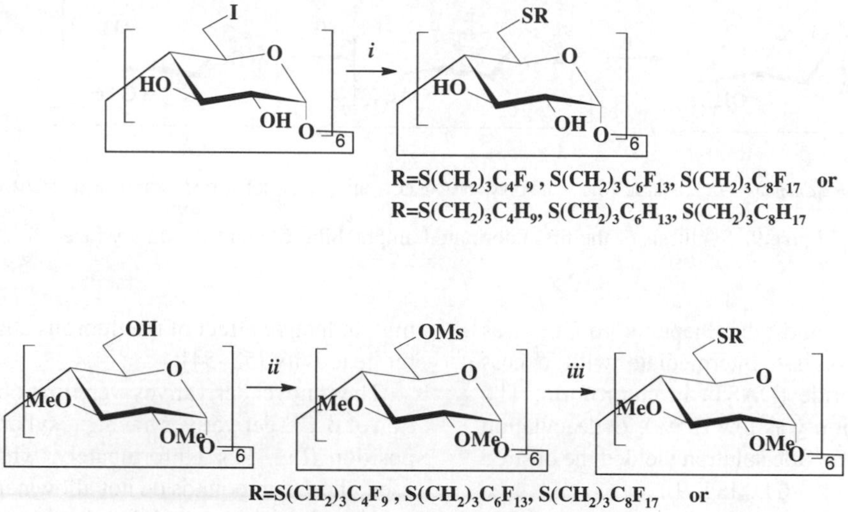

R=S(CH$_2$)$_3$C$_4$F$_9$, S(CH$_2$)$_3$C$_6$F$_{13}$, S(CH$_2$)$_3$C$_8$F$_{17}$ or
R=S(CH$_2$)$_3$C$_4$H$_9$, S(CH$_2$)$_3$C$_6$H$_{13}$, S(CH$_2$)$_3$C$_8$H$_{17}$

R=S(CH$_2$)$_3$C$_4$F$_9$, S(CH$_2$)$_3$C$_6$F$_{13}$, S(CH$_2$)$_3$C$_8$F$_{17}$ or
R=S(CH$_2$)$_3$C$_4$H$_9$, S(CH$_2$)$_3$C$_6$H$_{13}$, S(CH$_2$)$_3$C$_8$H$_{17}$

Reagents and conditions : *i*, [3-(perfluoroalkyl)propyl]isothiouronium salt or alkylisothiouronium salt, CS$_2$CO$_3$, DMF, 60°C, 4d; *ii*, MsCl, Pyridine, 5°C, 2h *iii*, [3-(perfluoroalkyl)propyl]isothiouronium salt or alkylisothiouronium salt, CS$_2$CO$_3$, DMF, 60°C, 4d

Figure 12. Preparation of hydrocarbonated and fluorinated α-CD amphiphiles.

alkylisothiouronium salts in the presence of cesium carbonate in DMF. Unfortunately, the molecules having free hydroxyls have shown very poor solubility in aqueous, organic, or fluorinated solvents. The O$_2$ and O$_3$ methyl ether analogs have shown higher solubilities in organic solvents and their interfacial properties have been studied. Due to the low solubility of the derivatives per-substituted at the primary face and nonmodified at the secondary face in common solvents, another strategy has been used by this same research group, to lead to more soluble α-CD derivatives, with the introduction of only two or four hydrophobic chains. These derivative syntheses are described in Section 3.2.2.

In Table 2 the different amphiphilic CDs belonging to this medusa-like family are described in the literature. This list is not exhaustive.

3.1.5. Skirt-Shaped CDs
Skirt-shaped CDs, persubstituted on the secondary face, are generally obtained in three steps

starting from native CDs, implying protection of the primary hydroxyls, acylation by an acid chloride or anhydride (or alkylation in the presence of a strong base) of the secondary hydroxyls, and then deprotection of the primary ones [20,21]. The secondary face is more crowded than the primary face due to the presence of twice the number of hydroxyl groups. Silyl ethers of CDs have been investigated widely because silyl groups are good protecting groups, due to their ease of removal [37,47,75]. A common synthetic approach to the synthesis of silyl ethers on the primary side is the treatment of *tert*-butyldimethylsilyl chloride (TBDMSCl) in pyridine with CDs at room temperature [76]. TBDMS is relatively more selective in pyridine or DMF at room temperature and proceeds to mostly the 6-positions, whereas at higher temperatures it starts reacting with the secondary side of CDs. Trimethylsilyl chloride (TMSCl) is less discriminatory than TBDMSCl and attacks all three positions of CDs in pyridine at room temperature [77,78]. An improved method of silylation

Table 2. Medusa-like Amphiphilic CDs and Their Corresponding Supramolecular Assemblies

CDs	R$_6$	Supramolecular Assemblies	Refs.
	Aminoalkyl		
α, β	NHC$_{12}$H$_{25}$	Air–water interface film	[40,57]
α, β, γ	NHC$_{16}$H$_{33}$	Air–water interface film	[58–60]
β	NH(C$_3$H$_7$)$_2$, NH(C$_4$H$_9$)$_2$, NH(C$_5$H$_{11}$)$_2$,	Nanospheres	[61]
	NH(C$_6$H$_{13}$)$_2$		
	Sulfoalkyls		
β	SOC$_4$H$_9$, SOC$_8$H$_{17}$	Air–water interface film	[15,62]
β	SOC$_{12}$H$_{25}$	Air–water interface film	[15,58,62]
	Thioalkyls/thiofluoroalkyls		
β	SCF$_3$	Air–water interface film	[51]
β	SC$_2$H$_5$, SC$_4$H$_9$		[41]
β	SC$_6$H$_{13}$	Air–water interface film	[63]
	SC$_6$H$_{13}$	Nanospheres	[64]
β	SC$_{10}$H$_{21}$	Air–water interface film	[41,63]
α, β, γ	SC$_{12}$H$_{25}$	Air–water interface film	[21]
β	SC$_{14}$H$_{29}$	Air–water interface film	[63]
α	SC$_{16}$H$_{33}$	Air–water interface film	[21]
β	SC$_{16}$H$_{33}$	Air–water interface film	[21,58]
		Micelles	[65]
γ	SC$_{16}$H$_{33}$	Air–water interface film	[21]
α	SC$_{18}$H$_{37}$	Air–water interface film	[62,66]
β	SC$_{18}$H$_{37}$	Air–water interface film	[29,41,62,63,65,67]
		Micelles	[65]
γ	SC$_{16}$H$_{37}$	Air–water interface film	[62,66]
β	SC$_3$H$_7$C$_6$F$_{13}$, SC$_3$H$_7$C$_8$F$_{17}$	Nanospheres	[64]
	Thioaryles		
α, β, γ	SPh, SPhBr(*p*), SPhOC$_4$H$_9$(*p*), SPhC5H$_{11}$(*p*),	Air–water interface film	[42,43,68]
	SPhNO$_2$(*p*)		
β	SPhNO$_2$(*o*), SPhNO$_2$(*m*)	Air–water interface film	[43]
β	Tetrathiafulvalene, triSMe, and derivatives	Air–water interface film	[69]
γ	Tetrathiafulvalene, triSMe	Air–water interface film	[69]
	Esters		
β	OCOC$_5$H$_{11}$	Air–water interface film	[70]
β	OCOC$_5$H$_{11}$	Nanospheres	[70]
		Nanocapsules	[45,70]
β	OCOC$_{11}$H$_{23}$	Nanospheres	[61]
	Amides		
β	NHCOC$_5$H$_{11}$,	Air–water interface film	[70]
		Nanospheres	[71,72]
		Nanocapsules	[45,71]
β	NHCOC$_5$H$_{11}$, NHCOC$_{13}$H$_{27}$, NHCOC(CH$_3$)$_3$	Nanocapsules	[45,71]
β	NHCOC$_{13}$H$_{27}$, NHCOCH-*t*Bu	Air–water interface film	[71]
		Air–water interface film	
β	NHCOC$_{11}$H$_{23}$, NHCOC$_{12}$H$_{25}$	Air–water interface film	[72,73]
		Air–water interface film	
	Pyridinyls		
α, β, γ	5′-Methyl-2,2′-bipyridinyle-ureido-5-methylene	Air–water interface film	[74]
	Tiazolyls		
β	5′-Methyl-2,2′-bithiazolyle-ureido-5-methylene	Air–water interface film	

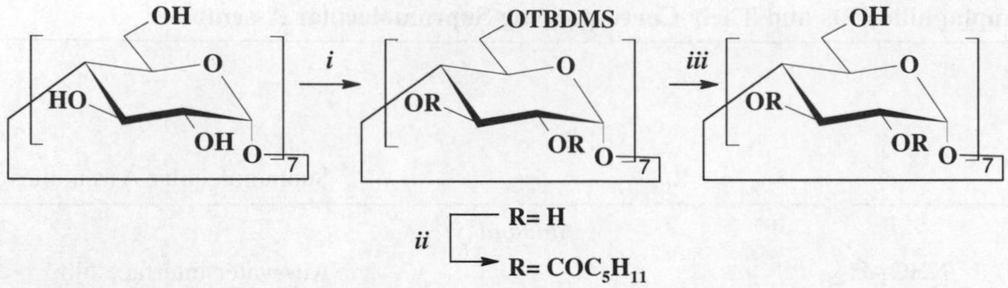

Reagents and conditions : *i*, TBDMSCl, Pyridine,r.t., 15h; *ii*, hexanoic anhydride, DMAP, Pyridine, 70°C, 48h; *iii*, TBAF, THF, r.t., 15h

Figure 13. Synthesis of acylated β-CD amphiphile.

in which the final product was purified by flash column chromatography has been published [75,79]. These silylated derivatives are key intermediates for most of the skirt-shaped CD derivative synthesis. After reaction of the secondary face with anhydride, the silyl ethers are removed by tetrabutylammonium fluoride (TBAF) in THF [80] or dioxane [47] refluxes. Furthermore, on the secondary face, the acidity of the hydroxyls at positions C_2 and C_3 are quite different. The first is the most acidic ($pK_a = 12.2$), whereas the last is the most inaccessible, then allowing a high selectivity in the modification of the secondary face directed toward C_2-hydroxyl [36].

The first method describing the per-2,3-di-*O*-acyl-β-CDs, was published in 1991 by Zhang et al. [20] and gave overacylated compound with 21 chains for the mean substitution degree. In 2000, Lesieur et al. reported conditions for grafting only 14 hexanoyl chains on β-CD in the presence of hexanoic anhydride and using *tert*-butyldimethylsilane to protect primary hydroxyls (Fig. 13) [18].

Duchêne et al. [10] also worked on esterification of the secondary face of the α-, β-, and γ-CD with hydrophobic

chains varying from C_2 to C_{14} [20,21]. By comparing these skirt-shaped β-CDs it has been shown that the most-surface-active molecule was the one with the C_6 chain length [56].

Wazynska et al. [81] also described the synthesis of skirt-shaped CDs by using these silyl-protected CDs which were then alkylated with iodo alcanes in DMF in presence of sodium hydride and then desilylated to give amphiphilic per-(2,3-di-*O*-alkyl)-α- and β-CD derivatives (Fig. 14).

The per-(6-thio)-α-CD analog was also prepared by the same group, and similar synthesis of per-(2,3-di-*O*-heptyl)-β and γ-CD has been reported by Badi et al. [82]. The characterization of those products is realized by ^1H and ^{13}C NMR in CDCl3 and mass spectrometries (MALDI-TOF MS and ESI MS).

Concerning fluorinated derivatives, the first β-cyclodextrin derivative, substituted on the secondary face by 14 C7F15 groups, has been described by Skiba et al. [83], who showed that these derivatives were able to self-organize in water. They also showed that nanocapsules formed from these derivatives can encapsulate oxygen because of the

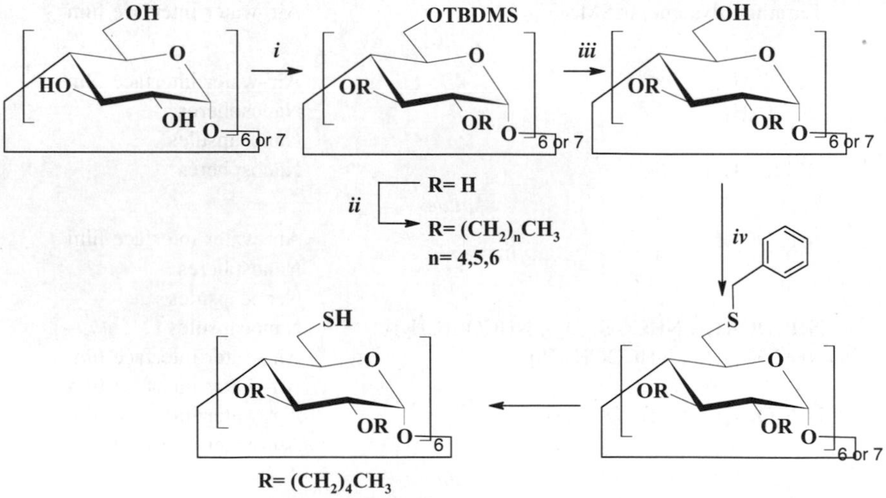

Reagents and conditions : *i*, TBDMSCl, Pyridine,r.t., 15h; *ii*, NaH, R-I, DMF, r.t. 3d, *iii*, TBAF, THF, reflux.*iv* PPh₃, I₂, imidazole, toluene then benzyl mercaptan, NaOMe, DMF, 60°C, 24h, *v* Na/NH₃, NH₄Cl, THF

Figure 14. Synthesis of alkylated CDs.

Table 3. Skirt-Shaped Amphiphilic CDs and Their Corresponding Supramolecular Assemblies

CDs	R$_2$ and R$_3$	Supramolecular Assemblies	Refs.
	Ester		
β	Ac, COC$_5$H$_{11}$, COC$_7$H$_{15}$, COC$_9$H$_{19}$, COC$_{11}$H$_{23}$	Air–water interface film	[21,56]
α, β, γ	COC$_{13}$H$_{27}$	Air–water interface film	[21,56,81]
β	21 chains COC$_5$H$_{11}$	Air–water interface film	[89]
β	COC$_5$H$_{11}$	Nanospheres	[4,5,45,71,86,88,90,92,93]
γ	COC$_5$H$_{11}$	Nanospheres	[10,91,99–101]
β	COC$_7$H$_{15}$, COC$_9$H$_{19}$, COC$_{11}$H$_{23}$, COC$_{13}$H$_{27}$	Nanospheres	[91,94,96,97,98]
α, γ	COC$_{13}$H$_{27}$	Nanospheres	[94,98,102]
β	COC$_7$F$_{15}$	Nanospheres	[103]
β, γ	COC$_5$H$_{11}$, COC$_{11}$H$_{23}$, COC$_{13}$H$_{27}$	Nanocapsules	[45,70,95,102,104–106] [91]
β	COC$_7$F$_{15}$	Nanocapsules	[83,84]
	Ethers		
β, γ	Bn	Air–water interface film	[107]
α, β	C$_5$H$_{11}$, C$_6$H$_{13}$, C$_7$H$_{15}$	Air–water interface film	[81]

fluorinated chains; these assemblies could have a potential role in oxygen delivery [83,84].

Among these skirt-shaped derivatives we can also site the work done by the group of Wouessidjewe [85], who shown that β-CD ester amphiphilic derivatives were found to self-organize into CD nanospheres [86] that can encapsulate drugs such as metronidazole [87] or tamoxifen used in breast cancer treatment [88]. Some of the skirt-shaped amphiphilic CDs are included in Table 3.

3.1.6. Bouquet-Shaped CDs

Bouquet-shaped CDs contain hydrocarbon chains on both sides of the cavity, increasing their hydrophobicity. All hydroxyl groups on the CD can be converted directly and in a nondiscriminatory way in ester or ether functions, using appropriate reactants. Esters are generally obtained by a carboxylic acid chloride in pyridine in the presence of a ternary amine as base. The size of the alkyl or aryl groups of the reactants generally have no effect on the substitution, and per-modified derivatives obtained are homogeneous, without side products. Complete acetylation [77,108,109] and benzoylation [18] of these CDs can be obtained with acetic anhydride and benzyl chloride in pyridine, respectively, with relatively long reaction times for total substitution.

Peralkyltions of CDs, which increase their solubility in organic solvents, are achieved by reacting alkyl halides with CD alkoxyde ions [110].

This type of amphiphilic CD can also result, for example, from the grafting of 14 polymethylene chains on 3-mono-methylated β-CDs, leading to an equal number of chains on the primary (C$_6$) and secondary (C$_2$) side [22]. Variants in

these series are per(2,6-di-*O*-alkyl)-cyclodextrins with alkyl chains of different length [23].

Lehn's group [111,112] has grafted one poly(oxyethylene), polymethylene, or *O*-alkyl chain on each side of the CD cavity (Fig. 15), and shown that these derivatives mimic transmembrane ion channels after insertion in lipid membranes.

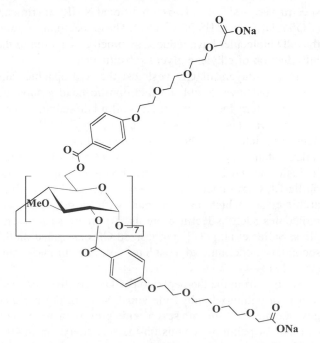

Figure 15. One of the amphiphilic CD designed by Lehn et al. [111,112] as an ion channel mimic.

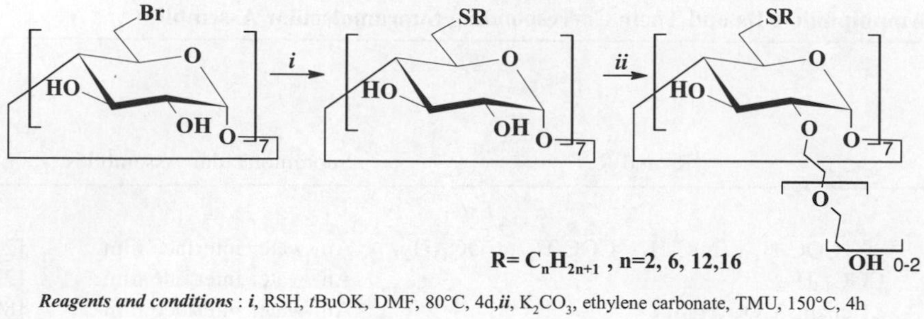

$R = C_nH_{2n+1}$, n=2, 6, 12,16

Reagents and conditions : *i*, RSH, *t*BuOK, DMF, 80°C, 4d,*ii*, K_2CO_3, ethylene carbonate, TMU, 150°C, 4h

Figure 16. Synthesis of oligo(ethylene oxide) β-CD.

In 2000, Ravoo, Darcy, and others [3,113,114] designed the first amphiphilic β-CD having the ability to form bilayer vesicles. 6-per-bromo-CD is first transformed by a nucleophilic substitution with the sodium or potassium salt of alkylthiols to per-alkylthio-β-CDs. These then reacted with an excess of ethylene carbonate only at the C_2 position, resulting in an average of two ethylene glycol units per grafted oligomers (Fig. 16).

These new compounds were characterized by microanalysis, electrospray mass spectrometry, and NMR spectroscopy. In positive-ion-mode electrospray spectroscopy, the peaks for the singly charged compounds are separated by 44 mass units, indicating a variable degree of substitution of the secondary face.

The *m/z* ratios in the spectra show that these compounds has been substituted with 9 to 14 ethylene glycol units, a conclusion that was confirmed by microanalysis and by integration of the ^1H NMR spectrum. The assignment of protons from the ^1H NMR and carbons from the ^{13}C NMR spectra were assisted by two-dimensional NMR experiments (COSY, DEPT, and HSQC spectra). These spectrums showed that CD molecules have reduced symmetry as a result of the introduction of ethylene glycol substituents.

In the bilayers of CD vesicles, the hydrophobic tails are directed inward and the hydrophilic head groups are facing water, increasing their colloidal stability. This concept was later extended to α- and γ-CD [3]. These neutral vesicles were shown to bind small guest molecules (adamantanes) and even polymers [115,116] by their inclusion in the CD cavity. Other bouquet-shaped amphiphilic CDs have been synthesized to make self-assembled architecture, which can interact with target biological molecules such as lectin proteins. To create such assemblies, Sallas et al. [117] synthesized CDs esterified on the secondary side and glycosylated on the primary side (Fig. 17).

Starting from the the perazido β-CD derivative, the reaction with palmitoyl anhydride and DMAP in dry pyridine gave an intermediate with seven azido groups on the primary rim and 14 palmitoyl chains on the secondary rim in 40% yield. The one-pot coupling reaction between these intermediates and an excess of the amino terminal unprotected

glucosamine derivative was carried out in pyridine in the presence of a large excess of triphenylphosphine and constantly bubbling CO_2. The first step is similar to a Staudinger reaction, after which the reaction proceeds through an isocyanate intermediate which is formed in situ and reacts with the nucleophilic amino derivative. In this derivative, the carbohydrate is linked to the CD by an urea spacer. The same kind of derivative has also been synthesized using another synthetic strategy, starting with the same azido derivative and an amide bond for linking the glycosylated residues to the CD core. It consisted of a simple coupling between an active ester and an amino group. The reaction was carried out in dry pyridine using a large excess of the active ester (Fig. 18).

All these compounds were characterized by NMR spectroscopy (CDCl3 or pyridine-*d*5), mass spectra (FAB and/or MALDI-TOF), and elemental analysis. The authors demonstrated that these amphiphilic CDs could then self-assemble, with polar glycosylated groups pointing outward, being able to bind to carbohydrate-specific lectin proteins.

The group of Stoddard has also synthesized CD-based carbohydrate clusters persustituted on the primary, secondary, or both faces, by using photochemical addition of sugar thiol to allylic groups [118]. One- and two-dimensional NMR spectroscopic investigations were performed on these compounds to determine their structures and establish that the products were homogeneous and contained no under-"substituted" products. Furthermore, MALDI-TOF mass spectrometry proved to be a useful technique for characterizing these cluster compounds. The mass ions expected, without fragmentation, have been observed for all of these compounds, and no peaks corresponding to undersubstituted compounds were observed (Table 4).

3.2. Selectively Modified Derivatives

The selective differentiation of two to five primary hydroxyl groups of a CD is a real synthetic challenge. Indeed, primary OH groups have a similar reactivity, implying selective synthetic methods and/or efficient purification methods. Before describing the amphiphilic CD derivatives, we have

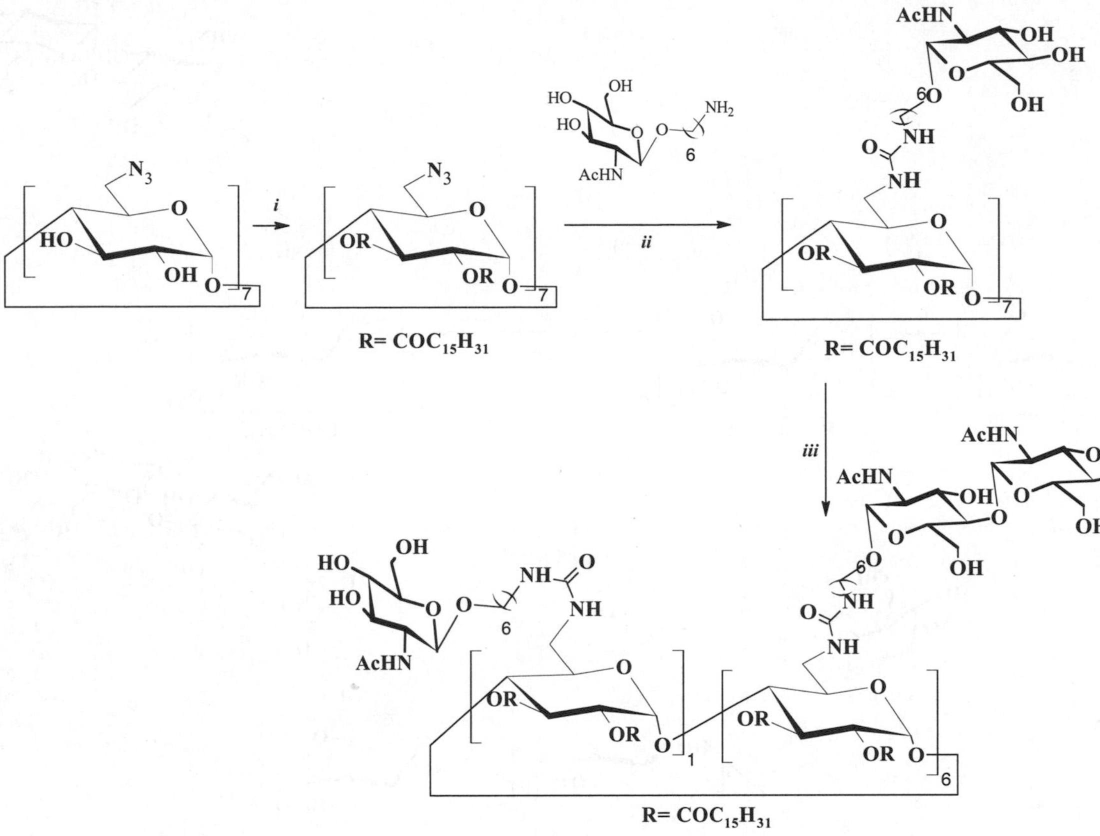

Reagents and conditions: *i* palmitoyl anhydride, DMAP, pyridine, 70 °C, 48 h; *ii* PPh$_3$, CO$_2$, pyridine, r.t., 24 h;

iii UDP-galactose, (β-1,4)galactosyl transferase, HEPES buffer, 50 mM, pH 6, 37 °C, 5 d.

Figure 17. Synthesis of glycosylated amphiphilic CDs via an urea spacer.

to focus our attention on the various methods for selective functionalizations leading to multisubstituted CD derivatives.

3.2.1. Monosubstituted Derivatives
The key step for the linkage of a functional group on only one of the primary hydroxyl groups is selective sulfonation, leading to the 6-*O*-tosylated derivative [124]. Two methods have been described in the literature for the synthesis of the well-used mono-6-*O*-tosyl-cyclodextrin. The "classical" method consists of the preparation of these monosulfonate derivatives by reacting 1 equivalent of benzene or *p*-toluenesulfonyl chloride with CD in pyridine or DMF containing a base, and is based on rigorous control of the progress of the reaction (i.e., temperature, pH, etc.) [125,126]. The major inconvenient of this synthesis is that di- and tri-tosylated derivatives are synthesized as side products, and thus long purification methods are required to obtain the pure mono-6-*O*-tosyl derivative. The yield of the final product is often reduced because the tosylate can undergo an exchange by chloride ions or an elimination process to give either the 3,6-anhydro compound or an alkene. The method of choice for the synthesis of mono-tosylcyclodextrin is to react cyclodextrin with tosyl chloride

in a 1:1 equivalent ratio in aqueous alkaline medium for a short time to give mono-6-tosylate in fairly good yield [127,128]. The product is obtained in reasonable purity either by repeated crystallization from water or by chromatography on a charcoal column. More recently, Brady et al. replaced tosyl chloride by *p*-toluenesulfonylimidazole, more stable toward hydrolysis [124].

Nucleophilic displacement of tosylate by nucleophilic groups as halides, azides, thiolates, hydroxylamines, or alkyl- or poly(alkylamine) produce the corresponding functionality: monohalogeno- [125,129], monoazido- [125], monothio- [130,131], mono(hydroxylamino) [132], or mono(alkylamino)-cyclodextrines [133], respectively. Monoazido-CDs are usually synthesized from monotosylated derivatives in the presence of lithium or sodium azide in refluxing DMF [125]. They can also be prepared directly from the native CD by a Vilsmeier–Haack type of reaction, in which the CD reacts with lithium azide in the presence of PPh3 in DMF [134].

Reduction of monoazide derivatives with PPh3 in the presence of NH$_3$ gives the monoamino-CD derivatives [134,135]. These aminated derivatives are precursors for the synthesis of a lot of monosubstituted CD derivatives,

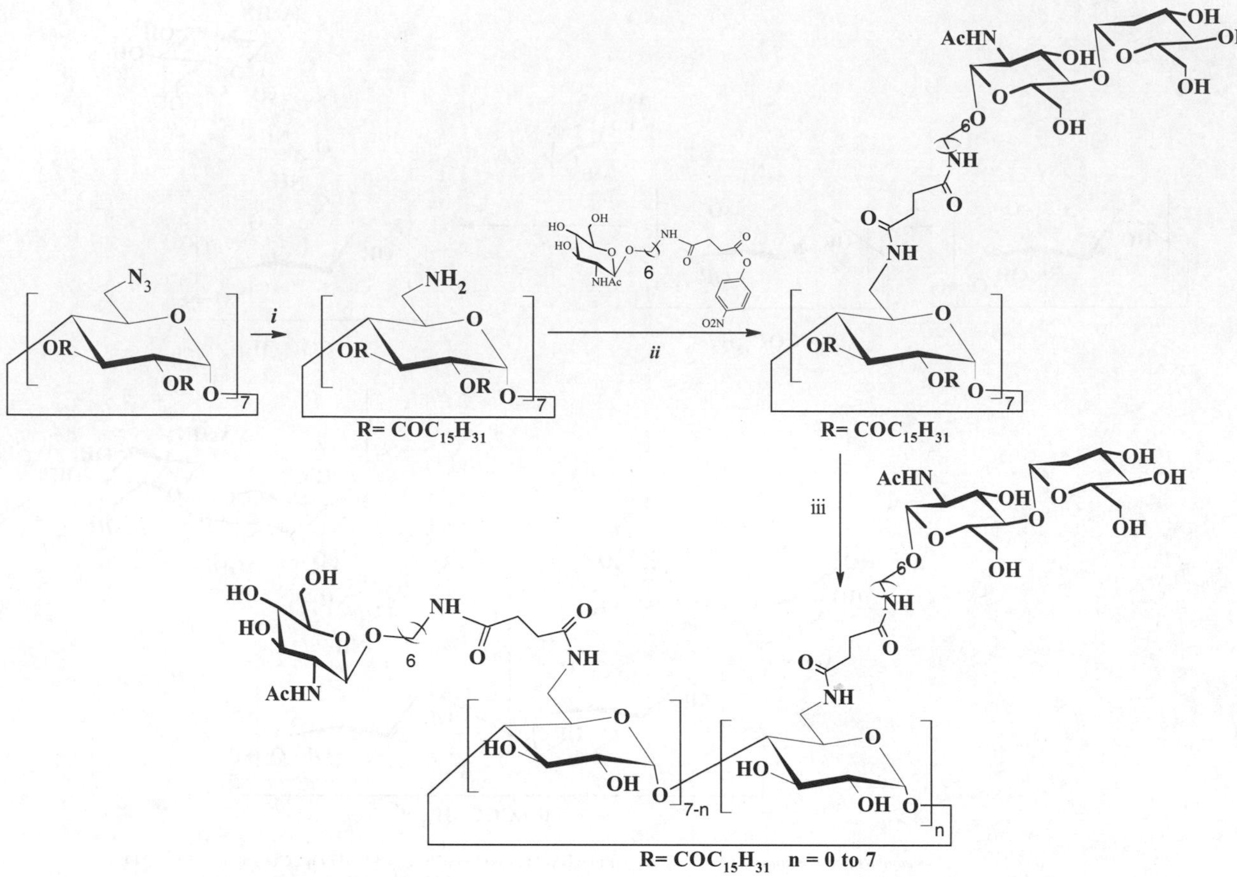

Reagents and conditions: *i* PPh$_3$, H$_2$O, THF, r.t., 7 d; ; *ii* pyridine, r.t., 7 d ; *iii* UDP-galactose, (β-1,4) galactosyl transferase, HEPES buffer, 50 mM, pH 6, 37 °C, 5 d.

Figure 18. Synthesis of glycosylated amphiphilic CDs via an amide bond.

especially for the anchorage of saccharides [136], peptides [137,138], or alkyl chains [24], by a peptide type of coupling. Monothio-CD derivatives have also been obtained from the monotosylated derivative [139]. Direct synthesis of monothio α-CD derivatives from native CDs and thioaryls has also been realized via a Mitsunobu reaction, giving a mixture of mono-, di-, and tri-substituted products, then purified by HPLC [140].

Among all the monosubstituted amphiphilic CDs described in the literature, we focus our attention on only few recent examples. Recently, the group of de Rossi [141] described the direct synthesis of monoacylated amphiphilic CD derivatives from the reaction of 3-(E)-dec-2-enyl)-dihydrofuran-2,5-dione with α-CD. ^{13}C NMR spectroscopy indicated that both isomers were formed (Fig. 19), but the exact amount of each of them could not be determined. ^{1}H NMR also indicated that the average substitution is one alkenyl succinic chain per CD molecule, but FAB analysis of the product indicates the presence of a small amount of unsubstituted and disubstituted product.

Nevertheless, the compound was an interesting surfactant that forms large aggregates. The self-inclusion of the chain in the cavity of CD as well as the intermolecular inclusion was demonstrated by ^{1}H NMR measurements that were able to detect methyl groups in three different environments. Besides, in the aggregates, the cavity is available to interact with external guests, such as phenolphthalein, 1-aminoadamantane, and prodan. The latter result was attributed to the fact that this probe interacts with the micelle in two binding sites: the cavity of the CD and the apolar heart of the micelle.

Djedaïni-Pilard et al. also described recently the synthesis of a monosubstituted amphiphilic CD by appending a single hydrophobic anchor, with the aim of improving the cell targeting of a drug containing CD cavities through their liposome transportation after insertion in the lipid bilayer [13]. They obtained the poorly soluble lollipop, in which the alkyl chain tends to loop back and enter the CD hydrophobic cavity, leading to intramolecular self-inclusion [24]. This self-inclusion can be weakened by adding a bulky Boc-amino protective group at the end of the alkyl chains, giving then a very water-soluble "cup and ball" derivative (Fig. 20) which can include sodium anthraquinone-2-sulfonate in the CD cavity, and which could be inserted in phospholipid membranes [144]. Furthermore, to increase the membrane

Table 4. Bouquet-Shaped Amphiphilic CDs and Their Corresponding Supramolecular Assemblies

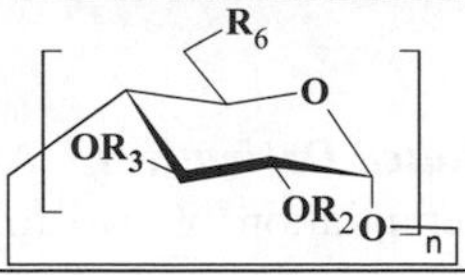

CDs	R_2	R_3	R_6	Supramolecular Assemblies	Refs.
β	Ac	Ac	SOC_4H_9, SOC_8H_{17}, $SOC_{12}H_{25}$, SC_4H_9, SC_8H_{17}, $SC_{12}H_{25}$	Air–water interface film	[15,62]
α, β, γ	$(C_2H_4O)_m$-H, $1 < m < 3$	H	$SC_{12}H_{25}$, $SC_{16}H_{33}$	Air–water interface film	[3]
α	Me	Me	SH	Air–water interface film	[81]
β	Me	Me	$NHCO(CH_2)_8C{\equiv}C{-}C{\equiv}C$ $(CH_2)_9CH_3$	Air–water interface film	[119]
β	Me	Me	$NHCO(CH_2)_{20}CH{=}CH_2$	Air–water interface film	[119]
β	C_8H_{17}	C_8H_{17}	C_8H_{17}	Air–water interface film	[120]
β	$C_{12}H_{25}$	Me	$C_{12}H_{25}$	Air–water interface film	[120]
β	$C_{12}H_{25}$	H	$C_{12}H_{25}$	Air–water interface film	[120]
β	C_8H_{17}	C_2H_5	C_8H_{17}	Air–water interface film	[120]
α, β, γ	Me	Me	5′-Methyl-2,2′-bipyridinyl-urei-do-5-methylen	Air–water interface film	[74]
β	$(C_2H_4O)_n$-C_2H_4S-β-D-glucose, $1 < n < 3$	H	SC_6H_{13}	Nanospheres	[121]
β	$(C_2H_4O)_n$-C_2H_4S-β-D-galactose, $1 < n < 3$	H	SC_6H_{13}	Nanospheres	[121]
β	$(C_2H_4O)_n$-$C_2H_4OC(S)$NH-α-D-mannose, $n_{moy} = 2$	H	SC_6H_{13}	Nanospheres	[122]
β	$(C_2H_4O)_n$-$C_2H_4OC(S)$NH-β-D-fucose, $n_{moy} = 2$	H	SC_6H_{13}	Nanospheres	[122]
β	$(C_2H_4O)_n$-H, $n_{moy} = 3$	H	$NHCOC_2H_4SS$-$C_2H_4COOC_8H_{17}$	Nanospheres	[123]

insertion, the hydrophobicity of the alkyl chain was increased by grafting a lipidlike anchor such as a cholesteryl or phospholipidyl moiety.

For the synthesis of cholesteryl-conjugated CDs, mono-amino-β-CD was reacted with cholesteryl chloroformate or with succinic anhydride, then with 3-α-aminocholesterol to give similar products, differentiated by the presence of a spacer link in the last case (Fig. 21). The synthesis of their methylated analogs has also been reported [12,142,143], and the authors showed that these conjugates could form monodisperse micelles and incorporate them into phospholipid membranes while retaining their included biological active compound [144].

Apart from these classical reactions (i.e., nucleophilic substitution, peptide-type coupling, etc.), the group of Parrot-Lopez prepared a series of novel primary face mono-substituted β-CD derivatives by using the olefin metathesis reaction (Fig. 22). Mono-6-allylamino-6-deoxy-β-cyclodextrin, easily synthesized by nucleophilic substitution of mono-6-tosyl-β-cyclodextrin, was the key synthon in the preparation in four steps of CD derivatives monofunctionalized at the primary face by alkyl, aryl, or perfluoroalkyl groups using Grubbs catalyst. For all these derivatives, the degree of the substitution was assessed by MALDI mass spectrometry and the structure confirmed by ^{1}H, ^{13}C, and HMBC NMR spectroscopies.

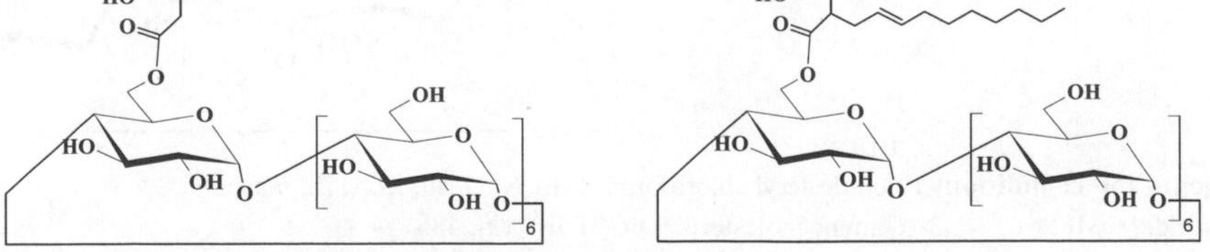

Figure 19. Monosubtituted β-CD directly prepared from native CD.

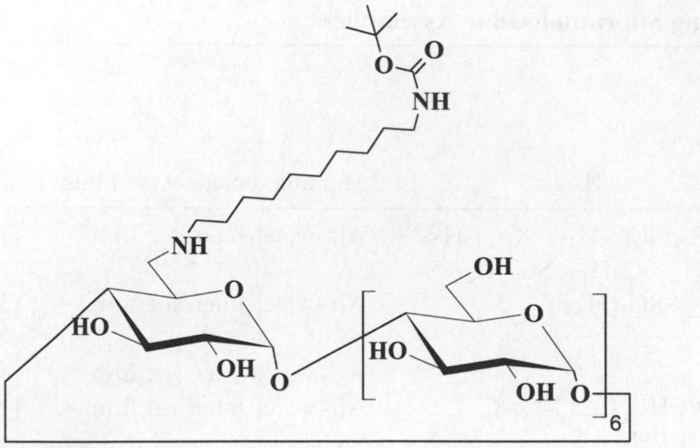

Figure 20. "Cup and ball" amphiphilic CD.

Finally, Cavalli et al. [145] described another type of amphiphilic CD derivatives, a β-CD/poly(4-acryloylmorpholine) monoconjugate, which is a tadpole-shaped polymer in which the β-CD ring is the hydrophilic head and the polymeric chain the amphiphilic tail (Fig. 23). The nanoparticles prepared with this derivative by the solvent injection technique showed good encapsulation properties. Particularly, the antiviral activity of acyclovir loaded into these nanoparticles against two clinical isolates of HSV-1 was evaluated

and found to be remarkably superior compared with that of the free drug.

3.2.2. Di-, Tri-, and Tetra-substituted Derivatives

As we have already seen, selective differentiation of two to five primary hydroxyls is a real synthetic challenge. Indeed, the primary hydroxyl functions have relatively the same reactivity, implying selective synthetic methods and/or efficient purification techniques for separating the different regioisomers (Fig. 24).

Even if the compound obtained is pure, the ^{1}H and ^{13}C NMR spectra become very complex because of the desymmetrization of the CD derivative compared to the native CD. Two-dimensional NMR is then required for the characterization of these derivatives, and mass spectroscopy is generally performed to confirm the structure.

Synthesis of Selectively Activated CDs Many strategies have been described in the literature for selectively differentiating the primary hydroxyl group in native CDs, and they can be grouped in four families: (1) efficient separation of a mixture of different substituted CDs, (2) use of bridged reactants, (3) use of an oversized reagent, and (4) use of the Sinaÿ reaction.

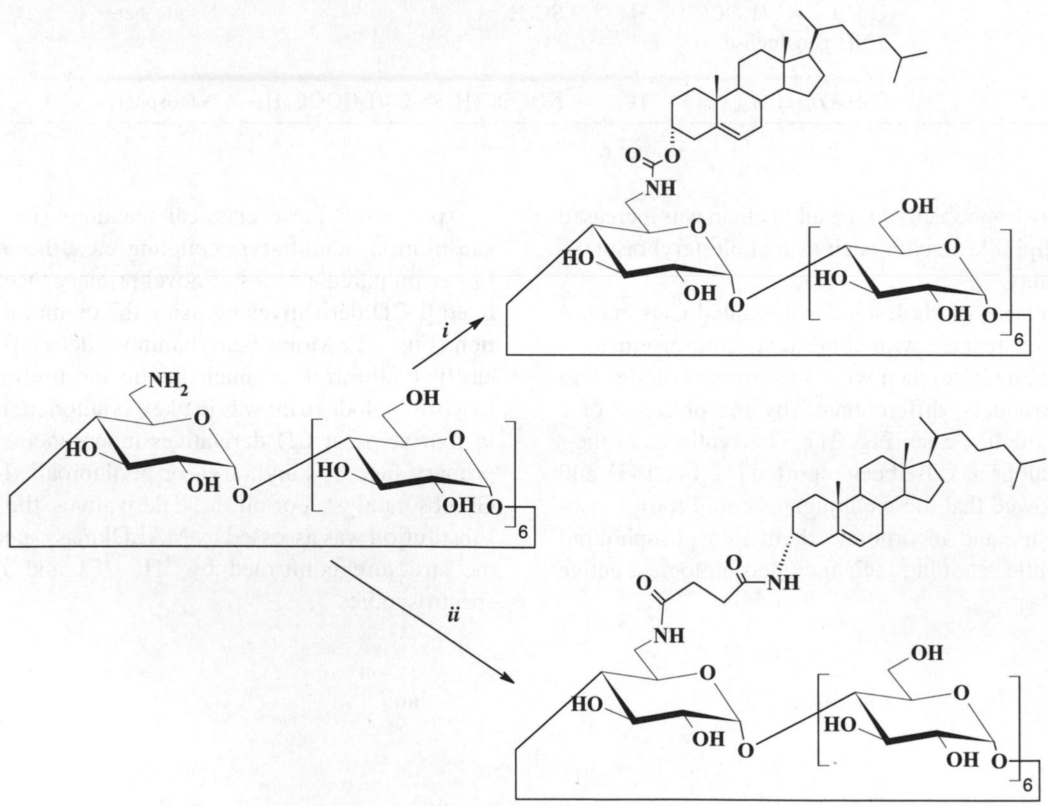

reagents and coonditions : *i* cholesteryl chloroformate, Et$_3$N, DMF, r.t., 17h; *ii* succinic anhydride, DMF, r.t., 3h, 3-α-aminocholesterol, DIC/HOBt, r.t., 48h

Figure 21. Synthesis of lipid-conjugated β-CDs.

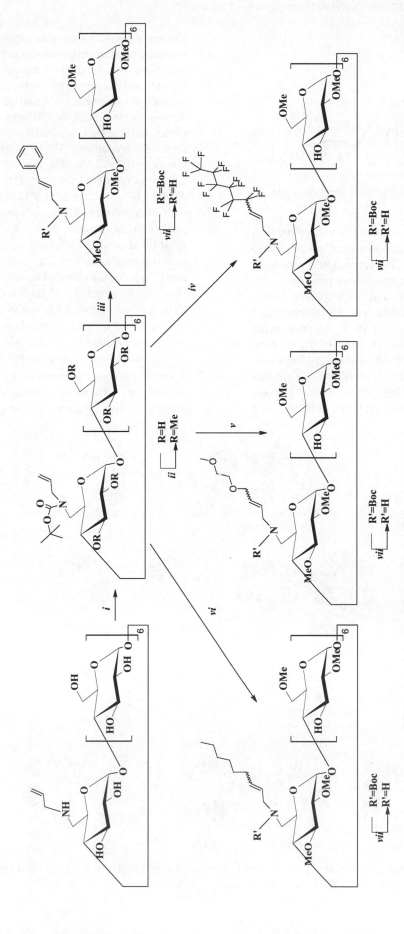

Reagents and conditions: *i* Boc₂O,NaHCO₃, MeOH, ultrasonication; *ii* NaH,CH₃I,DMF, 20 °C; *vi* oct-1-ene, Grubbs catalyst [Cl₂(PCy₃)₂Ru]CHPh], CH₂Cl₂, 55 °C; *iii* vinylbenzene, [Cl₂(PCy₃)₂Ru]CHPh], CH₂Cl₂, 55°C; *v* 3-(2-methoxyethoxy)-prop-1-ene, [Cl₂(PCy₃)₂Ru]CHPh], CH₂Cl₂, 55 °C; *iv* 1H,1H,2H-perfluoro-1-octene, Grubbs–Hoveyda catalyst (C₃₁H₃₈Cl₂N₂ORu), CH₂Cl₂, 55°C; *vii* TFA, 20 °C.

Figure 22. Synthesis of mono-functionalized derivatives by metathesis reaction.

215

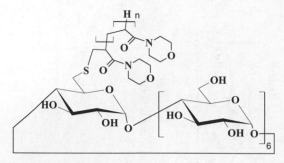

Figure 23. β-CD polyacryloylmorpholine monoconjugate.

1. Di-substituted CDs can be obtained by using more than 1 equiv of reagent with CD under conditions suitable to give a mixture of products. The formation of positional and regioisomers in these conditions requires extensive purification by HPLC and leads to poor yields. Fujita et al. [146] have synthesized the di-substituted α-CD in the A-B, A-C, and A-D positions using mesitylenesulfonyl chloride (8.7 equiv/CD) in pyridine (Fig. 25). The separation of these different isomers has been carried out by reversed-phase column chromatography, and the pure di-substituted isomers have been obtained with low yields (7 to 14%). Even if not mentioned in this paper, we can suppose the presence of other derivatives, more or less substituted in the reaction mixture. Tetra-substituted sulfonates derivatives have been also synthesized in the presence of an excess of reactant (18 equiv/CD) leading to three regioisomers: ABCD, ABCE, and ABDE. In addition to the purification described above, their purification requires a preparative HPLC, yields so far being very low, from 0.7 to 3.6%. Reactions of tosyl chloride with CDs were reported to give a mixture of di-O-6 [147], di-O-2 [148,149], or di-O-3 [150] derivatives along with other products. These were separated by reversed-phase column chromatography to give the desired di-substituted products. Despite all these difficulties, a variety of disulfonates are reported in the literature [151]. Di-sulfonated CDs are important intermediates for the synthesis of di-substituted amphiphilic CDs, the latter generally being synthesized by nucleophilic substitution. It is worth noting that the positional isomerism (AB, AC, and AD) of the disulfonates is retained in these di-substituted derivatives.

2. The use of bridging reagents is an elegant and efficient method of introducing selectively two sulfonate groups on the primary face (e.g., by reaction of

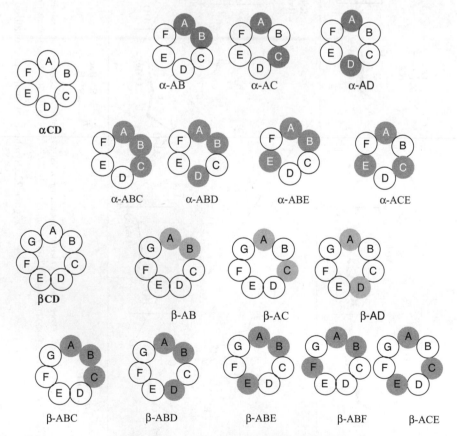

Figure 24. Possible positional isomers of di- and tri-modified α- and β-CDs (and tetra and penta by deduction. (*See insert for color representation of the figure.*)

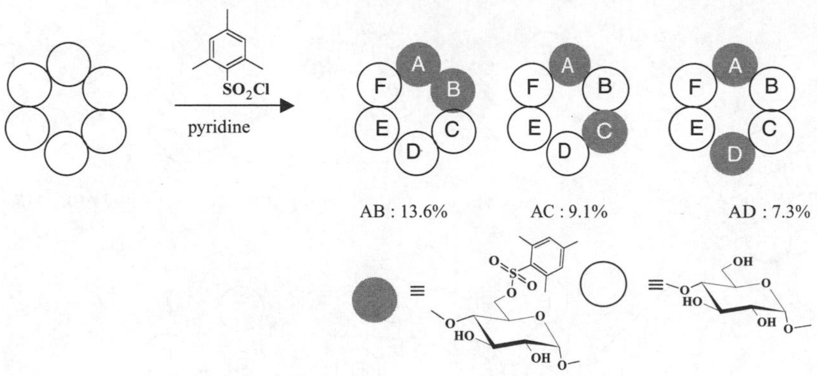

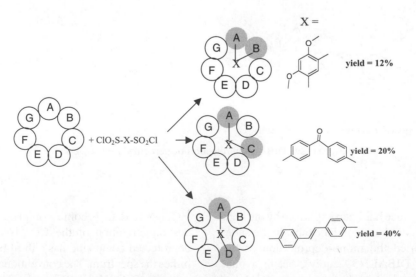

Figure 25. Synthesis of disulfonyl α-CDs.

arenedisulfonyl chlorides with CDs to give AB, AC, and AD isomers) [79,152,153]. Although these disulfonyl chlorides give a mixture of regioisomers, they show distinct regiospecificity based on their structures. An elegant method to control the regiospecificity to produce AB, AC, or AD isomers by the use of the geometry of the reagents has been described [152]. For example, as shown in Fig. 26, *trans*-stilbene and biphenyl-based capping reagents preferentially give AD isomers [79], benzophenone-based reagents give AC regioisomers, and 1,3-benzenedisulfonyl chlorides [155] gives the AB isomer. ACDF tetra-substituted derivatives can be synthesized in relative good yields (35%) by using an excess of benzophenone-based capping reagents (2.6 equiv/CD) [154].

3. The use of oversized reagents can limit the substitution of primary hydroxyl groups because of the steric hindrance thus generated on the primary face. Use of trityl chloride allows the formation of tri- or tetra-substituted CDs, depending of the ratio used

(Fig. 27) [156]. Purification is done by chromatography after methylation of all other free hydroxyl groups. In 2003, Poorters et al. [157] reported the use of a supertrityl protecting group, sTrCl or tris(4-*tert*-butyl-phenyl)methyl chloride, for the protection of four primary hydroxyl groups on a native α-CD (Fig. 28). After reaction of the free hydroxyl group with the appropriate reactant, deprotection of the supertrityl group with HBF_4 in acetonitrile afforded the corresponding tetrol in very good yields.

In 2000, Sinaÿ's group [158] described the selective deprotection of two diametrically opposed benzyl groups on a primary face of a perbenzylated CD (α, β, or γ), in the presence of diisobutylaluminium hydride in anhydrous toluene (Fig. 29). In the case of α-CD, the A-D di-deprotected regioisomer is obtained in 82% yield. The mechanism of this very nice reaction has been explained more recently, in 2004, following numerous studies [159]. The supposed mechanism implies two sequential reactions of de-*O*-benzylation, and

Figure 26. Use of the geometry of reagents to direct the regiospecificity in disubstitution of β-CDs.

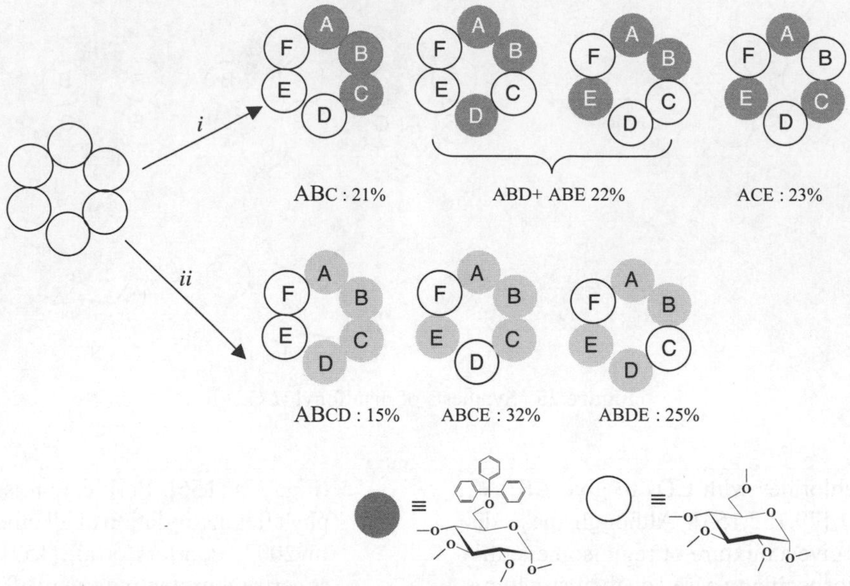

Reagents and conditions : ***i*** TrCl (3.3 eq.), pyridine, 70°C, 36h, NaH, MeI, DMF ; ***ii*** TrCl (4.5 eq.), pyridine, 70°C, 72h, NaH, MeI, DMF

Figure 27. Synthesis of tri- and tetra-substituted α-CD derivatives. (*See insert for color representation of the figure.*)

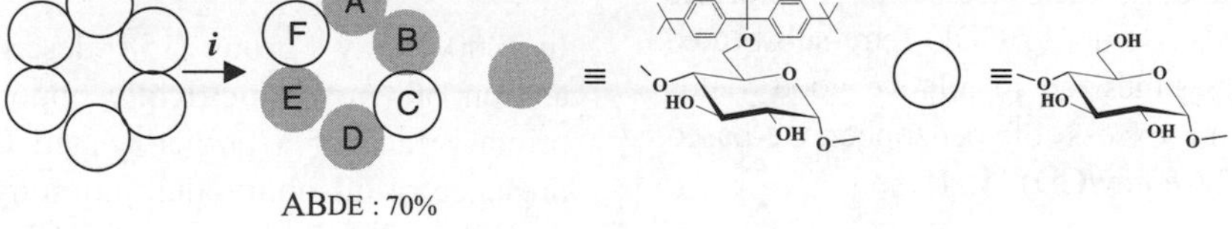

Reagents and conditions : ***i*** sTrCl (6 eq.), DMAP, pyridine, 70°C, 24h

Figure 28. Synthesis of the supertrityl tetra-protected α-CD.

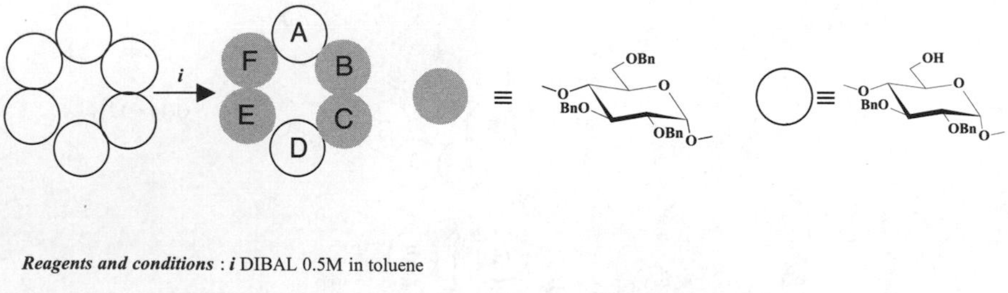

Reagents and conditions : ***i*** DIBAL 0.5M in toluene

Figure 29. DIBAL deprotection of perbenzylated α-CD.

selectivity for the primary face has been explained by the high density of benzyl groups on the secondary face. In the same paper, the authors reported the mono-deprotection of the perbenzylated CD with DIBAL (30 equiv), but in a final concentration of 0.1 M in toluene. The spectroscopic ^{1}H and ^{13}C NMR data become complexes (Fig. 30), due to the desymmetrization of the CD [160]. Nevertheless, the di-deprotected compound has a third-order symmetry that simplifies the spectrum. The conventional two-dimensional NMR analyses (COSY, HSQC, and HMBC) make possible the

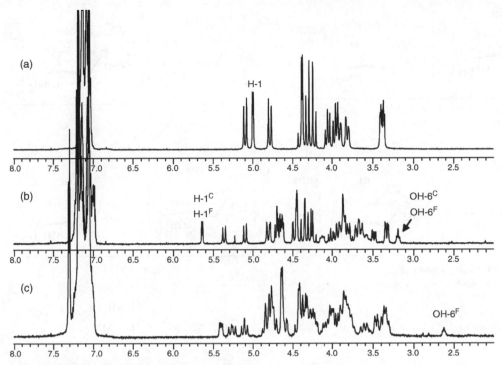

Figure 30. ^{1}H NMR (300 MHz, CDCl3) of perbenzylated (a), di- and mono-de-O-benzylated (b and c, respectively) CD derivatives.

differentiation of each type of proton (H-1, H-2, ..., H-6a and H-6b), failing to assign all protons to a unique glycoside unit. In these cases, mass spectroscopy experiments are useful for confirming the molecular weight of the CD derivatives.

To summarize, selective synthesis of di- and tetra-substituted CDs on the primary face is not trivial. To do that with a good yield, two techniques are most efficient: use of oversized reagents (TrCl or sTrCl) on a native CD or use of DIBAL in a perbenzylated CD. The latter seems to have had more and more success for the synthesis of amphiphilic CDs.

Synthesis of Di- and Tetra-functionalized Amphiphilic CDs Di-sulfonated or di-halogenated CDs react with alkanethiolates in an aqueous or DMF medium to give thioethers of CDs; thiolate ions act as good nucleophiles and do not produce elimination products or 3,6-anhydro compounds. Bis(methylthio)-, bis[(pyridinoalkyl)-thio]-, and bis(phenylthio)-α-cyclodextrin [147,161] have been prepared from ditosylates using the appropriate thiolate reagent.

More recently, Bertino-Ghera et al. [160] described the synthesis of di- and tetra-derivatized amphiphilic β-CD molecules having either alkylthio and perfluoroalkylpropanethio functions at the primary. These compounds had been synthesized in order to lead to more soluble β-CD derivatives, persubstituted derivatives being insoluble in most common solvents. The procedure of Sinaÿ for di-O-debenzylation of perbenzylated β-CDs, described in Fig. 29, has

been used and a new strategy of protection/deprotection has been developed for introducing the lipophilic chains, as shown in Fig. 31.

Activation of the free hydroxyl groups with methanesulfonyl chloride in anhydrous pyridine led to the desired compound in quantitative yield. Deprotection of all benzyl groups undertaken via catalytic hydrogenation in an AcOEt: MeOH 1:1 mixture using stoichiometric quantities led to 6A,6D-di-O-methylsulfonyl-α-cyclodextrin, with controlled regioselectivity in 75% yield starting from native α-CD. This intermediate compound is the precursor for the syntheses of all the di-substituted amphiphilic derivatives. For the introduction of perfluoroalkyl chains, authors have developed a new strategy using a polar reaction between a perfluoroalkylpropanethiol and this intermediate to give fluorinated derivatives (and their hydrocarbonated analogs) with good yields. Two-dimensional NMR spectroscopy (HSQC, HMBC two-dimensional NMR spectroscopy HSQC, HMBC, and TOCSY-HSQC) made possible the complete assignment of all proton and carbon signals, and the success of the coupling reaction in this series was confirmed by ^{13}C NMR spectroscopy. The structures of these derivatives were also confirmed by mass spectroscopy [160].

Tetra-substituted α-CDs have been synthesized from 6A,6B,6D,6E-tetra-O-benzyl-per-2,3-di-O-benzyl-α-cyclodextrin in four steps, as shown in Fig. 32. Free hydroxyl groups were first methylated in the presence of sodium

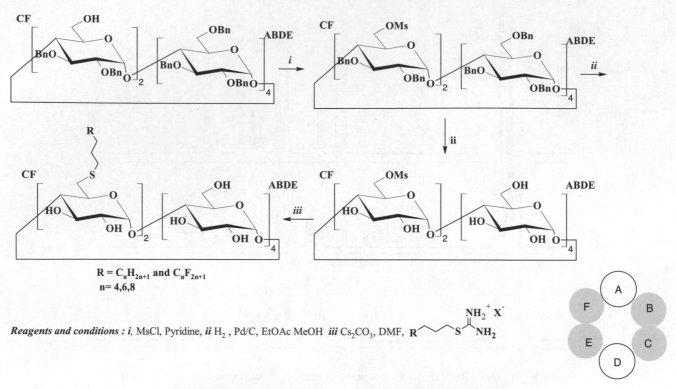

Reagents and conditions : *i*, MsCl, Pyridine, *ii* H$_2$, Pd/C, EtOAc MeOH *iii* Cs$_2$CO$_3$, DMF, R

Figure 31. Synthesis of disubstituted amphiphilic derivatives.

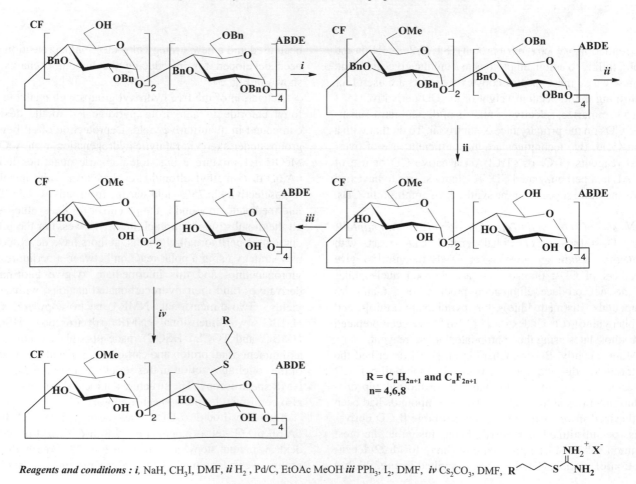

Reagents and conditions : *i*, NaH, CH$_3$I, DMF, *ii* H$_2$, Pd/C, EtOAc MeOH *iii* PPh$_3$, I$_2$, DMF, *iv* Cs$_2$CO$_3$, DMF, R

Figure 32. Synthesis of tetrasubstituted amphiphilic derivatives.

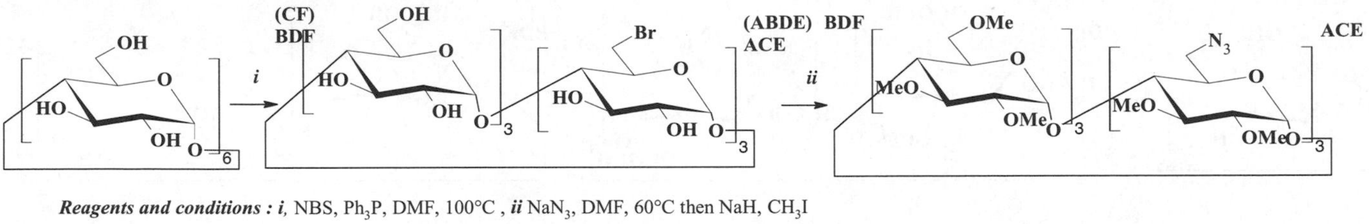

Reagents and conditions : *i*, NBS, Ph$_3$P, DMF, 100°C , *ii* NaN$_3$, DMF, 60°C then NaH, CH$_3$I

Figure 33. Synthesis of triazido α-CD derivative.

hydride and methyl iodide in DMF. Hydrogenolysis of the benzyl ether groups is then realized. Because of the selectivity of iodination for the primary face, the authors have chosen to introduce iodine atoms in the presence of PPh$_3$/I$_2$ for activating these positions. (6A,6B,6C,6D-tetradeoxy-6A,6B,6C,6D-tetraiodo)-(6C,6F-di-*O*-methyl)-α-cyclodextrin has been characterized by use of HSQC-TOCSY, HSQC, and HMBC. Then nucleophilic reagents obtained from the in situ basic hydrolysis of the alkylisothiouronium bromides or perfluoalkylropropane isothiouronium iodides are added to these iodo derivatives to afford the desired tetra-substituted amphiphilic α-CDs in good overall yields.

Synthesis of Tri-Functionalized Amphiphilic CDs As discussed previously, the literature contains well-characterized examples of tri-functionalized CDs derivatives, such as tri-acetylated α-CD [162] and tri-sulfonated β-CD [], contrary to trihalogenocyclodextrin, for which there is no example in the literature. Among the most used tri-functionalized intermediates are the triazido CDs. Reaction of sodium azide on permethylated-6-tri-mesylated cyclodextrins in DMF gives permethylated-6-triazidocyclodextrins [164]. These are also prepared directly from CD as a mixture of mono-, di-, and tri-azido derivatives and Ph$_3$P in DMF containing sodium azide. This mixture is then separated and purified by HPLC.

More recently, Marsura's group developed a new efficient and simple synthetic method making it possible to obtain triazido CDs derivatives on a large scale (Fig. 33) [165]. The first reaction consists of the direct bromination of freshly dried CD by NBS into DMF, giving after one crystallization the mixture of A,C,E-tribromo and A,B,D,E-tetrabromo isomer. The mixture of the crude bromomethyl CDs is readily transformed into the final 6A,6C,6E-triazido-6A,6C,6E-tri-deoxy-6B,6D,6F-tri-*O*-methylhexakis-(2,3-di-*O*-methyl)a-CD after an azidation/methylation "one-pot" reaction. The pure final product was obtained after extraction of the solid residue by ether and a simple crystallization in CH$_3$CN (overall yield 50%).

The tri-azidos are often reduced to triamino-cyclodextrins with Ph$_3$P in ammonia solution [134]. The amino functionalities behave as better nucleophiles than the

hydroxyl groups in these trifunctional cyclodextrins. Tris (alkyl or aryl) amino derivatives have been synthesized by displacing sulfonates groups by nucelophiles such as substituted amines [166].

One recent example of an amphiphilic CD functionalized with three hydrophobic arms has recently been described by Menuel et al. [167]. Indeed, they reported a four- or five-step original synthesis of three novel α-CD tripods bearing three ureido-bipyridyl tethers distributed symmetrically in the A, C, and E positions on the CD upper rim, or alternated with lauryl ester moieties grafted in the B, D, and F positions, using the tandem Staudinger–Aza–Wittig or phosphine imide strategy, starting with the triazido compound (Fig. 34).

All the compounds have been characterized by IR, UV-Vis, ^{1}H, ^{13}C NMR, and elemental analysis. The complexation properties toward different cations (i.e., EuIII, TbIII, CuII, etc.) were examined in solution, and it was shown that these complexes could be interesting as potent fluorescent tracers.

4. IONIC AMPHIPHILIC CDs

4.1. Cationic Amphiphilic CDs

Amphiphilic cationic derivatives have also been synthetized. We can note, for example, the amphiphilic derivative synthesized by Darcy's group [168], in which the amine groups have been added on oligoethylene oxide chains of bouquet-shaped amphiphilic CDs (Fig. 35). After iodination of the chain extremities at high temperature, the intermediate derivatives are then submitted to azidation. The per-azido compounds were then reduced to per-amino amphiphiles in PPh$_3$/NH$_3$, then protonated to their hydrochloride salts. The latter compounds have been used in gene delivery studies and proved to be at least five times more efficient again, that is, 20,000 times more efficient than uncomplexed DNA and comparable to commercial cationic vectors [169,170]. Positively charged nanoparticles obtained from the same derivatives were also shown to encapsulate an anionic porphyrin which could be used in photodynamic cancer therapy [171]. Furthermore, the amino derivative with a thio hexyl chain on

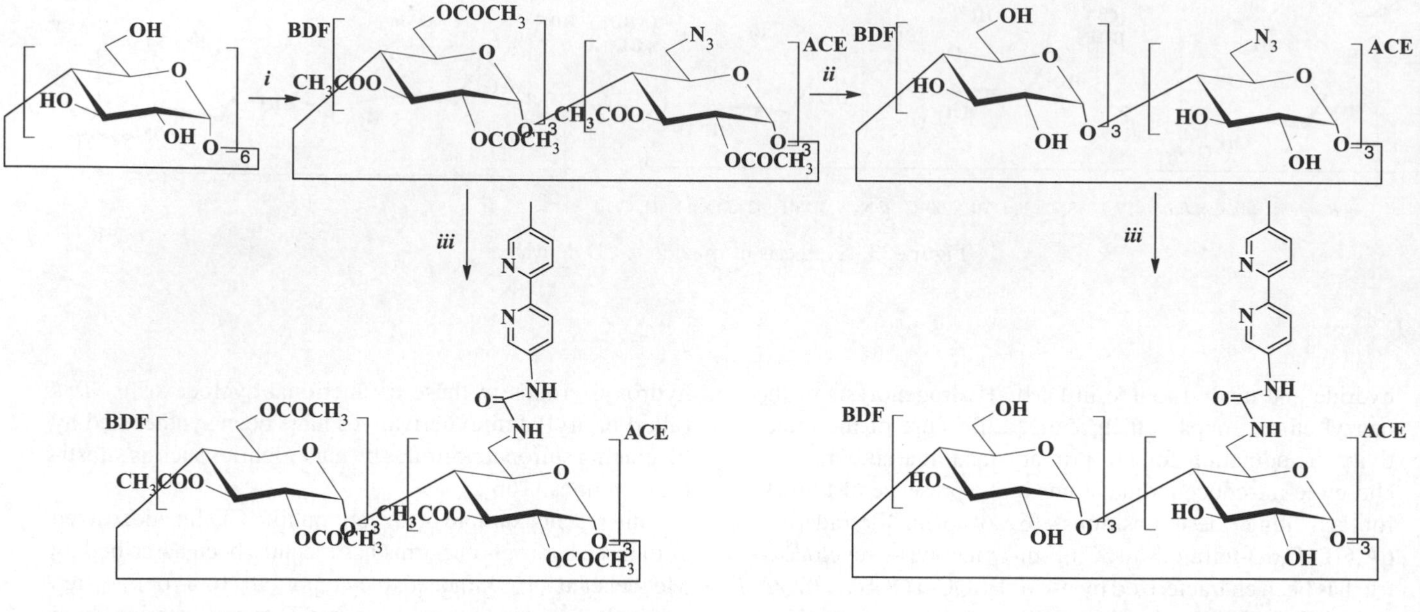

Reagents and conditions : *i*, NBS, Ph₃P, DMF, 100°C ; NaN₃, DMF, 60°C then pyridine, Ac₂O, *ii* MeOH, MeONa 0°C *iii* PPh₃, CO₂, DMF, 5-aminomethyl bipyridine

Figure 34. Amphiphilic tri-substituted α-CD derivatives.

the primary face, can form nanoparticules which have been used as an inhibitor against the photochemical degradation of diflusinal, a photosensitive commercial medicine. Pyridylamino amphiphilic derivatives have also been obtained directly from the iodo derivatives (Fig. 35).

More recently, in 2009, Byrne et al. described the synthesis of a series of amphiphilic CDs but containing cationic groups at the 6-position and alkyl or biolabile ester groups at the 2-positions (Fig. 36) [172]. The cationic groups are cysteamine derived, while the alkyl and ester groups are C_1 to C_{16} and benzyl ester groups. Two routes were first used for synthesis of the 2-O-allylated derivatives: introduction of the polar groups first on the primary side using an 6-deoxy-6-brominated intermediate or by protecting these positions and introducing first the allyl groups (which will be extended to form the lipophilic groups) on the secondary side. The following photochemical addition of lipophilic thiols allows solubility and self-assembly in water. The main advantages of polycationic CDs is their enhanced ability to interact with nucleic acid combined with self-organizational properties [170]. These derivatives are capable of acting as gene delivery vectors by condensing DNA and forming liquid crystalline complexes with oligonucleotides. Polyaminothiourea derivatives, derived from polyamino-β-CDs (Fig. 37) have also been described by Diaz-Moscoso et al. [173,174].

Gel electrophoresis revealed that these cationic CDs derivatives self-assemble in the presence of plasmid DNA to provide homogeneous and stable nanoparticles that fully protect plasmid DNA from the environment. The transfection efficiency of the resulting nanoparticles have been investigated on different cell lines and have been found to be intimately dependent on architectural features.

4.2. Anionic Amphiphilic CDs

Among these ionic amphiphilic CDs, anionic CDs have also been synthesized. Kraus et al. have used the allylated CDs described by Leydet et al. [175] and oxidized them in the presence of osmium tetroxyde [176]. The diastereoisomeric diols thus obtained were then oxidized in carboxylic acid to afford the desired carboxylated amphiphilic CDs (Fig. 38).

Another type of carboxylated amphiphilic derivative has been synthesized, but with anionic groups on the primary face of the CD. Roehri-Stoeckel et al. [177] used a click chemistry reaction between per(2,3-acetylated-6-azido-6-deoxy)cyclodextrin (prepared directly from acetylation of the perazido compound) and 2-butyne dicarboxylic acid ester at high temperatures in an aqueous alkaline medium to give a 1,3-dipolar cyclo-addition product (Fig. 39). This 1,2,3-triazole heterocyclic compound has been obtained in 91% yield and was highly soluble in water.

The most studied anionic amphiphilic CDs are those carrying a sulfate or sulfonate group in their structure. The first was prepared by Dubes et al. [178,179] via esterification of the O_2 and O_3 position with hexanoic anhydride of a CD protected on the primary face by silyl groups. Acylation by

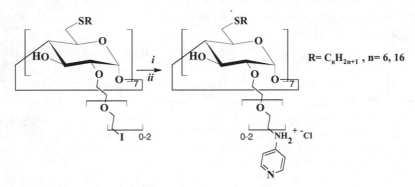

Reagents and conditions : *i*, DMF, PPh$_3$, NIS, 100°C, 4-5h ,*ii*, DMF, NaN$_3$, TMU, 100°C, 4-5d, *iii*,DMF, PPh$_3$, 2h,,NH$_3$ aq, then 1M HCl

Reagents and conditions : *i*, 4-aminopyridine, DMF, 100°C, 4d, *ii*, MeOH/H$_2$O or acetone, AgCl, r.t.

Figure 35. Synthesis of cationic amphiphilic CDs on the secondary face.

hexanoyl chloride led to suracylated derivatives, as shown by NMR, in which broad peaks and loss of the axial symmetry are observed. Removal of the silyl groups allows the sulfatation of the 6-hydroxyls with SO$_3$·pyridine complex (Fig. 40). These derivatives were able to form aggregates in an aqueous medium. It has also been shown that the sulfonated β-CD derivative with C$_{14}$ chains formed stable vesicles.

Sukegawa et al. [180] prepared sulfated amphiphiles using a similar procedure, and they showed that both sulfated and nonsulfated compounds form stable monolayers. However, the sulfated CDs showed higher monolayer collapse-pressure values. All these anionic derivatives formed 1:1 inclusion complexes with acyclovir, an antiviral drug. The complexes were characterized by UV-Vis spectrophotometry and electrospray ionization mass spectrometry (ESI-MS). Noncovalent interactions appear to involve only the hydrophobic region of the alkanoyl chains [179].

Schwinté et al. [181] synthesized sulfonated derivatices from acylated 6-per-bromo-β-CD via reaction with sulfonated thiols salts (Fig. 41). In contrast to the previous one, these sulfonated amphiphiles were water soluble and could form micellar aggregates in water that solubilized clofazimine, an antileprosy drug, better than did the native β-CD. Various charged CDs, such as sulfobutyl ether-β-cyclodextrin, are also well suited for pharmaceutical preparations [182].

5. CONCLUSIONS

Numerous methods have been described in the literature to modify selectively one or several specific positions on CDs. Due to the presence of 18 (α-CD), 21 (β-CD), or 24 (γ-CD) hydroxyl groups on the CD core, there are infinite

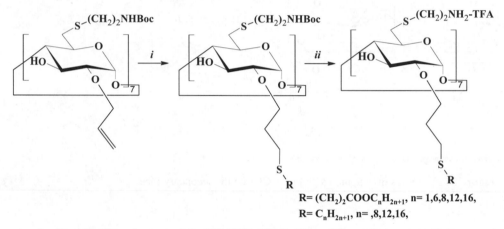

Reagents and conditions: Route 1: *i* tBuOK, BocNH(CH₂)₂SH, DMF, 80°C, 3 d; *ii* NaH, allyl bromide, r.t., 24 h. Route 2:

iii NaH, allyl bromide, r.t., 24 h; *iv* PPh₃, Br₂, CH₂Cl₂, r.t., 24 h; *v* NaH, BocNH(CH₂)₂SH, DMF, r.t., 16 h.

$R= (CH_2)_2COOC_nH_{2n+1}$, n= 1,6,8,12,16,

$R= C_nH_{2n+1}$, n= ,8,12,16,

Reagents and conditions: *i* R–SH (propionate esters 5–9), AIBN, MeOH; *ii* CH₂Cl₂, TFA, r.t. * Converted to hepta-hydrochloride.

Figure 36. Synthesis of cationic β-CD on the secondary face.

possibilities for substitution. Concerning amphiphilic CDs, numerous syntheses have been described in the literature, leading to medusa-like, bouquet-shaped or selectively substituted derivatives. Nevertheless, there is a lot of work to achieve on the methodology of synthesis in order to improve the efficiency of the reactions (better selectivity, better yields). Indeed, the majority of the amphiphilic CD derivatives reported will never have any industrial utilization because they involve complicated multistep syntheses, resulting in expensive products. The cost/benefit ratio often precludes their production and utilization. In order to be industrially produced and marketed, a CD derivative has to be produced by a simple, possibly "one-pot" reaction. For this

reason, the only CDs that are produced in ton quantities industrially are the methylated (RAMEB), hydroxyalkylated, acetylated, and branched (glucosyl- and maltosyl-α-CD) CDs. Furthermore, for pharmaceutical use these derivatives must be nontoxic, which is not always the case for amphiphilic deirvatives.

It is thus necessary to find more successful synthesis strategies for constructing target molecules with well-defined properties (i.e., increasing water solubilities of drugs, cell targeting, enzymatic activity, etc.) and low production cost. Amphiphilic cyclodextrins are original compounds that are still developed in many groups, primarily to form nanoparticulate drug delivery systems (nanospheres, nanocapsules,

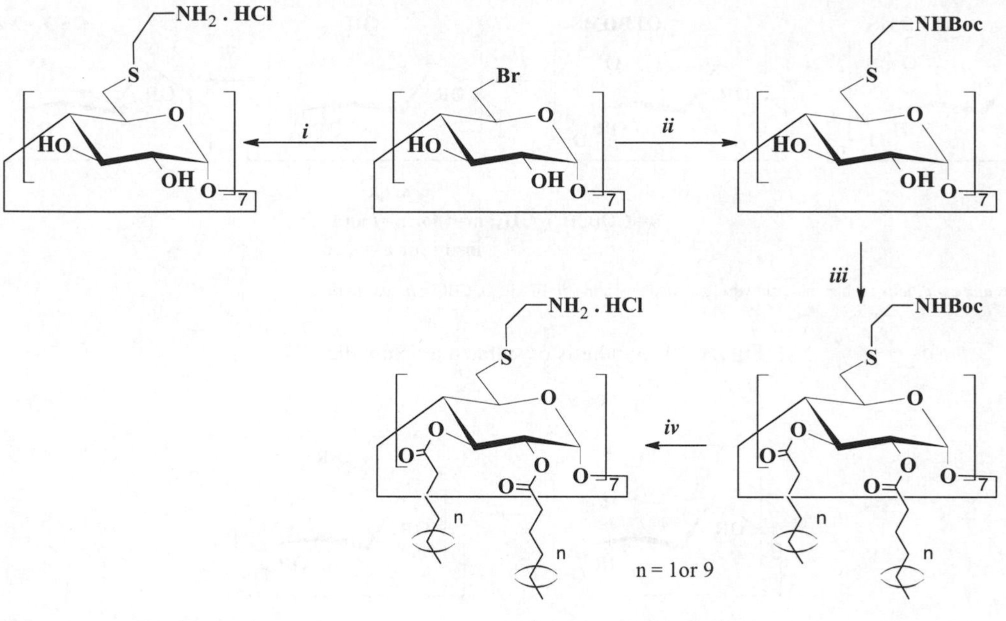

Reagents and conditions : *i* cysteamine hydrochloride, Et$_3$N, DMF, r.t.; *ii* N-Boc cysteamine , Cs$_2$CO$_3$ DMF, 70°C, *iii*, hexanoic (n=1) or myristic anhydride (n=9), DMAP, DMF, 45min, *iv* TFA/CH$_2$Cl$_2$ 1:1 r.t. 2h then HCl

Figure 37. Polyamino amphiphilic β-CDs.

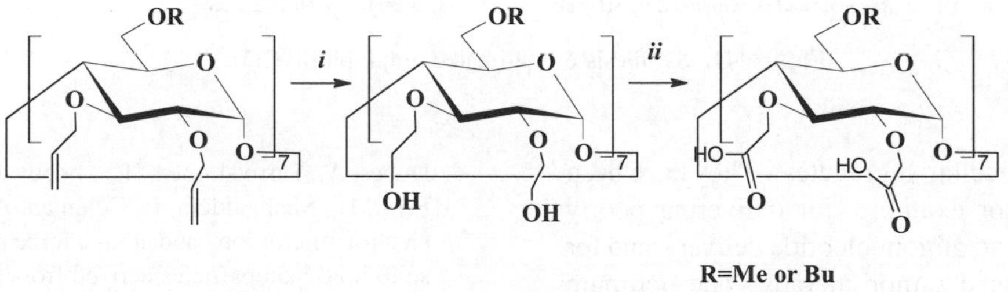

R=Me or Bu

Reagents and conditions : *i,* OsO$_4$, 4-methylmorpholine *N*-oxide, acetone / H$_2$O then NaIO$_4$, H$_2$O and NaBH$_4$, H$_2$O
ii, TEMPO, NaClO, KBr, H$_2$O, pH 10

Figure 38. Synthesis of O$_2$–O$_3$-carboxylated amphiphilic β-CD.

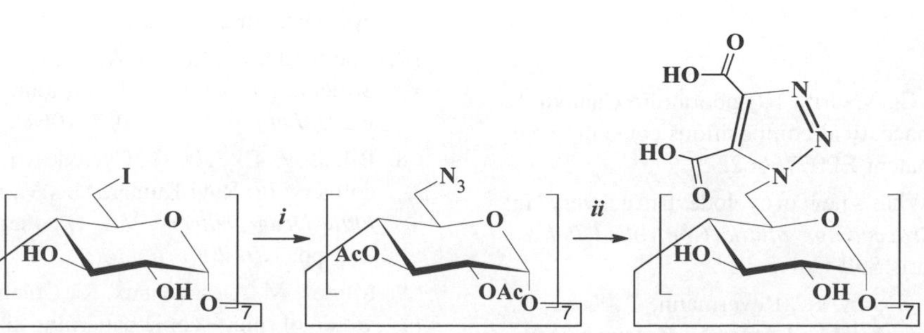

Reagents and conditions : *i,*NaN$_3$, DMF, 60°C then Ac$_2$O, Pyridine, 50°C *ii* MeOOC ⇌ COOMe toluene, 110°C then KOH, H$_2$O, MeOH, dioxane and H$^+$ (DOWEX 50W)

Figure 39. Synthesis of O$_6$-carboxylated amphiphilic β-CD.

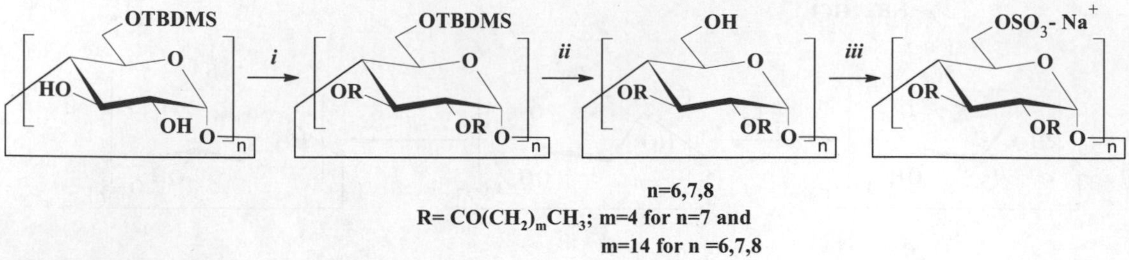

$n=6,7,8$

$R= CO(CH_2)_mCH_3$; $m=4$ for $n=7$ and

$m=14$ for $n =6,7,8$

Reagents and conditions : *i,* hexanoic anhydride, DMAP, pyridine *ii*, $BF_3.Et_2O$, $CHCl_3$, *iii* SO_3.Pyridine

Figure 40. Synthesis of sulfated amphiphilic CDs.

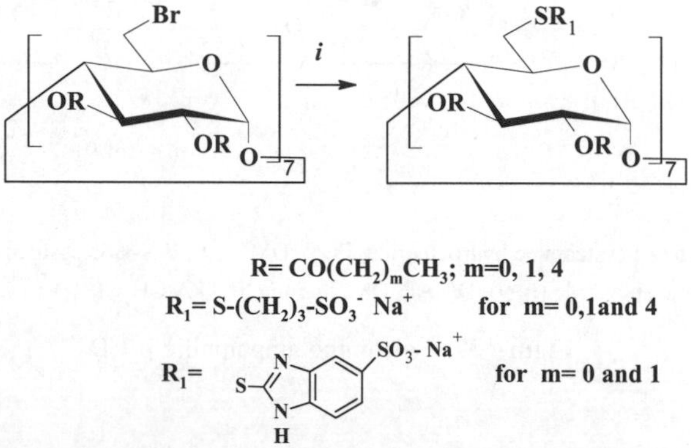

$R= CO(CH_2)_mCH_3$; $m=0, 1, 4$

$R_1= S-(CH_2)_3-SO_3^-$ Na^+ for $m= 0,1$ and 4

$R_1=$ [benzimidazole-$SO_3^-Na^+$ structure]   for $m= 0$ and 1

Reagents and conditions : *i,* R_1-SH, NaH or tBuOK, r.t., 30min, then 60-70°C, DMF, 2d

Figure 41. Synthesis of sulfonated amphiphilic CDs.

nanosize vesicles, and micellar aggregates). They have been shown to be efficient, for example, for delivering poorly soluble drugs, for DNA and oligonucleotide delivery, and for photodynamic and targeted tumor therapy. The optimum amphiphilic CD derivative for such a use should be nontoxic even in high doses, stable in physiological media, receptor specific, available in high purity, and should be produced on a large scale and at low cost. This ideal amphiphilic CD derivative does not yet exist.

REFERENCES

1. Coquelet, C., Latour, E., Maurin, F.(Laboratoire Chauvin S. A., Fr.) (1997). Pharmaceutical compositions containing mequitazine. European patent EP0776662.

2. Muller, B.W. (2000). What's new in cyclodextrin delivery? In: Duchêne, M., Ed., *Proceedings of the Controlled Release Society Workshop*, Paris Sud University.

3. Falvey, P., Lim, C.W., Darcy, R., Revermann, T., Karst, U., Giesbers, M., Marcelis, A.T.M., Lazar, A., Coleman, A.W., Reinhoudt, D.N., Ravoo, B.J. (2005). Bilayer vesicles of amphiphilic cyclodextrins: host membranes that recognize guest molecules. *Chem. Eur. J.*, *11*(4), 1171–1180.

4. Dubes, A., Parrot-Lopez, H., Abdelwahed, W., Degobert, G., Fessi, H., Shahgaldian, P., Coleman, A.W. (2003). Scanning electron microscopy and atomic force microscopy imaging of solid lipid nanoparticles derived from amphiphilic cyclodextrins. *Eur. J. Pharm. Biopharm.*, *55*, 279–282.

5. Memisoglu, E., Bochot, A., Ozalp, M., Sen, M., Duchêne, D., Hıncal, A.A. (2003). Direct formation of nanospheres from amphiphilic β-cyclodextrin inclusion complexes. *Pharm. Res.*, *20*, 117–125.

6. Ling, C.C., Darcy, R., Risse, W. (1993). Cyclodextrin liquid crystals: synthesis and self-organisation of amphiphilic thiocyclodextrins. *J. Chem. Soc. Chem. Commun.*, *5*, 438–440.

7. Lombardo, D., Longo, A., Darcy, R., Mazzaglia, A. (2004). Structural properties of non-ionic cyclodextrin colloids in water. *Langmuir*, *20*, 1057–1064.

8. Bilensoy, E. (2008). Cyclodextrin nanoparticles for drug delivery. In: Ravi Kumar, M.N.V., Ed., *Handbook of Particulate Drug Delivery*, Vol. 1, American Scientific, Valencia, CA, pp. 187–204.

9. Munoz, M., Deschenaux, R., Coleman, A.W. (1999). Observation of microscopic patterning at the air/water interface by mixtures of amphiphilic cyclodextrins: a compression isotherm and Brewster angle microscopy study. *J. Phys. Org. Chem.*, *12*(5), 364–369.

10. Duchêne, D., Wouessidjewe, D., Ponchel, G. (1999). Cyclo-dextrins and carrier systems. *J. Control. Release*, *62*(1–2), 263–268.

11. Sallas, F., Darcy, R. (2008). Amphiphilic cyclodextrins: advances in synthesis and supramolecular chemistry. *Eur. J. Org. Chem.*, *6*, 957–969.

12. Djedaïni-Pilard, F., Auzely-Velty, R., Perly, B. (2002). Recent developments in amphiphilic cyclodextrins: characterization and properties. *Recent Res. Dev. Org. Bioorg. Chem.*, *5*, 41–57.

13. Roux, M., Perly, B., Djedaïni-Pilard, F. (2007). Self-assemblies of amphiphilic cyclodextrins. *Eur. Biophys. J.*, *36*(8), 861–867.

14. Perly, B., Moutard, S., Djedaïni-Pilard, F. (2005). Amphiphilic cyclodextrins: from a general concept to properties and applications. *Pharm. Chem.*, *4*(1–2), 4–9.

15. Kawabata, Y., Matsumoto, M., Tanaka, M., Takahashi, H., Irinatsu, Y., Tamara, S., Tagaki, W., Nakahara, H., Fukuda, K. (1986). Formation and deposition of monolayers of amphiphilic β-cyclodextrin derivatives. *Chem. Lett.*, *15*(11), 1933–1934.

16. Djedaïni, F., Coleman, A.W., Perly, B. (1990). New cyclodextrin-based media for vectorization of hydrophobic drugs, mixed vesicles composed of phospholipids and lipophilic cyclodextrins. *Minutes, 5th International Symposium on Cyclodextrins*, pp. 328–331.

17. Liu, F.Y., Kildsig, D.O., Mitra, A.K. (1992). Complexation of 6-acyl-*O*-β-cyclodextrin derivatives with steroids: effects of chain length and substitution degree. *Drug Dev. Ind. Pharm.*, *18*(15), 1599–1612.

18. Lesieur, S., Charon, D., Lesieur, P., Ringard-Lefebvre, C., Muguet, V., Duchêne, D., Wouessidjewe, D. (2000). Phase behavior of fully hydrated DMPC–amphiphilic cyclodextrin systems. *Chem. Phys. Lipids* *106*(2), 127–144.

19. Memisoglu, E., Charon, D., Duchêne, D., Hincal, A.A. (1999). Synthesis of per(2,3-di-*O*-hexanoyl)-β-cyclodextrin and characterization of amphiphilic β-cyclodextrin nano-particles. *Proceedings of the 9th International Symposium on Cyclodextrins*, pp. 125–128.

20. Zhang, P., Ling, C.C., Coleman, A.W., Parrot-Lopez, H., Galons, H. (1991). Formation of amphiphilic cyclodextrins via hydrophobic esterification at the secondary hydroxyl face. *Tetrahedron Lett.*, *32*(24), 2769–2770.

21. Zhang, P., Parrot-Lopez, H., Tchoreloff, P., Baszkin, A., Ling, C.C., De Rango, C., Coleman, A.W. (1992). Self-organizing systems based on amphiphilic cyclodextrin diesters. *J. Phys. Org. Chem.*, *5*(8), 518–528.

22. Canceill, J., Jullien, L., Lacombe, L., Lehn, J.-M. (1992). Channel-type molecular structures. *Helv. Chim. Acta 75*, 791–812.

23. Wenz, G. (1991). Synthesis and characterisation of some lipophilic per(2,6-di-*O*-alkyl)cyclomalto-oligosaccharides. *Carbohydr. Res.*, *214*, 257–265.

24. Bellanger, N., Perly, B. (1992). NMR investigations of the conformation of new cyclodextrin-based amphiphilic transporters for hydrophobic drugs: molecular lollipops. *J. Mol. Struct.*, *273*, 215–226.

25. Dodziuk, H., Chmurski, K., Jurczak, J., Kozminski, W., Lukin, O., Sitkowski, J., Stefaniak, L. (2000). A dynamic NMR study of self-inclusion of a pendant group in amphiphilic 6-thiophenyl-6-deoxycyclodextrins. *J. Mol. Struct.*, *519*, 33–36.

26. Yamamura, H., Fujita, K. (1991). Preparation of heptakis[6-*O*-(*p*-tosyl)]-β-cyclodextrin and heptakis-[6-*O*-(*p*-tosyl)]-2-*O*-(*p*-tosyl)-β-cyclodextrin and their conversion to heptakis (3,6-anhydro)-β-cyclodextrin. *Chem. Pharm. Bull.*, *39*(10), 2505–2508.

27. Yamamura, H., Ezuka, T., Kawase, Y., Kawai, M., Butsugan, Y., Fujita, K. (1993). Preparation of octakis(3,6-anhydro)-β-cyclodextrin and characterization of its cation binding ability. *J. Chem. Soc. Chem. Commun.*, *7*, 636–637.

28. Fujita, K., Ohta, K., Masunari, K., Obe, K., Yamamura, H. (1992). Selective preparation of hexakis(6-*O*-arenesulfonyl)-β-cyclodextrin. *Tetrahedron Lett.*, *33*(38), 5519–5520.

29. Takeo, K., Kato, S., Kuge, T. (1974). Selective tritylation of phenyl α- and β-maltosides. *Carbohydr. Res.*, *38*, 346–351.

30. Berberan-Santos, M.N., Canceill, J., Brochon, J.C., Jullien, L., Lehn, J.M., Pouget, J., Tauc, P., Valeur, B. (1992). Multichromophoric cyclodextrins: 1. Synthesis of O-naphthoyl-β-cyclodextrins and investigation of excimer formation and energy hopping. *J. Am. Chem. Soc.*, *114*(16), 6427–6436.

31. Khan, A.R., D'Souza, V.T. (1994). Synthesis of 6-deoxychlorocyclodextrin via Vilsmeier–Haack-type complexes. *J. Org. Chem.*, *59*(24), 7492–7495.

32. Gadelle, A., Defaye, J. (1991). Selective halogenation of cyclic maltose oligosaccharides in the C-6 position and synthesis of per-(3,6-anhydro) cyclic maltose oligosaccharides. *Angew. Chem., Int. Ed. Engl.*, *1930*, 1978–1980.

33. Chmurski, K., Defaye, J. (1997). An improved synthesis of 6-deoxyhalo cyclodextrins via halomethylene-morpholinium halides: Vilsmeier–Haack type reagents. *Tetrahedron Lett.*, *38*(42), 7365–7368.

34. Chmurski, K., Defaye, J. (2000). An improved synthesis of per (6-deoxyhalo) cyclodextrins using *N*-halosuccinimides–triphenylphosphine in dimethylformamide. *Supramol. Chem.*, *12*(2), 221–224.

35. Yamamura, H., Kawase, Y., Kawai, M., Butsugan, Y. (1993). Preparation of polytosylated β-cyclodextrins. *Bull. Chem. Soc. Jpn.*, *66*(2), 585–588.

36. Khan, A.R., Forgo, P., Stine, K.J., D'Souza, V.T. (1998). Methods for selective modifications of cyclodextrins. *Chem. Rev.*, *98*(5), 1977–1996.

37. Takeo, K., Mitoh, H., Uemura, K. (1989). Selective chemical modification of cyclomalto-oligosaccharides via *tert*-butyldimethylsilylation. *Carbohydr. Res.*, *187*(2), 203–221.

38. Parrot-Lopez, H., Ling, C.C., Zhang, P., Baszkin, A., Albrecht, G., De Rango, C., Coleman, A.W. (1992). Self-assembling systems of the amphiphilic cationic per-6-amino-β-cyclodextrin 2,3-di-*O*-alkyl ethers. *J. Am. Chem. Soc.*, *114*(13), 5479–5480.

39. Guillo, F., Hamelin, B., Jullien, L., Canceill, J., Lehn, J.-M., De Robertis, L., Driguez, H. (1995). Synthesis of symmetrical cyclodextrin derivatives bearing multiple charges. *Bull. Soc. Chim. Fr.*, *132*(8), 857–866.

40. Parazak, D.P., Khan, A.R., D'Souza, V.T., Stine, K.J. (1996). Comparison of host–guest Langmuir–Blodgett multilayer formation by two different amphiphilic cyclodextrins. *Langmuir*, *12*(16), 4046–4049.

41. Ling, C.C., Darcy, R., Risse, W. (1993). Cyclodextrin liquid crystals: synthesis and self-organization of amphiphilic thio-β-cyclodextrins. *J. Chem. Soc. Chem. Commun.*, *5*, 438–440.

42. Chmurski, K., Coleman, A.W., Jurczak, J. (1996). Direct synthesis of amphiphilic α, β and γ cyclodextrins. *J. Carbohydr. Chem.*, *15*(7), 787–796.

43. Chmurski, K., Jurczak, J., Kasselouri, A., Coleman, A.W. (1994). Molecular building blocks: chromophoric amphiphilic cyclodextrin derivatives. *Supramol. Chem.*, *3*(3), 171–173.

44. Inamura, K., Ikeda, H., Ueno, A. (2002). Enhanced binding ability of β-cyclodextrin bearing seven hydrophobic chains each with a hydrophilic end group. *Chem. Lett.*, *2002*, 516–517.

45. Memisoglu, E., Bochot, A., Sen, M., Charon, D., Duchêne, D., Hincal, A.A. (2002). Amphiphilic β-cyclodextrins modified on the primary face: synthesis, characterization, and evaluation of their potential as novel excipients in the preparation of nanocapsules. *J. Pharm. Sci.*, *91*(5), 1214–1224.

46. Kruizinga, W.H., Strijtveen, B., Kellogg, R.M. (1981). Cesium carboxylates in dimethyl formamide: reagents for introduction of hydroxyl groups by nucleophilic substitution and for inversion of configuration of secondary alcohols. *J. Org. Chem.*, *46*, 4321–4323.

47. Takeo, K., Uemura, K., Mitoh, H. (1988). Derivatives of β-cyclodextrin and the synthesis of 6-*O*-α-D-glucopyranosyl-β-cyclodextrin. *J. Carbohydr. Chem.*, *7*(2), 293–308.

48. Gallois-Montbrun, D., Thiebault, N., Moreau, V., Le Bas, G., Archambault, J.-C., Lesieur, S., Djedaïni-Pilard, F. (2007). Direct synthesis of novel amphiphilic cyclodextrin. *J. Inclusion Phenom. Macrocyclic Chem.*, *57*(1–4), 131–135.

49. Ravey, J.C., Stebe, M.J. (1994). Properties of fluorinated non-ionic surfactant-based systems and comparison with non-fluorinated systems. *Colloids Surf. A*, *84*(1), 11–31.

50. Diakur, J., Zuo, Z., Wiebe, L.I. (1999). Synthesis and drug complexation studies with β-cyclodextrins fluorinated on the primary face. *J. Carbohydr. Chem.*, *18*(2), 209–223.

51. Granger, C.E., Felix, C.P., Parrot-Lopez, H.P., Langlois, B.R. (2000). Fluorine containing β-cyclodextrin: a new class of amphiphilic carriers. *Tetrahedron Lett.*, *41*(48), 9257–9260.

52. Bertino Ghera, B., Perret, F., Baudouin, A., Coleman, A.W., Parrot-Lopez, H. (2007). Synthesis and characterisation of *O*-6-alkylthio- and perfluoroalkylpropanethio-α-cyclodextrins and their O-2-, O-3-methylated analogues. *New J. Chem.*, *31*, 1899–1906.

53. Bertino Ghera, B., Perret, F., Chevalier, Y., Parrot-Lopez, H. (2009). Novel nanoparticles made from amphiphilic perfluoroalkyl α-cyclodextrin derivatives: preparation, characterisation and application to the transport of acyclovir. *Int. J. Pharm.*, *375*(1–2), 155–162.

54. Peroche, S., Parrot-Lopez, H. (2002). Novel fluorinated amphiphilic cyclodextrin derivatives: synthesis of mono-, di- and heptakis-(6-deoxy-6-perfluoroalkylthio)-β-cyclodextrins. *Tetrahedron Lett.*, *44*(2), 241–245.

55. Schalchli, A., Benattar, J.J., Tchoreloff, P., Zhang, P., Coleman, A.W. (1993). Structure of a monomolecular layer of amphiphilic cyclodextrins. *Langmuir*, *9*, 1968–1972.

56. Tchoreloff, P., Boissonade, M.M., Coleman, A.W., Baszkin, A. (1995). Amphiphilic monolayers of insoluble cyclodextrins at the water/air interface: surface pressure and surface potential studies. *Langmuir*, *11*, 191–196.

57. Tanaka, M., Ishizuka, Y., Matsumoto, M., Nakamura, T., Yabe, A., Nakanishi, H., Kawabata, Y., Takahashi, H., Tamura, S. (1987). Host–guest complexes of an amphiphilic β-cyclodextrin and azobenzene derivatives in Langmuir–Blodgett films. *Chem. Lett.*, *7*, 1307–1310.

58. Niino, H., Yabe, A., Ouchi, A., Tanaka, M., Kawabata, Y., Tamura, S., Miyasaka, T., Tagaki, W., Nakahara, H., Fukuda, K. (1988). Stabilization of a labile *cis*-azobenzene derivative with amphiphilic cyclodextrins. *Chem. Lett.*, *7*, 1227–1230.

59. Taneva, S., Ariga, K., Okahata, Y., Tagaki, W. (1989). Association between amphiphilic cyclodextrins and cholesterol in mixed insoluble monolayers at the air–water interface. *Langmuir*, *5*(1), 111–113.

60. Taneva, S., Ariga, K., Tagaki, W., Okahata, Y. (1989). Association of amphiphilic cyclodextrins with dipalmitoylphosphatidylcholine in mixed insoluble monolayers at the air–water interface. *J. Colloid Interface Sci.*, *131*(2), 561–566.

61. Terry, N., Rival, D., Coleman, A.W., Perrier, E. (2001). Novel non-hydroxylalkylated cyclodextrin derivatives and their use in the transport of active ingredients across skin tissue. British patent GB2362102, U.S. patent 6,524,595.

62. Kawabata, Y., Matsumoto, M., Nakamura, T., Tanaka, M., Manda, E., Takahashi, H., Tamura, S., Tagaki, W., Nakahara, H., Fukuda, K. (1988). Langmuir–Blodgett films of amphiphilic cyclodextrins. *Thin Solid Films*, *159*, 353–358.

63. Kobayashi, K., Kajikawa, K., Sasabe, H., Knoll, W. (1999). Monomolecular layer formation of amphiphilic cyclodextrin derivatives at the air/water interface. *Thin Solid Films*, *349*(1–2), 244–249.

64. Peroche, S., Degobert, G., Putaux, J.-L., Blanchin, M.-G., Fessi, H., Parrot-Lopez, H. (2005). Synthesis and characterization of novel nanospheres made from amphiphilic perfluoroalkylthio-β-cyclodextrins. *Eur. J. Pharm. Biopharm.*, *60*(1), 123–131.

65. Dey, J., Schwinte, P., Darcy, R., Ling, C.-C., Sicoli, F., Ahern, C. (1998). Amphiphilic 6-*S*-alkyl-6-thiocyclodextrins: unimolecular micellar and reverse micellar behavior. *J. Chem. Soc. Perkin Trans. 2*, *6*, 1513–1516.

66. Matsuzawa, Y., Matsumoto, M., Noguchi, S., Sakai, H., Abe, M. (2006). Hybrid Langmuir and LB films composed of amphiphilic cyclodextrins and hydrophobic azobenzene derivatives. *Mol. Cryst. Liq. Cryst.*, *445*, 139–147.

67. Kobayashi, K., Ishii, N., Sasabe, H., Knoll, W. (2001). Monomolecular layer formation of ferritin molecules on an amphiphilic cyclodextrin derivative at the air/water interface. *Biosci. Biotechnol. Biochem.*, *65*(1), 176–179.

68. Chmurski, K., Bilewicz, R., Jurczak, J. (1996). Monolayer behavior of [6-deoxy-6-*S*-phenyl]-α, β, and γ-cyclodextrins at the air–water interface. *Langmuir*, *12*(25), 6114–6118.

69. Le Bras, Y., Salle, M., Leriche, P., Mingotaud, C., Richomme, P., Moller, J. (1997). Functionalization of the cyclodextrin platform with tetrathiafulvalene units: an efficient access towards redox active Langmuir–Blodgett films. *J. Mater. Chem.*, *7*(12), 2393–2396.

70. Ringard-Lefebvre, C., Bochot, A., Memisoglu, E., Charon, D., Duchêne, D., Baszkin, A. (2002). Effect of spread amphiphilic β-cyclodextrins on interfacial properties of the oil/water system. *Colloids Surf. B*, *25*(2), 109–117.

71. Memisoglu, E., Bochot, A., Sen, M., Duchêne, D., Hincal, A. A. (2003). Non-surfactant nanospheres of progesterone inclusion complexes with amphiphilic β-cyclodextrins. *Int. J. Pharm.*, *251*(1–2), 143–153.

72. Kasselouri, A., Coleman, A.W., Baszkin, A. (1996). Mixed monolayers of amphiphilic cyclodextrins and phospholipids: I. Miscibility under dynamic conditions of compression. *J. Colloid Interface Sci.*, *180*(2), 384–397.

73. Coleman, A.W., Kasselouri, A. (1993). Supramolecular assemblies based on amphiphilic cyclodextrins. *Supramol. Chem.*, *1*(2), 155–161.

74. Badis, M., Van der Heyden, A., Heck, R., Marsura, A., Gauthier-Manuel, B., Zywocinski, A., Rogalska, E. (2004). Formation of Langmuir layers and surface modification using new upper-rim fully tethered bipyridinyl or bithiazolyl cyclodextrins and their fluorescent metal complexes. *Langmuir*, *20*(13), 5338–5346.

75. Pregel, M.J., Buncel, E. (1991). Cyclodextrin-based enzyme models: 1. Synthesis of a tosylate and an epoxide derived from heptakis(6-*O-tert*-butyldimethylsilyl)-Î²-cyclodextrin and their characterization using 2D NMR techniques: an improved route to cyclodextrins functionalized on the secondary face. *Can. J. Chem.*, *69*(1), 130–137.

76. Fugedi, P. (1989). Synthesis of heptakis(6-*O-tert*-butyldimethylsilyl)cyclomaltoheptaose and octakis(6-*O-tert*-butyldimethylsilyl)cyclomaltooctaose. *Carbohydr. Res.*, *192*, 366–369.

77. Cramer, F., Mackensen, G., Sensse, K. (1969). On ring compounds: XX. ORD-spectra and conformation of the glucose ring in cyclodextrins. *Chem. Ber.*, *102*(2), 494–508.

78. Beadle, J.B. (1969). Analysis of cyclodextrin mixtures by gas chromatography of their dimethylsilyl ethers. *J. Chromatogr.*, *42*(2), 201–206.

79. Tabushi, I., Kuroda, Y., Yokota, K., Yuan, L.C. (1981). Regiospecific A,C- and A, D-disulfonate capping of β-cyclodextrin. *J. Am. Chem. Soc.*, *103*(3), 711–712.

80. Khan, A.R., Barton, L., D'Souza, V.T. (1992). Heptakis-2,3-epoxy-β-cyclodextrin, a key intermediate in the synthesis of custom designed cyclodextrins. *J. Chem. Soc., Chem. Commun.*, *16*, 1112–1114.

81. Wazynska, M., Temeriusz, A., Chmurski, K., Bilewicz, R., Jurczak, J. (2000). Synthesis and monolayer behavior of amphiphilic per(2,3-di-*O*-alkyl)-α- and β-cyclodextrins and hexakis(6-deoxy-6-thio-2,3-di-*O*-pentyl)-α-cyclodextrin at an air–water interface. *Tetrahedron Lett.*, *41*(47), 9119–9123.

82. Badi, N., Jarroux, N., Guegan, P. (2006). Synthesis of per-2,3-di-*O*-heptyl-α and β-cyclodextrins: a new kind of amphiphilic molecules bearing hydrophobic parts. *Tetrahedron Lett.*, *47*(50), 8925–8927.

83. Skiba, M., Skiba-Lahiani, M., Arnaud, P. (2002). Design of nanocapsules based on novel fluorophilic cyclodextrin derivatives and their potential role in oxygen delivery. *J. Inclusion Phenom. Macrocyclic Chem.*, *44*, 151–154.

84. Skiba-Lahiana, M., Skiba, M. (2004). Nanocapsules based on perfluorinated cyclodextrins: potential role for oxygen carrier. *Formulation Composes Silicones Fluores*, *2004*, 192–200.

85. Choisnard, L., Geze, A., Putaux, J.-L., Wong, Y.-S., Wouessidjewe, D. (2006). Nanoparticles of β-cyclodextrin esters obtained by self-assembling of biotransesterified β-cyclodextrins. *Biomacromolecules*, *7*(2), 515–520.

86. Choisnard, L., Geze, A., Bigan, M., Putaux, J.-L., Wouessidjewe, D. (2005). Efficient size control of amphiphilic cyclodextrin nanoparticles through a statistical mixture design methodology. *J. Pharm. Pharm. Sci.*, *8*(3), 593–600.

87. Lahiani-Skiba, M., Bounoure, F., Shawky-Tous, S., Arnaud, P., Skiba, M. (2006). Optimization of entrapment of metronidazole in amphiphilic β-cyclodextrin nanospheres. *J. Pharm. Biomed. Anal.*, *41*(3), 1017–1021.

88. Memisoglu-Bilensoy, E., Vural, I., Bochot, A., Renoir, J.M., Duchêne, D., Hincal, A.A. (2005). Tamoxifen citrate loaded amphiphilic β-cyclodextrin nanoparticles: in vitro characterization and cytotoxicity. *J. Control. Release*, *104*(3), 489–496.

89. Dubes, A., Parrot-Lopez, H., Shahgaldian, P., Coleman, A.W. (2003). Interfacial interactions between amphiphilic cyclodextrins and physiologically relevant cations. *J. Colloid Interface Sci.*, *259*(1), 103–111.

90. Skiba, M., Wouessidjewe, D., Coleman, A., Fessi, H., Devissaguet, J.P., Duchêne, D., Puisieux, F. (1993). Preparation and use of novel cyclodextrin-based dispersible colloidal systems in the form of nanospheres. European patents EP0646003, EP19930913149.

91. Duchêne, D., Ponchel, G., Wouessidjewe, D. (1999). Cyclodextrins in targeting: application to nanoparticles. *Adv. Drug Deliv. Rev.*, *36*(1), 29–40.

92. Geze, A., Aous, S., Baussanne, I., Putaux, J., Defaye, J., Wouessidjewe, D. (2002). Influence of chemical structure of amphiphilic β-cyclodextrins on their ability to form stable nanoparticles, *Int. J. Pharm.*, *242*(1–2), 301–305.

93. Geze, A., Putaux, J.L., Choisnard, L., Jehan, P., Wouessidjewe, D. (2004). Long-term shelf stability of amphiphilic β-cyclodextrin nanosphere suspensions monitored by dynamic light scattering and cryo-transmission electron microscopy. *J. Microencapsul.*, *21*(6), 607–613.

94. Gulik, A., Delacroix, H., Wouessidjewe, D., Skiba, M. (1998). Structural properties of several amphiphile cyclodextrins and some related nanospheres: an x-ray scattering and freeze-fracture electron microscopy study. *Langmuir*, *14*(5), 1050–1057.

95. Memisoglu-Bilensoy, E., Sen, M., Hincal, A.A. (2006). Effect of drug physicochemical properties on in vitro characteristics of amphiphilic cyclodextrin nanospheres and nanocapsules. *J. Microencapsul.*, *23*(1), 59–68.

96. Skiba, M., Duchêne, D., Puisieux, F., Wouessidjewe, D. (1996). Development of a new colloidal drug carrier from chemically-modified cyclodextrins: nanospheres and influence of physicochemical and technological factors on particle size. *Int. J. Pharm.*, *129*(1–2), 113–121.

97. Skiba, M., Puisieux, F., Duchêne, D., Wouessidjewe, D. (1995). Direct imaging of modified β-cyclodextrin nanospheres by photon scanning tunneling and scanning force microscopy. *Int. J. Pharm.*, *120*(1), 1–11.

98. Skiba, M., Wouessidjewe, D., Puisieux, F., Duchêne, D., Gulik, A. (1996). Characterization of amphiphilic β-cyclodextrin nanospheres. *Int. J. Pharm.*, *142*(1), 121–124.

99. Lemos-Senna, E., Wouessidjewe, D., Duchêne, D. (1998). Amphiphilic cyclodextrin nanospheres: particle solubilization and reconstitution by the action of a nonionic detergent. *Colloids Surf. B*, *10*(5), 291–301.

100. Lemos-Senna, E., Wouessidjewe, D., Lesieur, S., Duchêne, D. (1998). Preparation of amphiphilic cyclodextrin nanospheres using the emulsification solvent evaporation method: influence of the surfactant on preparation and hydrophobic drug loading. *Int. J. Pharm.*, *170*(1), 119–128.

101. Lemos-Senna, E., Wouessidjewe, D., Lesieur, S., Puisieux, F., Couarraze, G., Duchêne, D. (1998). Evaluation of the hydrophobic drug loading characteristics in nanoprecipitated amphiphilic cyclodextrin nanospheres. *Pharm. Dev. Technol.*, *3*(1), 85–94.

102. Skiba, M., Nemati, F., Puisieux, F., Duchêne, D., Wouessidjewe, D. (1996). Spontaneous formation of drug-containing amphiphilic β-cyclodextrin nanocapsules. *Int. J. Pharm.* *145*(1–2), 241–245.

103. Skiba, M., Skiba, M., Duclos, R., Combret, J.-C., Arnaud, P. (2001). Cyclodextrins monosubstituted to persubstituted by fluoroalkyl groups, preparation and use thereof. Patent WO PCT/FR2000/002192.

104. Memisoglu-Bilensoy, E., Hincal, A.A. (2006). Sterile, injectable cyclodextrin nanoparticles: effects of gamma irradiation and autoclaving. *Int. J. Pharm.*, *311*(1–2), 203–208.

105. Skiba, M., Morvan, C., Duchêne, D., Puisieux, F., Wouessidjewe, D. (1995). Evaluation of gastrointestinal behavior in the rat of amphiphilic β-cyclodextrin nanocapsules, loaded with indomethacin. *Int. J. Pharm.*, *126*(1–2), 275–279.

106. Skiba, M., Wouessidjewe, D., Fessi, H., Devissaguet, J.P., Duchêne, D., Puisieux, F. (1998). Preparation and use of novel cyclodextrin-based dispersible vesicular colloidal systems in the form of nanocapsules. U.S. patent application 5,718,905.

107. Hamelin, B., Jullien, L., Laschewsky, A., Du Penhoat, C.H. (1999). Self-assembly of Janus cyclodextrins at the air–water interface and in organic solvents. *Chem. Eur. J.*, *5*(2), 546–556.

108. Fugedi, P., Nanasi, P., Szejtli, J. (1988). Synthesis of 6-*O*-α-D-glucopyranosylcyclomaltoheptaose. *Carbohydr. Res.*, *175*(2), 173–181.

109. Cottaz, S., Driguez, H. (1989). A convenient synthesis of 6-*S*-α and 6-*S*-β-D-glucopyranosyl-6-(thio)-maltodextrins. *Synthesis*, *10*, 755–758.

110. Szejtli, J., Liptak, A., Jodal, I., Fugedi, P., Nanasi, P., Neszmelyi, A. (1980). Synthesis and carbon-13 NMR spectroscopy of methylated β-cyclodextrins. *Starch Staerke*, *32*(5), 165–169.

111. Pregel, M.J., Jullien, L., Lehn, J.M. (1992). Toward artificial ion channels: transport of alkali metal ions through liposome membranes via bouquet-like molecules. *Angew. Chem. Int. Ed.*, *31*(12), 1637–1640.

112. Pregel, M.J., Jullien, L., Canceill, J., Lacombe, L., Lehn, J.-M. (1995). Channel-type molecular structures: 4. Transmembrane transport of alkali-metal ions by "Bouquet" molecules. *J. Chem. Soc. Perkin Trans. 2*, *3*, 417–426.

113. Ravoo, B.J., Darcy, R. (2000). Cyclodextrin bilayer vesicles. *Angew. Chem. Int. Ed.*, *39*(23), 4324–4326.

114. Mazzaglia, A., Donohue, R., Ravoo, B.J., Darcy, R. (2001). Novel amphiphilic cyclodextrins: graft synthesis of heptakis (6-alkylthio-6-deoxy)-β-cyclodextrin 2-oligo(ethylene glycol) conjugates and their halo derivatives. *Eur. J. Org. Chem.*, *9*, 1715–1721.

115. Crespo-Biel, O., Peter, M., Bruinink Christiaan, M., Ravoo Bart, J., Reinhoudt, D.N., Huskens, J. (2005). Multivalent host–guest interactions between β-cyclodextrin self-assembled monolayers and poly(isobutene-*alt*-maleic acid)s modified with hydrophobic guest moieties. *Chemistry*, *11*(8), 2426–2432.

116. Ravoo, B.J., Jacquier, J.-C., Wenz, G. (2003). Molecular recognition of polymers by cyclodextrin vesicles. *Angew. Chem. Int. Ed.*, *42*(18), 2066–2070.

117. Sallas, F., Niikura, K., Nishimura, S.-I. (2004). A practical synthesis of amphiphilic cyclodextrins fully substituted with sugar residues on the primary face. *Chem. Commun.*, *5*, 596–597.

118. Fulton, D.A., Stoddart, J.F. (2001). Synthesis of cyclodextrin-based carbohydrate clusters by photoaddition reactions. *J. Org. Chem.*, *66*, 8309–8319.

119. Niino, H., Miyasaka, H., Ouchi, A., Kawabata, Y., Yabe, A., Miyasaka, T., Tagaki, W., Nakahara, H., Fukuda, K. (1989). Photopolymerization in a Langmuir–Blodgett film of an amphiphilic cyclodextrin derivative containing a diacetylene group. *Thin Solid Films*, *179*, 53–57.

120. Greenhall, M.H., Lukes, P., Kataky, R., Agbor, N.E., Badyal, J.P.S., Yarwood, J., Parker, D., Petty, M.C. (1995). Monolayer and multilayer films of cyclodextrins substituted with two and three alkyl chains. *Langmuir*, *11*(10), 3997–4000.

121. Mazzaglia, A., Forde, D., Garozzo, D., Malvagna, P., Ravoo, B.J., Darcy, R. (2004). Multivalent binding of galactosylated cyclodextrin vesicles to lectin. *Org. Biomol. Chem.*, *2*(7), 957–960.

122. McNicholas, S., Rencurosi, A., Lay, L., Mazzaglia, A., Sturiale, L., Perez, M., Darcy, R. (2007). Amphiphilic

N-glycosyl-thiocarbamoyl cyclodextrins: synthesis, self-assembly, and fluorimetry of recognition by lens culinaris lectin. *Biomacromolecules*, 8(6), 1851–1857.

123. Nolan, D., Darcy, R., Ravoo, B.J. (2003). Preparation of vesicles and nanoparticles of amphiphilic cyclodextrins containing labile disulfide bonds. *Langmuir*, 19(10), 4469–4472.

124. Brady, B., Lynam, N., O'Sullivan, T., Ahern, C., Darcy, R. (2000). 6A-*O-p*-toluenesulfonyl-β-cyclodextrin. *Org. Synth.*, 77, 220–224.

125. Melton, L.D., Slessor, K.N. (1971). Synthesis of monosubstituted cyclohexaamyloses. *Carbohydr. Res.*, 18(1), 29–37.

126. Hao, A.Y., Tong, L.H., Zhang, F.S., Gao, X.M. (1995). Convenient preparation of monoacylated β-cyclodextrin (cyclomaltoheptaose) on the secondary hydroxyl side. *Carbohydr. Res.*, 277(2), 333–337.

127. Petter, R.C., Salek, J.S., Sikorski, C.T., Kumaravel, G., Lin, F.T. (1990). Cooperative binding by aggregated mono-6-(alkylamino)-β-cyclodextrins. *J. Am. Chem. Soc.*, 112(10), 3860–3868.

128. Petter, R.C., Salek, J.S. (1987). Cooperative binding by an amphipathic host. *J. Am. Chem. Soc.*, 109(25), 7897–7899.

129. Ueno, A., Breslow, R. (1982). Selective sulfonation of a secondary hydroxyl group of β-cyclodextrin. *Tetrahedron Lett.*, 23(34), 3451–3454.

130. Laine, V., Coste-Sarguet, A., Gadelle, A., Defaye, J., Perly, B., Djedaïni-Pilard, F. (1995). Inclusion and solubilization properties of 6-*S*-glycosyl-6-thio derivatives of $\hat{1}^2$-cyclodextrin. *J. Chem. Soc. Perkin Trans. 2*, 7, 1479–1487.

131. Defaye, J., Perly, B., Gadelle, A., Descamps, V., Coste, S.A. (1995). Method for solubilizing antitumor agents from the taxol family in an aqueous medium, and branched cyclodextrins therefor. Patent WO/1995/019994.

132. Fikes, L.E., Winn, D.T., Sweger, R.W., Johnson, M.P., Czarnik, A.W. (1992). Preassociating α-nucleophiles. *J. Am. Chem. Soc.*, 114(4), 1493–1495.

133. Lin, J., Creminon, C., Perly, B., Djedaini-Pilard, F. (1998). New amphiphilic derivatives of cyclodextrins for the purpose of insertion in biological membranes: the "cup and ball" molecules. *J. Chem. Soc. Perkin Trans. 2*, 12, 2639–2646.

134. Hanessian, S., Benalil, A., Laferriere, C. (1995). The synthesis of functionalized cyclodextrins as scaffolds and templates for molecular diversity, catalysis, and inclusion phenomena. *J. Org. Chem.*, 60(15), 4786–4797.

135. Hanessian, S., Hocquelet, C., Jankowski, C.K. (2008). Synthesis of aminocyclodextrin carboxylic acids. *Synlett*, 5, 715–719.

136. Yockot, D., Moreau, V., Demailly, G., Djedaini-Pilard, F. (2003). Synthesis and characterization of mannosyl mimetic derivatives based on a β-cyclodextrin core. *Org. Biomol. Chem.*, 1(10), 1810–1818.

137. Pean, C., Creminon, C., Wijkhuisen, A., Grassi, J., Guenot, P., Jehan, P., Dalbiez, J.-P., Perly, B., Djedaïni-Pilard, F. (2000). Synthesis and characterization of peptidyl-cyclodextrins dedicated to drug targeting. *J. Chem. Soc. Perkin Trans. 2*, 4, 853–863.

138. Djedaïni-Pilard, F., Desalos, J., Perly, B. (1993). Synthesis of a new molecular carrier: *N*-(Leu-enkephalin)yl 6-amino-6-deoxy-cyclomaltoheptaose. *Tetrahedron Lett.*, 34(15), 2457–2460.

139. Nelles, G., Weisser, M., Back, R., Wohlfart, P., Wenz, G., Mittler-Neher, S. (1996). Controlled orientation of cyclodextrin derivatives immobilized on gold surfaces. *J. Am. Chem. Soc.*, 118(21), 5039–5046.

140. Sallas, F., Leroy, P., Marsura, A., Nicolas, A. (1994). First selective synthesis of thio-β-cyclodextrin derivatives by a direct Mitsunobu reaction on free β-cyclodextrin. *Tetrahedron Lett.*, 35(33), 6079–6082.

141. Silva, O.F., Fernandez, M.A., Pennie, S.L., Gil, R.R., de Rossi, R.H. (2008). Synthesis and characterization of an amphiphilic cyclodextrin, a micelle with two recognition sites. *Langmuir*, 24(8), 3718–3726.

142. Moutard, S., Perly, B., Gode, P., Demailly, G., Djedaïni-Pilard, F. (2003). Novel glycolipids based on cyclodextrins. *J. Inclusion Phenom. Macrocyclic Chem.*, 44(1–4), 317–322.

143. Roux, M., Moutard, S., Perly, B., Djedaïni-Pilard, F. (2007). Lipid lateral segregation driven by diacyl cyclodextrin interactions at the membrane surface. *Biophys. J.*, 93(5), 1620–1629.

144. Auzely-Velty, R., Djedaïni-Pilard, F., Desert, S., Perly, B., Zemb, T. (2000). Micellization of hydrophobically modified cyclodextrins: 1. Micellar structure. *Langmuir*, 16(8), 3727–3734.

145. Cavalli, R., Donalisio, M., Civra, A., Ferruti, P., Ranucci, E., Trotta, F., Lembo, D. (2009). Enhanced antiviral activity of acyclovir loaded into β-cyclodextrin-poly(4-acryloylmorpholine) conjugate nanoparticles. *J. Control. Release*, 137(2), 116–122.

146. Fujita, K., Yamamura, H., Matsunaga, A., Imoto, T., Mihashi, K., Fijioka, T. (1990). Specific preparation and structure determination of 3A,3C, 3E-tri-*O*-sulfonyl-β-cyclodextrin. *J. Org. Chem.*, 55(3), 877–880.

147. Fujita, K., Matsunaga, A., Imoto, T. (1984). 6A6B, 6A6C, and 6A6D-ditosylates of β-cyclodextrin. *Tetrahedron Lett.*, 25(48), 5533–5536.

148. Fujita, K., Ishizu, T., Oshiro, K., Obe, K. (1989). 2A,2B-, 2A,2C-, and 2A,2D-bis-*O*-(*p*-tolylsulfonyl)-β-cyclodextrins. *Bull. Chem. Soc. Jpn.*, 62(9), 2960–2962.

149. Fujita, K., Nagamura, S., Imoto, T., Tahara, T., Koga, T. (1985). Regiospecific sulfonation of secondary hydroxyl groups of cyclodextrin: its application to preparation of 2A2B, 2A2C-, and 2A2D-disulfonates. *J. Am. Chem. Soc.*, 107(11), 3233–3235.

150. Fujita, K., Tahara, T., Imoto, T., Koga, T. (1986). Regiospecific sulfonation onto C-3 hydroxyls of β-cyclodextrin: preparation and enzyme-based structural assignment of 3A,3C and 3A3D disulfonates. *J. Am. Chem. Soc.*, 108(8), 2030–2034.

151. Szejtli, J., Osa, T., Eds. (1996). *Comprehensive Supramolecular Chemistry*. Elsevier, Oxford, UK.

152. Tabushi, I., Yamamura, K., Nabeshima, T. (1984). Characterization of regiospecific AC- and AD-disulfonate capping of

β-cyclodextrin: capping as an efficient production technique. *J. Am. Chem. Soc.*, *106*(18), 5267–5270.

153. Tabushi, I., Kuroda, Y., Yokota, K. (1982). A,C,D,F-tetrasubstituted β-cyclodextrin as an artificial channel compound. *Tetrahedron Lett.*, *23*(44), 4601–4604.

154. Tabushi, I., Yuan, L.C., Shimokawa, K., Yokota, K., Mizutani, T., Kuroda, Y. (1981). A,C;A',C'-doubly capped β-cyclodextrin: direct evidence for the capping structure. *Tetrahedron Lett.*, *22*(24), 2273–2276.

155. Tabushi, I., Nabeshima, T., Fujita, K., Matsunaga, A., Imoto, T. (1985). Regiospecific A,B capping onto β-cyclodextrin: characteristic remote substituent effect on carbon-13 NMR chemical shift and specific taka-amylase hydrolysis. *J. Org. Chem.*, *50*(15), 2638–2643.

156. Ling, C.C., Coleman, A.W., Miocque, M. (1992). Multiple tritylation: a convenient route to polysubstituted derivatives of cyclomaltohexaose. *Carbohydr. Res.*, *223*, 287–291.

157. Poorters, L., Armspach, D., Matt, D. (2003). Selective tetrafunctionalization of α-cyclodextrin using the supertrityl protecting group: synthesis of the first C2-symmetric tetraphosphane based on a cavitand (α-TEPHOS). *Eur. J. Org. Chem.*, *8*, 1377–1381.

158. Pearce, A.J., Sinaÿ, P. (2000). Diisobutylaluminum-promoted regioselective de-O-benzylation of perbenzylated cyclodextrins: a powerful new strategy for the preparation of selectively modified cyclodextrins. *Angew. Chem. Int. Ed.*, *39*(20), 3610–3612.

159. Lecourt, T., Herault, A., Pearce, A.J., Sollogoub, M., Sinaÿ, P. (2004). Triisobutylaluminium and diisobutylaluminium hydride as molecular scalpels: the regioselective stripping of perbenzylated sugars and cyclodextrins. *Chem. Eur. J.*, *10*(12), 2960–2971.

160. Bertino-Ghera, B., Perret, F., Fenet, B., Parrot-Lopez, H. (2008). Control of the regioselectivity for new fluorinated amphiphilic cyclodextrins: synthesis of di- and tetra(6-deoxy-6-alkylthio)- and 6-(perfluoroalkylpropanethio)-α-cyclodextrin derivatives. *J. Org. Chem.*, *73*(18), 7317–7326.

161. Breslow, R., Canary, J.W., Varney, M., Waddell, S.T., Yang, D. (1990). Artificial transaminases linking pyridoxamine to binding cavities: controlling the geometry. *J. Am. Chem. Soc.*, *112*, 5212–5219.

162. Cottaz, S., Apparu, S., Driguez, H. (1991). Chemoenzymatic approach to the preparation of regioselectively modified cyclodextrins: the substrate specificity of the enzyme cyclodextrin glucosyltransferase (CGTase). *J. Chem. Soc. Perkin Trans. 19*, 2235–2241.

163. Fujita, K., Tahara, T., Koga, T. (1989). Regioisomeric 6A,6X,6Y-tri-O-sulfonylated β-cyclodextrin. *Chem. Lett.*, *1989*, 821–824.

164. Boger, J., Brenner, D.G., Knowles, J.R. (1979). Symmetrical triamino-per-O-methyl-α-cyclodextrin: preparation and characterization of primary trisubstituted α-cyclodextrins. *J. Am. Chem. Soc.*, *101*, 7630–7631.

165. Heck, R., Jicsinszky, L., Marsura, A. (2003). Synthesis of symmetrically modified α-cyclodextrins: an efficient and easy method. *Tetrahedron Lett.*, *44*, 5411–5413.

166. Coleman, A.W., Lin, C.C., Miocque, M. (1992). Synthese und Komplexierungsverhalten eines auf Cyclodextrin beruhenden Siderophors. *Angew. Chem.*, *104*, 1402–1404.

167. Menuel, S., Corvis, Y., Rogalska, E., Marsura, A. (2009). Upper-rim alternately tethered β-cyclodextrin molecular receptors: synthesis, metal complexation and interfacial behavior. *New J. Chem.*, *33*(3), 554–560.

168. Donohue, R., Mazzaglia, A., Ravoo, B.J., Darcy, R. (2002). Cationic β-cyclodextrin bilayer vesicles. *Chem. Commun.*, *23*, 2864–2865.

169. Cryan, S.A., Donohue, R., Ravoo, B.J., Darcy, R., O'Driscoll, C.M. (2004). Cationic cyclodextrin amphiphiles as gene delivery vectors. *J. Drug Deliv. Sci. Technol.*, *14*(1), 57–62.

170. Cryan, S.-A., Holohan, A., Donohue, R., Darcy, R., O'Driscoll, C.M. (2004). Cell transfection with polycationic cyclodextrin vectors. *Eur. J. Pharm. Sci.*, *21*(5), 625–633.

171. Sortino, S., Mazzaglia, A., Scolaro, L.M., Merlo, F.M., Valveri, V., Sciortino, M.T. (2006). Nanoparticles of cationic amphiphilic cyclodextrins entangling anionic porphyrins as carrier-sensitizer system in photodynamic cancer therapy. *Biomaterials*, *27*(23), 4256–4265.

172. Byrne, C., Sallas, F., Rai, D.K., Ogier, J., Darcy, R. (2009). Poly-6-cationic amphiphilic cyclodextrins designed for gene delivery. *Org. Biomol. Chem.*, *7*(18), 3763–3771.

173. Diaz-Moscoso, A., Balbuena, P., Gomez-Garcia, M., Ortiz, M.C., Benito, J.M., Le Gourrierec, L., Di Giorgio, C., Vierling, P., Mazzaglia, A., Micali, N., Defaye, J., Garcia Fernandez, J.M. (2008). Rational design of cationic cyclooligosaccharides as efficient gene delivery systems. *Chem. Commun.*, *17*, 2001–2003.

174. Diaz-Moscoso, A., Le Gourrierec, L., Gomez-Garcia, M., Benito, J.M., Balbuena, P., Ortega-Caballero, F., Guilloteau, N., Di Giorgio, C., Vierling, P., Defaye, J., Ortiz, M.C., Fernandez, J.M.G. (2009). Polycationic amphiphilic cyclodextrins for gene delivery: synthesis and effect of structural modifications on plasmid DNA complex stability, cytotoxicity, and gene expression. *Chem. Eur. J.*, *15*(46), 12871–12888.

175. Leydet, A., Moullet, C., Roque, J.P., Witvrouw, M., Pannecouque, C., Andrei, G., Snoeck, R., Neyts, J., Schols, D., De Clercq, E. (1998), Polyanion inhibitors of HIV and other viruses. 7. Polyanionic compounds and polyzwitterionic compounds derived from cyclodextrins as inhibitors of HIV transmission, *J. Med. Chem.*, *41*, 4927–4932.

176. Kraus, T., Budesinsky, M., Zavada, J. (2001). General approach to the synthesis of persubstituted hydrophilic and amphiphilic β-cyclodextrin derivatives. *J. Org. Chem.*, *66*(13), 4595–4600.

177. Roehri-Stoeckel, C., Dangles, O., Brouillard, R. (1997). A simple synthesis of a highly water soluble symmetrical β-cyclodextrin derivative. *Tetrahedron Lett.*, *38*(9), 1551–1554.

178. Dubes, A., Bouchu, D., Lamartine, R., Parrot-Lopez, H. (2002). An efficient regio-specific synthetic route to multiply

substituted acyl-sulfated β-cyclodextrins. *Tetrahedron Lett.*, *42*(52), 9147–9151.

179. Dubes, A., Degobert, G., Fessi, H., Parrot-Lopez, H. (2003). Synthesis and characterisation of sulfated amphiphilic α-, β- and γ-cyclodextrins: application to the complexation of acyclovir. *Carbohydr. Res.*, *338*(21), 2185–2193.

180. Sukegawa, T., Furuike, T., Niikura, K., Yamagishi, A., Nishimura, S.-I. (2002). Erythrocyte-like liposomes prepared by means of amphiphilic cyclodextrin sulfates. *Chem. Commun.*, *5*, 430–431.

181. Schwinte, P., Ramphul, M., Darcy, R., O'Sullivan, J.F. (2003). Amphiphilic cyclodextrin complexation of clofazimine. *J. Inclusion Phenom. Macrocyclic Chem.*, *47*(3–4), 109–112.

182. Rajewski, R.A., Stella, V.J. (1996). Pharmaceutical applications of cyclodextrins: 2. In vivo drug delivery. *J. Pharm. Sci.*, *85*(11), 1142–1169.

12

GENE DELIVERY WITH CYCLODEXTRINS

Véronique Wintgens and Catherine Amiel
Institut de Chimie et Matériaux Paris Est, Thiais, France

1. INTRODUCTION

In this chapter we update recent progress made in developing cyclodextrin (CD)-based gene delivery systems. In the following, the principles of gene therapy [1–3] are presented briefly, and the criteria that a system should meet to be an efficient DNA vector for transfection are emphasized. The peculiarity of CD-based gene delivery systems is then introduced.

Gene therapy is the treatment of human genetic-based diseases by the transfer of genes into specific cells of the patient. Due to its broad potential, gene therapy has been investigated largely during the last 20 years. The first clinical trial was reported in 1990 [4] and the first clinical success in 2000 [5], but the number of successes is still small compared to the number of clinical trials.

Gene therapy has to fulfill the following challenges: cell targeting specificity, gene transfer efficiency, gene expression regulation, and vector safety. Gene delivery carriers can be divided into two categories:

1. *Viral vectors* have been used in the majority of gene delivery studies and for around 70% of ongoing clinical trials. Safety concerns limit their use; the possible side effects are induced cancer, immune reactions, and others.
2. *Synthetic vectors* can improve safety and show greater flexibility and easier manufacturing than can viral vectors. Usually, they are based on materials that bind DNA or RNA electrostatically, condensing the genetic materials into particles of a few tens to several hundred nanometers, and making possible DNA protection and

cellular entry. Plasmid DNA complexes with cationic lipids and polymers are called *lipoplexes* and *polyplexes*, respectively. The first reports of cationic lipid use for gene delivery were in 1987 [6], and recent reviews [7–9] update the state of the art in this field. Cationic polymers are numerous and varied, and chemistry allows the specific design of the polymer to provide the multiple functions required for efficient gene delivery. Therefore, they have a great potential, but their relatively poor gene-transfer efficiency has limited their clinical application.

Figure 1 summarizes the various biological barriers that face efficient gene delivery vectors (see, e.g., [2,3] and references cited herein), which should also remain biocompatible. The important steps, illustrated for polyplexes, are the following:

- *Gene packaging.* The vector should neutralize the negative charge of DNA to counterbalance the electrostatic repulsion against the anionic cell membrane, condense DNA to the appropriate length scale (nanometer scale for receptor-mediated endocytosis), and protect DNA from extracellular and intracellular nuclease degradation (e.g., stability and survival in the bloodstream). Mixing of DNA and cationic polymers leads spontaneously to polyplexes. Each polyplex particle often comprises many polymer chains and several DNA molecules. The number of cationic moieties has a strong effect on polymer–DNA interactions.
- *Cellular entry.* Most of the polyplexes have been designed to gain cellular entry via receptor-mediated

Cyclodextrins in Pharmaceutics, Cosmetics, and Biomedicine: Current and Future Industrial Applications, First Edition. Edited by Erem Bilensoy.
© 2011 John Wiley & Sons, Inc. Published 2011 by John Wiley & Sons, Inc.

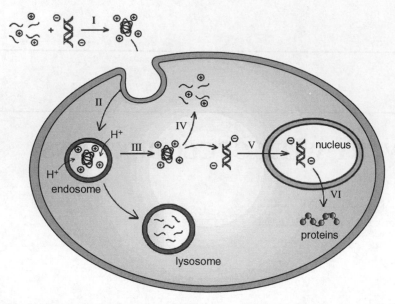

Figure 1. Barriers to gene delivery. Design requirements for gene delivery systems include the ability to (I) package therapeutics genes, (II) gain entry into the cells, (III) escape the endolysosomal pathway, (IV) effect DNA/vector release, (V) traffic through the cytoplasm and into the nucleus, and (VI) enable gene expression. (From [2].)

endocytosis. Polymer chemistry allows attachment of targeting molecules to induce cell uptake increase and cell specificity.

- *Endolysosomal escape.* Polyplexes are sequestered within endosomal vesicles. Only DNA that escapes into the cytoplasm can go on to reach the nucleus; otherwise, it undergoes degradation within the lysosomal compartment. Acidification occurs from endosomes to lysosomes. This influx of protons is used to disrupt the vesicle membrane, through materials known as *proton sponge*, therefore releasing DNA.

- *Transport through the cytoplasm and nuclear entry.* Once released from the endosomal vesicles, polyplexes must move through the cytoplasm to the nucleus. Cytosolic milieu is a physically and metabolically hostile environment, and the residence time should be minimized. Nuclear import of polyplexes is the last and probably most difficult step; various ligands have been proposed to promote cytosolic transport and nuclear import, such as nuclear localization signals (NLSs; short and cationic peptides) and carbohydrates. Subsequent to DNA–vector dissociation into the cytosol, or into the nucleus, genes released should be transcribed and translated into the therapeutic proteins.

- *DNA–vector dissociation.* Incorporation in a polyplex protects DNA from enzymatic degradation, but also prevents binding of the proteins required for gene expression. Expression could be enhanced if dissociation occurs within the nucleus. Polymers should be designed to incorporate a mechanism for nonspecific

or environmentally responsive release of genes; it has been suggested that thermoresponsive properties, hydrolytically degradable bonds, or reducible linkages be introduced into the vector.

Many systems involving CD compounds have been designed especially for gene delivery. The main properties of CDs—the biocompatibility, the ability to accommodate hydrophobic molecules within the cavity while remaining soluble, and the plurifunctionality enabling a large number of chemical modifications—make them suitable for the elaboration of biomimetic supramolecular architectures for gene therapy. A way to classify the different systems is to distinguish the structures of the CD compounds: cationic CD derivatives, polyrotaxanes, and CD-containing polymers or dendrimers. These classes constitute the various parts of the chapter.

Far beyond the objectives of this chapter, it should be noted that the ability of CDs to complex and interact with drugs led to their wide use for controlled drug delivery. However, interactions of CDs with nucleotides were explored (study of interactions between hydroxypropyl-β-CD and 26 nucleotides [10]) and then CD derivatives were studied with the aim of peptide, protein, and oligodeoxynucleotide (ODN) delivery [11,12]. The first ODN delivery studies were conducted with hydroxypropyl- and hydroxyethyl-β-CD [13], and other CD derivatives [14]. Delivery of therapeutic antisense ODN, or siRNA, involves a peculiar mechanism, but the synthetic vectors used have to meet almost the same criteria as in gene delivery; therefore, the same systems have been suggested as vectors (nanocarriers) for both ODN and gene delivery. Several reviews report progress in and development of synthetic polymer nanocarriers [15,16], in which linear cationic CD polymer has been proposed [17].

2. SYSTEMS BASED ON CD DERIVATIVES

2.1. Cationic CDs

Per-cationic CDs, usually amino-CD derivatives protonated at the physiological pH, were proposed to condense plasmid DNA. Stable complexes were observed between DNA and heptakis(6-amino-6-deoxy)-β-CD [18], and also with derivatives for which the amino groups are replaced by pyridylamino or imidazolyl groups. Complexes were characterized by size and zeta-potential measurements; electron microscopy experiments revealed toroid-shaped particles with sizes ranging from 200 to 300 nm. A relatively high ratio N/P (nitrogen/phosphate) was necessary to neutralize the negative charge of DNA and to produce charge inversion; therefore, in vitro transfection could be obtained only for N/P > 200. Transfection efficiencies deduced from luciferase expression were about 100 times lower than that obtained with DOTAP but were 20 times larger than that obtained with 2,6-dimethyl-β-CD. Indeed, neutral CDs are known to increase transfection levels due to their capacity to remove cholesterol from various cell types and thus to induce membrane instability [13]. Levels were improved 10- to 400-fold by the use of chloroquine, an endosomolytic agent, implying that the cationic CD–DNA complexes are endocytosed and that release from the endosome may be rate limiting.

The affinity of CD molecules for DNA is largely influenced by the type of substituent. It was shown that permodified α-, β-, and γ-CDs by guanidino group [19] had a strong affinity toward phosphorylated guests, and these CDs achieved mobility inhibition of calf thymus DNA during gel electrophoresis for an N/P ratio around 10 times lower than with heptakis(6-amino-6-deoxy)-β-CD. Recently, β- and γ-CDs bearing guanidinoalkylamino (Fig. 2) and aminoalkylamino groups [20] were designed as biomimetic structures of cell-penetrating peptides. The introduction of seven or eight branches resembling arginine or lysine side chains on the primary side of a CD combined with its cyclic rigid structure was considered as a favorable feature to achieve spatial organization of the positive charges. The compounds were found to cross the membranes of HeLa cells, the faster penetrations being observed with the guanidinylated CDs. Gel electrophoresis experiments have shown that the most active compounds in reducing calf thymus DNA mobility were guanidino derivatives bearing a propyl spacer group with the largest number of functional groups (γ-CD derivatives more active than β-CD derivatives) to enable multiple interactions with DNA. Additionally, ternary and quaternary

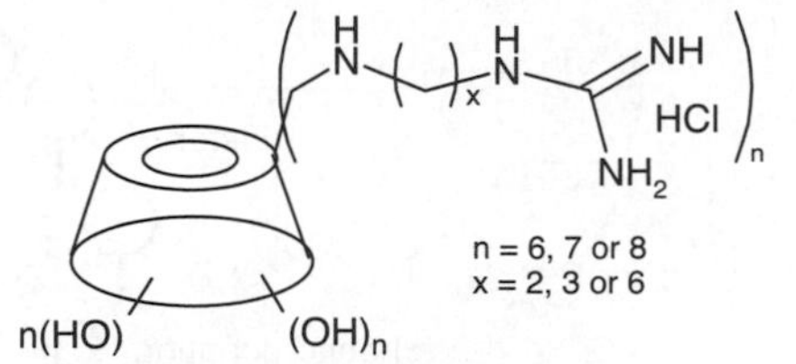

Figure 2. Chemical structure of per-guanidinoalkylamino α-, β-, and γ-CD (corresponding to $n = 6$, 7, or 8).

amino per-substituted CDs were also used in formulation with an adenovirus to enhance adenoviral transduction efficiency in two models of the intestinal epithelium [21]. CDs enhanced both viral binding and internalization.

Monosubstitued CDs were also used to condense DNA, as 6-monodeoxy-6-monoamino-β-CD that could bind and compact plasmid DNA [22] (revealed by atomic force microscopy images at an N/P ratio of 1). The strategy adopted for DNA delivery was to dissociate the role of condensation and the one of transport through the cell, which was ensured by anionic and pH-sensitive liposomes. DNA was first condensed by addition of cationic CD and was then encapsulated into the liposomes. Loaded vesicles of diameter around 200 nm with nonnegligible DNA encapsulation (10 μg/mL) were shown to be more stable and monodisperse than naked vesicles.

Cationic CDs [pyridylamino-β-CD (pCD)] have also been used in material coating based on polyelectrolyte multilayers containing adsorbed DNA and pCD [23]. The layer-by-layer (LBL) buildup was made from polyelecrolytes, poly(L-glutamic acid) (PLGA), and poly(L-lysine) (PLL) and was followed by quartz crystal microbalance. The final architectures comprised (PLL–PLGA)$_5$–pCD–DNA–pCD–PLGA–PLL and were put in contact with a cell culture medium. The biological activities of the films were tested by means of induced production of specific proteins (analyzed by fluorescence microscopy) into the cytoplasm or into the nucleus of three different types of cells. The authors showed that the films could act as an efficient gene delivery system, pCDs playing the role of transfection enhancer, and that multiple and sequential biological activity could be obtained depending solely on the level at which the DNA was embedded in the multilayer architectures.

2.2. Amphiphilic Cationic CDs

The strategy for incorporation of different functional elements on both CD sides has been developed to enhance DNA complexing and delivery properties. Such CD derivatives are constructed to perform specific tasks far beyond the simple formation of inclusion complexes. The synthesis of C$_7$-symmetric CD derivatives was realized giving amphiphilic cationic CDs. Seven or 14 fatty acyl chains are incorporated on either the primary [24–26] or the secondary

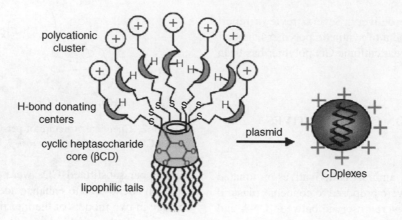

Figure 3. Amphiphilic β-CDs used as DNA vectors. (From [27].) (*See insert for color representation of the figure.*)

CD side [27], respectively, the other side being fully substituted by cationic groups. Figure 3 is a schematic representation of such CD derivatives that induce DNA condensation. These amphiphilic molecules self-associate in water into different types of architectures, such as vesicles or nanoparticles [28].

Designed specifically as gene delivery vectors, each substituent of the amphiphilic CDs has a definite task: The lipophilic chains are introduced to enhance membrane disruption activity, and the polycationic cluster should promote DNA condensation as mentioned in studies with cationic CDs [18–22]. The lipophilicity of the CDs was shown to be an important parameter for promoting gene transfection as transgene expression increased with the lipophilic chain length in transfection experiments of Hep G2 hepatocytes [26].

Two different types of cationic clusters have been reported. The first are made of seven short oligoethyleneglycol chains terminated by primary amino groups. The spacer role was to improve solubility and to reduce potential immunogenicity [24,26]. The second, described in Fig. 3, are made of 7 or 14 short chains bearing terminal primary amino groups and thiourea spacer groups [25,27]. The role of the thiourea groups was to promote cooperative binding with DNA by adding hydrogen-bond anchoring points.

It is quite difficult to compare quantitatively the performances of the CDs bearing oligoethylene glycol spacers to those bearing the thiourea spacers, as the experiments have not been performed with the same cells. Anyhow, the amphiphilic CDs show large in vitro transfection efficiencies which overpass that of the cationic poly(ethylenimine) (PEI; 25 kDa [27]) in the case of CDs bearing thiourea and that of DOTAP in the case of CDs bearing oligoethylene glycol [26], together with low-toxicity profiles.

2.3. Star-Shaped Cationic Polymers with a CD Core

Star-shaped cationic polymers were synthesized by conjugating multiple oligoethylenimine (OEI) arms onto an α-CD [29] or β-CD [30] core. An example is given in Fig. 4A for α-CD-OEI star polymers. Recently, star-shaped cationic polymers [31] consisting of β-CD core and poly[2-(dimethylamino)ethyl methacrylate] [p(DMAEMA)] and p(DMAEMA)-*block*-poly[poly(ethylene glycol)ethyl ether methacrylate] [p(PEGEEMA)] were prepared via atom transfer radical polymerization (Fig. 4B). All these star-shaped cationic polymers with an α- or β-CD core exhibited good efficiencies to condense plasmid DNA, already at low N/P ratios (around 2), as evidenced by gel electrophoresis. The polyplexes formed spherical nanoparticles with sizes

(A)

n = 5.8, x = 1
n = 6.8, x = 5
n = 3.4, x = 9
n = 5.0, x = 14

(B)

n = 4
x = 16, 56, 99
y = 0, 15

Figure 4. Chemical structure of (A) α-CD-OEI and (B) β-CD-[p(PDMAEMA)-*b*-p(PEGEEMA)] star polymers.

ranging from 80 to 200 nm and positive surface charges, at N/P ratios of 8 or higher. The nanoparticles could be observed by electron and atomic force microscopies.

The star polymers showed lower in vitro toxicity than that of branched PEI. Results also reported in vitro transfection efficiencies which are comparable or even higher than that of PEI. Additionally, it was shown that the stability of the polyplexes increased with their arm lengths, together with their transfection efficiency.

3. SYSTEMS BASED ON POLYPSEUDOROTAXANES AND POLYROTAXANES

The biomedical applications of polyrotaxanes span from biologically friendly degradable hydrogels, calcium-binding agents, and protein targeting agents to DNA transfection reagents [32,33]. As the properties and applications of polyrotaxanes are also described in other chapters, we essentially report only works related to gene delivery application. The gene carriers (or polyplexes) are formed through charge interactions between the phosphate anions of DNA and the cations borne by the polyrotaxanes. There are two ways to introduce a net positive charge in the polyrotaxanes: Either the polymer axle of polypseudorotaxane or polyrotaxane is charged with neutral CDs threaded onto the polymer axle (see Section 3.1), or the polymer axle is neutral with positively charged CDs threaded onto the polymer axle (see Section 3.2).

3.1. Polypseudorotaxanes Involving Neutral CDs

Poly(ethylenimine) (PEI) has attracted considerable interest due to its high positive charge density. PEI presents strong DNA binding, and high in vitro transfections are obtained. PEI with the secondary amines fulfilled the proton sponge requirement, which should facilitate endosomal and lysosomal escape of DNA into the cytoplasm. Unfortunately, PEI and PEI–DNA complexes show relatively high cytotoxicity. Reducing the cationic charge density is a key parameter for decreasing the cytotoxicity of the polyplexes. Yamashita et al. [34] tried to reduce the charge density of linear PEI (22 kDa) by using a polypseudorotaxane structure in which

the cationic groups are included in the threaded γ-CDs. Figure 5 reports schematic representation of PEI/γ-CD polypseudorotaxane inducing DNA condensation.

A 2.5-fold higher N/P (nitrogen/phosphate) ratio in the PEI/γ-CD polyplexes (pH 9) than in PEI is necessary to achieve mobility inhibition of DNA during gel electrophoresis. Yet PEI/γ-CD improved the cellular uptake (evaluated by flow cytometry and confocal-laser scanning microscopy), and the cell viability was largely increased even at high N/P ratios (N/P = 50). The transfection efficiency of the polypseudorotaxane was comparable to or greater than that of PEI, especially at high N/P ratios. It was suggested that γ-CDs could reduce the cytotoxicity by hindering the interaction of the secondary amines through γ-CDs complexation.

Copolymers of PEI were suggested to lower the charge density of PEI, but in most cases, if the cytotoxicity is reduced, the transfection efficiency is also decreased. Shuai et al. [35] proposed triblock copolymers PEI–PCL–PEG. α-CDs form inclusion complexes with both poly(ethylene glycol) (PEG) and poly(ε-caprolactone) (PCL) chains, and threading α-CDs largely improves the copolymer solubility. The copolymer/α-CDs complexes condensed DNA at N/P ratios similar to that of PEI (25 kDa). A reduced toxicity about 100 times lower than that of PEI was observed with copolymer/α-CD complexes. Confocal-laser scanning microscopy showed an efficient internalization of polyplexes, and it was suggested that DNA and copolymers were co-localized in the lysosomal compartment. The transfection efficiencies of the copolymer/α-CDs complexes were greater than that of PEI at higher N/P ratios.

3.2. Polypseudorotaxanes and Polyrotaxanes Involving Cationic CDs

The neutral polymer axle used primarily to synthesize the polypseudorotaxanes and polyrotaxanes involving cationic CDs are poly(ethylene oxide) (PEO), poly(propylene oxide) (PPO), or random copolymers (PEO–PPO). Ooya et al. [36] and Yamashita et al. [37] designed biocleavable polyrotaxanes based on dimethylaminoethyl-modified α-CDs threaded onto a PEO ($M_n = 4000$) chain end-capped with benzyloxycarbonyl tyrosine via disulfide linkages. Figure 6 reports the polyplex formation and its DNA release by supramolecular dissociation. Cleavage of the disulfide

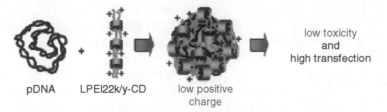

pDNA   LPEI22k/γ-CD   low positive charge   low toxicity and high transfection

Figure 5. PEI/γ-CD polypseudorotaxane inducing DNA condensation. (From [34].) (*See insert for color representation of the figure.*)

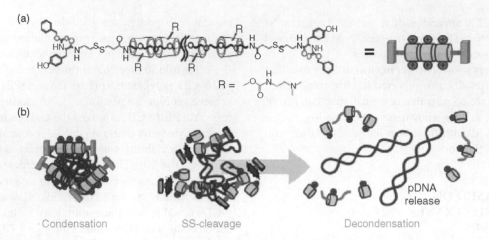

(a)

(b)

Condensation   SS-cleavage   Decondensation

pDNA
release

Figure 6. (a) Chemical structure of biocleavable polyrotaxane; (b) polyplex formation, terminal cleavage-triggered decondensation of DNA. (From [36].) (*See insert for color representation of the figure.*)

linkages decreases the molecular weight of the polycations inducing the dissociation of polyplexes—therefore the DNA release in the cytosolic milieu. The authors tried to optimize the transfection activity by varying the number of α-CDs and amino groups [37].

The DNA compaction, evaluated by the ethidium bromide displacement assay, is not influenced by the number of α-CDs but by the number of amino groups. The stability of the polyplexes has been determined quantitatively using a competitive displacement method against a counter-polyanion, such as dextran sulfate. The polyplexes formed with the polyrotaxanes showed a higher stability than the PEI polyplexes. In the reductive conditions (cleavage of the disulfide linkages), DNA release increased while decreasing the numbers of α-CDs and amino groups. High transfection efficiencies were observed with low dependence on the N/P values, particularly for low numbers of α-CDs and amino groups in the polyrotaxanes.

Similarly, Pérès et al. designed hydrolyzable polyaminorotaxanes based on PEO and spermine grafted α-CD. They used two pathways to build the polyaminorotaxanes, either by grafting spermine on a pre-made polyrotaxane or by grafting spermine on α-CD prior to threading along the PEO chains. They have shown that these tools made it

possible to adjust either the structure of the polyrotaxane chain or its charge density [38]. Yang et al. synthesized polyrotaxanes based on oligoethylenimine (OEI) grafted α-CDs [39] or β-CDs [40] threaded onto a random PEO–PPO copolymer chain end-capped with a bulky stopper (2,4,6-trinitrobenzene). Due to their cavity size, α- and β-CD are preferentially complexed with EO and PO units, respectively. Figure 7 shows the structure of one of these cationic polyrotaxanes.

The ability of the cationic polyrotaxanes to condense DNA was confirmed by gel electrophoresis, particle size analysis, and zeta-potential measurements. The polyrotaxanes had a similar or slightly better DNA condensation ability than that of PEI (25 kD). All the cationic polyrotaxanes with linear OEI exhibited less toxicity than did the PEI control. The transfection efficiency was dependent on the chain length of the OEI grafted onto CDs. The authors suggested that the best results obtained with one of the polyrotaxanes (linear OEI-grafted α-CDs threaded onto a random PEO–PPO copolymer) are related to the mobility increase in the cationic α-CD rings and to the polyrotaxane flexibility, which enhanced the interaction of α-CDs with DNA and/or the cellular membrane. Polypseudorotaxanes based on cationic β-CDs bearing anthryl groups and threaded onto a PPO chain showed good binding ability to DNA, evidenced by fluorescence titration experiments [41].

4. SYSTEMS BASED ON CD OLIGOMERS AND POLYMERS

In this section we report on CD oligomers and polymers used for gene delivery. Different polymer structures are described here: in the first part, polymers with a dendritic architecture; in the second part, linear cationic polymers containing CDs in the polymer backbone; in the third part, cationic polymers

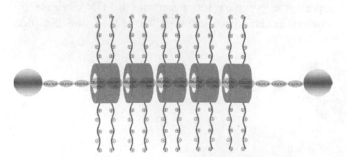

Figure 7. Structure of a cationic polyrotaxane. (From [40].) (*See insert for color representation of the figure.*)

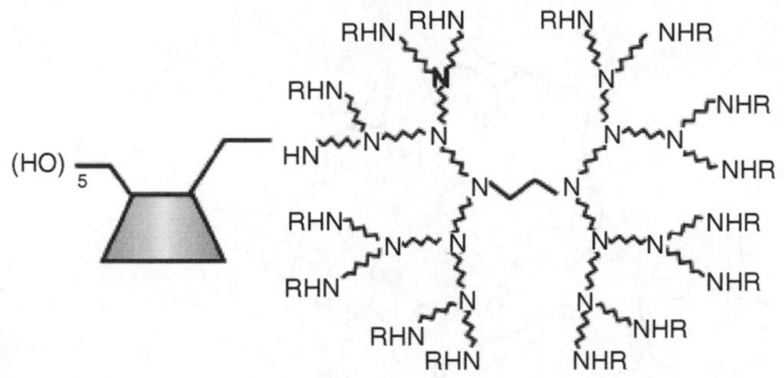

Figure 8. Chemical structure of polyamidoamine dendrimer conjugates (G2) with α-CD, R = H [45], or CS-NH-C$_6$H$_4$-mannose [46,47], or CS-NH-C$_6$H$_4$-galactose [48].

with CDs as a pendant of the chain; and in the last part, neutral CD polymers with a branched structure.

4.1. Dendrimer–CD Conjugates

The use of dendrimers in biomedical applications such as drug delivery devices and gene transfection vectors has been reviewed (see, e.g., [42–44] and references cited therein). Starburst poly(amidoamine) (PAMAM) dendrimers (Fig. 8) have been proposed as nonviral vectors. However, dendrimers with a low generation (generations 1 to 3) do not show efficient gene transfer activity, whereas those with a higher generation have sufficient gene transfer activity but exhibit too high cytotoxicity. To improve the transfection efficiency, Arima et al. proposed PAMAM dendrimer (generation 2, G2) conjugates with α-, β- or γ-CDs [45] (CDE conjugates with molar ratio CD/dendrimer 1 : 1). The results of gel electrophoresis indicated that all CDE conjugates formed complexes with plasmid DNA in the same manner as PAMAM dendrimers; the CD moiety is not involved in the complexation.

In any case, α-CDE conjugates showed the greatest transfection efficiency (around 100 times higher than those of PAMAM dendrimer alone and of the physical mixture PAMAM/CD), also superior to that of the commercial agent Lipofectin. At a high charge ratio, α-CDE conjugates affect the intracellular trafficking, and confocal fluorescence microscopy experiments showed that α-CDE conjugates change the distribution of DNA into the cells (probably due to an increase in the DNA release from the endosomes into the cytoplasm). Uekama's group also prepared α-CDE conjugates (G2) bearing mannose [46,47] (man-α-CDE) or galactose [48] (gal-α-CDE) with various degrees of substitution of the sugar moiety, since mannose and galactose moieties were proposed to target specific cells. The complexation ability of man-α-CDE and gal-α-CDE conjugates with plasmid DNA decreases with an increase in the substitution degree, due to a decrease in the number of primary amino groups. Therefore, there is an optimal degree of substitution of mannose residues for sufficient gene

transfer [46,48]. Man-α-CDE conjugate (with a substitution degree of 3.3) provided gene transfer activity higher than dendrimer or α-CDE conjugate in in vivo experiments, even though man-α-CDE did not show efficient DNA compaction, as revealed by ethidium bromide fluorescence experiments [47]. Additionally, Kihara et al. studied α-CDE conjugates with dendrimers of different generations (G3 and G4) [49]. α-CDE conjugate (G3) possesses the greatest gene transfer activity with the lowest cytotoxicity among the α-CDE conjugates. The degree of CD substitution of an α-CDE conjugate also affects the gene transfer activity, and it was shown that an average of 2.4 CD for a α-CDE conjugate of generation 3 gave the best results in terms of gene transfer efficiency and cytotoxicity [50], both in vitro and in vivo.

Finally, Uekama's group proposed α-CDE conjugate (G3) with a degree of CD substitution of 2.2 and with various degrees of substitution of mannose moieties (5 to 20) [51]. The most promising α-CDE conjugate has a degree of CD and a mannose substitution of 2.2 and 10, respectively. The enhancing effect of this vector on gene transfer activity in four different cell lines (deduced from luciferase expression at a charge ratio of 50 : 1) is unlikely to be dependent on cell surface mannose-binding receptors. On the other hand, this vector complexed plasmid DNA at a charge ratio of 1 : 1 (gel electrophoresis experiments) and showed a nearly zero value of the zeta potential even at a charge ratio of 50 : 1. This is probably at the origin of the serum-resistant gene transfer activity observed because the serum components bind preferentially to cationic complexes. No plasmid DNA compaction was observed (revealed by ethidium bromide fluorescence experiments), and additional studies suggested efficient endosome-escaping abilities, but also the nuclear translocation ability to account for the efficient gene transfer activity.

Regarding the remarkable aspects as a gene delivery carrier, Tsutsumi et al. recently proposed the use of one of the α-CDE conjugates (G3, degree of CD substitution of 2.4) as a novel carrier for small interfering RNA (siRNA) [52] and short hairpin RNA(shRNA) [53] expressing plasmid DNA.

Other dendrimers, such as polypropylenimine dendrimers (DABs), have also been proposed as nonviral vectors. DAB dendrimers with generations above 3, as PAMAM dendrimers, exhibit too high a level of cytotoxicity, limiting their transfection efficiency significantly. Zhang et al. [54] synthesized low-generation DAB (G2, eight terminal amino groups) linked to an average of 1.7 β-CD. This new vector provided a great capacity for DNA binding (as shown by gel electrophoresis, zeta potential, and TEM experiments), low cytotoxicity, and much higher transfection efficiency than those of the parent DAB dendrimer.

A dendrimer-like oligo(ethylenediamino)-β-CD-modified gold nanoparticle has been synthesized and used to complex DNA [55]. An average of 37 β-CDs surrounded the gold nanoparticle (around 4 nm in size), and DNA

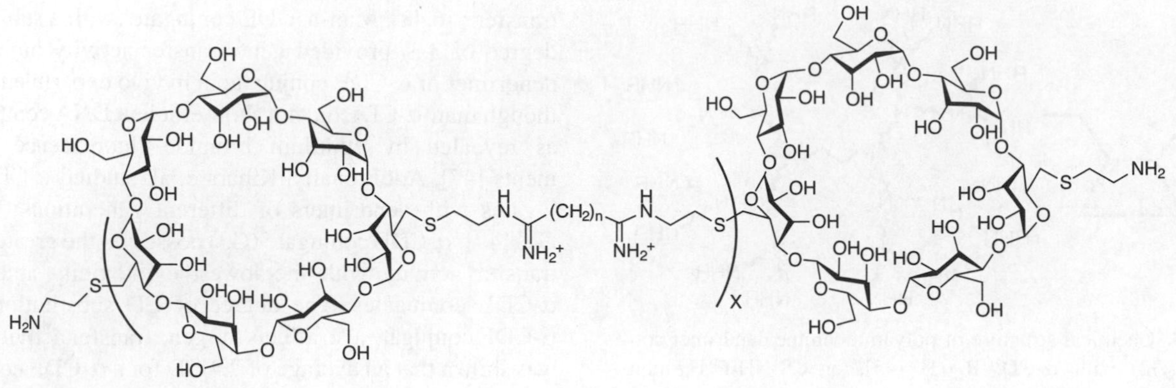

Figure 9. Chemical structure of a linear cationic β-CD copolymers, β-CDPn. (From [56].)

aggregation was shown by absorption, circular dichroism, and TEM experiments. Compared to the commercial transfection agent lipofectin, the cytotoxicity is lower but the transfection efficiency has to be improved.

4.2. Linear Cationic CD Oligomers and Polymers

The first example of linear cationic polymers containing β-CD in the polymer backbone for gene delivery applications was reported by Gonzalez et al. [56] in 1999. Copolymers with different structures were synthesized by copolymerizing difunctional β-CDs with other difunctionalized monomers. One of these structures is shown Fig. 9. The effect of the nature of the spacers [56] was studied, and the effect of the length of the spacer [57] was detailed for a series of β-CDP*n* copolymers, with *n* varying between 4 and 10. The spacer length slightly influences the size of DNA/β-CDP*n* polyplexes, but strongly affects the transfection efficiencies (determined by the luciferase protein activity) and the cytotoxicity. At a high N/P ratio of 50, β-CDP8 showed almost no cytotoxicity, whereas β-CDP6 showed the highest in vitro transfection efficiency among the vectors series. It was suggested that at this high N/P ratio, the free copolymers in solution are mainly responsible for the toxicity. On the other hand, heparin sulfate displacement studies revealed that β-CDP6 demonstrated the highest binding constant with DNA. A possible rationalization is that the spacing between the cationic amidine groups in β-CDP6 is optimal for DNA binding.

Davis's group also studied the effects of the carbohydrate size and its distance from the charge center [58,59] and the structural effect of the charge center type [60]. They confirmed that the toxicity increases as the DNA-binding charge center is farther removed from the carbohydrate moiety within the polycation backbone. Increasing the size of the carbohydrate moiety decreases the cytotoxicity, the β-CD polycations display lower toxicity than the trehalose polycations, and the absence of a carbohydrate moiety produces high toxicity. Additionally, the authors showed that β- and

γ-CD-containing polycations give an almost similar level of gene expression, with slightly lower toxicity for the γ-CD polycations. A change of the amidine group to a quaternary ammonium in similar polycations does not influence the toxicity. However, the amidine polycations exhibit higher gene expression than the quaternary ammonium analogs, due to the inability of the quaternary ammonium–based polyplexes to escape from endosomes (results based on in vitro experiments done in the presence of chloroquinine).

Modification of β-CDP6 with terminal imidazole groups (CDPim) improves the transfection efficiency at a low charge ratio without an increase in toxicity, the imidazole group introducing intracellular pH-buffering activity to the gene delivery vector [61], as reported for other systems. In any case, after investigation by various methods [62], it is still unclear if the improved transfection efficiency is a result of enhanced endosomal escape, as CDPim also generates greater amounts of unpackaged intracellular DNA than does β-CDP6.

An important feature of these gene delivery systems is that the polyplexes resulting from the association of the CD-containing cationic polymer and DNA might be modified further by inclusion complex formation. Davis's group reported postcomplexation pegylation of the β-CDP6/DNA polyplexes using poly(ethylene oxide)–adamantane conjugate. Targeting ligands such as galactose [63] or transferrin [64] were also added to the adamantane conjugates. Figure 10 depicts schematically surface modification of the polyplexes, involving the use of an adamantane group as an anchor.

Interestingly, post-DNA-complexation pegylation (through PEG–AD) introduced salt and serum stability to the polyplexes, as shown by DLS and electron microscopy experiments [63]. The salt-induced aggregation is prevented by a steric layer surrounding the particles. Additionally, an anionic galactosylated adamantane compound was used to target β-CDP6-based polyplexes to hepatocytes. Galactose-mediated targeting of the particles was demonstrated without any increase in cytotoxicity [63]. Among several conjugates, monofunctionalized transferrin adamantane conjugate

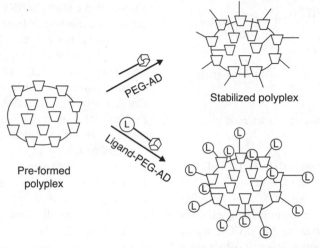

Figure 10. Post-DNA-complexation pegylation by inclusion complex formation. (From [63].)

(Tf–PEG–AD), with a lysine conjugation, has been shown to retain high receptor binding efficiency (tested on PC-3 and K562 cell lines). Transferrin-modified particles were therefore formulated by mixing β-CDP6–Imid/PEG–AD/Tf-PEG–AD/DNA. The ratio of the different polymers was optimized in order to get nanoparticles still positively charged, stable in the presence of salt, and leading to efficient in vitro transfection (determined by luciferase protein activity). These multiple components systems were proposed to deliver not only plasmid DNA but also siRNA and DNAzymes both in vitro and in vivo [65,66].

Recently, another family of linear polymers containing β-CDs in the main chain was obtained via "click reaction" with high molecular weight (up to 331 kDa) and was proposed as a gene delivery vector [67] (Fig. 11). Two series of polymers were tested to determine the effect of amine stoichiometry and polymer length on delivery efficiency and toxicity. All the polymers could condense pDNA into nanoparticles at a low N/P ratio of 2, but efficient protection of pDNA against DNAse occurred only with polymers that had a high amine stoichiometry. The transfection efficiency (determined by luciferase protein activity on HeLa cells) increased with the amine stoichiometry, and showed an optimum with the polymer length; this is due partly to the

increase in toxicity with polymer length. But even if all the polymers revealed higher cellular uptake than jet-PEI, their transfection efficiencies were always lower. The authors suggested that endosomal escape and/or transfection kinetics could influence the transgene expression.

Menuel et al. recently proposed a bis(guanidinium)tetrakis(β-CD) tetrapod [68], the first example of a new host family as a potential gene delivery system. The tetrapod has β-CD cavities regularly distributed around a central skeleton possessing a predefined number of cationic guanidinium centers (Fig. 12). Supramolecular 1 : 1 complexes were formed between the tetrapod and single-stranded DNA and siRNA, as shown by capillary electrophoresis. The efficiency of siRNA transfection in MRC-5 cells was comparable to that of PEI, with lower toxicity.

4.3. CD-Modified Poly(ethylenimine) and Polylysine

Poly(ethylenimine) (PEI) with high molecular weight is a cationic polymer in wide use in nonviral gene delivery. Its high in vitro efficiency, which is often used for comparison with other systems, is speculated to be due partly to enhanced endosomal escape via the proton sponge effect. Nevertheless, the significant cytotoxicity of PEI limited its use. Pun

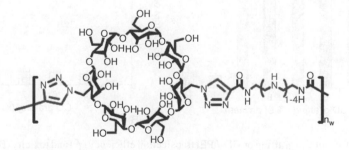

Figure 11. Structure of CD polymers (Cdn_w). (From [67].) (*See insert for color representation of the figure.*)

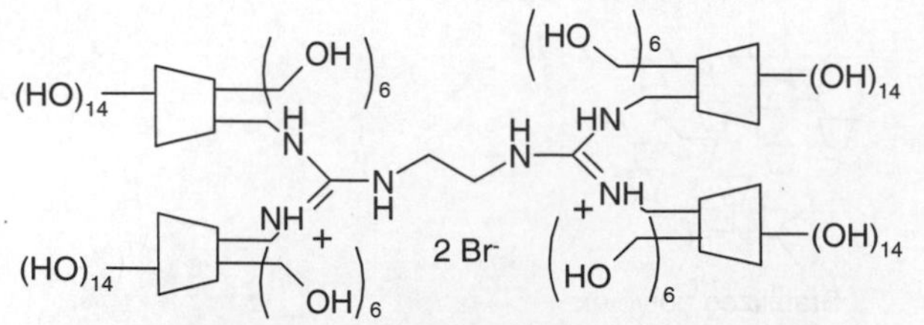

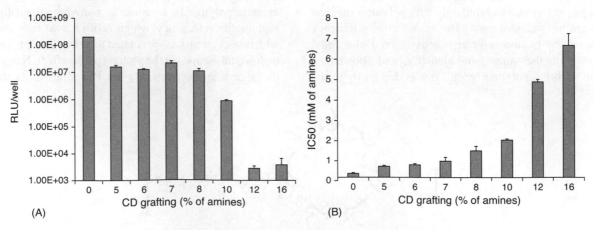

Figure 12. Chemical structure of CD tetrapod.

et al. [69] modified linear (*l*) and branched (*b*) PEI (25 kDa) to merge the beneficial qualities of the CD-based polymers (low toxicity and modification of the polyplexes by complex formation, as described earlier) with the interesting efficient transfection property of PEI. Increasing the β-CD grafting ratio of PEI reduces the transfection efficiency (determined by luciferase protein activity) but decreases the toxicity (determined by MTT cell viability assay), as shown in Fig. 13.

The best compromise activity, toxicity, was obtained for CD–*b*PEI and CD–*l*PEI with 8 and 12% CD grafting, respectively. The ability of these copolymers to complex and deliver DNA was compared with the parent *b*PEI and *l*PEI at an N/P ratio of 10. CD–PEI polymers delivered plasmids with higher efficiency than PEI, as shown by flow cytometry analysis using labeled plasmids. The transfection efficiencies were compared using the expression of EGFP in an analysis by flow cytometry. The transfection efficiencies were similar (CD–*b*PEI) or higher (CD–*l*PEI) than that of the parent polymer and were enhanced in the presence of chloroquinine (an endosomal buffering agent), whereas choroquinine did not enhance the transfection efficiency of PEIs. The authors could conclude that there were differences in the intracellular behaviors of CD–PEIs and PEIs. Formulation of the polyplexes in the presence of PEG–AD led to stable particles in physiological salt solutions. Then, in vivo experiments were run in mice. Blood analysis and histological evaluation

showed that CD–*l*PEI/PEG–AD/DNA was well tolerated and provided a method of liver gene expression.

Similarly, Forrest et al. [70] synthesized and used β-CD polyethylenimine conjugates for targeted in vitro gene delivery. They found that CD–*b*PEI polymers induced higher transfection efficiency through luciferase protein activity determinations on HEK293 cells (around fourfold) and lower cytotoxicity than that of unmodified *b*PEI (25 kD). Human insulin was derivatized with a hydrophobic palmitate group (pal-HI) which could bind to the polyplexes and target the insulin receptor of many cell groups. The addition of pal-HI to the polyplexes enhanced gene expression by more than one order of magnitude compared to unmodified *b*PEI, possibly by facilitating internalization via receptor-mediated endocytosis.

As low-molecular-weight PEIs displayed much less toxicity (but poor transfection efficiency), several groups have tried to cross-link low-molecular-weight PEI via CDs to get high-molecular-weight copolymers. A β-CD containing PEI polymer [71] (61 kDa) was synthesized by linking low-molecular-weight PEI (0.6 kDa) with β-CDs activated by 1,1′-carbonyldiimidazole. This copolymer has been proposed for gene transfer into the nervous system. Besides its degradability in the physiological medium, the CD–PEI copolymer showed much lower toxicity (measured by cell viability assays in neurons) than that of PEI (25 kDa). Size, zeta potential, and AFM measurements indicated that the copolymer could condense plasmid DNA at an N/P ratio higher than 20. In vitro transfection efficiency close to that offered by PEI (25 kDa) was determined by luciferase protein activity, and in vivo tests indicated that after intrathecal injection in the rat, transfected cells were mostly neuroectodermal cells, astrocytes, and microglia.

Similarly, low-molecular-weight PEI (0.6 kDa) was also cross-linked with (2-hydroxypropyl)-α- [72], -β-, and -γ-CD [73]. All three polymers exhibited lower toxicity than that of PEI (25 kDa), and in vitro transfection efficiency was optimal for N/P ratios of 50, 300, and 40 for α-, β-, and γ-CD/

Figure 13. Effect of CD grafting on CD-*b*PEI transfection efficiency (A) and toxicity (B) to PC3 cells. (From [69].)

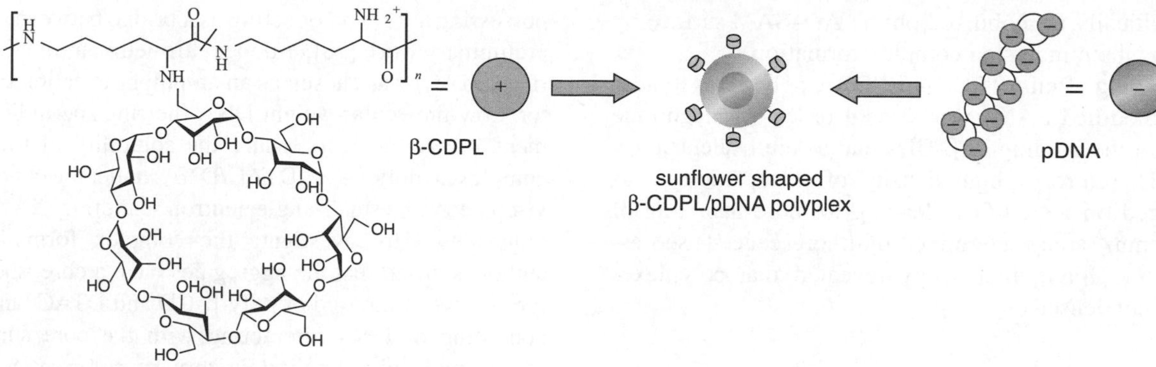

β-CDPL

sunflower shaped
β-CDPL/pDNA polyplex

pDNA

Figure 14. Sunflower-shaped polyplex formation. (From [74].) (*See insert for color representation of the figure.*)

PEI, respectively, and was close or even higher (for α-CD/PEI in bovine serum) than that of PEI (25 kDa).

Choi et al. [74] modified poly(ε-lysine) by a coupling reaction with 6-deoxy-6-monoaldehyde-β-cyclodextrin. The copolymer obtained (β-CDPL) could condense plasmid DNA, and confocal-laser scanning microscopy suggested a "sunflower"-shaped polyplex, the β-CD cavities being located at the outer surface of the polyplex (Fig. 14). At an N/P ratio of 10, the size and surface charge of the polyplexes depend on pH. At pH 6.0, polyplexes are tightly packed and slightly positive (the secondary amines being protonated), whereas polyplexes are negative at pH 7.4. This should suggest that the condensed polyplexes in the acidic endosomes will be weakened after release into the cytoplasm, affecting the intracellular trafficking of the polyplexes within cells. Efficient cellular uptake of β-CDPL/pDNA polyplexes (nearly one order of magnitude compared to free pDNA) was evidenced by both microscopy and fluorescence-based flow cytometry. Confocal-laser scanning microscopy also showed that the polyplexes escaped extensively from the endosome or lyosome. Transfection efficiency (determined by luciferase protein activity) was one order of magnitude higher than that of *l*PEI (25 kDa), with significantly lower cell viability. The authors suggested that outward-facing β-CD in the polyplex should promote the removal of cholesterol from the cell membrane via CD complex formation, inducing local membrane disturbances and assisting pDNA transfer into cells. Furthermore, the secondary amines could promote a proton-sponge effect, enhancing pDNA transfection.

Polyelectrolyte multilayers films (PEMs) have been used as vectors for polymer-precomplexed DNA. Recently, a β-CD-modified poly(ε-lysine) copolymer (PLL–CD) was reported [75] to complex pDNA, and the polyplexes formed were further incorporated into multilayered films made of hyaluronic acid (HA) and poly-(L-lysine) (PLL). PLL/pDNA and PLL–CD/pDNA polyplexes were prepared at an N/P ratio of 3. Formation of the multilayered films, (PLL–HA)$_5$–pDNA complexes–(PLL–HA)$_5$, was followed by quartz crystal microbalance. Transfection efficiencies obtained with

the polyplexes embedded in the multilayers were higher than those determined in solution, with a slightly better cell viability. Using fluorescent markers to label the pDNA and endosomes, the authors showed by confocal microscopy that the PLL–CD/pDNA polyplexes were not internalized by endocytosis when they were delivered from the multilayer system. Moreover, the green fluorescence of labeled pDNA was observed in the cytoplasm and in the nucleus. The nonendocytic intracellular pathway should contribute to the higher transfection efficiency.

Park et al. [76] developed a method for immobilizing gene delivery vehicles onto solid surfaces by inclusion complexes. Polyplexes were made of β-CD-modified linear PEI (CD–PEI) and DNA at an N/P ratio of 5; the nanoparticles formed were characterized by size and zeta-potential measurements. Amine-terminated, self-assembled monolayers (SAMs) on a gold surface were converted to adamantane-modified SAMs and characterized by x-ray photoelectron microscopy. The interaction of the polyplexes with SAMs was investigated by surface plasmon resonance spectroscopy. As represented schematically in Fig. 15, the CD–PEI/DNA nanoparticles

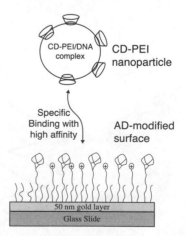

Figure 15. Cyclodextrin nanoparticle immobilization on adamantane-modified surfaces by inclusion complex formation. (From [76].)

were specifically immobilized on the AD–SAM surface by CD–adamantane inclusion complex formation.

The binding affinity of CD–PEI/DNA polyplexes to adamantane-modified SAMs was several orders of magnitude higher than the binding of β-CDs due to multivalent interactions. Therefore, a high density of nanoparticles was immobilized on the surface, leading to more than 1 ng of DNA per mm^2 area (determined by fluorescence-based assay). Atomic force microscopy revealed that polyplexes remained condensed.

4.4. Neutral CD Polymer

Among different systems described in a patent [77], Jon Wolff proposed the use of a polyion formed by an inclusion complex between amphipatic molecules, positively or negatively charged, and neutral β-CD polymer either to compact DNA or to be added to particles of DNA and poly-L-lysine. Examples were given with 4-*t*-butylbenzoic acid, 1-adamantanamine, and oleylamine. Particles around 100 nm in size were characterized, and in vitro transfection and in vivo expression were reported briefly.

Amiel's group [78–81] developed a DNA vector based on a macromolecular association obtained by inclusion complex formation between a neutral epichlorohydrin/β-CD copolymer (poly-β-CD) and an amphiphilic cationic connector. The resulting reversible polycation forms a polyplex with DNA by electrostatic interactions. The charge density of the vector can be controlled easily by the simple addition of a connector, and the polyplex characteristics may be modified by changing the connector. The interaction scheme is presented in Fig. 16.

The poly-β-CD used in these works was obtained by polycondensation of β-CD with epichlorohydrin. The copolymer has a branched structure where β-CDs are modified by poly(2-hydroxypropyl)ether sequences of different lengths,

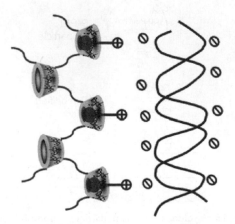

Figure 16. Ternary association between polyβCD, an amphiphilic cationic connector and DNA.

possessing a free end or acting as a bridge between CDs. In a preliminary work [78], *n*-dodecyltrimethylammonium chloride (DTAC) was chosen as an amphiphilic cationic connector. Low-molecular-weight DNA (herring sperm DNA fragments) were taken to ensure the solubility of the ternary complexes (poly-β-CD/DTAC/DNA) under the conditions of viscometry and small-angle neutron scattering (SANS) techniques used to investigate the complex formation. The authors showed that the aggregates have a core–shell structure: a core composed of poly-β-CD and DTAC, and a shell consisting of DNA interacting with the core surface via DTAC molecules implied in ternary complex formation. DTAC acts as a link between poly-β-CD and DNA, through inclusion (DTAC/poly-β-CD) and electrostatic (DTAC–DNA) interactions.

Then a series of nonsurfactant connectors were synthesized and tested [79–81]. They consist of a hydrophobic part [an adamantyl group (Ada)], a spacer with different chemical structure and length, and a polar head group, mono- or bicationic, as shown in Fig. 17. The affinity of the connectors to poly-β-CD, determined by fluorimetric titration, was shown to be strongly dependent on the steric hindrance generated by the spacer and polar head groups, which limit the entrance of the adamantyl groups into the CD cavities already embedded in a polymer structure. The affinities are thus in the order Ada4 < Ada2 < Ada3 < Ada5. The poly-β-CD/Ada/DNA polyplexes were usually prepared at a constant quantity of CD cavities (20 equiv) and DNA (1 equiv), with various amounts of Ada (ranging from 1 to 10 equiv). Size and zeta-potential measurements showed that all the poly-β-CD/Ada vectors could complex plasmid DNA. Moreover, agarose gel electrophoresis indicated that full retardation was obtained at low N/P ratios (∼5 for Ada2 and Ada3, ∼2 for Ada4, ∼1 for Ada5). The results were corroborated by surface-enhanced Raman spectroscopy (SERS) experiments, which monitored the accessibility of adenyl residues to silver colloids. The SERS signal intensity decreased with N/P increase, showing that adenyl residues have a reduced accessibility as DNA is bound to the vector. The bicationic head group of Ada4 and Ada5 led to polyplexes that are less sensitive to the ionic strength of the medium than those formed with Ada2 and Ada3, as evaluated by agarose gel electrophoresis and complemented with SANS measurements. In vitro transfection experiments were performed on two cell lines, HepG2 and HEK293, and transfection efficiency was determined by luciferase protein activity. The best efficiencies were obtained with Ada4 [80] and Ada5 [81], with polyplexes formed at a poly-β-CD/Ada/DNA ratio of 30 : 60 : 1 and 30 : 40 : 1, respectively. Addition of the fusogenic JTS1 peptide led to a noticeable increase in the transfection efficiency, reaching values close to those of DOTAP for Ada4 polyplexes and even more than 10 times higher for Ada5 polyplexes on the HEK293 cell line.

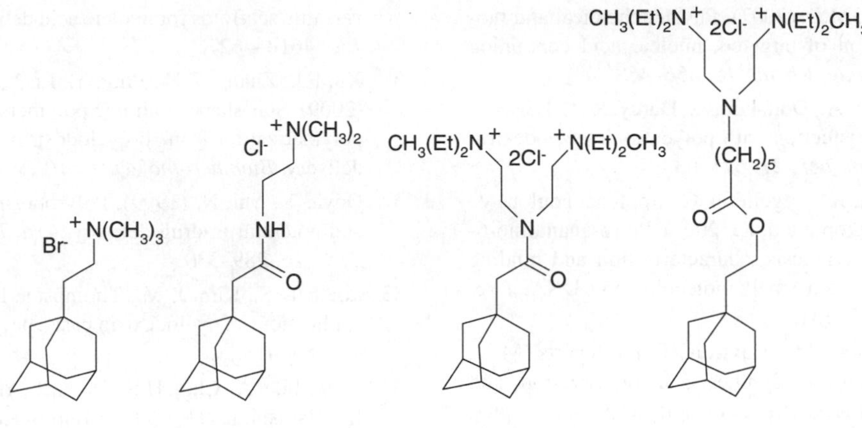

Figure 17. Chemical structure of the Ada connectors.

5. CONCLUSIONS

A large variety of architectures involving CDs have been designed to fulfill the requirements of gene delivery. Among them, it has been shown that CDs impart to these systems lower toxicities together with comparable or larger transfection efficiencies than comparable synthetic vectors free of CDs. A key feature of CD-containing vectors is that inclusion complex formation gives high flexibility to the system and allows surface modification of the vector (targeting ligands, PEGylation of the vectors).

REFERENCES

1. Li, J., Loh, X.J. (2008). Cyclodextrin-based supramolecular architectures: syntheses, structures, and applications for drug and gene delivery. *Adv. Drug Deliv. Rev.*, *60*, 1000–1017.

2. Wong, S.Y., Pelet, J.M., Putnam, D. (2007). Polymer systems for gene delivery: past, present, and future. *Prog. Polym. Sci.*, *32*, 799–837.

3. Pack, D.W., Hoffman, A.S., Pun, S., Stayton, P.S. (2005). Design and development of polymers for gene delivery. *Nat. Rev. Drug Discov.*, *4*, 581–593.

4. Blaese, R.M., Culver, K.W., Miller, A.D., Carter, C.S., Fleisher, T., Clerici, M., Shearer, G., Chang, L., Chiang, Y., Tolstoshev, P., Greenblatt, J.J., Rosenberg, S.A., Klein, H., Berger, M., Mullen, C.A., Ramsey, W.J., Muul, L., Morgan, R.A., Anderson, W.F. (1995). T lymphocyte–directed gene therapy for ADA SCID: initial trial results after 4 years. *Science*, *270*, 475–480.

5. Cavazzana-Calvo, M., Hacein-Bey, S., Basile, G.d.S., Gross, F., Yvon, E., Nusbaum, P., Selz, F., Hue, C., Certain, S., Casanova, J.-L., Bousso, P., Deist, F.L., Fischer, A. (2000). Gene therapy of human severe combined immunodeficiency (SCID)-X1 disease. *Science*, *270*, 669–672.

6. Felgner, P.L., Gadek, T.R., Holm, M., Roman, R., Chan, H.W., Wenz, M., Northrop, J.P., Ringold, G.M., Danielsen, M. (1987). Lipofection: a highly efficient, lipid-mediated DNA-transfection procedure. *Proc. Natl. Acad. Sci. USA*, *84*, 7413–7417.

7. Li, W., Szoka, F.C., Jr. (2007). Lipid-based nanoparticles for nucleic acid delivery. *Pharm Res.*, *24*, 438–449.

8. Martin, B., Sainlos, M., Aissaoui, A., Oudrhiri, N., Hauchecorne, M., Vigneron, J.-P., Lehn, J.-M., Lehn, P. (2005). The design of cationic lipids for gene delivery. *Curr. Pharm. Des.*, *11*, 375–394.

9. Aissaoui, A., Oudrhiri, N., Petit, L., Hauchecorne, M., Kan, E., Sainlos, M., Julia, S., Navarro, J., Vigneron, J.-P., Lehn, J.-M., Lehn, P. (2001). Progress in gene delivery by cationic lipids: guanidinium–cholesterol-based systems as an example. *Curr. Drug Targets*, *3*, 1–16.

10. Cserháti, T., Forgács, E., Szejtli, J. (1996). Inclusion complex formation of antisense nucleotides with hydroxypropyl-β-cyclodextrin. *Int. J. Pharm.*, *141*, 1–7.

11. Irie, T., Uekama, K. (1999). Cyclodextrins in peptide and protein delivery. *Adv. Drug Deliv. Rev.*, *36*, 101–123.

12. Redenti, E., Petra, C., Gerloczy, A., Szente, L. (2001). Cyclodextrins in oligonucleotide delivery. *Adv. Drug Deliv. Rev.*, *53*, 235–244.

13. Zhao, Q., Temsamani, J., Agrawal, S. (1995). Use of cyclodextrin and its derivatives as carriers for oligonucleotide delivery. *Antisense Res. Dev.*, *5*, 185–192.

14. Abdou, S., Collomb, J., Sallas, F., Marsura, A., Finance, C. (1997). β-Cyclodextrin derivatives as carriers to enhance the antiviral activity of an antisense oligonucleotide directed toward a coronavirus intergenic consensus sequence. *Arch. Virol.*, *142*, 1585–1602.

15. Chirila, T.V., Rakoczy, P.E., Garrett, K.L., Lou, X., Constable, I.J. (2002). The use of synthetic polymers for delivery of therapeutic antisense oligodeoxynucleotides. *Biomaterials*, *23*, 321–342.

16. deMartimprey, H., Vauthier, C., Malvy, C., Couvreur, P. (2009). Polymer nanocarriers for the delivery of small fragments of nucleic acids: oligonucleotides and siRNA. *Eur. J. Pharm. Biopharm.*, *71*, 490–504.

17. Bartlett, D.W., Davis, M.E. (2007). Physicochemical and biological characterization of targeted, nucleic acid–containing nanoparticles. *Bioconjug. Chem.*, *18*, 456–468.

18. Cryan, S.-A., Holohan, A., Donohue, R., Darcy, R., O'Driscoll, C.M. (2004). Cell transfection with polycationic cyclodextrin vectors. *Eur. J. Pharm. Sci.*, *21*, 625–633.

19. Mourtzis, N., Eliadou, K., Aggelidou, C., Sophianopoulou, V., Mavridis, I.M., Yannakopoulou, K. (2007). Per(6-guanidino-6-deoxy)cyclodextrins: synthesis, characterisation and binding behaviour toward selected small molecules and DNA. *Org. Biomol. Chem.*, *5*, 125–131.

20. Mourtzis, N., Paravatou, M., Mavridis, I.M., Roberts, M.L., Yannakopoulou, K. (2008). Synthesis, characterization, and remarkable biological properties of cyclodextrins bearing guanidinoalkylamino and aminoalkylamino groups on their primary side. *Chem. Eur. J.*, *14*, 4188–4200.

21. Croyle, M.A., Roessler, B.J., Hsu, C.-P., Sun, R., Amidon, G.L. (1998). β-Cyclodextrins enhance adenoviral-mediated gene delivery to the intestine. *Pharm. Res.*, *15*, 1348–1355.

22. Tavares, G.D., Viana, C.M., Araújo, J.G.V.C., Ramaldes, G.A., Carvalho, W.S., Pesquero, J.L., Vilela, J.M.C., Andrade, M.S., de Oliveira, M.C. (2006). Development and physico-chemical characterization of cyclodextrin–DNa complexes loaded with liposomes. *Chem. Phys. Lett.*, *429*, 507–512.

23. Jessel, N., Oulad-Abdelghani, M., Meyer, F., Lavalle, P., Haîkel, Y., Schaaf, P., Voegel, J.C. (2006). Multiple and time-scheduled in situ DNA delivery mediated by β-cyclodextrin embedded in a polyelectrolyte multilayer. *Proc. Natl. Acad. Sci. USA*, *103*, 8618–8621.

24. Cryan, S.-A., Donohue, R., Ravoo, B.J., Darcy, R., O'Driscoll, C.M. (2004). Cationic cyclodextrin amphiphiles as gene delivery vectors. *J. Drug Deliv. Sci. Technol.*, *14*, 57–62.

25. Ortega-Caballero, F., Mellet, C.O., Le Gourrierec, L., Guilloteau, N., Di Giorgio, C., Vierling, P., Defaye, J., Fernandez, J.M.G. (2008). Tailoring β-cyclodextrin for DNA complexation and delivery by homogeneous functionalization at the secondary face. *Org. Lett.*, *10*, 5143–5146.

26. McMahon, A., Gomez, E., Donohue, R., Forde, D., Darcy, R., O'Driscoll, C.M. (2008). Cyclodextrin gene vectors: cell trafficking and the influence of lipophilic chain length. *J. Drug Deliv. Sci. Technol.*, *18*, 303–307.

27. Diaz-Moscoso, A., Balbuena, P., Gomez-Garcia, M., Mellet, C.O., Benito, J.M., Le Gourrierec, L., Di Giorgio, C., Vierling, P., Mazzaglia, A., Micali, N., Defaye, J., Fernandez, J.M.G. (2008). Rational design of cationic cyclooligosaccharides as efficient gene delivery systems. *Chem. Commun.*, 2001–2003.

28. Donohue, R., Mazzaglia, A., Ravoo, B.J., Darcy, R. (2002). Cationic β-cyclodextrin bilayer vesicles. *Chem. Commun.*, 2864–2865.

29. Yang, C., Li, H., Goh, S.H., Li, J. (2007). Cationic star polymers consisting of α-cyclodextrin core and oligoethylenimine arms as nonviral gene delivery vectors. *Biomaterials*, *28*, 3245–3254.

30. Srinivasachari, S., Fichter, K.M., Reineke, T.M. (2008). Polycationic β-cyclodextrin "click clusters": monodisperse and versatile scaffolds for nucleic acid delivery. *J. Am. Chem. Soc.*, *130*, 4618–4627.

31. Xu, F.J., Zhang, Z.X., Ping, Y., Li, J., Kang, E.T., Neoh, K.G. (2009). Star-shaped cationic polymers by atom transfer radical polymerization from β-cyclodextrin cores for nonviral gene delivery. *Biomacromolecules*, *10*, 285–293.

32. Ooya, T., Yui, N. (1999). Polyrotaxanes: synthesis, structure, and potential in drug delivery. *Crit. Rev. Ther. Drug Carrier Syst.*, *16*, 289–330.

33. Loethen, S., Kim, J.-M., Thompson, D.H. (2007). Biomedical applications of cyclodextrin based polyrotaxanes. *Polym. Rev.*, *47*, 383–418.

34. Yamashita, A., Choi, H.S., Ooya, T., Yui, N., Akita, H., Kogure, K., Harashima, H. (2006). Improved cell viability of linear polyethylenimine through γ-cyclodextrin inclusion for effective gene delivery. *ChemBioChem*, *7*, 297–302.

35. Shuai, X., Merdan, T., Unger, F., Kissel, T. (2005). Supramolecular gene delivery vectors showing enhanced transgene expression and good biocompatibility. *Bioconjugate Chem.*, *16*, 322–329.

36. Ooya, T., Choi, H.S., Yamashita, A., Yui, N., Sugaya, Y., Kano, A., Maruyama, A., Akita, H., Ito, R., Kogure, K., Harashima, H. (2006). Biocleavable polyrotaxane–plasmid DNA polyplex for enhanced gene delivery. *J. Am. Chem. Soc.*, *128*, 3852–3853.

37. Yamashita, A., Kanda, D., Katoono, R., Yui, N., Ooya, T., Maruyama, A., Akita, H., Kogure, K., Harashima, H. (2008). Supramolecular control of polyplex dissociation and cell transfection: efficacy of amino groups and threading cyclodextrins in biocleavable polyrotaxanes. *J. Control. Release*, *131*, 137–144.

38. Pérès, B., Richardeau, N., Jarroux, N., Guégan, P., Auvray, L. (2008). Two independent ways of preparing hypercharged hydrolyzable polyaminorotaxane. *Biomacromolecules*, *9*, 2007–2013.

39. Yang, C., Wang, X., Li, H., Goh, S.H., Li, J. (2007). Synthesis and characterization of polyrotaxanes consisting of cationic α-cyclodextrins threaded on poly[(ethylene oxide)-*ran*-(propylene oxide)] as gene carriers. *Biomacromolecules*, *8*, 3365–3374.

40. Li, J., Yang, C., Li, H., Wang, X., Goh, S.H., Ding, J.L., Wang, D.Y., Leong, K.W. (2006). Cationic supramolecules composed of multiple oligoethylenimine-grafted β-cyclodextrins threaded on a polymer chain for efficient gene delivery. *Adv. Mater.*, *18*, 2969–2974.

41. Liu, Y., Yu, L., Chen, Y., Zhao, Y.-L., Yang, H. (2007). Construction and DNA condensation of cyclodextrin-based polypseudorotaxanes with anthryl grafts. *J. Am. Chem. Soc.*, *129*, 10656–10657.

42. Boas, U., Heegaard, P.M.H. (2004). Dendrimers in drug research. *Chem. Soc. Rev.*, *33*, 43–63.

43. Svenson, S., Tomalia, D.A. (2005). Dendrimers in biomedical applications: reflections on the field. *Adv. Drug Deliv. Rev.*, *57*, 2106–2129.

44. Kubasiak, L.A., Tomalia, D.A. (2004). Cationic dendrimers as gene transfection vectors. In: Amiji, M.M. Ed., *Polymeric Gene Delivery: Principles and Applications*. CRC Press, Boca Raton, FL.

45. Arima, H., Kihara, F., Hirayama, F., Uekama, K. (2001). Enhancement of gene expression by polyamidoamine dendrimer conjugates with α-, β-, and γ-cyclodextrins. *Bioconjugate Chem.*, *12*, 476–484.

46. Arima, H., Wada, K., Kihara, F., Tsutsumi, T., Hirayama, F., Uekama, K. (2002). Cell-specific gene transfer by α-cyclodextrin conjugates with mannosylated polyamidoamine dendrimers. *J. Inclusion Phenom. Macrocyclic Chem.*, *44*, 361–364.

47. Wada, K., Arima, H., Tsutsumi, T., Chihara, Y., Hattori, K., Hirayama, F., Uekama, K. (2005). Improvement of gene delivery mediated by mannosylated dendrimer/α-cyclodextrin conjugates. *J. Control. Release*, *104*, 397–413.

48. Wada, K., Arima, H., Tsutsumi, T., Hirayama, F., Uekama, K. (2005). Enhancing effects of galactosylated dendrimer/α-cyclodextrin conjugates on gene transfer efficiency. *Biol. Pharm. Bull.*, *28*, 500–505.

49. Kihara, F., Arima, H., Tsutsumi, T., Hirayama, F., Uekama, K. (2002). Effects of structure of polyamidoamine dendrimer on gene transfer efficiency of the dendrimer conjugate with α-cyclodextrin. *Bioconjugate Chem.*, *13*, 1211–1219.

50. Kihara, F., Arima, H., Tsutsumi, T., Hirayama, F., Uekama, K. (2003). In vitro and in vivo gene transfer by an α-cyclodextrin conjugate with polyamidoamine dendrimer. *Bioconjugate Chem.*, *14*, 342–350.

51. Arima, H., Chihara, Y., Arizono, M., Yamashita, S., Wada, K., Hirayama, F., Uekama, K. (2006). Enhancement of gene transfer activity mediated by mannosylated dendrimer/α-cyclodextrin conjugate (generation 3, G3). *J. Control. Release*, *116*, 64–74.

52. Tsutsumi, T., Hirayama, F., Uekama, K., Arima, H. (2007). Evaluation of polyamidoamine dendrimer/α-cyclodextrin conjugate (generation 3, G3) as a novel carrier for small interfering RNA (siRNA). *J. Control. Release*, *119*, 349–359.

53. Tsutsumi, T., Hirayama, F., Uekama, K., Arima, H. (2008). Potential use of polyamidoamine dendrimer/α-cyclodextrin conjugate (generation 3, G3) as a novel carrier for short hairpin RNA-expressing plasmid DNA. *J. Pharm. Sci.*, *97*, 3022–3034.

54. Zhang, W., Chen, Z., Song, X., Si, J., Tang, G. (2008). Low generation polypropylenimine dendrimer graft β-cyclodextrin: an efficient vector for gene delivery system. *Technol. Cancer Res. Treat.*, *7*, 103–108.

55. Wang, H., Chen, Y., Li, X.-Y., Liu, Y. (2007). Synthesis of oligo (ethylenediamino)-β-cyclodextrin modified gold nanoparticle as a DNA concentrator. *Mol. Pharm.*, *4*, 189–198.

56. Gonzalez, H., Hwang, S.J., Davis, M.E. (1999). New class of polymers for the delivery of macromolecular therapeutics. *Bioconjugate Chem.*, *10*, 1068–1074.

57. Hwang, S.J., Bellocq, N.C., Davis, M.E. (2001). Effects of structure of β-cyclodextrin-containing polymers on gene delivery. *Bioconjugate Chem.*, *12*, 280–290.

58. Reineke, T.M., Davis, M.E. (2003). Structural effects of carbohydrate-containing polycations on gene delivery: 1. Carbohydrate size and its distance from charge centers. *Bioconjugate Chem.*, *14*, 247–254.

59. Popielarski, S.R., Mishra, S., Davis, M.E. (2003). Structural effects of carbohydrate-containing polycations on gene delivery: 3. Cyclodextrin type and functionalization. *Bioconjugate Chem.*, *14*, 672–678.

60. Reineke, T.M., Davis, M.E. (2003). Structural effects of carbohydrate-containing polycations on gene delivery: 2. Charge center type. *Bioconjugate Chem.*, *14*, 255–261.

61. Davis, M.E., Pun, S.H., Bellocq, N.C., Reineke, T.M., Popielarski, S.R., Mishra, S., Heidel, J.D. (2004). Self-assembling nucleic acid delivery vehicles via linear, water-soluble, cyclodextrin-containing polymers. *Curr. Med. Chem.*, *11*, 179–197.

62. Mishraa, S., Heidela, J.D., Webster, P., Davis, M.E. (2006). Imidazole groups on a linear, cyclodextrin-containing polycation produce enhanced gene delivery via multiple processes. *J. Control. Release*, *116*, 179–191.

63. Pun, S.H., Davis, M.E. (2002). Development of a nonviral gene delivery vehicle for systemic application. *Bioconjugate Chem.*, *13*, 630–639.

64. Bellocq, N.C., Pun, S.H., Jensen, G.S., Davis, M.E. (2003). Transferrin-containing, cyclodextrin polymer–based particles for tumor-targeted gene delivery. *Bioconjugate Chem.*, *14*, 1122–1132.

65. Kulkarni, R.P., Mishra, S., Fraser, S.E., Davis, M.E. (2005). Single cell kinetics of intracellular, nonviral, nucleic acid delivery vehicle acidification and trafficking. *Bioconjugate Chem.*, *16*, 986–994.

66. Bartlett, D.W., Davis, M.E. (2006). Insights into the kinetics of siRNA-mediated gene silencing from live-cell and live-animal bioluminescent imaging. *Nucleic Acids Res.*, *34*, 322–333.

67. Srinivasachari, S., Reineke, T.M. (2009). Versatile supramolecular pDNA vehicles via "click polymerization" of β-cyclodextrin with oligoethyleneamines. *Biomaterials*, *30*, 928–938.

68. Menuel, S., Fontanay, S., Clarot, I., Duval, R.E., Diez, L., Marsura, A. (2008). Synthesis and complexation ability of a novel bis-(guanidinium)-tetrakis-(β-cyclodextrin) dendrimeric tetrapod as a potential gene delivery (DNA and siRNA) system: study of cellular siRNA transfection. *Bioconjugate Chem.*, *19*, 2357–2362.

69. Pun, S.H., Bellocq, N.C., Liu, A., Jensen, G., Machemer, T., Quijano, E., Schluep, T., Wen, S., Engler, H., Heidel, J., Davis, M.E. (2004). Cyclodextrin-modified polyethylenimine polymers for gene delivery. *Bioconjugate Chem.*, *15*, 831–840.

70. Forrest, M.L., Nathan, G., Daniel, W.P. (2005). Cyclodextrin–polyethylenimine conjugates for targeted in vitro gene delivery. *Biotechnol. Bioeng.*, *89*, 416–423.

71. Tang, G.P., Guo, H.Y., Alexis, F., Wang, X., Zeng, S., Lim, T.M., Ding, J., Yang, Y.Y., Wang, S. (2006). Low molecular weight polyethylenimines linked by β-cyclodextrin for gene transfer into the nervous system. *J. Gene Med. 8*, 736–744.

72. Huang, H., Yu, H., Li, D., Liu, Y., Shen, F., Zhou, J., Wang, Q., Tang, G. (2008). A novel co-polymer based on hydroxypropyl α-cyclodextrin conjugated to low molecular weight polyethylenimine as an in vitro gene delivery vector. *Int. J. Mol. Sci.*, *9*, 2278–2289.

73. Huang, H., Tang, G., Wang, Q., Li, D., Shen, F., Zhou, J., Yu, H. (2006). Two novel non-viral gene delivery vectors: low molecular weight polyethylenimine cross-linked by (2-

hydroxypropyl)-β-cyclodextrin or (2-hydroxypropyl)-γ-cyclodextrin. *Chem. Commun.* 2382–2384.

74. Choi, H.S., Yamashita, A., Ooya, T., Yui, N., Akita, H., Kogure, K., Ito, R., Harashima, H. (2005). Sunflower-shaped cyclodextrin-conjugated poly(ε-lysine) polyplex as a controlled intracellular trafficking device. *ChemBioChem*, *6*, 1986–1990.

75. Zhang, X., Sharma, K.K., Boeglin, M., Ogier, J., Mainard, D., Voegel, J.-C., Mély, Y., Benkirane-Jessel, N. (2008). Transfection ability and intracellular DNA pathway of nanostructured gene-delivery systems. *Nano Lett.*, *8*, 2432–2436.

76. Park, I.-K., von Recum, H.A., Jiang, S., Pun, S. H. (2006). Supramolecular assembly of cyclodextrin-based nanoparticles on solid surfaces for gene delivery. *Langmuir*, *22*, 8478–8484.

77. Wolff, J.A. (1999). Compositions and methods for drug delivery using amphiphile binding molecules. Patent WO 01/37665 A1.

78. Galant, C., Amiel, C., Auvray, L. (2005). Ternary complex formation in aqueous solution between a β-cyclodextrin polymer, a cationic surfactant and DNA. *Macromol. Biosci.*, *5*, 1057–1065.

79. Burckbuchler, V., Wintgens, V., Lecomte, S., Percot, A., Leborgne, C., Danos, O., Kichler, A., Amiel, C. (2006). DNA compaction into new DNA vectors based on cyclodextrin polymer: surface enhanced Raman spectroscopy characterization. *Biopolymers*, *81*, 360–370.

80. Burckbuchler, V., Wintgens, V., Leborgne, C., Lecomte, S., Leygue, N., Scherman, D., Kichler, A., Amiel, C. (2008). Development and characterization of new cyclodextrin polymer–based DNA delivery systems. *Bioconjugate Chem.*, *19*, 2311–2320.

81. Wintgens, V., Leborgne, C., Baconnais, S., Burckbuchler, V., Le Cam, E., Scherman, D., Kichler, A., Amiel, C.Inclusion of an amphipathic imidazole connector improves the transfection efficiency of a cyclodextrin polymer–based system (to be published).

13

TARGETED CYCLODEXTRINS

STEFANO SALMASO
Department of Pharmaceutical Sciences, University of Padua, Padua, Italy

FABIO SONVICO
Department of Pharmacy, University of Parma, Parma, Italy

1. INTRODUCTION

Among the different approaches aimed at drug delivery, targeting has always exerted a potent fascination. The idea of delivering a bioactive substance in a specific way to the site of action, the cells affected by the disease, or even to the subcellular compartment where the drug reaches its molecular target comes from the "magic bullet" concept of Paul Ehrlich. At the very dawn of the era of medicinal chemistry, he envisaged a specific targeting of the therapeutically active substance to the body region affected as the ideal treatment for any disease [1].

The successful targeting of the pharmaceutical active ingredient by conjugation to a targeting moiety allows for higher efficacy, milder side effects, and dose reduction as a consequence of the improved specificity of the treatment. From the very beginning, targeting has been proposed and investigated for the improvement and optimization of anticancer chemotherapies, generally associated with severe side effects [2].

Issues related to targeting are far more complex and challenging than the simplifications used for describing the mechanisms, in particular because of random distribution of the targeted drug delivery system and of the probability of interaction with its target [3]. However, once targeting has been achieved, the delivery system may penetrate cells, providing cellular accumulation that does not take place for untargeted systems. In addition, the discovery of new biological targets is expected to provide

further control on the efficiency and specificity of the delivery process [4].

Since the 1970s, cyclodextrins (CDs) and their derivatives, in virtue of their peculiar inclusion properties, have been proposed as pharmaceutical excipients and have found application in several marketed products. They are deemed unique products since natural and semisynthetic CDs can modulate the physical, chemical, and biopharmaceutical properties of guest molecules, eventually ameliorating critical drug properties such as water solubility, dissolution rate, stability, and bioavailability [5]. In addition, CDs have been studied as functional excipients in controlled-release systems [6]. The advances in chemical strategies for CDs derivatization, along with the exciting results obtained with site-selective colloidal drug carriers (i.e., liposomes and nanoparticles) has led to the development of targeted CDs. In recent years, CDs have been regarded not only as simple functional excipients but rather as multifunctional supramolecular carriers. Two requirements must be fulfilled to obtain such targeted CDs: (1) conjugation of targeting moieties on the CD scaffold, without impairing their specific recognition of cellular targets; and (2) adequate complexation of the guest molecule, which has to be transported efficiently to the specific disease site [7].

In this chapter, a critical overview of some of the most interesting approaches aimed at developing targeted CDs is presented. Particular emphasis is dedicated to tumor targeting, a field where selective treatments are required and expectations from targeted therapies are extremely high.

Cyclodextrins in Pharmaceutics, Cosmetics, and Biomedicine: Current and Future Industrial Applications, First Edition. Edited by Erem Bilensoy.
© 2011 John Wiley & Sons, Inc. Published 2011 by John Wiley & Sons, Inc.

We also review a number of other modified CDs developed as molecular carriers designed to attain tissues, organs, and body regions where more specific and efficient treatment is required.

Taking into account the latest evolution of semisynthetic CDs, applications of targeted CDs will increase in the near future. The new opportunities offered by biotechnology, in terms of both biological targets and potent but fragile bioactive molecules, are expected to sustain evolution in the pharmaceutical field, with targeted delivery systems playing a pivotal role.

2. TUMOR TARGETING

In recent years, significant advances have been made in cancer therapy [8], with nanotechnology and biotechnology playing a pivotal role in this evolution. Scientists have investigated two major fields: the development of new molecules addressed toward new pharmacological targets [9], and the site-selective delivery of drug-loaded nanosystems by targeting receptors or antigens overexpressed in a tumor. A few studies have shown interesting results by combining both strategies. Limitations in the therapeutic approach are mainly ascribable to an inability to design adequate drugs that would specifically interfere with a selected molecular or genetic target.

Site-selective drug delivery has been extensively investigated to target tumor cells displaying specific receptors and antigens on their surface. Enhanced site specificity and internalization can enhance the efficacy of a therapeutic treatment and decrease the secondary effects that are typical of chemotherapy. Some research groups have emphasized the pivotal role of active tumor cell targeting and uptake aimed at accessing molecular targets [10]. Drug delivery systems with cell-penetrating properties are often required to deliver effectively anticancer drugs that poorly permeate the cell membrane, such as oligonucleotides and biotherapeutics [11].

Cell-specific expression or overexpression of defined targets are essential for exploitation in selective drug delivery. So far, a few biological targets have been investigated for addressing drug delivery systems to tumor tissues:

1. Vascular endothelial growth factor receptor (VEGFR), a cytokine receptor overexpressed during neoangiogenesis in tumor tissues. Few delivery systems have been developed for VEGFR targeting.
2. $\alpha_v\beta_3$ integrin, highly expressed in neovasculature and involved in the signaling pathway leading to endothelial cell migration [12]. $\alpha_v\beta_3$ integrin has been targeted using nanosystems decorated with an arginine–glycine–aspartic acid sequence (RGD) or chemical entities mimicking its structure [13].
3. Vascular cell adhesion molecule-1 (VCAM-1), a transmembrane glycoprotein, absent on normal vasculature and expressed mostly during neoangiogenesis. Anti-VCAM-1 immunodecorated delivery systems have been developed.
4. Metalloproteineases (MMPs), a family of endopeptidases enrolled in degradation of the extracellular matrix [14]. These endopeptidases are extremely active in tissues undergoing neomorphogenesis as tumors. Metalloproteinases have been targeted by anti-MMP monoclonal antibodies to direct drug-loaded nanosystems [15].
5. Epidermal growth factor receptor (EGFR) and human epidermal receptor-2 (HER-2), overexpressed in 30% of solid tumors. These cytokines participate in the signaling pathway of growth and proliferation in tissues [16]. EGFR has been targeted by the monoclonal antibody Cetuximab (Imclone Systems, New York, NY and Bristol-Myers Squibb, Princeton, NJ) currently in the market as a therapeutic agent and under investigation as targeting tool for addressing nanosystems [17]. Anti-HER-2 antibody trastuzumab (Herceptin, Genentech, South San Francisco, CA) is used for the delivery of a variety of nanoparticulate systems [18].
6. Transferrin receptor, overexpressed on metastatic and drug-resistant cells compared to normal cells.
7. Folate receptor, upregulated on a variety of cancer cells compared to normal tissue.
8. Luteinizing hormone–releasing hormone (LHRH) receptor, a target that has been widely investigated for the diagnosis and treatment of localized and metastasized breast cancer.
9. Prostate-specific membrane antigen (PSMA), identified as the main target for prostate cancer treatment [19]. Recently, polymeric biodegradable nanoparticles decorated with aptamers recognizing the PSMA, have shown promising results in xenograft tumor models [20].

So far, only some of the biological targets mentioned above have been considered for obtaining CDs with active targeting properties aimed at site-specific delivery of chemotherapeutics or biotechnological drugs for the treatment of cancer.

2.1. Folate-Mediated Targeting

Folic acid is a vitamin essential for cell function and contributes a biosynthetic cycle of purines and pyrimidines. Folate has very low cell membrane permeability; thus, cells have developed two transport mechanisms for this vitamin. In normal cells, a low-affinity ($K_D \sim 1$ to $5\,\mu M$) membrane-spanning protein transports reduced folates directly into the cell cytosol and remains anchored to the cell membrane while shuttling folate. However, this transmembrane transporter is unable to translocate folate conjugates [21]. Malignant cells transport folate and folate conjugates in an oxidized form (i.e., folic acid) through a high-affinity ($K_D \sim 100\,\mathrm{pM}$) folate receptor by a mechanism called *potocytosis*. The folate receptor is a 38-kDa glycosyl-phosphatidylinositol-anchored glycoprotein expressed on the membrane surface of several cell types. It is located in caveolin-like clathrin-independent invaginations on the cell surface [22]. Upon receptor binding, the folate–folate receptor complex is taken up by cells and moves through the many organelles involved in endocytic trafficking [23]. The receptor is overexpressed by many types of tumor cells, including ovarian, endometrial, colorectal, breast, lung, renal, and neuroendocrine carcinomas and brain metastases [24]. By virtue of its ability to be taken up by tumor cells overexpressing the folate receptor, folic acid has been widely investigated as a targeting molecule for the selective delivery of anticancer drugs, because it allows cellular internalization of macromolecules and colloidal systems decorated with folic acid.

Proper synthetic procedures have been developed to conjugate folic acid to drug carriers. In particular, it has been demonstrated that the chemical derivatization of the glutamate γ-carboxyl group does not result in a significant loss in affinity of folic acid for its receptor, thereby allowing the conjugated folic acid to maintain its biological activity. Folate-based carriers have been produced by conjugating folic acid to radionuclide deferoxamine complexes aimed at radiopharmaceutical imaging, liposome-encapsulated drugs and DNA, as well as cytotoxins [25–28].

CDs have been modified successfully in order to achieve selective targeting of anticancer drugs to folate receptor–overexpressing tumor cells. A defined chemical protocol was established for folic acid conjugation to β-CD through a poly(ethylene glycol) (PEG) spacer [29]. The structure of the bioconjugate CD–PEG–folic acid (CD–PEG–FA) is reported in Scheme 1.

This supramolecular system has been designed to allow drug inclusion into the CD hydrophobic cavity, has a high solubility conferred by PEG, and possesses targeting properties via folic acid. It was hypothesized that the simultaneous presence of folic acid and drug-loaded β-CDs would guarantee uptake of the carrier into the target cells, where the drug can be released by environmental changes. For example, the pH-induced ionization of CD–loaded drug can take place in lysosomes or other subcellular compartments that have an acidic pH and can induce an affinity decrease for the CD cavity with consequential drug release.

The preparation of the monosubstituted CD–PEG–folate conjugate (CD–PEG–FA) was carried out by coupling a 700-Da diamino-PEG chain to 6′-monotosylated-β-CD. The resulting CD–PEG–NH$_2$ was reacted with an excess of succinimidyl ester–activated folic acid. PEG was used as a

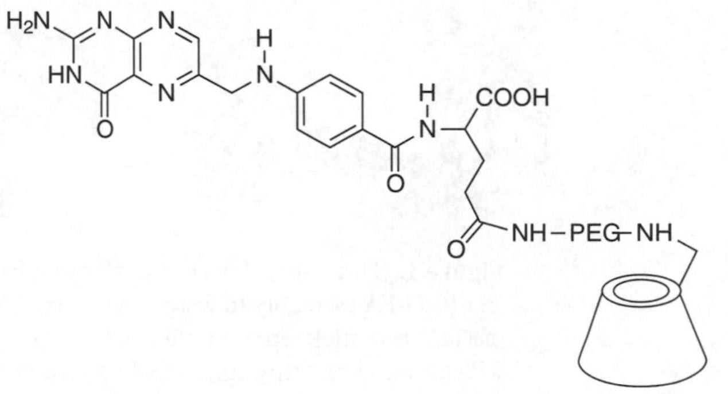

Scheme 1. Structure of the bioconjugate cyclodextrin-PEG-folic acid. (In part, from [29], with permission. Copyright © 2003 American Chemical Society.)

spacer arm because it confers flexibility to folic acid, favoring receptor interactions that may be hindered by the CD unit. CD–PEG–FA was found to be about three times more soluble than the unmodified β-CDs. The ability of the CD–PEG–FA conjugate to form inclusion complexes with model drugs was investigated using estradiol [30]. The solubility of β-estradiol complexed with CD–PEG–FA was about 540 times higher than that of drug in plain buffer and nine times that obtained with β-CD. The higher solubility of β-estradiol in the presence of CD–PEG–FA was attributed to the high solubility of the carrier. The CD–PEG–FA/estradiol inclusion constant (K_c) was about five times lower than the CD/estradiol K_c. The decreased β-estradiol affinity for the targeted CDs as compared to the native ones was partially ascribed to folic acid competition for the β-CD cavity. In addition, the modification of the epta-glucopyranose ring with the mobile and hydrophilic PEG chain may prevent the inclusion of estradiol, which is a molecule with a flat but large sterically hindered structure. *In silico* studies derived from the dissolution profiles of β-estradiol with molecular dynamic modeling showed that β-estradiol combines with CD–PEG–FA at a 2 : 1 stoichiometry for the CD–PEG–FA/β-estradiol complex, while CD/β-estradiol undergoes a 1 : 1 stoichiometry (Fig. 1).

Preservation of the included drug from chemical degradation was investigated using chlorambucil, which is a fast-degrading model drug. The study showed that the CD-based conjugate partially prevented drug degradation, although the protective effect was significantly lower than that displayed by native β-CDs. However, it should be noted that the chlorambucil/CD–PEG–FA affinity was significantly lower than that calculated for β-estradiol.

The systemic biocompatibility of CD–PEG-FA was investigated in vitro by the hemolysis test. Indeed, the parenteral administration of natural CDs is known to induce undesirable hemolysis as a consequence of cell membrane cholesterol extraction, and nephrotoxicity due to deposition of CD–cholesterol crystals in the kidneys [31]. CD–PEG-FA

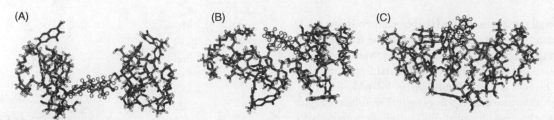

Figure 1. Three snapshots of the MD simulation performed for 2 ns at 310 K on the 1 : 2 β-estradiol/CD-PEG-FA assembly in water: (A) $t = 0$; (B): $t = 1$ ns; (C): $t = 2$ ns. For clarity, β-estradiol is shown in a ball-and-stick representation; CD-PEG-FA is shown in a stick representation. (Adapted from [29], with permission. Copyright © 2003 American Chemical Society.) (*See insert for color representation of the figure.*)

displayed negligible hemolytic characteristics (15%) up to a concentration of 20 mM compared to native β-CDs, which was ascribed to the lower affinity of the former for erythrocytic cholesterol.

The affinity of the bioconjugate CD–PEG–FA for immobilized folate-binding protein (FBP) and for the cell membrane folate receptor was investigated by BIAcore analysis and by cultured KB tumor cells overexpressing the folate receptor, respectively [32]. BIAcore studies showed that CD–PEG–FA bioconjugates interacted specifically with the FBP immobilized on the sensor chip and displayed a 400-fold lower affinity constant for the folate-binding protein compared to the affinity of free folic acid ($K_a = 50$ pM) [33]. However, it is important to note that a decreased affinity of the bioconjugated folate for the folate-binding protein was expected because about a 20% molar ratio of folic acid was conjugated to PEG through the α-carboxylic group of glutamate, which is involved in receptor recognition.

The binding of CD–PEG–FA to the folate receptor on KB cells was investigated by competition studies using radiolabeled folic acid. The study demonstrated that the bioconjugate competes and displaces [3H]folic acid from the cell surface receptor as the CD–PEG–FA concentration increased (Fig. 2). These results showed that CD–PEG–FA maintains 7% of the native affinity for the cell folate receptor in vitro.

Selective drug delivery to folate receptor overexpressing cells was also investigated. The interaction with the folic acid receptor is known to promote cell internalization of folic acid through a potocytosis pathway. CD–PEG–FA loaded with the fluorescent probe rhodamine-B was incubated with KB cells. The cell uptake of rhodamine-B delivered by CD–PEG–FA was 19% higher than that elicited by rhodamine-B-loaded β-CDs. The bioconjugate selectivity for targeting folate receptor overexpressing cells was confirmed by using MCF7 cells, a human breast cancer cell line that lacks the folate receptor, as a control [34]. Low levels of cell-associated fluorescence were observed after incubation with rhodamine-B-loaded CD–PEG–FA and untargeted CDs. Confocal laser-scanning microscopy visualized that CD–PEG–FA loaded with rhodamine-B allowed active transport of the fluorophore into KB cells and promoted its localization into tubular endosomal structures (Fig. 3).

Multivesicular bodies with a predominantly perinuclear fluorophore disposition were also observed. Cell trafficking studies on KB cells showed that early endosomes were formed as a consequence of folate-mediated selective uptake of the rhodamine-loaded CD–PEG–FA. Larger cytosolic fluorescent areas were observed over time, indicating that the endosomes migrate and are localized in proximity of the Golgi apparatus.

A second generation of folate-targeted CDs for selective drug delivery was obtained by expanding the hydrophobic cavity of the cyclic polysaccharide in order to enhance the drug-binding properties of the epta-glucopyranose carrier. The new derivative was synthesized by the conjugation of hexamethylene alkyl chains to the β-CD ring to obtain a bouquet-like CD arrangement [35]. The new CD-based carrier was obtained by a three-step procedure.

1. The CD cavity was expanded by reaction with an excess of hexamethylene diisocyanate to obtain

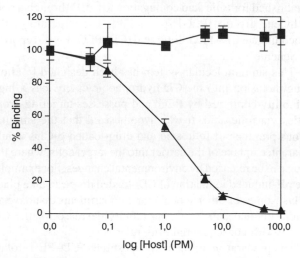

Figure 2. [3H]folic acid competition binding profiles of cellular folate receptor obtained with monolayer KB cells incubated with CD–PEG 1 : 1 molar ratio (■) and CD–PEG–FA (▲). (From [32], with permission. Copyright © 2004 American Chemical Society.)

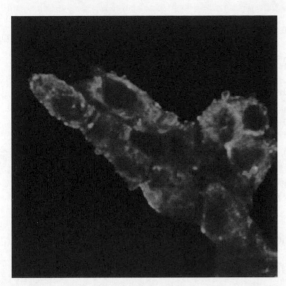

Figure 3. Confocal images obtained by incubation of KB cells with rhodamine-B-loaded CD–PEG–FA. (In part, from [32], with permission. Copyright © 2004 American Chemical Society.) (*See insert for color representation of the figure.*)

CD–(C$_6$-NCO)$_5$. A key point in the synthetic process was to avoid all cross-linking by using activated alkyl chains in excess and adequate reaction conditions. The ^{1}H NMR analysis showed that five out of seven primary hydroxyls of β-CDs were modified with hexamethylene.

2. The distal isocyanate groups of CD-(C$_6$-NCO)$_5$ underwent PEGylation with a 700-Da diamino-PEG polymer. PEG was essential to confer flexibility and solubility to the alkylated CDs.

3. CD–(C$_6$-PEG)$_5$–FA was synthesized by reacting CD–(C$_6$-PEG–NH$_2$)$_5$ with succinimidyl ester–activated folic acid. A derivative with a mean of 1.3 folic acid residues per CD moiety was obtained.

The ability of the new bioconjugate to form soluble inclusion complexes with hydrophobic molecules was investigated using β-estradiol and curcumin because of their poor solubility in water: 11 μM and 30 nM, respectively. Estradiol solubility was increased 320-fold when bioconjugated. The inclusion constant of CD–(C$_6$-PEG)$_5$–FA/β-estradiol was similar to that reported for CD–PEG–FA/β-estradiol discussed above ($K_c = 12975$ M^{-1}) [29]. Promising results were obtained with curcumin, a natural diarylheptanoid extracted from *Curcuma longa* L. Curcumin was chosen because it has a variety of biological activities, such as antiangiogenic, anti-inflammatory, and cytotoxic effects on many cancer cells [36]. Clearly, curcumin has a great potential as therapeutic agent, but suffers from poor solubility and stability in water. Curcumin solubility was increased 1.16×10^5 times when included in CD–(C$_6$-PEG)$_5$–FA, and the dissociation constant was calculated to be ~1 μM. By comparing the inclusion affinities of estradiol and curcumin for the carrier, it was concluded that small linear molecules display a higher affinity for these specific bouquet-like CDs, as they can penetrate deeper into the binding cavity and be surrounded by the newly introduced alkyl chains. In the case of curcumin, the increased drug solubility was due to both a higher carrier solubility and high complex stability.

The curcumin stability in free solution compared to being associated with β-CD and CD–(C$_6$-PEG)$_5$–FA was investigated in buffers at pH 6.5 and 7.2. These two pHs were selected to simulate the plasma (pH 7.2) and tumor tissue (pH 6.5) conditions. The drug in free solution undergoes pH-dependent degradation according to a first-order kinetic. CD–(C$_6$-PEG)$_5$–FA was found to enhance the drug stability significantly at both pHs. In addition, the drug stability observed with CD–(C$_6$-PEG)$_5$–FA was much higher than in the presence of β-CD (Fig. 4).

The selective targeting property of CD–(C$_6$-PEG)$_5$–FA was demonstrated with the KB and MCF7 cell lines. When

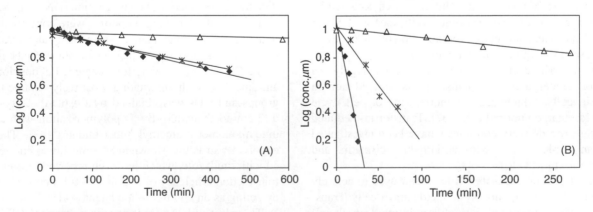

Figure 4. Curcumin hydrolysis profile in 0.1 M phosphate buffer, pH 6.5 (A) and pH 7.2 (B) at 30°C: free curcumin (◆), in the presence of β-cyclodextrins (∗), in the presence of CD–(C$_6$-PEG)$_5$–FA (△). (Adapted from [35], with permission. Copyright © 2007 Informa UK, Ltd.)

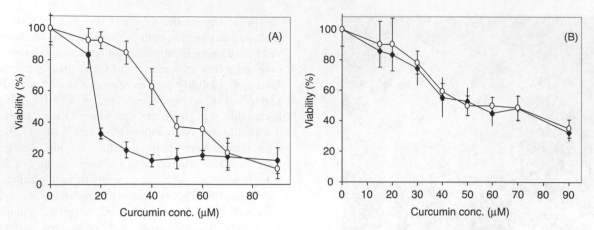

Figure 5. KB (A) and MCF7 (B) cell viability after incubation with curcumin/CD–$(C_6$-PEG–NH$_2)_5$ (○) or curcumin/CD–$(C_6$-PEG)$_5$–FA (◆).

KB cells were tested, curcumin-loaded CD–$(C_6$-PEG–NH$_2)_5$ exhibited a 58-μM ED$_{50}$, whereas curcumin-loaded CD–$(C_6$-PEG)$_5$–FA displayed a 21-μM ED$_{50}$, indicating that the physicochemical properties of the carrier are paramount for cell targeting (Fig. 5A). The study performed using MCF7 cells showed that curcumin loaded into targeted and non-targeted CDs induced similar cell toxicities (ED$_{50}$ over 60 μM) (Fig. 5B).

The nonspecific drug delivery to MCF7 cells confirmed that the macromolecular system is effective in addressing cell lines expressing the folate receptor. Notably, the limited beneficial effect on the specific delivery of curcumin to KB cells was ascribed partially to insufficient cell uptake due to the competition of the carrier free drug for the folate receptor.

2.2. Transferrin-Mediated Targeting

Transferrin is an 80-kDa iron-binding glycoprotein responsible for the transport of iron to proliferating cells. It is composed of a single polypeptide chain of 679 amino acids. In human serum, the transferrin concentration is about 2.5 mg/mL (35 mM), with a 30% binding sites occupied by Fe^{2+}. Iron-coupled transferrin binds to its specific receptor on the cell surface, undergoes internalization by endocytosis, and releases iron into acidic compartments.

The transferrin receptor has been investigated largely as a target for cancer therapy because of its upregulation in metastatic cells, which require more iron for cell cycle progression than do normal cells [37]. Transferrin-mediated anticancer drug delivery conjugates have been investigated in human clinical trials with adriamycin, cisplatin, and diphtheria toxin [38–40].

In recent decades, transferrin has been used to actively deliver colloidal therapeutic systems to cancer cells. Transferrin coupled to liposome through a poly(ethylene glycol) spacer was shown to specifically target C6 glioma cells in vitro and enhanced the uptake of liposome-encapsulated

doxorubicin via a receptor-mediated mechanism [41]. In a chronic myelogenous leukemia cell line (K562 cells), the cellular toxicity of transferrin-decorated doxorubicin/verapamil-loaded liposomes was 5.2 times higher than that of nontargeted drug-loaded liposomes. Importantly, competition binding studies confirmed the transferrin-mediated cell uptake. Thus, a combination therapy with transferrin-decorated liposomes has proven to effectively overcome multidrug resistance [42].

CDs-based self-assembled polymers have also been decorated with transferrin to endow a drug delivery system with targeting properties. These supramolecular constructs have been developed as nonviral vectors for the delivery of therapeutic macromolecules: namely, oligonucleotides. Aimed at developing an innovative, physically assembled supramolecular system for active macromolecular drug delivery, Mark Davis and his collaborators synthesized a new class of CD-based polycationic polymers that can be surface modified by modular attachment of stabilizing and targeting agents, thus representing a very versatile core platform. The novel cationic supramolecular system was designed to provide for oligonucleotides complexation (DNA, siRNA, etc.). In addition, the targeting properties were conferred by conjugating transferrin to adamantane, which can interact physically with the polymer. In fact, adamantane can be included in the CD cavity, allowing for transferrin exposition. Poly-amidine–CD with alternating positively charged amino groups and CDs was obtained by 6A,6D-dideoxy-6A,6D-di(2-aminoethanethio)-β-CD polymerization with a diamidine monomer (dimethyl suberimidate) [43]. The linear copolymer structure, shown in Scheme 2, was end-terminated with imidazole, which has been reported to improve the transfection efficiency of oligonucleotides delivery systems by acting as an endosomolitic agent [44].

This class of CD-based cationic polymer (CDP-Im) was found to bind DNA above the 1.5 polymer/DNA charge ratio. Furthermore, the DNA-binding efficiency depends on the

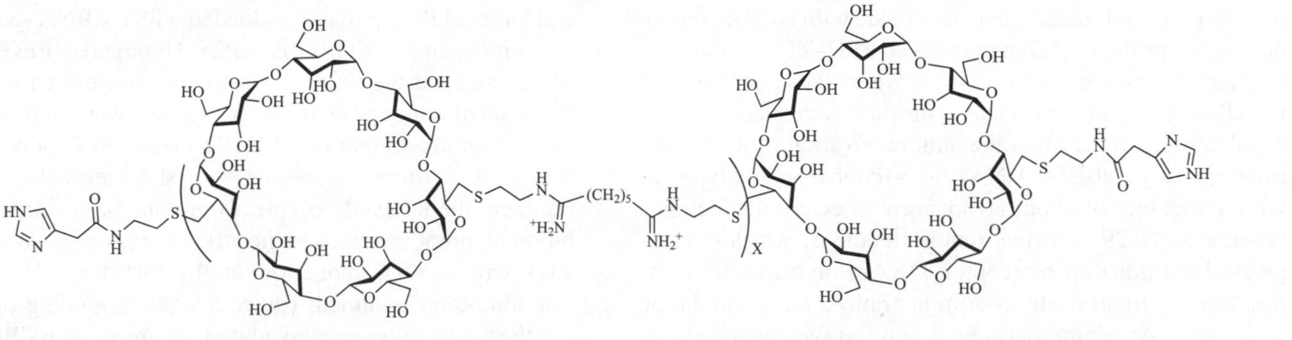

Scheme 2. Di(2 aminoethanethio)-β-CD/diamidine copolymer structure. (From [45], with permission. Copyright © 2003 American Chemical Society.)

distance of the amino groups from the CD moiety. Polymers containing a mean of 10 cationic charges per chain can condense DNA into discrete particles of 150 nm (polyplexes). As observed with DNA complexed with polyethylenimine, the stability and particle dimensions of these new polyplexes are strongly dictated by the polymer–DNA charge ratio.

To evaluate the gene delivery performance of this polycation, in vitro studies were performed using BHK-21 and CHO-K1 cells, which are cell lines known for their low transfection efficiency with diverse techniques. BHK-21 and CHO-K1 cells underwent successful transfection with a CD-based carrier complexed with a luciferase encoding plasmid. The polymer transfection increased luminescence, similarly to PEI and lipofectamine, and induced a two-order-of-magnitude higher luminescence than that of the commercial SuperFect transfection reagent. Preliminary in vitro studies showed low toxicity on BHK-21 cells, which was ascribed to the polycationic nature of the polymer, and in vivo studies in mice displayed no mortalities after a single injection of up to 200 mg/kg. In contrast, PEI showed a LD_{50} value of approximately 30 mg/kg in the same animal model.

The targeting moiety adamantane–poly(ethylene glycol)–transferrin (AD–PEG–Tf) was obtained by conjugating adamantane–PEG–vinylsulfone to the lysines of transferrin [45]. To characterize the targeting properties of the AD–PEG–Tf itself, competition studies using fluorescein-labeled transferrin were performed. The study was performed using PC-3 cells, a human prostate carcinoma cell line that expresses elevated levels of the transferrin receptor. The results showed

that AD–PEG–Tf maintained 94% of the cell recognition characteristics of the native protein.

The supramolecular system was obtained by mixing a solution containing AD–PEG–Tf, adamantane–PEG (AD–PEG), and CDP-Im to a solution of selected oligonucleotide sequences, which resulted in the spontaneous formation of targeted polyplexes. A scheme of the components and polyplex structure is reported in Fig. 6.

The CD cavities were occupied by AD–PEG and AD–PEG–Tf at 98 and 2% molar ratios, respectively. AD–PEG was used as a stabilizing and shielding agent. The colloidal polyplexes exhibit an average size of ~100 to 150 nm. The forces maintaining the physical identity of these polyplexes result from cooperative components of tightly packed PEG chains supported by the adamantane–CD inclusion enthalpy. Furthermore, PEGylation confers an elevated stability in biological media, reduced aggregation, and negligible erythrocyte aggregation, which is triggered by cationic polymers. The supramolecular system does not elicit complement activation, which may be ascribed to the low molecular weight of the CD-based copolymer [46]. Studies performed with siRNA showed that the polyplexes preserved the oligonucleotides from nuclease degradation in serum [47].

The cooperative effect of the targeting moieties on the carrier surface enhanced nanoparticle binding to the transferrin receptor–expressing cells. In vitro studies performed with K562 cells, which are notoriously difficult to transfect, showed that luciferase encoding plasmid-loaded polyplexes elicited a fourfold increase in transfection efficiency compared

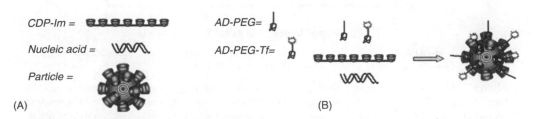

Figure 6. Formation of nucleic acid loaded particles using CD-based cationic polymer. (A) nanoparticle components; (B) particle assembly. (In part, from [47], with permission. Copyright © 2007 American Chemical Society.)

to untargeted polyplexes that were made with similar components, except the targeting agent AD–PEG–Tf.

Transferrin-targeted polyplexes were investigated in vivo for DNAzyme delivery. DNAzymes are sequences of DNA capable of catalytic cleavage and inactivation of RNA [48]. Fluorescently labeled DNAzymes condensed polyplexes with a diameter of about 50 nm were injected in nude mice bearing a HT-29 cell line xenograft tumor, which overexpressed the transferrin receptor. DNAzyme disposition was found to be time, route of administration, and formulation dependent. After intraperitoneal and intravenous injection, unformulated DNAzymes were cleared from the body completely, while transferrin–PEG polyplexes–formulated DNAzymes showed an elevated distribution in the tumor, kidneys, and liver. On the contrary, subcutaneous administration of native or formulated DNAzymes revealed that all material remained at the injection site. Histological analysis of tumor microsections showed that only intravenous injection of DNAzyme-loaded targeted polyplexes produced intracellular localization, while intraperitoneal administration did not result in tumor cell uptake. Therefore, intravenous administration is the preferred method of treatment for CD-based polyplexes and allows for their intracellular delivery to tumors only when specifically targeted. In vivo, CD-based polyplexes displayed a suitable pharmacokinetic profile, as they exhibit reduced renal clearance, adequate bloodstream circulation, DNA protection, and large payload delivery.

The CD-based targeted polyplexes were employed to deliver specific siRNA sequences for the therapeutic treatment of mice inoculated with a Ewing's family tumor (EFT) [49]. NOD/scid mice bearing EFT tumor overexpressing the transferrin receptor were injected with untargeted

and targeted PEG polyplexes loaded with a siRNA sequence for tumor suppression (siEFBP2). Untargeted PEG polyplexes delayed the tumor engraftment; however, following the onset of tumor growth, the growth rate was unaffected by further administration of siEFBP2-loaded PEG polyplexes. When administered according to a short-term therapeutic regimen, the targeted polyplexes provided a transient inhibition of tumor growth, and the effect lasted only two to three days with a 60% reduction in the targeted RNA level. The long-term treatment (twice a week beginning the first day the tumor cells were inoculated into mice) with siEFBP2-loaded transferrin/PEG polyplexes induced a significant inhibition against engraftment of malignant cells (Fig. 7). This study confirmed that targeted siRNA polyplexes are able to protect genetic material from nuclease degradation in vivo, and allow for receptor-mediated cellular uptake of biologically active siRNA.

Some reports claim that siRNA can trigger immuno-responses and alter the IFN levels. Nevertheless, a single administration of siEFBP2-loaded transferrin–PEG polyplexes to immunocompetent mice did not elicit any alteration of the levels of IL-12, IFN-α platelets, liver enzymes, or creatinine.

The biodistribution and pharmacokinetic profiles of siRNA-loaded transferrin–PEG polyplexes were evaluated by multimodal imaging. Positron emission tomography/computer tomography (PET/CT) for three-dimensional time-related biodistribution and bioluminescence correlates the disposition of a therapeutic agent and its biological effect [50]. NOD/scid mice bearing luciferase-transfected Neuro2A tumor cells were injected with transferrin–PEG polyplexes loaded with ^{64}Cu-labeled siRNA targeting

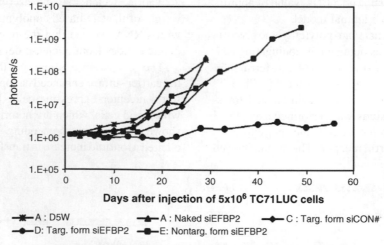

Figure 7. Growth curves for engrafted EFT tumor on mice receiving long-term treatment of siRNA formulations. The median integrated tumor bioluminescent signal (photons/s) for each treatment group is plotted versus time after cell injection. Treatment groups: *A*, 5% w/v glucose only (D5W); *B*, naked siEFBP2; *C*, targeted, formulated siCON1; *D*, targeted, formulated siEFBP2; *E*, nontargeted, formulated siEFBP2. (In part, from [49], with permission. Copyright © 2005 American Association for Cancer Research.)

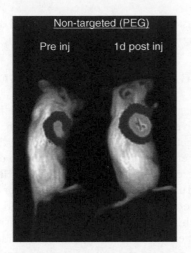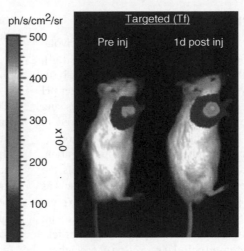

Figure 8. In vivo functional bioluminescence imaging of mice before injection and 1 day after injection of targeted and nontargeted siRNA nanoparticles. (In part, from [50], with permission. Copyright © 2007 The National Academy of Sciences of the USA.) (*See insert for color representation of the figure.*)

luciferase. MicroPET/CT animal scanning over 1 h after injection showed that transferrin-targeted polyplexes had a similar liver accumulation (26%) compared to naked ^{64}Cu-labeled siRNA, while the kidney excretion was delayed 10 min compared to free oligonucleotides. Both free and polyplex-loaded siRNAs displayed a $t_{1/2}$ of about 3 min and rapid kidney and liver distribution. It is relevant to note that targeted and nontargeted cationic polyplexes displayed very similar tumor dispositions 1 h after injection (about 1% ID/cm^3). Thus, it appears that the targeted system did not show any advantage over the untargeted PEG polyplexes. However, clear advantages of targeted nanosystems emerged by functional bioluminescence imagining, which assessed the higher therapeutic activity of the targeted nanoparticles compared to untargeted activity. In fact, transferrin–PEG polyplexes reduced the luciferase expression increase by 50% compared to nontargeted polyplexes, as shown by in vivo bioluminescence imaging (Fig. 8). Despite the fact that targeted and nontargeted polyplexes distribute in the tumor tissue to the same extent, targeted nanoparticles were able to deliver more functional siRNA inside tumor cells than untargeted polyplexes.

The best dosing schedule was also investigated for this family of CD-based targeted polyplexes [51]. Transferrin–PEG polyplexes loaded with antiproliferating siRNA were injected into A/J mice bearing a subcutaneous Neuro2A tumor cell line. Tumor growth inhibition was achieved when the siRNA concentration threshold inside the cells was reached. Thus, the efficacy of the siRNA-based therapeutic treatment was affected more by the dose schedules rather than the dose, and should be tailored according to the cell splitting time. In vivo pharmacokinetic studies showed that the best therapeutic performance can be obtained by three single administrations over three consecutive days of 2.5 mg/kg siRNA formulated with the transferrin-targeted polyplexes.

Recently, Mark Davis's group demonstrated the safety of siRNA-loaded transferrin–PEG polyplexes by administration to cynomolgus monkeys [52]. Targeted polyplexes were loaded with tumor antiproliferating siRNA and administered to healthy animals at 3 to 27 mg/kg siRNA dose, which was 100 times higher than the efficacy demonstrated in the mouse model. The animals treated did not exhibit loss of weight or alteration in food consumption. In addition, the serum markers and coagulation parameters remained in the physiological range. Overall, the targeted polyplexes were tolerated by the animals and only the highest dose showed mild kidney and liver toxicity. Although oligonucleotides are reported to have species-specific-related responses in terms of complement activation, siRNA-loaded transferrin–PEG polyplexes did not show this effect.

Specific interleukins, interferons, and TNFα were also evaluated in blood samples taken at various time points for each dose administered. The lowest doses did not elicit immunostimulation, while the 27-mg/kg siRNA dose induced a slight increase in Il-6. The absence of immunostimulation can be ascribed mainly to the polymeric carrier. Indeed, lipid-based siRNA delivery systems are reported to trigger an immune response only after intravenous administration. For comparison, a lipid-based formulation of siRNA showed a ninefold-higher liver toxicity and more evident organ dysfunction than those of the carrier proposed by Davis et al. In conclusion, the targeted siRNA-loaded CD-based carrier can be administered safely over a multiple-dose schedule, and it allows for a wide therapeutic index of siRNA [53]. The pharmacokinetic profile of siRNA in monkeys was in agreement with previous results obtained with mice. Since no antitransferrin antibodies are raised by animal exposure, after multiple administrations of targeted polyplexes, the oligonucleotides pharmacokinetic profile was not altered by the presence of antibodies.

2.3. Targeting Cathepsin B

Cathepsin B is a 31-kDa cysteine protease that plays a pivotal role in the protein turnover taking place in lysosomes [54]. Cathepsin B is primarily an exopeptidase at the acidic lysosome pH, but it displays endopeptidase activity at higher pH values. It is involved in pathologies such as arthritis and cancer. Cathepsin B overexpression has been observed in many human tumors, including gliomas and melanomas, prostate, breast, colorectal, esophageal, gastric, lung, ovarian, and thyroid carcinomas [55]. High cathepsin B levels are predictive of a poor prognosis in several tumors, such as colon [56] and ovarian [57] carcinomas. Cathepsin B expression and excretion are triggered by interactions of cells with extracellular matrix proteins. Defects in physiologic trafficking pathways lead to increased cellular secretion of procathepsin B, the cathepsin B precursor. Interestingly, despite cathepsin B not having transmembrane sequence, it has been shown to be closely associated with the cell surface of many types of tumors, and in a few tumors, membrane binding was found to take place through a macroglobulin and its receptor [58]. Because of its enzymatic activity, cathepsin B participates in invasive tumor progression in the following ways: (1) by direct degradation of extracellular matrices, and (2) by activating proteases that contribute to the degradation of extracellular matrices. Because cathepsin B is associated with the cell membrane of many tumors, extracellular cathepsin B has been considered an interesting target for selective tumor therapy. Several studies have been carried out to develop drug carriers for targeting cathepsin B positive tumor cells.

With the goal of endowing CDs with targeting properties toward cathepsin B–expressing tumor cells, an endo-epoxysuccinyl peptide (MeO–Gly–Gly–Leu–tEps–Leu–Pro–OH) with selective cathepsin B inhibitory activity was conjugated to β-CDs [59]. The conjugate was obtained by monoammino β-CDs reaction with BOC-protected ε-aminohexanoic acid. After deprotection, the derivative was coupled to the peptide sequence MeO–Gly–Gly–Leu–tEps–Leu–Pro–OH. Methotrexate was found to associate with the CD cavity at a 1 : 1 ratio. Biological investigation showed that conjugation to CDs did not dramatically alter the biological activity of the peptide sequence, and its inhibitory potency was reduced by only a factor of 0.7 upon conjugation, while the selectivity toward cathepsin B was increased. The non-membrane-permeable CD-endo-epoxysuccinyl peptide-conjugate was tested with MCF7 human breast cancer cells, which contain membrane-associated cathepsin B. The CD-endo-epoxysuccinyl peptide-conjugate displayed a high inhibitory activity for endogenous cathepsin B released from cell lysates, whereas limited inhibition was measured in living cells where cathepsin B is coupled with the membrane. Therefore, it can be concluded that the conjugate can selectively target the cathepsin B on tumor

cells overexpressing the enzyme and deliver the therapeutic drug included in the CD cavity.

3. GASTROINTESTINAL TARGETING

The most interesting site-specific delivery of biologically active molecules to the gastrointestinal tract is by far colon targeting. This type of delivery is particularly appealing for the effective therapy of colon-related diseases such as colon cancer and irritable bowel disease, including Crohn's disease and ulcerative colitis [60]. On the other hand, the colon shows several interesting physiological and chemical properties, such as nearly neutral pH, long transit time, and relatively low enzymatic activity, making it an ideal absorption site for various drug molecules, including peptides and proteins [61].

For these reasons, the application of CD for a site-specific delivery in the gastrointestinal tract after oral administration is probably the first type of targeted therapy attempted using CDs. Initially, CDs have been incorporated as an excipient in solid-dosage forms such as tablets, pellets, or microspheres [62–64], in order to complex the drug, to modulate the release, and eventually to act as an absorption enhancer [65]. However, in these cases, CDs are not directly responsible for the colon targeting, assured by polymeric components of the modified-release system [66].

In other approaches, CDs act not only as carriers able to solubilize, protect, and transport the drug by means of inclusion complexes, but because they are poorly absorbed from the GI tract and are fermented only in the colon by Bacteroides of the intestinal microflora [67,68], the oligosaccharides become the key component for colon delivery. In fact, to exploit these unique properties of CDs, several studies reported their use in the synthesis of prodrugs to achieve site-specific delivery of the drug in the colon.

In the treatment of inflammatory bowel diseases (IBDs), specific delivery to the colon of the drugs commonly used to treat this disease (i.e., aminosalicylates, corticosteroids, antibiotics and immunosuppressors) is expected to provide a more selective action on the principal sites of inflammation while decreasing the side effects, which for the most part are due to systemic absorption [69,70].

The anti-inflammatory drug biphenylylacetic acid (BPAA) has been selectively conjugated to one of the primary hydroxyl groups of α-, β-, and γ-CDs through esteric or amidic bonds and the cleavage of the prodrug studied in vitro [71] and in vivo [72,73]. In particular, in vitro studies performed using rat biological fluids showed that conjugates were stable in blood, stomach, and small intestinal fluid and in the homogenates of tissues such as liver and small and large intestine. On the contrary, BPAA was released efficiently when the ester conjugates were put in contact with cecal and colonic contents, suggesting a colon-specific enzymatic degradation of the conjugates. Interestingly, release

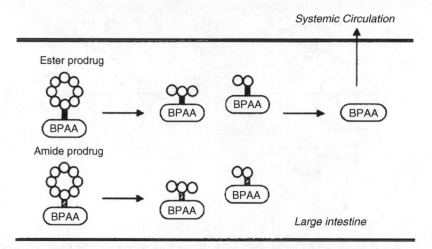

Figure 9. Suggested release mechanism of biphenylylacetic acid (BPAA) from BPAA/γ-cyclodextrin conjugates in rat cecum and colon as a result of the action of colon bacterioides (From [73], with permission. Copyright © 1998 John Wiley & Sons, Inc.)

from the β-CD ester conjugate was slower than for those with α- and γ-CD; this difference was attributed to the lower solubility of the β-CD conjugate. Amide conjugates showed no BPAA release in any of the media tested. When administered to rats, using as control the free drug and a BPAA/β-CD inclusion complex as controls, prodrugs obtained by esterification with α- and γ-CD were able to provide a delayed plasma concentration peak for the drug and higher bioavailabilities. The ester conjugate with β-CD and all the amide conjugates showed low or nondetectable plasma concentrations, suggesting that they passed through all gastrointestinal sections without being absorbed. It was elucidated that for ester conjugates a ring opening of CDs is followed by the enzymatic hydrolysis of the conjugate, leading to release of the free drug. The peculiar behavior of the β-CD ester conjugate was ascribed to its extremely low solubility in physiologic media, not allowing efficient transformation of the prodrug. On the contrary, for amide conjugates after CD ring opening, the small saccharide conjugates formed were not subjected to further cleavage, leading to stable hydrophilic derivatives such as maltose and triose amide conjugates, which were not absorbed and remained in the intestinal content (Fig. 9).

A similar approach to colon delivery was proposed for the corticosteroid prednisolone. This drug has been used for the treatment of inflammatory bowel disease; however, when administered orally, it is largely absorbed from the upper gastrointestinal tract, decreasing the therapeutic efficacy of the drug and producing severe systemic side effects, such as adrenosuppression, hypertension, and osteoporosis.

Also in this case, amide and ester conjugates with CD were prepared, but using a dicarboxylic acid, succininc acid, as a linker between the drug and the CD. Amide conjugates were abandoned, however, because unexpectedly they were found not to be stable in aqueous media. In fact, studies conducted on the 21-prednisolone/hemisuccinate/β-CD amide conjugate showed that whereas for the precursor 21-prednisolone/hemisuccinate ester, hydrolysis took up to 69 h to release 50% of the drug at 37°C in pH 7.0 buffer, the β-CD amide conjugate had a half-life of 6.50 min at even a lower temperature (25°C). This evidence was explained by an intramolecular nucleophilic catalysis of the hydrolysis of the ester linkage of the drug, involving the amide bond between the spacer and the CD and leading to the simultaneous release of prednisolone and mono(6-deoxy-6-succimino)-β-CD [74].

Ester conjugates with α-, β-, and γ-CDs were obtained by selective esterification of one of the secondary hydroxyl groups of CDs with prednisolone 21-hesuccinate using carbonyl diimidazol as a coupling agent. Aqueous solubility of conjugates was found to be more than 50% w/v, much higher than that for prednisolone and prednisolone 21-hemisuccinate, both poorly soluble. This high solubility was in contrast with that found for ester conjugates on the primary hydroxyl groups of β-CDs with biphenylyl acetic acid and n-butyric acid, where strong intermolecular interactions actually decreased the water solubility compared to parent compounds. In the case of prednisolone, however, it was demonstrated that the formation of an intramolecular inclusion complex between the lipophilic drug and the CD cavity (see Fig. 10) was at the base of the high water solubility observed for the conjugate. A steric hindrance to the action of ester-hydrolizing enzymes has also been suggested. In any case, at 37°C in phosphate buffer pH 7.4, the hydrolysis of ester conjugate and the release of the drug form of prednisolone and prednisolone succinate was slow and dependent on the CD appended (i.e., 49, 57, and 85% hydrolysis at 24 h for α-, β-, and γ-CD, respectively) [75]. Prednisolone appended to α-CD was

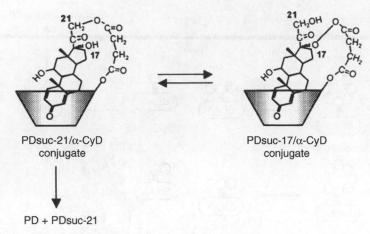

PDsuc-21/α-CyD
conjugate

PDsuc-17/α-CyD
conjugate

PD + PDsuc-21

Figure 10. Intramolecular inclusion complex of prednisolone 21–hemisuccinate/α-CD conjugate and its proposed behavior in aqueous media. (From [77], with permission. Copyright © 2002 Elsevier, Ltd.)

chosen for in vivo studies because of its slower hydrolysis kinetics compared to β- and γ-CDs. In vivo studies were performed in rats in which colitis was induced via 2,4,6-trinitrobenzensulfonic acid. Intracolonic administration of the drug was used to elucidate the in vivo colonic degradation of the prodrug and the extent of the anti-inflammatory effect of the conjugate administered locally. Interestingly, even if was evidenced that the conjugate was absorbed from the colon in the case of severe colitis, the conjugate was rapidly eliminated from blood by renal excretion and was found in urine unmodified.

These data were confirmed further after oral administration of the prednisolone/α-CD conjugate. In fact, in both administration routes, oral and intracolonic, not only was the anti-inflammatory effect, expressed in terms of a colon damage score, colon/body weight ratio, and myeloperoxidase activity, comparable to that obtained with the drug alone or its inclusion complex with hydroxypropyl-β-CD (HP-β-CD), but it succeeded in avoiding the side effects. No thymus atrophy was evidenced after administration of the prednisolone/α-CD conjugate, while the thymus/body weight ratio decreased significantly after treatment with prednisolone or its inclusion complex with HP-β-CD [76,77].

More recently, the colon-specific delivery of 5-aminosalicylic acid (5-ASA) was investigated. This drug is an active ingredient used in the long-term maintenance therapy to prevent relapses in Crohn's disease and ulcerative colitis. The 5-ASA was conjugated with α-, β-, and γ-CDs protecting the amino group by formylation, forming the intermediate imidazolide using an excess of carbonyldiimidazole and by successive conjugation with the CDs in the presence of triethylamine as catalyst. Different degrees of substitution were obtained using CD–drug ratios ranging from 1 : 1 to 1 : 10. In vitro studies showed that conjugates were stable at various pHs and that drug release in rat cecal and colonic content was obtained only with CD conjugates, whereas conjugates with other polysaccharides, such as hydroxypropylcellulose and

chitosan, were not able to release 5-ASA. It was also found that the release rate of the drug in the case of CD conjugates depended on both the type of CD and the degree of substitution. Faster release of 5-ASA from the prodrug was obtained for γ-CD, followed by α- and β-CD, while a higher degree of substitution corresponded to lower release rates, probably due to the reduction in water solubility of the prodrug [78]. When 5-ASA was administered to rats orally, its adsorption in the upper GI tract led to a peak in plasma and urine concentration of the drug 4 h after administration. On the contrary, after administration of the 5-ASA/CD conjugates, no drug was absorbed from the GI tract but was recovered almost completely in cecum, colon, and feces [79].

Other nonsteroidal anti-inflammatory drugs (NSAIDs), such as naproxen, sulindac, and flurbiprofen, have been conjugated to CDs in order to obtain a specific targeting to the colon. Flurbiprofen conjugate with α- and β-CD have been tested in vitro and in vivo. Both prodrugs were efficiently hydrolyzed by rat colon homogenate, except when clindamycin pretreatment impaired the intestinal microflora responsible for the specific degradation of CDs [80].

In another more general approach, the ketoprofen/α-CD conjugate 6A-O-[2-(3-benzoylphenyl)propionyl]-α-CD was used in combination with ketoprofen/HP-β-CD complex or ketoprofen–ethylcellulose solid dispersion, to obtain modified-release drug delivery systems. In fact, simultaneous oral administration of the CD conjugate with the fast-drug-releasing ketoprofen/HP-β-CD complex produced two peaks in rat ketoprofen plasma concentrations at 1–2 and 8–12 h, respectively. In the case of the combination between the ketoprofen/α-CD conjugate and the slow-releasing solid dispersion of the drug with ethylcellulose, after oral administration to rats, sustained plasma levels were obtained for at least 24 h. The plasma levels corresponded to repeated or prolonged anti-inflammatory effects, as demonstrated by measuring, at different times after the oral administration of the anti-inflammatory drug formulations, the swelling of the paws of rats induced by injection of 1% w/v carrageenan. While a constant swelling reduction against control was evidenced for the formulation containing the α-CD conjugate and the solid dispersion, discontinuous behavior was shown when ketoprofen conjugate plus the inclusion complex with CD were administered [81].

Even if in recent years the combination of surgical ablation with adjuvant chemotherapy has improved patient survival, colorectal cancer is still one of the leading causes of death in the United States and Europe. In particular, current methods of chemotherapy, administered at high doses and with a lot of undesirable side effects, are not optimal, and an oral site-specific delivery system would provide greater local exposure of the tumor to the drug, sparing healthy tissues [82]. As a consequence, inclusion complexes with CDs of substances having chemopreventive properties, such as geranyloxy–ferulic acid and auraptene [83], or anticancer properties, as in the case of platinum-based compounds [84], have been investigated to optimize their oral administration.

For specific colon delivery in the treatment of colorectal cancer, only a prodrug linking to β-CD has been proposed. In fact, among the numerous biological properties of short-chain fatty acids (SCFAs) such as *n*-butyric acid, an effect on tumors and, particularly, colorectal cancer was reported in some studies. In any case, study of β-CD/butyric acid conjugates have focused on the colon-specific delivery, and no experiments on in vivo animal tumor models were reported [85]. Nevertheless, some other CD conjugates with anticancer drugs have recently been proposed as prodrugs in tumor treatments. To our knowledge, studies involving those molecules actually concentrate only on the parenteral administration route [86,87].

Also in the case of oral delivery of peptides and proteins, the use of CDs has been suggested. In fact, CDs were shown to be able to interact and form inclusion complexes with side chains or amino acids of several peptides and proteins, providing several advantageous effects for drug delivery, such as higher aqueous solubility, decreased aggregation [88], reduced denaturation [89] and protection from enzymatic degradation [90], all this preserving biological activity [91]. However, targeting to the colon of peptides and proteins exploiting CD properties exclusively does not seem achievable. Actually, CDs have often been used as functional excipients in the preparation of microparticles [92,93] or nanoparticles [94] for the oral delivery of polypeptides, but their contribution, even if relevant or essential for the increased bioavailability, was not due to a specific targeting of an absorption site in the gastrointestinal tract.

4. TARGETING THE LIVER

Since the discovery that glycosylated lipids and proteins and carbohydrate-binding proteins (i.e., lectins) are heavily

involved in cell recognition, interaction, and adhesion, drug targeting via saccharide-linked moieties has become particularly appealing [95]. To obtain a targeted drug delivery system toward endogenous lectins, a number of different approaches have been proposed for CD conjugation with saccharides, such as glucose [96], galactose [97], mannose [98], and fucose [99,100], applying different linking strategies and spacer molecules. In fact, a point that has been investigated by several authors is the opportunity of linking to the same CD, multiple sugar molecules synthesizing simple polysubstituted structures or complex dendrimers (Fig. 11). This is expected to improve the efficacy of targeted CD because of the cluster effect: a multiple and optimized binding of the sugar moieties and cell surface protein [101,102].

Even though a lot of attention has been devoted to lectin-based targeting to the GI tract [103] and some interesting studies have shown enhancement in the oral absorption of phenytoin complexed with 6-*O*-α-D-glucosyl- or 6-*O*-α-D-maltosyl-β-CD [104], explicit CD targeting by an oral route using this approach has never been proposed.

Among the other applications, galactose-mediated targeting of the asialoglycoprotein receptor expressed by

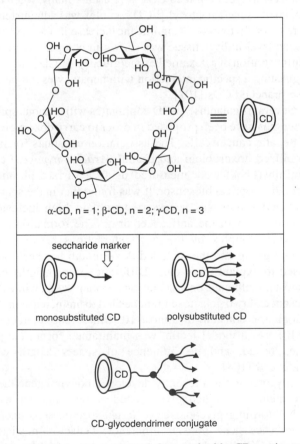

Figure 11. Structures proposed for saccharide–CD conjugates. (From [102], with permission. Copyright © 2004 American Chemical Society.)

hepatocytes seems more interesting for CDs. The specific uptake of the inclusion complex of *all trans*-retinoic acid with galactosyl β-CD by receptor-mediated endocytosis in primary mice hepatocytes and in a human hepatocellular carcinoma cell line (HepG$_2$) has been studied. Fluorescence microscopy and flow cytometry, using FITC-marked galactosylated CDs, showed a time-dependent uptake of the targeted inclusion complex [105].

Shinoda and co-workers proposed galactose-branched β- and γ-CDs synthesized by an enzymatic method using *Bacillus circulans* β-galactosidase. The specific interaction with hepatocytes was demonstrated by inhibition of the interaction between primary rat hepatocytes and FITC-labeled lactosyl polystyrene, a polymer demonstrated to have high affinity for the asialoglycoprotein receptor. The inhibition was obtained with galactose-branched CDs used as controls, but not with glucose-branched CDs [106]. The saccharide-targeted β-CDs were then used to form inclusion complexes with steroidal drugs, such as dexamethasone, hydrocortisone, triamcinolone, and prednisolone. Saccharide-branched CDs showed more stable complexation of the steroidal drugs than did the parent CDs: cholesterol was not able to replace the included drug in aqueous solution only when carbohydrate-linked CDs were used. Finally, when the dexamethasone/galactosyl-β-CD complex was administered intravenously to mice, a significant increase in the dexametasone level in liver tissue was shown in comparison to the administration of a dexamethasone/glucosyl-β-CD complex, suggesting a specific interaction with hepatocytes for galactose-branched CDs [107].

Several D-galactose/β-CD conjugates with various spacer lengths have been proposed in order to carry doxorubicin to hepatic cancer cells. The association constants for immobilized doxorubicin and peanut (*Arachis hypogea*) agglutinin (PNA) were measured using a surface plasmon resonance optical biosensor. It was found that in the spacer a phenyl group was necessary to obtain high inclusion association with the anticancer drug. The formation of a stacking complex by the π–π interactions between the phenyl group and the included doxorubicin has been proposed to explain these data. Different spacer lengths had almost no influence on association constants (K_a), while the presence of two galactose branches led to an increase in the association constant of almost 10^2 times. The interaction with PNA, a model lectin, was maintained for all conjugates tested, and no influence of spacer length was evidenced [108].

Glycofection is a recently developed nonviral gene therapy method in which glycosylated carriers interact with pDNA-forming glycoplexes to target-specific cells and/or to enhance gene transfer efficiency. Galactose-mediated targeted CDs have been proposed to enhance gene transfer efficiency. In a recent study, a polyamidoamine (PMMA) starburst dendrimer was used to link multiple galactose moieties to α-CD. The CD conjugates were able to condense pDNA and to protect it from the degradation of DNase I. The cytotoxicity and transfection efficiency of galactose-targeted α-CD complexed with a luciferase-expressing pDNA was evaluated on three different cell lines: HepG2, a human hepatocellular carcinoma cell line; NIH3T3, a mouse fibroblast cell line; and A549, a human lung epithelium cell line. Galactose-targeted complexes showed no cytotoxicity, and enhanced transfection efficiency in all three cell lines compared to pDNA complexes with the dendrimer alone or the dendrimer conjugated with α-CD but without the galactose-targeting moiety. This suggested an asialoglycoprotein receptor–independent enhancement of transfection efficiency. The absence of relevant galactose-mediated targeting was proved further by the very low inhibition effect on transfection shown in the presence of competitors, such as galactose and asialofetuin. However, to explain the enhanced transfection efficiency evidenced for galactose-conjugated α-CD/pDNA complexes, their nuclear translocation after interaction with intacellular galactose-binding lectins was suggested [109].

5. ROTAXANES: MULTIVALENT BINDING TO BIOLOGICAL TARGETS

Multivalent binding describes recognition events taking place in biochemistry and supramolecular chemistry. It refers to multiple noncovalent interactions between scaffolds endowed with multiple ligands and selective targets with recognition sites. In nature, many examples of such processes can be referred to as multivalent binding, including the immunological response, cell membrane adherence, signal-transduction interactions, and enzymatic machineries. Multivalent binding enforces the binding affinity between two entities by cooperative interactions. The multivalent concept is related to the effective concentration of a defined ligand, which accounts for its probability of docking the receptor. [110]. To mimic natural processes, supramolecular chemistry has lately exploited the multivalent concept by designing new nanomaterials and molecular devices. The following unique features play a key role in governing multivalent binding: (1) flexibility and length of spacers connecting the ligands, (2) strength of noncovalent binding, (3) number of ligands, and (4) flexibility of the scaffold.

Poly- and pseudopolyrotaxanes seem to possess suitable requisites for multivalent binding. These supramolecular systems are necklace-like structures in which cyclic components are interlocked mechanically to end-capped polymeric chains. These assemblies guarantee adequate flexibility and mobility of the macromolecular structure and the ligand moieties. CDs are commonly used as cyclic components that can be threaded onto a linear polymer chain and decorated with biorecognition moieties. Ligand-bearing CDs

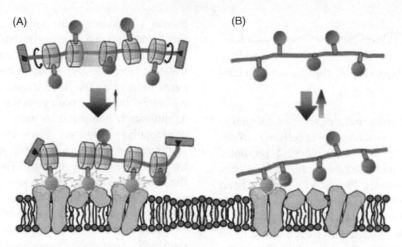

Figure 12. Effect of "mobile" motion of the cyclic compounds in polyrotaxanes on receptor binding in a multivalent manner: (A) ligand–polyrotaxane conjugate and receptor sites; (B) ligand-immobilized-polymer and receptor sites. (From [111], with permission. Copyright © 2006 Wiley-VCH Verlag.) (*See insert for color representation of the figure.*)

can slide and rotate in the rotaxane structure, thereby minimizing spatial mismatching between ligands conjugated to the cyclic polysaccharides and receptor sites. On the contrary, rigidity is a drawback in multivalent binding to specific receptors, as depicted in Fig. 12.

The binding benefits obtained with multivalent targeting can be described thermodynamically by taking into account the enthalpic and entropic energies of the binding process [111]. Entropic energy loss due to multiple binding interactions of a polyligand-bearing polymer is counterbalanced by a relevant enthalpic energy gain obtained only when a mismatching of ligands and receptors is minimized. This event is guaranteed by polyrotaxanes.

Polyrotaxanes decorated with biotin have been developed to exploit the intestinal transport of biotin-linked therapeutic agents, allowing for increased oral delivery of drugs in patients with compromized absorption [112,113]. Preliminary studies have shown that multivalent valine–lysine dipeptide-decorated polyrotaxanes selectively inhibit the human peptide transporter hPEPT1 on HeLa cells [114]. Carbonyldiimidazole-activated CD hydroxyls groups were reacted with biotin hydrazide. Polyrotaxanes for targeted delivery were obtained by threading 20 to 22 biotinilated α-CDs into a PEG chain end-capped with Z-L-Phe [115]. Surface plasmon resonance (SPR) analysis on a streptavidin immobilized sensor chip showed that rotaxanes derivatized with 11, 35, and 78 biotin molecules have high biotin/streptavidin affinity, although it was unaffected by the number of biotin molecules. This was explained by a rising mismatching effect: as the number of ligands increased on the rotaxane, they reduced the sliding opportunity along the PEG chain. On the other hand, the number of biotins in the rotaxanes directly affected the dissociation profile of the nanosystems. The dissociation constants decrease drastically as the number of biotins in the conjugates increased. SPR competition studies performed by biotin immobilization on the sensor chip and a 78-biotin-polyrotaxane/streptavidin or 1-biotin-CD/streptavidin mixture showed that the polyrotaxane carrier exhibits a four- to fivefold higher affinity for streptavidin than for 1-biotin-cyclodextrin ($K_{\text{inhibition}}$ 2.13 and 9.48 nM, respectively). The higher affinity of the multivalent biotinylated rotaxanes can result from noncovalent cross-linking of streptavidin molecules, which can take place due to the multivalent properties of the supramolecular system.

Furthermore, multivalent binding has been investigated for glycobiology applications. High-avidity interactions take place between membrane-associated carbohydrate-binding proteins (lectins) and their specific ligands, despite the low affinity provided by a single interaction between the two biological partners. These events allow for cell recognition and adhesion in both physiological and pathological processes. Flexibility, architecture, and density of polysaccharides influence the avidity of membrane carbohydrate antennas for their lectins [116]. Multiple binding sites to lectins are deemed a valuable strategy to enhance the avidity for lectins and the poly- and pseudopolyrotaxanes that have been used as scaffolds to mimic the natural mechanisms of binding. In fact, they can span large distances and modulate the density of conjugated ligands, thereby mimicking the fluidity of the cellular membrane. Glycosylated pseudopolyrotaxanes were prepared for galectin targeting. Galectin is a soluble galactosyl-binding lectin able to noncovalently cross-link lactose or *N*-acetyl lactosamine residues exposed on the cell surface, which is due to its two binding sites at opposite sides of the protein. Galectins are important mediators in cell adhesion, signaling, and death. Galectin-1 is overexpressed in many tumors as a self-defense mechanism. For example, by

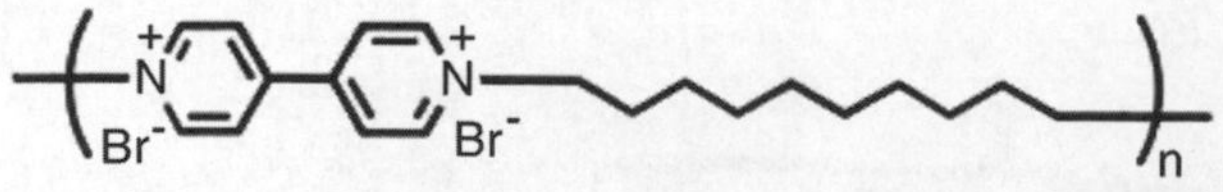

Scheme 3. Decamethylene-bipyridinium copolymer (polyviologen) structure.

releasing galectin-1, tumor cells emit a proapoptotic signal to induce T-cells and other immunocompetent cells toward the apoptotic pathway [117]. In addition, galectin-1 promotes tumor cell agglutination, which increases malignant cell survival in circulation and metastasis. Thus, it is clearly beneficial to study multivalent systems aimed at targeting galectin-1 for cancer diagnostics and delivery of therapeutics [118].

Multivalent and flexible systems have been found to be adequate for binding galectin-1. For example, 17α-cyclodextrin–lactosides were threaded onto a 7.76-kDa decamethylene and $4,4'$-bipyridinium copolymer [119] (polyviologen, Scheme 3).

The biological activity of the pseudopolyrotaxane was compared to four lactosylated derivatives with different structural architectures and flexibilities (one lactosylated chitosan and three trivalent lactoside glycoclusters). Chitosan was lactosylated by reductive amination, and 25% monomer substitution was achieved. The three glycoclusters were obtained by coupling three lactosylated arms via reductive amination to a central core through spacers of variable length and flexibility (Scheme 4).

The binding efficiency of the lactosylated polypseudorotaxane was investigated by the agglutination test, where

human T-leukemia cells were treated with galectin-1 preincubated with polypseudorotaxane. Lactosylated chitosan had a 1.7-fold higher affinity for galectin-1, and the three trivalent lactoside glycoclusters had a similar affinity compared to lactose. The three trivalent glycoclusters exhibited equal inhibitory effects on galectin-1-mediated leukemia cell agglutination compared to native lactose. In contrast, the pseudopolyrotaxane was very effective in inhibiting cell agglutination, due to its 10-fold higher affinity for galectin-1 with respect to lactose. These results indicate that flexibility and relative motion of the ligands along the polymer backbone in the pseudopolyrotaxane dictate the binding properties of the supramolecular system.

The relevance of CD sliding along the polymer backbone on the binding properties of pseudopolyrotaxanes was investigated using polyviologen copolymers with 8, 17, and 21 repeating units threaded with lactosylated CDs [120]. Pseudopolyrotaxane assemblies with different 1-lactosil-α-CD/diblock monomer molar ratios were obtained as summarized in Table 1.

Galectin-1 precipitation and inhibition of T-cell galectin-1-mediated agglutination studies were carried out to evaluate the biological and targeting properties of the polyrotaxanes. Pseudopolyrotaxanes [21:21] rapidly and efficiently coprecipitate galectin-1. Using a high galectin concentration, the precipitate was found to have a 1:1 lactose/galectin-1 molar ratio. Galectin-1 co-precipitation was proportional to the lactosylated-CDs ratio in the assemblies. Namely, assemblies [10:21] and [5:21] were less efficient in the precipitation process than was the [21:21] assembly.

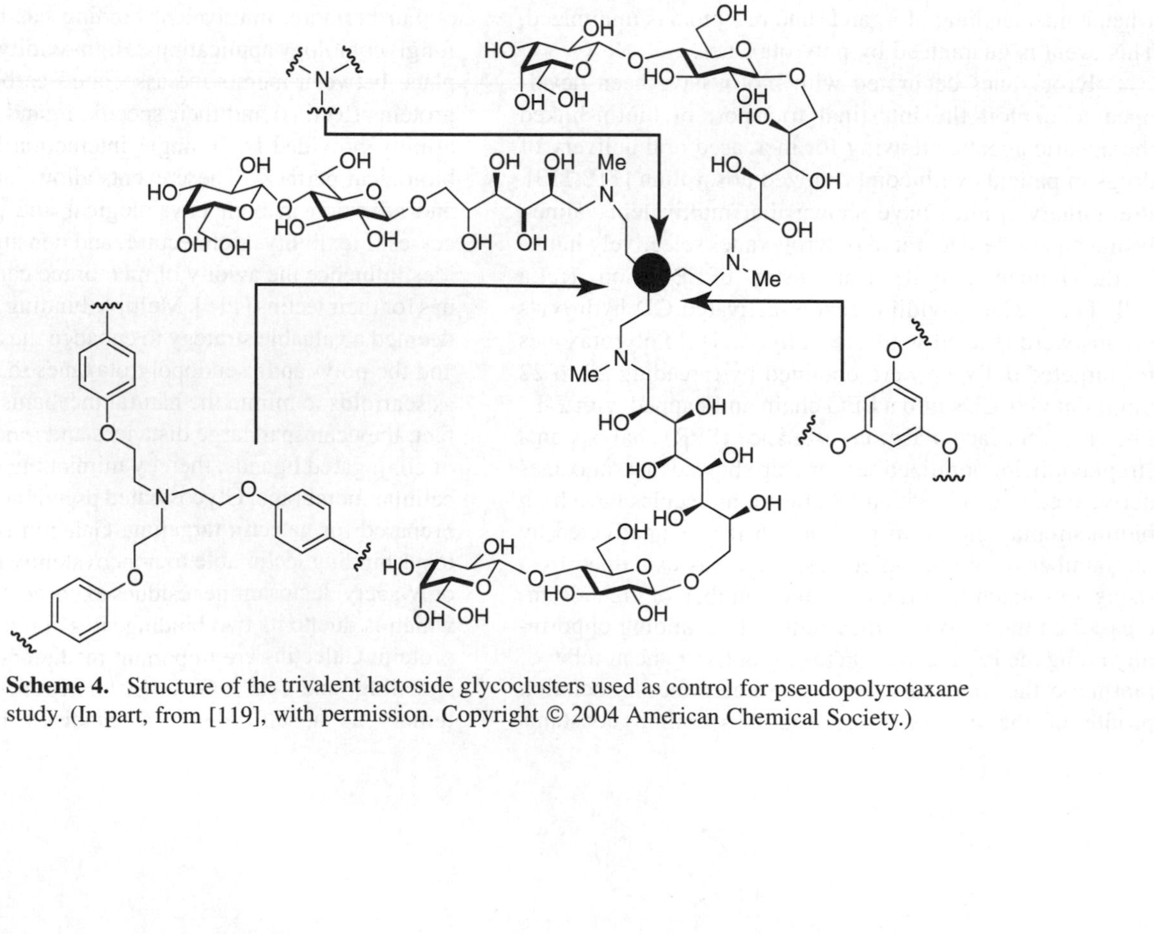

Scheme 4. Structure of the trivalent lactoside glycoclusters used as control for pseudopolyrotaxane study. (In part, from [119], with permission. Copyright © 2004 American Chemical Society.)

Table 1. Degree of Threading for a Series of Pseudopolyrotaxane Assemblies

1-Lactosil-α-CD Units	Diblock Monomers	Assembly Name
2	8	[2 : 8]
4	8	[4 : 8]
8	8	[8 : 8]
17	17	[17 : 17]
5	21	[5 : 21]
10	21	[10 : 21]
21	21	[21 : 21]

Inhibition of the T-cell galectin-1-mediated agglutination assay showed that both [21 : 21] and [10 : 21] assemblies inhibited agglutination at 100 μM, while the [5 : 21] displayed a twofold higher inhibition effect than the former two, as well as a 30-fold higher effect than lactose and a 20-fold higher effect than free lactosylated-CD. The [5 : 21] assembly was the best performing pseudopolyrotaxane, demonstrating that fully threaded pseudopolyrotaxanes, with limited sliding of the targeting moieties, are not favored for multivalent binding. Furthermore, the polyviologen backbone length affected the co-precipitation of galectin-1. The [8 : 8] assembly was less effective in precipitating the lectin than was the [21 : 21] assembly. The assembly with eight viologen units inhibited agglutination at 200 μM regardless of the number of lactosylated CDs threaded. Indeed, longer polymers have more connected ligands with a higher chance to cross-link different galectin-1 molecules. The effect of the polymer length and number of ligands threaded on the pseudopolyrotaxanes for multivalent binding may be due mainly to statistical effects. A linear correlation was found between the biological activity of these pseudopolyrotaxanes and the number of lactoside units. Only the [5 : 21] assembly, the most potent pseudopolyrotaxane, deviated from this correlation, showing that it binds galectin-1 by a mechanism in addition to cross-linking, as for the other pseudopolyrotaxanes. In fact, because of the distance between threaded lactoside–CDs, only the [5 : 21] assembly can chelate galectin-1 with lactoside units threaded on the polymer chain. This result confirms that multivalent systems require an adequate design in terms of the component ratio in order to have optimal biological performance.

The correlation between the molecular motion of each rotaxane component and the binding affinity to the biological target was investigated using mannosylated polyrotaxanes and concanavalin A, which is a model lectin [121,122]. Three polyrotaxanes were obtained by assembling 50 (Mal-PRX1), 85 (Mal-PRX2), and 120 (Mal-PRX3) α-CDs with 20-kDa PEG end-capped with Z-L-tyrosine. Concanavalin A selectivity was conferred by reacting carboxyethyl CDs with β-maltosylamine. The three polyrotaxanes had the same number of maltoses (230 to 240). The ^{1}H NMR spin-lattice

Table 2. Effect of Molecular Motion on the ^{1}H-NMR Relaxation Time of PEG Threaded Maltose-α-CD

	T_1	T_2
Slow molecular motion	↑	↓
Fast molecular motion	↑	↑

relaxation time (T_1) and spin-spin relaxation time (T_2) of the C(1)H of maltosyl groups, and the C(1)H of α-CDs and methylenes of PEG, were assessed to obtain information about the polyrotaxane components' relative motion. Table 2 shows the effect of molecular motion on the relaxation time within the magnetic field.

The time difference ($T_1 - T_2$) is diagnostic for molecular motion and can be correlated with the biological behavior of supramolecular systems. Smaller ($T_1 - T_2$) values indicate a faster motion of the corresponding groups. Both maltose and CD in Mal-PRX1 and Mal-PRX3 exhibited longer T_1 values than those of Mal-PRX2 and the free maltose–CD used as the "mobile" reference. This indicates that in MAL-PRX2 the threaded maltose–CD units undergo faster motion. The T_2 relaxation time accounts for the heterogeneous condition of the magnetic field, which was influenced by association, motion, and solubility. Mal-PRX2 showed longer T_2 values than those of the other two polyrotaxanes and was similar to that of free maltose–CD. Accordingly, it was deduced that the molecular motion of the maltosyl groups is dictated by the CD motion along the PEG backbone in the Mal–PRX2 assembly. The increased number of threaded maltose–CDs in Mal-PRX3 reduced the motion of maltose and CDs, due to intramolecular hydrogen bonding, whereas the reduced number of maltose–CDs in Mal-PRX1 promoted intermolecular association of Z-L-Tyr, thereby decreasing the motion of threaded moieties.

The affinity constants of the maltosylated polyrotaxanes for concanavalin A were assessed by incubating fluorescein-labeled concanavalin A with maltosylated conjugates. As expected, the association constant (K_a) of polyrotaxanes was higher than that measured for free maltose–CDs. Mal-PRX2 exhibited a 19- and twofold higher affinity than those of Mal-PRX1 and Mal-PRX3, respectively, which correlated well with data obtained from ^{1}H NMR spectroscopy. Furthermore, the local density of maltose sugars on each CD cannot be claimed as the driving force for the multivalent binding to concanavalin A. In fact, studies performed with polyrotaxanes with a fixed number of threaded CDs and increasing maltosyl residues conjugated per CD did not correlate with the binding potency to concanavalin A. This indicates that local ligand density plays only a marginal role in determining the binding potency of targeted polyrotaxanes.

The relevance of CDs motion on polyrotaxanes has been demonstrated to provide the main contribution to multivalent binding efficiency, which was determined by measuring the rate of aggregation of concanavalin A with the polyrotaxanes

Man-PRX1, Man-PRX2, and Man-PRX3 (the same number of maltosyl units). Mal-PRX2 exhibited the fastest aggregation rate of the three assemblies. Notably, the aggregation rate correlates nicely with the relaxation time T_2 of CDs and maltose. This result confirms that the motion of threaded CDs is the main factor affecting the spatial rearrangement of ligands on polyrotaxanes, thereby allowing for faster binding to the biological target.

6. OTHER TARGETING APPROACHES

Several other tissues have been proposed as potential targets for CD-based drug delivery systems. Methylated β-CDs conjugated with δ-opioid receptor agonist peptides, such as N-leucine-enkephalin or its cyclic analog, were proposed as a drug carrier for brain targeting. Even if in vitro experiments evidenced a decrease receptor binding, in vivo tests performed after intracerebroventricular or intravenous administration demonstrated that the β-CD conjugates maintained antinociceptive properties. Furthermore, the inclusion properties of N-leucine-enkephalin/β-CD conjugates toward a model neurotropic drug such as dothiepine were studied [123,124].

In another approach, β- and γ-CDs were conjugated to the neuropeptide substance P or its shorter but active C-terminal fragment. Substance P is an endogenous neuropeptide involved in numerous physiological or pathological processes in the central and/or peripheral nervous system by interaction with the G-protein-coupled receptor NK1. The structural integrity of the CD and of the neuropeptide forming the conjugates was demonstrated through a two-sided immunometric assay based on the simultaneous use of two antibodies specifically directed against both parts of the molecule. The ability of the conjugates to interact with NK1 receptors was demonstrated both by in vitro experiments with the CHO cell line transfected in order to express NK1 human receptor, and with ex vivo experiments on rat spinal cord. Even if a significant decrease in affinity was noted in cell culture experiments for CD conjugates in comparison to substance P, the conjugates were able to competitively displace ^{125}I-radiolabeled substance P from NK1 receptors in rat spinal cord sections autoradiography. Moreover, it was evidenced that substance P–CD conjugates were considerably more stable than the neuropeptide alone in biological media, probably because of a steric hindrance to the binding of proteolytic enzymes [125].

Mannosylated polyamidoamine (PMMA) starburst dendrimer conjugated with α-CD has been proposed for improve gene delivery to the kidneys. As in the case of the galactosylated dendrimer, in vitro experiments on various cell lines evidenced an improved transfection efficiency compared to the dendrimer or the nonmannosylated conjugate, even if independent from mannose receptor surface expression. However, after intravenous injection in mice, the mannosylated CD conjugate provided significantly higher luciferase activity in kidneys than that of administration of the nude plasmid or its complexes with the PMMA dendrimer or a nontargeted CD conjugate [126].

Bone-targeting drug delivery systems based on CD conjugates have been developed for the treatment of skeletal disease. An alendronate/β-CD conjugate was proposed for specific delivery of the bone anabolic agent prostaglandin E1 (PGE_1). The conjugate was demonstrated to have a strong affinity to hydroxyapatite and was able to form an inclusion complex with PGE_1. In vivo studies on rat mandibles, modeling the repair of craniofacial bony defects, showed a strong bone anabolic effect for the alendronate/β-CD/PGE_1 complex, but surprisingly, the CD conjugate alone caused significantly higher new bone formation while avoiding prostaglandin's side effects. A proposed explanation is in situ complexation by alendronate/β-CD conjugates of endogenous bone-active lipophilic compounds [127].

In a recent study, β-CD was transformed into its carboxylic derivative and conjugated to the nonapeptide oxytocin to target the uterus. After conjugation with β-CD, the oxytocin retained the ability to stimulate the contractions of myometrium isolated from rats, even if exhibiting a significantly lower potency than that of the original peptide. In any case, this targeted drug delivery system could be proposed for the administration of lipophilic labor-inducing drugs or of anticancer drugs in the treatment of cervix or endometrial cancer [128].

REFERENCES

1. Witkop, B. (1999). Paul Ehrlich and his magic bullets: revisited. *Proc. Am. Philos. Soc.*, *143*, 540–557.

2. Byrne, J.D., Betancourt, T., Brannon-Peppas, L. (2008). Active targeting schemes for nanoparticles systems in cancer therapeutics. *Adv. Drug Deliv. Rev.*, *60*, 1615–1626.

3. Bae, Y.H. (2009). Drug targeting and tumor heterogeneity. *J. Control. Release*, *133*, 2–3.

4. Hoffman, A.S. (2008). The origin and evolution of "controlled" drug delivery systems. *J. Control. Release*, *132*, 153–163.

5. Loftsson, T., Duchêne D. (2007). Cyclodextrins and their pharmaceutical applications. *Int. J. Pharm.*, *329*, 1–11.

6. Hirayama, F., Uekama K. (1999). Cyclodextrin-based controlled drug release. *Adv. Drug Deliv. Rev.*, *36*, 125–141.

7. Defaye, J., Garcia-Fernandez, J.M., Ortiz Mellet, C. (2007). Les cyclodextrines en pharmacie: perspectives pour le ciblage d'actifs thérapeutiques et le contrôle d'interactions membranaires. *Ann. Pharm. Fr.*, *65*, 33–49.

8. Geffen, D.B., Man, S. (2002). New drugs for the treatment of cancer 1990–2001. *Israeli Med. Assoc. J.*, *4*, 1124–1131.

9. Dassonville, O., Bozec, A., Fischel, J.L., Milano, G. (2007). EGFR targeting therapies: monoclonal antibodies versus

tyrosine kinase inhibitors: similarities and differences. *Crit. Rev. Oncol./Hematol.*, *62*, 53–61.

10. Kirpotin, D.B., Drummond, D.C., Shao, Y., Shalaby, M.R., Hong, K., Nielsen, U.B., Marks, J.D., Benz, C.C., Park, J.W. (2006). Antibody targeting of long-circulating lipidic nanoparticles does not increase tumor localization but does increase internalization in animal models. *Cancer Res.*, *66*, 6732–6740.

11. Atobe, K., Ishida, T., Ishida, E., Hashimoto, K., Kobayashi, H., Yasuda, J., Aoki, T., Obata, K., Kikuchi, H., Akita, H., Asai, T., Harashima, H., Oku, N., Kiwada, H. (2007). In vitro efficacy of a sterically stabilized immunoliposomes targeted to membrane type 1 matrix metalloproteinase (MT1-MMP). *Biol. Pharm. Bull.*, *30*, 972–978.

12. Nisato, R.E., Tille, J.C., Jonczyk, A., Goodman, S.L., Pepper, M.S. (2003). $\alpha_v\beta_3$ and $\alpha_v\beta_5$ integrin antagonists inhibit angiogenesis in vitro. *Angiogenesis*, *6*, 105–119.

13. Nasongkla, N., Shuai, X., Ai, H., Weinberg, B.D., Pink, J., Boothman, D.A., Gao, J. (2004). cRGD-functionalized polymer micelles for targeted doxorubicin delivery. *Angew. Chem. (Int. Ed. Eng.) 43*, 6323–6327.

14. Vihinen, P., Ala-aho, R., Kähäri, V.M. (2005). Matrix metalloproteinases as therapeutic targets in cancer. *Curr. Cancer Drug Targets*, *5*, 203–220.

15. Hatakeyama, H., Akita, H., Ishida, E., Hashimoto, K., Kobayashi, H., Aoki, T., Yasuda, J., Obata, K., Kikuchi, H., Ishida, T., Kiwada, H., Harashima, H. (2007). Tumor targeting of doxorubicin by anti-MT1-MMP antibody-modified PEG liposomes. *Int. J. Pharm.*, *342*, 194–200.

16. Laskin, J.J., Sandler, A.B. (2004). Epidermal growth factor receptor: a promising target in solid tumours. *Cancer Treat. Rev.*, *30*, 1–17.

17. Pan, X., Wu, G., Yang, W., Barth, R.F., Tjarks, W., Lee, R.J. (2007). Synthesis of cetuximab-immunoliposomes via a cholesterol-based membrane anchor for targeting of EGFR. *Bioconjugate Chem.*, *18*, 101–108.

18. Nobs, L., Buchegger, F., Gurny, R., Allemann, E. (2006). Biodegradable nanoparticles for direct or two-step tumor immunotargeting. *Bioconjugate Chem.*, *17*, 139–145.

19. Silver, D.A., Pellicer, I., Fair, W.R., Heston, W.D., Cordon-Cardo, C. (1997). Prostate-specific membrane antigen expression in normal and malignant human tissues. *Clin. Cancer Res.*, *3*, 81–85.

20. Farokhzad, O.C., Cheng, J., Teply, B.A., Sherifi, I., Jon, S., Kantoff, P.W., Richie, J.P., Langer, R. (2006). Targeted nanoparticle–aptamer bioconjugates for cancer chemotherapy in vivo. *Proc. Natl. Acad. Sci. USA*, *103*, 6315–6320.

21. Antony, A.C. (1992). The biological chemistry of folate receptors. *Blood*, *79*, 2807–2820.

22. Anderson, R.G.W., Kamen, B.A., Rothberg, K.G., Lacy, S.W. (1992). Potocytosis: sequestration and transport of small molecules by caveolae. *Science*, *255*, 410–411.

23. Turek, J.J., Leamon, C.P., Low, P.S. (1993). Endocytosis of folate–protein conjugates: ultrastructural localization in KB cells. *J. Cell Sci.*, *106*, 423–430.

24. Garin-Chesa, P., Campbell, I., Saigo, P.E., Lewis, J.L., Jr., Old, L.J., Rettig, W.J. (1993). Trophoblast and ovarian cancer antigen LK26: sensitivity and specificity in immunopathology and molecular dentification as a folate-binding protein. *Am. J. Pathol.*, *142*, 557–567.

25. Wang, S., Lee, R.J., Mathias, C.J., Green, M.A., Low, P.S. (1996). Synthesis, purification, and tumor cell uptake of 67Ga-deferoxamine-folate, a potential radiopharmaceutical for tumor imaging. *Bioconjugate Chem.*, *7*, 56–62.

26. Lee R.J., Low P.S. (1994). Delivery of liposomes into cultured KB cells via folate receptor-mediated endocytosis. *J. Biol. Chem.*, *269*, 3198–3204.

27. Gottschalk, S., Cristiano, R.J., Smith, L.C., Woo, S.L. (1994). Folate receptor mediated DNA delivery into tumor cells potosomal disruption results in enhanced gene expression. *Gene Ther.*, *1*, 185–191.

28. Leamon, C.P., Low, P.S. (1994). Selective targeting of malignant cells with cytotoxin–folate conjugates. *J. Drug Target.*, *2*, 101–112.

29. Caliceti, P., Salmaso, S., Semenzato, A., Carofiglio, T., Fornasier, R., Fermeglia, M., Ferrone, M., Pricl, S. (2003). Synthesis and physicochemical characterization of folate–cyclodextrin bioconjugate for active drug delivery. *Bioconjugate Chem.*, *14*, 899–908.

30. Higuchi, T., Connors, K.A. (1965). Phase-solubility techniques. *Adv. Anal. Chem. Instrum.*, *4*, 117–212.

31. Ohtani, Y., Irie, T., Uekama, K., Fukunaga, K., Pitha, J. (1989). Differential effects of α-, β- and γ-cyclodextrins on human erythrocytes. *Eur. J. Biochem.*, *186*, 17–22.

32. Salmaso, S., Semenzato, A., Caliceti, P., Hoebeke, J., Sonvico, F., Dubernet, C., Couvreur, P. (2004). Specific antitumor targetable β-cyclodextrin–poly(ethylene glycol)–folic acid drug delivery bioconjugate. *Bioconjugate Chem.*, *15*, 997–1004.

33. Salter, D.N., Scott, K.J., Slade, H., Andrews, P. (1981). The preparation and properties of folate-binding protein from cow's milk. *Biochem. J.*, *193*, 469–476.

34. Sonvico, F., Dubernet, C., Marsaud, V., Appel, M., Chacun, H., Stella, B., Renoir, J.-M., Colombo, P., Couvreur, P. (2005). Establishment of an in vitro model expressing the folate receptor for the investigation of targeted delivery systems. *J. Drug Deliv. Sci. Technol.*, *15*, 407–410.

35. Salmaso, S., Bersani, S., Semenzato, A., Caliceti, P. (2007). New cyclodextrin bioconjugates for active tumour targeting. *J. Drug Target.*, *15*, 379–390.

36. Ishida, J., Ohtsu, H., Tachibana, Y., Nakanishi, Y., Bastow, K. F., Nagai, M., Wang, H.K., Itokawa, H., Lee, K.H. (2002). Antitumor agents: Part 214. Synthesis and evaluation of curcumin analogues as cytotoxic agents. *Bioorg. Med. Chem.*, *10*, 3481–3487.

37. Singh, M. (1999). Transferrin as a targeting ligand for liposomes and anticancer drugs. *Curr. Pharm. Des.*, *5*, 443–451.

38. Faulk, W.P., Taylor, C.G., Yeh, C.J., McIntyre, J.A. (1990). Preliminary clinical study of transferrin–adriamycin conjugate for drug delivery to acute leukemia patients. *Mol. Biother.*, *2*, 57–60.

39. Head, J.F., Wang, F., Elliott, R.L. (1997). Antineoplastic drugs that interfere with iron metabolism in cancer cells. *Adv. Enzyme Regul.*, *37*, 147–169.

40. Rainov, N.G., Soling, A. (2005). Technology evaluation: TransMID, KS Biomedix/Nycomed/Sosei/PharmaEngine. *Curr. Opin. Mol. Ther.*, *7*, 483–492.

41. Eavarone, D.A., Yu, X., Bellamkonda, R.V. (2000). Targeted drug delivery to C6 glioma by transferrin-coupled liposomes. *J. Biomed. Mater. Res.*, *51*, 10–14.

42. Wu, J., Lu, Y., Lee, A., Pan, X., Yang, X., Zhao, X., Lee, R.J. (2007). Reversal of multidrug resistance by transferrin-conjugated liposomes co-encapsulating doxorubicin and verapamil. *J. Pharm. Sci.*, *10*, 350–357.

43. Gonzalez, H., Hwang, S.J., Davis, M. E. (1999). New class of polymers for the delivery of macromolecular therapeutics. *Bioconjugate Chem.*, *10*, 1068–1074.

44. Pack, D., Putnam, D., Langer, R. (2000). Design of imidazole-containing endosomolytic biopolymers for gene delivery. *Biotechnol. Bioeng.*, *67*, 217–223.

45. Bellocq, N.C., Pun, S.H., Jensen, G.S., Davis, M.E. (2003). Transferrin-containing cyclodextrin polymer-based particles for tumor-targeted gene delivery. *Bioconjugate Chem.*, *14*, 1122–1132.

46. Ward, C.M., Read, M.L., Seymour, L.W. (2001). Systemic circulation of poly(L-lysine)/DNA vectors is influenced by polycation molecular weight and type of DNA: differential circulation in mice and rats and the implications for human gene therapy. *Blood*, *97*, 2221–2229.

47. Bartlett, D.W., Davis, M.E. (2007). Physicochemical and biological characterization of targeted nucleic acid-containing nanoparticles. *Bioconjugate Chem.*, *18*, 456–468.

48. Pun, S.H., Tack, F., Bellocq, N.C., Cheng, J., Grubbs, B.H., Jensen, G.S., Davis, M.E., Brewster, M., Janicot, M., Janssens, B., Floren, W., Bakker, A. (2004). Targeted delivery of RNA-cleaving DNA enzyme (DNAzyme) to tumor tissue by transferrin-modified, cyclodextrin-based particles. *Cancer Biol. Ther.*, *3*, 641–650.

49. Hu-Lieskovan, S., Heidel, J.D., Bartlett, D.W., Davis, M.E. and Triche, T. J. (2005). Sequence-specific knockdown of EWS-FLI1 by targeted, nonviral delivery of small interfering RNA inhibits tumor growth in a murine model of metastatic Ewing's sarcoma. *Cancer Res.*, *65*, 8984–8992.

50. Bartlett, D.W., Su, H., Hildebrandt, I.J., Weber, W.A., Davis, M.E. (2007). Impact of tumor-specific targeting on the biodistribution and efficacy of siRNA nanoparticles measured by multimodality in vivo imaging. *Proc. Natl. Acad. Sci. USA*, *104*, 15549–15554.

51. Bartlett, D.W., Davis, M.E. (2008). Impact of tumor-specific targeting and dosing schedule on tumor growth inhibition after intravenous administration of siRNA-containing nanoparticles. *Biotechnol. Bioeng.*, *99*, 975–985.

52. Heidel, J.D., Yu, Z., Yi-Ching Liu, J., Rele, S.M., Liang, Y., Zeidan, R.K., Kornbrust, D.J., Davis, M.E. (2007). Administration in non-human primates of escalating intravenous doses of targeted nanoparticles containing ribonucleotide

reductase subunit M2 siRNA. *Proc. Natl. Acad. Sci. USA*, *104*, 5715–5721.

53. Zimmermann, T.S., Lee, A.C., Akinc, A., Bramlage, B., Bumcrot, D., Fedoruk, M.N., Harborth, J., Heyes, J.A., Jeffs, L.B., John, M., Judge, A.D., Lam, K., McClintock, K., Nechev, L.V., Palmer, L.R., Racie, T., Röhl, I., Seiffert, S., Shanmugam, S., Sood, V., Soutschek, J., Toudjarska, I., Wheat, A.J., Yaworski, E., Zedalis, W., Koteliansky, V., Manoharan, M., Vornlocher, H.P., MacLachlan, I. (2006). RNAi-mediated gene silencing in non-human primates. *Nature*, *441*, 111–114.

54. Mort, J.S., Buttle, D. J (1997). Cathepsin B. *Int. J. Biochem. Cell Biol.*, *29*, 715–720.

55. Frosch, B.A., Berquin, I., Emmert-Buck, M.R., Moin, K., Sloane, B.F. (1999). Molecular regulation, membrane association and secretion of tumor cathepsin B. *Acta Pathol., Microbiol., Immunol. Scand.*, *107*, 28–37.

56. Campo, E., Munoz, J., Miquel, R., Palacin, A., Cardesa, A., Sloane, B.F., Emmert-Buck, M.R. (1994). Cathepsin B expression in colorectal carcinomas correlates with tumor progression and shortened patient survival. *Am. J. Pathol.*, *145*, 301–309.

57. Scorilas, A., Fotiou, S., Tsiambas, E., Yotis, J., Kotsiandri, F., Sameni, M., Sloane, B.F., Talieri, M. (2002). Determination of cathepsin B expression may offer additional prognostic information for ovarian cancer patients. *Biol. Chem.*, *383*, 1297–1303.

58. Arkona, C., Wiederanders, B. (1996). Expression, subcellular distribution and plasma membrane binding of cathepsin B and gelatinases in bone metastatictissue. *Biol. Chem.*, *377*, 695–702.

59. Schaschke, N., Assfalg-Machleidt, I., Machleidt, W., Lassleben, T., Sommerhoff, C.P., Moroder, L. (2000). β-Cyclodextrin/epoxysuccinyl peptide conjugates: a new drug targeting system for tumor cells. *Bioorg. Med. Chem. Lett.*, *10*, 677–680.

60. Kosaraju, S.L. (2005). Colon targeted delivery systems: review of polysaccharides for encapsulation and delivery. *Crit. Rev. Food Sci. Nutr.*, *45*, 251–258.

61. Soares, A.F., Carvalho, R.A., Veiga, F. (2007). Oral administration of peptides and proteins: nanoparticles and cyclodextrins as biocompatible delivery systems. *Nanomedicine*, 2, 183–202.

62. Piao, Z.Z., Lee M.K., Lee B.J. (2008). Colonic release and reduced intestinal tissue damage of coated tablets containing naproxen inclusion complex. *Int. J. Pharm.*, *350*, 205–211.

63. Villar-Lopez, M.E., Nieto-Reyes, L., Anguiano-Igea, S., Otero-Espinar, F.J., Blanco-Méndez, J. (1999). Formulation of triamcinolone acetonide pellets suitable for coating and colon targeting. *Int. J. Pharm.*, *179*, 229–235.

64. Maestrelli, F., Zerrouk, N., Cirri, M., Mennini, N., Mura, P. (2008). Microspheres for colonic delivery of ketoprofen-hydroxypropyl-β-cyclodextrin complex. *Eur. J. Pharm. Sci.*, *34*, 1–11.

65. Sharma, P., Varma, M.V.S., Chawla, H.P.S., Panchagnula, R. (2005). Absorption enhancement, mechanistic and toxicity

studies of medium chain fatty acids, cyclodextrins and bile salts as peroral absorption enhancers. *Farmaco*, *60*, 884–893.

66. Mundargi, R.C., Patil, S.A., Agnihotri, S.A., Aminabhavi, T.M. (2007). Development of a polysaccharide-based colon targeted delivery systems for the treatment of amoebiasis. *Drug Dev. Ind. Pharm.*, *33*, 255–264.

67. Fourié, B., Molis, C., Achour, L., Cupas, H., Hatat, C., Rambaud, J.C. (1993). Fate of β-cyclodextrin in the human intestine. *J. Nutr.*, *123*, 676–680.

68. Antenucci, R., Palmer, J.K. (1984). Enzymatic degradation of α- and β-cyclodextrins by bacterioides of the human colon. *J. Agric. Food Chem.*, *32*, 1316–1321.

69. Meissner, Y., Lamprecht, A. (2008). Alternative drug delivery approaches for the therapy of inflammatory bowel disease. *J. Pharm. Sci.*, *97*, 2878–2891.

70. Friend, D.R. (1998). Review article: issues in oral administration of locally acting glucocorticosteroids for treatment of inflammatory bowel disease. *Aliment. Pharmacol. Ther.*, *12*, 591–603.

71. Hirayama, F., Minami, K., Uekama, K. (1996). In vitro evaluation of biphenylyl acetic acid–β-cyclodextrin conjugates as colon-targeting prodrugs: drug release behaviour in rat biological media. *J. Pharm. Pharmacol.*, *48*, 27–31.

72. Uekama, K., Minami, K., Hirayama, F. (1997). 6^A-O-[(4-Biphenylyl)acetyl]-α-, -β-, and -γ-cyclodextrins and 6^A-deoxy-6A-[[(4-biphenylyl)acetyl]amino]- α-, -β-, and -γ-cyclodextrins: potential prodrugs for colon-specific delivery. *J. Med. Chem.*, *20*, 255–276.

73. Minami, K., Hirayama, F., Uekama, K. (1998). Colon-specific drug delivery based on α-cyclodextrin prodrug release behaviour of biphenylylacetic acid from its cyclodextrin conjugates in rat intestinal tracts after oral administration. *J. Pharm. Sci.*, *87*, 715–720.

74. Yano, H., Hirayama, F., Arima, H., Uekama, K. (2000). Hydrolysis behaviour of prednisolone 21-hemisuccinate/β-cyclodextrin amide conjugate: involvement of intramolecular catalysis of amide group in drug release. *Chem. Pharm. Bull.* 48, 1125–1128.

75. Yano, H., Hirayama, F., Arima, H., Uekama, K. (2001). Preparation of prednisolone-appended α-, β-, and γ-cyclodextrins: substitution at secondary hydroxyl groups and in vitro hydrolysis behaviour. *J. Pharm. Sci.*, *90*, 493–503.

76. Yano, H., Hirayama, F., Arima, H., Uekama, K. (2001). Prednisolone-appended α-cyclodextrin: alleviation of systemic adverse effect of prednisolone after intracolonic administration in 2,4, 6-trinitrobenzensulfonic acid-induced colitis rats. *J. Pharm. Sci.*, *90*, 2103–2112.

77. Yano, H., Hirayama, F., Arima, H., Uekama, K. (2002). Colon specific delivery of prednisolone-appended α-cyclodextrin: alleviation of systemic adverse effect after oral administration. *J. Control. Release*, *79*, 103–112.

78. Zou, M., Okamoto, H., Cheng, G., Hao, X., Sun, J., Cui, F., Danjo, K. (2005). Synthesis and properties of polysaccharide prodrugs of 5-aminosalicylic acid as potential colon-specific delivery systems. *Eur. J. Pharm. Biopharm.*, *59*, 155–160.

79. Zou, M., Cheng, G., Okamoto, H., Hao, X., An, F., Cui, F., Danjo, K. (2005). Colon-specific drug delivery systems based on cyclodextrin prodrugs: in vivo evaluation of 5-aminosalicylic acid from its cyclodextrin conjugates. *World J. Gastroenterol.*, *11*, 7457–7460.

80. El-Kamel, A.H., Abdel-Aziz, A.A.-M., Fatani, A.J., El-Subbagh, H.I. (2008). Oral colon targeted delivery systems for treatment of inflammatory bowel diseases: synthesis, in vitro and in vivo assessment. *Int. J. Pharm.*, *358*, 248–255.

81. Kamada, M., Hirayama, F., Udo, K., Yano, H., Arima, H., Uekama, K. (2002). Cyclodextrin conjugate–based controlled release system: repeated- and prolonged-releases of ketoprofen after oral administration in rats. *J. Control. Release*, *82*, 407–416.

82. Ciftci, K. (1996). Alternative approaches to the treatment of colon cancer. *Eur. J. Pharm. Bipharm.*, *42*, 160–170.

83. Tanaka, T., de Azevedo, M.B., Durán, N., Alderete, J.B., Epifano, F., Genovese, S., Tanaka, M., Tanaka, T., Curini, M. (2010). Colorectal cancer chemoprevention by 2 β-cyclodextrin inclusion compounds of auraptene and 4′-geranyloxyferulic acid. *Int. J. Cancer*, *126*, 830–840.

84. Krause-Heuer, A.M., Wheate, N.J., Tilby, M.J., Pearson, D.G., Ottley, C.J., Aldrich-Wright, J.R. (2008). Substituted β-cyclodextrin and calix[4]arene as encapsulatory vehicles for platinum(II)-based DNA intercalators. *Inorg. Chem.*, *47*, 6880–6888.

85. Hirayama, F., Ogata, T., Yano, H., Arima, H., Udo, K., Takano, M., Uekama, K. (2000). Release characteristics of a short-chain fatty acid, *n*-butyric acid, from its β-cyclodextrin ester conjugate in rat biological media. *J. Pharm. Sci.*, *89*, 1489–1495.

86. Chen, J., Khin, K.T., Davis, M.E. (2004). Antitumor activity of β-cyclodextrin polymer conjugates. *Mol. Pharm.*, *1*, 183–193.

87. Shluep, T., Gunawan, P., Ma, L., Jensen, G.S., Duringer, J., Hinton, S., Richter, W., Hwang, J. (2009). Polymeric tubulysin–peptide nanoparticles with potent antitumor activity. *Clin. Cancer Res.*, *15*, 181–189.

88. Charman, S.A., Mason, K.L., Charman, W.L. (1993). Techniques for assessing the effects of pharmaceutical excipients on the aggregation of porcine growth hormone. *Pharm. Res.*, *10*, 954–962.

89. Kang, F., Singh, J. (2003). Conformational stability of a model protein (bovine serum albumin) during primary emulsification process of PLGA microspheres synthesis. *Int. J. Pharm.*, *260*, 149–156.

90. Garcia-Fuentes, M., Trapani, A., Alonso, M.J. (2006). Protection of the peptide glutathione by complex formation with α-cyclodextrin: NMR spectroscopic analysis and stability study. *Eur. J. Pharm. Biopharm.*, *64*, 146–153.

91. Kamphorst, A.O., Mendes de Sá, I., Faria, A.M.C., Sinisterra, R. D. (2004). Association complexes between ovalbumin and cyclodextrins have no effect on the immunological properties of ovalbumin. *Eur. J. Pharm. Biopharm.*, *57*, 199–205.

92. Jerry, N., Anitha, Y., Sharma, C.P., Sony, P. (2001). In vivo absorption studies of insulin from an oral delivery system. *Drug Deliv.*, *8*, 19–23.

93. Trapani, A., Laquintana, V., Denora, N., Lopedota, A., Cutrignelli, A., Franco, M., Trapani, G., Liso, G. (2007). Eudragit RS-100 microparticles containing 2-hydroxypropyl-β-cyclodextrin and glutathione: physicochemical characterization, drug release and transport studies. *Eur. J. Pharm. Sci.*, *30*, 64–74.

94. Sajeesh, S., Sharma, C.P. (2006). Cyclodextrin–insulin complex encapsulated polymethacrylic acid based nanoparticles for oral insulin delivery. *Int. J. Pharm.*, *325*, 147–154.

95. Bies, C., Lehr, C.-M., Woodley, J.F. (2004). Lectin-mediated drug targeting: history and applications. *Adv. Drug Deliv. Rev.*, *56*, 425–435.

96. Oda, Y., Kobayashi, N., Yamanoi, T., Katsuraya, K., Takahashi, K., Hattori, K. (2008). β-Cyclodextrin conjugates with glucose moieties designed as drug carriers: their syntheses, evaluations using concanavalin A and doxorubicin, and structural analyses by NMR spectroscopy. *Med. Chem.*, *4*, 244–255.

97. Ikuta, A., Mizuta, N., Kitahata, S., Murata, S., Usui, T., Koizumi, K., Tanimoto, T. (2004). Preparation and characterization of novel branched β-cyclodextrins having β-D-galactose residues on the nonreducing terminal of the side chains and their specific interactions with peanut (*Arachis hypogea*) agglutinin. *Chem. Pharm. Bull.*, *52*, 51–56.

98. Kubota, Y., Sanbe, H., Kizumi, K. (2002). Absorption, distribution and excretion of galactosyl-β-cyclodextrin and mannosyl-β-cyclodextrin in rats. *Biol. Pharm. Bull.*, *23*, 472–476.

99. Nishi, Y., Yamane, N., Tanimoto, T. (2007). Preparation and characterization of 6I-6n-di-*O*-(L-fucopyranosyl)-β-cyclodextrin (*n* = II–IV) and investigation of their functions. *Carbohydr. Res.*, *342*, 2173–2181.

100. Nishi, Y., Tanimoto, T. (2009). Preparation and characterization of branched β-cyclodextrin having α-L-fucopyranose and a study of their functions. *Biosci. Biotechnol. Biochem.*, *73*, 562–569.

101. Vargas-Berenguel, A., Ortega-Caballero, F., Santoyo-Gonzales, F., Garcia-Lopez, J.J., Gimenez-Martinez, J.J., Garcia-Fuentes, L., Ortiz-Salmeron, E. (2002). Dendritic galactoside based on a β-cyclodextrin core for the construction of site specific molecular delivery systems: synthesis and molecular recognition studies. *Chem. Eur. J.*, *8*, 812–827.

102. Benito, J.M., Gomez-Garcia, M., Ortiz-Mellet, C., Baussanne, I., Defaye, J., Garcia Fernandez, J.M. (2004). Optimizing saccharide-directed molecular delivery to biological receptors: design synthesis and biological evaluation of glycodendrimer–cyclodextrin conjugates. *J. Am. Chem. Soc.*, *126*, 10355–10363.

103. Woodley, J.F. (2000). Lectins for gastrointestinal targeting: 15 years on. *J. Drug Target.*, *7*, 325–333; Minko, T. (2004). Drug targeting to the colon with lectin and neoglycoconjugates. *Adv. Drug Deliv. Rev.*, *56*, 491–509.

104. Tanino, T, Ogiso, T., Iwaki, M. (1999). Effect of sugar-modified β-cyclodextrins on dissolution and absorption characteristics of phenytoin. *Biol. Pharm. Bull.*, *22*, 298–304.

105. Seo, S.J., Kim, S.H., Sasagawa, T., Choi, Y.J., Akoike T., Cho, C.S. (2004). Delivery of *all trans*-retinoic acid (RA) to

hepatocyte cell line from RA/galactosyl α-cyclodextrin inclusion complex. *Eur. J. Pharm. Biopharm.*, *58*, 681–687.

106. Shinoda, T., Maeda, A., Kagatani, S., Konno, Y., Sonobe, T., Fukui, M., Hashimoto, H., Hara, K., Fujita, K. (1998). Specific interaction between galactose branched-cyclodextrins and hepatocytes in vitro. *Int. J. Pharm.*, *167*, 147–154.

107. Shinoda, T., Kagatani, S., Maeda, A., Konno, Y., Hashimoto, H., Hara, K., Fujita, K., Sonobe, T. (1999). Sugar-branched-cyclodextrins as injectable drug carriers in mice. *Drug Dev. Ind. Pharm.*, *25*, 1185–1192.

108. Oda, Y., Yanagisawa, H., Maruyama, M., Hattori, K., Yamanoi, T. (2008). Design, synthesis and evaluation of D-galactose-β-cyclodextrin conjugates as drug-carrying molecules. *Bioorg. Med. Chem.*, *16*, 8830–8840.

109. Wada, K., Arima, H., Tsutsumi, T., Hirayama, F., Uekama, K. (2005). Enhancing effects of galactosylated dendrimer/a–cyclodextrin conjugates on gene transfer efficiency. *Biol. Pharm. Bull.*, *28*, 500–505.

110. Mandolini, L. (1986). Intramolecular reactions of chain molecules. *Adv. Phys. Org. Chem.*, *22*, 1–111.

111. Yui, N., Ooya, T. (2006). Molecular mobility of interlocked structures exploiting new functions of advanced biomaterials. *Chem. Eur. J.*, *12*, 6730–6737.

112. Ooya, T., Kawashima, T., Yui, N. (2001). Synthesis of polyrotaxanes–biotin conjugate and surface plasmon resonance analysis of streptavidin recognition. *Biotechnol. Bioprocess Eng.*, *6*, 293–300.

113. Ramanathan, S., Pooyan, S., Stein, S., Prasad, P.D., Wang, J., Leibowitz, M.J., Ganapathy, V., Sinko, P.J. (2001). Targeting the sodium-dependent multivitamin transporter (SMVT) for improving the oral absorption properties of a retro-inverso Tat nonapeptide. *Pharm. Res.*, *18*, 950–956.

114. Yui, N., Ooya, T., Kawashima, T., Saito, Y., Tamai, I., Sai, Y., Tsuji, A. (2002). Inhibitory effect of supramolecular polyrotaxane–dipeptide conjugates on digested peptide uptake via intestinal human peptide transporter. *Bioconjugate Chem.*, *13*, 582–587.

115. Ooya, T., Yui, N. (2002). Multivalent interactions between biotin–polyrotaxane conjugates and streptavidin as a model of new targeting for transporters. *J. Control. Release*, *80*, 219–228.

116. Gestwicki, J.E., Cairo, C.W., Strong, L.E., Oetjen, K.A., Kiessling, L.L. (2002). Influencing receptor-ligand binding mechanisms with multivalent ligand architecture. *J. Am. Chem. Soc.*, *124*, 14922–14933.

117. Rubinstein, N., Alvarez, M., Zwirner, N.W., Toscano, M. A., Ilarregui, J.M., Bravo, A., Mordoh, J., Fainboim, L., Podhajcer O.L., Rabinovich, G. A. (2004). Targeted inhibition of galectin-1 gene expression in tumor cells results in heightened T cell–mediated rejection. *Cancer Cell*, *5*, 241–251.

118. Perillo, N.L., Marcus, M.E., Baum, L.G. (1998). Galectins: versatile modulators of cell adhesion, cell proliferation, and cell death. *J. Mol. Med. 76*, 402–412.

119. Nelson, A., Belitsky, J.M., Vidal, S., Joiner, C.S., Baum, L.G., Stoddart, J.F. (2004). A self-assembled multivalent

pseudopolyrotaxane for binding galectin-1. *J. Am. Chem. Soc.*, *126*, 11914–11922.

120. Belitsky, J.M., Nelson, A., Hernandez, J.D., Baum, L.G., Stoddart, J.F. (2007). Multivalent interactions between lectins and supramolecular complexes: galectin-1 and self-assembled pseudopolyrotaxanes. *Chem. Biol.*, *14*, 1140–1151.

121. Ooya, T., Eguchi, M., Yui, N. (2003). Supramolecular design for multivalent interaction: maltose mobility along polyrotaxane enhanced binding with concanavalin A. *J. Am. Chem. Soc.*, *125*, 13016–13017.

122. Ooya, T., Utsunomiya, H., Eguchi, M., Yui., N. (2005). Rapid binding of concanavalin A and maltose–polyrotaxane conjugates due to mobile motion of α-cyclodextrins threaded onto a poly(ethylene glycol). *Bioconjugate Chem.*, *16*, 62–69.

123. Djedaïni-Pilard, F., Desalos, J., Perly, B. (2003). Synthesis of a new molecular carrier: *N*-(Leu-enkephalin)yl 6-amido-6-de-oxy-cyclomaltoheptaose. *Tetrahedron Lett.*, *34*, 2457–2460.

124. Hristova-Kazmierski, M.K., Horan, P., Davis, P., Yamamura, H.I., Kramer, T., Horvath, R., Kazmierski, W.M., Porreca, F., Hruby, V. (1993). A new approach to enhance bioavailability of biologically active peptides: conjugation of a δ opioid agonist to β-cyclodextrin. *Bioorg. Med. Chem. Lett.*, *3*, 831.

125. Péan, C., Wijkhuisen, A., Djedaïni-Pilard, F., Fisher, J., Doly, S., Conrath, M., Couraud, J.Y., Grassi, J., Perly, B., Créminon, C. (2001). Pharmacological in vitro evaluation of new substance P-cyclodextrin derivatives designed to drug targeting towards NK1-receptor bearing cells. *Biochim. Biophys. Acta*, *1541*, 150–160.

126. Wada, K., Arima, H., Tsutsumi, T., Chihara, Y., Hattori, K., Hirayama, F., Uekama, K. (2005). Improvement of gene delivery mediated by mannosylated dendrimer/α-cyclodextrin conjugates. *J. Control. Release*, *104*, 397–413.

127. Liu, X.M., Wiswall, A.T., Rutledge, J.E., Akhter, M.P., Cullen, D.M., Reinhardt, R.A., Wang, D. (2008). Osteotropic β-cyclodextrin for local bone regeneration. *Biomaterials*, *29*, 1686–1692.

128. Bertolla, C., Rolin, S., Evrard, B., Pochet, L., Masereel, B. (2008). Synthesis and pharmacological evaluation of a new targeted drug carrier system: β-cyclodextrin coupled with oxytocin. *Bioorg. Med. Chem. Lett.*, *18*, 1855–1858.

14

CYCLODEXTRINS AND BIOTECHNOLOGICAL APPLICATIONS

Amit Singh, Abhishek Kaler, Vachan Singh, Rachit Patil, and Uttam C. Banerjee

Department of Pharmaceutical Technology (Biotechnology), National Institute of Pharmaceutical Education and Research, SAS Nagar, India

1. INTRODUCTION

Any process that involves the use of a living system, including microorganisms, as whole cells or enzymes, to synthesize specific products useful for humanity is included in biotechnology. Biotechnology has a profound effect on the quality of life as well as on the world's environment. Biotechnology comprises various subjects, such as genetics, cell and molecular biology, tissue culture (both plant and animal), biochemistry, microbiology, biochemical engineering, and bioinformatics. On the basis of applications, biotechnology is divided into blue (marine and aquatic), green (agricultural), red (medical), and white (industrial). Cyclodextrins (CDs) are used in all the applications of biotechnology. CDs have emerged as a vital tool to improve the solubility of hydrophobic substrates in aqueous media, by forming stable inclusion complexes. Being of natural origin, organic and biocompatible, CDs cause no harm to microbial cells. CDs are cyclic oligosaccharides of α-D glucose, joined through 1,4-bonds. Major CDs are of three types: α-cyclodextrin (six glucopyranose units), β-cyclodextrin (seven glucopyranose units), and γ-cyclodextrin (eight glucopyranose units). β-Cyclodextrin is the most useful, most easily accessible, and lowest priced. Formation of CDs with fewer than six glucopyranose units is sterically hindered, whereas CDs with more than eight glucopyranose units are difficult to separate in crystalline form, due to higher solubility, and are of less practical importance [1].

CDs are truncated cone (doughnut-shaped) molecules which have a hydrophilic external surface and a hydrophobic internal cavity. Because of this cavity, wide varieties of hydrophobic molecules can form inclusion complexes without forming covalent bonds. Further, physical properties of CDs can be altered by chemical modifications to improve the solubility, stability, and chemical activity of guest molecules. This chapter covers with the use of CDs in various fields of biotechnology.

2. APPLICATION OF CDs IN BIOTECHNOLOGY

The application of CDs in biotechnology is enormous. Biocatalysis and biotransformation, fermentation and downstream processing, bioseparations, bioanalysis, biodegradation or bioremediation, and preparation of food and flavors are the major biotechnological fields in which the CDs are frequently used.

2.1. Biocatalysis and Biotransformation

Industrial biotechnology is largely confined to the use of microbial cells or the enzymes thereof as catalysts for the synthesis of chirally pure chemicals required in agriculture, pharmaceuticals, and other industrial sectors. This technology, termed *biocatalysis*, has gained popularity over chemical synthesis because of the inherent selective catalytic properties of the enzymes in terms of stereo-, regio-, and chemoselectivity [2]. Further, the biocatalysts work in aqueous media at mild temperature and pressure; hence, biocatalysis is an ecofriendly and greener process. Besides various

Cyclodextrins in Pharmaceutics, Cosmetics, and Biomedicine: Current and Future Industrial Applications, First Edition. Edited by Erem Bilensoy.
© 2011 John Wiley & Sons, Inc. Published 2011 by John Wiley & Sons, Inc.

advantages of biocatalysis, there are a few limitations, such as instability of enzymes at desired reaction conditions, process cost, reaction variability, and low substrate or product tolerance of the enzymes, but the major limitation is the low solubility of most chemical compounds in aqueous reaction media, which is required for most enzymatic activity. Various remedies are available to overcome this problem: (1) the solubilization of hydrophobic substrate using solubilizers [3], emulsifiers [4], and so on and (2) use of organic solvents as cosolvents [5], of biphasic reaction media [6], or of microcrystalline substrates [7].

Each of these agents has some degree of detrimental effect on microorganisms, due to unfavorable interaction with the cell membrane, affecting cell integrity. These interactions become very significant when a high substrate concentration must be used or the substrate is extensively hydrophobic, since to make them soluble, a large amount of these solubility enhancers must be added in the biocatalysis medium.

The insolubility of hydrophobic substrates in an aqueous reaction medium is the greatest hindrance to the industrial application of biocatalysis for the synthesis of fine chemicals, since the vast majority of organic substrates are lipophilic and sparingly soluble in water. Further, the substrate or product formed may be toxic to the enzyme system. Low solubility is generally overcome by the addition of solubilizing agents or cosolvents. However, if the substrate or product is toxic for the enzyme, batch addition to the substrate and in situ recovery of the product formed is a more suitable approach. Here, the inhibitory effect of the substrate and product can be encountered along with less exposure of solubilizing agents or cosolvents to the microbes, resulting in less toxicity to the cell wall. The addition of CDs in biocatalytic reaction media provides an impressive alternative. CDs form stable inclusion complexes with lipophilic substrates and exist in aqueous solutions. The CD is termed the *host* and the other molecule, the *guest*. Interaction between CDs and guest molecules depends on size; the smaller the guest, the more easily it can be fitted in to the cavity. With large molecules only partial interaction is possible. One or two guest molecules can be complexed with one, two, or three CDs.

Displacement of enthalpy-rich water molecules from the cavity by more hydrophobic guest molecules present in the solution is the driving force for complex formation and leads to decreased CD ring strain, resulting in a more stable lower-energy state [8]. The interaction between a CD and a guest molecule is a dynamic equilibrium exchange. This complexation–decomplexation equilibrium is very fast, and in practice does not influence the rate of bioconversion. The binding strength depends on how well the host–guest complex fits together and on the specific local interactions between surface atoms. The ability of CDs to form stable inclusion complexes with lipophilic compounds influences biocatalytic reactions through:

1. Increased solubility and hence availability of organic substrates to the enzyme [9].
2. Reduced substrate or product inhibition of enzyme by complexing the effective free concentration [10].
3. Increased enantioselectivity of the product [11].
4. Biocompatibility, causing no damage to the biocatalyst [12].

Table 1 include examples of CDs has been used to increase bioconversion through increased substrate solubility or decreased inhibitory effect of substrate or product.

It is reported in the literature [17] that microbial transformation of cholesterol to androst-4-ene-3,17-dione was increased up to 90% using β-CD, which not only increased the solubility of cholesterol and the bioconversion rate, but also decreased the product inhibition and steroid nucleus degradation. In the absence of CDs, only 40% biotransformation of cholesterol was achieved. In the biotransformation of aromatic aldehydes to alcohols by *Saccharomyces cerevisiae*, CDs increased the aqueous solubility of the substrate (aldehydes) and improved the inhibitory influence [18]. It is reported that [19,20] the hydrolytic activity of lipases improved in the presence of CDs and their derivatives. Several natural and chemically modified CDs, especially γ-CDs, have been shown to increase the conversion rate due to improved solubility of substrates. It is known that the

Table 1. Examples of CDs Used in Fermentation and Conversion Systems

Substrate	Microorganism/Enzyme	Product	Yield (%)		CD Used
			Without CD	With CD	
Vitamin D$_3$ [13]	*Amycolata autotrophica*	1α,25-Dihydroxy vitamin D$_3$	—	7-fold higher	Partiallymethylated-β- CD
Androstenedione [10]	*Saccharomyces cerevisiae*	Testosterone	27	78	Hydroxypropyl β-CD
3β,7α-Cholest-5-ene -3,7-diol [14]	Cholesterol oxidase	7α-Hydroxy cholest-4-en-3-one	<20	90	Hydroxypropyl-β-CD
17β-estradiol [15]	Mushroom tyrosinase	2-Hydroxy estradiol	No reduction	30	β-CD
Starch [16]	*A. oryzae*	6'-Galactosyl-lactose	37	65	β-CD

Table 2. Examples of CDs Used in Biocatalysis

Substrate	Biocatalyst	Product	Enantiomeric Excess (%)		CDs Used
			Without CD	With CD	
(R,S)-Ketoprofen ethyl ester [11]	Lipase	(S)- Ketoprofen	40	90	Hydroxypropyl-β-CD
(R,S)-O-Butyryl propranolol [25]	*Rhizopus niveus* lipase	(S)-Propanolol	—	90	Hydroxypropyl-β-CD
1-(2-Furyl) ethane [26]	*Pseudomonas cepacia* lipase	(R)- Acetic acid 1-furan-2-yl-ethyl ester	28	100	α- and γ-CDs
(R,S)-Arylpropionic esters [24]	Bovine serum albumin	Aryl acids	50–81	80–99	β- CD

hydrolysis of (S)-2-O-(N-benzoylglycyl)-β-phenyl lactate was limited, due to substrate inhibition of the enzyme carboxypeptidase A. This was partially overcome by addition of β-CD and hydroxypropyl-β-CD, which increased the substrate concentration in the reaction but did not change the reaction rate [21]. Biotransformation of cortisone acetate to prednisone acetate by *Arthrobacter simplex* was increased by the addition of hydroxypropyl-β-CD. Here, CD improved the solubility of cortisone acetate and prednisone acetate by 36.7- and 19.8-fold, decreasing product inhibition [22]. Addition of CD in a fermentation medium significantly increased the biotransformation of cholesterol to testosterone by *Lactobacillus bulgaricus* [23].

CDs have also been reported to improve enantioselectivity in biocatalytic reactions. In the inclusion complex of prochiral substrate with CD, only one face of the substrate is available for the enzyme to act. This preferential enzymatic attack resulted in higher enantioselectivity [24]. Table 2 includes examples of the use of Cds to increase the enantioselectivity of bioconversion reactions.

2.2. Fermentation and Downstream Processing

CDs are also useful in fermentation. They increase the aqueous solubility, hence there is greater availability of lipohilic substrates to microbial cells, which in turn results in increased product formation. Decreased toxicity of the substrates and products in a fermentation medium also contributes to improved fermentation. CDs can also be used as fermentation medium ingredients. It is reported [27] that solid and liquid media were prepared for the growth of *Helicobacter pylori*, replacing blood or its derivatives by CDs. This may be a suitable medium for the primary isolation of the bacterium from biopsy samples and routine laboratory growth. In another study [28] it was mentioned that *Candida albicans* was grown in liquid media containing 1% tripcasine

as the sole nitrogen source with 1.8% CDs (α-, β-, and γ-). Tripcasine alone was inefficient for the induction of *C. albicans* hyphae, but addition of CDs, especially β-CD, induced the mycelial transition, and 25 to 30% septate hyphae formation was achieved by the use of β-CD. This may be due to the existence of an uptake system for CDs in *C. albicans*, which uptakes CDs as linear oligosaccharides. An increased rate of formation of penicillin-G from phenylacetic acid and 6-aminopenicillanic acid was reported when penicillin acylase was complexed in β-methyl γ-CD [29]. Biotransformation of cholesterol to testosterone was increased significantly by the addition of CDs in the fermentation medium. The addition of 0.1% CD to the growth medium of *Lactobacillus bulgaricus* significantly increased the production of testosterone, indicating facilitated transport of the steroid substrate through the microbial cell wall. Similarly, production of lankacidin C group of antibiotics (lankacidin and lankacidinol) by *Streptomyces* sp. was improved when the cells were grown in the presence of β-CD. Also, the production of lankamycin and other by-products in lankacidin C fermentation was completely inhibited in the presence of CDs. An inclusion complex is formed between CD and the lankacidin, which prevented the inhibitory influence of lankacidin on its production [30].

Separation of a fermentation product from the broth is of great interest since it contributes largely to the production cost. The insoluble complex formation ability of CDs with products can be explored for the selective separation of the product from the fermentation broth. In the aqueous medium, inclusion complexes of CDs with different compounds have a diverse solubility profile, which easily can be separated. The complexation of CD with the product formed depends on the chemical structure and properties of the product, type of CD used, their relative concentrations, temperature, pH, and other parameters. To select the optimum conditions and CD to be used, different types of cyclodextrin–guest phase solubility relationships need to be examined. CDs were shown to offer an exciting possibility for the downstream processing of low concentration fermentation products [31].

2.3. Catalytic Agents

CDs are also being used as enzymes in catalytic reactions. The enzymes basically catalyze the reaction by molecular recognition of substrate and then stabilization of the transition state compared to the ground state. To catalyze the reaction, the enzyme must stabilize the transition state of the reaction more than it stabilizes the ground state of the substrate. Several modifications of CDs are possible by chemically modifying naturally occurring CDs through substituting various functional compounds on the primary or secondary face of the molecule or by attaching reactive groups to the primary or secondary –OH groups present [8]. These modified CDs can mimic molecular recognition through hydrogen, hydrophobic, and ionicbondings. The modifications are so designed that it binds favorably to the transition state of the substrate, and at the same time, the product is released in a thermodynamically favorable process [32]. Such modified CDs can enhance the rate of reaction by almost 1000-fold compared to the free solution. Due to their steric effect, the enantiomeric purity of the product formed was also found to be excellent. This is due to the preferential attack by the reagent on one of the enantioselective faces of the prochiral guest molecule.

The first chymotrypsin mimic was produced by coupling tripeptide Ser-His-Asp, the catalytic triad found in chymotrypsin, to the β-CD. This derivative enhanced the hydrolysis rate of activated esters and formation of amine bonds 3.4-fold [33]. β-CD was modified for the selective hydroxyethylation and hydroxymethylation of phenol and used to catalyze the conversion of phenylpyruvic acid to phenylalanine. It was reported [34] that chemical modification greatly promoted the catalytic activity. A β-CD derivative, 2-[(7-α-O-10-methyl-7-isoalloazino)methyl]-β-CD, increased the rate of conversion of benzyl alcohol to aldehyde 647-fold. The riboflavin was used as a catalyst for this study, and this is the highest increase in reaction rate with flavoenzyme mimics [35].

It is also reported [36] that hydrolysis of racemic arylpropionic esters by BSA resulted in low enantioselectivity (50 to 81%). Addition of β-CD to this reaction increased the enantioselectivity (80 to 99%) and the rate of hydrolysis. Enantioselectivity of baker's yeast as a chiral catalyst was improved up to 70% using CDs in cycloaddition reaction of nitriloxides or amines to the carbon–carbon triple bond.

Inclusion of catalyst with CDs also affects the catalytic activity in some reactions. This may result in either increased or decreased activity of the catalyst. CDs were found to strongly inhibit the intramolecular catalysis of amide hydrolysis. This is due to the changed geometry of the substrate,

resulting in changed interaction of the carboxylic and/or the amide groups of the substrate with the hydroxide groups [37]. It was observed that a nondistorted cavity with a secondary imidazole exhibited much greater catalytic activity in the ester hydrolysis than did its isomer with a distorted cavity [38].

Methyl-β-cyclodextrin and OH-propyl-β-cyclodextrin promoted cholesterol release from mouse sperm plasma membrane along with increased protein tyrosine phosphorylation, providing an alternative to replace the BSA from the media to support the signal transduction that leads to sperm capacitation [39]. It is demonstrated that the rate of cholesterol transfer between lipid vesicles was accelerated by β- and γ-CDs [40]. Similarly, activation of acyl-CoA cholesterol acyltransferase, its redistribution in microsomal fragments of cholesterol, and its facilitated diffusion by methyl-β-CD are also reported [41]. This study established that methyl-β-CD enhanced cholesterol transfer between liposomes and microsomes. This makes CDs very useful agents in in vitro studies of the transport of apolar molecules.

Hydrolysis of RNA is mediated through enzyme ribonuclease A. This enzyme contains two histidine residues, His12 and His119, at the catalytic site, which serve as principal catalytic groups. Breslow and Schmuck [42] prepared mimic of this enzyme by attaching two imidazole rings to the primary face of β-CD. This mimic catalyzed the hydrolysis of cyclic phosphate with $k_{cat} = 120 \times 10^{-5}\,s^{-1}$ compared to $k_{uncat} = 1 \times 10^{-5}\,s^{-1}$ for the uncatalyzed reaction (hydrolysis in basic solution). Also, the enantioseletivity increased from 1 : 1 to 99 : 1 for the product with this mimic compared to the uncatalyzed reaction. Further studies revealed that the relative positions of the imidazole groups on the β-CD ring were crucial. Maximum enanatioselectivity (100%) was obtained when the imidazole groups were attached to adjacent sugars.

2.4. Degradation and Remediation Process

CDs can play a major role in environmental science in terms of the solubilization of organic contaminants and the enrichment and removal of organic pollutants and heavy metals from soil, water, and atmosphere. The environmental pollution details are given below.

2.4.1. Wastewater Treatment
In biological wastewater treatment, toxic wastes present in water are degraded by microorganisms present in the biological sludge. Oxidation and hydrolysis reactions catalyzed by enzymes present in the diverse flora of yeast, bacteria, and fungi in the biological sludge are responsible for degradation of the toxic wastes. Biological waste includes organic and inorganic substances that are often toxic. Contaminants from the food industry are relatively easy to decompose, but those from the organic chemical industry, containing pesticides, drugs, and drug

intermediates, for example, are resistant to such biological treatment. On surpassing the tolerance level of toxic contaminants, the microbial flora is damaged and the biological activity of the sludge decreases more or less irreversibly. To prevent metabolism overload, the amount of toxicant has to be reduced either by diluting wastewater with toxicant-free water or by enhancing the metabolic capacity of microbes. Without doubt, neither process is easy.

Biodegradation of 2-nitrobiphenyl by CDs and modified CDs such as HPCD (hydroxypropyl-β-CD) and CMCD (carboxymethyl-β-CD) represents another example in which CDs increase the apparent solubility of 2--nitrobiphenyl and predispose it to microbial destruction by *Acinetobacter* sp. [43]. Moreover, CDs are not toxic to microbes and do not affect normal growth. CD-mediated complex formation is also employed to treat wastewater from industrial effluent containing pesticides such as karathane, folpet, and catpan [44].

Mechanism of Action One hypothesis suggests that toxicity reduction may be due to the reduced affinity to the cell membranes of the relatively large hydrophobic complexes formed. According to other hypotheses toxic moiety is more susceptible to biological decomposition in complex form than that in free form, resulting in rapid elimination from an activated sludge system. The former hypothesis seems to be a crucial factor since complex formed between a CD and a toxic moiety is degraded at a slower rate, which is contrary to the second hypothesis. It seems that microbial floras are better adapted than toxicants to toxic substances in the presence of β-CD or the β-CD resistant to metabolism remains longer in activated sludge, protecting it from toxic effects and maintaining its detoxicating capacity [45].

2.4.3. Soil Remediation

Human activities often result in the creation of organic micropollutants that contaminate soil. Heavily contaminated soil is found near industrial plants that use petroleum and coal. Pollutants includes polychlorinated biphenyls, heavy metals (Cr, Cu, Mn, Ni, Pb, Zn), and polycyclic aromatic hydrocarbons, which are highly toxic carcinogens often produced by incomplete combustion of carbon compounds, and many others [46]. Classical techniques for their removal from soil include chemical oxidation, thermal desorption, and washing techniques. These techniques are associated with the following inherent drawbacks:

1. Strong oxidizing agents such as $KMnO_4$ or H_2O_2 in acidic solution are required for chemical oxidation [47]. The generation of toxic metabolites by incomplete oxidation is the drawback associated with this technique that nullifies its beneficial effect.

2. Thermal desorption requires heating the contaminated soil to more than 150°C for resistant substances such as

polyaromatic hydrocarbons that have more than two rings [48].

3. Washing techniques use supercritical carbon dioxide [49], treatment with which requires special plants.

4. Differences in the properties of contaminants pose a problem in concurrent removal of bioburden by a single technique. For example, polychlorinated biphenyls are bound to soil by nonspecific hydrophobic interactions, whereas metal ions undergo weaker electrostatic forces. A single remediation process that can overcome both of these types of binding mechanisms simultaneously is not very easy to identify.

5. Ethylenediaminetetraacetic acid (EDTA) has a higher complexation capacity but poor selectivity, which results in the chelation of such metal ions as Ca^{2+}, Mg^{2+}, and Fe^{3+}. More consumption requires more EDTA [50].

6. Surfactants used for soil washing have a tendency to get adsorbed into the soil [51].

Small molecules such as CDs are known to trap efficiently many organic molecules into their rings. Binding constants of CD complexes in water can be higher than 10^4. The unique property of having a hydrophilic exterior and a hydrophobic interior makes CDs an interesting potential candidate for use in bioremediation [52]. Polycyclic aromatic hydrocarbons have limited bioavailability because of solubility constraints, which limit their biodegradation [53]. Low-polarity organic compounds having a size and shape complementary to the hydrophobic cavity can form inclusion complexes having higher water solubility. Based on the inclusion complex–forming ability of CDs or modified CDs, many of the hydrophobic organic pollutants in soil [e.g., polyaromatic hydrocarbons (PAHs), PCBs, chlorinated phenols, dioxins, and furans] are suitable guests for complex formation with the host molecule (i.e., CDs). This unique property provides CDs with a capacity to increase significantly the apparent solubility of low-polarity organic compounds. CDs increase the apparent solubility and, consequently bioavailability of organic compounds to degrading microorganisms and in this way are useful in soil remediation.

CDs as a bioremediation agent offer some advantages:

1. Complex formed between a CD and a contaminant is sometime more soluble than are micelles formed by surfactants. Hence, such problems as sorption and retention by soil are fewer than those with many surfactants.

2. Precipitation, phase separation, and foaming problems do not occur when CDs are used for bioremediation.

3. CDs are ecofriendly, as they are easily biodegradable and nontoxic.

4. CDs can be modified to complement a particular contaminant (i.e., a guest), so that inclusion complex formation can take place efficiently. Examples are hydroxypropyl-cyclodextrin (HPCD) methyl-cyclodextrin (MCD), and randomly methylated β-cyclodextrin (RAMEB) [54].

The target contaminants with CDs were polyaromatic hydrocarbons [55], pentachlorophenol [56], hexahydro-1,3,5-trinitrotriazine (RDX) and its transformation products [57], and polychlorinated biphenyl compounds (PCBs) [58]. Other trials have been directed to the desorption of trichloroethylene (TCE) and perchloroethylene (PCE) from diesel, transformer oil, and mazout [59].

2.4.4. Waste Gas Treatment

Volatile organic compounds (VOCs) are released from chemical, petrochemicals, and allied industries into the air and provoke environmental and health concerns [60]. Air pollutants include most solvent thinners, degreasers, and lubricants. Classical approaches rely on the destruction and recovery-based techniques, each associated with particular advantages and disadvantages [61]. In absorption, VOCs are removed from gas streams by contacting the contaminated air with a liquid solvent [62]. Several potential absorbents, such as vegetable, mineral, or silicon oil, polyglycols, alkylphthalates, and alkyladipates, are used.

CDs can immobilize VOCs into their low-polarity cavity in the aqueous phase, hence increasing the apparent solubility of low-polarity compounds. In this way they are expected to improve the transfer of gas into aqueous phase during the absorption process. Various forms of CDs used for this process include α-CD, β-CD, hydroxypropyl-β-CD (HP-β-CD), randomly methylated β-CD (RAMEB), a low methylated β-CD (CRYSMEB), and a sulfobutyl ether-β-CD (4-SBE-β-CD) [63].

2.5. Role of CDs in Vaccine Production and Gene Delivery

Bordetella pertussis, the human pathogen causing pertussis (whooping cough), has often been described as a fastidious organism. The difficulty encountered due to the puzzling in vitro growth behavior of *B. pertussis* has impeded the progress of large-scale vaccine production as well as clinical diagnostics [64]. An earlier assumption suggested the use of whole-blood and a potato extract of the first isolation medium (BG medium) of Bordet and Gengou (1906), satisfying the bacterium's complex nutritional requirements. Later it was found that the nutrient requirements of *B. pertussis* are quite simple. There are many growth inhibitors that can be present in growth media such as peptone, sulfur, iron, peroxides, and fatty acids. Of these, fatty acids may be the most critical inhibitor of *B. pertussis* growth. Even at concentrations as low as 1 ppm, growth is inhibited by these long-chain fatty acids. A further complication is the production of free fatty acids (FFAs) by *B. pertussis* itself, which could be autoinhibitory. Palmitic, palmitoleic, and stearic acids were also found in the extracellular medium during growth. FFA levels in the spent culture supernatant were significantly higher than those found in the uninoculated medium, indicating that they are released by the cells. This autoinhibition hypothesis explains why it was not possible previously to achieve maximum cell concentration with conventional media and deceleration of growth in batch culture much before the onset of the stationary growth phase.

Thus, one reason for the successful growth of the bacterium in BG medium can be attributed to the presence of substances that neutralized the effect of these fatty acids. The mechanism seems to be the uncoupling of oxidative phosphorylation. These inhibitors are adsorbed by certain components of media, such as starch and albumin. These adsorbents have been replaced by compounds with similar properties, such as charcoal, anion-exchange resins, poly(vinyl acetate) and poly(vinyl alcohol), methylcellulose, and CDs. Linear carbohydrate chains of starch are known to bind fatty acids by helical encapsulation; similarly, CDs bind fatty acids by forming inclusion complexes. The degree of inhibition should also depend on the partitioning of the fatty acids between the cell membrane and the bulk medium. A corresponding shift in the distribution of FFAs from the cells to the extracellular medium demonstrated that dimethyl-β-CD (M-β-CD) sequesters FFAs. Of all the adsorbents tried previously with *B. pertussis*, CD seems to be the best in terms of growth, foam control, and antigen production. M-β-CD was shown to have a high binding affinity for FFAs and to rapidly reverse FFA-induced inhibition.

The role of CDs as carriers of low-molecular-weight drugs and drugs of protein origin is well known [65]. Novel approaches include delivery of oligonucleotides by modifying CDs. Cationic polymers serve as excellent carriers for delivery of oligonucleotides. CDs were complexed with cationic agents [66,67] such as polyethylimine, oligoethylenimine [68], and polyamidoamine for gene delivery. Such complexes have shown excellent transfection ability and enjoy low toxicity.

2.6. Role of CDs in the Food Industry

β-CD was declared under the category "generally recognized as safe" (GRAS) since 1998 for the protection of sensitive components of food and as a flavor carrier. Following are the few properties of the CDs that makes them a suitable candidate for use in the food industry. CDs are nontoxic, are not absorbed in the upper gastrointestinal tract, and are completely metabolized by the colon microflora. The price of a β-CD-encapsulated flavor product is not much higher than that of a microencapsulated flavor formulation.

In the food industry, CDs can be used to play any of the following roles:

- Protection of active ingredients that are susceptible to thermodecomposition, photodecomposition, microbiological contamination, hygroscopicity, and those that undergo oxidation, loss by volatility, sublimation, etc.
- Masking undesired tastes and odors.
- Technological advantages, including operational ease from raw materials to finished product economically.

2.6.1. CDs as Flavor Carriers

Most natural and artificial flavors are volatile oils or liquids, and complexation with CDs provides a promising alternative to the conventional encapsulation technologies used for flavor protection. Molecular encapsulation is an encapsulation process on a molecular scale in which CD form complexes with flavors. Molecular encapsulation has an edge over traditional encapsulation techniques, as it ensures effective protection for every flavor constituent present in a multicomponent food system [69].

2.6.2. CDs as a Protectant of Food Ingredients

Photosensitive components such as citral were shown to be protected against the degrading effects of light when encapsulated in CDs which otherwise undergo a cyclization reaction in the presence of ultraviolet radiations. Decomposition products alter the taste of juices that have a citral flavor. The beneficial effect of CDs in the protection of light-sensitive volatile components of aromatic oils has already been reported (Table 3). CDs protect sensitive components such as citral and cinnamaldehyde from the deleterious effect of ultraviolet (UV) radiation. Certain food components such as unsaturated fatty acids are sensitive to oxidation. Partial or complete entrapment of such components inside CDs improves their resistance to oxidation [70]. It has already been reported and shown experimentally that oxidation has a deleterious effect on certain sensitive components, such as benzaldehyde and cinnamaldehyde. These components are vulnerable to oxidation in the free or naked state. Encapsulation inside CDs following complexation greatly enhanced their resistance to oxidation, as shown in Table 3 [71]. The stability of flavors in high-temperature extrusion cooking has already been reported [72]. CDs have been shown to provide better protection than traditional adsorbents such as lactose. Encapsulated compounds show a higher increase in resistance toward higher temperature and humidity conditions [70]. Excessive bitterness of flavors is undesirable. The efficacy of CDs to protect the bitter components of grapefruit juice, even at a pilot-plant scale, has been reported. The bitter component of grapefruit juice consists of limonin and naringin. β-CD was used to reduce these active components using a fluidized-bed process [73].

2.6.3. Miscellaneous Applications of CDs in the Food Industry

The resistance offered to environmental conditions has opened a new area for CD use: as packaging components [74]. They can also be employed to form a complex with antimicrobial agents, providing strong protection against microbial attack of food ingredients. Allyl isothiocyanate (AITC), a strong antimicrobial agent, was reported to be complexed with α- and β-CDs. This complex permits the controlled release of AITC and offers many advantages such as masking of strong odor and prolonging the antimicrobial effect [75].

2.7. Role of CDs as a Processing Aid

In juices, polyphenol-oxidase causes the browning of juice by converting the colorless polyphenols to colored compounds. The addition of CDs removes polyphenoloxidase from juices by complexation [76]. CD-mediated

Table 3. Protective Role of CDs Against Stress Conditions

Protection Against:	Sensitive Compounds	Complex Formed	Remarks
Oxidation	Benzaldehyde	Benzaldehyde/β-CD	Oxygen consumption is 10 to 11 times less in complex formed as in free component
	Cinnamaldehyde	Cinnamaldehyde/β-CD	
Photosensitivity	Citral	Citral/β-CD	80 to 90% increased resistance to photosensitivity in the complex formed
	Cinnamaldehyde	Cinnamaldehyde/β-CD	
Heat	Garlic	Garlic/β-CD	Less thermal degradation in the complex formed than in free component
	Onion	Onion/β-CD	
	Dill	Dill/β-CD	
Heat and humidity	Garlic oil	Garlic oil/β-CD	70 to 80% increased resistance to high temperature and humidity in the complex formed
	Lemon oil	Lemon oil/β-CD	

complexation is employed to remove cholesterol from products such as milk, butter, and eggs. The texture-improving effect on pastry and meat products may be due to complexation with free fatty acids [8,77].

2.8. Cosmetics, Toiletries, Personal care, and Miscellaneous Applications

The use of CDs-complexed fragrances in skin preparations such as talcum powder stabilizes the fragrance against loss by evaporation and oxidation over a long period. It also improves the antimicrobial efficacy. The major benefits of CDs in this sector are stabilization, odor control, and process improvement upon conversion of a liquid ingredient to a solid form, flavor protection and flavor delivery in lipsticks, water solubility, and enhanced thermal stability of oils [78]. O-Methoxycinnamaldehyde acts as an antimicrobial agent. The β-CD complex of o-methoxycinnamaldehyde has been incorporated into shoe insoles to inhibit microbial growth and foul odors. Fragrant paper or paper containing protective substances of perfumes, insecticides, rust inhibitors, mold- and mildew-proofing agents, fungicides, and bactericides can be prepared using CD complexes by incorporating suitable active agents [79].

3. CDs IN SEPARATION TECHNIQUES

Currently, chiral separations are one of the most important areas of the use of CDs and their derivatives [80]. Furthermore, CDs are also used extensively in high-performance liquid chromatography (HPLC) as stationary phases bonded to solid support or as mobile-phase additives in HPLC and in capillary electrophoresis for the separation of chiral compounds [81]. The size, shape, and selectivity of CDs confer the ability to discriminate between positional isomers, enantiomers, functional groups, and homologs [82]. Rivastigmine is a well-known agent for the treatment of Alzheimer's disease [83]. Native or derivatized cyclodextrins, such as heptakis (2,3,6-tri-O-methyl)-cyclodextrin, hydroxypropyl-cyclodextrin, and sulfated cyclodextrin are the most widely used chiral selectors in capillary electromigration methods, with the purpose of improving the enantiomeric resolution of racemic compounds. Resolution of rivastigmine, an anticholinesterase inhibitor, has been reported using the aforementioned CDs [84].

BTEX refers to benzene, toluene, ethylbenzene and o-, m-,and p-xylene. These are aromatic hydrocarbons used widely in the chemical industry. BTEX are small hydrophobic aromatic hydrocarbons with a great affinity for the nonpolar CD cavity. Selectivity is ensured, due to the electron sharing of the aromatic methylene groups with those of the glucoside oxygens. Due to the inclusion complex formation ability of CDs, they can detect and differentiate between

structural isomers such as o-, m-, and p-xylene. β-CD has been used as a mobile phase-component in reversed-phase HPLC and also in stationary phases, in both liquid and gas chromatography [85]. Hydrophilic CDs have frequently been used in capillary electrophoresis as buffer modifiers to effect chiral separation of drugs and chemicals [86].

4. CONCLUSIONS AND FUTURE PROSPECTS

From the discussion above it is evident that CDs play a major role in biotechnology-related products and processes. Their complex chemical structure makes them very useful in product and process development. The presence of hydrophobic and hydrophilic character in a single molecule makes them prone to practical use. The enthalpy-rich water molecules makes them unstable, and stabilization takes place in pulling guest molecules inside the cavity. In an appropriate environment, the guest molecules come out and deliver the product. Use of nano-CDs may be a future trend for the better use of CDs. Starting with media formulation and going fermentation, bioprocessing, and downstream processing, the use of CDs is everywhere. Chemical and biochemical transformation of CDs to a more useful complexing agent may be a research area where scientists should look into. Better complexing agents than CDs themselves may be generated from CDs through chemical and/or biochemical transformations of the sugar moieties, addition or deletion of functional groups, and so on.

REFERENCES

1. Villalonga, R., Cao, R., Fragoso, A. (2007). Supramolecular chemistry of cyclodextrins in enzyme technology. *Chem. Rev.*, *107*, 3088–3116.
2. Caner, H., Groner, E., Levy, L., Agranat, I. (2004). Trends in the development of chiral drugs. *Drug Discov. Today*, *9*, 105–110.
3. Ballesteros, A., Bornscheuer, U., Capewell, A., Combes, D., Condoret, J.S., Koenig, K., Kolisis, F.N., Marty, A., Menge, U., Scheper, T. (1995). Enzymes in non-conventional phases. *Biocatal. Biotransform.*, *13*, 1–42.
4. Rouse, J.D., Sabatini, D.A., Suflita, J.M., Harwell, J.H. (1994). Influence of surfactants on microbial degradation of organic compounds. *Crit. Rev. Environ. Sci. Technol.*, *24*, 325–370.
5. Klibanov, A.M. (2003). Asymmetric enzymatic oxidoreductions in organic solvents. *Curr. Opin. Biotechnol.*, *14*, 427–431.
6. Monot, F., Borzeix, F., Bardin, M., Vandecasteele, J.P. (1991). Enzymatic esterification in organic media: role of water and organic solvent in kinetics and yield of butyl butyrate synthesis. *Appl. Microbiol. Biotechnol.*, *35*, 759–765.
7. Schmitz, G., Franke, D., Stevens, S., Takors, R., Weuster-Botz, D., Wandrey, C. (2000). Regioselective oxidation of terfenadine with *Cunninghamella blakesleeana*. *J. Mol. Catal. B*, *10*, 313–324.

8. Szejtli, J. (1998). Introduction and general overview of cyclo-dextrin chemistry. *Chem. Rev.*, *98*, 1743–1754.

9. Szente, L., Szejtli, J. (1999). Highly soluble cyclodextrin derivatives: chemistry, properties, and trends in development. *Adv. Drug Deliv. Rev.*, *36*, 17–28.

10. Singer, Y., Shity, H., Bar, R. (1991). Microbial transformations in a cyclodextrin medium: Part 2. Reduction of androstenedione to testosterone by *Saccharomyces cerevisiae*. *Appl. Microbiol. Biotechnol.*, *35*, 731–737.

11. Kim, S.H., Kim, T.K., Shin, G.S., Lee, K.W., Shin, H.D., Lee, Y. H. (2004). Enantioselective hydrolysis of insoluble (*R,S*)-ke-toprofen ethyl ester in dispersed aqueous reaction system induced by chiral cyclodextrin. *Biotechnol. Lett.*, *26*, 965–969.

12. Bardi, L., Martini, C., Opsi, F., Bertolone, E., Belviso, S., Masoero, G., Marzona, M., Marsan, F.A. (2007). Cyclodex-trin-enhanced in situ bioremediation of polyaromatic hydro-carbons-contaminated soils and plant uptake. *J. Inclusion Phenom. Macrocyclic Chem.*, *57*, 439–444.

13. Takeda, K., Asou, T., Matsuda, A., Kimura, K., Okamura, K., Okamoto, R., Sasaki, J., Adachi, T., Omura, S. (1994). Appli-cation of cyclodextrin to microbial transformation of vitamin D$_3$ to 25-hydroxyvitamin D$_3$ and 1-α,25-dihydroxyvitamin D$_3$. *J. Ferment. Bioen.*, *78*, 380–382.

14. Alexander, D.L., Fisher, J.F. (1995). A convenient synthesis of 7-α-hydroxycholest-4-en-3-one by the hydroxypropyl-β-cy-clodextrin-facilitated cholesterol oxidase oxidation of 3-β,7-α-cholest-5-ene-3,7-diol. *Steroids*, *60*, 290–294.

15. Woerdenbag, H.J., Pras, N., Frijlink, H.W., Lerk, C.F., Mal-ingre, T.M. (1990). Cyclodextrin-facilitated bioconversion of 17-estradiol by a phenoloxidase from *Mucuna pruriens* cell cultures. *Phytochemistry*, *29*, 1551–1554.

16. Srisimarat, W., Pongsawasdi, P. (2008). Enhancement of the oligosaccharide synthetic activity of β-galactosidase in organic solvents by cyclodextrin. *Enzyme Microb. Technol.*, *43*, 436–441.

17. Bar, R. (1989). Cyclodextrin-aided bioconversions and fermen-tations. *Trends Biotechnol.*, *7*, 2–4.

18. Bar, R. (1989). Cyclodextrin-aided microbial transformation of aromatic aldehydes by *Saccharomyces cerevisiae*. *Appl. Micro-biol. Biotechnol.*, *31*, 25–28.

19. Ghanem, A. (2003). The utility of cyclodextrins in lipase-catalyzed transesterification in organic solvents: enhanced reaction rate and enantioselectivity. *Org. Biomol. Chem.*, *1*, 1282–1291.

20. Mine, Y., Zhang, L., Fukunaga, K., Sugimura, Y. (2005). Enhancement of enzyme activity and enantioselectivity by cyclopentyl methyl ether in the transesterification catalyzed by *Pseudomonas cepacia* lipase co-lyophilized with cyclodex-trins. *Biotechnol. Lett.*, *27*, 383–388.

21. Easton, C.J., Harper, J.B., Head, S.J., Lee, K., Lincoln, S.F. (2001). Cyclodextrins to limit substrate inhibition and alter substrate selectivity displayed by enzymes. *J. Chem. Soc. Perkin Trans. 1*, *2001*, 584–587.

22. Wang, M., Zhang, L., Shen, Y., Ma, Y., Zheng, Y., Luo, J. (2009). Effects of hydroxypropyl-β-cyclodextrin on steroids

23. Kumar, R., Dahiya, J.S., Singh, D., Nigam, P. (2001). Biotrans-formation of cholesterol using *Lactobacillus bulgaricus* in a glucose-controlled bioreactor. *Bioresource Technol.*, *78*, 209–211.

24. Kamal, A., Ramalingam, T., Venugopal, N. (1991). Enantio-selective hydrolysis of aryloxypropionic esters by bovine serum albumin: enhancement in selectivity by β-cyclodextrin. *Tetra-hedron Asymm.*, *2*, 39–42.

25. Ávila-González, R., Pérez-Gilabert, M., García-Carmona, F. (2005). Lipase-catalyzed preparation of *S*-propranolol in pres-ence of hydroxypropyl β-cyclodextrins. *J. Biosci. Bioeng.*, *100*, 423–428.

26. Ghanem, A., Schurig, V. (2001). Peracetylated β-cyclodextrin as additive in enzymatic reactions: enhanced reaction rate and enantiomeric ratio in lipase-catalyzed transesterifications in organic solvents. *Tetrahedron Asymm.*, *12*, 2761–2766.

27. Marchini, A., d'Apolito, M., Massari, P., Atzeni, M., Copass, M., Olivieri, R. (1995). Cyclodextrins for growth of *Helico-bacter pylori* and production of vacuolating cytotoxin. *Arch. Microbiol.*, *164*, 290–293.

28. Fekete-Forgacs, K., Szabo, E., Lenkey, B. (1997). The forma-tion of hyphae of *Candida albicans* induced by cyclodextrins. *Mycoses*, *40*, 451–453.

29. Prabhu, K.S., Ramadoss, C.S. (2000). Penicillin acylase cata-lyzed synthesis of penicillin-G from substrates anchored in cyclodextrins. *Indian J. Biochem. Biophys.*, *37*, 6.

30. Sawada, H., Suzuki, T., Akiyama, S., Nakao, Y. (1990). Mech-anism of the stimulatory effect of cyclodextrins on lankacidin-producing *Streptomyces*. *Applied Microbiol. and Biotechnol.*, *32*, 556–559.

31. Szejtli, J., Schmid, G. (1989). Downstream processing using cyclodextrins. *Biophys. J.*, *7*, 170–174.

32. Motherwell, W.B., Bingham, M.J., Six, Y. (2001). Recent progress in the design and synthesis of artificial enzymes. *Tetrahedron*, *57*, 4663–4686.

33. Ekberg, B., Andersson, L.I., Mosbach, K. (1989). The synthesis of an active derivative of cyclomaltoheptaose for the hydrolysis of esters and the formation of amide bonds. *Carbohydr. Res.*, *192*, 111–114.

34. Morozumi, T., Uetsuka, H., Komiyama, M., Pitha, J. (1991). Selective synthesis using cyclodextrins as catalysts: VI. Cy-clodextrin modification for para-selective hydroxymethylation and hydroxyethylation of phenol. *J. Mol. Catal.*, *70*, 399–406.

35. Ye, H., Tong, W., D'Souza, V.T. (1992). Efficient catalysis of a redox reaction by an artificial enzyme. *J. Am. Chem. Soc.*, *114*, 5470–5472.

36. Rao, K.R., Bhanumathi, N., Srinivasan, T.N., Sattur, P.B. (1990). A regioselective enzyme catalysed cycloaddition. *Tet-rahedron Lett.*, *31*, 892–899.

37. Granados, A.M., de Rossi, R.H. (2001). Effect of cyclodextrin on the intramolecular catalysis of amide hydrolysis. *J. Org. Chem.*, *66*, 1548–1552.

38. Chen, W.H., Hayashi, S., Tahara, T., Nogami, Y., Koga, T., Yamaguchi, M., Fujita, K. (1999). The dependence of catalytic activities of secondary functional cyclodextrins on cavity structures. *Chem. Pharm. Bull.*, *47*, 588–589.

39. Visconti, P.E., Galantino-Homer, H., Ning, X.P., Moore, G.D., Valenzuela, J.P., Jorgez, C.J., Alvarez, J.G., Kopf, G.S. (1999). Cholesterol efflux-mediated signal transduction in mammalian sperm cyclodextrins initiate transmembrane signaling leading to an increase in protein tyrosine phosphorylation and capacitation. *J. Biol. Chem.*, *274*, 3235–3242.

40. Leventis, R., Silvius, J.R. (2001). Use of cyclodextrins to monitor transbilayer movement and differential lipid affinities of cholesterol. *Biophys. J.*, *81*, 2257–2267.

41. Cheng, D., Tipton, C.L. (1999). Activation of acyl-CoA cholesterol acyltransferase: redistribution in microsomal fragments of cholesterol and its facilitated movement by methyl-β-cyclodextrin. *Lipids*, *34*, 261–268.

42. Breslow, R., Schmuck, C. (1996). Goodness of fit in complexes between substrates and ribonuclease mimics: effects on binding, catalytic rate constants, and regiochemistry. *J. Am. Chem. Soc.*, *118*, 6601–6605.

43. Cai, B., Gao, S., Lu, G. (2006). β-Cyclodextrin and its derivatives-enhanced solubility and biodegradation of 2-nitrobiphenyl. *J. Environ. Sci.*, *18*, 1157–1160.

44. Oláh, J., Cserháti, T., Szejtli, J. (1988). β-Cyclodextrin enhanced biological detoxification of industrial wastewaters. *Water Res.*, *22*, 1345–1351.

45. Szejtli, J. (1990). The cyclodextrins and their applications in biotechnology. *Carbohydr. Polym.*, *12*, 375–392.

46. Mackay, D., Shiu, W., Ma, K. (1996). *Illustrated Handbook of Physical–Chemical Properties and Environmental Fate for Organic Chemicals*, Vol. 1–4 Lewis Publishers, Boca Raton, FL.

47. Anipsitakis, G.P., Dionysiou, D.D. (2003). Degradation of organic contaminants in water with sulfate radicals generated by the conjunction of peroxymonosulfate with cobalt. *Environ. Sci. Technol.*, *37*, 4790–4797.

48. Smith, M.T., Berruti, F., Mehrotra, A.K. (2001). Thermal desorption treatment of contaminated soils in a novel batch thermal reactor. *Ind. Eng. Chem. Res.*, *40*, 5421–5430.

49. Fu, H., Matthews, M.A. (1999). Separation processes for recovering alloy steels from grinding sludge: supercritical carbon dioxide extraction and aqueous cleaning. *Sep. Sci. Technol.*, *34*, 1411–1427.

50. Kedziorek, M.A.M., Bourg, A.C.M. (2000). Solubilization of lead and cadmium during the percolation of EDTA through a soil polluted by smelting activities. *J. Contam. Hydrol.*, *40*, 381–392.

51. West, C.C., Harwell, J.H. (1992). Surfactants and subsurface remediation. *Environ. Sci. Technol.*, *26*, 2324–2330.

52. Khan, A.R., Forgo, P., Stine, K.J., D'Souza, V.T. (1998). Methods for selective modifications of cyclodextrins. *Chem. Rev.*, *98*, 1977–1996.

53. Guthrie, E.A., Pfaender, F.K. (1998). Reduced pyrene bioavailability in microbially active soils. *Environ. Sci. Technol.*, *32*, 501–508.

54. Brusseau, M.L., Wang, X., Wang, W.Z. (1997). Simultaneous elution of heavy metals and organic compounds from soil by cyclodextrin. *Environ. Sci. Technol.*, *31*, 1087–1092.

55. Cuypers, C., Pancras, T., Grotenhuis, T., Rulkens, W. (2002). The estimation of pah bioavailability in contaminated sediments using hydroxypropyl-β-cyclodextrin and triton X-100 extraction techniques. *Chemosphere*, *46*, 1235–1245.

56. Fenyvesi, É., Szemán, J., Szejtli, J. (1996). Extraction of PAHs and pesticides from contaminated soils with aqueous cyclodextrin solutions. *J. Inclusion Phenom. Macrocyclic Chem.*, *25*, 229–232.

57. Sheremata, T.W., Hawari, J. (2000). Cyclodextrins for desorption and solubilization of 2,4,6-trinitrotoluene and its metabolites from soil. *Environ. Sci. Technol.*, *34*, 3462–3468.

58. Fava, F., Bertin, L., Fedi, S., Zannoni, D. (2003). Methyl-β-cyclodextrin enhanced solubilization and aerobic biodegradation of polychlorinated biphenyls in two aged-contaminated soils. *Biotechnol. Bioeng.*, *81*, 381–390.

59. Molnar, M., Fenyvesi, E., Gruiz, K., Leitgib, L., Balogh, G., Muranyi, A., Szejtli, J. (2002). Effects of RAMEB on bioremediation of different soils contaminated with hydrocarbons. *J. Inclusion Phenom. Macrocyclic Chem.*, *44*, 447–452.

60. Ware, J.H., Spengler, J.D., Neas, L.M., Samet, J.M., Wagner, G. R., Coultas, D., Ozkaynak, H., Schwab, M. (1993). Respiratory and irritant health effects of ambient volatile organic compounds: the Kanawha County health study. *Am. J. Epidemiol.*, *137*, 1287–1301.

61. Khan, F.I., Kr. Ghoshal, A. (2000). Removal of volatile organic compounds from polluted air. *J. Loss Prev. Process Ind.*, *13*, 527–545.

62. Rahbar, M.S., Kaghazchi, T. (2005). Modeling of packed absorption tower for volatile organic compounds emission control. *Int. J. Environ. Sci. Technol.*, *2*, 207–215.

63. Blach, P., Fourmentin, S., Landy, D., Cazier, F., Surpateanu, G. (2008). Cyclodextrins: a new efficient absorbent to treat waste gas streams. *Chemosphere*, *70*, 374–380.

64. Fröhlich, B.T., d'Alarcao, M., Feldberg, R.S., Nicholson, M.L., Siber, G.R., Swartz, R.W. (1996). Formation and cell-medium partitioning of autoinhibitory free fatty acids and cyclodextrin's effect in the cultivation of *Bordetella pertussis*. *J. Biotechnol.*, *45*, 137–148.

65. Li, J., Loh, X.J. (2008). Cyclodextrin-based supramolecular architectures: syntheses, structures, and applications for drug and gene delivery. *Adv. Drug Deliv. Rev.*, *60*, 1000–1017.

66. Davis, M.E., Brewster, M.E. (2004). Cyclodextrin-based pharmaceutics: past, present and future. *Nat. Rev. Drug Discov.*, *3*, 1023–1035.

67. Gonzalez, H., Hwang, S.J., Davis, M.E. (1999). New class of polymers for the delivery of macromolecular therapeutics. *Bioconjugate. Chem.*, *10*, 1068–1074.

68. Yang, C., Li, H., Goh, S.H., Li, J. (2007). Cationic star polymers consisting of β-cyclodextrin core and oligoethylenimine arms as nonviral gene delivery vectors. *Biomaterials*, *28*, 3245–3254.

69. Szente, L., Szejtli, J. (2004). Cyclodextrins as food ingredients. *Trends Food Sci. Technol.*, *15*, 137–142.

70. Szente, L., Szejtli, J. (1988). Stabilization of flavors by cyclodextrins. *Flavour Encapsulation*. ACS Symposium series, Vol. 370 American Chemical Society, Washigton, DC, pp. 148–157.

71. Szejtli, J., Szente, L., Banky-Elod, E. (1979). Molecular encapsulation of volatile, easily oxidizable labile flavor substances by cyclodextrins. *Acta Chim. Acad. Sci. Hung.*, *101*, 27–46.

72. Bhandari, B., D'Arcy, B., Young, G. (2001). Flavour retention during high temperature short time extrusion cooking process. *Int. J. Food Sci. Technol.*, *36*, 453.

73. Wagner, C.J., Wilson, C.W., Shaw, P.E. (1988). Reduction of grapefruit bitter components in a fluidized β-cyclodextrin polymer bed. *J. Food Sci.*, *53*, 516–518.

74. Hashimoto, H. (2002). Present status of industrial application of cyclodextrins in japan. *J. Inclusion Phenom. Macrocyclic Chem.*, *44*, 57–62.

75. Ohta, Y., Takatani, K., Kawakishi, S. (2000). Kinetic and thermodynamic analyses of the cyclodextrin-allyl isothiocyanate inclusion complex in an aqueous solution. *Biosci. Biotechnol. Biochem.*, *64*, 190–193.

76. Sojo, M.M., Nunez-Delicado, E., Garcia-Carmona, F., Sanchez-Ferrer, A. (1999). Cyclodextrins as activator and inhibitor of latent banana pulp polyphenol oxidase. *J. Agric. Food Chem.*, *47*, 518–523.

77. Hedges, A.R. (1998). Industrial applications of cyclodextrins. *Chemical Reviews*, *98*, 2035–2044.

78. Buschmann, H.J., Schollmeyer, E. (2002). Applications of cyclodextrins in cosmetic products. *J. Cosmetic Sci.*, *53*, 185–192.

79. Szejtli, J. (1997). Utilization of cyclodextrins in industrial products and processes. *J. Mater. Chem.*, *7*, 575–587.

80. Lu, X., Chen, Y. (2002). Chiral separation of amino acids derivatized with fluoresceine-5-isothiocyanate by capillary electrophoresis and laser-induced fluorescence detection using mixed selectors of β-cyclodextrin and sodium taurocholate. *J. Chromatogr. A*, *955*, 133–140.

81. Zarzycki, P.K., Kulhanek, K.M., Smith, R. (2002). Chromatographic behaviour of selected steroids and their inclusion complexes with β-cyclodextrin on octadecylsilica stationary phases with different carbon loads. *J. Chromatogr. A*, *955*, 71–78.

82. Han, S.M. (1997). Direct enantiomeric separations by high performance liquid chromatography using cyclodextrins. *Biomed. Chromatogr.*, *11*, 259–271.

83. Polinsky, R.J. (1998). Clinical pharmacology of rivastigmine: a new-generation acetylcholinesterase inhibitor for the treatment of Alzheimer's disease. *Clin. Ther.*, *20*, 634–647.

84. Lucangioli, S.E., Tripodi, V., Masrian, E., Scioscia, S.L., Carducci, C.N., Kenndler, E. (2005). Enantioselective separation of rivastigmine by capillary electrophoresis with cyclodextrines. *J. Chromatogr. A*, *1081*, 31–35.

85. Campos-Candel, A., Llobat-Estellés, M., Mauri-Aucejo, A. (2009). Comparative evaluation of liquid chromatography versus gas chromatography using a β-cyclodextrin stationary phase for the determination of BTEX in occupational environments. *Talanta*, *78*, 1286–1292.

86. Fanali, S. (2000). Enantioselective determination by capillary electrophoresis with cyclodextrins as chiral selectors. *J. Chromatogr. A*, *875*, 89–122.

15

CYCLODEXTRINS AND CELLULAR INTERACTIONS

JUSTIN M. DREYFUSS AND STEVEN B. OPPENHEIMER

Center for Cancer and Developmental Biology, California State University–Northridge, Northridge, California

1. INTRODUCTION

Cyclodextrins (CDs) are a class of cyclic oligosaccharides made up of six, seven, eight, or more glucopyranose units linked together [38]. They are of particular biochemical importance, due to their possession of a hydrophilic outer covering paired with their hydrophobic inner core, properties that allow them to greatly increase the solubility of nonpolar or hydrophobic molecules in an aqueous solution. CDs have also become a favorite tool of researchers, due to the very low toxicity dangers they present [38]. The majority of biological research involving CDs has involved this ability to increase the solubility of important drugs and other macromolecules, and their capacity to draw cholesterol out of the plasma membrane. This research has covered a vast array of topics, ranging from cellular adhesion interactions, the fertilization process, bacterial and viral infections, and even their involvement in neurodegenerative disease. As the following sections illustrate, CDs have been extremely important, and will continue to be important, to an increasing multitude of biological subjects.

2. CELL MEMBRANE CHOLESTEROL EFFLUX

As mentioned earlier, and as will become exceedingly clear throughout this chapter, the majority of research involving CDs has revolved around their ability to extract cholesterol from the plasma membrane. While plenty of studies have focused on the consequences of this phenomenon, there has been quite a bit of research into the dynamics of this process.

Numerous studies have examined the kinetics of CD-mediated cellular cholesterol efflux in many different cell lines, including murine fibroblasts [26], human red blood cells [66], and rat cerebellar neurons [50]. The results of these studies have indicated that CDs are able to induce an efflux of cholesterol, sphingolipids, and phophatidylcholine from the plasma membrane in a manner that is both time and dose dependent [60]. This efflux usually results in the loss of very large amounts of membrane cholesterol, often in the range of 50 to 90% of the original amount. CDs are not only able to extract large amounts of membrane cholesterol, they are able to do so at an incredibly rapid rate.

Studies with human red blood cells showed that the rate of membrane cholesterol efflux created by treatment with CDs is approximately three orders of magnitude more rapid than the rate of cholesterol transfer from red blood cells to other lipid acceptors, and approximately five orders of magnitude faster than the rate of transfer from cultured cells to synthetic vesicles [66]. Collectively, these studies show that CDs are able to extract large amounts of membrane cholesterol in a rapid and efficient manner, making them excellent tools for studying cholesterol efflux and its associated effects.

3. CELLULAR ADHESION

It has only been within the past few years that CDs have started being used as tools to study the mechanisms that mediate cellular adhesion interactions. However, their use has already provided researchers with insights into these occurrences, which had previously remained a mystery.

Cyclodextrins in Pharmaceutics, Cosmetics, and Biomedicine: Current and Future Industrial Applications, First Edition. Edited by Erem Bilensoy.
© 2011 John Wiley & Sons, Inc. Published 2011 by John Wiley & Sons, Inc.

Platelets are anuclear cells that attach to blood vessel walls to form clots during hemostasis. This adhesive interaction is mediated by the platelet glycoprotein (Gp) Ib-IX-V which binds to the von Willebrand factor (VWF), a glycoprotein found in the blood. Gp Ib-IX-V has been shown to be localized to lipid rafts, cholesterol-enriched microdomains, within the platelet plasma membrane [62]. Depletion of membrane cholesterol with CDs inhibits platelet aggregation by as much as 70% [63]. Similar results have been obtained with respect to the healing of wounds in the intestinal mucosa, which occurs through the migration of intestinal epithelial cells. The adhesion of these cells to the wound area involves the use of integrins, receptors that mediate the attachment of cells to their surrounding tissues. In migrating intestinal epithelial cells these receptors are internalized via endocytosis involving membrane lipid rafts [74]. In this instance, depletion of membrane cholesterol by CD was able to slow the wound-healing process drastically.

CDs have also been shown to interfere with adhesive interactions during developmental processes. Gastrulation in sea urchin embryos involves an invagination of part of the developing embryo and the attachment of the archenteron tip to the blastocoel roof. CD treatment was able to block this attachment effectively, resulting in unattached archenterons in approximately 70% of the treated embryos [58]. In these experiments it may be that CDs bound to lectin-type receptors involved in this cellular interaction.

Finally, CD treatment was able to inhibit adhesion between chicken embryonic retina cells and several lines of human cancer cells. In a study involving embryonic retina cells, CDs suppressed the adhesive interactions in a dose-dependent manner, with a concentration of 8 mM reducing cell adhesion by about 92% compared to controls [23]. Experiments have also been carried out using human breast cancer cells, cervix epitheloid carcinoma cells, and hepatocarcinoma cells. These studies indicated that CDs were able to disrupt adhesion between the different cell lines by disrupting the lipid raft–associated integrin signaling pathway in the same way as emodin, a well-known anticancer agent [21]. These results, combined with the data for urchin embryos, platelets, and intestinal epithelial cells, beautifully demonstrate the ability of CDs to inhibit cellular adhesion interactions.

4. MEMBRANE PROTEINS AND RECEPTORS

Much of the research involving the use of CDs has focused on specific cellular interactions (cholesterol efflux, cell–cell adhesion, etc.), but there have also been a number of studies that have examined CD effects on specific, isolated membrane proteins and receptors. Although the majority of these studies cannot really be placed into any broader categories, it is still important to elaborate on the results that this research has produced.

- CDs are able to block effectively the channels formed by connexin proteins, which compose gap junctions and are involved in the intercellular movement of signaling molecules [35].
- Membrane lipid rafts play an important role in the activation of nuclear factor κB, a transcription factor that is used in osteoclast functioning [19].
- Kappa opioid receptors, which elicit effects such as water diuresis, analgesia, and dysphoria, are localized to lipid rafts in human placental tissue [81].
- Voltage-gated K^+ channels, voltage-gated Ca^{2+} channels, caveolin-2 (a lipid-raft structural protein), and soluble N-ethylmaleimide-sensitive factor attachment protein receptor proteins (SNARE proteins) are localized to lipid rafts in pancreatic α-cells and function in the stimulation of glucagon secretion [79].
- The α1a-adrenergic receptor (a catecholamine-targeted G-protein-coupled receptor) occupies membrane lipid rafts in rat fibroblasts [40].
- The function of the dihydropyridine receptor (which controls excitation–contraction coupling) is modulated by cholesterol levels of the plasma membrane in mice fetal skeletal muscle cells [52].
- Ras proteins, which play a critical role in the transduction of extracellular signals, are located in membrane lipid rafts and cholesterol-independent microdomains [53].
- The inhibition of endocytosis by CDs induces the accumulation of aquaporin-2 (a protein that regulates the flow of water) in rat renal epithelial cells [56].
- The glutamate transport-associated protein GTRAP3-18 is a membrane-bound protein that interacts with the excitatory amino acid carrier-1 cysteine transporter (EAAC), which regulates the intracellular glutathione content. Treatment with CDs has been shown to increase the amount of GTRAP3-18 in the plasma membrane of human embryonic kidney cells, thus decreasing the intracellular amount of glutathione and rendering the cell vulnerable to oxidative stress [76].
- Epidermal growth factor (EGF) is a protein that plays an important role in several cellular functions, including cell growth and differentiation. Depletion of membrane cholesterol levels caused increased binding of the EGF receptor (EGFR) [55] as well as ligand-independent activation of the EGFR-associated MAP kinase pathway [9]. This suggests that cellular cholesterol levels possibly play important roles in both the inhibition and the activation of cell growth and differentiation.

Although there are certainly other cholesterol-associated membrane-bound proteins and receptors whose location and function remain to be determined, the research that has already been performed clearly demonstrates the usefulness of CDs in elucidating the role of membrane proteins in cellular interactions.

5. VIRAL INFECTIONS

Cholesterol levels in the plasma membrane have been found to be extremely important in many parts of the viral infection process, including the entry and release of virions from the host cell and the transport of various viral proteins. The use of CDs to decrease membrane cholesterol and impede viral entry into the host cell has been demonstrated for foot-and-mouth disease [48], duck hepatitis B virus [13], and murine corona virus [71]. Conversely, treatment with CDs increased the rate of release but decreased the relative infectivity of the virions for the influenza A virus [4], and the Newcastle virus [29,30], while decreasing the observed viral titer for the bluetongue virus [5]. Finally, cholesterol depletion greatly disrupted the surface transport of influenza virus hemagglutinin [20,25]. These results demonstrate a clear affect of CDs on the pathogenicity of several different viruses, and provide evidence of the potential potent antiviral properties of CD-based treatments.

5.1. HIV Infections

While extensive research has been performed on the effects of CDs on viral infections, no virus has been more studied in terms of the effects of CDs than human immunodeficiency virus (HIV). Initial studies in the early 1990s revealed that sulfated cyclodextrins were quite affective at inhibiting HIV infection [61,77], but the mechanisms of this process had not yet been determined. Since then, further research has revealed that lipid raft–based receptors are necessary for HIV entry into CD^+ T-cells [51] and that membrane cholesterol is required for the infectivity of HIV virions [7,17]. However, the most current studies have shown that despite the known ability of CDs to affect HIV virions, the constantly changing nature of the virus still makes it an extremely difficult target to eradicate. A 2008 study demonstrated the usefulness of CDs in creating a lentivirus vector that was able to selectively kill HIV positive macrophages (a feat that had previously proven to be extremely difficult) [83], but that same year also saw the publication of a paper demonstrating that the use of CDs as a topical antiviral was only able to provide partial protection against the transmission of the simian form of HIV [1]. Nevertheless, the evidence is clear that CDs still present a viable basis for anti-HIV treatment, and further research should continue despite the recent setbacks.

6. BACTERIAL INFECTIONS

Even though there has been significant amounts of research aimed at identifying the effects of CDs on viral infections, there has been even more research into the effects of CDs on bacterial infections. In addition to the research involving the ability of CDs to solubilize antibacterial agents [12,36], the CD-based studies on bacterial infections have covered a broad spectrum of bacterial species and toxins. It has been shown that lipid rafts play a vital role in the binding of and entry into host cells by *Francisella tularensis* [70], *Listeria monocytogenes* [62], cholera toxin [79], *Bacillus thuringiensis* Cry1A toxin [87], *Pseudomonas aeruginosa* [82], *Escherichia coli* and *E. coli* shiga toxin [3,24,27], and *Clostridium perfringens* ι-toxin [44]. The loss of membrane cholesterol has even been able to prevent the cytoskeletal rearrangement that occurs during *E. coli* infection [54]. The *Clostridium* species have been the subjects of especially intensive studies, the results of which have indicated that cholesterol depletion is able to block the binding of and/or the pore formation caused by the *C. perfringens* α- and β-toxins [43,47] and *C. difficile* toxin A [15], but does not impair the ATP depletion and subsequent cell death caused by *C. perfringens* ε-toxin [8]. These findings provide a clear demonstration of the potential antibacterial properties of CDs, and when combined with their previously discussed antiviral properties, paint a clear picture of CDs as being potent weapons in the war against disease.

7. ORGANELLES AND INTRACELLULAR TRANSPORT

Membrane cholesterol levels can have important effects on the activities of various organelles and movement of molecules within the cell. Consequently, the extraction of cholesterol by CDs can have negative effects on these processes as well. Removal of cholesterol inhibits the intra-Golgi transport of proteins [67], as well as effecting the movement of the Golgi along the cytoskeleton in cells that contained mutant kinesin proteins [80]. Transcytosis is the process of moving molecules across the interior of a cell, and CDs have been shown to affect this process greatly as well. Indeed, depletion of cholesterol inhibits the efflux of proteins from endosomes [46] as well as the postendocytic sorting and trafficking of said molecules [32,46].

8. LECTINS

Lectins are sugar-binding proteins that have been widely used in research, especially in the investigation of polysaccharide receptors in cellular adhesion interactions. CDs have been used increasingly as part of these studies, due to their ability to effect the binding of lectins to their target receptors. These effects are due mostly to the ability of CDs to bind lectins directly [14] and to disrupt lipid rafts where the target receptors may reside [6]. However, the lectin-binding ability of CDs has also been utilized to allow researchers to determine the topological differences between different lectins [2]. Additionally, recent advancements have indicated that an increase in cellular lectin binding is a prognostic indicator for lung cancer [68]. From the research already conducted, it can surely be concluded that CDs will continue to be used to study the effects brought about by lectin binding.

9. IMMUNE SYSTEM

The various components of the immune system have proven to be particularly sensitive to cholesterol depletion by CDs. These effects could prove to be especially important to researchers as they search for novel ways to manipulate the bodies' immune response. The depletion of cholesterol from macrophages regulates cholesterol efflux and cholesteryl ester clearance from the cells [34], and also increases the expression of the SR-BI receptor [85], all of which could greatly affect the ability of macrophages to help fight off infection. T-cells have also shown to be highly affected by CDs, as their activation by the major histocompatability complex appears to be facilitated by lipid raft formation [30]. However, T-cells are not the only part of the innate immune system that relies on lipid rafts for activation. Several receptor molecules that function in the lipopolysaccharide-activated secretion of TNFα and other antibacterial chemokines, such as CD14, hsp70 and 90, and chemokine receptor 4, are all present in lipid raft microdomains as well [72]. B-cells have also shown to be sensitive to CD treatment as CDs were able to inhibit the ligand-mediated internalization of the CD22 receptor [22], as well as reversing the inhibition of K^+ channels caused by phosphatidylinositol 4,5-bisphosphate [45]. Finally, there is evidence that CDs could potentially be used to help prevent gene transfer between bacteria and their mammalian hosts. The human antimicrobial peptide LL-37 functions to bring about the lysis of invading microbes during bacterial infections. However, this same peptide is also able to transfer the free DNA plasmids from lysed bacteria into the nuclear compartment of mammalian host cells via interactions with lipid rafts [60]. This ability to protect bacterial DNA and bind to host cell DNA could help explain the mutations and periods of dormancy observed in bacterial and viral infections, as well as the initiation of autoimmune disorders. Thus, while certain aspects of the immune system appear to be quite sensitive to CD-mediated cholesterol depletion, there are certain phenomena that could be avoided by proper CD usage.

10. NERVOUS SYSTEM

Like the cells of the immune system and other body systems that will be discussed later, the components of the nervous system are also susceptible to CD-mediated cholesterol depletion. Similar to what was observed in human placental tissue, neurons also possess the GTRAP3-18 and EAAC1 proteins that function in the synthesis of glutathione, a protein that helps protect cells from oxidative stress. And as was observed in the placental tissue, CDs increased the interaction between GTRAP3-18 and EAAC1, thus increasing glutathione synthesis [76]. The protective effects of glutathione are mirrored by statin drugs, which are used to reduce both the incidence of strokes and infarct volume. In studies examining the link between cellular cholesterol and the protective properties of statins, it was found that their ability to protect cells from excitotoxicty were reversed by cotreatment with cholesterol, and were mimicked by treatment with CDs [85].

Alzheimer's disease is a neurodegenerative disorder that is characterized by the formation of amyloid plaques within the brain. Although Alzheimer's is very often associated with decreased levels of glutathione and other neuroprotective enzymes, the presence of the amyloidogenic peptide AB_{42} is the most prominent risk factor for developing the disease. The production of AB_{42} is mediated by a pair of neuronal enzymes, β- and γ-secretase, and cholesterol depletion by CDs was able to inhibit the enzymatic activities of the two secretases both additively and independently [18]. There is also additional evidence demonstrating the importance of cholesterol in other neurological disorders for which impaired muscle movement is a symptom, as CD-mediated cholesterol depletion of axons at crayfish neuromuscular junctions effectively blocked synaptic transmission [86]. These studies suggest a potential role for CDs in the treatment of Alzheimer's disease as well as other neurodegenerative disorders as they could be used to either extract cholesterol from the neurons exhibiting Alzheimer's symptoms or deliver cholesterol to the neuromuscular junctions of patients exhibiting decreased motor functioning.

Finally, neural cell adhesion molecule (NCAM) is a glycoprotein that is expressed on the membrane surface of neurons. This molecule is extremely important for mediating the outgrowth of neurites (neuronal cell body projections), but to do so it must form a complex with two other important neuronal proteins, spectrin and PKC-$β_2$. Treatment of cells with CDs was able to disrupt the binding of NCAM120 with spectrin, suggesting the interaction involves lipid rafts [31], and demonstrating the importance of membrane cholesterol in the formation of neural connections.

CD-based research on neurons has provided us with some very interesting results regarding the possible use of CDs as a means for protection against excitotoxicity and as a treatment for Alzheimer's disease. Hopefully, future research will provide us with further inside into these possibilities.

11. ENDOCRINE SYSTEM

The endocrine system is responsible for mediating intercellular communication over long distances via the release of hormones. Studies have shown that both the release and binding of these molecules is often dependent on cellular interactions involving membrane cholesterol. The hormone insulin regulates the cellular uptake of glucose from the bloodstream and is one of the best known biological signaling molecules, due to the widespread prevalence of diabetes. Insulin is synthesized and secreted from the β-cells of the islets of Langerhans within the pancreas via exocytosis. Cholesterol depletion by CDs inhibits the docking of the secretory granules with the plasma membrane with laboratory experiments showing that this can decrease insulin secretion by as much as 50% [49,75]. Conversely, luteinizing hormone (LH) is essential for triggering ovulation in women and testosterone production in men. LH receptors normally translocate to the plasma membrane after binding of human chorionic gonadotropin, but this process was severely inhibited in cells that had been depleted of membrane cholesterol [65]. Considering the number of patients worldwide who suffer from endocrine-related conditions, these results suggest a very important role for CDs and membrane cholesterol in the pathogenesis and possible treatment of these disorders.

12. FERTILIZATION AND EMBRYOGENESIS

CDs have recently gained popularity in research focused on fertilization and the reproductive process. The most recent evidence suggests that CDs may be able to play a very effective role at improving the techniques currently used for in vitro fertilization and artificial insemination. Cryopreservation of sperm is widely used, as it provides an economical and relatively reliable means of maintaining lines of genetically engineered laboratory animals. Of the lines of genetically engineered mice that are routinely used as research subjects, the C57BL/6 line is the most important. However, the observed rate of fertilization from frozen C57BL/6 mice is quite low compared to other strains. Frozen sperm from this strain that were treated with CD exhibited a fertilization rate that was nearly three times that of control samples [69]. However, these results were not replicated in similar studies performed on the frozen sperm of rainbow trout. This study showed that neither the depletion of cholesterol by up to 50%, nor the doubling of the baseline cholesterol level, had any effect on the fertilization rate [41]. This suggests that the ability of CDs to improve fertilization from cryopreserved sperm is limited to species in which the sperm possess an acrosome.

Other interesting findings have come from studies on *Xenopus laevis* oocytes. Incubation of *X. laevis* oocytes with CDs greatly accelerated the rate of meiotic maturation, the first such observation in vertebrate oocytes [57]. Although these experiments have yet to be carried out on mammalian oocytes, similar results could potentially lead to dramatically improved rates of success for artificial insemination in humans or possibly even further progress in the field of cloning.

Similar to the attempts that have already been made in trying to use CDs to prevent the transmission of HIV, CDs have also been considered as possible topical spermicides [10]. Although it is believed that their ability to extract high amounts of membrane cholesterol would allow them to be affective spermicidal agents, their inability to prevent HIV transmission has prevented the initiation of any sort of clinical trials. Nevertheless, CDs have still provided researchers with a potential avenue to explore in terms of fertilization and embryology, and further research in these topics could lead to vastly improved methods for artificial insemination and the maintenance of important strains of laboratory animals.

13. CARDIOVASCULAR SYSTEM

Heart disease is currently the leading cause of death in the United States, and its connection to elevated levels of cholesterol has been known for quite some time now. Considering the ability of CDs to remove cellular cholesterol, it would only seem prudent to investigate their potential uses in combating this deadly condition.

The recruitment of monocytes which then differentiate into macrophages and ingest low-density lipoproteins is a critical step in the formation of atherosclerotic plaques [42]. Activation of monocyte adhesion molecules such as CD11b is required for this step to proceed. CDs, but not CDs complexed with cholesterol, were able to prevent the activation of CD11b [42], suggesting that the extraction of cholesterol from serum monocytes could be an effective means of inhibiting the development of atherosclerotic plaques.

The membrane-bound receptors and channel proteins of various cell types have been found to be either located in or directly associated with lipid rafts, and the membrane proteins of cardiac cells are no different. Of the cardiac membrane proteins that have been found to be associated with the lipid rafts, the most important ones are the Ca^{2+} channels. Depletion of cholesterol from cardiac myocytes produces profound inhibitory effects on the ion flux mediated by the Ca^{2+} channels [37,73]. This suggests that the depletion of cholesterol from cardiac myocytes could potentially have grave consequences if it were to occur in a human being.

Ion channels important to the cardiovascular system are not just located within the heart itself; they are also present in the smooth muscle that mediates the contraction of arteries. CD-mediated cholesterol depletion has not only shown that ATP-sensitive K^+ channels and adenyl cyclase both localize to lipid rafts in the membrane of arterial smooth muscle [59], it also elicited an impaired arterial contraction in response to the binding of 5-hydroxytryptamine (5-HT), vasopressin, and endothelin [11].

The ability of CD to elicit drastic changes in the ion flow and the contractility of arterial smooth muscle indicate that lipid rafts and cellular cholesterol levels are extremely important factors in the functioning of the cardiovascular system. Continued research on this front could potentially lead to major advancements in our ability to fight heart disease.

14. MUSCLE

In addition to the smooth and cardiac muscle of the cardiovascular system, CDs have been very useful in studying smooth and skeletal muscle throughout the rest of the body. Depletion of membrane cholesterol by CDs decreases the membrane capacitance of rat uterine myocytes [64] and inhibits rat penile erection as well [33]. But CDs do not just exhibit inhibitory effects on muscle function and growth. Cholesterol depletion has also been shown to enhance myoblast fusion and induce the formation of myotubes in chick skeletal muscle, although the resulting tissue did exhibit disorganized nuclei [39]. CDs have also been implicated in the development of chronic airway diseases. Conditions like these, such as asthma, are associated with increased smooth muscle mass in the airway. Disruption of the caveolae (invaginations of the plasma membrane that are usually associated with high levels of lipids) with CDs caused spontaneous activation of the p42/p44 MAP kinase pathway, leading to localized smooth muscle proliferation [16]. This suggests that caveolae play an important role in inhibition of excessive smooth muscle proliferation, and combined with other examples presented, shows that membrane cholesterol levels are very important for maintaining proper muscle functioning.

15. CONCLUSIONS

CDs are cyclic oligosaccharides that due to their unique physical properties have proven to be very useful tools in biological research. Their ability to extract cholesterol from the plasma membrane has allowed researchers to shed new light on a variety of cellular interactions, some of which could be extremely useful in a variety of areas, including the prevention of viral and bacterial infections, the prevention of

neurodegenerative diseases, and increased efficiency for sperm storage and in vitro fertilization. Hopefully, CD-based research will continue into the future, as the results already obtained suggest that such studies could prove to be extremely beneficial to humankind.

Acknowledgments

Work from the authors' laboratory was supported by NIH, MBRS, SCORE (S0648680), RISE, and MARC programs, and the Sidney Stern Memorial Trust.

REFERENCES

1. Ambrose, Z., Compton, L., Piatak, M., Lu, D., Alvord, G.W., Lubomirski, M.S., Hildreth, J.E.K., Lifson, J.D., Miller, C.J., KewalRamani, V.N. (2008). Incomplete protection against simian immunodeficiency virus vaginal transmission in rhesus macaques by a topical antiviral agent revealed by repeat challenges. *J. Virol., 82,* 6591–6599.

2. Andre, S., Kaltner, H., Furuike, T., Nishimura, S., Gabius, H. (2004). Persubstituted cyclodextrin-based glycoclusters as inhibitors of protein-carbohydrate recognition using purified plant and mammalian lectins and wild-type and lectin-gene-transfected tumor cells as targets. *Bioconjugate Chem., 15,* 87–98.

3. Baorto, D.M., Gao, Z., Malaviya, R., Dustin, M.L., van der Merwea, A., Lublin, D.M., Abraham, S.N. (1997). Survival of FimH-expressing enterobacteria in macrophages relies on glycolipid traffic. *Nature, 389,* 636–639.

4. Barman, S., Nayak, D.P. (2007). Lipid raft disruption by cholesterol depletion enhances influenza A virus budding from MDCK cells. *J. Virol., 81,* 12169–12178.

5. Bhattacharya, B., Roy, P. (2008). Bluetongue virus outer capsid protein VP5 interacts with membrane lipid rafts via a SNARE domain. *J. Virol., 82,* 10600–10612.

6. Brewer, C.F., Miceli, M.C., Baum, L.G. (2002). Clusters, bundles, arrays and lattices: novel mechanisms for lectin-saccharide-mediated cellular interactions. *Curr. Opin. Struct. Biol., 12,* 616–623.

7. Campbell, S., Gaus, K., Bittman, R., Jessup, W., Crowe, S., Mak, J. (2004). The raft-promoting property of virion-associated cholesterol, but not the presence of virion-associated Brij 98 rafts, is a determinant of human immunodeficiency virus type 1 infectivity. *J. Virol., 78,* 10556–10565.

8. Chassin, C., Bens, M., de Barry, J., Courjaret, R., Bossu, J.L., Cluzeaud, F., Mkaddem, S.B., Gibert, M., Poulain, B., Popoff, M.R., Vandewalle, A. (2007). Pore-forming epsilon toxin causes membrane permeabilization and rapid ATP depletion-mediated cell death in renal collecting duct cells. *Am. J. Physiol. Renal Physiol., 293,* 927–937.

9. Chen, X., Resg, M. (2002). Cholesterol depletion from the plasma membrane triggers ligand independent activation of the EGF receptor. *J. Biol. Chem., 277,* 49631–49637.

10. Doncel, G.F. (2006). Exploiting common target in human fertilization and HIV infection: development of novel contraceptive microbicides. *Hum. Reprod. Update*, *12*, 103–117.

11. Dreja, K., Voldstedkund, M., Vinten, J., Tranum-Jensen, J., Hellstrand, P., Sward, K. (2002). Cholesterol depletion disrupts caveolae and differentially impairs agonist-induced arterial contraction. *Arterioscler. Thromb. Vasc. Biol.*, *22*, 1267–1275.

12. Duan, M.S., Zhao, N., Ossurardotti, I.B., Thorsteinsson, T., Loftsson, T. (2005). Cyclodextrin solubilization of the antibacterial agents triclosan and triclocarbon: formation of aggregates and higher-order complexes. *Int. J. Pharm.*, *297*, 213–222.

13. Funk, A., Mhamdi, M., Hohenberg, H., Heeren, J., Reimer, R., Lambert, C., Prange, R., Sirma, H. (2008). Duck hepatitis B virus required cholesterol for endosomal escape during virus entry. *J. Virol.*, *82*, 10532–10542.

14. Garcia-Lopez, J.J., Hernandez-Mateo, F., Isac-Garcia, J., Kim, J. M., Roy, R., Santoyo-Gonzalez, F., Vargas-Berenguel, A. (1999). Synthesis of per-glycosylated β-cyclodextrins having enhanced lectin binding affinity. *J. Org. Chem.*, *64*, 522–531.

15. Giesemann, T., Jank, T., Gerhard, R., Maier, E., Just, I., Benz, R., Aktories, K. (2006). Cholesterol-dependent pore formation of *Clostridium difficile* toxin A. *J. Biol. Chem.*, *281*, 10808–10815.

16. Gosens, R., Stelmack, G., Dueck, G., McNeil, K.D., Yamasaki, A., Gerthoffer, W.T., Unruh, H., Gounni, A.S., Zaagsma, J., Halayko, A.J. (2006). Role of caveolin-1 in p42/p44 MAP kinase activation and proliferation of human airway smooth muscle. *Am. J. Physiol. Lung Cell Mol. Physiol.*, *291*, L523–L534.

17. Graham, D.R.M., Chertova, E., Hilburn, J.M., Arthur, L.O., Hildreth, J.E.K. (2003). Cholesterol depletion of human immunodeficiency virus type 1 and simian immunodeficiency virus with β-cyclodextrin inactivates and permeabilizes the virions: evidence for virion-associated lipid rafts. *J. Virol.*, *77*, 8237–8248.

18. Grimm, M.O.W., Grimm, H.S., Tomic, I., Beyreuther, K., Hartmann, T., Bergmann, C. (2008). Independent inhibition of Alzheimer disease β-and γ-secretase cleavage by lowered cholesterol levels. *J. Biol. Chem.*, *283*, 11302–11311.

19. Ha, H., Kwak, H.B., Lee, S.K., Na, D.S., Rudd, C.E., Lee, Z.H., Kim, H. (2003). Membrane rafts play a crucial role in receptor activator of nuclear factor ϰB (RANK) signaling and osteoclast function. *J. Biol. Chem.*, *273*, 18573–18580.

20. Hess, S.T., Kumar, M., Verma, A., Farrington, J., Kenworthy, A., Zimmerberg, J. (2005). Quantitative electron microscopy and fluorescence spectroscopy of the membrane distribution of influenza hemagglutinin. *J. Cell Biol.*, *169*, 965–976.

21. Huang, Q., Shen, H., Shui, G., Wenk, M., Ong, C. (2006). Emodin inhibits tumor cell adhesion through disruption of the membrane lipid raft-associated integrin signaling pathway. *Cancer Re.*, *66*, 5807–5815.

22. John, B., Herrin, B., Raman, C., Wang, Y., Bobbitt, K., Brody, B., Justement, L.B. (2003). The B cell coreceptor CD22 associates with AP50, a clathrin-coated pit adapter protein, via tyrosine-dependent interaction. *J. Immunol.*, *170*, 3534–3543.

23. Kanee, A. http://www.mawhiba.org.sa/NR/rdonlyres/2923D2E4-2B92-4632-BF5C-5284E93E33D8/0/Abdullah-KaneeRSI2007.pdf.

24. Kansau, I., Berger, C., Hospital, M., Amsellem, R., Nicolas, V., Servin, A., Bernet-Camard, M. (2004). Zipper-like internalization of Dr-positive *Escherichia coli* by epithelial cells is preceded by an adhesin-induced mobilization of Raf-associated molecules in the initial step of adhesion. *Infect. Immun.*, *72*, 3733–3742.

25. Keller, P., Simons, K. (1998). Cholesterol is required for surface transport of influenza virus hemagglutinin. *J. Cell Biol.*, *140*, 1357–1367.

26. Kilsdonk, E.P.C., Yancey, P.G., Stoudt, G.W., Bangerter, F.W., Johnson, W.J., Phillips, M.C., Rothblat, G.H. (1995). Cellular cholesterol efflux mediated by cyclodextrins. *J. Biol. Chem.*, *270*, 17250–17256.

27. Kovbasnjuk, O., Edidin, M., Donowitz, M. (2001). Role of lipid rafts in shiga toxin 1 interaction with the apical surface of Caco-2 cells. *J. Cell Sci.*, *114*, 4025–4031.

28. Laliberte, J.P., McGinnes, L.W., Peeples, M.E., Morrison, T.G. (2006). Integrity of membrane lipid rafts is necessary for the ordered assembly and release of infectious Newcastle disease virus particles. *J. Virol.*, *80*, 10652–10662.

29. Laliberte, J.P., McGinnes, L.W., Morrison, T.G. (2007). Incorporation of functional HN-F glycoprotein-containing complexes into Newcastle disease virus is dependent on cholesterol and membrane lipid raft integrity. *J. Virol.*, *81*, 10636–10648.

30. Larbi, A., Douziech, N., Khalil, A., Dupuis, G., Gherairï, S., Guérard, K.P., Fülöp, T. Jr. (2004). Effects of methyl-beta-cyclodextrin on T lymphocytes lipid rafts with aging. *Exp. Gerontol.* *39*, 551–558.

31. Leshchyns'ka, I., Sytnky, V., Morrow, J.S., Schachnew, M. (2003). Neural cell adhesion molecule (NCAM) association with PKC-β_2 via β-I spectrin is implicated in NCAM-mediated neurite outgrowth. *J. Cell Biol.*, *161*, 625–639.

32. Leyt, J., Melamed-Book, N., Vaerman, J., Cohen, S., Weiss, A. M., Aroeti, B. (2007). Cholesterol-sensitive modulation of transcytosis. *Mol. Biol. Cell*, *18*, 2057–2071.

33. Linder, A.E., Leite, R., Lauria, K., Mills, T.M., Webb, R.C. (2005). Penile erection requires association of soluble guanylyl cyclase with endothelial caveolin-1 in rat corpus cavernosum. *Am. J. Physiol. Integr. Compar. Physiol.*, *290*, 1302–1308.

34. Liu, S.M., Cogny, A., Kockx, M., Deat, R.T., Gaus, K., Jessup, W., Kritharides, L. (2003). Cyclodextrins differentially mobilize free and esterified cholesterol from primary human foam cell macrophages. *J. Lipid Res.*, *44*, 1156–1166.

35. Locke, D., Koreen, I.V., Liu, J.Y., Harris, A. (2004). Reversible pore block of connexin channels by cyclodextrins. *J. Biol. Chem.*, *279*, 22883–22892.

36. Loftsson, T., Leeves, N., Bjornsdóttir, B., Duffy, L., Másson, M. (1999). Effect of cyclodextrins and polymers on triclosan availability and substantivity in toothpastes in vivo. *J. Pharm. Sci.*, *88*, 1254–1258.

37. Maguy, A., Hebert, T.E., Nattel, S. (2005). Involvement of lipid rafts and caveolae in cardiac ion channel function. *Cardiovasc. Res.*, *69*, 798–807.

38. Martin Del Valle, E.M. (2003). Cyclodextrins and their uses: a review. *Process Biochem.*, *39*, 1033–1046.

39. Mermelstein, C.S., Portilho, D.M., Medeiros, R.B., Matos, A. R., Einicker-Lama, M., Tortelote, G.G., Vieyra, A., Costa, M.L. (2005). Cholesterol depletion by methyl-β-cyclodextrin enhances myoblast fusion and induces the formation of myotubes with disorganized nuclei. *Cell Tissue Res.*, *319*, 289–297.

40. Morris, D.P., Lei, B., Wu, Y., Michelotti, G.A., Schwinn, D. (2007). The α_{1a}-adrenergic receptor occupies membrane rafts with its G protein effectors but internalizes via clathrin-coated pits. *J. Biol. Chem.*, *283*, 2973–2985.

41. Muller, K., Muller, P., Pincemy, G., Kurtz, A., Labbe, C. (2008). Characterization of sperm plasma membrane properties after cholesterol modification: consequences for cryopreservation of rainbow trout spermatozoa. *Biol. Reprod.*, *78*, 390–399.

42. Murphy, A.J., Woollard, K.J., Hoang, A., Mukhamedova, N., Stirzaker, R.A., McCormick, S.P.A., Remaley, A.T., Sviridov, D., Chin-Dustin, J. (2008). High density lipoprotein reduces the human monocyte inflammatory response. *Arterioscler. Thromb. Vasc. Biol.*, *28*, 2071–2077.

43. Nagahama, M., Hayashi, S., Morimitsu, S., Sakurai, J. (2003). Biological activities and pore-formation of Clostridium perfringens β-toxin in HL 60 cells. *J. Biol. Chem.*, *278*, 36934–36941.

44. Nagahama, M., Yamaguchi, A., Hagiyama, T., Ohkubo, N., Kobayashi, K., Sakurai, J. (2004). Binding and internalization of *Clostridium perfringens* iota- toxin in lipid rafts. *Infect. Immun.*, *72*, 3267–3275.

45. Nam, J.H., Lee, H.S., Nguren, Y.H., Kang, T.M., Lee, S.W., Kim, H.Y., Kim, S.J., Earm, Y.E., Kim, S.J. (2007). Mechanosensitive activation of K⁺ channel via phospholipase C–induced depletion of phosphatidylinositol 4,5-bisphosphate in B lymphocytes. *J. Physiol.*, *582* (3), 977–990.

46. Nyasae, L.K., Hubbarb, A.L., Tuma, P.L. (2003). Transcytotic efflux from early endosomes is dependent on cholesterol and glycosphingolipids in polarized hepatic cells. *Mol. Biol. Cell*, *14*, 2689–2705.

47. Oda, M., Matsuno, T., Shiihara, R., Ochi, S., Yamauchi, R., Imagawa, H., Nagahama, M., Nishizawa, M., Sakurai, J. (2008). The relationship between the metabolism of sphingomyelin species and hemolysis of sheep erythrocytes by *Clostridium perfringens* α-toxin. *J. Lipid Res.*, *49*, 1039–1047.

48. O'Donnell, V., LaRocco, M., Baxt, B. (2008). Heparan sulfate binding foot-and-mouth disease virus enters cells via caleola-mediated endocytosis. *J. Virol.*, *82*, 9075–9085.

49. Ohara-Imaizumi, M., Nishiwaki, C., Kikuta, T., Kumakura, K., Nakamichi, Y., Nagamatsu, S. (2004). Site of docking and fusion of insulin secretory granules in live MIN6 β cells analyzed by TAT-conjugated anti-syntaxin 1 antibody and total internal reflection fluorescence microscopy. *J. Biol. Chem.*, *279*, 8403–8408.

50. Ottico, E., Prinetti, A., Prioni, S., Giannotta, C., Basso, L., Chigorno, V., Sonnino, S. (2003). Dynamics of membrane lipid domains in neuronal cells differentiated in culture. *J. Lipid Res.*, *44*, 2142–2151.

51. Popik, W., Alce, T.M., Au, W. (2002). Human immunodeficiency virus type 1 uses lipid raft–colocalized CD4 and chemokine receptors for productive entry into CD4⁺ T cells. *J. Virol.*, *76*, 4709–4722.

52. Pouvreau, S., Berthier, C., Blaineau, S., Amsellem, J., Coronado, R., Strube, C. (2004). Membrane cholesterol modulates dihydropyridine receptor function in mice fetal skeletal muscle cells. *J. Physiol.*, *552* (2), 365–381.

53. Prior, I.A. Muncke. C., Parton, R.G., Hancock, J.F. (2003). Direct visualization of Ras proteins in spatially distinct cell surface microdomains. *J. Cell Biol.*, *160*, 165–170.

54. Riff, J.D., Callahan, J.W., Sherman, P.M. (2005). Cholesterol-enriched membrane microdomains are required for inducing host cell cytoskeleton rearrangements in response to attaching-effacing *Escherichia coli. Infect. Immun.*, *73*, 7113–7125.

55. Ringerike, T., Blystad, F.D., Levy, F.O., Madshus, I.H., Stang, E. (2002). Cholesterol is important in control of EGF receptor kinase activity but EGF receptors are not concentrated in caveolae. *J. Cell Sci.*, *115*, 1331–1340.

56. Russo, L.M., McKee, M., Brown, D. (2006) Methyl-β-cyclodextrin induces vasopressin-independent apical accumulation of aquaporin-2 in the isolated, perfused rat kidney. *Am. J. Physiol. Renal Physiol.*, *291*, F246–F253.

57. Sadler, S.E., Jacobs, N.D. (2004). Stimulation of *Xenopus laevis* oocyte maturation by methyl-β-cyclodextrin. *Biol. Reprod.*, *70*, 1685–1692.

58. Sajadi, S., Rojas, P., Oppenheimer, S.B. (2007). Cyclodextrin, a probe for studying adhesive interactions. *Acta Histochem.*, *109*, 338–342.

59. Sampson, L.J., Hayabuchi, Y., Standen, N.B., Dart, C. (2004). Caveolae localize protein kinase A signaling to arterial ATP-sensitive potassium channels. *Circ. Res.*, *95*, 1012–1018.

60. Sandgren, S., Wittrup, A., Cheng, F., Jonsson, M., Eklund, E., Busch, S., Beltig, M. (2004). The human antimicrobial peptide LL-37 transfers extracellular DNA plasmid to the nuclear compartment of mammalian cells via lipid rafts and proteoglycan-dependent endocytosis. *J. Biol. Chem.*, *279*, 17951–17956.

61. Schols, D., De Clercq, E., Witvrouw, M., Nakashima, H., Snoeck, R., Pauwels, R., Van Schepdael, A., Claes, P. (1991). Sulphated cyclodextrins are potent anti-HIV agents acting synergistically with 2,3-dideoxynucleoside analogues. *Antiviral Chem. Chemother.*, *2*, 45–53.

62. Seveau, S., Bierne, H., Giroux, S., Prevost, M., Cossart, P. (2004). Role of lipid rafts in E-cadherin- and HGF-R/Met-mediated entry of *Listeria monocytogenes* into host cell. *J. Cell Biol.*, *166*, 743–753.

63. Shrimpton, C.N., Borthakur, G., Larrucea, S., Cruz, M.A., Dong, J., Lopez, J. (2002). Localization of the adhesion receptor glycoprotein Ib-IX-V complex to lipid rafts is required for platelet adhesion and activation. *J. Exp. Med. 196*, 1057–1066.

64. Shmygol, A., Noble, K., Wray, S. (2007). Depletion of membrane cholesterol eliminates the Ca²⁺-activated component of outward potassium current and decreases membrane capacitance in rat uterine myocytes. *J. Physiol.*, *581* (2), 445–456.

65. Smith, S.M.L., Lei, Y., Liu, J., Cahill, M.E., Hagen, G.M., Barisas, B.G., Roess, D.A. (2006). Luteinizing hormone

receptors translocate to plasma membrane microdomains after binding of human chorionic gonadotropin. *Endocrinology*, *147*, 1789–1795.

66. Steck, T.L., Ye, J., Lange, Y. (2002). Probing red cell membrane cholesterol movement with cyclodextrin. *Biophys. J.*, *83*, 2118–2125.

67. Stuven, E., Porat, A., Shimron, F., Fass, E., Kaloyanova, D., Brugger, B., Wieland, F.T., Elazar, Z., Helms, J.B. (2003). Intra-Golgi protein transport depends on a cholesterol balance in the lipid membrane. *J. Biol. Chem.*, *278*, 53112–53122.

68. Szoke, T., Kayser, K., Baumhakel, J., Trojan, I., Furak, J., Tiszlavicz, L., Horvath, A., Szluha, K., Gabius, H., Andre, S. (2005). Prognostic significance of endogenous adhesion/growth-regulatory lectins in lung cancer. *Oncology*, *69*, 167–174.

69. Takeo, T., Hoshii, T., Kondo, Y., Toyodome, H., Arima, H., Yamamura, K., Irie, T., Nakagata, N. (2008). Methyl-β-cyclo-dextrin improves fertilizing ability of C57BL/6 mouse sperm after freezing and thawing by facilitating cholesterol efflux from the cells. *Biol. Reprod.*, *78*, 546–551.

70. Tamilselvam, B., Daefler, S. (2008). *Francisella* targets cholesterol-rich host cell membrane domains for entry into macrophages. *J. Immunol.*, *180*, 8262–8271.

71. Thorp, E.B., Gallagher, T.M. (2004). Requirements for CEACAMs and cholesterol during murine coronavirus cell entry. *J. Virol.*, *78*, 2682–2692.

72. Triantafilou, M., Miyake, K., Golenbock, D.T., Triantafilou, K. (2002). Mediators of innate immune recognition of bacteria concentrate in lipid rafts and facilitate lipopolysaccharide-induced cell activation. *J. Cell Sci.*, *115*, 2603–2611.

73. Tsujikawa, H., Song, Y., Watanabe, M., Masumiya, H., Gupte, S., Ochi, R., Okada, T. (2007). Cholesterol depletion modulates basal L-type Ca^{2+} current and abolishes its β-adrenergic enhancement in ventricular myocytes. *Am. J. Physiol. Heart Circ. Physiol.*, *294*, H285–H292.

74. Vassilieva, E.V., Gerner-Smidt, K., Ivanov, A.I., Nusrat, A. (2008). Lipid rafts mediate internalization of $β_1$-integrin in migrating intestinal epithelial cells. *Am. J. Physiol. Gastrointest. Liver Physiol.*, *295*, G965–G976.

75. Vikman, J., Jimenez-Feltstrom, J., Nyman, P., Thelin, J., Eliasson, L. (2009). Insulin secretion is highly sensitive to desorption of plasma membrane cholesterol. *FASEB J.*, *23*, 58–67.

76. Watabe, M., Aoyama, K., Nakaki, T. (2008). A dominant role of GTRAP3-18 in neuronal glutathione synthesis. *J. Neurosci.*, *28*, 9404–9413.

77. Weiner, D.B., Williams, W.V., Weisz, P.B., Greene, M.I. (1992). Synthetic cyclodextrin derivatives inhibit HIV infection in vitro. *Pathobiology*, *60*, 206–212.

78. Wolf, A.A., Fukinaga, Y., Lencer, W.I. (2002). Uncoupling of the cholera toxin G_{M1} ganglioside-receptor complex from endocytosis, retrograde Golgi trafficking, and downstream signal transduction by depletion of membrane cholesterol. *J. Biol. Chem.*, *277*, 16249–16256.

79. Xia, F., Leung, Y.M., Gaisano, G., Gao, X., Chen, Y., Manning, J.E., Fox, M., Bhattacharjee, A., Wheeler, M.B., Gaisano, H.Y., Tsushima, R.G. (2007). Targeting of voltage-gated K^+ and Ca^{2+} channels and soluble *N*-ethylmaleimide-sensitive factor attachment protein receptor proteins to cholesterol-rich lipid rafts in pancreatic α-cells: effects on glucagon stimulus-secretion coupling. *Endocrinology*, *148*, 2157–2167.

80. Xu, Y., Takeda, S., Nakata, T., Noda, Y., Tanaka, T., Hirokawa, N. (2002). Role of KIFC3 motor protein in Golgi positioning and integration. *J. Cell Biol.*, *158*, 293–303.

81. Xu, W., Yoon, S., Huang, P., Wang, Y., Chen, C., Chong, P.L., Liu-Chen, L. (2006). Localization of the ϰ opioid receptor in lipid rafts. *J. Pharmacol. Exp. Ther.*, *317*, 1295–1306.

82. Yamamoto, Na., Yamamoto, No. Petroll, M.W., Cavanagh, H.D., Jester, J.V. (2005). Internalization of *Pseudomonas aeruginosa* is mediated by lipid rafts in contact lens-wearing rabbit and cultured human corneal epithelial cells. *Invest. Ophthalmol. Vis. Sci.*, *46*, 1348–1355.

83. Young, J., Tang, Z., Yu, Q., Yu, D., Wu, Y. (2008). Selective killing of HIV-I-positive macrophages and T cells by the Rev-dependent lentivirus carrying anthrolysin O from *Bacillus anthracis*. *Retrovirology*, *5*, doi: 10.1186/1742-4690-5-36.

84. Yu, L., Cao, G., Repa, J., Stangl, H. (2004). Sterol regulation of scavenger receptor class B type I in macrophages. *J. Lipid Res.*, *45*, 889–899.

85. Zacco, A., Togo, J., Spence, K., Ellis, A., Lloyd, A., Furlong, S., Piser, T. (2003). 3-Hydroxy-3-methylglutaryl coenzyme a reductase inhibitors protect cortical neurons from excitotoxicity. *J. Neurosci.*, *23*, 11104–11111.

86. Zamir, O., Charlton, M. (2006). Cholesterol and synaptic transmitter release at crayfish neuromuscular junction. *J. Physiol.*, *571* (1), 83–99.

87. Zhuang, M., Oltean, D., Gomez, I., Pullikuth, A.K., Soberon, M., Bravo, A., Gill, S.S. (2002). *Heliothis virescens* and *Manduca sexta* lipid rafts are involved in Cry 1A toxin binding to the midgut epithelium and subsequent pore formation. *J. Biol. Chem.*, *277*, 13863–13872.

16

CYCLODEXTRIN-BASED HYDROGELS

CARMEN ALVAREZ-LORENZO
Departamento de Farmacia y Tecnologia Farmaceutica, Universidad de Santiago de Compostela, Santiago de Compostela, Spain

MARIA D. MOYA-ORTEGA AND THORSTEINN LOFTSSON
Faculty of Pharmaceutical Sciences, University of Iceland, Reykjavik, Iceland

ANGEL CONCHEIRO AND JUAN J. TORRES-LABANDEIRA
Departamento de Farmacia y Tecnologia Farmaceutica, Universidad de Santiago de Compostela, Santiago de Compostela, Spain

1. HYDROGELS IN DRUG DELIVERY

1.1. Concept, Structure, and General Features

The term *gel* was introduced late in the nineteenth century in an attempt to classify semisolid substances according to their physical characteristics rather than their molecular composition [1]. Structurally, a gel consists of a relatively small fraction of solid materials, mainly entangled polymers, dispersed in quite a large volume of liquid in which the polymers form a three-dimensional structure. The liquid prevents the network from collapsing into a compact mass, and the network prevents the liquid from flowing away. Thus, gel can be thought of as an intermediate between solid and liquid states [2].

The appearance and properties of gels come from the links among the polymer chains [3]. Depending on the strength of the interactions, the cross-linking degree and the ratio between the number of cross-linking points and the length of the polymer chains, gels vary in consistency from viscous fluids to fairly rigid solids. Physically cross-linked gels are formed by simple entanglement of polymer chains and are, in general, easily spreadable. Simple addition of more solvent induces dilution of the polymer and the system loses its gel behavior. Chemically cross-linked gels, in which the polymer chains are bound to each other through relatively strong bonds (principally, electrostatic or covalent bonds), may be used as highly viscoelastic macroscopic platforms or as individualized microgel particles. Cross-linked gels swell in water without solution, keeping their cross-linking points (Fig. 1). Comprehensive reviews on gel synthesis and properties may be found elsewhere [3,4].

Any gel in which the liquid phase is water is called a *hydrogel*, although the term is used mainly to design chemically cross-linked systems. Hydrogels are of great interest as food components, water superabsorbents, chemical traps, artificial organs and scaffolds, agents able to immobilize enzymes, and in the pharmaceutical field, as contact lenses and drug vehicles or carriers [5,6]. High water content, soft consistency, and viscoelastic behavior make hydrogels resemble natural living tissues more than any other class of synthetic biomaterials. Hydrogels can be made to resist the physiological stress caused by skin flexion, blinking, and mucociliar movement, to adopt the shape of the application area without damage, and even to provide support to the surrounding tissues [7,8]. Despite hydrogels having long been known, they have in recent years generated increasing interest as drug delivery, systems owing to advances in polymer physics and chemistry that have led to resourceful synthetic methods and to myriads of network structures and performances [9].

Cyclodextrins in Pharmaceutics, Cosmetics, and Biomedicine: Current and Future Industrial Applications, First Edition. Edited by Erem Bilensoy.
© 2011 John Wiley & Sons, Inc. Published 2011 by John Wiley & Sons, Inc.

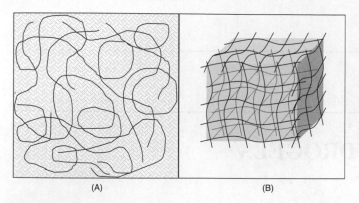

Figure 1. Structure of a (A) physically cross-linked gel and (B) a continuous chemically cross-linked gel.

1.2. Pharmaceutical and Biomedical Applications

Hydrogels offer important advantages over conventional dosage forms. The versatility of hydrogel-based drug carriers makes them able to be adapted to practically all delivery routes [10–13]. Furthermore, these outstandingly patient-friendly delivery systems enable a precise release of drugs for more or less prolonged periods of time. Two of the critical parameters determining the kinetics and the mechanism of drug release, the permeability (mesh size) of the polymer network and its swelling properties, are easily adjustable by choosing an adequate polymer composition and cross-linking degree. A hydrogel may act as a simple reservoir of drugs

that are released by diffusion or erosion, as targeting drug delivery systems, or as triggered drug release devices [14–16]. Physically cross-linked hydrogels control drug release primarily through the effect of viscosity on drug diffusion [17]. The drug diffusion barrier through the gel is not as high as predicted by the apparent macroviscosity, but depends on the viscosity of the microenvironment through which the drug has to pass [18,19]. In general, the greater the hydrodynamic size of the diffusible species, the slower the diffusion. Thus, attempts to control drug release by incorporation of enlarged structures, such as cyclodextrin complexes, have been successful under certain circumstances [20–24].

Chemically cross-linked hydrogels can control drug delivery through changes in mesh size (e.g., swelling or eroding) or through changes in the binding strength of the drug to some chemical groups of the network that respond to modifications of some physiological variables (Fig. 2). If the hydrodynamic diameter of the drug is greater than the mesh size, the drug is not released; if it is lower, the release rate depends on the tortuousness of the diffusion path [5]. Once in contact with the physiological fluids, some hydrogels suffer deep structural changes, and their volume, mesh size, and even integrity may be reversibly (smart behavior) or irreversibly altered. Specific conditions of the environment, such as pH, temperature, light wavelength, nature and concentration of ions, or enzymatic activity, determine the intensity of such changes and, in consequence, may affect the drug

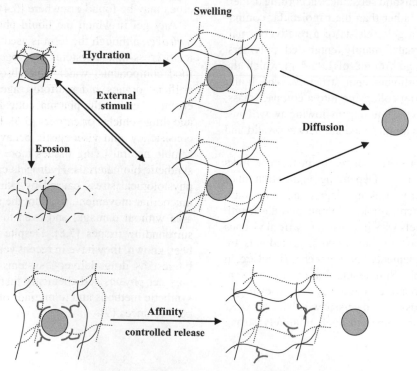

Figure 2. Mechanisms potentially involved in the drug release process from a hydrogel-based formulation.

release process [16,25]. Obviously, enhanced knowledge of the effects of these external variables will increase the possibilities of designing hydrogels with adequate drug release profiles through swelling- or erosion-controlled mechanisms.

The highly hydrophilic character of hydrogels, which gives gels their biocompatibility and appropriate mechanical properties, is, however, a handicap for performance as a drug delivery device. Hydrophobic drugs that are poorly soluble in water can be difficult to load into hydrogels. When immersed in a drug solution, equilibrium between the aqueous phase of the hydrogel and the surrounding loading solution is achieved in such a way that the drug concentration in both phases becomes the same [26]. The amount of drug loaded is expected to be proportional to the volume of water in the hydrogel [27]:

$$\text{drug in aqueous phase (w/w dry hydrogel)} = \frac{V_s}{W_p} C_0 \quad (1)$$

where V_s is the volume of the aqueous phase in the hydrogel, W_p the weight of dried hydrogel, and C_0 the drug concentration in the loading solution. According to this equation, the lower the concentration of drug in the loading solution, the smaller the amount that can be loaded by the hydrogel. This is an important concern since most drugs considered essential by the World Health Organization are poorly water soluble, and nearly 90% of the new chemical entities regarded as drug candidates also lack sufficient aqueous solubility [28–30]. Therefore, the amount of hydrophobic drug loaded by hydrogels is usually therapeutically insufficient.

On the other hand, hydrophilic drugs may exhibit rapid and great uptake by the hydrogel, but their release profiles typically show a burst delivery. The rate of drug release from a hydrogel obeys the diffusion laws and is mainly dependent on thickness, degree of hydration, and drug concentration in the hydrogel. The following equation can be used to predict, under *sink* conditions, the release of a drug by diffusion through hydrophilic chemically cross-linked networks [26]:

$$\frac{dM}{dt} = \frac{8DM_\infty}{l^2}\exp\left(\frac{-\pi^2 Dt}{l^2}\right) \quad (2)$$

In this equation, M_∞ represents the total amount of drug released, l the hydrogel thickness, and D the coefficient of drug diffusion, which is expected to remain constant if no changes occur in the degree of swelling of the hydrogel during drug release. For a given drug, the diffusion time decreases as the water content of the hydrogel increases; and for a given hydrogel, the lower the molecular weight of the drug, the shorter the release time [26]. Bearing this in mind, if the drug diffuses passively through the aqueous phase of the network without interacting with the polymeric structure,

both the amount loaded and the control of the release will be deficient. Therefore, efficient mechanisms of drug retention in the hydrogel are being searched.

The development of hydrogels that interact effectively with a drug through ionic or hydrophobic bonds has been shown to improve the loading and release performance of ionizable drugs [15,31,32]. When the chemical groups and even the conformation of the polymer network match the features of the drug molecules, specific binding sites for the drug into the network can be obtained [33,34]. In such a case, the drug is not only loaded as drug molecules dissolved in the aqueous phase of the hydrogel, but also as drug molecules effectively interacting with the binding sites. This situation enhances remarkably the affinity of the drug to the hydrogel, increasing the total amount of drug that can be loaded and offering an affinity-dependent release rate [31,35,36].

The concerns about the loading of nonionizable poorly soluble drugs are more demanding. In this regard, inclusion of the drug into colloidal carriers, such as liposomes or surfactant micellar structures, occupies a prominent place in the design of both physically and chemically cross-linked networks [37–41]. Drug-loaded liposomes or micelles are dispersed in the aqueous phase of the chemically cross-linked hydrogel during synthesis. Then the drug is released at a rate that depends on the diffusion of the colloidal structures, on the drug partition between the colloids and the aqueous phase of the hydrogel, and on the diffusion coefficient of the free drug [42]. The main limitations of this approach refer to the stability of the colloidal structures during the synthesis and storage of the hydrogels and to the changes that the colloids can induce in the optical and mechanical properties of the network.

In recent years, the capability of cyclodextrins (CDs) to form complexes with relatively hydrophobic molecules is being explored as a way for overcoming the loading and release limitations of common hydrogels. In this context, functionalization with CDs that can be chemically attached to the network and that retain the capability to host drug molecules is gaining raising attention [43]. It is worth mentioning that CDs can also interact with the components of physically cross-linked hydrogels, such as certain blocks of in situ gelling copolymers (forming polypseudorotaxanes) or hydrophobic chains (acting as reversible cross-linkers), enabling remarkable control of the gelling process, the viscoelastic properties, and the performance for pharmaceutical and biomedical applications. The latter aspects are covered partially in other chapters of this book and have also been the aim of recent comprehensive reviews and papers [44–48]. Therefore, in the next sections we focus on the benefits of using CDs as components of chemically cross-linked hydrogels and show relevant examples of their key role in the performance of quite diverse networks as drug delivery systems.

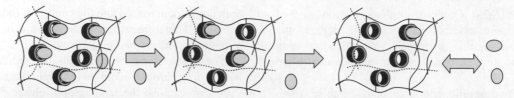

Figure 3. Drug release from a chemically cross-linked CD network. The CDs behave as dimples and the drug has to escape the dimples to be released.

1.3. Benefits of Incorporating CDs

With the exception of some specially designed CD derivatives (e.g., sugammadex), administration of drug–CD solutions leads to a practically instantaneous decomplexation when the complexes are diluted in aqueous fluids [49,50]. Therefore, in general, drug–CD solutions do not result in sustained drug release and may even lead to drug precipitation, due to complete and rapid drug release upon dilution of the drug–CD complex solution [51]. Covalent attachment of CDs to chemically cross-linked networks may enable CDs to fully display their complexation capability and, at the same time, prevent the dilution phenomenon that occurs when solutions and physical gels are administered. Once in contact with the physiological fluids, the hydrogel swells but the volume of water taken up is limited by the polymer network and, consequently, the polymeric chains do not dissolve and move apart. This creates a microenvironment rich in cavities available to interact with the guest drug molecules. In such a network, the affinity of the drug molecules for the CD cavities control the delivery. It is important to note that covalent attachment of CDs to a polymeric structure does not decrease their complexation ability but may even improve it, particularly in the case of large molecules that require more than

one CD to fulfill the complexation [52–58]. Decomplexation of a drug molecule from one cavity makes the drug available to form complexes with a neighboring empty cavity, the likelihood of recomplexation also being dependent on the drug–CD affinity. Therefore, drug release from the hydrogel can be seen as successive escape of the drug molecules from the CD cavities until the molecules reach the surface (Fig. 3). The higher affinity the drug has for the CD cavity, the slower the drug release will be by decomplexation. Drug release is faster when most CD cavities are occupied and the likelihood of recomplexation is lower. Oppositely, as the hydrogel delivers the drug, the number of empty CD cavities that are available for hosting the just-passing-through drug molecules increases. Furthermore, if the drug molecules previously released accumulate around the hydrogel, it could again recapture the drug toward the network. Drug affinity for the CD network is expected to be sensitive to the environmental conditions or to the presence of certain competitors for the CDs, opening the possibility of developing hydrogels that self-regulate drug delivery. All these features provide CD hydrogels with unique potential as controlled-release devices. The screening of such a potential is the focus of the steadily increasing number of publications (mainly journal articles but also patents) that appeared in the last decade (Fig. 4).

Hydrogels in which CDs are forming part of the structure of chemically cross-linked networks can be obtained by direct cross-linking of CDs (condensation with a cross-linker), by copolymerization of CDs with vinyl or acrylic co-monomers, or by first preparing the network and then anchoring the CDs to it. In the next section we review the different strategies pursued to develop CD cross-linked networks useful as drug delivery systems.

2. SYNTHESIS AND APPLICATIONS OF HYDROGELS WITH CDs

2.1. Hydrogels Obtained by Cross-Linking of CDs

CD polymers and hydrogels were first obtained through condensation reactions of the hydroxyl groups of natural CDs or of the amine or carboxylic acid groups of functionalized CDs with di- or multifunctional cross-linking agents, such as aldehydes, ketones, isocyanates, or epoxides,

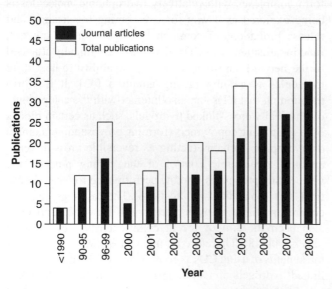

Figure 4. Evolution of the publications related to "cyclodextrin and hydrogel," indexed in the SciFinder Scholar database. No references before 1985 were found.

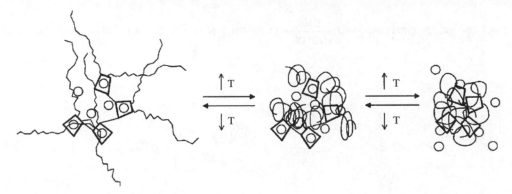

Figure 5. Temperature-responsive inclusion of a probe (o) by a cross-linked β-CD network bearing poly(*N*-isopropylacrylamide) chains. (From [80], with permission of John Wiley & Sons, Inc.)

particularly epichlorhydrin (EPI) [59,60]. Under alkaline conditions, the two reactive functional groups of EPI can react with the hydroxyl groups of CDs or with other EPI molecules. This results in a mixture of cross-linked CDs joined by repeating glyceryl units of polymerized EPI [61,62]. A careful control of the reaction process makes it possible to obtain water-soluble CD polymers [63] or microgel particles. The EPI/β-CD weight ratio also determines the fraction of CD cavities that are available to host guest molecules, the maximum being observed for hydrogels made with 50% β-CD [64]. EPI–CD microgels have been shown useful as selective traps for removal of components from water or food [65–71], bioremediation [72], and in separation science [73–77].

With the aim of achieving a greater hydrophilicity and more versatile mechanical properties for biomedical applications, studies pioneered by Szejtli et al. [78] focused on mixed networks of CDs and hydrophilic polymers [e.g., poly (vinyl alcohol), PVA] obtained using EPI and ethylene glycol bis(epoxypropyl)ether as cross-linking agents. The hydrogels were then modified with carboxymethyl and acetyl groups, rendering the networks even more hydrophobic. This approach ensured that CDs retain their capability to form complexes and the hydrogels demonstrated a high loading capacity for disinfecting drugs, such as ethacridine lactate, brilliant green, fuchsin acid, or cetylpyridinium chloride [79].

Temperature-sensitive EPI–CD networks have been prepared including poly(*N*-isopropyl acrylamide) (PNIPA) as the responsive component. PNIPA was grafted directly to previously cross-linked β-CDs, leading to hydrogels that maintain the transition temperature of PNIPA; being swollen at room temperature and shrunken at 37°C. Below the transition temperature, the cross-linked β-CDs easily formed complexes with fluorescent probes; the affinity constant being 100 times larger than that recorded for free β-CDs in solution. This finding is explained by the hydrophobic microenvironment that the PNIPA chains provide around the

CDs (Fig. 5). By contrast, the association constant decreased sharply above the transition temperature due to steric hindrance of the collapsed PNIPA chains, which constrained the access of the fluorescent probe to the CD cavities [80].

Interpenetrating (IPN) or semi-interpenetrating (semi-IPN) networks of EPI–CD and PNIPA have also been prepared. PNIPA hydrogels synthesized in the presence of the EPI/β-CD network maintained the transition temperature of PNIPA. The main feature of these IPNs was that even at the swollen state they were able to control the release of ibuprofen, owing to the complexation of the drug with the CDs [81]. Similar results were found for semi-IPNs made of β-CDs grafted to poly(ethyleneimine) and cross-linked PNIPA hydrogels and that were loaded with propranolol [82].

pH-sensitive microgels have been obtained by interpenetrating EPI–CD–PVA networks with poly(methacrylic acid) (PMAA) [83]. As expected, the microgels collapsed at pH 1.4 and swelled at pH 7.4. Nevertheless, they exhibited an unexpected release pattern of methyl orange, the release rate being much faster at the collapsed state. This finding was due to the fact that the affinity of methyl orange for β-CD was one order of magnitude larger at neutral pH, when the probe was not ionized, than at acid pH. Therefore, the pH-responsive delivery is afforded by the effect of pH on the host–guest interactions and not by the macroscopic swelling of the IPN. These results clearly highlight the key role of the CDs in controlling drug release.

Electric-responsive systems have also been developed starting from EPI–CD networks. The large content of CDs in hydroxyl groups enables the preparation of biocompatible smart fluids that undergo rapid and reversible changes of the rheological properties under small voltage electric fields. EPI–CD networks cannot endure a high electric field for a long time, and the polarization is restricted, owing to the rigidity and the high density in CDs [84]. Including starch during the cross-linking led to particulate networks that once mixed with silicon oil showed improved electrorheological properties [85].

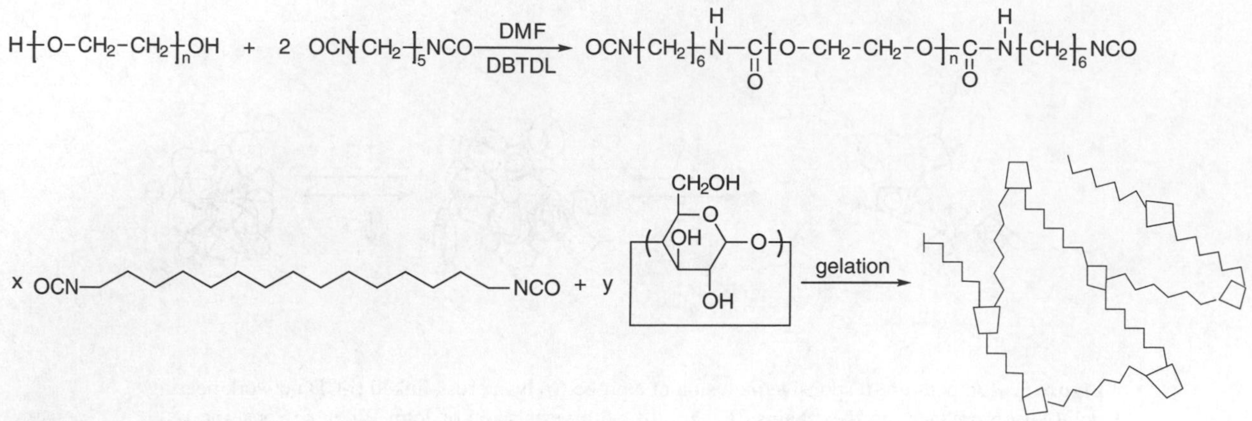

Figure 6. End-capping process of PEG with hexamethylene diisocyanate (HDMI) in *N,N*-dimethyl-formamide (DMF), using dibutyltin dilaurate (DBTDL) as catalyst, and reaction among the end-capped PEG and β-CD, which results in a hydrogel. (From [91], with permission of Wiley-VCH Verlag GmbH.)

EPI–CD hydrogels can also uptake substances by forming noninclusion complexes, taking advantage of the numerous free hydroxyl groups of CDs. For example, hydrogels able to recognize creatinine selectively were prepared in the presence of creatinine molecules at alkaline pH [86]. In this medium, the OH groups at C_6's are ionized and can interact electrostatically with amine groups of creatinine. The arrangement of CDs around creatinine molecules was maintained after polymerization and removal of creatinine. Molar ratios of β-CD/creatinine 3 : 2 and β-CD/EPI 1 : 10 were found as adequate to achieve a high rebinding effect. EPI–CD networks have also been functionalized with alkyl quaternary ammonium groups to obtain traps for biliar salts. A low degree of cross-linking and the presence of ammonium groups notably enhanced the loading capability and the affinity to sodium salts of cholic, glycocholic, and chenodeoxycholic acids [87].

Although EPI–CD hydrogels have demonstrated a great potential for pharmaceutical and biomedical applications, the relatively high toxicity of EPI and its pollutant character have motivated an intense search for alternative cross-linking agents. Diisocyanates have received a significant attention for preparing CD hydrogels or beads [88–90]. This approach enables the use of poly(ethylene glycol) (PEG), which is a highly hydrophilic and biocompatible polymer, as a main component of the hydrogels. PEG chains previously end-capped with isocyanate groups react with β-CD-forming urethane links. The CDs can be bonded to both ends of each PEG chain, acting as tie junctions among several chains (Fig. 6) [91]. The molecular weight of PEG and the PEG/β-CD molar ratio during reaction determine the structure of the hydrogel, its swelling properties, and its capability to load naphthol, adsorbed physically onto the polymer network, and forming inclusion complexes with β-CD.

PEG-based hydrogels can also be obtained by first activating the β-CD with hexamethylene diisocyanate (HDMI) in anhydrous dimethyl sulfoxide (DMSO) in order to obtain a mean of five isocyanate moieties attached to each β-CD. Addition of the activated β-CDs to PEG–diamine yields to instantaneous hydrogel formation [92]. These hydrogels exhibited high hydrophilicity and biocompatibility as well as the capability to load and to sustain the delivery of estradiol, quinine, and lysozyme [92]. It is important to note that lysozyme, which is hydrophilic and has no tendency to penetrate into the β-CD cavities, was better loaded by loosely cross-linked hydrogels prepared with low proportions of β-CD than by highly β-CD cross-linked hydrogel. Lysozyme release occurred rapidly from the aqueous phase of these hydrogels and was completed within few hours. By contrast, loading of estradiol and quinine was found to increase and their release rate to decrease (dramatically in the case of estradiol) as the content in β-CD was increased. The behavior observed for estradiol has been explained by estradiol/β-CD complex formation and sorption of drug molecules onto hydrophobic clusters within the matrix.

Following a more sophisticated approach, PEG–diamine was used to create networks in which the tie junctions were polyrotaxanes having isocyanate-activated β-CDs [93]. The hydrogels thus obtained were microporous and excellent scaffolds for chondrocytes. The polyrotaxanes were designed to have hydrolyzable ester linkages at the terminals and therefore eroded slowly without causing significant pH changes. Biodegradable polyrotaxane-based hydrogels have also been prepared by threading of α-CD onto thiolated four-arm PEG and subsequent oxidation of thiol groups (Fig. 7) [94]. The α-CDs protected the disulfide bonds to some extent from reduction degradation caused by

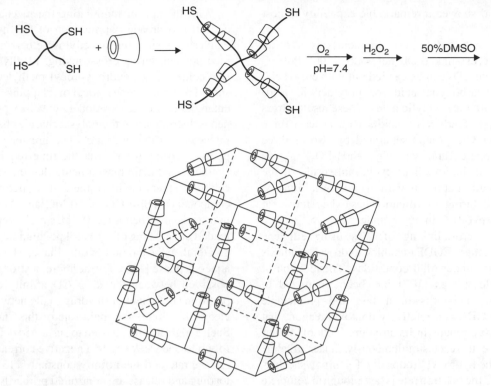

Figure 7. Supramolecular self-assembly and oxidation cross-linking processes. (From [94], with permission of The Royal Society of Chemistry.)

glutathione and, consequently, modulating the content in α-CDs enabled the control of the degradation rate.

CD-based hydrophilic hyperbranched polymers that show high ability to complex guest molecules [95], and nanoporous CD particles that rapidly retain solutes from aqueous environments and release them to organic phases [96], have also been prepared using diisocyanates. In this context, molecular imprinting is a quite attractive technology to improve the yield and selectivity of the loading and to achieve a better control of the release from the hydrogels [33,34]. Asanuma and Komiyama and co-workers have extensively explored the potential of molecular imprinting for making β-CD networks cross-linked with toluene-2,4-diisocyanate and capable to selectively recognize biologically relevant molecules or to remove pollutants from water streams [97]. Cholesterol and stigmasterol were used as templates during synthesis of the network. The presence of these templates induced β-CDs to arrange as dimers and trimers to bind the steroids cooperatively. The cross-linking made the arrangement permanent, and once the templates were removed, molecularly imprinted networks (MIP) were obtained. MIPs selectively bind cholesterol and stigmasterol, showing much lower affinity for other structurally related steroids [98,99].

Comparative studies of the performance of β-CD hydrogels cross-linked with EPI, succinyl chloride, HDMI, or toluene-2,4-diisocyanate revealed the importance of the

nature of the cross-linker regarding the affinity of guest molecules for the CD cavities. The results suggest that diisocyanates lead to networks of smaller mesh size and a lower degree of swelling in water compared to EPI, which self-polymerizes and provides longer bridges between the CDs. Nonspecific hydrophobic interactions with the network are more likely to occur in the case of diisocyanate cross-linked networks, while complexation with CDs predominates in the EPI–CD hydrogels. Compared to other commercially available pollutant sorbents, CD networks showed higher sorption capacities, particularly at high phenol concentrations [100,101]. Such a good performance, together with the environmental-friendly and recyclable character, make CD networks a cheap and efficient resource for removing organic pollutants and heavy metals from aqueous solutions [66,102].

Interfacial cross-linking of β-CDs with diacyl chlorides can lead to microcapsules with walls made of cross-linked CDs [103]. The CD cavities in the capsules were readily accessible to guest molecules, enabling complete loading within 5 min. Furthermore, the microcapsules resulted in controlled release of propranolol for up to several hours [104]. Recently, biodegradable β-CD hydrogels have been prepared using diacylated PEG. Mixing diacylated PEG with β-CD in DMF at 50°C for 48 h resulted in formation of ester bonds that led to hydrogels with maximum stability at pH 4. The networks degraded by hydrolysis in acidic and alkaline media and showed a remarkable capability to host 1-naphthol [105].

Condensation with poly(carboxylic acid)s is a clean method to obtain cross-linked CD networks. [106]. Polyesterification of native CDs can be carried out with citric acid, 1,2,3,4-butanetetracarboxylic acid, or poly(acrylic acid) (PAA), but fails with dicarboxylic acids. These results stress the need to use poly(carboxylic acid)s with at least three neighboring carboxylic groups separated by two or three carbons. A phosphate catalyzer (e.g., NaH_2PO_4) is also required to form an intermediate cyclic anhydride of the poly(carboxylic acid) that reacts with the CD. Additionally, the water produced during esterification has to be removed at temperatures above 140°C in air or under vacuum.

One-step direct cross-linking of CDs using ethylene glycol diglycidyl ether (EGDE) enables hydrogel synthesis in aqueous medium under mild conditions. The reaction of CDs with epoxide groups [107] has been optimized to render viscoelastic networks in a fast and predictable way [108–110]. EGDE is a relatively nontoxic reagent that possesses two epoxy groups in its structure, both of similar reactivity and able to react simultaneously, under alkaline conditions, with the hydroxyl groups of CDs or polysaccharides [111]. At 50°C the reaction rate is fast enough to complete the cross-linking process in a few hours without compromising the stability of the CDs and the other polysaccharides [110]. Most glycidyl ether groups of EGDE are consumed in the reaction, but if any remain in the hydrogel, washing with diluted HCl solutions opens the rings to give hydroxyl groups, resulting in highly biocompatible hydrogels [112]. Three different types of hydrogels have been prepared so far using EGDE: (1) CD solely hydrogels; (2) CD-*co*-polysaccharide hydrogels, and (3) IPNs of CD network and acrylic network. Regarding the first option, an intense search of the best conditions for obtaining hydrogels in short time and mild conditions has been carried out. A minimum concentration of 10 wt% HP-β-CD or M-β-CD and at least 14.28 wt% EGDE is required for preparing hydrogels with β-CD derivatives. It was shown that using those proportions, two-thirds of the hydroxyl groups of each CD can react with EGDE, rendering transparent and superabsorbent hydrogels that retain the capability of CD cavities to host drug molecules [110]. Both HP-β-CD and M-β-CD hydrogels were able to host twice the predicted amount of diclofenac, assuming that the loading exclusively took place in the aqueous phase of the hydrogel and exhibited a drug partition coefficient between the polymer network and the loading solution of almost 10. It is interesting to note that no hydrogels could be obtained with sulfobutyl ether-β-cyclodextrin (SBE-β-CD), probably because the high degree of substitution of the variety tested (hepta-substituted, M.S. 1) led to ionic repulsions among the CD molecules. On the other hand, γ-CD and HP-γ-CD did render hydrogels when cross-linked with EGDE [109]. The features of γ-CD hydrogels are currently under study.

With the aim of modulating the mechanical properties, and also to broaden application of CD hydrogels as drug delivery systems, linear cellulose ethers and dextran were incorporated during cross-linking. Ideally, the polysaccharide content (e.g., hydroxypropyl methylcellulose, HPMC) should be between the critical overlapping (0.2 wt%) and the entanglement concentration (1 wt%) to promote homogeneous distribution of the polysaccharide chain and of the CD molecules in the hydrogel. The greater the proportion of polysaccharide, the shorter the time required for hydrogel synthesis, the main reason being that the long chains of the cellulose ethers facilitate the contact between the different components. HP-β-CD-*co*-HPMC and M-β-CD-*co*-HPMC hydrogels absorbed 4 to 10 times their weight in water and loaded up to 24 mg of estradiol per gram, which is 500 times greater than the amount of drug that can be dissolved in their aqueous phase [113]. Furthermore, a strong correlation was observed between the drug–CD affinity constant and the partition coefficient of the drug to the network (Fig. 8). This clearly highlights the main role of the CDs in the loading. Such a high affinity also led to sustained delivery of estradiol for up to one week and to a negative correlation between the release rate and the affinity constant [113]. Although drug loading and release are controlled primarily by the drug–CD affinity constant, the cellulose ethers contribute to the improvement of the physical properties and to modulate the release rate (Fig. 9).

A comparative study of the performance of HP-β-CD (20%) hydrogels prepared in the presence of 0.4 or 0.8% HPMC, methylcellulose (MC), hydroxypropyl cellulose (HPC), carboxymethyl cellulose (CMCNa), or dextran, as carriers of sertaconazole was recently carried out [114]. Sertaconazole is an antifungal agent very effective for treatment of *Candida albicans* infections; however, its poor aqueous solubility is still a challenging issue for developing suitable formulations. HP-β-CD-based hydrogels provided a microenvironment that is very rich in CD cavities responsible for hosting the drug and controlling its release rate. All the hydrogels were superabsorbents, although those containing MC, CMCNa, or HPC were the ones with the lowest degree of swelling.

This phenomenon can be attributed to the concomitance of two effects: (1) a less hydrophilic character compared to HP-β-CD and (2) a higher degree of cross-linking due to an easier reaction of EGDE with the unsubstituted hydroxyl groups of cellulose. The hardness and compressibility of hydrogels prepared with HPMC or dextran were similar to those of the HP-β-CD sole hydrogel (4.2 N and 3.1 N mm). The addition of other polysaccharides caused, in general, an increase in these parameters (up to 9.4 N and 8.7 N mm), which confirms the greater apparent cross-linking density of HP-β-CD-*co*-MC, HP-β-CD-*co*-CMCNa, and HP-β-CD-*co*-HPC hydrogels. Sertaconazole loading was carried out by immersion of hydrogels in drug suspensions. Application of

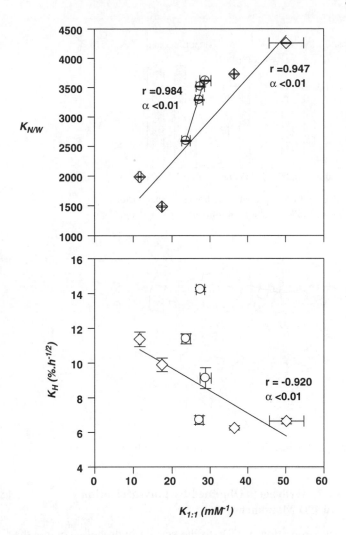

Figure 8. Dependence of network/water partition coefficient ($K_{N/W}$) and of the release rate constant (K_H) of estradiol for HP-β-CD (◇) and M-β-CD-based hydrogels (○) on the estradiol–cyclodextrin stability constants ($K_{1:1}$). (From [113], with permission of Elsevier.)

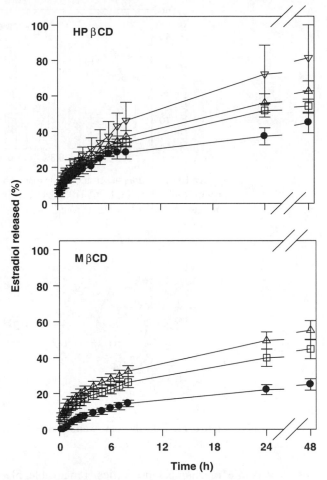

Figure 9. Estradiol release profiles from HP-β-CD or M-β-CD hydrogels prepared with 20% CD (●), 20% CD–0.25% HPMC (□), 25% CD–0.25% HPMC (△), and 30% CD–0.25% HPMC (▽), which were previously loaded in aqueous suspensions of estradiol. (From [113], with permission of Elsevier.)

autoclaving during hydrogel loading revealed that this thermal treatment still promoted the drug uptake by the CDs (Fig. 10) and did not cause relevant changes in the mechanical properties of the hydrogels. Nontreated HP-β-CD hydrogels loaded 21.7 mg/g and, when autoclaved, 18.7 mg/g. Hydrogels containing MC or HPMC loaded similar amounts or even greater. By contrast, nonautoclaved hydrogels made with HPC, CMCNa or dextran showed a significantly lower loading capability. Once autoclaved, HP-β-CD/HPC hydrogels reached values similar to those obtained with HP-β-CD sole hydrogels. In the case of CMCNa or dextran hydrogels, autoclaving enhanced the loading to such a high extent that these hydrogels became those with the greatest loading capability. All hydrogels showed a relatively fast delivery of drug in the first 24 h, followed by a more sustained release step up to 4 days. The antifungal effectiveness of the

sertaconazole-loaded hydrogels was verified using *Candida albicans* cultures in exponential phase of growth with positive results. Therefore, the EGDE cross-linked CD–polysaccharide hydrogels have a great potential as efficient carriers of topically applied antifungal drugs on mucosal surfaces.

IPNs of HP-β-CD and poly(acrylic acid) (PAAc) microgels have been prepared by one-step cross-linking of HP-β-CD with EGDE in the presence of Carbopo lmicrogels aqueous solutions [115]. As the chains of cross-linked CDs growth, they thread the micronetworks of carbopol. The individualized microgels are trapped in the continuous CD network, preventing their leaking from the system (Fig. 11). Thus, unlike common IPNs in which both networks are continuous, the HP-β-CD/carbopol systems present microdomains of IPN (microscale IPN). Such a unique microstructure facilitates the relative movement of both networks and

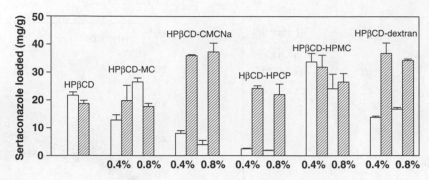

Figure 10. Amounts of sertaconazole loaded by HPβCD hydrogels prepared without polysaccharides or with MC, CMCNa, HPC, HPMC, or dextran at 0.4 or 0.8%. Hashed columns identify hydrogels autoclaved during loading. (From [114].)

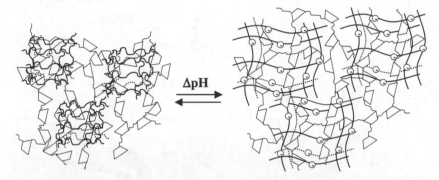

Figure 11. pH-sensitive swelling of HP-β-CD/carbopol microscale IPNs. (From [115], with permission of Elsevier.)

leads to versatile mechanical properties, remarkable pH-responsive degree of swelling, and bioadhesion.

The microscale IPNs loaded estradiol and ketaconazole with a remarkably high affinity; the amounts loaded being up to 200-fold greater than those expected if the drug molecules were only taken up in the aqueous phase of the hydrogels. Drug loading was enhanced even by the presence of carbopol, owing to the increase in the degree of swelling, which results in greater mesh size. The microscale IPNs sustained the release for several days, the rate also being dependent on carbopol content and pH (Fig. 12). Therefore, an adequate design of the HP-β-CD/carbopol IPNs may provide a single material with tunable mechanical properties, in which the complexation ability of CDs is combined with the bioadhesive and pH-responsive features of PAA.

Just recently, 3-(glycidoxypropyl)trimethoxysilane has been used to prepare CD microparticles containing anionic polysaccharides (carboxymethyl or sulfopropyl pullulan). This cross-linker agent acted both through grafting with the epoxy end on the hydroxyl groups of the CD and the polysaccharide, and through hydrolysis and condensation of the methoxy silane groups at the other end. The microparticles showed the ability to retain water pollutants (phenol and benzoic acid derivatives, β-naphthol), drugs (salicylic acid, indomethacin), and proteins (lysozyme) [116].

2.2. Hydrogels Obtained by Polymerization of CD Monomers

Incorporation of CDs to the acrylic hydrogels requires the previous synthesis of copolymerizable monomers of CD with reactivity similar to that of the other monomers used as components of the hydrogels. Several synthetic routes of CD monomers have been developed. The presence of a high number of equally reactive hydroxyl groups in the CD structure makes the preparation of monofunctionalized monomers particularly challenging, and in most publications, multifunctional monomers are reported.

Monofunctional acryloyl monomers of CDs (Fig. 13a) have been prepared by the reaction of α- and β-CD with *m*-nitrophenyl ester in alkaline medium at room temperature for 5 min. The nitrophenyl esters form complexes with the CDs and lead to selective transesterification at one of the secondary hydroxyl groups [117]. If no cross-linker is added, the acryloyl monomers of CD render water-soluble polymers that show a lower-affinity constant for small molecules (e.g., *m*-chlorobenzoic acid and cinnamic acid) but higher for large substrates with two aromatic rings (e.g., methyl red and orange I), owing to a cooperative binding effect [118]. Copolymerization of acryloyl-β-CDs with *N*-isopropylacrylamide (NIPA) has been used to prepare porous hydrogels

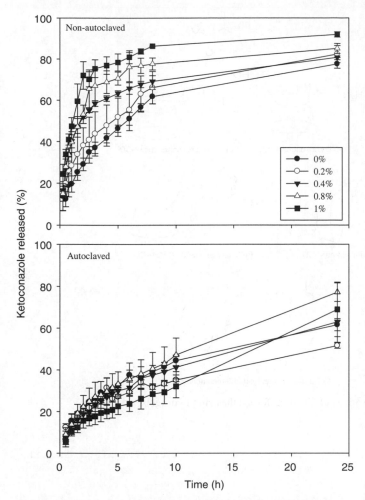

Figure 12. Ketoconazole release profiles in 0.3% SDS solution (pH 7.8) from HP-β-CD/carbopol microscale IPNs that were prepared with different proportions in carbopol (see legend) and that were previously loaded by immersion in a drug suspension applying autoclaving or not. (From [115], with permission of Elsevier.)

that undergo volume-phase transitions very rapidly when immersed in aqueous medium at 37°C [119]. Acryolyl-α-CD and acryloyl-(6-O-α-D-glucosyl)-β-CD have also been tested for creating imprinted networks for certain molecules that form complexes with two or more CD units simultaneously. These monomers cross-linked in the presence of various guest molecules (vancomycin, cefazolin, phenethicillin, and some dipeptides) rendered rigid particles with a microstructure able to fit the complexation preferences of each guest molecule [120]. In average, the imprinted networks were able to load twice the amount of drug loaded by the CD networks prepared in the absence of the guest molecule. Bisacryloyl-β-CD monomers combined with 2-acryloylamido-2,2′-dimethylpropane sulfonic acid led to networks with an enhanced affinity for amphiphilic molecules, such as phenylalanine, particularly one of its enantiomers [121,122].

Monotosyl derivatives of β-CD (Fig. 13b) can be obtained by reaction between one C_6 primary hydroxyl group of β-CD

and tosyl chloride [123]. These derivatives have been used for functionalizing polyvinylamine for chromatographic purposes [124], for improving the performance of natural polymers as drug carriers [125], and for preparing new monofunctionalized monomers, such as ethylenediamine (EDA)/β-CD or 1,6-hexanediamine (HAD)-β-CD (Fig. 13c) [126]. The amino group of alkylenediamine CDs can react with glycidyl methacrylate (GMA) to render monomethacrylate β-CD monomers (Fig. 13d). The use of these monomers enabled the preparation of linear copolymers with NIPA that maintain the temperature responsiveness of PNIPA. Hydrogels were obtained first by polymerizing NIPA with GMA and then coupling with EDA/β-CD because, using these monomers, the rate of conjugation between EDA and GMA is slower than the cross-linking among NIPA-GMA chains [127]. The resulting hydrogels were temperature sensitive and loaded methyl orange with an affinity similar to that exhibited by EDA/β-CD in solution. A modification of this method to render a hydrogel in one step involved the mixing of PNIPA, β-CD, and GMA (10 : 2 : 1 mole ratio) and the addition of $K_2S_2O_8$ [128]. However, since GMA forms inclusion complexes with β-CD, the structure of the network is still unclear (Fig. 14).

Acrylic hydrogels have also been obtained by free-radical copolymerization of 2-hydroxyethyl acrylate (HEA; 87 to 100 mol%) with GMA–EDA/β-CD (0 to 13 mol%) in aqueous solution. Nevertheless, since HEA homopolymerization occurred faster than the copolymerization with GMA–EDA/β-CD, the actual content in CD in the hydrogels was much lower than expected. In general, the presence of GMA–EDA/β-CD led to a much higher glass transition temperature (enhanced network rigidity) and a lower degree of swelling, but also to a slower release of melatonin (from a 90% to a 70% release at 120 min) [129].

A monotosyl derivative of β-CD has recently been used to prepare mono-(6-N-allylamino-6-deoxy)-β-CD (Fig. 13e) that was copolymerized with NIPA via free-radical polymerization in a DMF/water medium [130]. PNIPA-co-β-CD hydrogels showed greater affinity for 8-anilino-1-naphthalenesulfonic acid ammonium salt (ANS) than did PNIPA hydrogel. The affinity increased notably as the temperature was raised from 21°C to 60°C, which was attributed to the adsorption of ANS on hydrophobic domains of the collapsed PNIPA network (Fig. 15).

Acrylamidomethyl-CD (Fig. 13f) can be prepared quickly in an aqueous environment from the acid-catalyzed reaction of N-methylolacrylamide (NMA) and CD [131]. NMA/β-CD with one to three acrylamidomethyl groups per CD molecule has been used to functionalize cotton fibers [131] and to improve the drug loading and release properties of pH-responsive sodium acrylate hydrogels [132]. Regardless of the NMA/γ-CD ratio, swollen hydrogels were highly flexible and transparent. The glass transition temperature of the acrylate network (130°C) was not modified when

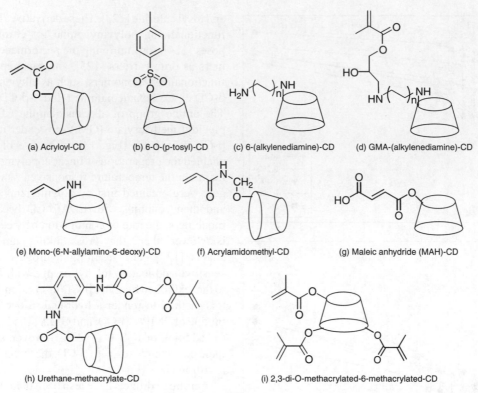

(a) Acryloyl-CD   (b) 6-O-(p-tosyl)-CD   (c) 6-(alkylenediamine)-CD   (d) GMA-(alkylenediamine)-CD

(e) Mono-(6-N-allylamino-6-deoxy)-CD   (f) Acrylamidomethyl-CD   (g) Maleic anhydride (MAH)-CD

(h) Urethane-methacrylate-CD   (i) 2,3-di-O-methacrylated-6-methacrylated-CD

Figure 13. Structure of monomeric derivatives of CD used for synthesizing hydrogels.

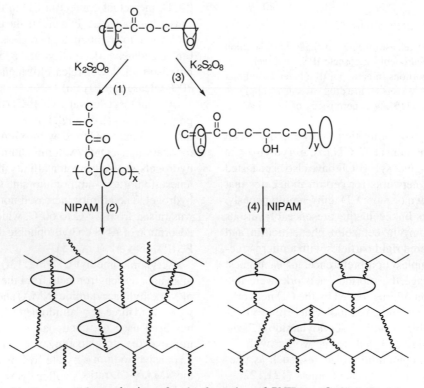

Figure 14. Two feasible mechanisms for the formation of PNIPA-*co*-β-CD hydrogel networks obtained with GMA as a cross-linker. (From [128], with permission of Wiley-VCH Verlag GmbH.)

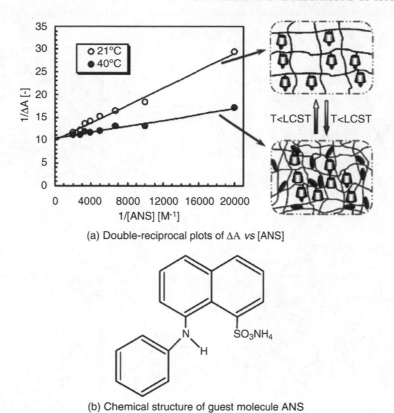

(a) Double-reciprocal plots of ΔA *vs* [ANS]

(b) Chemical structure of guest molecule ANS

Figure 15. Double-reciprocal plots of the change in absorbance of ANS solution upon loading by P (NIPA-*co*-CD) gel at temperatures below LCST (21°C) and above (40°C). (From [130], with permission of the American Chemical Society.)

copolymerized with NMA/γ-CD, which means that the CD monomer did not act as a plasticizer and was effectively attached to the network. The presence of γ-CD notably enhanced the loading of triamcinolone acetonide and led to sustained delivery for 24 h, disregarding the pH of the medium, indicating that the drug–CD affinity governs the delivery process (Fig. 16). Such a hydrogel might be useful for the treatment of inflammatory processes in colon.

In this context, it is worth stressing that the position of the acrylamide group in the glucopyranose ring may strongly determine the functionality of the network. Studies carried out with acrylamide monomers of β-CD that tether the reactive double bond on either the wider or the smaller rim of the truncated cone indicate that the monomers adopt a different spatial arrangement in the presence of amino acid derivatives and oligopeptides that were used as templates during polymerization [133]. Since molecules able to be hosted in the CD cavity take a specific orientation, the position of the reactive double bond on the CD monomer affects enormously the microstructure of imprinted networks. The reactive double bond determines the distance between the template and the polymerizable moiety. Most molecules that form complexes with β-CD enter through the larger rim [134]. This is also observed for protected amino

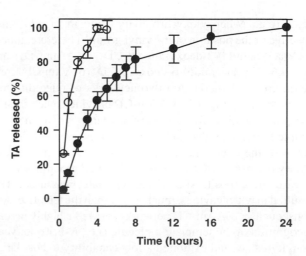

Figure 16. Triamcinolone release profiles in pH 6.8 phosphate buffer from hydrogels synthesized with 3 M sodium acrylate and 39 mM BIS without (open symbol) or with 9.4% of γ-CD/NMA (filled symbol). The content in TA of hydrogels prepared without γ-CD/NMA was significantly lower (0.6 mg/g dried gel) than that of hydrogels prepared with 9.4% γ-CD/NMA (32 mg/g dried gel). (From [132], with permission of Elsevier.)

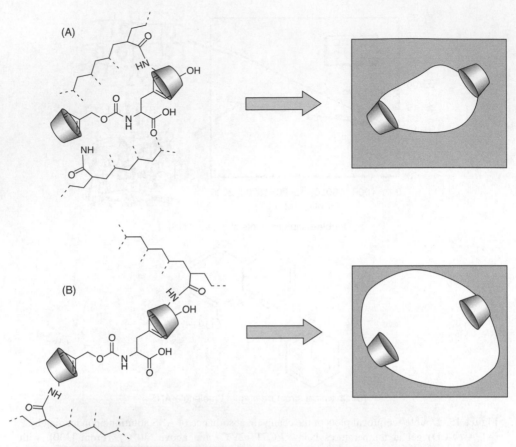

Figure 17. Arrangement of (A) mono-3-(N-acrylamido)-3-deoxy-altro-β-CD or (B) mono-6-(N-acrylamido)-6-deoxy-β-CD during polymerization and cross-linking with $N,\acute{N}$-methylenebis(acrylamide) in the presence of N-benzyloxycarbonyltyrosine. (From [133], with permission of the American Chemical Society.)

acids (e.g., N-benzyloxycarbonyltyrosine in Fig. 17) irrespective of the position of the vinyl group. Therefore, mono-3-(N-acrylamido)-3-deoxy-altro-β-CD (3-AAm-CD) and mono-6-(N-acrylamido)-6-deoxy-β-CD (6-AAm-CD) rendered after polymerization two networks of quite different microstructure (Fig. 17). 3-AAm-CD provided much smaller receptor cavities for N-benzyloxycarbonyltyrosine and adapted to the molecular shape of this amino acid. By contrast, the copolymerization occurred far from the template when 6-AAm-CD, which has the vinyl group protruding toward the opposite site of the template, was used. This resulted in wider cavities, much larger than the template. As a consequence, 3-AAm-CD networks showed a highly precise recognition of the template molecule (i.e., N-benzyloxycarbonyltyrosine) and exhibited a low capability to host larger amino acids and peptides. The opposite was true for 6-AAm-CD [133]. Recently, 6-AAm-CD has been found useful to create networks with artificial receptors that clearly distinguished angiotensin I and angiotensin II, despite the apparent similarity of amino acid sequences of these two oligopeptides. The excellent results obtained when used, as the

stationary phase of HPLC indicates that the molecular imprinting is not directly associated with the primary structure of the oligopeptide template, but with the conformation of the oligopeptide in solution [135].

Smart monomers that combine the ability of CDs to host guest molecules and the pH sensitiveness of carboxylic groups have been obtained by condensation of CDs with maleic anhydride (MAH) (Fig. 13g). The number of vinyl and carboxylic acid groups per CD can be controlled effectively by tuning the MAH/β-CD mole ratio [136]. Highly biocompatible and biodegradable hydrogels have been prepared through photo cross-linking of hyaluronic acid (HA) with MAH/β-CD [137]. The hydrogels showed an improved ability to uptake hydrocortisone through complex formation with β-CD, as confirmed in competitive binding studies carried out in the presence of adamantine carboxylic acid. Hydrogels sensitive to both pH and ionic strength have also been obtained by electron beam irradiation of MAH/β-CD: acrylic acid (AA) aqueous solutions at 1:3, 1:3.5, 1:4, 1:4.5, and 1:5 w/w [138]. The hydrogels collapsed at acid pH and showed an abrupt swelling at pH 7.4, particularly

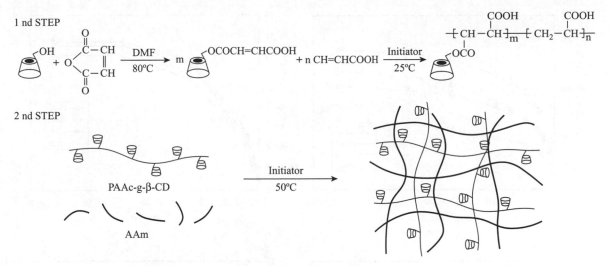

Figure 18. Preparation of IPNs of PAAc-*grafted*-MAH-β-CD and PAAm. (From [140], with permission of Wiley-VCH Verlag GmbH.)

when the ionic strength of the medium was low. The changes in degree of swelling were reversible after several collapsed-swollen cycles.

Multiresponsive hydrogels sensitive to temperature, pH, and ionic strength were achieved by free-radical copolymerization of NIPA (91 to 64%w/w) and MAH/β-CD (9 to 36% w/w) [136]. The changes in the degree of swelling induced by these variables were reversible and reproducible after several cycles, exhibiting truly intelligent behavior. PNIPA-*co*-MAH/β-CD hydrogels have been proposed to uptake chlorambucil and to release it at a pH-dependent rate [139]. Recently, IPNs of PAAc-grafted-MAH/β-CD and poly(acrylamide) (PAAm) have been prepared (Fig. 18) and exhibited differential features compared to IPNs prepared without β-CD [140]. PAAc and PAAm from intermolecular complexes via hydrogen bonding at temperatures lower than the upper critical solution temperature (UCST, ca. 33°C), whereas they dissociate at higher temperatures. Thus, the swelling pattern of PAAc/PAAm hydrogels (shrunken at temperature below 33°C and swollen at greater temperature) is opposite to PNIPA hydrogels. Grafting of β-CD onto PAAc chains caused minor changes in the UCST (ca. 35°C) and in the degree of swelling (from 25.6% to 22.5%) at temperatures above UCST, but notably enhanced the loading of ibuprofen and delayed the drug release rate at both 25 and 37°C. Furthermore, the IPNs showed pulsed release of ibuprofen when subjected to temperature cycles, exhibiting a truly smart behavior (Fig. 19).

MAH/β-CD has also been copolymerized with macromonomers of Pluronic F68 and poly(ε-caprolactone) with the aim of obtaining temperature-sensitive biodegradable networks. UV-induced copolymerization of acryloyl monomers of Pluronic F68-g-poly(ε-caprolactone) (5 to 30%) with MAH/β-CD (0 to 25%) was carried out in water (70%). Different from polypseudorotaxane formation, the vinyl

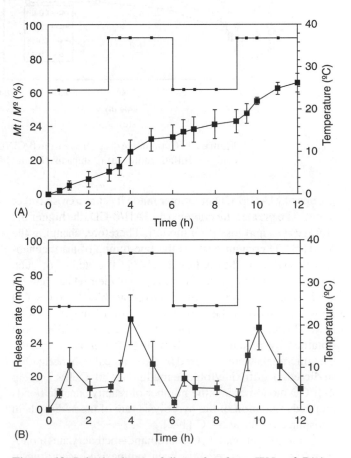

Figure 19. Pulsed release of ibuprofen from IPNs of PAAc-*grafted*-MAH-β-CD and PAAm when subjected to temperature cycles: (A) cumulative release; (B) release rate. The ibuprofen load was 16.63 mg/g. (From [140], with permission of Wiley-VCH Verlag GmbH.)

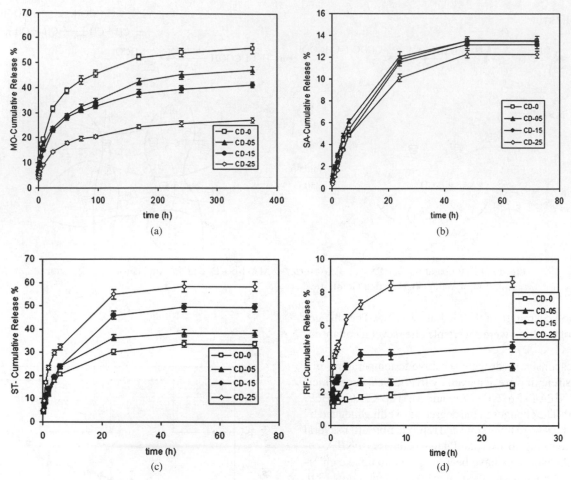

Figure 20. Cumulative drug release from β-CD/UM-based hydrogels: (a) methyl orange; (b) salicylic acid; (c) sulfathizaole; (d) rifampicin. (From [143], with permission of John Wiley & Sons, Inc.)

groups of MAH/β-CD monomer made it act as a cross-linker agent. The greater the content of MAH/β-CD, the higher the storage (G′) and loss (G″) moduli. Therefore, changing the MAH/β-CD content enables the fine tuning of the mechanical behavior of the hydrogels [141]. The ability of such hydrogels to take up drugs and to control their release has not yet been tested. Another attempt to create biodegradable CD networks involved the copolymerization of poly(D,L-lactic acid) (PLA) macromonomer with a β-CD derivative (both obtained by reaction with 1-allyloxy-2,3-epoxy propane). The microgels were prepared by free-radical polymerization in dimethyl sulfoxide/toluene medium at 70°C. The content in β-CD monomer and the number of reactive double bonds enabled regulation of the hydrolysis rate of the microgels in phosphate buffer at 37°C [142].

A multifunctional β-CD urethane–methacrylate monomer (Fig. 13h) has recently been synthesized according to a two-step addition mechanism [143]. Hydrogels were prepared by ultraviolet (UV) irradiation of HEMA (87.5 to 90 mol%), urethane/methacrylate/β-CD monomer (0 to 2.5 mol%) and cross-linker poly(ethylene glycol) diacrylate. The

presence of the β-CD monomer at 2.5 mol% caused a remarkable increase in the degree of swelling of the hydrogel (from 34 to 50%) and on the amount of salicylic acid, sulfathiazole, rifampicine, and methyl orange that the hydrogels were able to take up. Nevertheless, the effect on drug release was not homogeneous, delaying slightly the delivery of the hydrophilic methyl orange and salicylic acid but accelerating the release of the hydrophobic sulfathiazole (Fig. 20). Such intricate effect may be related to the different affinity constants of the drugs for the CDs and to the different solubility of the drugs in the buffer used as the release medium. Both variables should be taken into account when the release from CD–hydrogels is interpreted.

Methacrylic monomers of CDs (Fig. 13i) have attracted a great interest for a wide range of purposes and even the monomer 2-hydroxy-3-methacryloyloxy-propyl-β-CD (βW7MAHP from Wacker-Chemie GmbH) with an average number of 2.5 double bonds per HP-β-CD unit was commercially available for some time. This monomer can be prepared by reaction of glycidyl methacrylate (GMA) with HP-β-CD in an alkaline medium [144]. Cross-linked βW7MAHP and copolymers of βW7MAHP with 2-hydroxyethyl methacrylate (HEMA) have been shown useful for the removal of pollutants from water [145]. In particular, copolymerization with HEMA notably enhanced the degree of swelling in water and thus the sorption capacity of the βW7MAHP networks, due to better accessibility of the CD cavities. Copolymerization of βW7MAHP with 1-vinyl-2-pyrrolidinone rendered soluble polymers when the content in βW7MAHP was below 60 mol%, while greater proportions led to cross-linked hydrogels [144].

Methacrylic monomers of CDs with one to six double bonds per CD unit can also be obtained via a single-step reaction of β-CD with methacrylic anhydride using sodium hydroxide as catalyst [146]. Direct photopolymerization of 6% methacrylic-β-CD solutions enabled the formation of hydrogels, while concentrations above 8% produced brittle and white networks. CD monomers having methacrylic groups only at positions 2 and 3 can be obtained by acetylation of primary hydroxyl groups and esterification of secondary hydroxyl groups with methacrylic anhydride [147]. Methacrylic/β-CD monomers have been used successfully as templates during polymerization of other methacrylate monomers in order to achieve degrees of polymerization of 7 or 14 [148–150]. Methacrylated monomers have been explored in detail for preparing dental fillings. Changes in the degree of conversion and the flexural strength of the fillings have been evidenced to occur when the photoinitiators (e.g., camphorquinone or ethyl-4-dimethylaminobenzoate) form inclusion complexes with the CD monomers [151,152]. Optimization of the ratio of polymerizable methacrylate–hydroxyl groups enabled the preparation of adhesive CD monomers that promote the bonding of dental composites to dentin [153]. Resins prepared with 33% methacrylated β-CD, 30% HEMA, and 37% acetone showed the maximum shear bond strength (16 MPa), with values similar to those exhibited by commercial products.

Regarding drug delivery, foldaway, and viscoelastic loosely cross-linked hydrogels made of HEMA and (2,3-di-*O*-methacrylated-6-methacrylated)-β-CD [with all primary (7) and secondary (14) hydroxyl groups substituted with the polymerizable moiety] have been tested for obtaining medicated soft contact lenses (SCLs). The limited ocular bioavailability achieved with conventional ophthalmic formulations has prompted the search of alternatives, among which SCLs able to combine the capability to correct optic deficiencies with the capability of loading a drug and to control its release are one of the most promising candidates [154]. Nevertheless, most SCLs cannot uptake therapeutic doses of hydrophobic drugs or control the delivery of the hydrophilic ones. Differently from β-CD, which is not soluble in the liquid HEMA, (2,3-di-*O*-methacrylated-6-methacrylated)-β-CD enabled the synthesis of hydrogels without using solvents. The presence of CDs in the network (0.23 to 1.82 mol%) led to transparent hydrogels that showed a high cytocompatibility and did not induce

macrophage response [155]. The greater the methacrylated β-CD content, the higher the glass transition temperature, the lower the degree of swelling and free water proportion, and the greater the storage and loss moduli of the swollen hydrogels. These findings were related directly to the increase in cross-linking degree caused by the methacrylated β-CD. The capability of CD to form complexes was tested using 3-methyl benzoic acid (3-MBA), which has a high affinity for β-CD $(1.3 \times 10^7 \, \text{M}^{-1})$, as a probe. pHEMA-co-βCD hydrogels were able to uptake greater amounts of 3-MBA than were pHEMA hydrogels, although a progressive decrease in the drug/β-CD molar ratio was observed as the proportion of β-CD monomer increased. Hydrogels with a low content in β-CD loaded more than one 3-MBA molecule per β-CD, owing to the formation of 3-MBA dimmers. Oppositely, in hydrogels prepared with a high β-CD monomer proportion (>0.167 g/mL) the complexation capability of the CDs was not fulfilled, which can be attributed to a smaller mesh size (i.e., a greater cross-linking degree and a lower content in water) that hinders the diffusion of 3-MBA. Additionally, the contribution of steric impediments owing to the proximity of the β-CD units in the network cannot be discarded. Drug-loading studies were carried out with hydrocortisone and acetazolamide, both of practical interest for the local treatment of ocular pathologies and the ability to form complexes with CDs in water and in lachrymal fluid. Hydrocortisone loading decreased progressively as the content in methacrylated β-CD rose, due to a decrease in the volume of aqueous phase of the hydrogel. Acetazolamide loading showed a maximum for an intermediate content in β-CD (0.125 to 0.167 g/mL), owing to a balance between complexation with β-CD and hydrogel mesh size (Fig. 21). In fact, these hydrogels showed a two-fold (three-fold when autoclaved) increase in acetazolamide loading compared to the hydrogels prepared without β-CD. The hydrogels sustained hydrocortisone delivery for 7 days. The acetazolamide release rate was dependent on the β-CD content and could be prolonged for 24 days (Fig. 22). In sum, an adequate selection of the content in β-CD makes it possible to obtain pHEMA-*co*-β-CD hydrogels suitable for specific biomedical applications.

Click chemistry is now being explored as a way to prepare CD hydrogels. It was observed that when solutions of azide-PNIPA and of alkyne-β-CD are mixed together, a cross-linking reaction takes places through a Huisgeńs 1,3-dipolar azide-alkyne cycloaddition catalyzed by Cu(I) [156]. The hydrogels exhibited a porous structure and were able to load and to control the delivery of DNA.

2.3. Functionalization with CDs of Preformed Hydrogels

Functionalization with CDs of preformed materials is been explored for a broad range of purposes. In general, materials have to combine appropriate bulk and surface properties for being suitable for a certain application. The bulk properties, such as strength, toughness, and chemical and mechanical

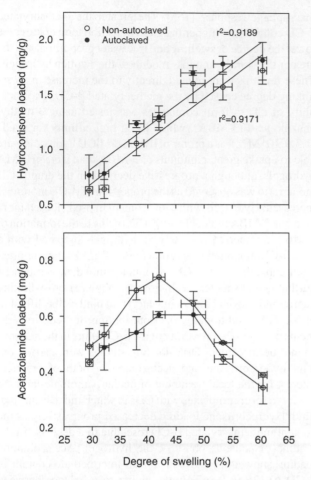

Figure 21. Dependence of the amount of drug loaded on the degree of swelling of pHEMA hydrogels prepared with 80 mM EGDMA and different amounts of methacrylated-β-CD. (From [155], with permission of Elsevier.)

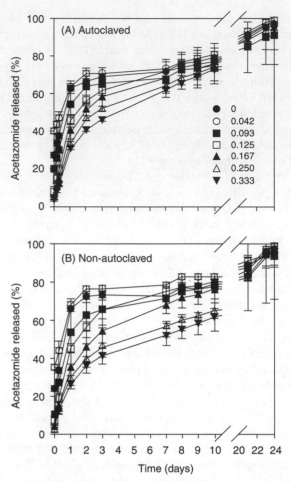

Figure 22. Acetazolamide release profiles from pHEMA hydrogels prepared with 80 mM EGDMA and different amounts of metha-crylated-β-CD: 0 (●), 0.042 (■), 0.083 (□), 0.125 (○), 0.167 (▲), 0.250 (△), 0.333 (▼) g/mL of monomeric solution. (From [155], with permission of Elsevier.)

stability, influence the long-term durability of the material, while the surface properties govern the interfacial interactions and performance when it enters into contact with foreign compounds or surfaces (e.g., other materials or the living tissues). Surface functionalization with CDs opens up the possibility of modulating the affinity of the surface (particularly when highly hydrophilic) toward certain molecules. For example, textile materials are coupled with CDs with the aim to retain colors, fragrances, insect repellents, or even antimicrobial substances [157–159]. Cotton with immobilized CDs shows a slower volatilization of fragrances and can even stand up to 15 washes retaining the fragrance [160]. In the biomedical field, the surface modification of polymeric medical devices with CDs resulted in a lower adsorption of proteins and enhanced blood compatibility [161].

In the particular case of hydrogels, functionalization with CDs was motivated by the aim of maintaining the bulk properties of networks that had been shown adequate for

specific purposes. As mentioned in previous sections, CD monomers usually have more than one reactive double bond, and therefore they act as cross-linking points, altering the viscoelastic, mechanical, and swelling characteristics of the hydrogels. Thus, attachment of CDs to preformed hydrogels is envisioned as a way to keep the favorable bulk properties of the networks and to provide them with new functionalities. Two recent examples of this successful approach are described below.

Hydrogels able to undergo autonomous shrinking and swelling phase transitions accompanied by the complexation or decomplexation of a guest molecule were prepared using β-CD as the sensing moiety and NIPA as the actuating moiety [162]. NIPA (20 g) was copolymerized with *p*-nitrophenyl acrylate (3.4 g) in DMF. Then, amine-substituted CDs were incorporated in the *p*-nitrophenyl acrylate moieties using an ester exchange reaction. CD complexation of a guest molecule changed the hydrophilic–hydrophobic

(a)   (b)

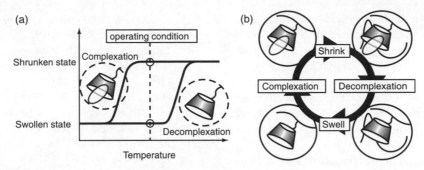

Figure 23. (a) Operating temperature for autonomous oscillatory phenomenon, which would be realized in a nonequilibrium open system of poly(NIPAM-*co*-CD)/8-anilino-1-naphthalene-sulfonic acid. (b) Autonomous oscillatory phenomenon in the open system: polymer shrinking (upper process), decomplexation of 8-anilino-1-naphthalene-sulfonic acid (right-hand process), polymer swelling (lower process), and complexation of 8-anilino-1-naphthalene-sulfonic acid (left-hand process). (From [162], with permission of the American Chemical Society.)

balance of the network and, consequently, the transition temperature. Particularly, complexation of 8-anilino-1-naphthalene-sulfonic acid creates a hydrophobic microenvironment that decreases the transition temperature, while decomplexation reestablishes the transition temperature. At the same time, the phase transition alters the complexation; shrinking makes the complex unstable and leads to decomplexation. Coordinating these two mutual effects at a temperature in between that of the hydrogel transition, when CDs are forming complexes, and that observed when CDs are free, an autonomous oscillatory phenomenon was achieved (Fig. 23). These hydrogels have potential as components of sensors.

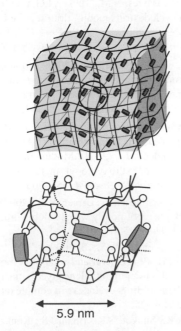

Figure 24. pHEMA-*co*-GMA hydrogel with pendant β-CDs. (From [164], with permission of Elsevier.)

Attachment of CDs to acrylic hydrogels enabled the development of SCLs that maintain the mechanical properties, the swelling degree, the oxygen permeability, and the biocompatibility of the starting hydrogels but show notable improvement in their ability to load drugs and to control their release rate [163]. PHEMA hydrogels were prepared by copolymerization with GMA at various proportions, and then β-CD was grafted to the network by reaction with the glycidyl groups under mild conditions. This led to networks where the β-CDs form no part of the structural chains but are hanging on two to three ether bonds through the hydroxyl groups (Fig. 24). The pendant β-CDs enhanced diclofenac loading by 1300% and drug affinity 15-fold and provided the hydrogels with the ability to sustain drug delivery in lachrymal fluid for two weeks [164].

3. CONCLUSIONS AND PERSPECTIVES

Hydrogels consist of a three-dimensional network of polymeric chains in water or aqueous solution that may be over 99% of the total weight of the gel. A chemically cross-linked polymeric network results in the formation of relatively firm hydrogel systems that do not dissociate, even when submersed in aqueous solutions. CDs are hydrophilic cyclic carbohydrates that possess the unique ability to form inclusion complexes with many lipophilic drugs or drug moieties. Thus, CDs are able to solubilize and stabilize lipophilic water-insoluble drugs through formation of inclusion complexes. Incorporation of CD molecules into hydrogels, either through the cross-linking process or through attachment of individual CD molecules to the polymeric chains, results in the formation of hydrogels that possess unique drug delivery properties. The CD-containing hydrogels are able to dissolve amphiphilic and hydrophobic drugs within the aqueous matrix, as well as much larger molecules such as peptides and proteins, and hamper their chemical and physical decomposition. Due to the swelling-limited structure of cross-linked hydrogels, the drug molecules are not released through the dilution of drug–CD complex, which is the general means of drug release from drug–CD complexes. Instead, the release is a diffusion-controlled process where the rate is controlled by the affinity of the diffusing molecule (i.e., drug, peptide, or protein) for the CD cavities and the polymeric network.

The variety of CD-containing polymers and polymer matrixes is also increasing rapidly. New CD-cross-linking methods are being developed as well as novel reactive CD monomers for the synthesis of functionalized CD polymers. Tailored biomaterials that load therapeutically relevant amounts of drugs and control their release rate according to specific requirements are under development. Thus, CD-based hydrogels will take a prominent place among innovative drug delivery systems.

Acknowledgments

This work was financed by Ministerio de Ciencia e Innovación (SAF2008-01679), FEDER, and Xunta de Galicia (PGIDT07CSA002203PR), Spain.

REFERENCES

1. Klech, C.M. (1999). Gels and jellies. In: Swarbrick, J., Boilan, J.C. Eds. *Encyclopedia of Pharmaceutical Technology*, Vol. 6. Marcel Dekker, New York; pp. 415–439.

2. Tanaka, T. (1981). Gels. *Sci. Am.*, 124–138.

3. Osada, Y. (2001). Polymer gels: crosslink formations. In: Osada, Y., Kajinara, K. Eds. *Gels Handbook*, Vol. 1. Academic Press, San Diego; CA.

4. Mathur, A.M., Moorjani, S.K., Scranton, A.B. (1996). Methods for synthesis of hydrogel networks: a review. *Polym. Rev.*, 36, 405–430.

5. Peppas, N.A., Bures, P., Leobandung, W., Ichikawa, H. (2000). Hydrogels in pharmaceutical formulations. *Eur. J. Pharm. Biopharm. 50*, 27–46.

6. Lin, C.C., Metters, A.T. (2006). Hydrogels in controlled release formulations: network design and mathematical modeling. *Adv. Drug Deliv. Rev.*, 58, 1379–1408.

7. Kashyap, N., Kumar, N., Ravi Kumar, M.N.V. (2005). Hydrogels for pharmaceutical and biomedical applications. *Criti. Rev. Ther. Drug Carrier Syst.* 22, 107–150.

8. Spanoudaki, A., Fragiadakis, D., Vartzeli-Nikaki, K., Pissis, P., Hernandez, J.C.R., Pradas, M.M. (2006). Nanostructured and nanocomposite hydrogels for biomedical applications. In: Blitz, J.P., Gun'ko, V. Eds. *Surface Chemistry in Biomedical and Environmental Science*. NATO Science Series, Vol. 228. Springer-Verlag, Dordrecht, The Netherlands, pp. 229–240.

9. Kopecek, J. (2007). Hydrogel biomaterials: A smart future? *Biomaterials*, 28, 5185–5192.

10. Knuth, K., Amiji, M., Robinson, J.R. (1993). Hydrogel delivery systems for vaginal and oral applications. *Adv. Drug Deliv. Rev.*, 11, 137–167.

11. Hoare, T.R., Kohane, D.S. (2008). Hydrogels in drug delivery: progress and challenges. *Polymer*, 49, 1993–2007.

12. Kim, S., Kim, J.H., Jeon, O., Kwon, I.C., Park, K. (2009). Engineered polymers for advanced drug delivery. *Eur. J. Pharm. Biopharm. 71*, 420–430.

13. Raemdonck, K., Demeester, J., De Smedt, S. (2009). Advanced nanogel engineering for drug delivery. *Soft Matter*, 5, 707–715.

14. Yuk, S.H., Bae, Y.H. (1999). Phase transition polymer for drug delivery. *Crit. Rev. Ther. Drug Carrier Syst.*, 16, 385–423.

15. Alvarez-Lorenzo, C., Concheiro, A. (2002). Reversible adsorption by a pH- and temperature-sensitive acrylic hydrogel. *J. Control. Release*, 80, 247–257.

16. Alvarez-Lorenzo, C.; Concheiro, A. (2008). Intelligent drug delivery systems: polymeric micelles and hydrogels. *Mini-Rev. Med. Chem.*, 8, 1065–1074.

17. Amsden, B. (1998). Solute diffusion within hydrogels: Mechanisms and models, *Macromolecules*, 31, 8382–8395.

18. Alvarez-Lorenzo, C.; Gomez-Amoza, J.L.; Martínez-Pacheco, R.; Souto, C.; Concheiro, A. (1999). Microviscosity of hydroxypropylcellulose gels as a basis for prediction of drug diffusion rates. *Int. J. Pharm.*, 180, 91–105.

19. Barreiro-Iglesias, R.; Alvarez-Lorenzo, C.; Concheiro, A. (2001). Incorporation of small quantities of surfactants as a way to improve the rheological and diffusional behavior of carbopol gels. *J. Control. Release*, 77, 59–75.

20. Bibby, D.C.; Davies, N.M.; Tucker, I.G. (2000). Mechanisms by which cyclodextrins modify drug release from polymeric drug delivery systems. *Int. J. Pharm.*, 197, 1–11.

21. Quaqlia, F.; Varrichio, G.; Miro, A.; La Rotonda, M.I.; Larobina, D.; Mensitieri, G. (2001). Modulation of drug release from hydrogels by using cyclodextrins: the case of nicardipine/β-cyclodextrin system in cross-linked polyethylenglycol. *J. Control. Release*, 71, 329–337.

22. Pose-Vilarnovo, B.; Rodriguez-Tenreiro, C.; Rosa dos Santos, J.F.; Vazquez-Doval, J.; Concheiro, A.; Alvarez-Lorenzo, C.; Torres-Labandeira, J.J. (2004). Modulating drug release with cyclodextrins in hydroxypropyl methylcellulose gels and tablets. *J. Control. Release*, 94, 351–363.

23. Woldum, H.S.; Madsen, F.; Larsen, K.L. (2008). Cyclodextrin controlled release of poorly water-soluble drugs from hydrogels. *Drug Deliv.*, 15, 69–80.

24. Tomic, K.; Veeman, W.S.; Boerakker, M.; Litvinov, V.M.; Dias, A.A., (2008). Lateral and rotational mobility of some drug molecules in a poly(ethylene glycol) diacrylate hydrogel and the effect of drug–cyclodextrin complexation. *J. Pharm. Sci.*, 97, 3245–3256.

25. Bajpai, A.K.; Shukla, S.K.; Bhanu, S.; Kankane, S. (2008). Responsive polymers in controlled drug delivery. *Prog. Polym. Sci.*, 33, 1088–1118.

26. Wajs, G., Meslard, J.C. (1986). Release of therapeutic agents from contact lenses. *Crit. Rev. Ther. Drug Carrier Syst.*, *2*, 275–289.

27. Kim, S.W., Bae, Y.H., Okano, T. (1992). Hydrogels: swelling, drug loading, and release. *Pharm. Res.*, *9*, 283–290.

28. Lipinski, C.A. (2000). Drug-like properties and the cause of poor solubility and poor permeability. *J. Pharmacol. Toxicol. Methods*, *44*, 235–249.

29. Lindenberg, M., Kopp, S., Dressman, J.B. (2004). Classification of orally administered drug on the World Health Organization model list of essential medicines according to the biopharmaceutics classification system. *Eur. J. Pharm. Biopharm.*, *58*, 265–278.

30. Ku, M.S. (2008). Use of the biopharmaceutical classification system in early drug development. *AAPS J.*, *10*, 208–212.

31. Sen, M., Yakar, A. (2001). Controlled release of antifungal drug terbinafine hydrochloride from poly(*N*-vinyl 2-pyrrolidone/itaconic acid) hydrogels. *Int. J. Pharm.*, *228*, 33–41.

32. Jimenez-Kairuz, A.F., Allemandi, D.A., Manzo, R.H. (2003). Equilibrium properties and mechanism of kinetic release of metoclopramide from carbomer hydrogels. *Int. J. Pharm.*, *250*, 129–136.

33. Alvarez-Lorenzo, C.; Concheiro, A. (2004). Molecularly imprinted polymers for drug delivery. *J. Chromatogr. B*, *804*, 231–245.

34. Alvarez-Lorenzo, C.; Concheiro, A. (2006) Molecularly imprinted materials as advanced excipients for drug delivery systems. In: El-Gewely, M.R. Ed., *Biotechnology Annual Review*, Vol. 12. Elsevier, Amsterdam, pp. 225–268.

35. Paulsson, M., Edsman, K. (2002). Controlled drug release from gels using lipophilic interactions of charged substances with surfactants and polymers. *J. Colloid Interface Sci.*, *248*, 194–200.

36. Andrade-Vivero, P., Fernández-Gabriel, E., Alvarez-Lorenzo, C., Concheiro, A. (2007). Improving the loading and release of NSAIDs from pHEMA hydrogels by copolymerization with functionalized monomers. *J. Pharm. Sci.*, *96*, 802–813.

37. Paulsson, M., Edsman, K. (2001). Controlled drug release from gels using surfactant aggregates: I. Effect of lipophilic interactions for a series of uncharged substances. *J. Pharm. Sci.*, *90*, 1216–1225.

38. Paulsson, M., Edsman, K. (2001). Controlled drug release from gels using surfactant aggregates: II. Vesicles formed from mixtures of amphiphilic drugs and oppositely charged surfactants. *Pharm. Res.*, *18*, 1586–1592.

39. Barreiro-Iglesias, R., Alvarez-Lorenzo, C., Concheiro, A. (2001). Incorporation of small quantities of surfactants as a way to improve the rheological and diffusional behavior of carbopol gels. *J. Control. Release 77*, 59–75.

40. Alvarez-Lorenzo, C., Concheiro, A. (2003). Effects of surfactants on gel behavior: design implications for drug delivery systems. *Am. J. Drug Deliv. 1*, 77–101.

41. Mulik, R., Kulkarni, V., Murthy, R.S.R. (2009). Chitosan-based thermosensitive hydrogel containing liposomes for sustained delivery of cytarabine. *Drug Dev. Ind. Pharm.*, *35*, 49–56.

42. Kapoor, Y., Thomas, J.C., Tan, G., John, V.T., Chauhan, A. (2009). Surfactant-laden soft contact lenses for extended delivery of ophthalmic drugs. *Biomaterials*, *30*, 867–878.

43. Crini, G., Morcellet, M. (2002). Synthesis and applications of adsorbents containing cyclodextrins. *J. Sep. Sci.*, *25*, 789–813.

44. Ooya, T., Yui, N. (1999). Polyrotaxanes: synthesis, structure, and potential in drug delivery. *Crit. Rev. Ther. Drug Carrier Syst.*, *16*, 289–330.

45. Loethen, S., Kim, J.M., Thompson, D.H. (2007). Biomedical applications of cyclodextrin based polyrotaxanes. *Polym. Rev.*, *47*, 383–418.

46. Li, J., Loh, X.J. (2008). Cyclodextrin-based supramolecular architectures: Syntheses, structures, and applications for drug and gene delivery. *Adv. Drug Deliv. Rev.*, *60*, 1000–1017.

47. Wu, D.Q., Wang, T., Lu, B., Xu, X.D., Cheng, S.X., Jiang, X.J., Zhang, X.Z., Zhuo, R.X. (2008). Fabrication of supramolecular hydrogels for drug delivery and stem cell encapsulation. *Langmuir*, *24*, 10306–10312.

48. Alvarez-Lorenzo, C., Rosa dos Santos, F., Sosnik, A., Torres-Labandeira, J.J., Concheiro, A. (2009). Hydrogels with cyclodextrins as highly versatile drug delivery systems. In: Stein, D.B. Ed., *Handbook of Hydrogels: Properties, Preparation and Applications*. Nova Publishers, New York.

49. Uekama, K., Hirayama, F., Irie, T. (1994). Application of cyclodextrins in pharmaceutical preparations. *Drug Target. Deliv.*, *3*, 411–456.

50. Stella, V.J., Rao, V.M., Zannou, E.A., Zia, V. (1999). Mechanisms of drug release from cyclodextrin complexes. *Adv. Drug Deliv. Rev.*, *36*, 3–16.

51. Hirayama, F., Uekama, K. (1999). Cyclodextrin-based controlled drug release system. *Adv. Drug Deliv. Rev.*, *36*, 125–141.

52. Szemán, J., Fenyvesi, E., Szejtli, J., Ueda, H., Machida, Y., Nagai, T. (1987). Water soluble cyclodextrin polymers: their interaction with drugs. *J. Inclusion Phenom. Macrocyclic Chem.*, *5*, 427–431.

53. Crini, G., Bertini, S., Torri, G., Naggi, A., Sforzini, D., Vecchi, C., Janus, L., Lekchiri, Y., Morcellet, M. (1998). Sorption of aromatic compounds in water using insoluble cyclodextrin polymers. *J. Appl. Polym. Sci.*, *68*, 1973–1978.

54. Layre, A.M., Gosselet, N.M., Renard, E., Sebille, B., Amiel, C. (2002). Comparison of the complexation of cosmetical and pharmaceutical compounds with γ-cyclodextrin, 2-hydroxypropyl-β-cyclodextrin and water-soluble β-cyclodextrin-co-epichlorhydrin polymers. *J. Inclusion Phenom. Macrocyclic Chem.*, *43*, 311–317.

55. Li, J., Xiao, H., Li, J., Zhong, Y.P. (2004). Drug carrier systems based on water-soluble cationic β-cyclodextrin polymers. *Int. J. Pharm.*, *278*, 329–342.

56. Liu, Y., Yang, Y.W., Yang, E.C., Guan, X.D. (2004). Molecular recognition thermodynamics and structural elucidation of interactions between steroids and bridged bis(β-cyclodextrins)s. *J. Org. Chem.*, *69*, 6590–6602.

57. Qian, L., Guan, Y., Xiao, H. (2008). Preparation and characterization of inclusion complexes of a cationic β-cyclodextrin

polymer with butylparaben or triclosan. *Int. J. Pharm.*, *357*, 244–251.

58. Gazpio, C., Sanchez, M., Isasi, J.R., Velaz, I., Martin, C., Martinez-Oharriz, C., Zornoza, A. (2008). Sorption of pindolol and related compounds by a β-cyclodextrin polymer: isosteric heat of sorption. *Carbohydr. Polym. 71*, 140–146.

59. Crini, G., Morcellet, M. (2002). Synthesis and applications of adsorbents containing cyclodextrins. *J. Sep. Sci.*, *25*, 789–813.

60. Xu, W.L., Liu, J.D., Sun, Y.P. (2003). Preparation of a cyclomaltoheptaose (β-cyclodextrin) cross-linked chitosan derivative via glyoxal or glutaraldehyde. *Chin. Chemi. Lett.*, *14*, 767–770.

61. Kobayashi, N., Shirai, H., Hojo, N. (1989). Virtues of a polycyclodextrin for the de-aggregation of organic molecules in water. *J. Polym. Sci. B*, *27*, 191–195.

62. Crini, G., Cosentino, C., Bertini, S., Naggi, A., Torri, G., Vecchi, C., Janu, L., Morcellet, M. (1998). Solid state NMR spectroscopy study of molecular motion in cyclomaltoheptaose (β-cyclodextrin) crosslinked with epichlorohydrin. *Carbohydr. Res.*, *308*, 37–45.

63. Li, J., Xiao, H., Li, J., Zhong, Y.P. (2004). Drug carrier systems based on water-soluble cationic β-cyclodextrin polymers. *Int. J. Pharm.*, *278*, 329–342.

64. Velaz, I., Isasi, J.R., Sánchez, M., Uzqueda, M., Ponchel, G. (2007). Structural characteristics of some soluble and insoluble β-cyclodextrin polymers. *J. Inclusion Phenom. Macrocyclic Chem.*, *57*, 65–68.

65. Orprecio, R., Evans, C.H. (2003). Polymer-immobilized cyclodextrin trapping of model organic pollutants in flowing water streams. *J. Appl. Polym. Sci.*, *90*, 2103–2110.

66. Crini, G. (2005). Recent developments in polysaccharide-based materials used as adsorbents in wastewater treatment. *Prog. Polym. Sci.*, *30*, 38–70.

67. Crini, G. (2008). Kinetic and equilibrium studies on the removal of cationic dyes from aqueous solution by adsorption onto a cyclodextrin polymer. *Dyes Pigments Int. J.*, *77*, 415–426.

68. Su, C.H., Yang, C.P. (1991). Partial removal of various food components from aqueous-solution using cross-linked polymers of cyclodextrins with epichlorohydrin. *J. Sci. Food Agric.*, *54*, 635–643.

69. Astray, G., Mejuto, J.C., Rial-Otero, R., González-Barreiro, C., Simal-Gándara, J. (2009). A review on the use of cyclodextrins in foods. *Food Hydrocolloids*, doi: 10.1016/j.foodhyd.2009.01.001.

70. Wiedenhof, N., Lammers, J.N.J.J., Van Panthaleon van Eck, C.L. (1969). Properties of cyclodextrins: III. Cyclodextrin–epichlorhydrin resins: preparation and analysis. *Starch Stärke*, *21*, 119–123.

71. Hoffman, J.L. (1973). Chromatography of nucleic-acids on crosslinked cyclodextrin gels having inclusion-forming capacity. *J. Macromole. Sci. Chem.*, *A7*, 1147–1157.

72. Sevillano, X., Isasi, J.R., Peñas, F.J. (2008). Feasibility study of degradation of phenol in a fluidized bed bioreactor with a cyclodextrin polymer as biofilm carrier. *Biodegradation*, *19*, 589–597.

73. Harada, A., Furue, M., Nozakura, S. (1978). Optical resolution of mandelic-acid derivatives by column chromatography on crosslinked cyclodextrin gels. *J. Polym. Sci.- Polym. Chem. Ed.*, *16*, 189–196.

74. Thuaud, N., Sebille, B., Renard, E. (2002). Insight into the chiral recognition of warfarin enantiomers by epichlorohydrin/β-cyclodextrin polymer-based supports: determination of stoichiometry and stability of warfarin/β-cyclodextrin polymer complexes. *J. Biochem. Biophys. Methods.*, *54*, 327–337.

75. Wang, H.D., Chu, L.Y., Song, H., Yang, J.P., Xie, R., Yang, M. (2007). Preparation and enantiomer separation characteristics of chitosan/β-cyclodextrin composite membranes. *J. Membrane Sci.*, *297*, 262–270.

76. Schneiderman, E., Stalcup, A.M. (2000). Cyclodextrins: a versatile tool in separation science. *J. Chromatogr. B*, *745*, 83–102.

77. Scriba, G.K.E. (2008). Cyclodextrins in capillary electrophoresis enantioseparations recent developments and applications. *J. Sep. Sci.*, *31*, 1991–2011.

78. Szejtli, J., Fenyvesi, E., Zsadon, B. (1978). Cyclodextrin polymers. *Starch Stärke*, *30*, 127–131.

79. Fenyvesi, E., Ujhazy, A., Szejtli, J., Pütter, S., Gan, T.G. (1996). Controlled release of drugs from CD polymers substituted with ionic groups. *J. Inclusion Phenom. Macrocyclic Chem.*, *25*, 443–447.

80. Nozaki, T., Maeda, Y., Kitanao, H. (1997). Cyclodextrin gels which have a temperature responsiveness. *J. Polym. Sci. A*, *35*, 1535–1541.

81. Zhang, J.T., Huang, S.W., Gao, F.Z., Zhuo, R.X. (2005). Novel temperature-sensitive, β-cyclodextrin-incorporated poly(*N*-isopropylacrylamide) hydrogels for slow release of drugs. *Colloid Polym. Sci.*, *283*, 461–464.

82. Zhang, J.T., Xue, Y.N., Gao, F.Z., Huang, S.W., Zhuo, R.X. (2008). Preparation of temperature-sensitive poly(*N*-isopropylacrylamide)/β-cyclodextrin-grafted polyethylenimine hydrogels for drug delivery. *J. Appl. Polym. Sci.*, *108*, 3031–3037.

83. Liu, Y.Y., Fan, X.D., Kang, T., Sun, L. (2004). A cyclodextrin microgel for controlled release driven by inclusion effects. *Macromol. Rapid Commun.*, *25*, 1912–1916.

84. Gao, Z.W., Zhao, X.P. (2003). Enhancing electrorheological behaviors with formation of β-cyclodextrin supramolecular complex. *Polymer*, *44*, 4519–4526.

85. Gao, Z., Zhao, X. (2004). Preparation and electrorheological characteristics of β-cyclodextrin–epichlorohydrin–starch polymer suspensions. *J. Appl. Polym. Sci.*, *93*, 1681–1686.

86. Tsai, H.A., Syu, M.J. (2005). Synthesis of creatinine-imprinted poly(β-cyclodextrin) for the specific binding of creatinine. *Biomaterials*, *26*, 2759–2766.

87. Baille, W.E., Huang, W.Q., Nichifor, M., Zhu, X.X. (2000). Functionalized β-cyclodextrin polymers for the sorption of bile salts. *J. Macromol. Sci. A*, *37*, 677–690.

88. Mocanu, G., Vizitiu, D., Carpov, A. (2001). Cyclodextrin polymers. *J. Bioact. Compat. Polym.*, *16*, 315–342.

89. Yamasaki, H., Makihata, Y., Fukunaga, K. (2008). Preparation of crosslinked β-cyclodextrin polymer beads and their application as a sorbent for removal of phenol from wastewater. *J. Chem. Technol. Biotechnol.*, *83*, 991–997.

90. Ozmen, E.Y., Sezgin, M., Yilmaz, A., Yilmaz, M. (2008). Synthesis of β-cyclodextrin and starch based polymers for sorption of azo dyes from aqueous solutions. *Bioresource Technol.*, *99*, 526–531.

91. Cesteros, L.C., Ramized, C.A., Peciña, A., Katime, I. (2007). Synthesis and properties of hydrophilic networks based on poly(ethylene glycol) and β-cyclodextrin. *Macromol. Chem. Phys.*, *208*, 1764–1772.

92. Salmaso, S., Semenzato, A., Bersani, S., Matricardi, P., Rossi, F., Caliceti, P. (2007). Cyclodextrin/PEG based hydrogels for multi-drug delivery. *Int. J. Pharm.*, *345*, 42–50.

93. Ooya, T., Ichi, T., Furubayashi, T., Katoh, M., Yui, N. (2007). Cationic hydrogels of PEG crosslinked by a hydrolysable polyrotaxane for cartilage regeneration. *React. Func. Polym.*, *67*, 1408–1417.

94. Yu, H., Feng, Z., Zhang, A., Sun, L., Qian, L. (2006). Synthesis and characterization of three-dimensional crosslinked networks based on self-assembly of α-cyclodextrins with thiolated 4-arm PEG using a three-step oxidation. *Soft Matter*, *2*, 343–349.

95. Chen, L., Zhu, X., Yan, D., He, X. (2003). A straightforward method to synthesize cyclodextrin-based hyperbranched polymer from natural cyclodextrin. *Polym. Prepr.*, *44*, 669–670.

96. Ma, M., Li, D.Q. (1999). New organic nanoporous polymers and their inclusion complexes. *Chem. Mater.*, *11*, 872–874.

97. Asanuma, H., Hishiya, T., Komiyama, M. (2004). Efficient separation of hydrophobic molecules by molecularly imprinted cyclodextrin polymers. *J. Inclusion Phenom. Macrocyclic Chem.*, *50*, 51–55.

98. Hishiya, T., Shibata, M., Kakazu, M., Asanuma, H., Komiyama, M. (1999). Molecularly imprinted cyclodextrins as selective receptors for steroids. *Macromolecules*, *32*, 2265–2269.

99. Hishiya, T., Asanuma, H., Komiyama, M. (2002). Spectroscopic anatomy of molecular-imprinting of cyclodextrin. evidence for preferential formation of ordered cyclodextrin assemblies. *J. Am. Chem. Soc.*, *124*, 570–575.

100. Garcia-Zubiri, I.X., Gonzalez-Gaitano, G., Isasi, J.R. (2007). Isosteric heats of sorption of 1-naphthol and phenol from aqueous solutions by β-cyclodextrin polymers. *J. Colloids Interface Sci.*, *307*, 64–70.

101. Romo, A., Penas, F.J., Isasi, J.R., Garcia-Zubiri, I.X., Gonzalez-Gaitano, G. (2008). Extraction of phenols from aqueous solutions by β-cyclodextrin polymers: comparison of sorptive capacities with other sorbents. *React. Func. Polym.*, *68*, 406–413.

102. Crini, G. (2003). Studies on adsorption of dyes on β-cyclodextrin polymer. *Bioresource Technol.*, *90*, 193–198.

103. Pariot, N., Edwards-Levy, F., Andry, M.C., Levy, M.C. (2000). Cross-linked β-cyclodextrin microcapsules: preparation and properties. *Int. J. Pharm.*, *211*, 19–27.

104. Pariot, N., Edwards-Levy, F., Andry, M.C., Levy, M.C. (2002). Cross-linked β-cyclodextrin microcapsules: II. Retarding effect on drug release through semi-permeable membranes. *Int. J. Pharm.*, *232*, 175–181.

105. Cesteros, L.C., Gonzalez-Teresa, R., Katime, I. (2009). Hydrogels of β-cyclodextrin cross-linked by acylated poly(ethylene glycol): synthesis and properties. *Eur. Polym J.*, *45*, 674–679.

106. Martel, B., Ruffin, D., Weltrowski, M., Lekchiri, Y., Morcellet, M. (2005). Water-soluble polymers and gels from the polycondensation between cyclodextrins and poly(carboxylic acid)s: a study of the preparation parameters. *J. Appl. Polym. Sci.*, *97*, 433–442.

107. Komiyama, M., Hirai, H. (1987). Preparation of immobilized β-cyclodextrins by use of alkanediol diglycidyl ethers as cross-linking agents and their guest binding abilities. *Polym. J.*, *19*, 773–775.

108. Zhang, X., Wang, Y., Yi, Y. (2004). Synthesis and characterization of grafting β-cyclodextrin with chitosan. *J. Appl. Polym. Sci.*, *94*, 860–864.

109. Alvarez-Lorenzo, C., Rodríguez-Tenreiro, C., Torres-Labandeira, J.J., Concheiro, A. (2006). Method of obtaining hydrogels of cyclodextrins with glycidyl ethers, compositions thus obtained, and applications thereof. *PCT international patent application* WO 2006089993.

110. Rodríguez-Tenreiro, C., Alvarez-Lorenzo, C., Rodriguez-Perez, A., Concheiro, A., Torres-Labandeira, J.J. (2006). New cyclodextrin hydrogels cross-linked with diglycidylethers with a high drug loading and controlled release ability. *Pharm. Res.*, *23*, 121–130.

111. Yui, N., Okano, T., Sakurai, Y. (1992). Inflammation responsive degradation of crosslinked hyaluronic acid gels. *J. Control. Release*, *22*, 105–116.

112. Huang, L.L.H., Lee, P.C., Chen, L.W., Hsieh, K.H. (1998). Comparison of epoxides on grafting collagen to polyurethane and their effects on cellular growth. *J. Biomed. Mater. Res.*, *39*, 630–636.

113. Rodríguez-Tenreiro, C., Alvarez-Lorenzo, C., Rodriguez-Perez, A., Concheiro, A., Torres-Labandeira, J.J. (2007). Estradiol sustained release from high affinity cyclodextrin hydrogels. *Euro. J. Pharm. Biopharm.*, *66*, 55–62.

114. Lopez-Montero, E., Alvarez-Lorenzo, C., Rosa dos Santos, J.F., Torres-Labandeira, J.J., Concheiro, A. (2009). Sertaconazole-loaded cyclodextrin–polysaccharide hydrogels for improved antifungal activity. *Open Drug Deliv. J.*, *3*, 1–9.

115. Rodríguez-Tenreiro, C., Diez-Bueno, L., Concheiro, A., Torres-Labandeira, J.J., Alvarez-Lorenzo, C. (2007). Cyclodextrin/carbopol micro-scale interpenetrating networks (ms-IPNs) for drug delivery. *J. Controll. Release*, *123*, 56–66.

116. Mocanu, G., Mihai, D., LeCerf, D., Picton, L., Moscovici, M. (2009). Cyclodextrin–anionic polysaccharide hydrogels: synthesis, characterization, and interaction with some organic

molecules (water pollutants, drugs, proteins). *J. Appl. Polym. Sci. 112*, 1175–1183.

117. Harada, A., Furue, M., Nozakura, S. (1976). Cyclodextrin-containing polymers: 1. Preparation of polymers. *Macromolecules, 9*, 701–704.

118. Harada, A., Furue, M., Nozakura, S. (1976). Cyclodextrin-containing polymers: 2. Cooperative effects in catalysis and binding. *Macromolecules, 9*, 705–710.

119. Zhang, J.T., Huang, S.W., Zhuo, R.X. (2004). Preparation and characterization of novel temperature sensitive poly(*N*-isopropylacrylamide-co-acryloyl-β-cyclodextrin) hydrogels with fast shrinking kinetics. *Macromol. Chem. Phys., 205*, 107–113.

120. Asanuma, H., Akiyama, T., Kajiya, K., Hishiya, T., Komiyama, M. (2001). Molecular imprinting of cyclodextrin in water for the recognition of nanometer-scaled guests. *Anal. Chim. Acta, 435*, 25–33.

121. Piletsky, S.A., Andersson, H.S., Nicholls, I.A. (1999). Combined hydrophobic and electrostatic interaction-based recognition in molecularly imprinted polymers. *Macromolecules, 32*, 633–636.

122. Piletsky, S.A., Andersson, H.S., Nicholls, I.A. (2005). On the role of electrostatic interactions in the enantioselective recognition of phenylalanine in molecularly imprinted polymers incorporating β-cyclodextrin. *Polym. J., 37*, 793–796.

123. Seo, T., Kajihara, T., Iijima, T. (1987). The synthesis of poly (allylamine) containing covalently bound cyclodextrin and its catalytic effect in the hydrolysis of phenyl-esters. *Macromol. Chem., 188*, 2071–2082.

124. Crini, G., Torri, G., Guerrini, M., Martel, B., Lekchiri, Y., Morcellet, M. (1997). Linear cyclodextrin–poly(vinylamine): synthesis and NMR characterization. *Eur. Polym. J., 33*, 1143–1151.

125. Ramirez, H.L., Cao, R., Fragoso, A., Torres-Labandeira, J.J., Dominguez, A., Schacht, E.H., Baños, M., Villalonga, R. (2006). Improved anti-inflammatory properties for naproxen with cyclodextrin-grafted polysaccharides. *Macromol. Biosci., 6*, 555–561.

126. Liu, Y.Y., Fan, X.D., Gao, L. (2003). Synthesis and characterization of β-cyclodextrin based functional monomers and its copolymers with *N*-isopropylacrylamide. *Macromol. Biosci. 3*, 715–719.

127. Liu, Y.Y., Fan, X.D., Zhao, Y.B. (2005). Synthesis and characterization of a poly(*N*-isopropylacrylamide) with ß-cyclodextrin as pendant groups. *J. Polym. Sci. A, 43*, 3516–3524.

128. Deng, J., He, Q., Wu, Z., Yang, W. (2008). Using glycidyl methacrylate as cross-linking agent to prepare thermosensitive hydrogels by a novel one-step method. *J. Polym. Sci. A, 46*, 2193–2201.

129. Liu, Y.Y., Fan, X.D. (2005). Synthesis, properties and controlled release behaviors of hydrogel networks using cyclodextrin as pendant groups. *Biomaterials, 26*, 6367–6374.

130. Wang H.D., Chu, L.Y., Yu, X.Q., Xie, R., Yang, M., Xu, D., Zhang, J., Hu, L. (2007). Thermosensitive affinity behavior of poly(*N*-isopropyl acrylamide) hydrogels with β-cyclodextrin moieties. *Ind. Eng. Chem. Res. 46*, 1511–1518.

131. Lee, M.H., Yoon, K.J., Ko, S.H. (2001). Synthesis of a vinyl monomer containing β-cyclodextrin and grafting onto cotton fibber. *J. Appl. Polym. Sci., 80*, 438–446.

132. Siemoneit, U., Schmitt, C., Alvarez-Lorenzo, C., Luzardo, A., Otero-Espinar, F., Concheiro, A., Blanco-Méndez, J. (2006). Acrylic/cyclodextrin hydrogels with enhanced drug loading and sustained release capability. *Int. J. Pharm., 312*, 66–74.

133. Osawa, T., Shirasaka, K., Matsui, T., Yoshihara, S., Akiyama, T., Hishiya, T., Asanuma, H., Komiyama, M. (2006). Importance of the position of vinyl group on β-cyclodextrin for the effective imprinting of amino acid derivatives and oligopeptides in water. *Macromolecules, 39*, 2460–2466.

134. Bender, M.L., Komiyama, M. (1978). *Cyclodextrin Chemistry*. Springer-Verlag, Berlin.

135. Song, S., Shirasaka, K., Katayama, M., Nagaoka, S., Yoshihara, S., Osawa, T., Sumaoka, J., Asanuma, H., Komiyama, M. (2007) Recognition of solution structures of peptides by molecularly imprinted cyclodextrin polymers. *Macromolecules, 40*, 3530–3532.

136. Liu, Y.Y., Fan, X.D. (2002). Synthesis and characterization of pH- and temperature-sensitive hydrogel of *N*-isopropylacrylamide/cyclodextrin based copolymer. *Polymer, 43*, 4997–5003.

137. Zawko, S.A., Truong, Q., Schmidt, C.E. (2008). Drug-binding hydrogels of hyaluronic acid functionalized with β-cyclodextrin. *J. Biomed. Mater. Res., 87A*, 1044–1052.

138. Shan, T., Chen, J., Yang, L., Jie, S., Qian, Q. (2009). Radiation preparation and characterization of pH-sensitive hydrogel of acrylic acid/cyclodextrin based copolymer. *J. Radioanal. Nucl Chem., 279*, 75–82.

139. Liu, Y.Y., Fan, X.D., Hu, H., Tang, Z.H. (2004). Release of chlorambucil from poly(*N*-isopropylacrylamide) hydrogels with β-cyclodextrin moieties. *Macromol. Biosci., 4*, 729–736.

140. Wang, Q., Li, S., Wang, Z., Liu, H., Li, C. (2009). Preparation and characterization of a positive thermoresponsive hydrogel for drug loading and release. *J. Appl. Polym. Sci. 111*, 1417–1425.

141. Ma, D., Zhang, L.M., Yang, C., Yan, L. (2008). UV photopolymerized hydrogels with β-cyclodextrin moities. *J. Polym. Res., 15*, 301–307.

142. Lu, D., Yang, L., Zhou, T., Lei, Z. (2008). Synthesis, characterization and properties of biodegradable polylactic acid-β-cyclodextrin cross-linked copolymer microgels. *Eur. Polym. J., 44*, 2140–2145.

143. Demir, S., Kahraman, M.V., Bora, N., Apohan, N.K., Ogan, A. (2008). Preparation, characterization, and drug release properties of poly(2-hydroxyethyl methacrylate) hydrogels having β-cyclodextrin functionality. *J. Appl. Polym. Sci., 109*, 1360–1368.

144. Janus, L., Carbonnier, B., Morcellet, M., Ricart, G., Crini, G., Deratani, A. (2003). Mass spectrometric characterization of a new 2-hydroxypropyl-β-cyclodextrin derivative bearing methacrylic moities and its copolymerization with 1-vinyl-2-pyrrolidone. *Macromol. Biosci., 3*, 198–209.

145. Janus, L., Crini, G., El-Rezzi, V., Morcellet, M., Cambiaghi, A., Torri, G., Naggi, A., Vecchi, C. (1999). New sorbents

containing β-cyclodextrin: synthesis, characterization, and sorption properties. *React. Func. Polym.*, *42*, 173–180.

146. Zawko, S.A., Schmidt, C.E. (2006). Preparation of methacryoloyl-β-cyclodextrin monomer for hydrogel functionalization. *Polym. Mater. Sci. Eng.*, *95*, 1022–1023.

147. Saito, R., Okuno, Y., Kobayashi, H. (2001). Synthesis of polymers by template polymerization: I. Template polymerization of poly(methacrylic acid) with β-cyclodextrin. *J. Polym. Sci. A*, *39*, 3539–3576.

148. Saito, R., Kobayashi, H. (2002). Synthesis of polymers by template polymerization: 2. Effects of solvent and polymerization temperature. *Macromolecules*, *35*, 7207–7213.

149. Saito, R., Yamaguchi, K. (2003). Synthesis of bimodal methacrylic acid oligomers by template polymerization. *Macromolecules*, *36*, 9005–9013.

150. Saito, R., Yamaguchi, K. (2005). Effect of guest compounds on template polymerization of multivinyl monomer of cyclodextrins. *Macromolecules*, *38*, 2085–2092.

151. Hussain, L.A., Dickens, S.H., Bowen, R.L. (2004). Effects of polymerization initiator complexation in methacrylated β-cyclodextrin formulations. *Dent. Mater.*, *20*, 513–521.

152. Hussain, L.A., Dickens, S.H., Bowen, R.L. (2005). Properties of eight methacrylated β-cyclodextrin composite formulations. *Dent. Mater.*, *21*, 210–216.

153. Hussain, L.A., Dickens, S.H., Bowen, R.L. (2005). Shear bond strength of experimental methacrylated β-cyclodextrin-based formulations. *Biomaterials*, *26*, 3973–3979.

154. Alvarez-Lorenzo, C., Hiratani, H., Concheiro, A. (2006). Contact lenses for drug delivery: achieving sustained release with novel systems. *Am. J. Drug Deliv.*, *4*, 131–151.

155. Rosa dos Santos, J.F., Couceiro, R., Concheiro, A., Torres-Labandeira, J.J., Alvarez-Lorenzo, C. (2008). Poly(hydroxyethyl methacrylate-*co*-methacrylated-β-cyclodextrin hydrogels: synthesis, cytocompatibility, mechanical properties and drug loading/release properties. *Acta Biomater. 4*, 745–755.

156. Xu, X.D., Chen, C.S., Lu, B., Wang, Z.C., Cheng, S.X., Zhang, X.Z., Zhuo, R.X. (2009). Modular synthesis of thermosensitive P(NIPAAm-*co*-HEMA)/β-CD based hydrogels via click chemistry. *Macromol. Rapid Commun.*, *30*, 157–164.

157. Wang, C.X., Chen, S.L. (2004). Anchoring β-cyclodextrin to retain fragrances on cotton by means of heterobifunctional reactive dyes. *Color Technol.*, *120*, 14–18.

158. Romi, R., Lo Nostro, P., Bocci, E., Ridi, F., Baglioni, P. (2005). Bioengineering of a cellulosic fabric for insecticide delivery via grafted cyclodextrin. *Biotechnol. Prog.*, *21*, 1724–1730.

159. Hebeish, A., Fouda, M.M.G., Hamdy, I.A., El-Sawy, S.M., Abdel-Mohdy, F.A. (2008). Preparation of durable insect repellent cotton fabric: limonene as insecticide. *Carbohydr. Polym.*, *74*, 268–273.

160. Wang, C.X., Chen, S.L. (2006). Surface treatment of cotton using β-cyclodextrins sol-gel method. *Appl. Surf. Sci. 252*, 6348–6352.

161. Zhao, X., Courtney, J.M. (2007). Surface modification of polymeric biomaterials: utilization of cyclodextrins for blood compatibility improvement. *J. Biomed. Mater. Res. A*, *80*, 539–553.

162. Ohashi, H., Hiraoka, Y., Yamaguchi, T. (2006). An autonomous phase transition-complexation/decomplexation polymer system with a molecular recognition property. *Macromolecules*, *39*, 2614–2620.

163. Alvarez-Lorenzo, C., Rosa dos Santos, J.F., Torres-Labandeira, J.J., Concheiro, A. (2008). Hidrogeles acrílicos con ciclodextrinas colgantes, su preparación y su aplicación como sistemas de liberación y componentes de lentes de contacto. Patent ES 200802364.

164. Rosa dos Santos, J.F., Alvarez-Lorenzo, C., Silva, M., Balsa, L., Couceiro, J., Torres-Labandeira, J.J., Concheiro, A. (2009). Soft contact lenses functionalized with pendant cyclodextrins for controlled drug delivery. *Biomaterials*, *30*, 1348–1355.

17

CYCLODEXTRIN NANOSPONGES AND THEIR APPLICATIONS

FRANCESCO TROTTA

Dipartimento di Chimica, Università di Torino, Torino, Italy

1. INTRODUCTION

Nanosponges may be made of many different organic or inorganic materials; their structure presents a nanometric or smaller dimension. Well-known examples are titanium or other metal oxide–based nanosponges [1–3], silicon nanosponge particles [4], carbon-coated metallic nanosponges [5], hyper-cross-linked polystyrene nanosponges [6], and of course, cyclodextrin-based nanosponges. The common characteristic of these materials is the presence of nanoscale pores, which give them particular properties.

Preparation of these different types also differs considerably; for example, gallium antimonide nanosponges are prepared by bombarding the surface with gallium ions [7]. Polystyrene nanosponges are obtained via a two-step cross-linking reaction. First, linear polystyrene is chloromethylated to the desired degree of substitution. In the second stage, a diluted solution (to avoid intramolecular cross-linking) of the chloromethylated polymer above is allowed to react in the presence of Friedel–Craft catalyst (i.e., tin tetrachloride).

Polystyrene is also involved in the preparation of titanium nanosponges. In this case, functionalized polystyrene nanoparticles are obtained by the copolymerization of styrene in the presence of a polymerizable surfactant. The polystyrene nanoparticles are then coated with titanium using *n*-butyl titanate. The core shell particles are turned into TiO_2 nanosponges by calcining the dried particles in a furnace. The resulting nanosponges find applications in highly sensitive gas sensors, separation of peptides with polyionic nanosponges, and for the purification of water from polycyclic aromatic hydrocarbons.

On the other hand, cyclodextrins (CDs) [8] are capable of including compounds whose geometry and polarity are compatible with those of the cavity. However, native CDs are incapable of forming inclusion compounds with certain molecules, such as hydrophilic or high-molecular-weight molecules. In addition, since the cheapest and most useful CD (i.e., β-CD) has low water solubility (1.85 wt% at room temperature) and is toxic when injected intravenously, many chemical modifications of CDs have been studied in order to overcome their drawbacks and to improve their technological characteristics. Some CD derivatives are well tolerated parenterally, especially hydroxypropyl CD, which is reported to have a toxicity approaching that of glucose. However, because individual CDs, and even individual CD derivatives, easily dissociate from the drug on dilution, many of the advantages of CDs are limited in parenteral treatment.

A possible solution is to synthesize dimers [9–11] or trimers [12–14] of CDs having the hydrophobic cavities positioned in such a way as to cooperate in forming the inclusion compound with the guest molecule. This leads to much higher binding constants in comparison with either native or modified CD monomers. Unfortunately, due to the presence of many reactive hydroxyl groups on CDs, complex reactions are required in order to prepare the desired monomer, generally with very low yields and high cost. On the other hand, the production of simple cross-linked CDs has long been possible. The best known network is generated using epichloridrine [15–17] as a cross-linking agent, and these CDs have been used for several purposes, including column packing for inclusion chromatography, elimination of bitter components from grapefruit juice, for copper

Cyclodextrins in Pharmaceutics, Cosmetics, and Biomedicine: Current and Future Industrial Applications, First Edition. Edited by Erem Bilensoy.
© 2011 John Wiley & Sons, Inc. Published 2011 by John Wiley & Sons, Inc.

analysis, and for cobalt determination in foods. Glutaraldehyde [18] is also reported to give a cross-linked CD. Highly cross-linked CDs have also been synthesized for molecular recognition purposes [19–21], but only after the work by Li and Ma [22] was the term *cyclodextrin nanosponges* introduced and the peculiarities of such compounds studied. CD-based nanosponges are generally used for decontamination processes.

At present, CD-based nanosponges can easily be obtained by reacting the selected CD with a suitable cross-linking agent; these include diisocianates [23], diarylcarbonates and carbonyl diimidazoles [24], carboxylic acid dianhydrides [25], and 2,2-bis(acrylamido)acetic acid [26]. More recently, through the work of Trotta and co-workers [24–26], many other applications have been found. In this chapter we describe the uses of CD nanosponges in different fields.

2. CD-BASED CARBAMATE NANOSPONGES

Researchers at the Los Alamos National Laboratory were the first to introduce the term *nanosponge* for CD cross-linked derivatives. The starting problem was to find an alternative to activated carbon for wastewater decontamination processes. Activated carbon has some affinity to apolar organic compounds but fails to remove the more polar compounds and is unable to remove contaminants to ppb levels. By reacting CDs with suitable diisocyanates [i.e., hexamethylene diisocyanate (HDI) or toluene-2,4-diisocyanate (TDI)] in DMF solution at 70°C for 16 to 24 h under a nitrogen atmosphere, after the usual workup and prolonged washing with a large excess of acetone to remove residual DMF, a powder of the cross-linked polymer was obtained (Scheme 1). The

chemical conversion was monitored using infrared spectrophotometry following the complete disappearance of the isocyanate group at 2270 cm^{-1}. Nanoporous polymers were obtained in the form of granular solids, powders, and even films, but in all cases the purified CD was allowed to react with diisocyanate in a 1:8 molar ratio. No other CD/cross-linking agent ratios have been reported.

The resulting nanoporous polymers can bind organic molecules; for example, nitrophenol is removed from its water solution even at very low concentrations. This compound shows a characteristic absorption at 400 nm (yellow), but at very diluted concentrations (i.e., 10^{-7} to 10^{-9} M), the yellow color of nitrophenol solutions cannot be seen by the naked eye. Nevertheless, when CD-based carbamate nanosponges is introduced into a colorless nitrophenol solution, the native white polymer gradually turns yellow due to absorption of nitrophenol, which is concentrated on the nanosponge powder.

It is an important characteristic of CD-based carbamate nanosponges that they have a very low surface area, 1 to 2 m^2/g compared with activated carbon (600 to 700 m^2/g). However, their loading capacities for organic molecules are very close to those of activated carbon, being in the range 20 to 40 mg/cm^3. It thus appears probable that the organic molecules are not simply adsorbed onto the surface of the nanosponges, but are transported into the bulk of the nanoporous polymer during the inclusion process. The intercalation of the guest molecule into CD-based carbamate nanosponges is reported to be a "downhill" process, that is, to be spontaneous, with a driving force, arising from its $\Delta G°$ at 298 K of about 10 to 13 kcal/mol [27].

Table 1 reports some chemical–physical data of the inclusion compounds of selected organic molecules with

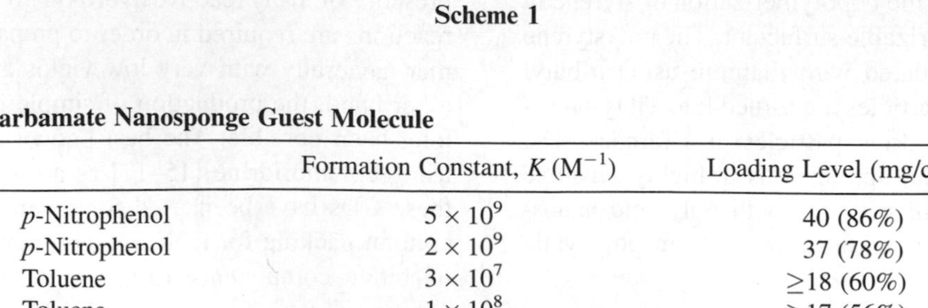

Scheme 1

Table 1. CD-Based Carbamate Nanosponge Guest Molecule

		Formation Constant, K (M^{-1})	Loading Level (mg/cm^3)	ΔG (kcal/mol)
β-CD/HDI	*p*-Nitrophenol	5×10^9	40 (86%)	−13.2
β-CD/TDI	*p*-Nitrophenol	2×10^9	37 (78%)	−12.7
β-CD/HDI	Toluene	3×10^7	≥18 (60%)	−10.2
β-CD/OMe-HDI	Toluene	1×10^8	≥17 (56%)	−10.9
β-CD/HDI	Trichloroethylene	1.8×10^8	≥38 (87%)	−11.2
β-CD/OMe-HDI	Trichloroethylene	2.2×10^9	≥35 (79%)	−12.7

several nanosponges. The high value of the formation constants is significant: 10^7 to $5 \times 10^9\,\mathrm{M}^{-1}$. This means that the affinity of nanosponges for the reported molecules is very high, being among the highest observed for noncovalent inclusion phenomena. Bearing in mind that native CDs show maximum formation constants no greater than $10^4\,\mathrm{M}^{-1}$ (e.g., β-CD with phenolphthalein), the formation of the nanosponge network greatly enhances the (apparent) stability constant of the inclusion compounds [28].

Los Alamos researchers Li and Ma reported numerous characterization experiments, but the only application claimed by the patent deals with water purification. Superior performances in comparison with activated carbon, zeolites, and reverse osmosis were reported. Because zeolites are good material for absorbing water from organics, but not vice versa, and because reverse osmosis fails for low-contaminant concentrations, nanosponges appear to be compared preferably with activated carbon. Although the latter is cheaper than CD nanosponges, it has drawbacks in water purification. For example, it is easily deactivated by moisture in the air and cannot function effectively once completely saturated with water; it is also inefficient at reducing low-concentration contaminants in wastewater (ppb level) and is not suitable for polar molecules. All these drawbacks are overcome by using CD carbamate nanosponges. The polymers developed can be used to bind trichloroethylene, toluene, some phenol derivatives, and many dye compounds.

Li and Ma also demonstrated that nanosponges form an inclusion compound with the guest molecules by using circular dichroism [27]. It is well known that a compound must have both chirality and suitable electron-optical absorption in order to show circular dichroic absorption. Although CDs possess chirality, they do not absorb in the ultraviolet/visible (UV/Vis) region. On the other hand, the guest molecule (e.g., nitrophenol) absorbs light but is not chiral. An induced circular dichroism will be observed only when CD and the guest molecule form a true inclusion compound that has both chirality and optical absorption. That is the case for nanosponges: The hosted molecule is not simply absorbed on the surface, but enters the CD cavity.

Another important feature of nanosponges is that the formation constant depends closely on the solvent used. In aqueous environments, formation of the inclusion compounds is greatly favored. On the contrary, the inclusion process is completely reversible in organic and less polar solvents, such as ethanol. Because of this reversibility, nanosponges are easily regenerated simply by washing with an ecofriendly solvent such as ethanol and can thus be recycled for indefinite numbers of cycles without weight loss and without the dangerous burning processes necessary for partial recovery of activated carbon.

The effectiveness of CD-based carbamate nanosponges in contaminant removal from wastewater was more recently also shown by Mamba et al. [29]. They report that nanosponges can remove as much as 84% of dissolved organic carbon (DOC) from wastewater. The same group demonstrated the possibility to remove undesirable taste and odor compounds such as geosmin [30] and 2-methylisoborneol by using appropriate CD polyurethane nanosponges.

More recently, Tang et al. applied CD-based carbamate nanosponges to adsorb aromatic amino acids [31] (AAA) from phosphate buffer. The adsorption efficiencies of AAA were in the order L-tryptophane > L-phenylalanine > L-tyrosine, and the adsoprtion process followed the Freundlich isotherm.

3. CD-BASED CARBONATE NANOSPONGES

Among the many difunctional compounds suitable for use as cross-linking agents, particularly interesting results have been obtained using active carbonyl compounds such as carbonidiimidazole, diphenylcarbonate, and trifosgene. The resulting CD nanosponges present carbonate bonds between two CD monomers, as shown in Scheme 2. Depending on the carbonylating agent used and on the required reaction time, the reaction can be carried out at room temperature (CDI) or at 80 to 100°C (DPC) in the presence of a solvent, but even in melted DPC or under ultrasound assistance [32]. A mixture of CD and linear dextrin is also used to tune the porous network and, consequently, affects the inclusion ability of CD-based nanosponges. In all cases, after the addition of water to eliminate any possible nonreacted carbonyl derivative, prolonged washing with water and ethanol, Soxhlet extraction with acetone or ethanol, and drying in an oven at 70 to 80°C, a white powder is obtained.

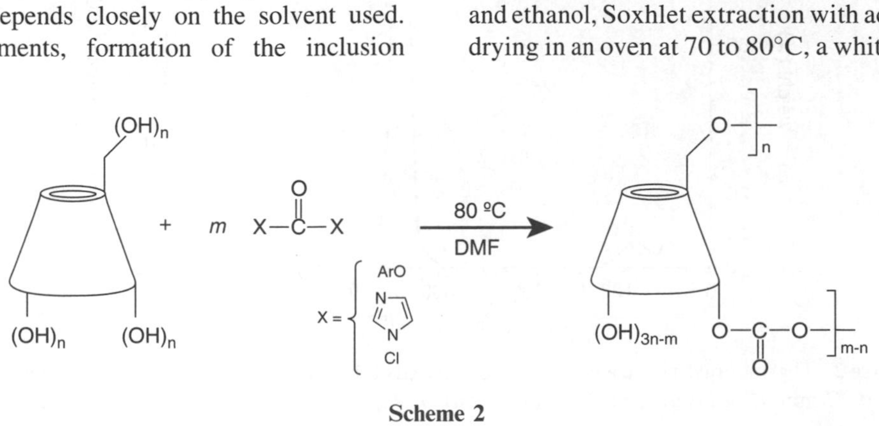

Scheme 2

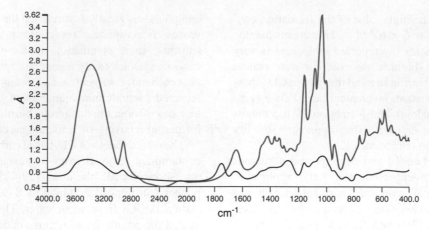

Figure 1. FTIR spectra of native β-CD- (blue line) and β-CD-based carbonate nanosponges (black line.) (*See insert for color representation of the figure.*)

The reaction is easily monitored by the appearance of a band at about $1750 \, \text{cm}^{-1}$ in FTIR spectra due to the carbonyl group of the carbonate bond. It is evident from the spectrum that the peak $1750 \, \text{cm}^{-1}$ is missing from the infrared spectrum of CD, which is a starting material for nanosponge synthesis (Fig. 1).

CD-based carbonate nanosponges are insoluble in water and all organic solvents and do not swell appreciably; they are nontoxic, porous, and stable above 300°C. The latter characteristic is evident from the gravimetric thermal analyses (TGA) reported in Fig. 2. The TGA shows no degradation until 340°C, indicating that the nanosponges have good thermal stability. This assumption is confirmed by differential scanning calorimetry (DSC), as reported in Fig. 3. It is evident that an exothermic peak is present at about 340°C, which could be ascribed to degradation of the nanosponge network. The endothermic peak below 100°C is related to residual moisture in the sample.

Carbonate nanosponges exhibit some interesting features:

- They can be formed in spherical particles of diameter as low as 1 μm (Fig. 4). They can be obtained in a wide range of dimensions, ranging from 1 μm to tens of micrometers.
- The cavities of the framework have a tunable polarity.
- Different functional groups can be linked to the structure.
- They can be given magnetic properties (by adding magnetic particles in the reaction mixture).
- They can be modified further.

The ability of carbonate nanosponges to remove organics from wastewater is evident in Fig. 5, where methyl red is completely removed from its water solution by adding a small amount of carbonate nanosponges. More interestingly,

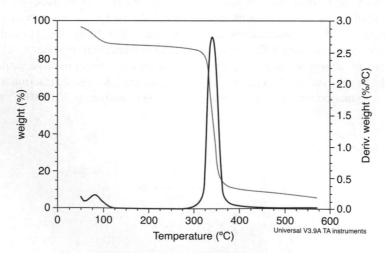

Figure 2. Thermogravimetric analysis of β-CD-based carbonate nanosponges. Nitrogen flow. Ramp rate 10°C/min. (*See insert for color representation of the figure.*)

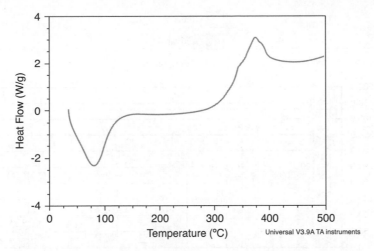

Figure 3. DSC analyses of β-CD-based carbonate nanosponges. Nitrogen flow. Ramp rate 10°C/min.

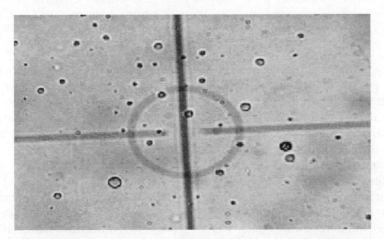

Figure 4. Microscope photograph of spherical β-CD-based carbonate nanosponges particles (<10 μm). [Reprinted with permission from *Journal of Inclusion Phenomena and Macrocyclic Chemistry* 56, 211 (2006).]

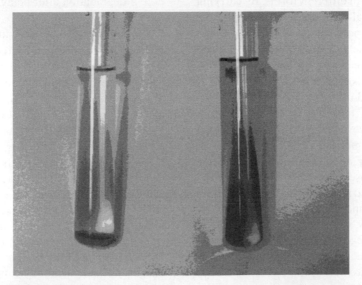

Figure 5. Discoloration of a methyl red solution using β-CD-based carbonate nanosponges. (*See insert for color representation of the figure.*)

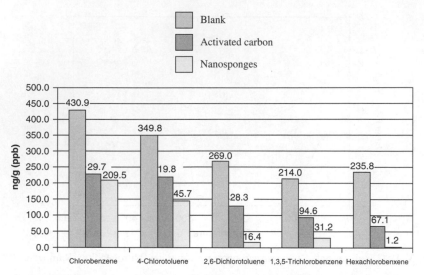

Figure 6. Comparison between β-CD-based carbonate nanosponges and activated carbon. Water decontamination from aromatic compounds after a 4-h batch treatment with 2 wt% adsorbent. [Reprinted with permission from *Composite Interfaces* 16, 46 (2009).]

carbonate nanosponges may be the ideal solution to purify water contaminated by persistent organic pollutants (POPs), such as chlorobenzenes, chlorotoluenes, or polychlorobiphenyls. Currently, the most widespread technique is based on the use of activated carbon. Thanks to their adjustable polarity and the changeable dimensions of their cavities, nanosponges are able to equal, and in some cases exceed, the performance of activated carbon. As reported in Fig. 6, this is particularly true for highly chlorinated aromatic compounds such as hexachlorobenzene, which is almost completely removed from wastewater using CD-based carbonate nanosponges [33].

Unlike other classical nanosponges, carbonate-CD-based nanosponges can be obtained in different forms. By carrying out the reaction under classical conditions in DMF solution, an amorphous cross-linked polymer is obtained. On the contrary, by reacting the CD with melted diphenylcarbonate under controlled reaction conditions, a rather crystalline product is synthesized. This is confirmed by XRD (Fig. 7) and TEM analyses (Fig. 8) [34].

Nanosponge samples are unstable under the electron beam of the microscope. However, TEM analysis has shown ordered areas of restricted dimensions (10 to 50 nm) on an amorphous phase. The *d*-spacing marked out in each TEM image perfectly matches those calculated from XRD patterns. The inclusion ability and the controlled release behavior may be influenced by the morphism of the nanosponges; this theme is discussed below.

Preliminary toxicity studies were carried out on nanosponges to assess their safety and in view of preclinical studies on laboratory animals. In vitro cell culture toxicity assays were carried out on different cell lines (i.e., HeLa, MCF7) using the MTT test. Cells were incubated for between 24 and

72 h and their viability determined. Figure 9 shows cell survival after incubation with nanosponges. Other in vitro methods to evaluate the toxicity of a new material include the determination of hemolytic properties. Nanosponges were incubated with human erythrocytes for 90 min; no hemolytic activity was evidenced up to a concentration of 6 mg/mL, showing the nanosponges to possess good blood compatibility.

In vivo toxicity testing is currently under way. Systemic acute toxicity was evaluated after injection in mice; nanosponges were found to be safe between 500 and 5000 mg/kg in Swiss albino mice; they did not show any sign of toxicity or adverse reactions. Nor did nanosponges show any aggregation or degradation after being incubated in plasma at 37°C for 3 h.

Carbonate nanosponges do not significantly affect the surface tension of water. Moreover, dynamic vapor sorption (DVS) studies confirmed the non-hygroscopic nature of nanosponges and their retention of crystal structure during absorption and desorpion of moisture. In FTIR spectra of nanosponges before and after sterilization, the peak at 1750 cm^{-1} does not disappear; thus, carbonate nanosponges are autoclavable without loss of their properties.

As reported for other nanosponges, CD-based carbonate nanosponges have a low surface area, no greater than 2 m^2/g, and they can be regenerated easily simply by washing with ethanol, acetone, or similar organic compounds without any loss of activity. Carbonate nanosponges have long-term stability and can be stored for six months at 40°C and 75% relative humidity with marginal loss of native CD (less than 3%). Finally, carbonate nanosponges are quite stable in neutral or acid solution, and even more so in a slightly basic water solution at room temperature (Fig. 10).

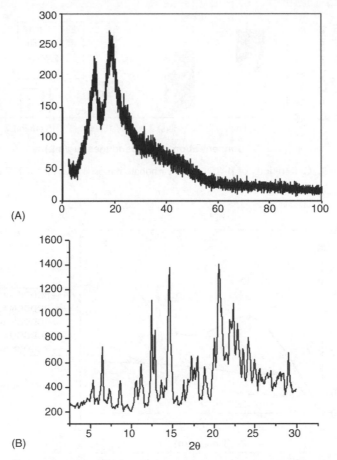

(A)

(B)

2θ

Figure 7. XRD of (A) amorphous β-CD-based carbonate nanosponges and (B) crystalline C β-CD-based carbonate nanosponges.

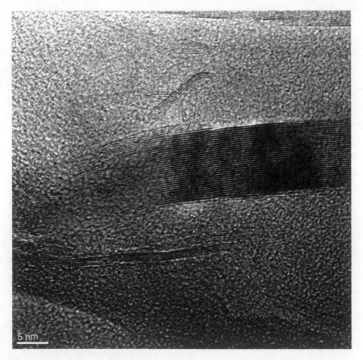

5 nm

Figure 8. TEM photograph of β-CD-based carbonate nanosponges. Magnification ×400,000.

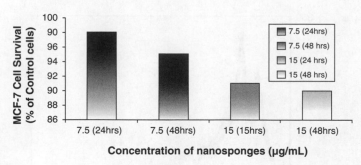

Figure 9. Cytotoxicity of β-CD-based carbonate nanosponges on MCF-7 cells.

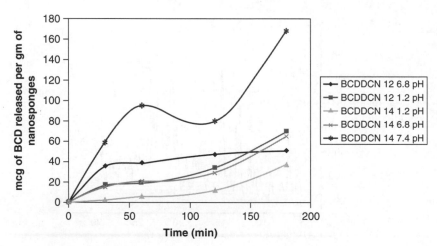

Figure 10. Stability of β-CD-based carbonate nanosponges with different degrees of cross-linking at different pH values.

3.1. CD-Based Carbonate Nanosponges in Agrochemistry

Postharvest flower life is of crucial importance in determining the value of a crop, and many species are extremely sensitive to the senescence process. Ethylene, a potent plant growth regulator, is widely implicated in postharvest senescence. For example, petal senescence in carnation flowers is associated with an increase in ethylene production. However, the initiation of senescence can also be hastened by exposure to exogenous ethylene. Silver nitrate, 2,5-norbornadiene (2,5-NBD), and 1-methylcyclopentene are used to prolong flower longevity but are not used commercially because of their toxicity or undesired smell. 1-Methylcyclopropene (1-MCP) can prevent some of the effects of ethylene without toxicity, but often does so only for a rather time. In recent years some commercial products based on CD inclusion compounds have appeared on the market (e.g., SmartFresh). However, very low antiethylenic compounds are included, and the stability of the inclusion compounds is not high.

Antiethylene compounds, interacting with putative ethylene receptors, can modulate ethylene responses, especially if included in nanosponges. The use of antiethylene–nanosponge complexes can thus improve the longevity of cut flowers, but also fruit and vegetables, thus becoming a significant tool for limiting world starvation. In carnation cultivars, nanosponges containing an antiethylenic molecule favored cut-flower longevity (23 days) and the maintenance of fresh weight, in particular using 1-methylcyclopentene (Fig. 11) [35].

CD-based carbonate nanosponges are also effective in root propagation. For example, *P. umbilicata* is usually recalcitrant to traditional propagation methods. The results obtained in micropropagation experiments showed that this species is also probably recalcitrant to micropropagation and that the use of nanosponges, either with indolbutyrric acid (IBA) or with naphthyl acetic acid (NAA), was not effective. In *P. cirrhiflora*, IBA improved the number of roots produced by each plant but showed no significant difference compared to controls without nanosponge inclusion. On the contrary,

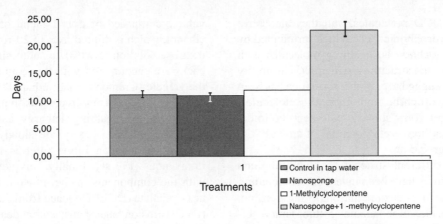

Figure 11. Longevity of *Dianthus caryophyllus* cut stems treated with 1-methylcyclopentene-β-CD-based carbonate nanosponges complex, β-CD-based carbonate nanosponges, and 1-methylcyclopentene compared to controls in tap water.

the NAA-nanosponge (0.2 mg/L) complex improved root development (a mean number of 6.32 roots/plant) of *P. cirrhiflora* explants significantly compared with controls containing nonincluded NAA (1.5 roots/plant) (Fig. 12). In this case, nanosponges were able to improve rooting, which is an essential step in in vitro propagation of ornamental species, probably facilitating control of phytoregulator release, stimulating root development and ensuring constant availability of the active principle [35].

This new nanocolloidal carrier system has been shown to provide several advantages in an important economic field, that of nursery gardening. The use of nanosponges to carry phytohormones and antiethylene substances favors the gradual release of these products over time, reducing the concentrations normally required, and increasing their bioavailability, with substantial reduction of production costs.

3.2. CD-Based Carbonate Nanosponges in Pharmaceuticals

Considering their biocompatibility and versatility, nanosponges have many possible applications in the pharmaceutical field. These applications were recently reported by Cavalli et al. [36]. Nanosponges can be used as excipients in preparing tablets, capsules, pellets, granules, suspensions, solid dispersions, or topical dosage forms. In particular, the design of new drug delivery carriers to improve administration is now under study. Many drugs show poor solubility, low permeability, short half-life, low stability and/or high molecular weight, and their formulation is thus challenging. Many bioavailability problems may be solved by enhancing the solubility and dissolution rate of a substance, and nanosponges can increase solubility. The interaction sites available for complexation in nanosponges are increased markedly

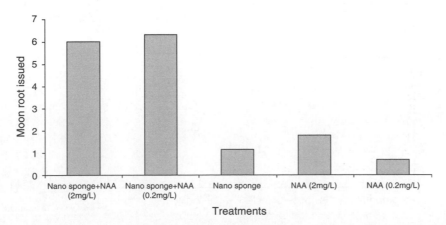

Figure 12. Influence of β-CD-based carbonate nanosponges on root propagation in *Passiflora cirrhiflora*.

compared to those in CD molecules, and they also have different polarities (hydrophobic CD cavities surrounded by rather hydrophilic nanochannels) enabling molecules with different lipophilicities and structures to interact. The ability of CD nanosponges to engage in complex formation has been evaluated for both hydrophilic and lipophilic molecules. Moreover, nanosponges form stable, opalescent, colloidal dispersion in water, and their zeta potential of around −30 mV prevents their aggregation.

Many drugs with different structures, solubilities, and pharmacological activities have been investigated, including paclitaxel, camptothecin, dexamethasone, flurbiprofen, doxorubicin hydrocloride, itraconazole, 5-fluorouracil, cilostazol, progesterone, oxcarbamazepine, nelfinavir mesylate, resveratrol, and tamoxifen. Nanosponges particularly improve the wetting and solubility of molecules with very poor aqueous solubility. For example, Swaminathan et al. [37] have studied the formulation of nanosponges with itraconazole, a drug with aqueous solubility of about 1 ng/mL at physiological pH. The presence of nanosponges improved the drug's solubility more than 27-fold; on adding PVP as an auxiliary component in the nanosponge formulation, this was enhanced to 55-fold. Moreover, the dissolution profiles of the drug from the two formulations was faster than from the marketed formulations. Nanosponges could thus increase the bioavailability of itraconazole. Nanosponges can also be used as carriers of more hydrophilic drugs, and they may protect the drug during passage through the stomach. This is the case of the well-known anticancer drug doxorubicin, which is released very slowly at pH 1.1, whereas release is faster if the pH is increased to 7.4 [35]. Particularly good results were obtained using carbonate nanosponges in the delivery of some anticancer drugs such as paclitaxel and campothecin [38].

Paclitaxel is a diterpenoid with high anticancer activity (in particular for breast cancer) but extremely low aqueous solubility. For this reason it is currently formulated in a vehicle composed of a 1:1 blend of Cremophor EL and ethanol, which is diluted five- to 20-fold in normal saline or dextrose solution (5 wt%) for administration. Many drawbacks are encountered with this administration route; first, the vehicle (Cremophor EL: ethanol) required for solubilization is toxic, causing hypersensitivity reactions, vasodilatation, labored breathing, lethargy, and hypotension. Paclitaxel also tends to precipitate slowly out of the aqueous media, and thus an inline filter is recommended in the intravenous (IV) set. Finally, this vehicle is incompatible with the components of the infusion sets, causing diethylhexylpthalate (DHEP) to leach from the poly(vinyl chloride) (PVC) infusion bags. Thus, a great deal of work is needed to develop a cremophore-free formulation of paclitaxel for intravenous (IV) use, and to enhance the drug's oral bioavailability. The use of nanosponges as a carrier for paclitaxel could overcome all these problems, leading to a Cremophor-free formulation. Paclitaxel-loaded nanosponges have been prepared using three types of nanosponges having different degrees of cross-linking (1:2, 1:4, and 1:8); the formulations have been characterized. The different nanosponge solubilization capacity is correlated with the degree of cross-linking. The solubility enhancement factor versus plain paclitaxel increased with the degree of cross-linking: nanosponges with a 1:8 ratio showed a high solubilization capacity, with an enhancement factor of more than 60, while that of β-CD is 9.

The significant change in XRD defractograms of complexes from nanosponges indicates complexation; this is confirmed by FTIR analyses, in particular by DSC (Fig. 13) and also by XRPD and NMR analyses. The cytotoxicity of paclitaxel-loaded 1:4 ratio nanosponges was evaluated on MCF-7 cells; paclitaxel carried by nanosponges was found to be significantly more cytotoxic than the plain drug (Fig. 14).

Pharmacokinetic profiles were tested in vivo for both IV and oral administration. Briefly, nanosponges tend to release

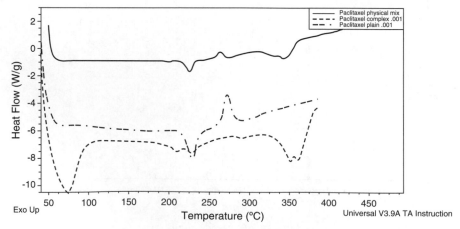

Figure 13. DSC analyses of paclitaxel complex with β-CD-based carbonate nanosponges compared with plain drug and neat nanosponges.

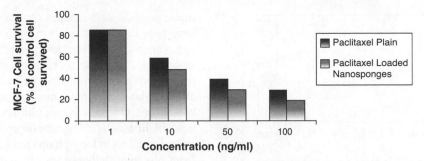

Figure 14. Cytotoxicity of paclitaxel-loaded β-CD-based carbonate nanosponges compared with plain drug on MCF-7 cells after 48 h.

paclitaxel faster than commercial formulation when administered IV, making the drug available for rapid distribution in the peripheral compartments, as is evident from the $t_{1/2}\alpha$ and K_{12} values obtained. Results were even better for oral administration: As reported in Fig. 15, the AUC of paclitaxel in the nanosponge formulation increased threefold. The percentage relative bioavailability of paclitaxel from the nanosponges was 256%. There was a 2.5-fold increase in the mean absolute bioavailability (AB) in nanosponges. The C_{max} value for nanosponges was $74.36 \pm 8.3\,\mu g/mL$ versus $60.76 \pm 7.2\,\mu g/mL$ for Taxol. There was no significant change in t_{max}, mean residence time, or terminal half-life ($t_{1/2}$).

On the other hand, camptothecin, originally isolated from the stem wood of *Camptotheca acuminata*, is a potent cytotoxic alkaloid active against a variety of tumors, which unfortunately is insoluble in water. Camptothecin delivery suffers from its poor solubility, which causes low bioavailability; its high systemic toxicity; the fact that the lactone ring of camptothecin opens in vivo, rendering it inactive; and from photo-instability. As reported in Fig. 16, nanosponges enhance the photo-stability of camptothecin [39]. This is particularly true for surface-modified nanosponges with succinic anhydrides and thus bearing carboxylic acid residues. Nanosponges also stabilize camptothecin in the plasma (Fig. 17), but although the results are in agreement with PBS stability results, the degradation is more pronounced in the

plasma, possibly due to the presence of albumin, known to play a role in the degradation of camptothecin. Again, cytotoxicity studies show that the complex is more than six times more cytotoxic than the plain drug. Cytotoxicity even after 72 h indicates stability of the lactone ring. Moreover, since the amount of drug released is higher after 24 h than it is after 15 min, the drug may be said to be stable in the plasma (Fig. 18).

An interesting feature of CD-based carbonate nanosponges is that the inclusion ability depends significantly on their degree of crystallinity: Different molecules can be solubilized to different extents depending on the nanosponges used. For example, the well-known anticancer drug dexametasone is fourfold better solubilized by crystalline nanosponges, whereas the widely used antiviral compound acyclovir is solubilized twice as much by amorphous nanosponges [40]. Thus, the degree of crystallinity of CD-based carbonate nanosponges is an important parameter to be considered in the correct formulation of drug carriers.

3.3. CD-Based Carbonate Nanosponges in Catalysis

The use of immobilized enzymes is advantageous for various reasons: It allows enzymes to be recycled, facilitates the separation and recovery of reaction products from the reaction environment, and enables continuous flow processes to

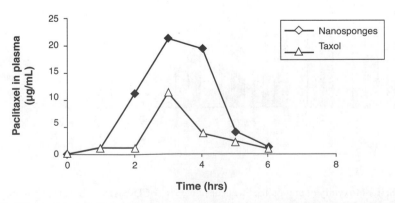

Figure 15. Plasma concentration of paclitaxel after oral administration. [Reprinted with permission from *Drug Delivery* 17, 422 (2010).]

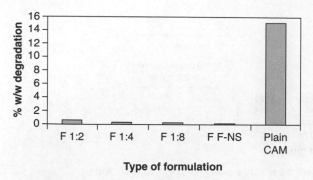

Figure 16. Photodegradation of camptothecin.

be applied. Exploitation of enzymes in industrial processes often relies on enzyme immobilization on solid supports, which facilitates working under continuous-flow conditions and has been reported to increase enzyme stability and to limit inhibition effects exerted by reaction products or components of the reaction mixture. As a consequence, the demand for economically advantageous and environmentally friendly supports is growing constantly.

Hydrolytic enzymes, such as proteases, amylases, esterases, and lipases, are among the enzymes used most widely for industrial purposes. Lipases are triacylglycerol hydrolases (EC 3.1.1.3) that either catalyze the hydrolysis of triacylglycerols (in aqueous media) or catalyze

transesterification reactions (in non-aqueous media). They have found widespread use in a growing range of industrial applications, including detergents, food processing, pharmaceutical, cosmetic, textiles, the leather and paper industries, biotransformations, waste treatment, and bioremediation. Lipases also have medical applications as drugs or biosensors. One of the most intriguing industrial applications of lipases is the synthetic process producing methyl and ethyl esters of long-chain fatty acids (biodiesel) starting from oils and methanol (or ethanol); however, it must be said that the performance of lipases is not yet comparable to that provided by basic catalysis, in the presence of NaOH or KOH.

A system obtained by adsorbing *Pseudomonas fluorescens* lipase on a newly synthesized CD-based carbonate nanosponge has lead to improved activity and better structural stability of this lipase, when adsorbed on this new support, compared with those of the free enzyme in solution. This new type of CD-based nanosponge stabilizes *Ps. fluorescens* lipase, the enzyme still being active after 66 days of incubation at $T \sim 18°C$. Moreover, unlike the solubilized enzyme, adsorbed lipase is active at $T > 40°C$, at pH 5 and after 24 h of incubation with 70% v/v methanol (13% residual activity) [41].

Catechol dioxygenases are enzymes containing iron that can convert the substrate catechol into the product *cis-cis*

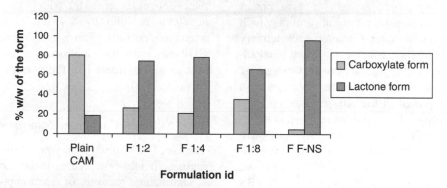

Figure 17. Plasma stability of camptotheticin complexes over 24 h.

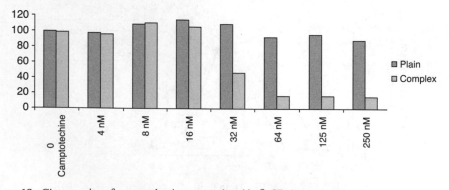

Figure 18. Citotoxycity of camptothecin encapsulated in β-CD-based carbonate nanosponges versus plain drug on HCPC1 cells after 72 h. [Reprinted with permission from *European Journal of Pharmaceutics and Biopharmaceutics* 74, 199 (2010).]

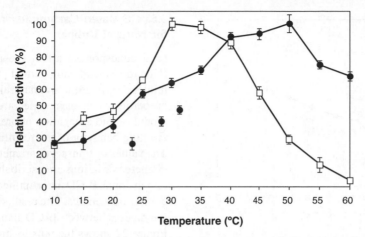

Figure 19. Stability of 1,2-dioxygenase to temperature: enzyme on β-CD-based carbonate nanosponges (□); free enzyme (●). [Reprinted with permission from *Dalton Transactions* 33, 6510 (2009).]

muconic acid, a precursor of the industrially important compound adipic acid. Catechol 1,2-dioxygenase from *Acinetobacter radioresistens* S13 was immobilized on β-CD-based carbonate nanosponges [42]. The enzyme was adsorbed on this matrix with a yield of 29 mg enzyme per gram of support. The activity profiles at different pH values and temperatures showed that the optimal pH was between 8.5 and 9.5, and the optimal temperature was 30°C in free and 50°C in immobilized protein, respectively. Kinetic parameters were calculated, and the K_M values were found to have increased to $2.0 \pm 0.3\,\mu M$ for the free form and to $16.6 \pm 4.8\,\mu M$ for the immobilized enzyme, whereas the k_{cat} values were found to be $32 \pm 2\,s^{-1}$ and $27 \pm 3\,s^{-1}$ for the free and immobilized forms, respectively. The immobilization process was also found to increase the thermostability of the enzyme, with 60% residual activity after 90 min at 40°C for the immobilized protein versus 20% for the free enzyme, and 75% residual activity after 15 min at 60°C for the immobilized enzyme versus a total loss of activity for the free form (Fig. 19). The immobilized enzyme retained its activity toward other substrates, such as 3- and 4-methylcatechol and 4-chlorocatechol. A small-scale bioreactor was constructed and was able to convert catechol into *cis-cis* muconic acid with high efficiency for 70 days.

3.4. CD-Based Carbonate Nanosponges as Gas Traps

The high cross-linked network of CD nanosponges could host even small molecules such as gases. The reversible encapsulation of gases in a solid matrix is of great interest for many applications, such as gas separation, sensors, gas storage, and fuel cells. It has long been possible to complex halogen and halogen halides with native cyclodextrins: α-CD can host Cl_2, Br_2, and I_2 equally well, β-CD can host Br_2 and I_2, while γ-CD forms an inclusion complex only with iodine. The inclusion complex is formed simply by bubbling the gas

through a CD aqueous solution at 10°C. The gas is liberated by placing the solid complex in water at a slightly higher temperature.

Many other gases can be entrapped in CDs: α-CD is generally the CD of choice due to the small size of its cavity, making it more suitable for low-molecular-weight compounds. The particular nanoporous network of nanosponges enables β-CD to host gas molecules. As mentioned in connection with agrochemistry applications, carbonate nanosponges can retain 1-methylcyclopropene and other similar antiethylenic gases. They can also host other molecules, such as CO_2 and oxygen. For example, β-CD carbonate nanosponges can entrap carbonic anhydride when previously outgassed. The isotherms obtained at 20°C are reported in Fig. 20. The stored CO_2 amounts to 0.4 ± 0.1 mass% for the sample treated at room temperature and 0.6 ± 0.1 mass% for that treated at 120°C. In both cases, the kinetics of CO_2 adsorption was extremely slow, requiring an equilibration time of 4 h for each point of the isotherm. Whereas for the former sample a good coincidence of the first and second runs indicates the good reversibility of the adsorption process, for the sample treated at 120°C, a partial hysteresis between the first and second cycle indicates that 0.1 mass% of CO_2 was irreversibly stored in the sample. The initial weight of the sample was restored only after heat treatment at 62°C for 3 h in vacuum [43].

Oxygen also shows good interactions with nanosponges. It is important to recall that gases play a very relevant role in medicine, for both diagnostic and treatment purposes. However, it can be difficult to deliver oxygen in an appropriate form and at an optimal dosage, making it necessary to develop carriers that provide patients with oxygen under optimal conditions. From a therapeutic standpoint, any compound encapsulating oxygen for a long time is a potential carrier. Local hypoxia is related to various important diseases, ranging from inflammation to cancerous lesions.

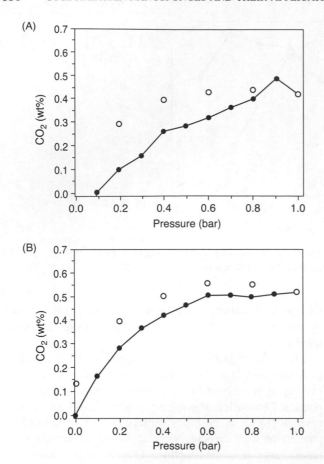

Figure 20. Excess CO_2 gravimetric isotherms recorded at 20°C on β-CD-based carbonate nanosponges sample after thermal treatment at (A) room temperature and (B) 120°C. Solid and ring scatters refer to the adsorption and desorption branches, respectively.

3.5. CD-Based Carbonate Nanosponges as Filler for Natural Rubber

Neat nanosponges affect the properties of natural rubber (NR) to a considerable extent [45] when they are added as a reinforcing phase to the natural rubber latex matrix. Tensile properties such as tensile strength, tear strength, tensile modulus, and elongation at break are altered, as well as the swelling behavior of the resulting nanocomposites in water. The values of transport parameters, relative weight loss, and diameter variations are ascribable to the presence of a three-dimensional β-CD nanosponge network within the nanocomposite samples. The results indicate that there is a strong interaction between β-CD nanosponges and the NR latex. Figure 22 shows the tensile strength of NR samples versus nanosponge content. Moreover, NS/NR nanocomposites, prepared by the water evaporation method, offer molecular transport of different solvents (i.e., toluene, benzene, and xylene); tests were undertaken at room temperature for neat NR film and for nanocomposites with 1-, 2-, and 3-g nanosponge content per 100 g of rubber. Transport parameters—diffusion coefficient, sorption coefficient, permeation coefficient, relative weight loss (RWL), and sol fraction—were determined. Uptake was slowest for xylene and fastest for benzene, as measured by these coefficients. A first-order kinetics model was used to investigate the transport kinetics (Fig. 23) [46].

3.6. CD-Based Carbonate Nanosponges in Cosmetics

γ-Orizanol (GO), extracted from rice bran, is a mixture of ferulate esters of triterpene alcohols and is known to be a powerful inhibitor of hydroxyl radical formation. Unfortunately, GO is a light-sensitive ingredient, and thus it would be advantageous to include it in a carrier system to give it adequate efficacy and stability. The permeation behavior of GO through an artificial silicon membrane has been studied under UVA and UVB radiation; photodegradation was found to be related to the initial GO concentration and to the nature of the vehicle. Sapino and co-workers [47] complexed GO with β-CD-based carbonate nanosponges and found that inclusion in the nanosponges led to a more photo-stable form than free GO (Fig. 24). The peroxidation of linoleic acid was investigated under UVA irradiation in the absence and presence of GO, either free or complexed with β-NS. Lipid peroxidation was assessed by malondialdehyde (MDA) determination through a colorimetric reaction with thiobarbituric acid (TBA). As Fig. 25 shows. inclusion of GO in β-NS does not limit its antioxidative activity. Using slices of pig-ear skin, it was shown that inclusion in nanosponges slightly increases γ-orizanol skin accumulation. CD-based nanosponges may thus be a promising carrier system for GO, as they improve photostability without limiting antioxidative properties or skin

Tumor hypoxia is a therapeutic problem because of its adverse impact on the effectiveness of radiotherapy and chemotherapy. Oxygen delivery to hypoxic tumor cells could be particularly important and has been found to be a powerful strategy in cancer therapy. Nanosponges could be used as a long-term oxygen delivery system.

The particle-size analysis of nanosponges encapsulating oxygen shows that the average diameter is between 400 and 600 nm. The particle-size distribution is unimodal with a narrow size range. The formulations show sufficiently high zeta potential, which prevents nanosponges from forming aggregates. The oxygen is released from the NS formulation over long periods [i.e., 60 min starting from a hypoxic condition (0.4 mg/L)]. The release profiles indicate that there is an initial burst effect, followed by constant release maintained for a long period. The same release behavior was observed in the presence of a silicon membrane. Figure 21 shows profiles of oxygen release. Ultrasound sonication produced an increase of approximately 30% in oxygen release [44].

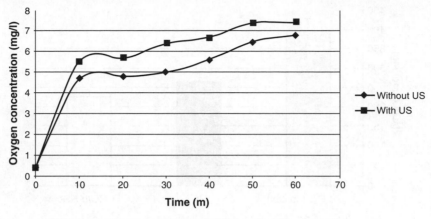

Figure 21. Oxygen release by β-CD-based carbonate nanosponges.

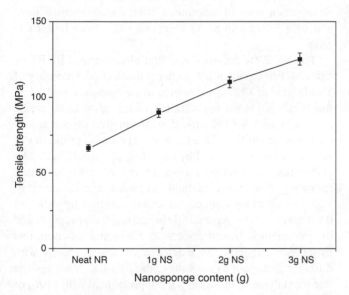

Figure 22. Effect of β-CD-based carbonate nanosponges on the tensile strength of natural rubber.

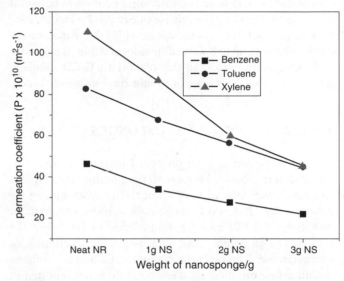

Figure 23. Effect of β-CD-based carbonate nanosponges on the permeation coefficient of nanocomposite samples in different solvents.

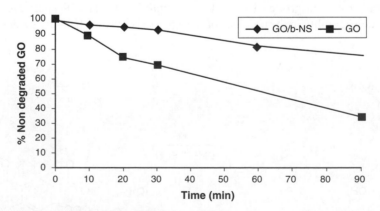

Figure 24. UVB-induced photodegradation of γ-orizanol (GO) (0.05 mM) in (2%) hydroxyethyl cellulose gel.

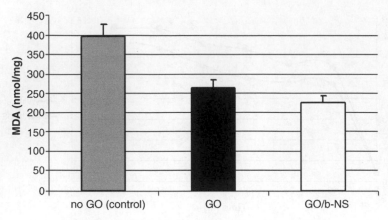

Figure 25. Antooxidative activity of γ-orizanol (GO).

accumulation. CD-based nanosponges could also be useful to entrap and prolong the release of essential oil molecules, such as linalool. It has been reported [48] that nanosponges can incorporate up to 8 wt% of linalool and that its release at room temperature was half that of a pristin β-CD complex, thus prolonging the effect of pristin fragrance over time.

4. CD-BASED ESTER NANOSPONGES

Nanosponges can also be produced using a suitable dianhydride as a cross-linking agent; pyromellitic anhydride is frequently the anhydride of choice. The cross-linking reaction is very fast and is carried out at room temperature, dissolving the CD and the dianhydride in DMSO in the presence of an organic base such as pyridine, collidine, or triethylamine. The reaction is exothermic and is complete within a few minutes; reaction conditions are reported in Scheme 3 [25]. It is significant that this type of nanosponge (NS-PYR) contains a polar free carboxylic acid group, and can thus host both apolar organic molecules and cations simultaneously. By using different amounts of dianhydride it is possible to vary the number of acid groups. Pyromellitic nanosponges also swell considerably in the presence of

water when a small amount of cross-linker agent is used and to a lesser extent on increasing the cross-linker/CD ratio.

The NS-PYR polymer was first characterized by FT-IR analysis. Principally it is significant thanks to the presence of a wide band at $1727\,cm^{-1}$ related to the ester carbonyl group that is missing in native β-CD. Characterization of the acidic properties of NS-PYR provides information on the anion-exchange capability of the material. The acidic properties of the sites were determined by modeling pH-metric data. The elaboration of titration curves of the NS-PYR polymer, following Soldatov's method, provides a pK_a value of 4.38. In view of the synthetic procedure, we may hypothesize the presence of the pyromellitic biacid residue as a spacer and the pyromellitic triacid residue as a terminal function; this hypothesis suggests the existence of acidic groups with different protonation constants in the polymer. If we assume that the polymer behaves as a monoprotic acid with a pK_H of 4.38, the trend of titration curves cannot be interpreted across the range of pH values. However, the pH-metric data can be explained exhaustively with a chemical model that assumes the presence of two independent sites: a biprotic site and a triprotic site. The IR spectroscopy data strengthen the hypothesis of this chemical model [49].

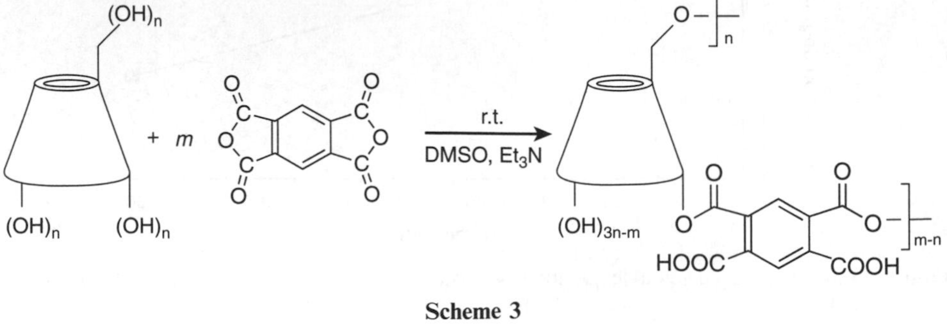

Scheme 3

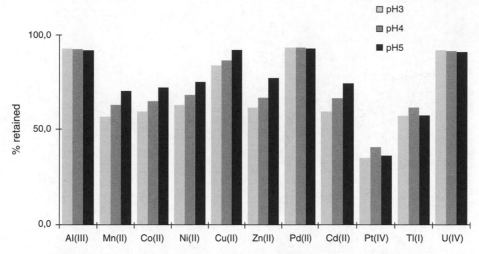

Figure 26. Retention of heavy metal cations on CD-based ester nanosponges.

The ionic moiety of NS-PYR can be exploited for complexation of heavy metal cations at different pH values (Fig. 26). Adsorptions have been reported to be above 70% for Al(III), Mn(II), Co(II), Ni(II), Cu(II), Zn(II), Cd (II), Pd(II), and U(IV). The new synthesized nanosponges present a particular structure, which confers cation exchange and coordinating properties, enhancing the complexing activity of the native CD toward metal ions. Cu(II) retention on the polymer has been characterized at different pH values and through pH-metric titrations [50]. Through the modeling of pH-metric data, the presence of four different complexes has been shown, one of which involves dissociated alcoholic groups present on the CD structure. This complex is present in the pH range 5.5 to 7. The retention of Cd(II) on NS-PYR is exploited in a flow system; the NS-PYR packed column shows the possibility of retaining and releasing Cd(II) simply by changing the pH conditions of the eluent. Furthermore, the polymer maintains its properties unchanged after 10 consecutive adsorption and desorption cycles. These data support the possibility of using NS-PYR in environmental applications.

Recently, deeper insight into the inclusion abilities of NS-PYR was achieved by Mele and co-workers using a high-resolution magic angle spinning (HR MAS) NMR technique that can be used to investigate semisolid samples such as swellable NS-PYR. The diffusivity of water and the interaction of fluorescein in the inner cavities were thus investigated [51].

5. POLYAMIDOAMINE NANOSPONGES

Polyamidoamine nanosponges are quite different from the other types of nanosponges described above. First, the reaction is carried out in water without using any organic solvents, working at room temperature in the presence of LiOH·H$_2$O. Under these reaction conditions, β-CD polymerizes with acetic acid 2,2'-bis(acrylamide) simply after long standing (i.e., 94 h at room temperature). A white powder is then recovered after the usual workup, as reported by Ferruti et al. [26]. A possible structure of polyamidoamine nanosponges is shown in Fig. 27. They swell in water and have both acid and basic residues; swelling and water uptake were found to be pH dependent. The polymer formed a translucent gel instantly on contact with water, and time-dependent swelling studies in biorelevant media confirmed the stability of the gel for up to 72 h. Model protein albumin gave a very high encapsulation efficiency, around 90%. In vitro drug release studies show that the protein release can be modulated up to 24 h. The conformational stability of the protein, examined using the SDS-PAGE technique, showed the formulation to be stable even for several months [52].

6. MODIFIED NANOSPONGES

As could be predicted, the presence of a large set of hydroxyl groups on the surface of nanosponges opens the way to further modification of this material and offers the possibility of fitting the performance more closely to the desired application. Among others, we mention the fluorescent derivative obtained by reacting the carbonate nanosponges with fluorescein isothiocyanate in DMSO at 90°C for a few hours [53]. After the usual workup, a yellow powder comprising nanosponges is obtained that is of interest for biological studies. For example, Fig. 28 is a photograph of human squamous cell carcinoma surrounded by fluorescent nanosponges. Another possibility is to carry out the same reaction, in the conditions reported above,

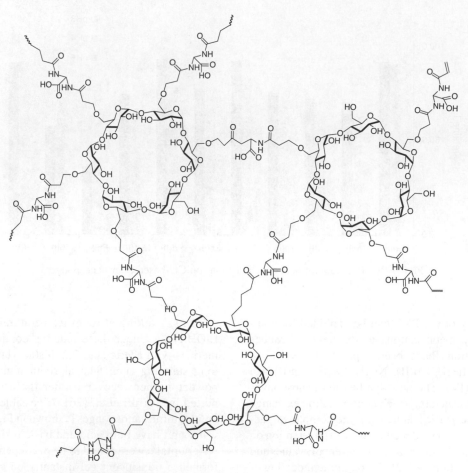

Figure 27. CD-based polyamidoamine nanosponge structure.

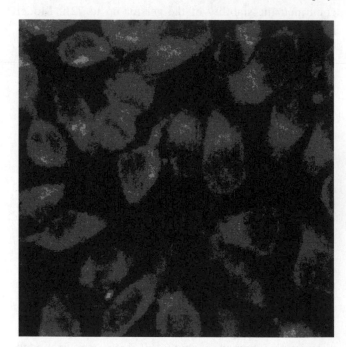

Figure 28. Human squamous cells carcinoma treated with fluorescent β-CD-based carbonate nanosponges. Confocal Microscope Zeiss Axiovert 100 M. (*See insert for color representation of the figure.*)

between carbonate nanosponges and a cyclic organic anhydride such as succinic anhydride or maleic anhydride. In this way, CD nanosponges bearing carboxylic acid residues can be obtained [54]. These novel functionalized nanosponges could react with biologically important carriers such as biotin, chitosan, or proteins, possibly providing promising carriers for targeting drugs to specific receptors. Titrimetry revealed the presence of amounts of acidic residues as high as 1 meq/g of nanosponge. XRPD sudies showed that these nanosponges are amorphous in nature; they are also nonhemolytic and noncytotoxic. They interact strongly with chitosan, as is shown by zeta-potential measurements. As reported above for camptothecin, carboxylated nanosponges appear to be a promising safe carrier for drug release.

7. CONCLUDING REMARKS

Nanosponges are a new type of biocompatible, versatile cross-linked polymer that greatly expand the performances of their parent CDs. In particular, thanks to very high inclusion stability constants, they find numerous applications in pharmaceutics, biomedicine, cosmetics, bioremediation processes, catalysis, agrochemistry, gas entrapping, and other fields. Their synthesis can be modulated by varying the reaction conditions to better fit the application selected. Moreover, thanks to the presence of residual oxydril groups on the surface, they can be further modified with suitable spacers, to increase their specificity and efficiency.

REFERENCES

1. Guo, L., Gao, G., Liu, X., Liu, F. (2008). Preparation and characterization of TiO$_2$ nanosponge. *Mater. Chem. Phys.*, *111*, 322–325.

2. Zuruzi, A.S., Kolmakov, A., MacDonald, N.C., Moskovits, M. (2006). Highly sensitive gas sensor based on integrated titania nanosponge arrays. *Appl. Phys. Lett.*, *88*, 102904-1-3.

3. Zuruzi, A.S., Kolmakov, A., MacDonald, N.C., Moskovits, M. (2007). Metal oxide nanosponges as chemical sensors: highly sensitive detection of hydrogen using nanosponge titania. *Angew. Chem. Int. Ed.*, *46*, 4298–4301.

4. Farrell, D., Limaye, S.Y., Subramanian, S. (2006). Silicon nanosponge particles. U. S. patent 0,251,561A1.

5. Lian, K., Wu, Q., Carbon encased metal nanoparticles and sponges, methods of synthesis and methods of use. IPC patent AB01D5394FI.

6. Dakankov, V.A., Llyin, M.M., Tsyurupa, M.P., Timofeeva, G.I., Dubronina, L.V. (1998). From a dissolved polystyrene coil to intramolecularly hyper cross linked nanosponges. *Macromolecules*, *29*, 8398–8403.

7. Kluth, S.M., Fitz Gerald, J., Ridgway, M.C. (2005). Ion-irradiation-induced porosity in GaSb. *Appl. Phys. Lett.*, *86*, 131920/1–3.

8. Szejtli, J. (1988). *Cyclodextrin Technology*. Kluwer, Dordrecht, The Netherlands.

9. Ishimaru, Y., Masuda, T., Iida, T. (1997). Synthesis of secondary face-to-face cyclodextrin dimers linked at each 2-position. *Tetrahedron Lett.*, *38*, 3743–3744.

10. Costes, J.H., Easton, C.J., van Eyk, S.J., Lincoln, F., May, B.L., Whalland, C.B., Williams, M.L. (1990). A new synthesis of cyclodextrin dimers. *J. Chem. Soc. Perkin Trans. 1*, 2619.

11. Breslow, R., Zhang, B. (1996). Cholesterol recognition and binding by cyclodextrin dimers. *J. Am. Chem. Soc.*, *118*, 8495–8496.

12. Leung, D.K., Atkins, J.H., Breslow, R. (2001). Synthesis and binding properties of cyclodextrin trimers. *Tetrahedron Lett.*, *42*, 6255–6258.

13. Nakajima, H., Sakabe, Y., Ikeda, H., Ueno, A. (2004). Cyclodextrin trimers as receptors for arranging ester and catalyst at optimized location to achieve enhancement of hydrolitic activity. *Biorg. Med. Chem. Lett.*, *5*, 1783–1786.

14. Luo, M.M., Chen, W.H., Yuan, D.Q., Gang Xie, R. (1998). Synthesis of novel cyclodextrin trimers. *Synth. Commun.*, *28*, 3845–3848.

15. Harada, A., Furue, M., Nozakura, A. (1981). Inclusion of aromatic compounds by a β-cyclodextrin epichlorohydrin polymer. *Polym. J.*, *13*, 777–781.

16. Xu, W., Wang, Y., Shen, S., Li, Y., Xia, S., Zhang, Y. (1988). Studies on the polymerization of β-cyclodextrin with epichlorohydrin. *Chin. J. Polym. Sci.*, *7*, 16–22.

17. Volkova, D.A., Bannikov, G.E., Lopatin, S.A., Il'in, M.M., Gabinskaya, K.N., Andryushina, V.A., Gracheva, I.M., Varlamov, V.P. (1999). Effect of β-cyclodextrin epichlorohydrin copolymer on the solubility of cortexolone. *Pharm. Chem. J.*, *33*, 30–32.

18. Xu, W.L., Liu, J.D., Sun, Y.P. (2003). Preparation of cyclomaltoheptaose (β-cyclodextrin) cross-linked chitosan derivative via glyoxal or glutaraldehyde. *Chin. Chem. Lett.*, *14*, 767–770.

19. Asanuma, H., Kakazu, M., Shibata, M., Hishiya, T., Komiyama, M. (1997). Molecularly imprinted polymer of β-cyclodextrin for the efficient recognition of cholesterol. *Chem. Commun.* 1971–1972.

20. Hishiya, T., Shibata, M., Kakazu, M., Asanuma, H., Komiyama, M. (1999). Molecularly imprinted cyclodextrins as selective receptors for steroids. *Macromolecules*, *32*, 2265–2269.

21. Yang, Y., Long, Y., Cao, Q., Li, K., Liu, F. (2008). Molecularly imprinted polymer using β-cyclodextrin as functional monomer for the efficient recognition of bilirubin. *Anal. Chem. Acta*, *606*, 92–97.

22. Li, D., Ma, M. (1998). Cyclodextrin polymer separation materials. Patent WO 9822197.

23. Li, D., Ma, M. (1999). New organic nanoporous polymers and their inclusion complexes. *Chem. Mater.*, *11*, 872–874.

24. Trotta, F., Tumiatti, V. (2003). Cross-linked polymers based on cyclodextrins for removing polluting agents. Patent WO 03/085002 A1.

25. Trotta, F., Tumiatti, V., Vallero, R. (2004). Nanospugne a base di ciclodestrine funzionalizzate con gruppi carbossilici: sìntesi ed utilizzo nélla contaminazióne da metàlli pesanti e da compósti organici, separazióni cromatogràfiche e veicolazióne di farmaci. Dom. Brev. It. MI2004A000614.

26. Ferruti, P., Ranucci, E., Trotta, F., Cavalli, R., Gilardi, G., Dinardo, G. (2008). Nanospugne a base di ciclodestrine cóme suppòrto per catalizzatóri biologíci e nélla veìcolazióne e rilàscio di enzimi, proteine, vaccini ed anticòrpi. Dom. Brev. It. MI2008A1056.

27. Li, D., Ma, M. (1999) Nanosponges: from inclusion chemistry to water purifying technology. *Chemtech*, May, pp. 31–37.

28. Li, D., Ma, M. (2000). Nanosponges for water purification. *Clean Prod. Processes*, *2*, 112–116.

29. Mamba, B.B., Krause, R.W., Malefetse, T.J., Sithhole, S.P. (2008). Cyclodextrin nanosponges in the removal of organic matter to produce water for power generation. *Water SA*, *34*, 657–660.

30. Mamba, B.B., Krause, R.W., Malefetse, T.J., Mhlanga, S.D., Sithhole, S.P., Salipira, K.L., Nxumalo, E.N. (2007). Removal of geosmin and 2-methylisoborneol (2-MIB) in water from Zuikerbosch water treatment plant (Rand Water) using β-cyclodextrin polyurethane. *Water SA*, *32*, 223–228.

31. Tang, S., Kong, L., Ou, J., Liu, Y., Li, X., Zou, H. (2006). Application of cross-linked β-cyclodextrin polymer for adsorption of aromatic amino acid. *J. Mol. Recogn. Macrocyclic Chem.*, *19*, 39–48.

32. Trotta, F., Cavalli, R., Tumiatti, V., Zerbinati, O., Roggero, C., Vallero, R. (2006). Ultrasound assisted synthesis of cyclodextrin based nanosponges. Patent WO 2006/002814 A1.

33. Trotta, F., Cavalli, R. (2009). Characterization and application of new hyper-cross-linked cyclodextrin. *Compos. Interface*, *16*, 39–48.

34. Tumbiolo, S., Bertinetti, L., Trotta, F., Colucia, S., Cavalli, R., Cravotto, G. (2008). Structural characterization of β-cyclodextrin nanosponges. *Proceedings of the 14th International Cyclodextrin Symposium*, Kyoto, Japan, pp. 336–338.

35. Seglie, L., Devecchi, M., Trotta, F., Chiavazza, P., Dolci, M., Scariot, V. (2008). Effect of nanosponges including new anti-ethylene compounds on post-harvest longevity of cut flowers and nanosponges including phytohormones on in vitro regeneration of ornamental species. *Proceedings of the 14th International Cyclodextrin Symposium*, Kyoto, Japan, pp. 332–335.

36. Cavalli, R., Trotta, F., Tumiatti, W. (2006). Cyclodextrin-based nanosponges for drug delivery. *J. Inclusion Phenom. Macrocyclic Chem.*, *56*, 209–213.

37. Swaminathan, S., Vavia, P.R., Trotta, F., Torne, S. (2007). Formulation of β-cyclodextrin based nanosponges of itraconazole. *J. Inclusion Phenom. Macrocyclic Chem.*, *57*, 89–94.

38. Trotta, F., Tumiatti, V., Cavalli, R., Roggero, C.M., Mognetti, B., Berta, G.N. (2007). Nanospugne a base di ciclodestrine cóme veìcolo per farmaci antitumorali. Dom. Brev. It. MI2007A1321.

39. Cavalli, R., Trotta, F., Vavia, P.R., Swaminathan, S., Trotta, M. (2008). Nanosponges based prolonged release injectable formulations of camptothecin. *Proceedings of the 14th International Cyclodextrin Symposium*, Kyoto, Japan, pp. 122–116.

40. Swaminathan, S. (2009). Ph.D. dissertation. University Institute of Chemical Technology, Mumbai, India.

41. Boscolo, B., Trotta, F., Ghibaudi, E. (2010). High catalytic performances of *Ps. fluorescens* lipase adsorbed on a new type of cyclodextrin-based nanosponges. *J. Mol. Catal. B: Enzymatic*, *22*, 155–161.

42. Di Nardo, G., Roggero, C., Campolongo, S., Valetti, F., Trotta, F., Gilardi, G. (2009). Catalytic properties of catechol 1,2-dioxygenase from *Acinetobacter radioresistens* S13 immobilized on nanosponges. *J. Chem. Soc. Dalton Trans.*, *33*, 6507–6512.

43. Trotta, F., Cavalli, R., a K., Vitillo, J., Bordiga, S., Swaminathan, S., Ansari, K., Vavia, P.R. (2009). Cyclodextrin nanosponges as effective gas carriers. Presented at the First European Cyclodextrin Conference, Aalborg, Denmark.

44. Ansari, K., Vavia, P.R., Trotta, F., Cavalli, R. (2009). Nanosponge formulations as long term oxygen delivery carrier. Presented at the CRS meeting, Copenaghen, Denmark.

45. Visakh, P.M., Thomas, S., Trotta, F., Tumiatti, V., Roggero, C.M. (2009). Tensile properties and swelling behaviour of β-cyclodextrin nanosponges (NS) reinforced natural rubber latex (NR) nanocomposites (unpublished results).

46. Visakh, P.M., Thomas, S., Trotta, F. (2009). Molecular transport characteristics of nanosponge (NS) reinforced natural rubber latex (NR) nanocomposites in different solvents (unpublished results).

47. Sapino, S., Carlotti, M.E., Trotta, M. (2008). Photostability and antioxidative acitivity of γ-orizanol included in cyclodextrin based nanosponges. Presented at the 7th Central European Symposium on Pharmaceutical Technology and Biodelivery Systems, Ljubljana, Slovenia.

48. Cavalli, R., Carlotti, M.E., Trotta, F., Thorne, S., Roggero, C., Trotta, M. (2006). Linalool-loaded nanosponges. Presented at XX Symposium ADRITELF, Catania, Italy.

49. Berto, S., Bruzzoniti, M.C., Cavalli, R., Perrachon, D., Prenesti, E., Sarzanini, C., Trotta, F., Tumiatti, W. (2007). Synthesis of new ionic β-cyclodextrin polymers and characterization. *J. Inclusion Phenom. Macrocyclic Chem.*, *57*, 631–636.

50. Berto, S., Bruzzoniti, M.C., Cavalli, R., Perrachon, D., Prenesti, E., Sarzanini, C., Trotta, F., Tumiatti, W. (2007). Highly cross-linked ionic β-cyclodextrin polymers and their interaction with heavy metals. *J. Inclusion Phenom. Macrocyclic Chem.*, *57*, 637–643.

51. Mele, A., Castiglione, F., Malpezzi, L., Ganazzoli, F., Raffini, G., Trotta, F., Rossi, B., Fontana, A., Giunchi, G. (2010). HR MAS NMR, powder XRD and Raman spectroscopy study of inclusion phenomena in β-CD nanosponges. *J. Inclusion Phenom. Macrocyclic Chem.* DOI: 10.1007/S10847-010-9772-x.

52. Swaminathan, S., Vavia, P.R., Trotta, F., Cavalli, R., Ferruti, P., Ranucci, E., Gerges, I. (2009). Release modulation and conformational stabilization of a model protein by use of swellable nanosponges of β-cyclodextrin. Presented at the First European Cyclodextrin Conference, Aalborg, Denmark.

53. Trotta, F. Unpublished results.

54. Trotta, F., Cavalli, R., Swaminathan, S., Sarzanini, C., Vavia, P.R. (2008). Novel funtionalized nanosponges: synthesis, characterization. Safety assessment, cytotoxicity testing and interaction studies. *Proceedings of the 14th International Cyclodextrin Symposium*, Kyoto, Japan, pp. 338–342.

Complex formation

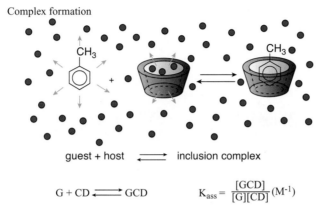

guest + host ⇌ inclusion complex

$$G + CD \rightleftharpoons GCD \qquad K_{ass} = \frac{[GCD]}{[G][CD]} \, (M^{-1})$$

Chapter 2, Figure 3. Complex formation with CDs.

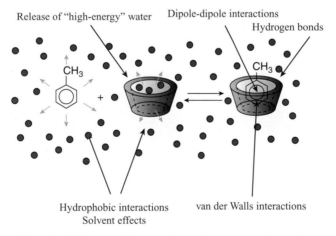

Release of "high-energy" water   Dipole-dipole interactions

Hydrogen bonds

Hydrophobic interactions   van der Walls interactions
Solvent effects

Chapter 2, Figure 4. Driving forces for complex formation.

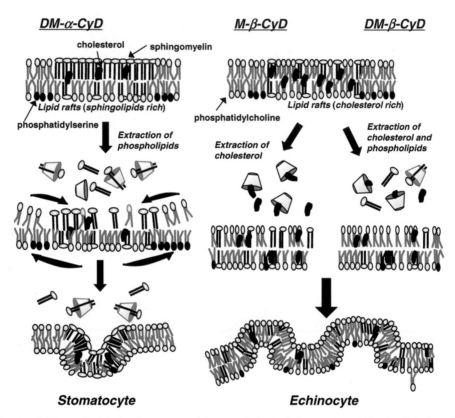

Chapter 5, Figure 3. Mechanism proposed for morphological changes in red blood cells induced by methylated CDs.

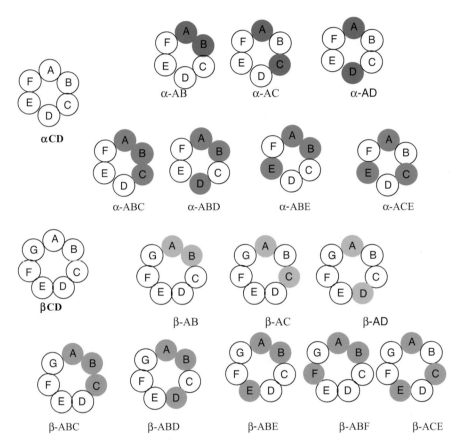

Chapter 11, Figure 24. Possible positional isomers of di- and tri-modified α- and β-CDs (and tetra and penta by deduction.

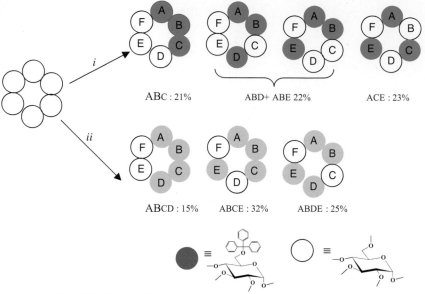

ABC : 21%　　　ABD+ ABE 22%　　　ACE : 23%

ABCD : 15%　　　ABCE : 32%　　　ABDE : 25%

Reagents and conditions : *i* TrCl (3.3 eq.), pyridine, 70°C, 36h, NaH, MeI, DMF ; *ii* TrCl (4.5 eq.), pyridine, 70°C, 72h, NaH, MeI, DMF

Chapter 11, Figure 27. Synthesis of tri- and tetra-substituted α-CD derivatives.

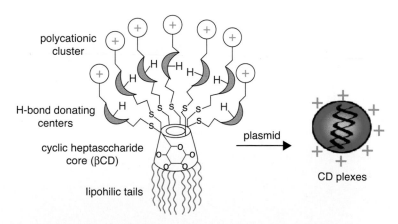

polycationic cluster

H-bond donating centers

cyclic heptasccharide core (βCD)

lipohilic tails

plasmid

CD plexes

Chapter 12, Figure 3. Amphiphilic β-CDs used as DNA vectors.

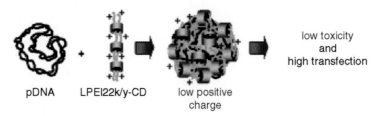

Chapter 12, Figure 5. PEI/γ-CD polypseudorotaxane inducing DNA condensation.

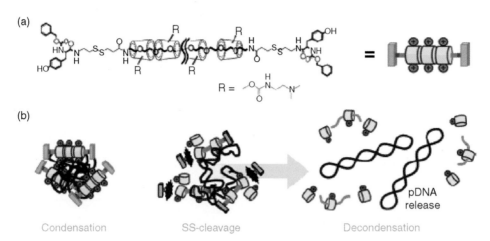

Chapter 12, Figure 6. (a) Chemical structure of biocleavable polyrotaxane; (b) polyplex formation, terminal cleavage-triggered decondensation of DNA.

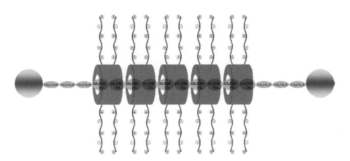

Chapter 12, Figure 7. Structure of a cationic polyrotaxane.

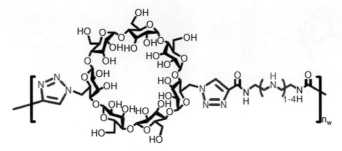

Chapter 12, Figure 11. Structure of CD polymers (Cdn$_w$).

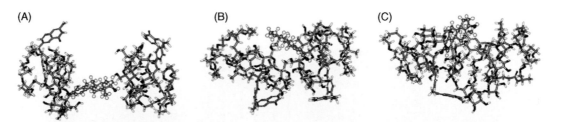

Chapter 12, Figure 14. Sunflower-shaped polyplex formation.

β-CDPL

sunflower shaped
β-CDPL/pDNA polyplex

pDNA

(A)   (B)   (C)

Chapter 13, Figure 1. Three snapshots of the MD simulation performed for 2 ns at 310 K on the 1 : 2 β-estradiol/CD-PEG-FA assembly in water: (A) $t = 0$; (B): $t = 1$ ns; (C): $t = 2$ ns. For clarity, β-estradiol is shown in a ball-and-stick representation; CD-PEG-FA is shown in a stick representation.

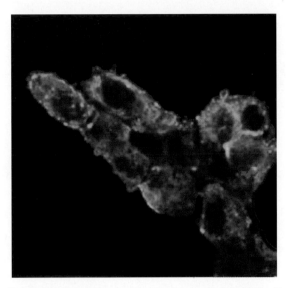

Chapter 13, Figure 3. Confocal images obtained by incubation of KB cells with rhodamine-B-loaded CD–PEG–FA.

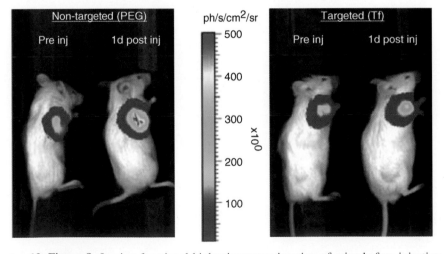

Chapter 13, Figure 8. In vivo functional bioluminescence imaging of mice before injection and 1 day after injection of targeted and nontargeted siRNA nanoparticles.

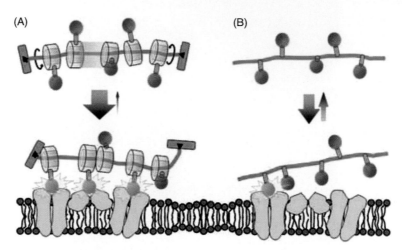

Chapter 13, Figure 12. Effect of "mobile" motion of the cyclic compounds in polyrotaxanes on receptor binding in a multivalent manner: (A) ligand–polyrotaxane conjugate and receptor sites; (B) ligand-immobilized-polymer and receptor sites.

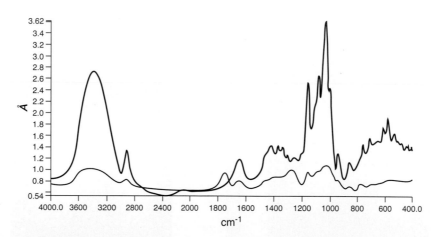

Chapter 17, Figure 1. FTIR spectra of native β-CD- (blue line) and β-CD-based carbonate nanosponges (black line).

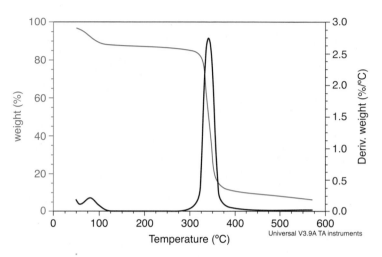

Chapter 17, Figure 2. Thermogravimetric analysis of β-CD-based carbonate nanosponges. Nitrogen flow. Ramp rate 10°C/min.

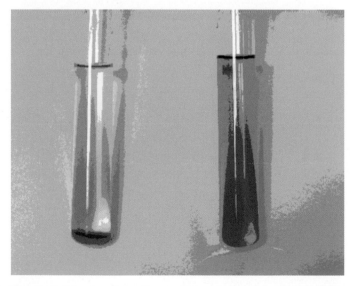

Chapter 17, Figure 5. Discoloration of a methyl red solution using β-CD-based carbonate nanosponges.

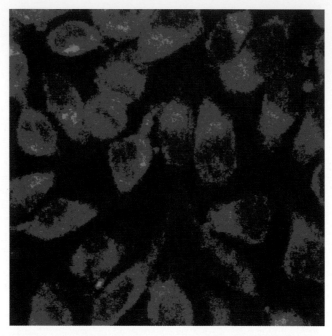

Chapter 17, Figure 28. Human squamous cells carcinoma treated with fluorescent β-CD-based carbonate nanosponges. Confocal Microscope Zeiss Axiovert 100 M.

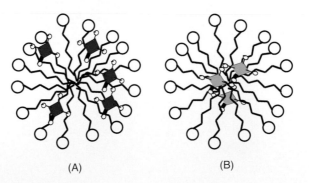

(A)   (B)

Chapter 18, Figure 3. PS entrapment in colloidal nanoassemblies: (A) hydrophylic PS in the outer casing of micelles; (B) hydrophobic PS in the core of micelles.

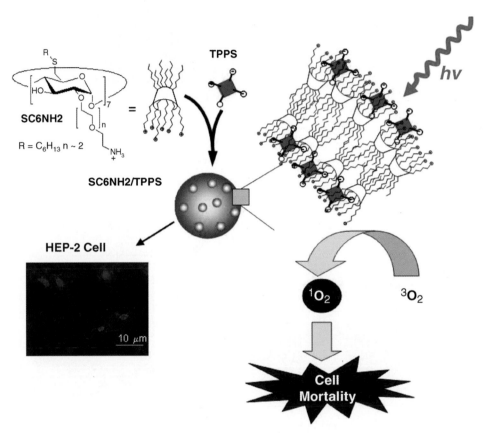

Chapter 18, Figure 6. PS delivery by CD nanoassemblies, HEP-2 intracellular delivery and generation of singlet oxygen.

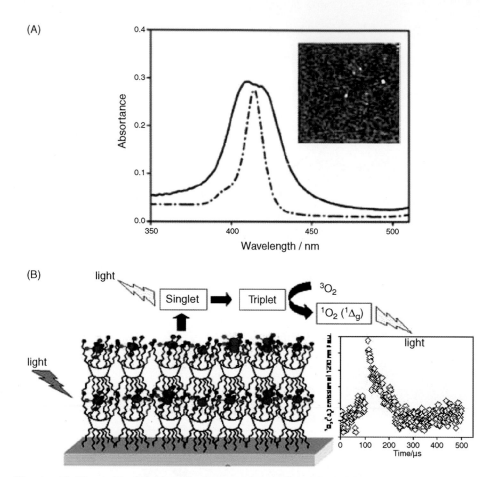

Chapter 18, Figure 10. (A) Absoption spectra of a multilayer $SC_{16}NH_2$–TPPS Langmuir–Schäfer film (solid line) and TPPS in water solution (dashed-dotted line). The inset shows a typical image (width = 1 μm) of the $SC_{16}NH_2$–TPPS photoactive nanoassemblies transferred onto hydrophobized silicon. (B) Potential arrangement of the $SC_{16}NH_2$–TPPS multilayer. The inset shows a representative decay kinetic of $^1O_2(^1\Delta_g)$ monitored at 1270 nm.

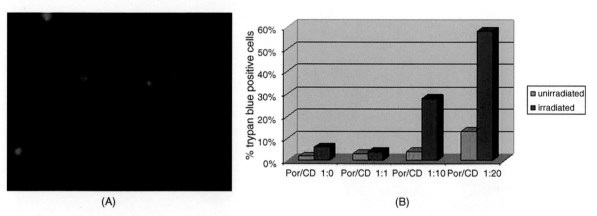

(A)   (B)

Chapter 18, Figure 12. (A) Fluorescence microscopy analysis of HeLa cells treated with TPPS–SC$_6$NH$_2$ nanoassemblies (TPPS 1 μM) at 1:10 molar ratio (merging of cell visualized with a rhodamine filter and DAPI filter). (B) Cell death percenteage in HeLa through trypan blue assay.

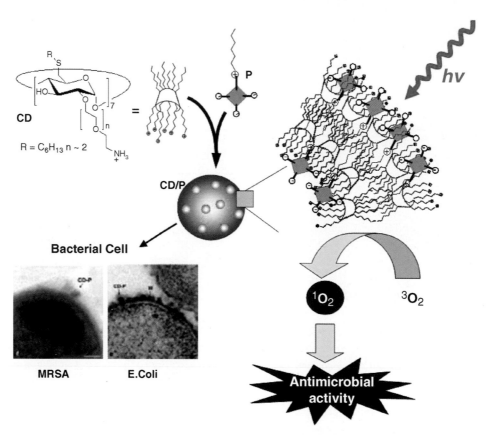

Chapter 18, Figure 13. Cationic porphyrin bearing a long hydrophobic chain (TDPyP) delivery entrapped in cationic amphiphilic CD (SC$_6$NH$_2$). TDPyP–SC$_6$NH$_2$ can be delivered to both gram-positive and gram-negative bacterial cells (MRSA and *E. Coli*, respectively) by photosensitising their inactivation.

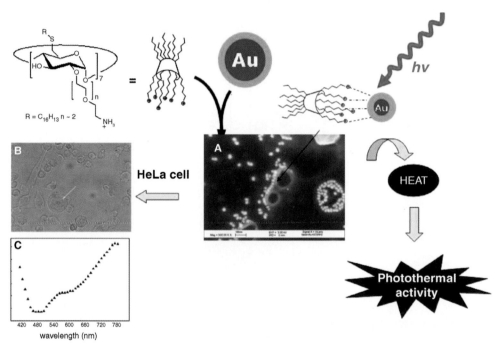

Chapter 18, Figure 14. Supramolecular hybrid nanoassemblies of Au/SC$_6$NH$_2$ delivery for Pt: (A) typical FEG–SEM image of Au/SC$_6$NH$_2$ system; (B) Optical microscope images of immobilized HeLa cells treated with Au/SC$_6$NH$_2$; (C) UV/Vis extinction spectra of (B) at the point marked.

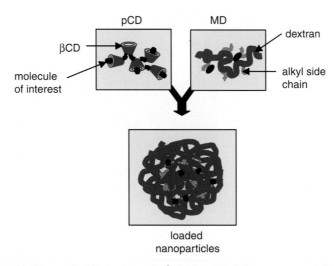

Chapter 20, Figure 2. Mechanism of pβ-CD/modified dextran nanogel formation.

18

PHOTODYNAMIC TUMOR THERAPY WITH CYCLODEXTRIN NANOASSEMBLIES

ANTONINO MAZZAGLIA

CNR-Istituto per lo Studio dei Materiali Nanostrutturati, Università di Messina, Messina, Italy

1. OVERVIEW ON THE PHOTODYNAMIC THERAPY OF TUMORS: STATE OF THE ART AND PERSPECTIVES

Photodynamic therapy (PDT) is an innovative modality for the treatment of localized tumors as an alternative or adjuvant to classical therapies such as radiotherapy, surgery, and chemotherapy; at the same time, several nononcological applications are also emerging for the treatment of a variety of dermatological, ophthalmic, cardiovascular, and infective diseases. PDT aims at selectively killing neoplastic lesions by the combined action of a photosensitizer (PS) and visible light: in most cases, wavelengths in the red or near-infrared spectral range are used, since they are characterized by a greater penetration depth into most human tissues and are not absorbed by normal tissue constituents, thereby avoiding the promotion of generalized photosensitivity. The selectivity of the PDT modality is achieved partly by the selective or at least preferential accumulation of the PS in malignant cells and tissue, and partly by restricting application of the incident light to the desired area. The latter measure is incapable of sparing peritumoral tissues completely, especially in the case of infiltrating tumors; thus, the degree of selectivity of PS accumulation is of utmost importance.

This therapeutic modality is based on the property of some porphyrin derivatives or analogs thereof to be preferentially accumulated and retained for prolonged periods of time by many types of solid tumors. Thus, irradiation of the neoplastic lesions with red light wavelengths, which are specifically absorbed by the photosensitising agent and are endowed with a relatively high penetration power (up to 2 cm) into most human tissues, generates cytotoxic species [i.e., reactive oxygen species (ROS)], mainly singlet oxygen, by energy or electron transfer from the electronically excited photosensitiser, inducing irreversible tumor damage [1–3]. At present, regulatory approval for PDT has been obtained for selected tumors using Photofrin, a complex mixture of hematoporphyrin derivatives, or Foscan, a chlorin derivative, as photosensitizers, while 5-aminolevulinic acid (ALA) and its methyl ester (which are converted metabolically to the photosensitizer protoporphyrin IX) are being widely used for the PDT of nonmelanoma skin cancer and other dermatological diseases. Several thousand patients worldwide have been treated by PDT for palliative or curative purposes with objectively positive results [4].

PS photodynamic action consists of a photooxidative process that induces multiple consecutive biochemical and morphological reactions [5]. In detail, PS from its ground state (PS^0) is brought to a short-lived excited state ($^1PS^*$) and from the latter may convert to a long-lived triplet state ($^3PS^*$). This state is the photoactive state, which may produce cytotoxic species via a type I or type II mechanism (Fig. 1). In the type I mechanism, $^3PS^*$ may interact with substrate molecules via hydrogen abstraction or electron transfer to give radical and ionic species which, in turn, can react with molecular oxygen to furnish cytotoxic ROS. Alternatively, $^3PS^*$ may react via energy transfer with ground-state molecular oxygen to yield singlet oxygen (type II mechanism) that can interact with a large number of biological substrates to begin oxidative damage via

Cyclodextrins in Pharmaceutics, Cosmetics, and Biomedicine: Current and Future Industrial Applications, First Edition. Edited by Erem Bilensoy.
© 2011 John Wiley & Sons, Inc. Published 2011 by John Wiley & Sons, Inc.

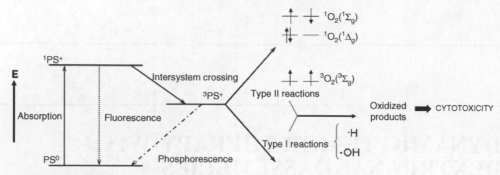

Figure 1. Mechanism of photodynamic action according to a modified Jablonski diagram (vibrational levels are omitted). Transitions between electronic states of oxygen in solution ($^3\Sigma_g \to {}^1\Delta_g$; $^3\Sigma_g \to {}^1\Sigma_g$) are reported.

peroxidative chain reactions. The triplet energy (E_T) is usually ≥ 94 J/mol for efficient energy transfer to ground-state dioxygen [2]. The type II mechanism is the most widely accepted, leading to cell damage [6,7]. The necessity of molecular oxygen in the environment of action can limit the potential of PDT in solid tumors, where poor vascularized tissues with insufficient levels of O_2 may exist [1]. There is, however, evidence that an oxygen-independent photodynamic effect is possible from the upper excited triplet state T_n of PS, such as rose bengal and a substituted magnesium phthalocyanine [8]. Also, there is an increasing interest in the application of concomitant biphotonic activation of photosensitizer molecules using two NIR photons [9].

Novel synthesized analogs of porphyrins displayed absorption bands peaking at longer wavelengths and possess larger molar extinction coefficients than do conventional porphyrins [1,10]. Such spectroscopic properties play a key role, since they guarantee a higher probability and efficiency of light absorption, hence a higher number of electronically excited states. For therapeutic applications, the triplet quantum yield, lifetime, and energy are the main parameters. There was interest in expanding the conjugation of the macrocycle ring in order to shift the absorption spectrum toward the far-red/near-infrared wavelength region (i.e., porphyricenes), or to prepare chlorines, which are characterized by the reduction via hydrogenation of one pyrrole ring; phthalocyanines and naphthalocyanines, where a benzene and a naphthalene ring, respectively, is condensed with each pyrrole moiety. Basic chemical structures of the most used photosensitisers and their precursors are displayed in Fig. 2.

Among several compounds of second-generation photosensitizers for PDT, phthalocyanines have been investigated widely, due to their powerful absorption ($\varepsilon \approx 10^5$ M^{-1}/cm) in the far-red region of the spectrum ($\lambda \approx 680$ nm), where tissue transparency is optimal [11,12]. For each class of compounds, it is possible utilize both the free base and selected metalloderivatives: several metal diamagnetic ions can be coordinated at the center of the macrocycle; therefore, it was possible to synthesize porphyrinoid derivatives with Mg(II), Zn(II), Al(III), Ge(IV), Si(IV), and Sn(IV) [13]. In a few cases, axial ligands are located in the fifth and sixth coordination positions of the central metal ion in order to modulate the physicochemical and (photo)biological features of the porphyrinoid molecules. Recently, two water-soluble sulfonated phthalocyanines containing paramagnetic central metal ions—copper phthalocanine tetrasulfonate (CuPcS$_4$) and nickel phthalocanine tetrasulfonate (NiPcS$_4$)—were studied. Such paramagnetic ions significantly shorten the lifetime of the triplet state ($\approx$ nanoseconds), making the complexes far less effective photosensitizers for PDT with respect to phthalocyanines chelated with diamagnetic metal ions (Al^{3+}, Ga^{3+}, Zn^{2+}), whose triplet lifetimes are much longer. For CuPcS$_4$ and NiPcS$_4$ it was possible to select satisfactorily the optimal parameter for two-photon PDT [14].

Also, scientists have focused on porphyrinoids substituted in the peripheral positions of the macrocycle by functional groups with well-defined chemical structure, in order to generate compounds differing in size, electric charge, hydrophilic or hydrophobic character, and bulkiness. One of the strategies was to introduce in the periphery of the macrocycle suitable hydrophobic or hydrophilic groups conferring amphiphilicity and improving selectivity in the tumor localization. Therefore, the PDT efficiency depends on the physicochemical features of the PS, its pharmaceutical formulations, the localization and quantity of PS in treated tissue, time activation with light, light intensity, and amount of oxygen.

2. PHYSICOCHEMICAL PROPERTIES AND MECHANISM OF PHOTOSENSITIZERS IN NANO- AND MICROENVIRONMENTS

The binding properties of complex systems formed by PS in colloidal nano- and microenvironments, such as liposomes, micelles, emulsions, and polymer particles, were actively investigated [15–17]. Over the last 20 years, liposomes

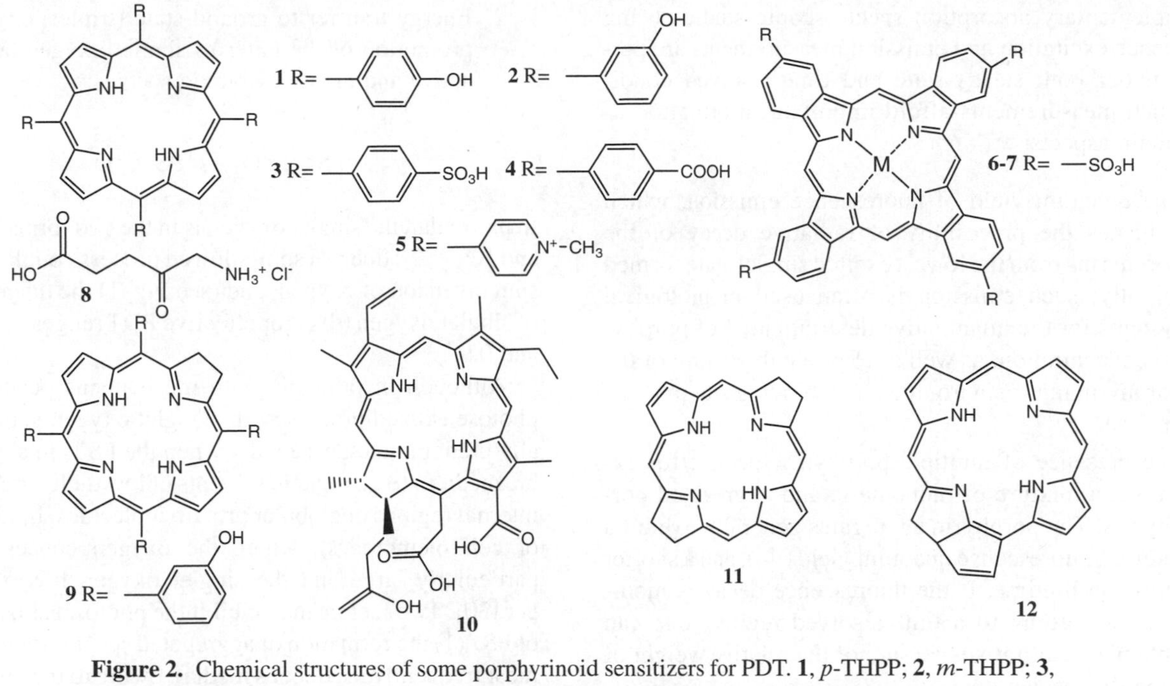

Figure 2. Chemical structures of some porphyrinoid sensitizers for PDT. **1**, *p*-THPP; **2**, *m*-THPP; **3**, TPPS; **4**, TPPC; **5**, TMPyP; **6**, M = Zn ZnPcS4; **7**, M = Al-X, AlPcS4; **8**, ALA; **9**, *m*-THPC (Foscan); **10**, Chlorin e6 (Fotolon); **11**, basic structures of Bacteriochlorin; **12**, Purpurin.

interacting with hydrophobic porphyrins and metalloporphyrins have been employed to mimic membrane proteins [18] and to investigate distribution of sensitizers in cell compartments [15,19] in order to improve the PDT efficacy substantially and to preserve the quantum yield of the generation of hypereactive species [20].

Studies on porphyrins incorporated in neutral and charged micelles have also been proposed [21] in terms of their equilibrium or kinetic behavior [22]. The mechanism of interaction with liposomes of photodynamic drugs such as hematoporphyrins, protoporphyrins, and glycoporphyrins [23,24], has been investigated widely and it has been established that photophysical and photochemical parameters are quite sensitive to the physicochemical features of the microenvironment. Moreover, chiral recognition of micelles versus a chiral functionalized porphyrin was observed [25]. Biopolymers and calixarenes constitute further microdomains where chromophores can be localized [26–28].

A combination of noncovalent and electrostatic interactions influences the partition of dye in a lipid bilayer; hydrophobic guests are incorporated in the lipid region, while hydrophilic sensitizers interact primarily with the aqueous interface at the hydrated internal core of liposomes. The entrapment of photosensitizers with different polarities in colloidal nanoassemblies is sketched in Fig. 3. As an example, liposomal porphyrins could provoke endocytoplasmatic damage, leading to a change in mitochondrial shape, while water-soluble hematoporphyrin photosensitized primarily the plasma membrane [29].

As a result of the localization of sensitizers inside vesicles, monomers and self-aggregates of chromophores can be formed. The presence of these supramolecular oligomers is due to a high local concentration of sensitizer. After being incorporated in the bilayer, the collision process of excited states and the rotational and diffusional freedom of the sensitizer molecule is consistently slowed down as a consequence of the increasing microviscosity. Finally, the incorporation of photosensitizer influences sensitively the photophysical properties of PS noncovalently bound to host molecules or host nanoassemblies. The general mutual interaction is outlined interestingly in a review of Lang et al. [15].

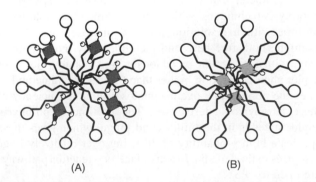

(A)   (B)

Figure 3. PS entrapment in colloidal nanoassemblies: (A) hydrophylic PS in the outer casing of micelles; (B) hydrophobic PS in the core of micelles. (*See insert for color representation of the figure.*)

Complementary absorption spectroscopic studies using fluorescence excitation and emission measurements are performed under both steady-state and time-resolved conditions. Such measurements afford important information as to two main aspects:

1. The quantum yield of fluorescence emission, which indicates the probability of radiative decay of the porphyrins from the lowest excited singlet state formed initially; such emission is often used in biological systems for the quantitative determination of porphyrin concentration, as well as to probe the nature of the porphyrin microenvironment and binding site in a cell or tissue.

2. The presence of multiple porphyrin species: for example, a mixture of monomeric and aggregated porphyrins. The porphyrin aggregates generally exhibit a reduced fluorescence quantum yield (Φ_f), and a shorter emission lifetime. If the fluorescence decay is monitored according to a time-resolved regime, one can obtain a quantitative measure of the relative weight of each PS species.

Determining the amount of species aggregated also plays a predictive role in the photosensitizing efficiency of porphyrins. Generally, the Φ_f value of PS is about 0.1. This value points out that a predominant fraction of the electronically excited PS molecules ($^1PS^*$) undergoes intersystem crossing to the lowest excited triplet state ($^3PS^*$); the latter plays a major role in photosensitized processes, owing to its particularly long lifetime (on the order of milliseconds) even in fluid media. The lowest excited triplet state represents the main reactive intermediate in the photophysical processes promoted by PS. The efficiency for the generation of such as electronically excited state can be measured quantitatively through a specific parameter, the triplet quantum yield (Φ_T), which is the ratio of the number of triplet species formed to the total number of photoexcited photosensitizer molecules. This parameter can be obtained by means of laser flash photolysis techniques, which can determine the triplet-state lifetime with a resolution level of microseconds. As an example, both free base and metalloporphyrinoids coordinated with diamagnetic metal ions exhibit quantum yields for triplet photogeneration higher than 0.5; such a value underlines the appreciable overall photosensitizing efficiency of these tetrapyrrolic derivatives. The lifetime of porphyrin triplet states is in the millisecond range, which gives these species a high probability of diffusing over relatively large distances in fluid media. In actual fact, two reaction pathways are open to $^3PS^*$:

1. Direct interaction with a nearby substrate molecule with the formation of radical derivatives, which in turn reacts with oxygen to yield oxidized products.

2. Energy transfer to ground-state (triplet) oxygen with promotion of the latter to singlet oxygen, a hyperreactive and highly cytotoxic species:

$$^3PS^* + {}^3O_2 \rightarrow {}^1PS + {}^1O_2 \qquad (1)$$

In particular, the singlet oxygen is in the two forms $^1O_2\,(^1\Delta_g)$ and $^1O_2\,(^1\Sigma_g)$ due to a spin-allowed process coupled with the spin inversion of oxygen, and (see Fig. 1) the quantum yield of singlet oxygen (Φ_Δ) for effective PDT ranges between 0.7 and 0.9.

Although in principle both mechanisms occur in most photosensitized processes, the singlet oxygen pathway usually predominates, especially when the PS is in a hydrophobic medium (e.g., organic solvents of low dielectric constant, internal regions of globular protein molecules, lipid domains of cell membranes), where the oxygen concentration is particularly large and the singlet oxygen lifetime is longer [30]. Two factors may inhibit the photosensitizing action of PS: (1) the formation of aggregated species (which occurs most frequently in aqueous media), owing to the related drop in the lifetime of the triplet state; and (2) the insertion of paramagnetic metal ions [e.g., Fe(III), Co(III) ions] at the center of the macrocycle. Such derivatives possess a quantum yield of triplet generation close to unity; however, the triplet lifetime falls below the nanosecond level, preventing any interaction between the triplet and other molecules.

Finally, it was ascertained that a photoexcited PS may also generate syperoxide anion via electron transfer to oxygen:

$$^3PS^* + {}^3O_2 \rightarrow {}^3(^2PS^+ \ldots {}^2O_2^-) \leftrightarrow {}^2PS^+ + {}^2O_2^- \quad (2)$$

However, the efficiency of this process is about 1% of that typical of singlet oxygen generation; hence, superoxide does not really play an important role in PS photosensitized processes.

Finally, two main factors have to be considered in a PS for potential use in PDT: (1) the quantum yields of the excited triplet states and its singlet oxygen production from ground-state $^3O_2$, and (2) the stability relative to oxidative degradation (photobleaching).

3. PASSIVE AND ACTIVE PDT WITH COLLOIDAL NANOASSEMBLIES

A full exploitation of PDT potential is limited by factors such as the often limited selectivity of tumor targeting, the prolonged persistence of available PDT agents in the skin, causing a generalized cutaneous photosensitivity, and the chemical heterogeneity (due to the different chemical species present) and inefficient red light absorption typical of PS as Photofrin and ALA-derived protoporphyrin [1,4].

Novel approaches are being investigated to overcome the present limitations of PDT, including the development of second-generation photosensitizers with improved photochemical and tumor-localizing properties, and the combination of PDT with other therapeutic modalities. The encapsulation of PS in vehicles tagged with a receptor targeting group aims to enhance the selectivity of drug by preserving its efficacy.

3.1. Disaggregation Process of Supramolecular PS Oligomers

The aggregation of PS is disadvantageous in preserving a high level of PDT efficiency. The aggregation reflects the tendency of the flat tetrapyrrolic macrocycle to yield face-to-face dimers or higher oligomers due to hydrophobic interactions and is more pronounced in aqueous media, while generally its importance decreases upon moving to organic solvents or micellar/liposomal dispersions [17,21,31,32]. In the case of phthalocyanines, the disaggregation into individual molecules in water or a reduction in the extent of the aggregates extent is promoted by different strategies, including (1) the use of hydrophilic groups as axial ligands coordinated to a central metal such as Zn(II) or Ru(II) [33,34], (2) the use of surfactants to create a microheterogeneous environment [35–38], or (3) interaction with dendrimers, substituents on the phthalocyanine macrocycle, which sterically inhibit molecular aggregation and increase solubility [39]. To circumvent aggregation in water in the absence of surfactants or other disaggregating agents, synthesis and characterization of highly water-soluble phthalocyanines with terminal carboxylate functionalities as monomeric species in water was proposed [40].

3.2. Targeting of Photosensitizers Mediated by Nanoassemblies

Micelles, liposomes, oil dispersions, biodegradable polymeric particles, and hydrophilic polymer–PS conjugates and nanoparticles generally are used to delivery PS (Fig. 3). Nanoparticles are considered as passive targeting systems because they exploit both the natural distribution processes (passive diffusion and phagocytosis) and selective accumulation in target tissue (i.e., tumors or neovasculature) by an enhanced permeability and retention effect (EPR). Various (in most cases, hydrophobic) photosensitizers were bound to water-soluble polymers [mainly to poly(lactic-*co*-glycolic acid) and polylactic acid], to overcome the problems associated with solubilization of the photosensitizers (e.g., bacteriochlorophyll-a [41], verteporfin [42], mesotetra(4-hydroxyphenyl)porphyrin [43], methylene blue [44], hypericine [45], and various phthalocyanines [46]). Passive targeting was illustrated by Konan et al. [47] using data on biodistribution, pharmacokinetics, and phototoxicity both

for PS encapsulated in liposomes [20,48,49], oil dispersion or micellar systems [50–52], polymer particles [53], and hydrophilic polymer–PS conjugates [32,54]. Silica nanoparticles have been used extensively as well, loaded with typical PDT sensitizers [55,56] or with less usual ones, such as fullerene [57]. Gold nanoparticles as carriers of PS [58] were also proposed, since a surface-bound photosensitizer with gold nanoparticles may represent an advantage, as singlet oxygen does not need to diffuse out of the polymeric structure as it does for encapsulated photosensitizers. Later it was shown that these conjugates are indeed effective in PDT [59].

Tumor targeting of the photosensitizers has been a long-sought-after goal in PDT research. Using as PDT agent a lipophylic photosensitizer [60], by its nature it passively targets malignant tissue, since the hydrophobic compound is bound in the plasma mainly to low-density lipoproteins (LDLs) [61]. Indeed both neovascular endothelial cells and actively proliferating tumor cells usually have a high expression of LDL receptors. Thus, hydrophobic photosensitizers are preferentially accumulated by malignant cells through the LDL-receptor-mediated endocytosis pathway.

The idea of active targeting of a photosensitizer toward malignant cells by attaching them to various molecule-recognizing proteins, peptides, and receptors overexpressed on the tumor cell membrane has a long history [62,63] but without reaching a real breakthrough in PDT efficacy. Selective delivery of PS to the site of action can occur toward covalent grafting of receptor targeting group directly on the PS [64,65] or by tagging nanoparticles with receptor-binding moieties and PS [66]. Therefore, active targeting may represent a significant improvement. Effective PDT requires the preferential accumulation of relatively large concentrations of the photosensitizer in the targeted cells. In conjugates composed of nanoparticles equipped with PDT sensitizer and targeting units, the latter should facilitate the delivery of several photosensitizer molecules by increasing efficiency and selectivity. These targeting molecules (units) can be small molecules such as short peptides [67] or the RGD sequence, antibodies, or fractions thereof [68,69].

4. PDT WITH PHOTOSENSITIZERS ENCAPSULATED IN CD NANOASSEMBLIES

Due to the high resistance to thermal and oxidative degradation, diminution of side effects and increase in the solubility and bioavailability of complexed drug, CDs are a promising nontoxic carrier for PS in PDT and photodynamic antimicrobial chemotherapy (PACT) [70]. Cyclodextrins (CDs) present a hydrophobic cavity which excludes water molecules and can include a variety of PS. The binding mode of PS on CDs depends strongly on the size of the cavity and on functional groups. As an example, anionic *meso-*

tetrapheyl-substituted porphyrins can bind CD in the following modes [71–73]:

1. Inclusion through the lower rim of CD. This mode is characteristic for β-CD and modified β-CD.
2. Inclusion through the primary face of CD. This typology of binding is typical for γ-CD, which possesses a secondary face wider than that of β-CD.
3. External binding between the exterior binding sites of CDs and PS monomers or aggregates.

Therefore, the mode of binding influences the photophysical features of the complexes by modulating the amount of singlet oxygen produced, of which the quantum yield is crucial for a highly effective PDT.

4.1. Supramolecular Complex of CDs and Photosensitizers

Systems of CDs and porphyrin (Por) have been considered extensively as models for hemoproteins (74,75), and some of these investigations have focused on the metalloporphyrin environments which were designed to mime the microdomains of myoglobin and hemoglobin, in which an iron-containing coordination site is involved (71,76). Hydrophilic Por/β-CD conjugates [77] and porphyrins bearing four

covalently bound permethylated β-CDs [78] were synthesized and characterized as soluble hosts having the benefit of multiple interactions with different substrates. Supramolecular interaction between CDs and green plant pigments [79] has also been reported. Most of these topics concerning systems of CD–PS complexes have been reported in the literature [15,80,81]. The strong affinity of anionic porphyrin 5,10,15,20-tetrakis(4-sulfonatophenyl)-21H,23H-porphine (TPPS) toward the heptakis (2,3,6-tri-O-methyl)-β-cyclodextrin (TM-β-CD) host to form stable inclusion complexes has been exploited [82]. It was found that in water, a TPPS (TM-β-CD) inclusion complex (see Fig. 4) behaves as a highly efficient and resistant supramolecular sensitizer with high photo-oxygenation turnover numbers and a low extent of bleaching. Also, supramolecular complexes of hydroxypropyl-β-cyclodextrin (HP-β-CD), TM-β-CD, and heptakis (2,6-di-O-methyl)-(β-cyclodextrin) (DM-β-CD) with a neutral porphyrin (TPyP) have been investigated as potential agents for PDT [83].

In the recent past, a poorly soluble second-generation photosensitizer (purpurin) was administrated to animals in the presence of γ-CD or in dipalmitoyl phosphatidylcholine liposomes [84]. Natural curcumin has been evaluated as a potential photosensitizer for oral applications. Phototoxicity was studied in an aqueous dispersion of nonionic micelles, cyclodextrin, liposomes, and so on. It was assessed that the

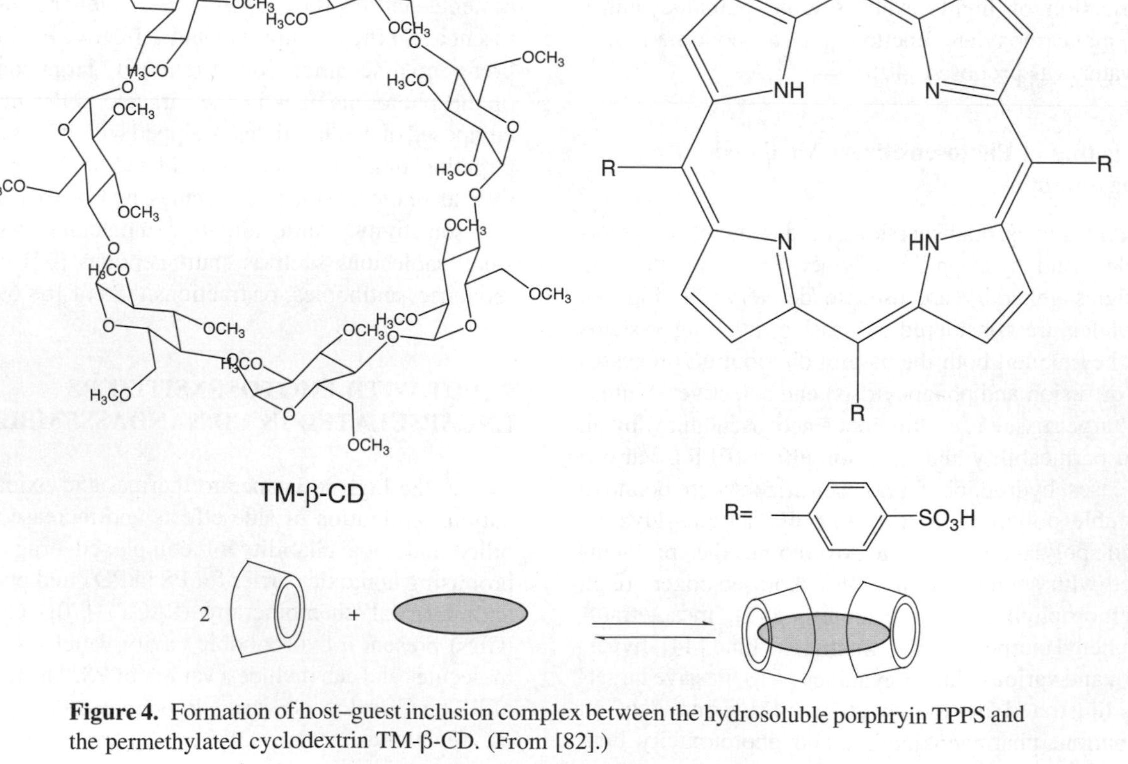

Figure 4. Formation of host–guest inclusion complex between the hydrosoluble porphryin TPPS and the permethylated cyclodextrin TM-β-CD. (From [82].)

phototoxic effect induced by curcumin is highly dependent on the type of preparation [85].

Recently, it was demonstrated that photoexcitation of fullerene derivatives efficiently produces an excited triplet state and, through energy and electron transfer to molecular oxygen, produces both singlet molecular oxygen and superoxide. It was found that γ-cyclodextrin bicapped C_{60} [$(\gamma\text{-CD})_2/C_{60}$] is a very efficient singlet oxygen producer and therefore can be used in PDT to kill tumor cells [86].

4.2. Supramolecular Aggregates of Amphiphilic CD and Porphyrins

One of the main challenges is to achieve targeted PDT, that is, to obtain selective delivery of a photodynamically active drug directly to action sites [64]. Our group is interested on the design of supramolecular colloidal systems formed by porphyrin and amphiphilic CDs in order to characterize nanomaterials as "smart" carriers for specific recognition and cellular internalization [87,88].

Several chemical modifications have been carried out on CDs by grafting substituent groups to different positions (narrow rim, wide rim, or both rims), thus providing nonionic or cationic derivatives able to form supramolecular nanoassemblies potentially useful in pharmaceutical applications [89]. However, the ability of these molecules in engineering nanocarriers to deliver and especially target a drug is largely unknown. In fact, supramolecular nanoassemblies, either obtained spontaneously or designed through a proper preparation method, are of a size compatible with intravenous injection and could be utilized to guarantee drug distribution in the body. Strategies involving nanocarriers are highly appropriate for the delivery of highly toxic anticancer drugs to solid tumors, using their features to extravasate at the level of the tumor porous capillary bed and deliver the drug at the site of action. The most significant studies in this sense have been produced from Bilensoy and co-workers, who demonstrated that 6-*N*-CAPRO-β-CD and β-CDC6 (modified on the primary and secondary face, respectively, with 6C aliphatic chains) could furnish nanoparticles or nanocapsules able to load anticancer drugs such as tamoxifen citrate and paclitaxel with good efficiency [90]. Furthermore, nonhemolytic and noncytotoxic nanoparticles based on amphiphilic cyclodextrins were demonstrated [91].

In this context, our group has recently investigated several neutral and cationic amphiphilic derivatives of β-CD (Fig. 5) which form supramolecular assemblies such as micelles and vesicles, and generally, nanoparticle dispersions showing different colloidal stability in water according to the CD component and preparation method [92–94]. This new generation of amphiphilic CDs provides more water-soluble and adaptable nanoparticles through modulation of the balance between hydrophobic and hydrophilic chains at both CD

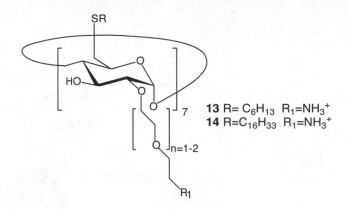

13 R= C_6H_{13} R_1=NH_3^+
14 R=$C_{16}H_{33}$ R_1=NH_3^+

Figure 5. Structures of amphiphilic cationic CDs (**13**, SC_6NH_2; **14**, $SC_{16}NH_2$) investigated as nanocarriers of PS for PDT. (From [87,100].)

sides. The grafting of small portions of poly(ethylene glycol) at the secondary rim of CD can increase drug bioavailability and potentially decrease the immunogenicity [95].

Furthermore, the use of long hydrophobic chains (C_{16} or C_{12}) at a CD's primary side may make it possible to preserve the affinity for a biological membrane. Amphiphilic CDs can be tailored conveniently by covalently appending receptor-targeting glycosyl groups [88,96] in order to build nanocarriers with increased drug selectivity toward specific cell lines. Supramolecular aggregates of amphiphilic cyclodextrins are thereby versatile systems toward the encapsulation of both hydrophobic and hydrophilic guests [97,98]. PS drugs embedded in cationic CD nanoassemblies were effective in inducing photodynamic damage in cancer cells. In particular, the entanglement process between water-soluble TPPS and the amphiphilic cationic CD, heptakis(2-ω-amino-*O*-oligo (ethylene oxide)-6-hexylthio-β-CD, (SC_6NH_2) and the occurrence of various species at different porphyrin–CD ratios were studied by a combination of ultraviolet/visible (UV/Vis) absorption, fluorescence anisotropy, time-resolved fluorescence, resonance light scattering, and circular dichroism. The encapsulating process of TPPS on the mean vesicle diameter was investigated over a wide concentration range by quasielastic light-scattering techniques (QELS) [99]. The experimental results indicate that the presence of PS in this colloidal system promotes some morphological rearrangements, due essentially to charge interactions, which are responsible for a sensitive change in vesicle dimensions. In the range of porphyrin–CD molar ratios between 1:10 and 1:50, the porphyrin is solubilized in monomeric form and photosensitizes the production of singlet oxygen. At the same molar ratio, this amphiphilic cyclodextrin is able to transport porphyrins into tumoral cells [87] (Fig. 6).

Elastic light scattering experiments provided the first structural characterization of these cationic nanoassemblies in both the absence and presence of TPPS porphyrin,

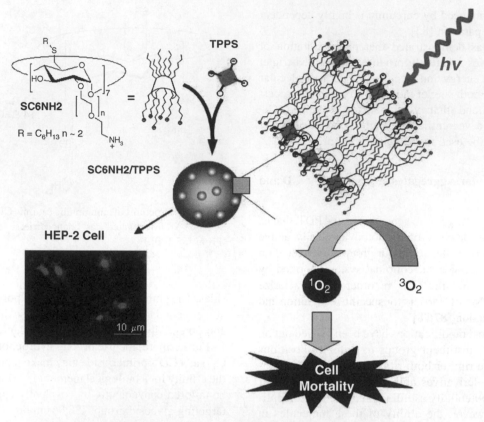

Figure 6. PS delivery by CD nanoassemblies, HEP-2 intracellular delivery and generation of singlet oxygen. (From [87,100].) (*See insert for color representation of the figure.*)

modeling the system as a spherical particle described by a single thin-shell form factor [99]. The structure of mixed heteroaggregates is modulated by the charge and size of the two components as a function of different porphyrin–cyclodextrin molar ratios. Scanning near-field optical microscopy (SNOM) of the samples evaporated on glass surfaces gave further insights on the morphology and optical properties of these systems, confirming the embedding of TPPS on the nanoassemblies and evidencing the role of the solvent (Fig. 7).

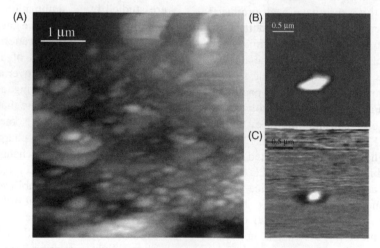

Figure 7. (A) SNOM topography on a typical sample of TPPS/SCNH$_2$ aggregates (1:10 porphyrin–/CD molar ratio; transmission mode, $\lambda_{exc} = 457$ nm); (B) Detail of topography; (C) relative fluorescence SNOM images. (From [99].)

4.3. Photophysics and Photochemistry of Porphryrins and CD Nanoassemblies

As a consequence of the previous results, in this section we report on photophysical and photochemical behavior of PS encapsulated in CD nanoassemblies. These properties can shed light on the correlation structure–activity relationships of the colloidal complex systems.

Lang et al. [28] show in detail some photophysical peculiarities of anionic and cationic porphyrins bound to CDs in aqueous solution as model sensitizers. It was evidenced that the binding constants (in the range 10^3 to $10^5 M^{-1}$ range) depend strongly on the cavity size and on the substituent groups grafted on the hydrophilic rims of CDs. Furthermore, the phototoxicity in vitro of these supramolecular sensitizers was described. The noncovalent interaction between CDs and anionic porphyrins affects the spectroscopic and photophysical properties of photosensitizers, leading to the red shift of the Soret band and an increase in triplet lifetime in both the absence and presence of oxygen. On the other hand, the quantum yield of singlet and triplet emission of the sensitizer and singlet oxygen formation does not undergo sensitive changes. These results suggest that stable complexes of CDs and porphyrins can prevent extensive aggregation (i.e., the formation of J-aggregates), which would decrease the photodynamic efficacy.

In TPPS–SC$_6$NH$_2$ systems the steady-state anisotropy can be related to the molecular motion by time-resolved fluorescence analysis. The physical–chemical parameters such as size, lifetimes, anisotropy, and rotation correlation time are strictly dependent on the TPPS–SC$_6$NH$_2$ molar ratio. The full photophysical behavior of this system has described by Mazzaglia et al. [81,99].

The quantum yield of singlet oxygen photogeneration by TPPS relies strictly on the CD concentration. Although Φ_Δ drops down significantly at low concentrations of nanoparticles (see Fig. 8A), it rises up again as the carrier amount increases (at a TPPS–CD molar ratio of 1:20) to a limiting value comparable to that of the free PS. The dependence of Φ_Δ on the molar ratio is in excellent agreement with the behavior exhibited by Φ_Δ (see Fig. 8A and 8B for a comparison). This point clearly suggests that the capability of the encapsulated TPPS to photogenerate $^1O_2(1\Delta_g)$ at the different ratios is modulated almost exclusively by the population of the triplet state rather than by the efficacy of the energy transfer to oxygen. The combination of the photophysical and photobiological results lead to useful insights into the photodynamic activity of the TPPS–SC$_6$CNH$_2$ nanoassemblies [100]. Despite the high TPPS–SC$_6$CDNH$_2$ molar ratio (i.e., 1:1, 1:2), the nanoparticles support PS aggregation with consequent loss of photodynamic activity and clearly avoid aggregation at the lowest ratios (i.e., 1:10 to 1:50), representing a valid alternative to other biovehicles in which self-aggregation of the PS in the encapsulated state represents a severe disadvantage [101,102]. On these bases we are inclined to think that the triplet signal observed under these conditions is generated primarily from residual free TPPS molecules still present in monomeric form and noninteracting with the nanoparticles (Fig. 9A). With a decreasing molar ratio (a higher CD concentration), porphyrins can interact statistically with more nanoparticles, and most of them are probably entangled in the amphiphilic phase (Fig. 9B).

These results are in agreement with the structural characterization shown by light scattering, microscopy, and photophysics. Therefore, at high carrier concentrations, a carrier–PS model system was proposed and the correlation supramolecular design and activity were investigated in detail. Some typical examples are reported below.

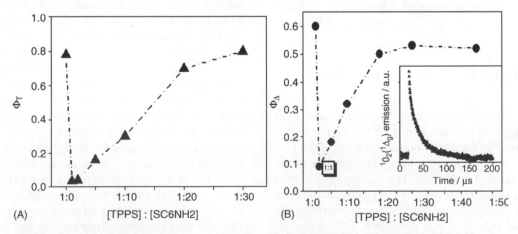

Figure 8. (A) Dependence of the triplet quantum yield of TPPS on the concentration of SC$_6$NH$_2$; (B) Dependence of the quantum yield for singlet oxygen production from TPPS (12 μM) on the SC$_6$NH$_2$ concentration in air-saturated solutions. The inset shows a representative decay kinetic of $^1O_2(^1\Delta_g)$ monitored at 1270 nm at the molar ratio 1:20. (From [100].)

(A)

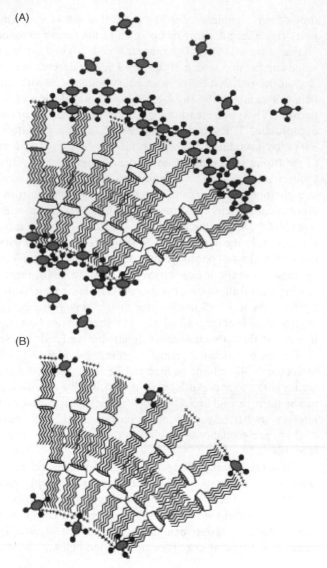

(B)

Figure 9. Model for the entangling process of TPPS into SC_6NH_2 at (A) low and (B) high CD concentration. (From [100].)

4.4. Nanostructured Molecular Film with CD and Porphyrins

Self-assembly of porphyrins onto the surface and their activation by light have been object of strong interest in recent years, not only for a better understanding of the life-related mechanism but also in designing optoelectronic and photonic molecular devices. The Langmuir–Blodgett (LB) technique is one of the smartest ways to control both packing and molecular orientation in two dimensions. We have shown that homogeneous multilayer films based on amphiphilic CDs entrapping hydrophilic porphyrins can be fabricated by exploiting interfacial electrostatic interactions between the two components. The presence of TPPS

absorption in the films (Fig. 10) clearly indicates a successful transfer procedure according to the formation of hybrid heptakis(2-ω-amino-oligo-(ethylene oxide)-6-hexadecyllthio-β-CD ($SC_{16}NH_2$)–TPPS multilayers [103]. A comparison of the TPPS absorption in the films with that shown in water solution allows insights to be gained into the possible organization of the porphyrin within the amphiphilic network. As illustrated in the figure, despite the presence of photochemically "silent" H-aggregates (see the contribution in the UV band at shorter wavelengths), these LS films reveal a good answer to light excitation. The characteristic shape of the amphiphilic $SC_{16}NH_2$ seems to play a role in this concern, entrapping a satisfactory amount of TPPS as a monomer and exerting protection against self-quenching and/or annihilation processes. It was stressed that the CD cavity is not involved in the binding with TPPS and therefore can be available for incorporation of additional guest molecules. This additional advantage contributes to making the present architectures intriguing platforms for the fabrication of more complex supramolecular assemblies on two-dimensional surfaces in which electron or energy transfer processes from suitable guests incorporated within the cavity to the entangled porphyrin, or vice versa, could be activated by photons of appropriate energy.

4.5. Other Applications

Light-induced energy and electron transfer between porphyrins and suitable donors and acceptors have been a topic of strong interest for many years. The literature abounds in descriptions of smart covalently linked systems. On the other hand, investigations in which the donor and acceptor are held together by noncovalent interactions either in solution or in a confined region remain quite limited [104]. The capability of SC_6NH_2 nanoparticles embedding TPPS to facilitate photo-induced energy and electron transfer with suitable donor and acceptor guest molecules was investigated using a combination of absorption, induced circular dichroism, and fluorescence spectroscopy. It is shown that anthracene (AN) and anthraquinone-2-sulfonate (AQS), chosen as an appropriate energy donor and electron acceptor, respectively, can be trapped within the nanoassembly network (Fig. 11). Fluorescence experiments performed in the presence and, for comparison, in the absence of nanoassemblies provide clear evidence that the amphiphilic nanoassemblies strongly promote singlet–singlet energy transfer from AN to TPPS as well as photo-induced electron transfer from TPPS to AQS. In view of the biocompatibility and drug-delivering properties of the CD nanoassemblies used, these results are of interest from the perspective of designing photo-activated carrier systems [105].

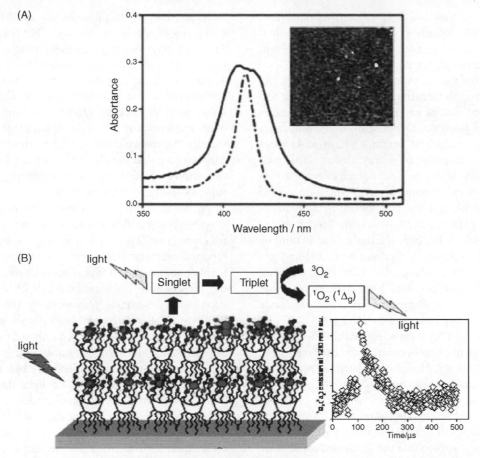

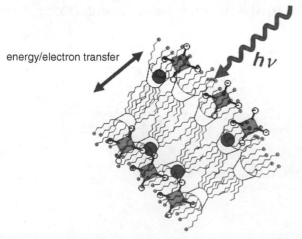

Figure 10. (A) Absoption spectra of a multilayer $SC_{16}NH_2$–TPPS Langmuir–Schäfer film (solid line) and TPPS in water solution (dashed-dotted line). The inset shows a typical image (width = 1 μm) of the $SC_{16}NH_2$–TPPS photoactive nanoassemblies transferred onto hydrophobized silicon. (B) Potential arrangement of the $SC_{16}NH_2$/TPPS multilayer. The inset shows a representative decay kinetic of $^1O_2(^1\Delta_g)$ monitored at 1270 nm. (From [103].) (*See insert for color representation of the figure.*)

Figure 11. Portion of spherical nanoaggregate of SC_6NH_2–TPPS–guest ternary system for energy and electron transfer. (From [105].)

5. CELLULAR INTERNALIZATION AND PHOTOTOXICITY

PDT aims at selectively targeting neoplastic lesions by the combined action of a photosensitizer and visible light, which generally produces ROS. A classical chemotherapeutic agent combined with light-activated therapy triggers a series of signaling pathways in the cells that activate not only the apoptotic machinery but also cell survival pathways [106]. Some of these pathways are also altered by genetic changes in specific types of tumors. As a consequence, cancer treatment, based on drugs of high efficacy, is facing several problems related to their unselective uptake by both healthy and tumor cells, as well as to the development of resistance due to modified cellular processes. Cross-resistance to a multitude of chemotherapeutic agents, called *multidrug resistance*, is believed to develop in a small subset of cancer cells and is the reason for tumor recurrence despite intensive chemotherapy.

The availability of new molecules, such as molecules interfering with survival pathways and the use of combinations of new nanotechnologies for the treatment of different tumors has a positive impact on tumor treatment. Nanoparticles designed for cancer treatment alter drug pharmacokinetics, improving the treatment's ability to target and kill cells of diseased tissues and organs while affecting as few healthy cells as possible. Depending on the properties of the carrier, large variations in drug pharmacokinetics with major clinical implications may occur. It is well known that sterically stabilized nanocarriers with a biomimetic coating (Stealth nanocarriers) show increased longevity in the circulation and a potential to accumulate predominantly in pathological sites with compromized leaky vasculature, which is a characteristic of solid tumors. Encouraging results have been obtained by recent findings pointing out that nanoparticle-loaded PDT agents are accumulated in significant amounts by a variety of tumor cells and are able to induce efficient cell photodamage once activated by suitable visible-light wavelengths [55,107]. In this scenario the use of CD–PS nanoassemblies for PDT is an exciting challenge in cancer research. On the other hand, exploitation of CD–PS complexes was restricted by various factors, including toxicity in the absence of irradiation, while in most cases of properly designed inclusion complexes, in which the formation of aggregated PS species is prevented, the photophysical features are not affected. As a consequence, it was observed that the PS drug does not lose its sensitizing properties (i.e., $\Phi\Delta$) after binding with CD. The shielding effect of CD can protect PS from aggregation and from interaction with other species in solution. In addition, complexes of CD dimers can avoid PS delivery by the lipoprotein pathway to healthy tissue and reduce unwanted targeting apart from tumors [108]. Selected supramolecular complexes of CD–PS were tested biologically to investigate their phototoxicity and their potential applications in PDT.

As an example, the phototoxicity of TPPS and ZnTPPS in a supramolecular complex with HP-β-CD was reported. These porphyrins may represent efficient PSs with high phototoxicity against G361 human melanoma cells [109]. The cytotoxic and phototoxic activity of complexes between chlorophyll a and different types of CDs toward human leukemia T-lymphocytes (Jurkat cells) were tested by means of experiments aimed at discriminating between the intrinsic toxicity and the toxicity induced by light [110]. The overall data indicate that the HP-β-CD is the CD that has the best characteristics to form, with chlorophyll a, a potential supramolecular system for PDT.

As noted in Section 4.1 the difference in generation of singlet oxygen and ocular toxicity between CD-complexed fullerene [$(\gamma$-CD$)_2$/C$_{60}$] and its aggregated derivatives was reported recently. It was determined that $(\gamma$-CyD$)_2$/C$_{60}$ is highly phototoxic to human lens epithelial cells HLE B-3 in the presence of UVA radiation. With aggregation, these compounds lose their phototoxicity [86]. In the case of TPPS–SC$_6$NH$_2$ nanoassemblies (at a molar ratio of 1:10 to 1:50) [87], the carrier entrapped only the monomeric form of the porphyrin, preserving almost entirely its capacity to generate a singlet oxygen effectively and, as a consequence, to kill tumor cells upon visible light illumination. It was demonstrated that the system can deliver TPPS in HEP-2 and HeLa cells (Fig. 12A). In the latter case, HeLa cells are damaged upon irradiation (Fig. 12B). In this respect, the molar ratio 1:10 seems to be the best compromise. In fact, under such conditions we find (1) a significantly high quantum yield of singlet oxygen photogeneration, (2) the highest efficiency of intracellular delivery, (3) the highest percentage of cells alive after treatment and before irradiation, and (4) a high percentage of cell death after irradiation. In view of the appropriate combination of these pecularities, the TPPS–SC$_6$NH$_2$ system represents a potential candidate for the useful application of photodynamic therapy, such as in vivo experiments on tumor-model animals [100].

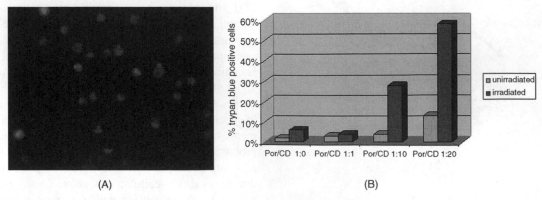

(A)   (B)

Figure 12. (A) Fluorescence microscopy analysis of HeLa cells treated with TPPS–SC$_6$NH$_2$ nanoassemblies (TPPS 1 μM) at 1:10 molar ratio (merging of cell visualized with a rhodamine filter and DAPI filter). (B) Cell death percenteage in HeLa through trypan blue assay. (For details see [100].) (*See insert for color representation of the figure.*)

5.1. Antimicrobial PDT with CD Nanoassemblies

The present increasing global occurrence of infectious diseases caused by multiantibiotic-resistant microbial pathogens represents a major challenge. Recent results suggest that PACT could represent a useful tool addressing this problem correctly, at least in the case of restricted infections [111]. Current confirmation [112] points out that the photosensitizing agents showing the maximal antimicrobial efficiency are represented by cationic dyes, including cationic phenothiazines, porphyrins, phthalocyanines, and fullerenes, which can promptly interact electrostatically with the array of negative charges that are present at the walls of bacterial cells [113–115].

It has been reported that the monocationic meso-substituted cationic porphyrin 5-[4-(dodecanoylpyridinium)] 10,15,20-tryphenylporphine (TDPyP) complexed into supramolecular aggregates of SC_6NH_2 appeared able (with favorable properties) to operate as a photosensitizing agent, including a very high quantum yield ($\Phi_\Delta = 0.90$) for the generation of $^1O_2$ (Fig. 13). Although the yield of $^1O_2$ generation was similar to that obtained after TDPyP

incorporation into cationic unilamellar liposomes of N-[1-(2,3-dioleoyloxy)propyl]-N,N,N-trimethylammonium chloride (DOTAP), SC_6NH_2-bound TDPyP was more active than DOTAP-bound TDPyP in photosensitizing the inactivation of the gram-positive methicillin-resistant bacterium *Staphylococcus aureus*. Conversely with respect to DOTAP-bound TDPyP, photoactivated SC_6NH_2-bound TDPyP was efficient also in photokilling Gram-negative bacterial pathogens such as *Escherichia coli* [70]. These remarks are in agreement with the well-known photobactericidal outcome of positively charged porphyrin derivatives, which can be improved markedly after incorporation into carriers with multiple positive charges. In addition, transmission electron microscopy studies revealed that potentiation of the TDPyP-mediated photobactericidal effect by incorporation into SC_6NH_2 is an effect of the carrier's ability to promote an efficient crossing of the very compactly organized three-dimensional architecture of the bacterial outer wall by the embedded porphyrin so that prompt interaction can take place between the short-lived photogenerated $^1O_2$ and close targets, whose integrity is critical for cell survival.

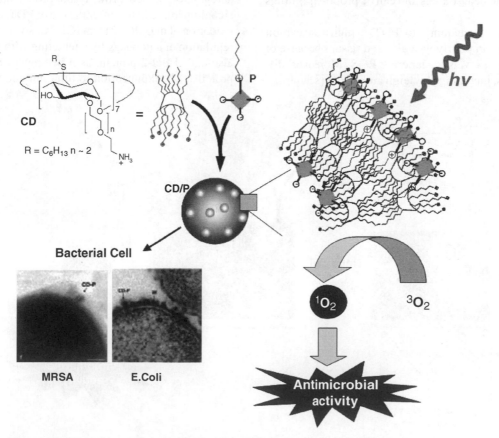

Figure 13. Cationic porphyrin bearing a long hydrophobic chain (TDPyP) delivery entrapped in cationic amphiphilic CD (SC_6NH_2). TDPyP–SC_6NH_2 can be delivered to both gram-positive and gram-negative bacterial cells (MRSA and *E. Coli*, respectively) by photosensitising their inactivation. (From [70].) (*See insert for color representation of the figure.*)

5.2. Therapies Combined using Multifunctional Nanoparticles

The increasing interest toward hybrid organic and inorganic multifunctional nanosystems is the foundation of consistent multidisciplinary research. In the field of PDT, small gold colloidal particles having an average diameter of 2 to 4 nm can successfully deliver photosensitizer drugs to cancer cells acting as carrier systems [59]. Recently, NO has proven to be an efficient anticancer agent that inhibits key metabolic pathways to block growth or kill cells outright [116]. Using light as an external trigger provides extreme rapidity of delivery and precise control over the NO amount released. Promising multifunctional nanoassemblies of amphiphiles and porphyrins for simultaneous delivery of nitric oxide and singlet oxygen have also been reported recently [117].

PDT combined with photothermal therapy (PT), another noninvasive treatment making use of the heat released upon light excitation (bimodal treatments), is known be promising for some cancer types. It has been shown that tumor cells are more sensitive to heat damage than are healthy cells. In the field of therapeutic applications, gold colloids of various shapes (spheres, rods, and shells) can enhance the sensitivity of cancer tissues to heat sources and can be proposed as drugs in PT [118].

PT can be complementary to PDT, which can exhibit decreased efficacy in poorly vascularized tissues because of insufficient oxygen levels to generate ROS. Moreover, different techniques, including Rayleigh and Raman scattering,

absorption, diffractometry, and microscopy (SNOM, TEM, etc.), have been exploited to reveal gold in biological matrices and to understand cellular component alteration following neoplastic diseases.

SC_6NH_2 can bind gold colloids, yielding Au–CD hybrid nanoassemblies with an average hydrodynamic radius (R_H) of 2 and 25 nm in water solution [119]. Under physiological conditions, the gold–aminoamphiphile system can internalize in HeLa cells and induce photothermal damage upon irradiation, doubling the cell mortality with respect to uncovered gold colloids. These findings, sketched in Fig. 14, can open useful perspectives to the use of these self-assembled systems in cancer photothermal therapy. Consequently, gold nanoparticles capped with amphiphilic CDs could also encapsulate drugs and therefore could be the starting point for developing effective systems for combined therapies (i.e., PT and PDT) [120].

5.3. PDT and Multimodal Cancer Therapy: Toward an Understanding of Cellular Mechanisms

Recently, data have shown in vitro a reduction in total mTOR, a regulator of cell growth and proliferation, and mTOR signaling immediately after the use of PDT, affording a direct oxidation of mTOR [121]. mTOR is also involved in negative regulation of autophagy by interacting with Beclin 1 protein, also called Bcl-2 protein, as it belongs to the Bcl-2 family proteins. The competition to limit amounts of Beclin 1 has

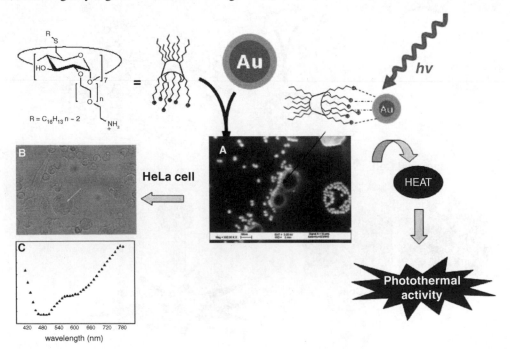

Figure 14. Supramolecular hybrid nanoassemblies of Au/SC₆NH₂ delivery for Pt: (A) typical FEG–SEM image of Au/SC₆NH₂ system; (B) Optical microscope images of immobilized HeLa cells treated with Au/SC₆NH₂; (C) UV/Vis extinction spectra of (B) at the point marked. (From [119].) (*See insert for color representation of the figure.*)

been shown to prevent overactivation of autophagy. In addition to its role in the degradation of proteins and organelles, autophagy plays a critical role in cellular survival by providing energy during periods of starvation. Interestingly, regulation of autophagy involves multiple signaling pathways which have also been implicated in tumorigenesis, suggesting the potential link between authophagy and cancer. The efforts of many group of researchers are headed in the direction of the use of novel multifunctional carrier systems for cancer therapy based on mixed organic–inorganic hybrid CD nanoparticles, simultaneously incorporating conventional and photoactive anticancer drugs as well as stimuli-responsive metal particles.

The inactivation of mTOR (autophagy) generated by photoactive drugs, the association with silencing therapy using silencing-RNA (si-RNA), and the use of conventional anticancer drugs which cause a mitotic catastrophe [94]—all combined in a CD nanocarrier—could represent a novel multifunctional tool to circumvent inefficient apoptosis and to optimize treatment of many malignant diseases. These research themes (PDT and, generally, photodynamic medicine), which represent a consistent contribution among the challenging strategies in the field of nanomedicine, are fascinating the scientific community as a result of their multidisciplinary approach and open perspectives for improvements in health.

Acknowledgments

I express my gratitude to N. Micali, L. Monsù Scolaro, M. T. Sciortino, and S. Sortino for their fruitful collaboration and encouragement to study multifunctional cyclodextrin systems according to a multidisciplinary approach. I dedicate this chapter to the memory of Professor R. Romeo (Department of Inorganic, Physical and Analytical Chemistry, University of Messina).

REFERENCES

1. Dougherty, T.J., Gomer, C.J., Henderson, B.W., Jori, G., Kessel, D., Korbelik, M., Moan, J., Peng, Q. (1998). Photodynamic therapy. *J. Natl. Cancer Inst.*, *90*, 889–905.

2. Bonnett, R. (1995). Photosensitizers of the porphyrin and phthalocyanine series for photodynamic therapy. *Chem. Soc. Rev.*, *24*, 19–33.

3. Ali, H., Van Lier, J. (1999). Metal complexes as photo- and radiosensitizers. *Chem. Rev.*, *99*, 2379–2450.

4. Huang, Z. (2005). A review of progress in clinical photodynamic therapy. *Technol. Cancer Res. Treat.*, *4*, 283–293.

5. Van Lier, J.E. (1991). Photosensitization: reaction pathways. In: Valenzeno, D.P. Ed., *Photobiological Techniques*. Plenum Press, New York, pp. 85–97.

6. Weishaupt, K.R., Gomer, C.J., Dougherty, T.J. (1976). Identification of singlet oxygen as the cytotoxic agent in photoinactivation of a murine tumor. *Cancer Res.*, *36*, 2326–2329.

7. Sharman, W.M., Allen, C.M., van Lier, J.E. (2000). Role of activated oxygen species in photodynamic therapy. *Methods Enzymol.*, *319*, 376–400.

8. Stiel, H., Teuchner, K., Paul, A., Freyer, W., Leupold, D. (1994). Two photon excitation of alkaly-substituted magnesium phthalocyanine: radical formation via higher excited states. *J. Photochem. Photobiol. A*, *80*, 289–298.

9. Goyan, R.L., Cramb, D.T. (2000). Near-infrared two-photon excitation of photoporphyrin: IX. Photodynamics and photoproduct generation. *Photochem. Photobiol.*, *72*, 821–827.

10. Kostron, H., Jori, G., Eds. (1996). Perspectives in Photodynamic Therapy. *J. Photochem. Photobiol., B*, *36*(2), 157–168.

11. Ali, H., van Lier, J.E. (1999). Metal complexes as photo- and radiosensitizers. *Chem. Rev.*, *99*, 2379–2450.

12. Allen, C.M., Sharman, W.M., van Lier, J.E. (2001). Current status of phthalocyanines in the photodynamic therapy of cancer. *J. Porphyrins Phthalocyanines*, *5*, 161–169.

13. Huang, Q., Pan, Z., Wang, P., Chen, Z., Zhang, X., Xub, H. (2006). Zinc(II) and copper(II) complexes of β-substituted hydroxylporphyrins as tumor photosensitizers. *Bioorg. Med. Chem. Lett.*, *16*, 3030–3033.

14. Fournier, M., Pépin, C., Houde, D., Ouellet R., van Lier, J.E. (2004). Ultrafast studies of the excited-state dynamics of copper and nickel phthalocyanine tetrasulfonates: potential sensitizers for the two-photon photodynamic therapy of tumors. *Photochem. Photobiol. 3*, 120–126.

15. Lang, K., Mosinger, J., Wagnerova, D.M. (2004). Photophysical properties of porphyrinoid sensitizers non-covalently bound to host molecules; models for photodynamic therapy. *Coord. Chem. Rev.*, *248*, 321–350, and references therein.

16. Li, G., Fudickar, W., Skupin, M., Klyszcz, A., Draeger, C., Lauer, M., Fuhrhop, J.-H. (2002). Rigid lipid membranes and nanometer clefts: motifs for the creation of molecular landscapes. *Angew. Chem. Int. Ed. 41*, 1828–1852.

17. Castriciano, M.A., Romeo, A., Villari, V., Angelini, A., Micali, N., Monsù Scolaro, L. (2005). Aggregation behavior of tetrakis(4-sulfonatophenyl)porphyrin in AOT/water/decane microemulsions. *J. Phys. Chem. B*, *109*, 12086–12092.

18. Feiters, M.C., Rowan, A.E., Nolte, R.J.M. (2000). From simple to supramolecular cytochrome P450 mimics. *Chem. Soc. Rev.*, *29*, 375–384.

19. Ricchelli, F., Jori, G., Gobbo, S., Tronchin, M. (1991). Liposomes as models to study the distribution of porphyrins in cell membranes. *Biochim. Biophys. Acta Biomembr.*, *1065*, 42–48.

20. Derycke, A.S.L., de Witte, P.A.M. (2004). Liposomes for photodynamic therapy. *Adv. Drug Deliv. Rev.*, *56*, 17–30.

21. Maiti, C., Mazumdar, S., Periasamy, N. (1998). J- and H-aggregates of porphyrin-surfactant complexes: time-resolved fluorescence and other spectroscopic studies. *J. Phys. Chem. B*, *102*, 1528–1538.

22. Simplicio, J., Schwenzer, K., Maenpa, F. (1975). Kinetics of cyanate and imidazole binding to hemin in micelles. *J. Am. Chem. Soc.*, *97*, 7319–7326.

23. Ricchelli, F., Gobbo, S., Jori, G., Salet, C., Moreno, G. (1995). Temperature-induced changes in fluorescence properties as a probe of porphyrin microenvironment in lipid membranes: 2. The partition of hematoporphyrin and protoporphyrin in mitochondria. *Eur. J. Biochem.*, *233*, 165–170.

24. Voszka, I., Galantái, R., Maillard, P., Csík, G. (1999). Interaction of glycosylated tetraphenyl porphyrins with model lipid membranes of different compositions. *J. Photochem. Photobiol. B*, *52*, 92–98.

25. Monti, D., Cantonetti, V., Venanzi, M., Ceccacci, F., Bombelli C., Mancini, G. (2004). Interaction of a chirally functionalised porphyrin derivatives with chiral micellar aggregates. *Chem. Commun.*, 972–973.

26. Pasternack, R.F., Fleming, C., Herring, S., Collings, P.J., de Paula, J., De Castro, G., Gibbs, E.J., (2000). Aggregation kinetics of extended porphyrin and cyanine dye assemblies. *Biophys. J.*, *79*, 550–560.

27. Costanzo, L.D., Geremia, S., Randaccio, L., Purrello, R., Lauceri, R., Sciotto, D., Gulino, F.G., Pavone, V. (2001). Calixarene–porphyrin supramolecular complexes: pH-tuning of the complex stoichiometry. *Angew. Chem. Int. Ed. 40*, 4245–4247.

28. Lang, K., Kubát, P., Lhoták, P., Mosinger, J., Wagnerová, D.M. (2001). Photophysical properties and photoinduced electron transfer within host–guest complexes of 5,10,15,20-tetrakis (4-*N*-methylpyridyl)porphyrin with water-soluble calixarenes and cyclodextrins. *Photochem. Photobiol.*, *74*, 558–565.

29. Milanesi, C., Sorgato, F., Jori, G. (1989). Photokinetic and ultrastructural studies on porphyrin photosensitization of HeLa cells. *Int. J. Rad. Biol.*, *55*, 59–69.

30. Jiménez-Banzo, A., Ragàs, X., Kapusta, P., Nonell, S. (2008). Time-resolved methods in biophysics: 7. Photon counting vs. analog time-resolved singlet oxygen phosphorescence detection. *Photochem. Photobiol. Sci.*, *7*, 1003–1010.

31. Barber, D.C., Freitag-Beeston, R.A., Whitten, D.J., (1991). Atroisomer-specific formation of premicellar porphyrin J-aggregates in surfactant solutions. *J. Phys. Chem.*, *95*, 4074–4086.

32. Castriciano, M.A., Romeo, A., Angelini, N., Micali, N., Longo, A., Mazzaglia, A., Monsù Scolaro, L. (2006). Structural features of meso-tetrakis(4-carboxyphenyl)porphyrin interacting with amino-terminated poly(propylene oxide). *Macromolecules*, *39*, 5489–5496.

33. Kimura, M., Nakada, K., Yamaguchi, Y., Hanabusa, K., Shirai, H., Kobayashi, N., (1997). Dendritic metallophthalocyanines: synthesis and characterization of a zinc(II) phthalocyanine[8] 3-arborol. *Chem. Commun.* 1215–1216.

34. Bossard, G.E., Abrams, M.J., Darkes, M.C., Vollano, J.F., Brooks, R.C. (1995). Convenient synthesis of water soluble, isomerically pure ruthenium phthalocyanine complexes. *Inorg. Chem.*, *34*, 1524–1527.

35. Li, X.-Y., Ng, A.C.H., Ng, D.K.P. (2000). Influence of surfactants on the aggregation behavior of water-soluble dendritic phthalocyanines. *Macromolecules, 22*, 2119–2123.

36. Gu, Z.W., Omelyanenko, V., Kopecková, P., Kopecek, J., Konák, C. (1995). Association of a substituted zinc(II) phthalocyanine-*N*-(2-hydroxypropyl)methacrylamide copolymer conjugate. *Macromolecules*, *28*, 8375–8380.

37. Dhami, S., Cosa, J.J., Bishop, S.M., Phillips, D. (1996). Photophysical characterization of sulfonated aluminum phthalocyanines in a cationic reversed micellar system. *Langmuir*, *12*, 293–300.

38. Howe, L., Zhang, J.Z. (1998). The effect of biological substrates on the ultrafast excited-state dynamics of zinc phthalocyanine tetrasulfonate in solution. *Photochem. Photobiol.*, *67*, 90–96.

39. Ng, A.C.H., Li, X.-Y., Ng, D.K.P. (1999). Synthesis and photophysical properties of nonaggregated phthalocyanines bearing dendritic substituents. *Macromolecules*, *32*, 5292–5298.

40. Liu, W., Jensen, T.J., Fronczek, F.R., Hammer, R.P., Smith, K.M., Vicente, M.G.H. (2005). Synthesis and cellular studies of nano-aggregated water-soluble phthalocyanines. *J. Med. Chem.*, *48*, 1033–1041.

41. Gomes, A.J., Lunardi, L.O., Marchetti, J.M., Lunardi, C.N., Tedesco, A.C. (2005). Photobiological and ultrastructural studies of nanoparticles of poly(lactic-*co*-glycolic acid)-containing bacteriochlorophyll-*a* as a photosensitizer useful for PDT treatment. *Drug Deliv.*, *12*, 159–164.

42. Konan-Kouakou, Y.N., Boch, R., Gurny, R., Allemann, E. (2005). In vitro and in vivo activities of verteporfin loaded nanoparticles. *J. Control. Release*, *103*, 83–91.

43. Konan, Y.N., Berton, M., Gurny, R., Allemann, E. (2003). Enhanced photodynamic activity of meso-tetra(4-hydroxyphenyl)porphyrin by incorporation into sub-200 nm nanoparticles. *Eur. J. Pharm. Sci.*, *18*, 241–249.

44. Tang, W., Xu, H., Kopelman, R., Philbert, M.A. (2005). Photodynamic characterization and in vitro application of methylene blue-containing nanoparticle platforms. *Photochem. Photobiol. 81*, 242–249.

45. Zeisser-Labouebe, M., Lange, N., Gurny, R., Delie, F. (2006). Hypericin-loaded nanoparticles for the photodynamic treatment of ovarian cancer. *Int. J. Pharm.*, *326*, 174–181.

46. Ricci-Júnior, E., Marchetti, J.M. (2006). Zinc(II) phthalocyanine loaded PLGA nanoparticles for photodynamic therapy use. *Int. J. Pharm.*, *310*, 187–195.

47. Konan, Y.N., Gurny, R., Allémann, E. (2002). State of the art in the delivery of photosensitizers for photodynamic therapy. *J. Photochem. Photobiol. B*, *66*, 89–106.

48. Damoiseau, X., Schuitmaker, H.J., Lagerberg, J.W.M., Hoebeke, M. (2001). Increase of photosensitizing efficiency of the bacteriochlorin *a* by liposomes-incorporation. *J. Photochem. Photobiol. B.*, *60*, 50–60.

49. Jori, G., Reddi, E., Cozzani, I., Tomio, L. (1986). Controlled targeting of different subcellular sites by porphyrin in tumour-bearing mice. *Brit. J. Cancer*, *53*, 615–621.

50. Wörle, D., Muller, S., Shopova, M., Mantareva, V., Spassova, G., Vietri, F., Ricchelli, F., Jori, G. (1999). Effect of delivery system on the pharmacokinetic and phototherapeutic properties of bis(methyloxyethyleneoxy)silicon–phthalocyanine in tumor-bearing mice. *J. Photochem. Photobiol. B, 50*, 124–128.

51. Kessel, D., Morgan, A., Garbo, G.M. (1991). Sites and efficacy of photodamage by tin etiopurpurin in vitro using different delivery systems. *Photochem. Photobiol.*, *54*, 193–196.

52. Bourdon, O., Mosqueira, V., Legrand Ph., Blais, J. (2000). A comparative study of the cellular uptake, localization and phototoxicity of *meta*-tetra(hydroxyphenyl)chlorin encapsulated in surface-modified submicronic oil/water carriers in HT29 tumor cells. *J. Photochem. and Photobiol. B*, *55*, 164–171.

53. Brasseur, N., Brault, D., Couvreur, P. (1991). Adsorption of hematoporphyrin onto polyalkylcyanoacrylate nanoparticles: carrier capacity and drug release. *Int. J. Pharm.*, *70*, 129–135.

54. Brasseur, N., Ouellet, R., La Madeleine, C., Van Lier, J. E. (1999). Water soluble aluminium phthalocyanine–polymer conjugates for PDT. *Brit. J. Cancer*, *80*, 1533–1541.

55. Roy, I., Ohulchanskyy, T.Y., Pudavar, H.E., Begey, E.J., Oseroff, A.R., Morgan, J., Dougherty, T.J., Prasad, P.N. (2003). Ceramic-based nanoparticles entrapping water-insoluble photosensitising anticancer drugs: novel drug-carrier system for photodynamic therapy. *J. Am. Chem. Soc.*, *125*, 7860–7865.

56. Yan, F., Kopelman, R. (2003). The embedding of *meta*-tetra (hydroxyphenyl)-chlorin into silica nanoparticle platforms for photodynamic therapy and their singlet oxygen production and pH-dependent optical properties. *Photochem. Photobiol.*, *78*, 587–591.

57. Davydenko, M.O., Radchenko, E.O., Yaschchuk, V.M., Dmitruk, I.M., Prylutskyy, Y.I., Matishevska, O.P., Golub, A.A. (2006). Sensibilization of fullerene C-60 immobilized at silica nanoparticles for cancer photodynamic therapy. *J. Mol. Liq.*, *127*, 145–147.

58. Hone, D.C., Walker, P.I., Evans-Gowing, R., FitzGerald, S., Beeby, A., Chambrier, I., Cook, M.J., Russell, D.A. (2002). Generation of cytotoxic singlet oxygen via phthalocyanine-stabilized gold nanoparticles: a potential delivery vehicle for photodynamic therapy. *Langmuir*, *18*, 2985–2987.

59. Wieder, M.E., Hone, D.C., Cook, M.J., Handsley, M.M., Gavrilovic, J., Russell, D.A. (2006). Intracellular hotodynamic therapy with photosensitizer-nanoparticle conjugates: cancer therapy using a "Trojan horse." *Photochem. Photobiol. Sci.*, *5*, 727–734.

60. Hahn, F., Schmitz, K., Balaban, T. S., Stefan Bräse, S., Schepers, U. (2008). Conjugation of spermine facilitates cellular uptake and enhances antitumor and antibiotic properties of highly lipophilic porphyrins. *ChemMedChem*, *3*, 1185–1188.

61. Jori, G., Reddi, E. (1993). The role of lipoproteins in the delivery of tumour-targeting photosensitisers. *Int. J. Biochem.*, *25*, 1369–1375.

62. Chen, B., Pogue, B.W., Hoopes, P.J., Hasan, T. (2006). Vascular and cellular targeting for photodynamic therapy. *Crit. Rev. Eukaryot. Gene Expr.*, *16*, 279–305.

63. Taquet, J.P., Frochot, C., Manneville, V., Barberi-Heyob, M. (2007). Phthalocyanines covalently bound to biomolecules: targeting photosensitizers for photodynamic therapy. *Curr. Med. Chem.*, *14*, 103–112.

64. Amessou, M., Carrez, D., Patin, D., Sarr, M., Grierson, D., S., Croisy, A., Tedesco, A.C., Maillard, P., Johannes, L. (2008). Retrograde delivery of photosensitizer (TPPp-*O*-β-GluOH)$_3$ selectively potentiates its photodynamic activity. *Bioconjugate Chem.*, *19*, 532–538.

65. Schneider, R., Schmitt, F., Frochot, C., Fort, Y., Lourette, N., Guillemin, F., Müller, J.-F., Barberi-Heyob, M. (2005). Design, synthesis, and biological evaluation of folic acid targeted tetraphenylporphyrin as novel photosensitisers for selective photodyanamic therapy. *Bioorg. Med. Chem.*, *13*, 2799–2808.

66. Reddi, E. (1997). Role of delivery vehicles for photosensitisers in the photodynamic therapy of tumors. *J. Photochem. Photobiol. B*, *37*, 189–195.

67. Schneider, R., Tirand, L., Frochot, C., Vanderesse, R., Thomas, N., Gravier, J., Guillemin, F., Barberi-Heyob, M. (2006). Recent improvements in the use of synthetic peptides for a selective photodynamic therapy. *Curr. Med. Chem. Anti-Cancer Agents Med. Chem.*, *6*, 469–488.

68. Vrouenraets, M.B., Visser, G.W.M., Stewart, F.A., Stigter, M., Oppelaar, H., Postmus, P.E., Snow, G.B., van Dongen, G.A.M. S. (1999). Development of *meta*-tetrahydroxyphenylchlorin-monoclonal conjugates for photoimmunotherapy. *Cancer Res.* *59*, 1505–1513.

69. Stanelloudi, C., Smith, K., Hudson, R., Malatesti, N., Savoie, H., Boyle, R. W., Greenman, J. (2007). Development and characterization of novel photosensitizer: scFv conjugates for use in photodynamic therapy of cancer. *Immunology*, *120*, 512–517.

70. Ferro, S., Jori, G., Sortino, S., Stancanelli, S., Nikolov, P., Tognon, G., Ricchelli, F., Mazzaglia, A. (2009). Inclusion of 5-[4-(1-dodecanoylpyridinium)]-10,15,20-triphenylporphine in supramolecular aggregates of cationic amphiphilic cyclodextrins: physicochemical characterization of the complexes and strengthening of the antimicrobial photosensitizing activity. *Biomacromolecules*, *10*, 2592–2600.

71. Kano, K., Nishiyabu, R., Asada, T., Kuroda, Y. (2002). Static and dynamic behaviour of 2:1 inclusion complexes of cyclodextrins and charged porphyrins in aqueous organic media. *J. Am. Chem. Soc. 124*, 9937–9944.

72. Mosinger, J., Kliment, V., Sejbal, J., Kubát, P., Lang, K. (2002). Host–guest complexes of anionic porphyrin sensitizers with cyclodextrins. *J. Porphyrins Phthalocyanines*, *6*, 514–526.

73. El-Hachemi, Z., Farrera, J.-A., García-Ortega, H., Ramírez-Gutiérrez, O., Ribó, J.M. (2001). Heteroassociation of meso-sulfonatophenylporphyrins with β- and γ-cyclodextrin. *J. Porphyrins Phthalocyanines*, *5*, 465–473.

74. Ribó, J. M., Farrera, J., Valero, M.L., Virgili, A. (1995). Self-assembly of cyclodextrins with meso-tetrakis(4-sulfonatophenyl)porphyrin in aqueous solution. *Tetrahedron*, *51*, 3705–3712.

75. Dick, D.L., Rao, T.V.S., Sukumaran, D., Lawrence, D.S. (1992). Molecular encapsulation: cyclodextrin-based analogues of heme-containing proteins. *J. Am. Chem. Soc.*, *114*, 2664–2669.

76. Kano, K., Kitagishi, H., Tamura, S., Yamada, A. (2004). Anion binding to a ferric porphyrin complexed with per-*O*-methylated-α-cyclodextrin in aqueous solution. *J. Am. Chem. Soc.*, *126*, 15202–15210.

77. Carofiglio, T., Fornasier, R., Lucchini, V., Simonato, L., Tonellato, U. (2000). Synthesis, characterization, and supramolecular properties of a hydrophilic porphyrin-β-cyclodextrin conjugate. *J. Org. Chem.*, *65*, 9013–9021.

78. Sasaki, K., Nakagawa, H., Zhang, X., Sakurai, S., Kano, K., Kuroda, Y. (2004). Construction of porphyrin–cyclodextrin self-assembly with molecular wedge. *Chem. Commun.* 408–409.

79. Agostiano, A., Catucci, L., Cosma, P., Fini, P. (2003). Aggregation processes and photophysical properties of chlorophyll *a* in aqueous solutions modulated by the presence of cyclodextrins. *Phys. Chem. Chem. Phys.*, *5*, 2122–2128.

80. Monti, S., Sortino, S. (2002). Photoprocesses of photosensitizing drugs within cyclodextrin cavities. *Chem. Soc. Rev.*, *31*, 287–300.

81. Mazzaglia, A., Micali, N., Monsù Scolaro, L. (2006). Structural characterization of colloidal cyclodextrins: molecular recognition by means of photophysical investigation. In: Douhal, A. Ed., *Cyclodextrin Materials Photochemistry, Photophysics and Photobiology.* Elsevier, *Amsterdam*, pp 203–222.

82. Bonchio, M., Carofiglio, T., Carraro, M., Fornasier, R., Tonellato, U. (2002). Efficient sensitized photooxygenation in water by a porphyrin–cyclodextrin supramolecular complex. *Org. Lett.*, *4*, 4635–4637.

83. Cosma, P., Catucci, L., Fini, P., Dentuto, P.L., Agostiano, A., Angelini, N., Monsù Scolaro, L. (2006). Tetrakis(4-pyridyl) porphyrin supramolecular complexes with cyclodextrins in aqueous solution. *Photochem. Photobiol.*, *82*, 563–569.

84. Sekher, P., Garbo, G.M. (1993). Spectroscopic studies of tin ethyl etiopurpurin in homogeneous and in heterogeneous systems. *J. Photochem. Photobiol. B*, *20*, 117–125.

85. Bruzell, E.M., Morisbaka, E., Tønnesen, H.H. (2005). Studies on curcumin and curcuminoids: XXIX. Photoinduced cytotoxicity of curcumin in selected aqueous preparations. *Photochem. Photobiol. Sci.*, *4*, 523–530.

86. Zhao, B., He, Y.-Y., Chignell, C.F., Yin, J.-J., Andley, U., Roberts, J.E. (2009). Difference in phototoxicity of cyclodextrin complexed fullerene [(γ-CyD)$_2$/C$_{60}$] and its aggregated derivatives toward human lens epithelial cells. *Chem. Res. Toxicol.*, *22*, 660–667.

87. Mazzaglia, A., Angelini, N., Darcy, R., Donohue, R., Lombardo, D., Micali, N., Villari, V., Sciortino, M.T., Monsù Scolaro, L. (2003). Novel heterotopic colloids of anionic porphyrins entangled in cationic amphiphilic cyclodextrins: spectroscopic investigation and intracellular delivery. *Chem. Eur. J.*, *9*, 5762–5769.

88. Mazzaglia, A., Valerio, A., Villari, V., Rencurosi, A., Lay, L., Spadaro, S., Monsù Scolaro, L., Micali, N. (2006). Probing specific protein recognition in controlled host nanoassemblies of glycosylated cyclodextrins. *New J. Chem.*, *30*, 1662–1668.

89. Sallas, F, Darcy, R. (2008). Amphiphilic cyclodextrins: advances in synthesis and supramolecular chemistry. *Eur. J. Org. Chem.* 957–969.

90. Bilensoy, R. (2008). Cyclodextrin nanoparticles for drug delivery. In: Ravi Kumar, M.N.V. Ed, *Handbook of Particulate Drug Delivery*, Vol. 1, American Scientific, Valencia, CA, pp. 187–204.

91. Memisoglu-Bilensoy, E., Dogan, A. L, Hincal, A.A. (2006). Cytotoxic evaluation of injectable cyclodextrin nanoparticles. *J. Pharm. Pharmacol.*, *58*, 585–589.

92. Ravoo, B.J., Darcy, R. (2000). Cyclodextrin bilayer vesicles. *Angew. Chem. Int. Ed. 39*, 4324–4326.

93. Lombardo, D., Longo, A., Darcy, R., Mazzaglia, A. (2004). Structural properties of non-ionic cyclodextrin colloids in water. *Langmuir*, *20*, 1057–1064.

94. Quaglia, F., Ostacolo, L., Mazzaglia, A., Villari, V., Zaccaria, D, Sciortino, M.T. (2009). The intracellular effects of non-ionic amphiphilic cyclodextrin nanoparticles in the delivery of anticancer drugs. *Biomaterials*, *30*, 374–382.

95. Mazzaglia, A., Donohue, R., Ravoo, B.J., Darcy, R. (2001). Novel amphiphilic cyclodextrins: graft synthesis of heptakis (6-alkylthio-6-deoxy)-β-cyclodextrin 2-oligo(ethylene glycol) conjugates and their ω-halo derivatives. *Eur. J. Org. Chem.* 1715–1721.

96. McNicholas, S., Rencurosi, A., Lay, L., Mazzaglia, A., Sturiale, L., Perez M., Darcy, R. (2007). Amphiphilic *N*-glycosylthiocarbamoyl cyclodextrins: synthesis, self-assembly, and fluorimetry of recognition by lens culinaris lectin. *Biomacromolecules*, *8*, 1851–1857.

97. Sortino, S., Petralia, S., Darcy, R., Donohue, R., Mazzaglia, A. (2003). Photochemical outcome modification of diflunisal by a novel cationic amphiphilic cyclodextrin. *New J. Chem.*, *27*, 602–608.

98. Ravoo B. J, Jacquier J.C., Wenz, G. (2003). Molecular recognition of polymers by cyclodextrin vesicles. *Angew. Chem. Int. Ed. 42*, 2066–2070.

99. Mazzaglia, A., Angelini, N., Lombardo, D., Micali, N., Patanè, S., Villari, V., Monsù Scolaro, L. (2005). Amphiphilic cyclodextrin carriers embedding porphyrins: charge and size modulation of colloidal stability in heterotopic aggregates. *J. Phys. Chem. B*, *109*, 7258–7265.

100. Sortino, S., Mazzaglia, A., Monsù Scolaro, L., Marino Merlo, F., Valveri, V., Sciortino, M.T. (2006). Nanoparticles of cationic amphiphilic cyclodextrins entangling anionic porphyrins as a "carrier-sensitizer" system in photodynamic cancer therapy. *Biomaterials*, *27*, 4256–4265.

101. Konan, Y.N., Gruny, R., Allemann, E.J. (2002). State of the art in the delivery of photosensitizers for photodynamic therapy. *J. Photochem. Photobiol. B*, *66*, 89–106.

102. Damoiseau, X., Schuitmaker H.J., Lagerberg J.W.M., Hoebeke M. (2001). Increase of the photosensitizing efficiency of the Bacteriochlorin a by liposome-incorporation. *J. Photochem. Photobiol. B*, *60*, 50–60.

103. Valli, L., Giancane, G., Mazzaglia, A., Monsù Scolaro, L., Conoci, S., Sortino, S. (2007). Photoresponsive multilayer films by assembling cationic amphiphilic cyclodextrins and anionic porphyrins at the air/water interface. *J. Mater. Chem.* 1660–1663.

104. Hayashi, H., Ogoshi, H. (1997). Molecular modelling of electron transfer systems by noncovalently linked porphyrin–acceptor pairing. *Chem. Soc. Rev.*, *26*, 355–364.

105. Callari, F.L., Mazzaglia, A., Monsù Scolaro, L., Valli, L., Sortino, S. (2008). Biocompatible nanoparticles of amphiphilic cyclodextrins entangling porphyrins as suitable vessels for light-induced energy and electron transfer. *J. Mater. Chem.*, *18*, 802–805.

106. Perona, R., Sánchez-Pérez, I. (2007). Signalling pathways involved in clinical responses to chemotherapy. *Clin. Transl. Oncol.*, *9*, 625–633.

107. Huang, Z. (2005). A review of progress in clinical photodynamic therapy. *Technol. Cancer Res. Treat.*, *4*, 283–294.

108. Moser, J.G. (2000). Porphyrins and phthalocyanines as model compounds for detoxification of tumor chemotherapeutic drugs. *J. Porphyrins Phthalocyanines*, *4*, 129–135.

109. Kolárová, H., Mosinger, J., Lenobel, R., Kejlová, K., Jírová, D., Strnad, M. (2003). In vitro toxicity testing of supramolecular sensitizers for photodynamic therapy. *Toxicol. in Vitro*, *17*, 775–778.

110. Cosma, P., Fini, P., Rochira, S., Catucci, L., Castagnolo, M., Agostiano, A., Gristina, R., Nardulli, M. (2008). Phototoxicity and cytotoxicity of chlorophyll *a*/cyclodextrins complexes on Jurkat cells. *Bioelectrochemistry*, *74*, 58–61.

111. Hamblin, M.R., Hasan, T. (2004). Photodynamic therapy: a new antimicrobial approach to infectious disease. *Photochem. Photobiol. Sci.*, *3*, 436–450.

112. Jori, G., Coppellotti, O. (2007). Inactivation of pathogenic microorganisms by photodynamic techniques: mechanistic aspects and perspective applications. *Anti-Infection Agents Med. Chem.*, *6*, 119–131.

113. Hamblin, M. R., O'Donnell, D. A., Murthy, N., Rajagopalan, K., Michaud, N., Sherwood, M. E., Hasan, T. (2002). Polycationic photosensitiser conjugates: effect of chain length and Gram classification on the photodynamic inactivation of bacteria. *J. Antimicrob. Chemother.*, *49*, 941–951.

114. Merchat, M., Bertoloni, G., Giacomoni, P., Villanueva, A., Jori, G., (1996). Meso-substituted cationic porphyrins as efficient photosensitisers of gram-positive and gram-negative bacteria. *J. Photochem. Photobiol. B*, *32*, 153–157.

115. Tegos, G.P., Demidova, T.N., Arcila-Lopez, D., Lee, H., Wharton, T., Gali, H., Hamblin, M.R. (2005). Cationic fullerenes are effective and selective antimicrobial photosensitisers. *Chem. Biol.*, *12*, 1127–1135.

116. Wang, P.G., Xian, M., Tang, X., Wu, X., Wen, Z., Cai, T., Janczuk, A.J. (2002). Nitric oxide donors: chemical activities and biological applications. *Chem. Rev.*, *102*, 1091–1134.

117. Caruso, E.B., Cicciarella, E., Sortino, S. (2007). A multifunctional nanoassembly of mesogen-bearing amphiphiles and porphyrins for the simultaneous photodelivery of nitric oxide and singlet oxygen. *Chem. Commun.* 5028–5030.

118. El-Sayed, I.H., Huang, X., El-Sayed, M.A. (2006). Selective laser photo-thermal therapy of epithelial carcinoma using anti-EGFR antibody conjugated gold nanoparticles. *Cancer Lett.*, *239*, 129–135.

119. Mazzaglia, A., Trapani, M., Villari, V., Micali, N., Marino Merlo, F., Zaccaria, D., Sciortino, M.T., Previti, F., Patanè, S., Monsù Scolaro, L. (2008). Amphiphilic cyclodextrins as capping agents for gold colloids: a spectroscopic investigation with perspectives in photothermal therapy. *J. Phys. Chem. C*, *112*, 6764–6769.

120. Mazzaglia, A., et al. Unpublished results.

121. Weyergang, A., Berg, K., Kaalhus, O., Peng, Q., Selbo, P.K. (2008). Photodynamic therapy targets the mTOR signaling network in vitro and in vivo. *Mol. Pharma.*, *6*, 255–264.

19

SUGAMMADEX: A CYCLODEXTRIN-BASED NOVEL FORMULATION AND MARKETING STORY

FRANÇOIS DONATI

Department of Anesthesiology, Université de Montréal, and Hôpital Maisonneuve–Rosemont, Montréal, Québec, Canada

1. INTRODUCTION

Most applications of cyclodextrins (CDs) are concerned with solubilizing a drug in a vial to make it water soluble. When injected in the body, dilution of the drug occurs, dissociation of the CD–drug complex takes place, and free drug is available for pharmacological action in the body. Anesthetic drugs are usually highly lipid soluble, because they must cross the blood–brain barrier to reach their site of action. Thus, CDs are useful in the formulation of such poorly water-soluble agents, such as propofol, sufentanil, etomidate, and ropivacaine [1]. To achieve adequate effectiveness, the dissociation constant must be less than the concentration of the drug in the vial and greater than the effective concentration in the body. A typical dissociation constant for this purpose is 10^{-4} M.

Sugammadex was developed with another purpose in mind: to remove a drug from the body. This required much lower dissociation constants. In essence, the drug had to be tightly bound to the CD; the dissociation constant needed to be lower than typical effective concentrations of the drug, so that encapsulation by a CD would decrease the free drug concentration below the level that corresponds to pharmacological activity. The drugs that are targeted are neuromuscular blocking agents, which are not lipid-soluble drugs, but whose removal from the body at the end of the anesthetic constitutes a serious clinical problem.

2. THE PROBLEM

2.1. Components of Anesthesia

Modern anesthesia is a judicious mixture of unconsciousness, analgesia (or freedom from pain) and muscle relaxation, adapted to the particular patient and the surgical procedure. Over the latter part of the twentieth century, drug development has focused on agents that are specific for each component. Unconsciousness is provided by inhalational drugs such as sevoflurane and desflurane, or intravenous drugs such as propofol. Opioid drugs such as fentanyl, sufentanil, and remifentanil are commonly given for analgesia. Muscle relaxation, provided by neuromuscular blocking agents, is required when an immmobile surgical field is needed, when muscle tone must be abolished, and when the patient must be ventilated mechanically. Neuromuscular blocking agents are thus required in the vast majority of surgical procedures involving the thoracic and abdominal cavities.

2.2. Neuromuscular Blocking Agents

Neuromuscular blocking agents, or curare-like drugs, have virtually no effect on the central nervous system. They are water-soluble drugs that do not cross the blood–brain barrier. Their site of action is the neuromuscular junction of skeletal muscle, where they bind to the nicotinic

Cyclodextrins in Pharmaceutics, Cosmetics, and Biomedicine: Current and Future Industrial Applications, First Edition. Edited by Erem Bilensoy.
© 2011 John Wiley & Sons, Inc. Published 2011 by John Wiley & Sons, Inc.

cholinergic receptor. The interaction between most neuromuscular blocking agents and the receptor is competitive. Transmission can be restored if an excess of the agonist, acetylcholine, is present. The only practical method to increase the acetylcholine level is to administer acetylcholinesterase inhibitors such as neostigmine. These agents inhibit acetylcholine breakdown, and this improves neuromuscular transmission. However, even high doses of acetylcholinesterase inhibitors have a limited effect when profound blockade is present [2]. This mechanism applies to all neuromuscular blocking agents available except succinylcholine, which belongs to the class of depolarizing agents. All other neuromuscular blocking agents are thus called *nondepolarizing agents*. They are the only agents that are used to maintain muscle relaxation during surgery. Succinylcholine is normally used only for tracheal intubation.

2.3. Restoration of Neuromuscular Function

At the end of surgery it is essential to make sure that the effects of nondepolarizing agents are no longer present, because these drugs have an effect on all skeletal muscles, including those concerned with respiratory function. Several strategies have been adopted to deal with the problem. There has been a trend in drug development to produce drugs with a short duration of action so that the effect would wear off before the end of surgery. Unfortunately, the shortest agents available—atracurium, cisatracurium, vecuronium, and rocuronium—when given in the usual doses, have a duration of action of 1 to 2 h, and in some cases up to 4 h [3], and do not offer the flexibility that is required. If the initial dose is reduced, the price to pay is poor muscle relaxation during surgery.

Another strategy is to administer agents that restore neuromuscular function. This is normally achieved by giving acetylcholinesterase agents, which inhibit the breakdown of acetylcholine at the neuromuscular junction, thus allowing more agonist to compete with the neuromuscular blocking agent and restoration of neuromuscular function. Acetylcholinesterase agents are used widely and can prevent signs of residual paralysis after emergence from anesthesia [4]. Unfortunately, they have limited effectiveness, and an anticholinergic drug has to be coadministered to prevent serious parasympathetic side effects, such as bradycardia and bronchoconstriction [2]. Despite the administration of anticholinesterase agents, residual paralysis after anesthesia is relatively common (5 to 10%), but is less frequent than if anticholinesterase agents are omitted (40 to 70%) [4]. Although transient, residual paralysis is an important problem, because it has been associated with an increased incidence of postoperative pulmonary complications [5].

3. THE PROPOSED SOLUTION: SPECIFIC BINDING

3.1. Classes of Neuromuscular Blocking Agents

Several classes of drugs have been recognized as having neuromuscular blocking properties. Most modern neuromuscular blocking drugs available today belong to one of two classes: the benzyl isoquinolines (atracurium, cisatracurium, mivacurium, doxacurium) and the steroid-based (rocuronium, vecuronium, pancuronium). Development of chlorofumarates (gantacurium), which have a short duration of action and can be antagonized by cysteine, is under way [6,7]. Rocuronium and vecuronium are devoid of cardiovascular effects and have become very popular in anesthesia. Rocuronium has a faster onset of action than vecuronium and is probably the most widely used neuromuscular blocking agent throughout the world. Pancuronium is longer acting, has some cardiovascular effects, and is more difficult to reverse [2,4]. Over the past few years, its use has declined steadily.

3.2. The Emergence of Sugammadex

Since the 1960s, the pharmaceutical industry has concentrated its efforts into the production of new compounds with a shorter and shorter duration of action [6]. The goal was to make the duration of action short enough so that administration of the drug would be flexible and recovery would be quick and reliable once administration of the agent is discontinued. The introduction of atracurium and vecuronium in the mid-1980s represented a significant improvement, but they were still too long-acting (30 to 45 min with anticholinesterase agents). A few years later, mivacurium became available. It duration of action (20 min) was shorter than that of agents previously available, but its slow onset of action and cardiovascular side effects were seen as negative aspects, so that mivacurium has been withdrawn in several countries, including the United States. Rocuronium was marketed in the 1990s. It had the same duration of action as vecuronium but became extremely popular because of its fast onset of action. Cisatracurium also became available in the 1990s but was less successful because of a slow onset of action [8].

In the late 1990s, attempts to produce a shorter-acting drug produced gantacurium, which has a duration of action of less than 10 to 12 min but some cardiovascular side effects at high doses [6]. Development has been hindered by successive acquisition of commercial rights by different firms. Another approach was taken by Organon, the manufacturer of rocuronium. Anton Bom, a chemist at Organon, was working on a series of cyclodextrin (CD) compounds with the intention of finding one with sufficient affinity for rocuronium to encapsulate the drug. The size of rocuronium, a molecule with a molecular weight of 672 Da, required a γ-CD, consisting of eight residues, for a tight fit [1]. Development resulted in a compound with a molecular weight of 2170 Da, initially called Org 25969, for which the name *sugammadex* was later coined. It features a pocket that is sufficiently deep to accommodate the hydrophobic end of the rocuronium molecule and negatively charged side chains attached to the sugar residues that interact with the hydrophilic, positively charged quaternary ammonium end of the rocuronium molecule to keep it tightly bound. Initial estimates of the dissociation constant were in the range $0.1\,\mu M$ ($10^{-7}\,M$), indicating a large affinity between sugammadex and rocuronium [9].

3.3. Interactions Between Neuromuscular Blocking Agents and Sugammadex

The concentrations of rocuronium associated with its pharmacological action (i.e., neuromuscular blockade) are in the μM range [10]. Thus, the addition of sugammadex, with a K_D value at least one order of magnitude lower, has the potential to lower the free concentration of rocuronium to levels that do not produce neuromuscular blockade. It is not known whether this dissociation constant is modified by pH, temperature, or blood constituents. Vecuronium reportedly has a lower affinity for sugammadex, which may be two to three times less [1]. Also, vecuronium is six times more potent than rocuronium [8], and concentrations corresponding to neuromuscular blockade are less than for rocuronium, typically in the range 0.1 to $0.2\,\mu M$. A lower potency means that fewer molecules have to be injected to produce an effect, so fewer molecules are needed to encapsulate the substance. However, this effect is compensated by the lower affinity, so that in practice, the sugammadex dose requirements are approximately the same for both vecuronium and rocuronium. It seems that pancuronium, which has a potency similar to that of vecuronium, has even less affinity for sugammadex, and little experimental work has been performed with the combination [11].

4. PRECLINICAL PHARMACOLOGY

4.1. Measurement of Effect

The effect of a neuromuscular blocking drug is typically evaluated by stimulating a nerve electrically and measuring the force of contraction of the corresponding muscle. As the concentration of neuromuscular blocking agent increases, competitive antagonism occurs at the neuromuscular junction and neurotransmission is impaired progressively. In other words, the twitch response of the muscle decreases from 100% of the preblockade value to 0% if the amount of neuromuscular blocking agent is sufficient. Complete reversal of blockade is accompanied by a return to baseline. Nondepolarizing blockade is also characterized by another phenomenon: The response is not sustained when stimulation

frequency is sufficiently high. This fade is produced at frequencies higher than 2 Hz, and this property serves as a useful tool to measure the degree of muscle relaxation. Fade is maximum after the fourth response, and contractions at 2 Hz are easily seen. Thus, the "train-of-four" pattern has been retained in clinical practice and research to assess the degree of blockade. The degree of neuromuscular blockade during surgery usually corresponds to 0 to 2 twitches visible in response to train-of-four stimulation. When no visible response is seen, another mode of stimulation is used, *post-tetanic count*, which relies on the application of a high-frequency stimulus (50 Hz for 5 s), followed by 15 to 20 stimulations at 1 Hz. The effect of the high-frequency, or tetanic, stimulation is to produce a temporary reversal of blockade. The depth of blockade depends inversely on the number of 1-Hz responses seen: 0 to 2 representing deep blockade, 15 to 20 moderate block. It is generally agreed that full neuromuscular recovery compatible with adequate respiratory function requires a train-of-four ratio of 90%; that is, the size of the fourth twitch is > 90% that of the first [4]. The muscle relaxation required for surgery is associated with a train-of-four ratio of 0%.

4.2. Binding Studies

The dissociation constant (K_D) value for the rocuronium–sugammadex complex was initially reported to be $0.1\,\mu M$ [9]. More recent articles mention values of $0.04\,\mu M$ for rocuronium–sugammadex and $0.1\,\mu M$ for vecuronium–sugammadex [12]. Dissociation constants for pancuronium have not been reported, but they are probably higher because sugammadex has less effect on a pancuronium-induced block in vitro than that produced by rocuronium or vecuronium [1].

4.3. In Vitro Studies

The effect of sugammadex was first evaluated in the mouse phrenic nerve–hemidiaphragm preparation, where the concept of reversal by encapsulating rocuronium, vecuronium, and pancuronium was proven. At rocuronium concentrations of $3.6\,\mu M$, the twitch height is abolished. With increasing concentrations of sugammadex, the twitch height is restored progressively, with maximum response occurring before the molar sugammadex/rocuronium ratio reaches 1.0 [12]. A ratio of exactly 1.0 is not required, probably because of the safety margin at the neuromuscular junction. The concept of safety margin refers to the need for neuromuscular blocking agents to occupy a large proportion of receptors before a blocking effect is seen. Vecuronium is more potent than rocuronium, and the twitch height is abolished at $1.0\,\mu M$. The sugammadex/vecuronium ratio required for the restoration of twitch height exceeds 1.0, probably because of the lesser affinity of vecuronium for sugammadex compared with rocuronium. Since there are wide discrepancies between species in their sensitivities to neuromuscular blocking agents [12], the actual numbers obtained in the phrenic nerve–hemidiaphragm preparation have little practical significance.

4.4. Animal Studies

Sugammadex has been given to guinea pigs, monkeys, and cats, and the efficacy of the drug was measured [1]. Again, the doses of rocuronium and vecuronium are different from humans. Nevertheless, these experiments showed that the train-of-four ratio reached the target value of 0.9 rapidly after sugammadex when complete blockade was produced by rocuronium or vecuronium, but there was no effect if blockade was produced by benzylisoquinoline compounds such as atracurium or mivacurium [11]. It was also established that the plasma concentration of rocuronium increased after injection of sugammadex [13]. This provides insights into the mechanism of action of sugammadex. Since both free and sugammadex-bound fractions are measured by the assay, it appears that sugammadex is effective in creating a movement of rocuronium from the peripheral to the central compartment.

5. MECHANISM OF ACTION IN HUMANS

5.1. Pharmacokinetics

Plasma concentrations of sugammadex were measured in humans, and the pharmacokinetic profile was shown to be linear with increasing dose. The drug has a rather small volume of distribution (10 to 12 L in adults), consistent with distribution in the extracellular fluid volume [14]. Its elimination half-life is approximately 2 h, and the rocuronium–sugammadex complexes are excreted by the kidney. In patients with impaired renal function (mean creatinine clearance of 12 mL/min), the clearance of sugammadex was markedly reduced, to 5.5 mL/min from 95 mL/min in normal subjects [15]. Elimination half-life was increased markedly in renal failure (35 h) compared with normal subjects (2.3 h) [15]. As in animals, the injection of sugammadex was also found to produce an increase in the measured free and bound plasma concentration of rocuronium. This suggests that the creation of rocuronium–sugammadex complexes in plasma reduces the free concentration of rocuronium, which in turn creates a concentration gradient between the extravascular tissue and plasma. Free rocuronium then moves from the periphery into plasma, down its concentration gradient, which causes even more rocuronium molecules to bind to sugammadex. The site of action of rocuronium is at the neuromuscular junction, which is located extravascularly, and termination of action of rocuronium occurs because of movement of rocuronium away from the neuromuscular junction. Sugammadex does not appear to move to the site of action of rocuronium.

5.2. Clinical Pharmacology

5.2.1. *Volunteers*   Sugammadex has been administered to unanesthetized volunteers [16–18] either alone or in combination with rocuronium or vecuronium given simultaneously. Injection of sugammadex with either rocuronium or vecuronium did not produce neuromuscular blockade [18], probably because of the rapid binding. Side effects are reported as mild, with pain on injection and a change in the sense of taste being the most frequent complaints. No QTc change has been reported on the electrocardiogram. The doses used have been as high as 32 mg/kg, which is considerably more than the usual doses of 2 to 4 mg/kg. There is one case of "hypersensitivity reaction" in one volunteer that resulted in discontinuation of administration of sugammadex but did not require any treatment [7]. Six volunteers also showed possible hypersensitivity after 32 mg/kg [7]. Other volunteers were anesthetized and received rocuronium or vecuronium followed by sugammadex. In these persons, rapid reversal of neuromuscular blockade was documented, with a dose– response relationship similar to that found in patients. Side effects may be masked when a general anesthetic is given. However, it was observed that although volunteers appeared to have an adequate depth of anesthesia before sugammadex was injected, signs of light anesthesia (such as movement, coughing, and grimacing) were often observed after sugammadex [7]. This probably has little clinical significance because sugammadex is intended to be given at the end of anesthesia and surgery.

5.2.2. *Efficacy*   In anesthetized patients, the efficacy of sugammadex has been evaluated by measuring recovery time, defined as the interval from injection until a train-of-four ratio > 90% is obtained at the adductor pollicis muscle. Secondary measures of efficacy were the absence of recurrence of blockade and clinical assessment of muscle power after emergence from anesthesia. It was expected that the response would depend on the intensity of blockade, so sugammadex was tested in three typical situations: moderate blockade, when two twitches are visible after train-of-four stimulation; deep blockade, when the post-tetanic count or number of visible responses after a 50-Hz, 5-s tetanus is 1 to 2; or rescue reversal, 3 to 5 min after the administration of rocuronium, to mimic the situation when attempts at intubation were unsuccessful.

5.2.3. *Moderate Blockade*   Not surprisingly, recovery time decreases as the dose of sugammadex increases, and for the same recovery time, the dose required increases with depth of blockade. For moderate rocuronium blockade in healthy adults (i.e., when two twitches are visible), doses of 0.5 to

Table 1. Mean Times (min) for Complete Recovery with Sugammadex Given for Moderate Blockade

Study	No Sugammadex	Sugammadex Dose			
		0.5 mg/kg	1 mg/kg	2 mg/kg	4 mg/kg
For Rocuronium					
Sorgenfrei et al. [18]	21	4.3	3.3	1.3	1.1
Shields et al. [19]	—	6	3	2	2
Suy et al. [20]	31.8	3.7	2.3	1.7	1.1
Sacan et al. [21]	—	—	—	—	1.8
Vanacker et al. [22]	—	—	—	1.8	
Flockton et al. [23]	—	—	—	1.9	
For Vecuronium					
Suy et al. [20]	48.8	7.7	2.5	2.3	1.5
Khuenl-Brady et al. [24]	—	—	—	2.7	

8 mg/kg have been tested and compared with control subjects that were not given sugammadex [18–20]. Spontaneous recovery takes approximately 40 min under those circumstances. With 2 mg/kg, recovery time is markedly reduced, to approximately 2 min, and plateaus for higher doses (Table 1). With 0.5 and 1 mg/kg, recovery is longer and more variable. For moderate vecuronium blockade, the dose–response relationship is approximately the same, with 2 mg/kg being the lowest effective dose [20]. This might seem surprising because the affinity of sugammadex for vecuronium is less than for rocuronium, but this effect is probably compensated for by the greater potency of vecuronium, which implies that for a given degree of blockade, fewer molecules of vecuronium are present in the body compared with rocuronium. Other studies supported the notion that 2 mg/kg was effective for moderate blockade [21–24].

5.2.4. Deep Blockade Sugammadex has been given to patients with deep blockade, defined as a post-tetanic count (PTC) of 1 to 2. With either rocuronium or vecuronium, it usually takes 15 to 30 min to recover from deep blockade, or a PTC of 1 to 2, to reach moderate blockade, when two twitches are visible in response to train-of-four stimulation. When the blockade is deep, recovery times are increased at all sugammadex doses less than, and including, 2 mg/kg [25,26]. Recovery times of approximately 2 min are observed with 4 mg/kg [25–28]. There is no further improvement, even with doses as high as 8 mg/kg, because the effect is limited by circulatory factors; adequate mixing of any drug within the circulation takes approximately 2 min. Again, these doses apply to rocuronium and vecuronium (Table 2). Neuromuscular recovery is usually sustained when a train-of-four ratio > 0.9 is attained, except when relatively low doses of sugamamdex (0.5 to 1.0 mg/kg) are given [26].

5.2.5. Rescue Reversal Vecuronium is not indicated for emergency tracheal intubation, so the ability of sugammadex to restore neuromuscular function soon after intubation has been investigated for rocuronium only. Here, the required dose of sugammadex depends on how much rocuronium is given. With a relatively modest dose (0.6 mg/kg), rapid reversal time (< 2 min after sugammadex) is obtained with sugammadex, 8 mg/kg [14]. Doses as high as 16 mg/kg are required for a rocuronium dose of 1.2 mg/kg [29] (Table 3).

Table 2. Time (min) for Complete Recovery with Sugammadex Given for Deep Blockade

Study	Sugammadex Dose				
	0.5 mg/kg	1 mg/kg	2 mg/kg	4 mg/kg	8 mg/kg
For Rocuronium					
Groudine et al. [25]	32.4	15.3	4.3	2.5	1.5
Duvaldestin et al. [26]	79.8	28	3.2	1.7	1.1
Rex et al. [27]	—	—	—	1.3	
Jones et al. [28]	—	—	—	2.9	
For Vecuronium					
Duvaldestin et al. [26]	68.4	15.1	9.1	3.3	1.7

Table 3. Time (min) for Complete Recovery with Sugammadex 3 to 5 Minutes After Rocuronium

Study	No Sugammadex	Sugammadex Dose				
		1 mg/kg	2 mg/kg	4 mg/kg	8 mg/kg	16 mg/kg
For Rocuronium, 0.6 mg/kg						
Sparr et al. [14]	52.4	25.0	6.9	4.3	1.6	
For Rocuronium, 1.2 mg/kg						
deBoer et al. [29]	122.1	—	56.5	15.8	2.8	1.9
Lee et al. [30]	—	—	—	—	—	3.2

The time required from injection of rocuronium until sugammadex-induced recovery is shorter than the interval between the administration of succinylcholine until full recovery [30]. This indicates that rocuronium, with sugammadex as a backup in case of failure to intubate, may be given instead of succinylcholine for emergency tracheal intubation.

5.2.6. Special Populations

Experience with sugammadex in special populations is still limited. A multicenter study in pediatric patients suggests that sugammadex is equally effective in children as in adults, using the same mg/kg dose [31]. In renal-failure patients, sugammadex has a longer half-life than in normal subjects, but the effect of sugammadex is essentially the same as in patients with normal renal function [15]. Concern arises when a patient requires another surgical procedure within hours, or perhaps days in the case of renal failure, of receiving sugammadex. Residual effects of the drug might still persist, and the blocking effect of rocuronium or vecuronium might be less. It is too early to formulate recommendations, and other neuromuscular blocking agents that do not interfere with sugammadex might be given. Sugammadex appears as a good choice in neuromuscular disease, such as myasthenia gravis, or muscle disease, such as the muscular dystrophies, but at the time of writing, no case reports describing these situations had been published.

5.2.7. Comparison with Anticholinesterase Drugs

The current method to accelerate reversal of neuromuscular blockade is to administer an anticholinesterase agent, such as neostigmine, pyridostigmine, or edrophonium. Of these, neostigmine is most frequently used and is considered to be the gold standard. Anticholinesterase agents have parasympatomimetic effects, and among these the dramatic slowing of heart rate is most important clinically. To counteract these side effects, an anticholinergic drug, atropine or glycopyrrolate, is given concurrently. Still, changes in heart rate are common after atropine–neostigmine or glycopyrrolate–neostigmine mixtures. Also, like other anticholinesterase drugs, neostigmine is not very effective when given during deep blockade [27]. At moderate blockade, when its effectiveness is optimal, recovery time is 10 to 30 min, longer than that of sugammadex [21,23,24,28]. Thus, sugammadex is much more effective than currently available agents (Table 4). For deeper levels of blockade, the current therapeutic option is to provide mechanical ventilation of the lungs and proper sedation until sufficient neuromuscular recovery returns for neostigmine reversal of blockade. Sugammadex has been found to be free of cardiovascular side effects, whereas there are marked variations in heart rate with the neostigmine–glycopyrrolate mixture [23]. Thus, sugammadex has a better efficacy and fewer side effects than do existing drugs.

6. THE FUTURE OF SUGAMMADEX IN CLINICAL PRACTICE

6.1. Approval Issues

Sugammadex was submitted for approval in the United States, the European Union, and several other countries. In

Table 4. Comparison of Sugammadex with Neostigmine

Study	Neuromuscular Blocking Agent	Level of Block	Sugammadex		Neostigmine	
			Dose (mg/kg)	Time (min)	Dose (mg/kg)	Time (min)
Sacan et al. [21]	Rocuronium	Moderate	4	1.8	0.07	17.4
Jones et al. [28]	Rocuronium	Deep	4	2.9	0.05	50.4
Khuenl-Brady et al. [24]	Rocuronium	Moderate	2	2.7	0.05	17.9
Flockton et al. [23]	Rocuronium	Moderate	2	1.9	0.05	9.0

mid-2008, the U.S. Food and Drug Administration (FDA) ruled that the drug was "not approvable," based on a few mild cases of hypersensitivity or allergy seen in volunteers [7]. The FDA required more clinical and safety data. Based on the same information, the European authorities approved the drug in 2008.

6.2. Cost Issues

In countries where sugammadex is available in clinical practice, its cost is markedly higher than that of any of the other drugs used in anesthesia, and probably greater than the combined cost of all other drugs used for a typical case. A 200-mg vial, which would be appropriate for reversal of moderate block, has a price of approximately 70 euros. Using it for all general anesthesia cases would in fact more than double the cost of anesthetic drugs, so a restricted use of sugammadex might be considered a prudent option. Still, most would agree that sugammadex is valuable and may be lifesaving in emergency situations. Unfortunately, these situations cannot be anticipated easily, so it is not possible to ask clinicians to request the drug only if they have an indication to use it. Anesthesia departments, individual anesthesiologists, and hospital administrations have to make tough decisions on the possible indications of the drug in their setting and on the ways to dispense it immediately should the need arise.

6.3. Possible Impact on Clinical Practice

At the time of writing, sugammadex had been available for just over a year, and only in a restricted number of countries. Cost has hindered its general use, so experience is limited. It can therefore be said with a reasonable degree of confidence that sugammadex has not yet changed anesthetic practice. However, a certain number of trends can be predicted. At present, anesthesiologists are reluctant to give more neuromuscular blocking agent than is required for the duration of the planned surgical procedure, because reversal of profound blockade is difficult with neostigmine and other anticholinesterase agents. Availability of sugammadex essentially removes that restriction. With the administration of large doses of neuromuscular blocking agents, some signs of inadequate anesthesia and analgesia such as movement are lost, so the possibility of patient awareness might be of concern. Another possible development could occur at the end of surgery. Recovery rooms were developed, at least in part, in response to the need for close monitoring of respiratory function, which might be impaired after anesthesia because of the residual effects of neuromuscular blocking agents. If such a concern is no longer present thanks to the judicious use of sugammadex, will recovery rooms be needed in the future?

7. CONCLUSIONS

Sugammadex is the first cyclodextrin with sufficient specific binding affinity to be used as an effective antagonist of the effects of another drug in the body. It is used against the neuromuscular blocking agents rocuronium and vecuronium when a rapid transition from complete muscle relaxation to full neuromuscular recovery is required at the end of anesthesia. Sugammadex has few side effects, is well tolerated, and has improved effectiveness compared with currently available methods of accelerating reversal of neuromuscular blockade.

REFERENCES

1. Booij, L.H.D.J. (2009). Cyclodextrins and the emergence of Sugammadex. *Anaesthesia, 64,* 31–37.
2. Bevan, D.R., Donati, F., Kopman, A.F. (1992). Reversal of neuromuscular blockade. *Anesthesiology, 77,* 785–805.
3. Debaene, B., Plaud, B., Dilly, M.P., Donati, F. (2003). Residual paralysis in the PACU after a single intubating dose of nondepolarizing muscle relaxant with an intermediate duration of action. *Anesthesiology, 98,* 1042–1048.
4. Naguib, M., Kopman, A.F., Ensor, J.E. (2007). Neuromuscular monitoring and postoperative residual curarisation: a meta-analysis. *Br. J. Anaesth., 98,* 302–316.
5. Berg, H., Roed, J., Viby-Mogensen, J., Mortensen, C.R., Engbaek, J., Skovgaard, L.T., Krintel, J.J. (1997). Residual neuromuscular block is a risk factor for postoperative pulmonary complications: a prospective, randomised, and blinded study of postoperative pulmonary complications after atracurium, vecuronium and pancuronium. *Acta Anaesthesiol. Scand., 41,* 1095–1103.
6. Belmont, M.R., Lien, C.A., Tjan, J., Bradley, E., Stein, B., Patel, S.S., Savarese, J.J. (2004). Clinical pharmacology of GW280430A in humans. *Anesthesiology, 100,* 768–773.
7. Naguib, M., Brull, S.J. (2009). Update on neuromuscular pharmacology. *Curr. Opin. Anaesthesiol., 22,* 483–490.
8. Donati, F. (2000). Neuromuscular blocking drugs for the new millenium: current practice, future trends–comparative pharmacology of neuromuscular blocking drugs. *Anesth. Analg., 90,* S2–S6.
9. Bom, A., Bradley, M., Cameron, K., (2002). A novel concept of reversing neuromuscular block: chemical encapsulation of rocuronium bromide by a cyclodextrin-based synthetic host. *Angew. Chem. Int. Ed. 41,* 265–270.
10. Kuipers, J.A., Boer, F., Olofsen, E., Bovill, J.G., Burm. A.G. (2001). Recirculatory pharmacokinetics and pharmacodynamics of rocuronium in patients: the influence of cardiac output. *Anesthesiology, 94,* 47–55.
11. Booij, L.H.D.J., van Egmond, J., Driessen, J.J., de Boer, H.D. (2009). In vivo animal studies with sugammadex. *Anaesthesia, 64,* 38–44.

12. Bom, A., Hopr, F., Rutherford, S., Thompson, K. (2009). Preclinical pharmacology of sugammadex. *J. Crit. Care*, 24, 29–35.

13. Epemolu, O., Bom, A., Hope, F., Mason, R. (2003). Reversal of neuromuscular blockade and simultaneous increase in plasma rocuronium concentration after the intravenous infusion of the novel reversal agent Org 25969. *Anesthesiology*, 99, 632–637.

14. Sparr, H.J., Vermeyen, K.M., Beaufort, A.M. (2007). Early reversal of profound rocuronium-induced neuromuscular blockade by sugammadex in a randomized multicenter study: efficacy, safety, and pharmacokinetics. *Anesthesiology*, 1065, 935–943.

15. Staals, L.M., Snoeck, M.M.J., Driessen, J.J., van Hamersvelt, H.W., Flockton, E.A., van den Heuvel, M.W., Hunetr, J.M. (2010). Reduced clearance of rocuronium and sugammadex in patients with severe to end-stage renal failure: a pharmacokinetic study. *Br. J. Anaesth.*, 104, 31–39.

16. Gijsenbergh, F., Ramael, S., Houwing, N., van Iersel, T. (2005). First human exposure of Org 25969, a novel agent to reverse the action of rocuronium bromide. *Anesthesiology*, 103, 695–703.

17. Cammu, G., De Kam, P.J., Demeyer, I., Decoopman, M., Peeters, P.A., Smeets, J.M., Foubert, L. (2008). Safety and tolerability of single intravenous doses of sugammadex administered simultaneously with rocuronium or vecuronium in healthy volunteers. *Br. J. Anaesth.*, 100 (3), 373–379.

18. Sorgenfrei, I.F., Norrild, K., Larsen, P.B., Stensballe, J., Ostergaard, D., Prins, M.E., Viby-Mogensen, J. (2006). Reversal of rocuronium-induced neuromuscular block by the selective relaxant binding agent sugammadex: a dose-finding and safety study. *Anesthesiology*, 104, 667–674.

19. Shields, M., Giovannelli, M., Mirakhur, R.K., Moppett, I., Adams, J., Hermens, Y. (2006). Org 25969 (sugammadex), a selective relaxant binding agent for antagonism of prolonged rocuronium-induced neuromuscular block. *Br. J. Anaesth.*, 96, 36–43.

20. Suy, K., Morias, K., Cammu, G., Hans, P., van Duijnhoven, W.G., Heeringa, M., Demeyer, I. (2007). Effective reversal of moderate rocuronium- or vecuronium-induced neuromuscular block with sugammadex, a selective relaxant binding agent. *Anesthesiology*, 106, 283–288.

21. Sacan, O., White, P.F., Tufanogullari, B., Klein, K. (2007). Sugammadex reversal of rocuronium-induced neuromuscular blockade: a comparison with neostigmine–glycopyrrolate and edrophonium–atropine. *Anesth. Analg.*, 104, 569–574.

22. Vanacker, B.F., Vermeyen, K.M., Struys, M.M., Rietbergen, H., Vandermeersch, E., Saldien, V., Kalmar, A.F., Prins, M.E. (2007). Reversal of rocuronium-induced neuromuscular block with the novel drug sugammadex is equally effective under maintenance anesthesia with propofol or sevoflurane. *Anesth. Analg.*, 104, 563–568.

23. Flockton, E.A., Mastronardi, P., Hunetr, J.M., Goamr, C., Mirakhur, R.K., Aguilera, L., Guinta, F.G., Mesitelman, C., Prins, M.E. (2008). Reversal of rocuronium-induced neuromuscular block with Sugammadex is faster than reversal of cisatracurium-induced block with neostigmine. *Br. J. Anaesth.*, 100, 622–630.

24. Khuenl-Brady, K.S., Wattwill, M., Vanacker, B.F., Lora-Tamayo, J.I., Rietbergen, H., Alvarez-Gomez, J.A. (2010). Sugammadex provides faster reversal of vecuronium-induced neuromuscular blockade compared with neostigmine: a multicenter, randomized, controlled trial. *Anesth. Analg.*, 110, 64–73.

25. Groudine, S.B., Soto, R., Lien, C., Drover, D., Roberts, K. (2007). A randomized, dose-finding, phase II study of the selective relaxant binding drug, sugammadex, capable of safely reversing profound rocuronium-induced neuromuscular block. *Anesth. Analg.*, 104, 555–562.

26. Duvaldestin, P., Kuizenga, K., Saldien, V., Claudius, C., Servin, F., Klein, J., Debaene, B., Heeringa, M. (2010). A randomized, dose-response study of sugammadex given for the reversal of deep rocuronium- or vecuronium-induced neuromuscular blcokade under sevoflurane anesthesia. *Anesth. Analg.*, 110, 74–82.

27. Rex, C., Wagner, S., Spies, C., Scholz, J., Rietbergen, H., Heeringa, M., Wulf, H. (2009) Reversal of neuromuscular blockade by sugammadex after continuous infusion of rocuronium in patients randomized to sevoflurane or propofol maintenance anesthesia. *Anesthesiology*, 111, 30–35.

28. Jones, K.R., Caldwell, J.E., Brull, S.J., Soto, R.G. (2008). Reversal of profound rocuronium-induced blockade with sugammadex: a randomized comparison with neostigmine. *Anesthesiology*, 109, 816–824.

29. de Boer, H.D., Driessen, J.J., Marcu, M.A.E., Kerkkaamp, H., Heeringa, M., Klimek, M. (2007). Reversal of rocuronium-induced (1.2 mg/kg) profound neuromuscular block by sugammadex: a multicenter, dose-finding and safety study. *Anesthesiology*, 107, 239–244.

30. Lee, C., Jahr, J.S., Candiotti, K.A., Warriner, B., Zornow, M.H., Naguib, M. (2009). Reversal of profound neuromuscular block by sugammadex administered three minutes after rocuronium. *Anesthesiology*, 110, 1020–1025.

31. Plaud, B., Meretoja, O., Hofmockel, R., Raft, J., Stoddart, P.A., van Kuijk, J.H.M., Hermens, Y., Mirakhur, R.K. (2009). Reversal of rocuronium-induced neuromuscular blockade with sugammadex in pediatric and adult surgical patients. *Anesthesiology*, 110, 284–294.

20

CYCLODEXTRINS AND POLYMER NANOPARTICLES

DOMINIQUE DUCHÊNE AND RUXANDRA GREF

UMR CNRS 8612, Physico-Chimie–Pharmacotechnie–Biopharmacie, Faculté de Pharmacie, Université Paris–Sud, Châtenay Malabry, France

1. INTRODUCTION

For many years cyclodextrins (CDs) have been proposed as valuable tools in the design of drug dosage forms to improve active ingredient stability, solubility, and bioavailability whatever the type of dosage form: solid, viscous, or liquid, or the administration route: oral, buccal, nasal, ophthalmic, dermal, or parenteral [1]. More recently, nanoparticulate vectors, after liposomes and microparticles, have become the necessary carriers of drugs such as fragile molecules, anticancer products, proteins, and genes. In this new type of formulation, CDs demonstrate their ability to be more than simple additives but, rather, intelligent parts of the systems themselves [2,3].

In this chapter we focus exclusively on nanosystems resulting from combinations of CDs with polymers (or oligomers) to create original drug delivery systems. The main objective is not to present all the literature references existing in this domain, but rather to show how CDs can participate in the creation of new intelligent nanoparticles. Two main types of particle are considered: the already classical nanoparticles obtained from a mixture of polymers and CDs, and, the more elaborate gene delivery systems represented by the polyplexes in which CDs are involved.

2. CDs IN POLYMER NANOPARTICLES

When compared with other nano- or microcarriers, nanoparticles present several advantages. They are more stable than liposomes, and because of their small size, they have a larger surface area than microparticles, leading to higher bioavailability due to better contact with biological membranes [2]. Unfortunately, polymer nanoparticles may have limited drug-loading capacity, especially for weakly water-soluble drugs. It is one of the reasons leading to the use of CDs in the preparation of nanoparticles.

The group of Ponchel and Duchêne was one of the first to work in that domain, using poly(alkylcyanoacrylate) polymers. They have been followed by many others groups working with different polymers or copolymers. Two types of nanoparticles have to be considered: that in which the CD is just physically associated to the (co)polymer without any chemical bond (polymer/CD), and that in which the CD is chemically linked to the polymer, leading to a copolymer (polymer–CD).

2.1. Nanoparticles of CD/Polymer Associations

2.1.1. CDs in Poly(alkylcyanoacrylate) Nanoparticles
Poly(alkylcyanoacrylates) are particularly interesting in the preparation of nanoparticles, not only because they are biodegradable, but also because of the simple emulsion polymerization process that occurs in an aqueous medium and leads simultaneously to the formation of nanoparticles. Of course, the loading capacity (amount of drug associated to a unit mass of polymer) is often limited and may result in the necessity to administer considerable amounts of polymer, limiting their usefulness. This drawback has been the starting point of a series of studies by Ponchel and Duchêne on the potential role of CDs added to the polymerization medium.

Feasibility of Poly(alkylcyanocrylate) Nanoparticles in the Presence of CDs Monza da Silveira et al. [4,5] investigated

Cyclodextrins in Pharmaceutics, Cosmetics, and Biomedicine: Current and Future Industrial Applications, First Edition. Edited by Erem Bilensoy.
© 2011 John Wiley & Sons, Inc. Published 2011 by John Wiley & Sons, Inc.

Table 1. Poly(isobutylcyanoacrylate)/HP-β-CD Nanoparticles: Influence of the CD Nature on the Nanoparticles' Main Characteristics

Cyclodextrin (5 mg/mL)	Size (nm ± SD)	Zeta Potential (mV ± SD)	CD Content[a] (μg CD/mg Particles)
α-CD	228 ± 69	−34.4 ± 4.0	ND
β-CD	369 ± 7	−24.7 ± 8.2	360
γ-CD	286 ± 9	−22.9 ± 0.6	240
HP-α-CD	244 ± 25	−27.0 ± 2.2	ND
HP-β-CD	103 ± 6	−8.6 ± 0.9	247
HP-γ-CD	87 ± 3	−2.6 ± 2.2	220
SBE-β-CD	319 ± 10	−45.4 ± 2.4	ND

Source: Adapted from [4].

[a] ND, not determined.

the feasibility of poly(isobutylcyanoacrylate) nanoparticles in the presence of a series of CDs: α-, β- and γ-cyclodextrins, their hydroxylpropyl derivatives, and the sulfobutylether of β-cyclodextrin (α-CD, β-CD, γ-CD, HP-α-CD, HP-β-CD, HP-γ-CD, and SBE-β-CD, respectively). Nanoparticles were prepared by anionic polymerization of 100 μL of poly (isobutylcyanoacrylate) in 10 mL of hydrochloric acid solution (pH 2) containing 1% poloxamer 188 in the presence (or not) of CD. After 6 h of stirring at room temperature, the suspension was filtered (2-μm prefilter) to eliminate the aggregates and the product was characterized.

The main characteristics of the particles obtained are reported in Table 1. The smallest particle size was obtained for HP-γ-CD (87 ± 3 nm) followed by HP-β-CD (103 ± 6 nm), the largest for β-CD (369 ± 7 nm) and SBE-β-CD (319 ± 10 nm). The zeta potential, which was around −40 mV in the absence of CDs, decreased only in the case of the negatively charged SBE-β-CD (ca. −45 mV), and increased in the presence of neutral natural or hydroxypropylated CDs (ca. −25 mV for β-CD, −9 mV for HP-β-CD, and −3 mV for HP-γ-CD) as a consequence of the CD hydroxyl groups, probably because of the presence of the hydrophilic CDs at the surface of the hydrophobic nanoparticles.

It was demonstrated with HP-β-CD that the CD concentration had a noticeable effect on the particle size and charge. When the HP-β-CD concentration varied from 0 to 12.5 mg/mL, the particle size decreased from 300 nm to less than 50 nm and the zeta potential increased from −40 mV to approximately 0. The amount of HP-β-CD that can be associated to the nanoparticles could be as high as 60%. In the presence of HP-β-CD in the polymerization medium, addition of poloxamer 188 was not essential to obtain nanoparticles. The large amount of CD associated with the nanoparticles suggests that they have a prominent role in their formation. CDs could intervene in two manners in the nanoparticle formation. First, due to their high number of hydroxyl

groups, they could initiate the polymerization process of isobutylcyanoacrylate; however, this hypothesis was not confirmed by further studies [5]. Second, the strong influence of CDs on the nanoparticle surface properties (zeta potential) and their very fast release [6] suggest that they are adsorbed at the nanoparticle surface and that they probably have a steric stabilizing effect on the nanoparticles obtained [5].

Loading of Nanoparticles In a first attempt, the previous poly(isobutylcyanoacrylate)/HP-β-CD nanoparticles were loaded with a model drug: progesterone [4,5]. For that, a progesterone/HP-β-CD complex was prepared and added to the polymerization medium in order to obtain HP-β-CD concentrations varying from 2.5 to 20.0 mg/mL, in the presence (or not) of poloxamer 188. The progesterone loading was increased dramatically in the presence of HP-β-CD. In the absence of HP-β-CD, the loading was 0.79 μg/mg, and in the presence of 20.0 mg/mL HP-β-CD, the loading increased up to 45 μg/mg, which represents a 50-fold increase. There were no significant differences between the particles prepared in the presence or absence of poloxamer.

The in vitro release of progesterone, in pH 8.4 phosphate buffer, occurred in two steps [5,6]. After a fast release, a plateau was observed at 30 or 50%, depending on the nanoparticle size (150 or 70 nm). Complete release was obtained only after degradation of nanoparticles by the presence of esterases in the dissolution medium. This result indicated that part of the progesterone was located at the nanoparticle surface (fast release) and in the nanoparticle core (release after degradation).

The loading mechanism was investigated on a series of steroids [7] (hydrocortisone, prednisolone, estradiol, spironolactone, testosterone, megestrol acetate, danazol, and progesterone) differing by their solubility in poloxamer 188 or HP-β-CD solutions, and their octanol–water partition coefficient (correlated with the polymer–water partition coefficient) (Table 2). The presence of HP-β-CD in the polymerization medium resulted in significant increases in the drug payload of all the steroids, except for estradiol, which showed a reduction in loading. The highest increase in the particle payload was 129-fold for prednisolone.

Whatever the presence of HP-β-CD, no direct correlation appeared between the steroid loading of nanoparticles and either the steroid solubility in polymerization medium or the steroid partition coefficient considered independently. On the other hand, in the presence of HP-β-CD, a combined influence of the initial drug concentration (solubility) and partition coefficient was observed.

A general mechanism for drug loading has been proposed. In the polymerization medium the drug/CD complex reversibly dissociates according to its stability constant (D/CD ↔ D + CD); this leads to free entities. The free hydrophobic drug can be displaced from the solution to the polymer network, driven by its partition coefficient in favor of the

Table 2. Poly(isobutylcyanoacrylate)/HP-β-CD Nanoparticles Loaded with Steroids: Physical Characteristics of Steroids and Loading Increase Due to the Presence of HP-β-CD in the Poly (isobutylcyanocrylate) Nanoparticles

Steroid	Octanol/ Water Partition Coefficient	Solubility in 1% Poloxamer (mM)	Solubility in 1% HP-β-CD (mM)	Loading Increase (Fold Number)
Hydrocortisone	1.86	0.086	0.267	7.0
Prednisolone	1.60	0.059	0.337	129.2
Estradiol	3.78	0.0011	0.136	0.6
Spironolactone	2.80	0.0156	0.314	6.9
Testosterone	3.35	0.0018	0.207	8.6
Megestrol acetate	3.90	0.0013	0.0065	5.6
Danazol	4.20	0.0052	0.0296	32.9
Progesterone	3.85	0.0019	0.197	27.7

Source: Adapted from [7].

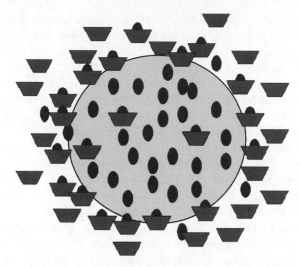

Figure 1. Localization of drug, CD, drug/CD inclusion molecules inside or on the surface of a poly(alkycyanoacrylate) nanoparticle.

polymer. The level of drug incorporation into the polymer network depends on the affinity between the drug and the polymer compared with the affinity of the drug for the CD cavity. The remaining drug/CD complex, as well as empty CDs can be adsorbed at the nanoparticle surface (Fig. 1).

Saquinavir-Increased Bioavailability by CD/ Poly(alkylcyanoacrylate) Nanoparticles Saquinavir is a potent HIV-1 and HIV-2 protease inhibitor. Unfortunately, saquinavir has low oral bioavailability, because of the hepatic first-pass effect, limited absorption due to low water solubility, and the effect of P-glycoprotein (P-Gp) responsible for an efflux mechanism. Boudad et al. tried to overcome these drawbacks by preparing CD/poly(alkylcyanoacrylate) nanoparticles intended for oral administration [8,9]. Two types of alkylcyanoacrylate were investigated: isobutylcyanoacrylate (IBCA) and isohexycyanoacrylate (IHCA), and two types of CD were used: HP-β-CD [8] and M-β-CD (methylated β-cyclodextrin) [8,9]. The nanoparticles were prepared as described above. Their main characteristics are reported in

Table 3 [9]. The drug content was increased considerably when using the HP-β-CD/ or M-β-CD/saqinavir complexes. These increases were independent of the polymer type. Even if M-β-CD was less associated with the particles than HP-β-CD was, the saquinavir loading was much higher with M-β-CD than with HP-β-CD. The association of saquinavir was slightly higher with PIHCA; this could be due to the higher hydrophobicity of PIHCA than with PIBCA.

The release of saquinavir studied on PIBCA or PIHCA/ HP-β-CD nanoparticles in reconstituted gastric (or intestinal) medium was very rapid and complete in less than 4 h [10,11]. Studied on Caco-2 cell monolayers, the transport of saquinavir alone from the basolateral to the apical compartment was higher than the transport existing from the apical to the basolateral side, this phenomenon being attributed to the expression of P-Gp efflux pump present at the apical side of Caco-2 cells. The simple inclusion of saquinavir in either HP-β-CD or M-β-CD did not improve the absorption of the drug. On the other hand, the association of

Table 3. Poly(alkylcyanoacrylate)/CD Nanoparticles Loaded with Saquinavir: Main Characteristics of Particles, and Drug Content

Formulation	Initial Concentration of Saquinavir (mg/mL)	Size (nm ± S.D.)	Zeta Potential (mV)	Drug Content (μg saq./mg Particles)
PIBCA/saq.	0.0385[a]	310.8 ± 83.6	−4.2	2.4
PIBCA/M-β-CD/saq.	1.2[b]	393.3 ± 34.9	9.7	75.6
PIBCA/HP-β-CD/saq.	1.2[b]	351.9 ± 86.6	14.5	38.5
PIHCA/saq.	0.0385[a]	322.7 ± 79.8	−10.3	2.9
PIHCA/M-β-CD/saq	1.2[b]	195.4 ± 61	10.2	85.0
PIHCA/HP-β-CD/saq.	1.2[b]	243.1 ± 56.3	10.2	42.5

Source: Adapted from [9].

[a] Saquinavir alone in the polymerization medium.

[b] Saquinavir in the form of inclusion in a CD.

saquinavir to PIBCA/HP-β-CD or PIBCA/Me-β-CD promoted faster transport of the drug whatever the direction of the transport. The addition of free M-β-CD (2.5%) to the complexes, or to the saquinavir-loaded PIBCA/CD nanoparticles, increased the amount of saquinavir recovered in the basolateral chamber. This was correlated with a decrease in the transport from the basolateral to the apical chambers [10,12].

Interestingly, either HP-β-CD or M-β-CD decreases poly (alkylcyanoacrylate) nanoparticle cytotoxicity measured on Caco-2 monolayers [9]. This could be due to the ability of CDs to shield the alkyl chains borne by the polymer when it is not yet degraded, thus lowering interactions with cell membranes. It could also be due to the complexation of fatty alcohols issued from degradation of the polymers, rendering the free alcohols less available to the cells.

2.1.2. CDs in Chitosan Nanoparticles

Chitosan is a polysaccharide well known to improve drug bioavailability not only because of its mucoadhesive properties, which can prolong the residence time at the absorption site, but also because of its ability to open the tight junctions. Alonso's group carried out a series of works on chitosan nanoparticles and investigated the value of adding CDs to these particles. Their main objective was to combine the advantages of chitosan with that of CDs which can both carry hydrophobic drugs and thus increase their bioavailability by increasing their apparent water solubility. Such systems would be of interest in the formulation of drugs belonging to classes 2 (low solubility, high permeability) and 4 (low solubility, low permeability) of the Biopharmaceutics Classification System [13]. In this domain they have been pioneers and were followed by other researchers.

Preparation of Chitosan/CD Nanoparticles The preparation of chitosan/CD nanoparticles was first described with HP-α- and HP-β-CD [13], but, as will be seen, it can be applied to other CDs. Nanoparticles were prepared by a modified ionotropic gelation method [14]. Aqueous solutions of different concentrations of HP-β-CD (0 to 25 mM) were incubated with chitosan (0.2% w/v) for 24 h under stirring, and then filtered (0.45-μm filter). The negatively charged cross-linking agent tripolyphosphate (TPP) was added (TPP/chitosan ratios varying from 1 : 3 to 1 : 6), leading to gelation in the form of nanoparticles. In preliminary studies, the optimal ratio of TPP/chitosan was demonstrated to be 1 : 4, hydrochloride chitosan was preferred to glutamate salt, and low molecular weight chitosan (110 kDa) was shown to lead to the smallest particles (around 450 to 600 nm). For nanoparticles prepared with HP-α-CD and 1.25 mg/mL TPP concentration, a slight reduction in the size (from 401 ± 29 nm to 361 ± 18 nm) was observed with an increase in the CD concentration (from 3.14 mM to 25 mM), but this effect was not detected

when using 2 mg/mL of TPP, this result probably being due to the broad size distribution observed in this case. The production yield increased with an increased amount of CDs [13].

Nanoparticles were loaded with poorly water-soluble drugs: triclosan and furosemide. The drugs were incubated for 24 h with chitosan water solutions containing different amounts of HP-β-CD, and after filtration the solution was used for the nanoparticle preparation as above. The association of the drugs to HP-β-CD facilitated their entrapment into the nanoparticles, increasing up to 4- and 10-fold (for triclosan and furosemide, respectively) the final loading. In any case, it is very difficult to give general conclusions because if there was a very positive effect of HP-β-CD on the solubility of the drugs (125-fold for triclosan, and 27-fold for furosemide), there was also a negative effect of chitosan on triclosan inclusion in HP-β-CD, when it was rather a synergistic effect for furosemide inclusion [13].

Whatever the drug investigated (triclosan or furosemide) and the CD nature (HP-α- or HP-β-CD), the in vitro release profile (pH 6.0 acetate buffer) was very similar and very characteristic: There was an initial rapid release followed by a plateau, the remaining amount of drug being released only in the presence of enzyme chitosanane [13]. This release behavior was rather similar to that observed by the group of Ponchel and Duchêne for the steroids loaded into poly (alkylcyanoacrylate)/HP-β-CD nanoparticles.

In further studies the same group prepared nanoparticles with anionic CDs such as the carboxymethyl-β-cyclodextrin (CM-β-CD) or the sulfobutyl ether of β-cyclodextrin (SBE-β-CD) [15–18]. One of the ideas was that these CDs could interact favorably with the cationic chitosan molecules in the same way as TPP. In the presence of either CM-β-CD or SBE-β-CD, the nanoparticles were prepared as previously using the ionotropic gelation method by mixing the cationic chitosan solution with a polyanionic phase containing either CM-β-CD or SBE-β-CD, or both either CM-β-CD or SBE-β-CD and TPP [16].

It appeared that the cross-linking agent TPP was not necessary for the formation of nanoparticles. For nanoparticles prepared without TPP, the incorporation efficiency (percentage of the starting material incorporated in the nanoparticles) of the CDs was variable and proportional to the amount of CD in the starting materials. The incorporation efficiency was higher with SBE-β-CD than with CM-β-CD; this was explained by the stronger affinity of SBE-β-CD for chitosan than that of CM-β-CD. In the presence of TPP the incorporation efficiency of the SBE-β-CD and the chitosan was higher than that observed without TPP. Finally, in vitro stability studies indicated that the presence of CDs in the nanoparticles can prevent their aggregation in simulated intestinal fluid [16].

Chitosan/CD Nanoparticles for the Delivery of Peptides, Polypeptides, and Genes Considering that glutathione, a tripeptide used in the treatment of psoriasis, diabetes, liver diseases, and alcoholism, cannot still be administered orally because of its degradation, Ieva et al. proposed to encapsulate it into chitosan/SBE-β-CD nanoparticles [19]. The preparation process, based on the ionic gelation process, was very simple. An aqueous solution of SBE-β-CD and glutathione was prepared and kept under stirring in nitrogen atmosphere for 24 h at room temperature, then the solution was poured in an acidic solution of chitosan and the nanoparticles formed spontaneously. The presence of TPP as cross-linking agent was not necessary, due to the use of the anionic SBE-β-CD. Chitosan/SBE-β-CD/glutathione nanoparticles prepared with a weight ratio of 4 : 3 : 3 had a mean diameter of 160 ± 30 nm and an encapsulation efficiency of 25 ± 3%.

To characterize the glutathione localization inside the nanoparticles, the authors carried out a very interesting chemical analysis. X-ray photoelectron spectroscopy was used for the surface characterization, and the depth-profile analysis was carried out using an ion gun. The technique evidenced the presence of thiol groups (belonging to glutathione) not only at the particle surface but also in the particle core. The authors concluded that these chitosan/SBE-β-CD nanoparticles acted as a protective coating for glutathione, encapsulated in their inner part [19].

Chen et al. [2] wanted to assess the potential role of the biodegradable and biocompatible polyanionic dextran sulfate in the presence (or not) of SBE-β-CD, for the formulation of dalargin chitosan nanoparticles [20]. Dalargin, a hydrophilic hexapeptide and a Leu-enkephalin analog with opioid activity, was chosen as model peptide because of its potential use as biomarker for peptide delivery to the brain. Nanoparticles were prepared by the complex coacervation technique.

Loaded or not with dalargin, chitosan/dextran sulfate nanoparticles presented a smaller diameter than chitosan/SBE-β-CD nanoparticles, and the use of a mixture of dextran sulfate and SBE-β-CD resulted in the smallest particles. The entrapment efficiency was much higher with dextran sulfate than with SBE-β-CD. This was explained by the high charge density of dextran sulfate, leading to nanoparticles with more negatively charged sulfate groups for interaction with positively charged dalargin, resulting in high peptide loading. When a mixture of SBE-β-CD and dextran sulfate was used, instead of SBE-β-CD, much smaller particles were obtained as the proportions of anionic components increased. The presence of both SBE-β-CD and dextran sulfate led to small nanoparticles with relatively good dalargin entrapment efficiency [20].

The in vitro release of dalargin was almost as slow as chitosan/dextran sulfate or chitosan/SBE-β-CD nanoparticles (20% in 72 h), when it was fast for simple chitosan nanoparticles for which more than 80% was released in 10 h [20].

The Alonso's group worked on insulin in chitosan/CD nanoparticles either as a macromolecular model drug [15] or in view of its nasal administration [18]. With the latter objective, they compared the role of two anionic cyclodextrins: SBE-β-CD and CM-β-CD [18]. From previous work [15], they selected two formulations: chitosan/SBE-β-CD/TPP 4 : 3 : 0.25 and chitosan/CM-β-CD/TPP 4 : 4 : 0.25. They calculated the loading efficiency and the association efficiency:

loading efficiency (%)

$$= \frac{\text{total amount of drug} - \text{amount of unbound drug}}{\text{nanoparticles weight}} \times 100$$

association efficiency (%)

$$= \frac{\text{total amount of drug} - \text{amount of unbound drug}}{\text{total amount of drug}} \times 100$$

The main nanoparticles characteristics are reported in Table 4. SBE-β-CD nanoparticles displayed the smallest size, the highest association efficiency and yield, and the lowest loading efficiency.

The nanoparticle cell interaction was assessed on Calu-3 cell monolayers by measuring the transepithelial electric resistance (TEER). Incubation with both nanoparticle formulations led to a significant decrease in the TEER of the cell monolayers, evidencing their ability to open the tight junctions between the epithelial cells. The interaction with nasal epithelium was studied on the rat by administration of fluorescein-labeled chitosan/CD nanoparticles. After sacrifice of the rats and excision of the mucosa, this one was submitted to confocal laser scanning microscopy. Not only was an intense fluorescence observed, but it was possible to visualize the intracellular localization of the nanoparticles, and thus the transcellular mechanism of transport. Finally, 5 IU/kg of insulin (solution or nanoparticles) were administered intranasally to rats. The maximum blood glucose level decrease obtained with the solution was only 14% at 30 min post-administration when it was about 35% at 1 h post-administration for the chitosan/CD nanoparticles. Whatever the CD nature and the corresponding nanoparticle characteristics, the

Table 4. Chitosan/SBE-β-CD/TPP and Chitosan/CM-β-CD/TPP Nanoparticles Loaded with Insulin: Main Characteristics of Particles and Drug Content

CD Type	Size (nm ± SD)	Association Efficiency (%)	Loading Efficiency (%)	Yield (%)
SBE-β-CD	327 ± 27	94.9 ± 0.1	23.3 ± 1.9	74.1 ± 6.1
CM-β-CD	436 ± 34	88.6 ± 0.8	46.7 ± 0.8	27.6 ± 2.3

Source: Adapted from [18].

hypocalcemic effect was similar. Compared with results obtained with insulin chitosan nanoparticles [22], it appeared that the main advantage of the formulations with cyclodextrins was that the amount of chitosan required to achieve a similar effect was much lower in the presence of SBE-β-CD or CM-β-CD (0.096 mg/kg with CD, 0.16 mg/kg without) [18].

After this work on nasal delivery, Alonso's group investigated the potential of chitosan/CD nanoparticles as gene delivery systems to the airway epithelium [17]. They worked on chitosan/SBE-β-CD/TPP and chitosan/CM-β-CD/TPP nanoparticles, and they used a plasmid DNA (pDNA) encoding the expression of secreted alkaline phosphatase (pSEAP). The ability of the nanoparticles to entrap pDNA was studied by agarose gel electrophoresis. It was shown that most of the DNA was associated with the nanoparticles since no migration of free DNA was observed [17].

Since in the design of a pDNA carrier the information on its ability to enter the target cells is often connected with cytotoxicity, this factor was assessed on Calu-3 cells. It appeared that the chitosan/CD nanoparticles exhibited a significantly lower cytotoxicity than those composed of solely chitosan. The IC_{50} values estimated were three-fold higher in the presence of CDs than in its absence. Furthermore, there was significantly lower toxicity with CM-β-CD than with SBE-β-CD. To study the interaction of nanoparticles with cells, Calu-3 cell monolayers were exposed to fluorescent nanoparticles and the cell nuclei were stained with propidium iodide to facilitate the nanoparticle localization. Confocal laser scanning microscopy indicated that irresponsive of their composition, the nanoparticles were effectively internalized by the cells [17].

Finally, the ability of the nanoparticles to transfect the monolayer was investigated. For that, the nanoparticles containing pSEAP were added to the monolayer and the amount of alkaline phosphatase produced by the cells, and then secreted to the culture medium, was determined for up to 6 days. It appeared that transfection with naked pSEAP produced very low gene expression, whereas all the nanoparticle formulations were able to elicit a significantly higher response. The transfection effect was higher when using low molecular weight chitosan than with higher molecular weight. Furthermore, with low molecular weight chitosan the efficacy of SBE-β-CD was similar to that of CM-β-CD, when it was lower with high-molecular-weight chitosan. The authors concluded that these nanocarriers represented a promising approach for gene therapy at the level of mucosa surfaces, in particular the respiratory mucosa [17].

2.1.3. CD in Poly(anhydride) Nanoparticles

Poly(anhydride) nanoparticle surface can easily be modified by incubation with molecules showing hydroxyl or amino groups, leading to "decorated" nanoparticles displaying new abilities of interest for drug delivery purposes. For example, the Irache's group worked on poly(methyl vinyl ether-*co*-maleic anhydride) (Gantrez AN) nanoparticles, at the surface of which they linked different molecules in order to increase their specific distribution and interactions with the gut: *Sambucus nigra* agglutinin [23], *Salmonella enteritidis* flagellin [24], mannosamine [25], thiamine [26] or vitamin B12 [27]. Having evidenced the bioadhesive properties of poly(anhydride) nanoparticles [28], the same group wanted to increase the incorporation of lipophilic drugs in these carriers by combining poly(anhydride) with CDs [29].

Preparation and Characteristics of Poly(anhydride)/CD Nanoparticles The poly(anhydride) used was the poly(methyl vinyl ether-*co*-maleic anhydride) (Gantrez AN 119) with a molecular weight of 200,000, and the CDs investigated were β-cyclodextrin (β-CD), 2-hydroxylpropyl-β-CD (HP-β-CD), and the 6-monodeoxy-6-monoamino-β-cyclodextrin (β-CD-NH$_2$). The CDs were dispersed in acetone and the poly(anhydride) dissolved in acetone was added to the suspension. The mixture was incubated under stirring at room temperature for different times. The nanoparticles were obtained by addition of an ethanol/water mixture and the organic solvents eliminated by evaporation under reduced pressure. The purification was carried out by centrifugation and dispersion of particles in an aqueous solution of sucrose (cryoprotector) repeated twice, and finally they were freeze-dried [29].

All the nanoparticles containing CDs, whatever their nature, displayed a slightly smaller size (140 to 150 nm) than the control nanoparticles without CD (180 nm). This decrease in particle size is in agreement with results described either with poly(isobutylcyanoacrylate) nanoparticles [4] or with chitosan [13]. The amount of CD associated to the nanoparticles was higher for β-CD than for HP-β-CD or β-CD-NH$_2$. This fact was probably due to the lower water solubility of β-CD compared with the two other CDs, giving it a better affinity for the polymer. Finally the surface of nanoparticles containing β-CD was smooth, whereas that of nanoparticles containing either HP-β-CD of β-CD-NH$_2$ was rough [29].

Gastrointestinal Transit of Poly(anhydride)/CD Nanoparticles The bioadhesive interactions of the nanoparticles through the gastrointestinal tract was investigated with nanoparticles loaded with the fluorescent rhodamine B isothiocyanate administered orally to rats. All the poly(anhydride)/CD nanoparticles displayed a higher bioadhesion than that of the control nanoparticles. They also showed a certain potential to concentrate in the stomach mucosa after 30 min. These results were more pronounced for HP-β-CD and β-CD-NH$_2$ than for β-CD itself. This is probably the consequence of the differences observed in particle surface characteristics, the rough surfaces providing larger surface areas to establish bioadhesive interactions [29]. Finally, from in vivo imaging biodistribution studies, carried out after oral administration to the rat of nanoparticles labeled with [99m]technetium, it appeared that the possibility for the poly(anhydride)/β-CD nanoparticles to be translocated and/or absorbed was negligible because no distribution to organs other than the gut was observed [29].

2.1.4. CD/Dextran Self-Assembled Nanogels Recently, Gref's group has proposed a mild procedure to obtain CD-containing spherical nanoassemblies in the absence of organic solvents, called nanogels because of the large amount of water retained in their structure [30].

Formation and Characterization of the CD/Dextran Nanogel Preparation of the nanogel is very simple. Spherical nanogels of 100 to 200 nm formed spontaneously in aqueous medium upon the association of two water-soluble polymers: a hydrophobically modified dextran obtained by grafting alkyl side chains and a β-CD polymer (pβ-CD) [30]. Part of the alkyl chains formed inclusion complexes with some CD cavities, leaving most of the CDs available to include hydrophobic molecules of interest such as therapeutic agents (Fig. 2) [31].

The pβ-CD was prepared by cross-linking of β-CD under strong alkaline conditions with epichlorohydrin [32]. The modified dextran was a dextran bearing hydrophobic lauryl side chains synthesized by coupling lauroyl chloride to dextran [33,34]. This method allowed varying the modified dextran substitution from 2 to 7% of glucose units.

The mechanism of pβ-CD/modified dextran nanogel formation, as well as its supramolecular architecture, were gained by ^{1}HNMR spectroscopy [23]. It is well established that alkyl chains can form inclusion with CDs [35,36]. It was

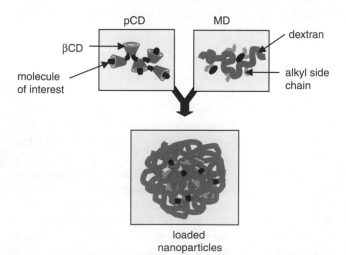

Figure 2. Mechanism of pβ-CD/modified dextran nanogel formation. (*See insert for color representation of the figure.*)

shown that the modified dextran alkyl chains were interacting with β-CD either in their monomeric or in their polymeric form, and that the complex obtained with pβ-CD was more stable than that obtained with β-CD [37]. Moreover, if the molar ratio CD/alkyl chains was ≤1, all the alkyl chains were interacting with pβ-CD cavities. This might be the principal reason for the stability of the nanogels: Sufficient physical cross-links between the chains were formed by means of inclusion complexes, thus stabilizing the supramolecular nanoassemblies.

The strong association between the two polymers was demonstrated by measuring the surface tension of pβ-CD, modified dextran and their mixture [37]. The amphiphilic-modified dextran adsorbed at the air–water interface, lowering its surface tension. The highly water-soluble pβ-CD did not adsorb and did modify the surface tension. When it was injected into the modified dextran solution, the surface tension increased and tended to that of pure water. This was the consequence of the progressive inclusion of the modified dextran lauryl chains into the pβ-CD cavities, and the progressive disappearance of modified dextran from the water surface. This phenomenon could be considered as surprising since the hydrophobic alkyl chains of the modified dextran normally were oriented toward the air and not toward the water, where the pβ-CD was, but it was probably sufficient for a few alkyl chains immersed in the water to be captured by the β-CD cavities to produce a chain reaction leading to complete depletion of the air–water interface by a "gear-wheel" process. It was shown that the use of high molar mass pβ-CD was crucial for the spontaneous interlocking of the two polymers.

Molecular modeling was employed to simulate the interaction between the modified dextran and pβ-CD in the presence of water molecules [37]. Two linear chains of pβ-CD (8 units) and modified dextran (16 units) were inserted within a cylinder containing 720 water molecules. The system was energy-minimized and subjected to molecular dynamics. Modified dextran tightly interlaced around pβ-CD, which tended to adopt a bent helical structure. Moreover, molecular dynamics clearly demonstrated the displacement of water molecules from β-CD cavities as a result of alkyl chain inclusion. Indeed, it was shown previously using molecular modeling that β-CD could easily accommodate up to 12 hydrogen-bonded water molecules within its hydrophobic core [37]. The resulting "free" water molecules were no longer at a distance compatible with the establishment of hydrogen bonds. As their number increased, it was inferred that polymer organization in nanoassemblies was essentially entropy-driven.

The nature of the modified dextran and especially the alkyl chain length had prominent influence. The study of different dextrans with alkyl chain length showed that there was an increased affinity between the chains and β-CD as the alkyl chain length increased (a factor of 2 to 3 from C_8 to C_{12}), but

Table 5. pβ-CD/Modified Dextran Nanogels Loaded with Benzophenone: Main Characteristics of Particles and Drug content[a]

Polymer Conc. (g/L)	MD/pβ-CD Ratio (w/w)	Mean Diameter (nm)			Entrapment Efficiency BL (%)	Loading BL (% w/w)
		UL	L	BL		
1	50/50	110 ± 20	135 ± 19	152 ± 23	26.6 ± 5.9	2.9 ± 0.8
	33/67	106 ± 15	141 ± 24	160 ± 20	22.0 ± 2.1	2.6 ± 0.3
2.5	50/50	136 ± 24	173 ± 30	192 ± 42	40.7 ± 7.0	2.5 ± 0.6
	33/67	140 ± 17	180 ± 28	188 ± 35	42.5 ± 5.0	2.9 ± 0.3

Source: Adapted from [30].

[a] MD, modified dextran; UL, unloaded nanogel; L, loaded gel; BL, bi-loaded gel.

the sensitivity was much lower than for small alkyl amphiphiles (a factor of 10 from C_8 to C_{12}) [38]. The short-term stability of modified dextran/pβ-CD (50:50) nanogel suspensions, having total polymer (modified dextran/pβ-CD) concentrations varying from 1 to 10 g/L, was assessed [31]. Whatever the polymer concentration, nanoassemblies smaller than 150 nm were formed initially. The most diluted suspensions were the most stable. For the 1-g/L suspensions the mean diameter remained below 200 nm over 24 h (from 125 ± 9 nm at $t = 0$ to 168 ± 5 at $t = 24$ h). In the case of high polymer concentrations, the diameter increased rapidly in the first 8 h and tended to reach a plateau. The higher the polymer concentration, the higher the size increase over time. For example, at $t = 12$ h, the mean diameter was 154 ± 4, 190 ± 2, 201 ± 18, and 387 ± 15 nm for 1-, 2.5-, 5- and 10-g/L polymer suspensions.

Because freeze-drying is one of the most convenient methods for long-time storage of colloidal systems, this method was tested on modified dextran/pβ-CD nanogels with a polymer concentration varying from 0.5 to 2.5 g/L [31]. All the nanogels were successfully freeze-dried except the 2.5-g/L suspension, and the dried cakes could easily be redispersed in milliQ water. The Tyndall effect was preserved and the size after freeze-drying was even slightly smaller than the initial size. At a 2.5-g/L polymer concentration, the rehydration of the cake led to the formation of a highly viscous phase.

Nanogels Loaded with Hydrophobic Drugs The entrapment of two model hydrophobic drugs, benzophenone and tamoxifen, into self-assembling nanogels was studied [39]. In a first attempt, nanogels were prepared by a one-step procedure. A solution of benzophenone or tamoxifen with pβ-CD (corresponding to solutions of inclusion complexes) was added to a solution of modified dextran. Nanogels were obtained instantaneously. Suspensions of tamoxifen nanogels were very instable and formed a gel deposit very rapidly, possibly as a consequence of the negative charge of the particles in contact with the positive charge of tamoxifen. Nanogels obtained with benzophenone were stable (more than 15 days at 4°C) and presented a mean diameter below 200 nm. Whatever the polymer concentrations and the ratios

tested, the benzophenone loading (amount of benzophenone associated per 100 mg of dried polymer) was less than or equal to 1% w/w.

A second loading method was proposed: the bi-loading method. Nanogels were prepared by mixing a benzophenone-pβ-CD solution and a benzophenone-modified dextran solution. The main characteristics of the nanogels are reported in Table 5. The particle size was independent of the modified dextran/pβ-CD ratio, but increased with the polymer concentration. It increased with the loading (unloaded < loaded < bi-loaded). The entrapment efficiency increased with the polymer content, while the loading remained constant. The bi-loading technique led to benzophenone loading approximately three-fold higher than the simple loading [40].

Benzophenone phase solubility diagrams carried out either with β-CD or pβ-CD showed that in the case of β-CD a B_s type of diagram (limited solubility of the inclusion) was obtained, whereas an A_l type was obtained with pβ-CD, with a 1:1 stoichiometry for both. Furthermore, the stability constants with β-CD or pβ-CD were not changed significantly (2680 and 2710 M^{-1} respectively). These results showed the advantage of using pβ-CD, rather than β-CD, because if the stability constants were similar, the solubility of the inclusion could be much higher with the polymer than with the monomer [40].

Nanogels Loaded with Contrast Agents Magnetic resonance imaging (MRI) is a powerful, non invasive diagnostic technique with high spatial resolution. The contrast in MRI is the result of a complex interplay between instrument parameters and intrinsic differences in the relaxation rates of tissue water protons. Gd^{3+} chelates, which reduce the proton relaxation times locally, are widely used as contrast agents. Gd^{3+}-loaded nanoparticles appear as promising candidates for molecular imaging and there is a growing need for more powerful new systems. In the pursuit of different in vivo delivery methods, one can change the size, charge, and surface properties of these carriers as well as the Gd^{3+} payloads. In this context, the host–guest type of interaction with CDs have been exploited to obtain large supramolecular structures [41], especially high-molecular-weight adducts of pβ-CD with suitably functionalized chelates, leading to an efficient relaxation enhancement [42,43].

Battistini et al. [44] described pβ-CD/modified dextran self-assembled nanogels loaded with Gd^{3+}. The preparation was the classical one in which Gd^{3+} was added to pβ-CD solution, and after stirring overnight, the solution was added to a modified dextran solution. The particles obtained had a diameter about 200 nm, a payload of 1.8×10^5 units of Gd^{3+}, and a relaxivity of 48.4 mM^{-1}/s at 20 MHz and 37°C. This macromolecular Gd-based system could be less toxic than that in which Gd^{3+} chelate is covalently bound to a polymeric matrix, since it would follow the elimination pathway of the free low-molecular-weight complex.

2.2. Nanoparticles of CD-Based Copolymers

CD-based copolymers have been used by different research groups for the formulation of nanoparticles combining the advantages of the hydrophilic CD to that of the hydrophobic polymer. When one CD is conjugated with one polymer chain, they form a "tadpole"-shaped copolymer. But there also exist copolymers in which more than one CD are combined.

2.2.1. Poly(lactide)-, Poly(glycolide)-, Poly(lactide-co-glycolide)-β-CD Nanoparticles
Because of their biocompatibility and biodegradability poly(lactide) (PLA), poly(glycolide) (PLG) and their copolymers, poly(lactide-*co*-glycolide) (PLGA) are among the most widely used polymers for drug encapsulation [45,46]. Their association to β-CD has been in detail studied by Ma's group [47–49].

In a first attempt, Gao et al. [47], considering the impossibility of encapsulating proteins in PLA nanospheres without deactivation, proposed to combine PLA to β-CD potentially capable of protecting the drug against deactivation. To graft only one PLA on one β-CD, they worked with a monoamino-substituted β-CD: the mono(6-(2-aminoethyl)amino-6-deoxy)-β-CD (β-CDen) prepared via the reaction of mono(6-*O*-(*p*-toluenesulfonyl))-β-CD [50,51] with freshly distilled ethyldiamine [52]. The coupling reaction with PLA of different molecular weights was carried out in the presence of *N,N'*-dicyclohexylcarbodiimide as catalyst.

Nanoparticles were prepared either by double emulsion [53] or by modified nanoprecipitation [54]. Bovine serum albumin (BSA) was used as a model protein. In both preparation methods the β-CDen-PLA nanoparticle diameter became smaller as the molecular weight of the copolymer decreased. When prepared by the double emulsion method, the nanoparticle diameters were in the range 300 to 500 nm; by the nanoprecipitation method the diameters were much smaller, in the range 150 to 250 nm. This could be interpreted by the sketch of the proposed mechanisms for the nanoparticle formations (Fig. 3): the double emulsion method leads to a multi-nanoreservoir system, when the nanoprecipitation results in single-layer nanospheres [47].

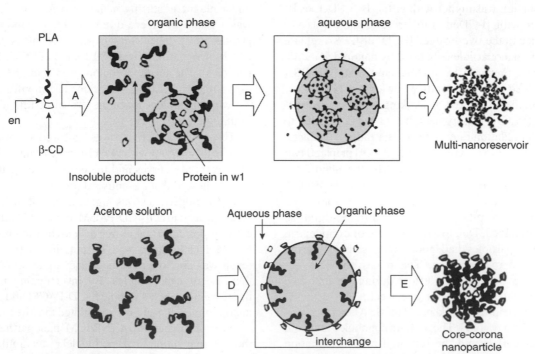

Figure 3. Supposed mechanisms of protein encapsulation into the CDen-PLA copolymer nanoparticles by the double emulsion (top) or nanoprecipitation method. (From [47], with permission of Elsevier.)

The encapsulation efficiency of BSA in β-CDen-PLA nanoparticles was much improved compared to that obtained in simple PLA nanoparticles. Different reasons could explain this phenomenon. The β-CD cavity is large enough to include aromatic and alkyl groups of BSA. Second, ionic interactions could occur between BSA and β-CDen-PLA: The amino group of the copolymer could exist in the ionic form $-NH_2{}^+-$ in neutral or acidic medium; this group may interact with $-COO^-$ in BSA. Third, the β-CDen-PLA copolymer being amphiphilic its hydrophilic part is accessible to the water-soluble BSA molecule [47]. Finally, the structure stability of BSA released in phosphate buffer saline (pH 7.4) was preserved [47].

In a second study, Gao et al. [48] worked with β-CD dimers, which could have better binding ability to BSA than β-CD itself because of the cooperative binding effect of the two adjacent cavities [55]. Two different types of β-CD dimers were prepared with either ethylenediamino or diethylenetriamino spacers at the upper rim of β-CD, leading to ethylenediamino-bridged bis-β-CD (B-β-CDen) and diethylenetriamino-bridged bis-β-CD (B-β-CDden). PLA was conjugated on them, leading to what could be considered as two-headed tadpoles. Nanoparticles were prepared as previously, either by double emulsion or nanoprecipitation methods [47].

The same conclusions as previously could be drawn: The nanoprecipitation method led to smaller nanoparticles with lower encapsulation efficiency than did the double emulsion method. It seemed that the encapsulation efficiency obtained by double emulsion was higher with either B-β-CDen or B-β-CDden than with β-CDen, furthermore, there was not much difference in the two bridged β-CD [48]. BSA release was faster from nanoparticles obtained by nanoprecipitation than from those prepared by double emulsion. The BSA released remained stable in solution for almost 14 days.

Considering the advantage of working with poly(lactide-co-glycolide) (PLGA) copolymers, because it is possible to control their degradation rate by the molar ratio of the lactic and glycolic acid in the polymer, Gao et al. [49] prepared two types of nanoparticles: β-CDen-PLGA [with the mono(6-(2-aminoethyl)amino-6-deoxy)-β-CD] and B-β-CDen-PLGA (with the ethylenediamino-bridged bis-β-CD). They used the repeated nanoprecipitation method and freeze-dried the nanoparticles obtained. Due to the existence of cyclodextrin corona, the nanoparticles had good steric stability during freeze-drying and resuspension. The release of BSA occurred in three phases: a burst effect during the first day, followed by a plateau for about one week, and sustained release of the protein over 14 days. The BSA structure was maintained for 21 days [49]. The authors concluded that both the novel biomaterials and the repeated nanoprecipitation technique were very promising for the delivery of hydrophilic proteins and the nanoparticle industrial production [49].

2.2.2. Poly(4-acryloylmorpholine)-β-CD Nanoparticles

Due to its capability to include a large variety of therapeutic molecules, β-CD is very often chosen in drug delivery design, but its low water solubility resulted in the search of chemical modifications to overcome this drawback. Among these modifications is the association to biocompatible hydrophilic synthetic polymers. The preparation of such β-CD-polymer derivatives has many advantages. When amphiphilic polymers are used, the derivative obtained may be soluble in both water and organic solvents, facilitating complex formation. The polymer can also cooperate with the β-CD moieties to stabilize the complex via secondary reactions.

A group with Cavalli, Trotta, Ferruti, Ranucci, et al. used a well-known amphiphilic biocompatible polymer, poly(4-acryloylmorpholine) (PACM), to prepare a water-soluble tadpole-like β-CD conjugate (β-CD-PACM) capable of forming nanoparticles for the delivery of acyclovir [56,57]. The β-CD-PACM polymer was prepared by radical polymerization of 4-acrylmorpholine in the presence of 6-deoxy-6-mercapto-β-CD (β-CD-SH) as chain transfer agent. These derivatives have an average molecular weight of 10^4 and carry a single β-CD moiety at one terminus. β-CD-PACM was able to solubilize a considerable amount of poorly water-soluble drugs such as acyclovir, and in vitro studies demonstrated that the polymer was noncytotoxic and nonhemolytic [56].

The solvent injection method [58] was used to prepare acyclovir/β-CD-PACM nanoparticles. Acetone solution of preloaded β-CD-PACM was added dropwise to filtered water under magnetic stirring maintained for 2 h. The dispersion was washed or freeze-dried to obtain nanoparticles as a powder [57]. The average diameter of acyclovir/β-CD-PACM nanoparticles was 180 ± 11 nm (135 ± 100 for blank particles). The acyclovir loading was about 13% w/w, due to the formation of stable inclusion. The in vitro release at pH 7.4 was rather slow, reaching 30% after 120 min without an initial burst effect [57].

The antiviral activity was assessed on monolayers of Vero cells infected with two clinical isolates of HSV-1. The dose–response curves and the corresponding IC_{50} and IC_{90} values showed that the antiviral potency of acyclovir/β-CD-PACM nanoparticles was higher than that of free acyclovir, whereas acyclovir/β-CD-PACM complex had an antiviral activity similar to that of free acyclovir, and unloaded β-CD-PACM nanoparticles had no antiviral activity [57]. This result was due to intracellular accumulation of acyclovir in the case of nanoparticles. By investigating the uptake of fluorescently labeled soluble β-CD-PACM and β-CD-PACM nanoparticles, the authors speculated that the higher antiviral activity of acyclovir/β-CD-PACM nanoparticles is due to the internalization and perinuclear accumulation of the nanoparticles, delivering the drug in the vicinity of the nucleus, which is the compartment where acyclovir exerts its antiviral activity [57].

2.2.3. Aminoethylcarbamoyl-β-CD–Based Nanoparticles

With the aim of creating a material for nanobiotechnology, Kodaka's group prepared nanoparticles composed of aminoethylcarbamoyl-β-CD (AEC-β-CD) and ethylene glycol diglycidyl ether (EGDGE) [59,60]. Nanoparticles were prepared by interfacial polyaddition reaction [61] of EGDGE and AEC-β-CD. An aqueous solution of AEC-β-CD was added to a surfactant (tetraglycerin monoester) cyclohexane solution and emulsified by ultrasonication. A toluene solution of EGDGE was mixed with the emulsion and the polymerization was carried out at 50°C for 4 h. After washing, the EGDGE-AEC-β-CD particles were suspended in water [59]. The particle size depends on both the water and the surfactant contents. It decreased with a water content increase, from 17.5 μm (water 0.25%/surfactant 5.0%) to 0.3 μm (water 5.0%/surfactant 5.0%), and to gel appearance for higher water concentrations. It increased with a surfactant content increase, from 0.3 μm (water 5.0%/surfactant 5.0%) to 26.5 μm (water 5.0%/surfactant 20.0%), for lower surfactant concentrations (1%), fiberlike product was obtained [60]. Working on 8-anilino-1-naphthalenesulfonic acid, the authors demonstrated that there is a synergistic effect of cavity inclusion and electrostatic interaction in the uptake of guest molecules by the particles [60].

2.2.4. Poly(γ-benzyl-L-glutamate)-β-CD Nanoparticles

The group of Ponchel and Fontaine, looking for "ideal" nanoparticles, were very much interested in families of polymers composed of the same backbone and bearing adapted chemical groups. The polymers should then be combined on demand for conferring the desired properties to the nanoparticles. Poly(γ-benzyl-L-glutamate) (PBLG), a synthetic polypeptide with a degradable amide bound in the backbone and a carboxylic acid function in the side chain, available after hydrolysis of the lateral benzyl ester, retained their attention. They carried out the synthesis of PBLG derivatives using initiators allowing the preparation of biocompatible nanoparticulate systems [62,63]. PBLG derivatives were synthesized by ring-opening polymerization of γ-benzyl-L-glutamate N-carboxyanhydride (γ-BLG-NCA) using different amine-terminated initiators among which the 6-monodeoxy-6-monoamino-β-cyclodextrin (β-CD-NH$_2$) prepared from its hydrochloride (β-CD-NH$_3^+$Cl$^-$) [64].

γ-BLG-NCA and β-CD-NH$_2$ were mixed in dimethylformamide at 30°C until disappearance of NCA bands from a Fourier transform infrared spectrum. The reaction product was poured in cold diethyl ether. The PBLG-β-CD precipitate was purified by methanol and diethyl ether and dried under vacuum at 35°C for 12 h. The estimated molecular weight of PBLG-β-CD was 46,000 Da [63]. Nanoparticles were prepared by a slightly modified nanoprecipitation method [65]. PBLG-β-CD was dissolved in tetrahydrofurane (THF) at 30°C for 18 h, and after gentle homogenization it was added to water in the presence (or not) of poloxamer 188. After 15 min of stirring, the solvent was evaporated and the nanoparticles were washed, their suspension in water can be freeze-dried [63]. Whatever the presence of poloxamer, the particle size was in the range 50 to 60 nm. The FTIR characterization suggested that PBLG-β-CD polymer presented an α-helix conformation before or after nanoparticle formation. In such a conformation the PBLG chain length is about 31 nm, corresponding to the nanoparticle radius. This suggested that a significant amount of β-CD could be exposed at the nanoparticle surface because of their hydrophilic nature, while the hydrophobic PBLG moieties could self-associate. This hypothesis was verified using isothermal microcalorimetry to characterize the accessibility of β-CDs within the nanoparticles by 1-adamantylamine: About 20% of the β-CDs remain functional within the nanoparticles [63]. The authors concluded that this system may be of interest when association of large amounts of hydrophobic drugs to nanoparticles is required.

3. CD SYSTEMS FOR GENE DELIVERY

Viral and antiviral systems can be used for gene delivery. Viral transfection systems can be very efficient but they have disadvantages: Adenoviruses have high immunogenicity, retroviruses produce random genomic integration, problematic scale-up manufacture exists, and so on. Nonviral transfection systems have low immunogenicity and the capacity to handle larger size of DNA than viruses; nevertheless, they also have some drawbacks, such as less gene expression and higher toxicity than that of viral vectors.

Among the nonviral systems, polyplexes (association of cationic polymer and nucleic acid) [66] are taking a prominent place. In the search of more efficient and less toxic systems, CDs have been employed either in linear copolymer-containing polyplex systems or in more elaborated structures, such as star, dendron, or polyrotaxane-containing polyplexes.

3.1. CDs in Linear Copolylmer-Containing Polyplexes

Cationic polymers can deliver plasmid DNA into cells by self-assembling with the anionic DNA via electrostatic interactions and condensation into nanoparticles which can easily be endocytosed. Whatever their nature, cationic polymers have similar mechanisms of delivery, but their transfection efficiency differs, and not only with their chemical nature, but also with their exact structure and their molecular weight. Considering the ability of CDs to form inclusion complexes exploited in drug formulation, and also their low toxicity and lack of immunogenicity, Davis's group undertake a series of works on CD-containing copolymer polyplexes for gene delivery.

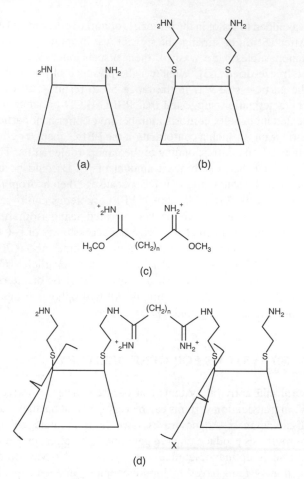

Figure 4. Products and synthesis of diamino-β-CD-NH₂-diimidate copolymer: (a) dideoxydiamino-β-CD; (b) dideoxy-diaminoethanethio-β-CD; (c) diimidate; (d) diamino-β-CD-diimidate copolymer. (From [67,68].)

3.1.1. Nature of the Linear Cationic β-CD-Containing Copolymer Polyplexes

β-CD-containing copolymers were prepared by the polymerization of a difunctionalized β-CD monomer (A) with a difunctionalized monomer (B) to give a ABAB copolymer. The difunctionalized β-CDs were either 6A,6D-dideoxy-6A,6D-diamino-β-CD or a 6A,6D-dideoxy-6A,6D-di(2-aminoethanethio)-β-CD (Fig. 4a and b). The difunctionalized monomers were either a series of diimidates differing by the number of methylene units (Fig. 4c) or dithiobis(succinimidyl propionate). The resulting cationic β-CD-diimidate copolymers (cpβ-CD) are schematized in Fig. 4d [67,68]. The copolymer yield was proportional to the diimidate length. The average polymerization degree was 4 to 5, with a cationic charge of 8 to 10. The cpβ-CDs formed polyplexes with different plasmids by rapid self-assembly in aqueous solutions. Polyplex sizes differed slightly with the cpβ-CD nature, from 120 to 150 nm. For effective gene delivery, the polycation must protect the plasmid from nuclease activity. The cpβ-CDs with even numbers of

methylenes ($n = 4$, 6, 8, 10 in Fig. 4) were those best protecting the plasmid DNA [68].

All the cpβ-CD were able to transfect BHK-21 cells, but with variable efficiency, depending on the monomer B length; the higher transfection was obtained for $n = 6$, the lowest ones for $n = 4$, 5, or 10. It could be explained by the fact that the spacing between cationic amidine groups in cpβ-CD with six methylenes was optimal for DNA binding. This seemed to be confirmed by the determination of relative binding constants by heparin sulfate displacement [68]. The toxicity associated with the transfection was evaluated by determination of the total protein concentration in cell lysates 48 h after transfection. No toxicity was observed for cpβ-CD with $n = 8$, and it increased for all the other cpβ-CDs. The IC$_{50}$ values of the cpβ-CDs to BHK-21 cells showed the same trend as the toxicities associated with polyplex transfection [68].

Polyethylenimine (PEI) has also been used for polyplex preparation, but its use as an in vitro and in vivo transfection agent is limited by its toxicity and difficulties in formulation. For this reason, different authors have tried to combine the qualities of CD-based polymers (among them, low toxicity) with those of linear or branched PEI [69–72]. As will seen later, PEI have been specially used in the preparation of dendrimers and β-CD-containing dendrimer polyplexes.

3.1.2. Influence of the Copolymer Structure on the Polyplex Characteristics

After demonstrating the value of cpβ-CDs in the formation of polyplexes, and the differences obtained when varying the copolymer structure, the same group of researchers decided to better define the structural effects of the different units of the copolymer [73–75]. They first studied the influence of the carbohydrate size and its distance from charge centers [73]. For that they synthesized a series of amidine-based polycations containing either β-CD or trehalose (among others). Because of the previous results, they kept six methylene units ($n = 6$ in Fig. 4) between the two charges (NH$_2^+$), and they varied the chain length between the charged part and the CD. The polymerization degree is affected by the monomer unit: Trehalose gives a higher polymerization degree than β-CD does.

Polyplexes were obtained with the different polycations, their diameters varied from 80 to 100 nm. All the polyplexes were able to transfect BHK-21 cells. As the charge center is removed farther from the carbohydrate unit, the toxicity is increased. As the size of carbohydrate moiety is enlarged from trehalose to β-CD, the toxicity is reduced [73].

In a second series of experiments, the influence of the charge center type was investigated by replacing the amidine charge center by different quaternary ammonium [74]. The charge center type did not greatly modify the toxicity since no significant difference was observed between the quaternary ammonium and the amidine families. On the other hand, the charge center had a significant influence on gene delivery since the amidine polyplexes exhibited a higher gene expression than that of their quaternary ammonium analogs.

Finally, the authors studied the influence of the cyclodextrin type (β and γ) and of their functionalization: 3A,3D-dideoxy-3A,3D-diamino-CDs compared with the 6A,6D-dideoxy-6A,6D-diamino-CDs studied previously. In each series, the length and character of the spacer between the CD and the amidine charge center were varied [69]. The nature of the spacer between the CD and the primary amines of each monomer influenced both the molecular weight and the polydispersity of the polycations. When these polycations were used to form polyplexes with plasmid, longer alkyl regions between the CD and the charge centers in the polycation backbone resulted in increased transfection efficiency and toxicity in BHK-21 cells. Increasing the hydrophilicity of the spacer (alkoxy or alkyl) resulted in lower toxicity. Finally, γ-CD-based polycations were less toxic than the β-CD-based polycations [69].

3.1.3. Tailored Polyplexes for Gene Delivery

Whatever their efficiency in vitro, polyplexes present a number of drawbacks when used in vivo. For example, when in the bloodstream, cationic polyplexes interact with serum proteins and are rapidly eliminated by phagocytosis; furthermore, they aggregate at physiological ionic strength. Finally, they should not be taken by Kupffer cells, but rather, they target specific tumor cells, thanks to specific ligands. For all these reasons, modified polyplexes have been developed by Davis and Pun, who grafted on the polyplexes moieties displaying specific biological activities, such as poly(ethylene glycol) (PEG) to obtain stealth particles, or ligands capable of affinity for defined molecules or cells. On the model of the complex formation between adamantane end-capped PEG and cyclodextrin polymers [76,77], they grafted

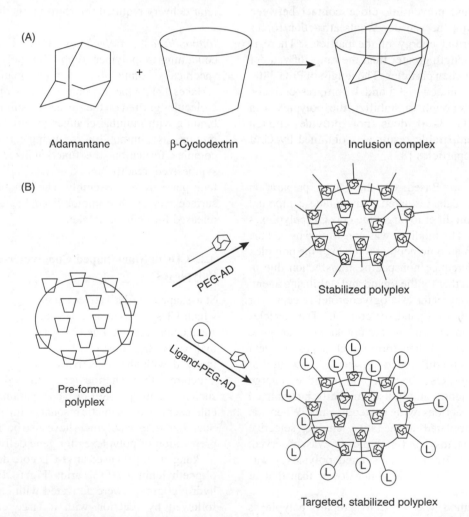

Figure 5. (A) Inclusion compound formation between adamantane and β-CD; (B) post-DNA-complexation PEGylation by inclusion compound formation by PEG-AD (adamantane) and ligand-PEG-AD. (From [78], with permission of the American Chemical Society.)

different modifiers on β-CD-based polyplexes [69,71,78–81]. The reaction mechanisms are represented in Fig. 5.

Stealth Polyplexes A very simple and efficient method to prepare stealth polyplexes was by self-assembly of already prepared polyplexes with adamantane–PEG at 100% adamantane to cyclodextrin (mol%). This is the "post-DNA-complexation PEGylation by inclusion compound formation" [78]. In salt solution (PBS) the average diameter of "nude" polyplexes increased from 58 to 272 nm; the presence of PEG 5000 in solution does not prevent aggregation (average diameter, 278 nm); polyplexes PEGylated with adamantane-PEG 3400 aggregated to 204 nm in diameter, those PEGylated with adamantane-PEG 5000 only increased to 95 nm. The PEGylated particles maintained small sizes in physiological salt concentration (150 mM) for hours. The difference in stabilization observed between PEG 3400 and PEG 5000 was explained by the fact that the PEG formed a steric layer around the particles, preventing close contact between particles and keeping the van der Waals attraction forces lower than the thermal energy of the particles. There is probably a critical length greater than the range of van der Waals attraction between particles. The length of PEG 3400 and PEG 5000 is estimated to 3.5 and 4.3 nm, respectively. PEG 5000 is long enough to stabilize the polyplexes in salt, whereas PEG 3400 does not provide enough stabilization. A comparable result was obtained by Gref et al. for PLA nanoparticles [82].

Targeted Polyplexes Based on the same preparation process, an anionic galactosylated adamantane compound was evaluated as a modifier to target the cpβ-CD polyplexes to hepatocytes [78]. The modifier was constituted by (1) the adamantane, (2) an anionic peptide to change the polyplex surface charge and reduce nonspecific transfection due to charge, (3) PEG as tether for the ligand and stabilizing agent, and (4) galactose to target the asialoglycoprotein receptor, or glucose as a low-affinity ligand for control. The particles were exposed to HepG2 cell line for uptake and transgene expression experiments. When formulated at positive zeta potentials, there was no difference between glucosylated or galatosylated polyplexes, because at positive charge polyplexes are efficiently and indiscriminantly internalized by nonspecific interactions with proteoglycans. When the particles were formulated at negative charge, selective transfection by galactosylated polyplexes was observed. The gene expression from galactosylated polyplexes was approximately one order of magnitude lower than that of unmodified polyplexes.

A different method was used to prepare polyplexes tailored with transferrin [79–81], an iron-binding glycoprotein ligand for tumor targeting. In this case, the following components in water solutions were mixed: (1) cpβ-CD

polymer, (2) adamantane–PEG, and (3) adamantane–PEG–transferring; then they were added to plasmid DNA solution. The polymers and DNA self-assembled in few seconds to form particles with diameters of 100 to 150 nm. At low transferrin modification, the particles remained stable in physiologic salt concentrations and transfected K 562 leukemia cells with increased efficiency over untargeted particles. The increase in transfection disappeared when transfection was carried out in the presence of an excess of free transferrin [79]. The transferrin polyplexes containing fluorescently labeled DNAzyme molecules were administered to tumor-bearing nude mice and their biodistribution and clearance were monitored. Four administration methods were compared: intraperitoneal injection, intraperitoneal infusion, subcutaneous injection, and intravenous injection. DNAzymes from polyplex formulations were concentrated and retained in tumor tissue and other organs, whereas unformulated DNAzyme was eliminated from the body within 24 h after injection. Tumor cell uptake was observed with intravenous injection only, and intracellular delivery required transferrin targeting [81].

Immobilized Polyplexes Cationic β-cyclodextrin-containing copolymer (cpβ-CD) polyplexes have been specifically immobilized on adamantane-functionalized surfaces [71]. One advantage of this method is that guest molecules grafted on the surfaces result in high nanoparticle binding with minimal changes in surface properties. In the present case, immobilization being carried out by inclusion complex formation, the functionalized surfaces could be synthesized readily and the nanoparticle immobilization took place by self-assembly. The nanoparticles held on the surface were stably immobilized but could be reversibly released for cellular uptake.

3.2. CDs in Star-Shaped Copolymer-Containing Polyplexes

Star-shaped CD derivatives are supramolecular assemblies in which CDs are used as cores for the synthesis of medium-sized conjugates obtained by per-substitution of the primary hydroxyls [83]. Most often such assemblies have been synthesized with the objective of coating the supramolecule structure with saccharide ligands toward biological receptors such as lectins [84–89] or Gd^{3+} paramagnetic chelates for enhancement of magnetic resonance imaging resolution [90]. Such star-shaped polymers have also been investigated in the preparation of polyplexes for gene delivery [91–93].

Yang et al. [91] used an α-CD core decorated with many oligoethylenimine (OEI) arms (Fig. 6). For that, the primary hydroxyl groups were activated with carbonyldiimidazole, followed by reaction with a large excess of linear or branched OEI of different lengths. These α-CD-OEI star polymers were capable of condensing pDNA (pRL-CMV encoding *Renilla* luciferase) into particulate structures.

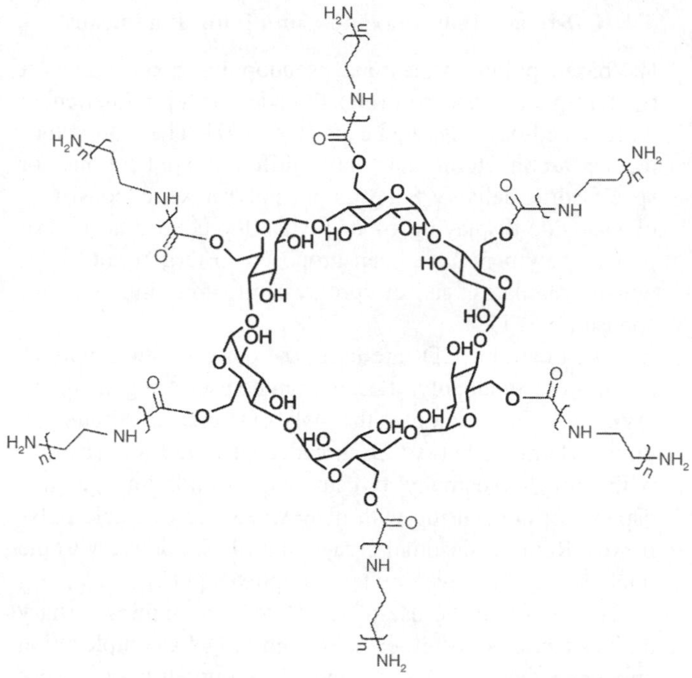

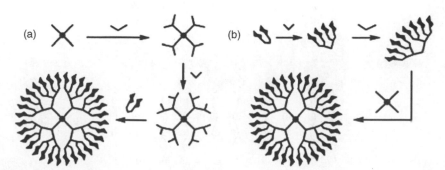

Figure 6. α-CD-oligoethylenimine star polymers. (From [91], with permission of Elsevier.)

The particle size depends on the nitrogen (in α-CD-OEI)-phosphate (in pDNA) ratio (N/P). Generally, the particle size decreased with an increase in N/P until 6 and 8; afterward, the particle size remained in the range 100 to 200 nm. The cytotoxicity of the α-CD-OEI star polymers, analyzed on Cos7 and HEK293 cells was much lower than that of PEI itself. Depending on the exact structure of the star polymers, the LD_{50} in Cos7 (the concentration of polymer resulting in 50% inhibition of cell growth), which is 12 μg/mL for PEI, varied from 88 to 560 μg/mL. The transfection efficiency of α-CD-OEI polyplexes in both cell lines increased with the chain length of the OEI grafted [91].

Other star-shaped polymers were prepared by Srinivasachari et al. [92] with β-CD decorated with alkine dendrons containing between zero and four secondary amines, and by Xu et al. [93] with poly(2-(dimethylamino)ethyl methacrylate) [P(DMAEMA)] and P(DMAEMA)-*block*-poly(ethylene glycol)ethyl ether methacrylate) [P(PEGEEMA)] arms. In both cases it was possible to prepare polyplexes by condensation of pDNA. The particle sizes were in the range 80 to 130 nm [92] or 100 to 200 nm [93] with an influence of the N/P ratios. In both cases, there was a decrease in cytotoxicity and an increase in gene transfection efficiency.

3.3. CDs in Dendrimer Copolymer-Containing Polyplexes

Dendrimers are a class of highly branched, monodisperse, globular macromolecules. They have a molecular architecture characterized by regular dendritic branching with radial symmetry. Their spherical structure is achieved by the ordered assembly of polymer subunits in concentric dendritic tiers around a central multifunctional core unit, terminated in many reactive groups around their peripheries [94,95]. There are two main processes for the synthesis of dendrimers: divergent and convergent growth (Fig. 7). In the divergent synthesis, the dendrimer is grown outward from the core, by doubling the number of reactive functionalities introduced at each new generation. In the convergent synthesis the synthesis of dendritic wedges is needed first, followed by the linking of these wedges to further branching components, and finally, by the attachment of these dendrons to the core [95].

Dendrimers have been prepared with different types of subunits: polyamidoamine (PAMAM) [94,96–98] and polypropylenimine [99–101]. They are characterized by high surface charge density and water solubility enabling electrostatic interactions with nucleic acids, allowing the formation of dendrimer–DNA complexes, which can be used for efficient transfection of cells [94]. However, dendrimers with low generations (first to third) have only low gene transfer activity [102,103], whereas dendrimers with higher generations exhibit cytotoxicity [104]. Similar to what was done for linear polyplexes, dendrimers have been conjugated to Cds [96–99].

Figure 7. (a) Divergent and (b) convergent syntheses of carbohydrate-coated dendrimers. (From [95], with permission of Elsevier.)

Different authors demonstrated the advantage of grafting β-CD to dendrimers to increase the transfection efficiency and decrease the cytotoxicity [94,99]. However, Uekama's group carried out a series of systematic studies on dendrimer-CD conjugates [96–98]. First, they investigated the role of the different natural cyclodextrins [96]. α-, β-, and γ-CDs were covalently bound to polyamidoamine dendrimers (second generation) in a molar ratio of 1 : 1. Substituted (or not) dendrimers were incubated with pDNA solution for 15 min to form the polyplexes (size around 700 nm). The dendrimer-CD conjugates formed complexes with pDNA and protected its degradation by DNase. They showed a potent luciferase gene expression, especially with the dendrimer conjugated with α-CD, which provided the greatest transfection efficiency: about 100-fold that of dendrimer without CD in NIH3T3 and RAW264.7 cells. The authors postulated that the enhanced gene transfer obtained with dendrimer-α-CD conjugate could be due not only to an increase in cellular association, but also to a change in the intracellular trafficking of DNA [96].

In a second study, the same group investigated the influence of the dendrimer generation, which varied from second to fourth (G2, G3, and G4) in dendrimer-α-CD conjugates [97]. They concluded that the gene transfer activity of the α-CD conjugate G3 was higher than that of dendrimers G2, G3, or G4 and also α-CD conjugate G2. On the other hand the cyctotoxicity of α-CD conjugate G3 was lower than that of α-CD conjugate G4. Therefore, they claimed that α-CD conjugate G3 might be a potent candidate for nonviral gene delivery.

Finally, in a third study, they investigated the influence of the dendrimer average substitution degree [98]. They prepared α-CD conjugates G3 containing different average numbers of α-CD: 1.1, 2.4, and 5.4. Their conclusion was that the α-CD conjugate G3 with a 2.4 degree of substitution was best because it provided good gene transfection in vitro and in vivo together with low toxicity.

3.4. CD-Based Polyrotaxane-Containing Polyplexes

CD-based polyrotaxanes and pseudopolyrotaxanes consist by multiple cyclodextrin rings threaded on a polymer chain with or without bulky end caps [105–107]. These supramolecular architectures may have different applications for specific drug delivery. For example, polyrotaxanes consisting of lactoside-displaying α- or β-CDs threaded onto hydrophobic polymers have been proposed to target galectin-1, which regulates cancer progression and immune response [108–111].

When cationic CDs are used, the corresponding polyrotaxanes or pseudopolyrotaxanes can be used for gene delivery [107]. For example, the polyrotaxane constituted by dimethylaminoethyl-α-CDs threaded on a PEG chain capped with benzyloxycarbonyl tyrosine via disulfide linkages was capable of condensing with pDNA, leading to stable polyplexes. Rapid endosomal escape and pDNA delivery to the nucleus were achieved by the polyplexes [112].

Thinking that the use of α-CD only containing tertiary amines could not be efficient enough in DNA complexation and gene delivery, Jun Li and his group decided to use oligoethylenimine-grafted α- or β-CDs (Fig. 8) [113–116].

α-CD was threaded either on a random copolymer, poly [(ethylene oxide)-*ran*-(propylene oxide) [114], or a homo poly(ethylene glycol) chain [115]. All the cationic polyrotaxanes could efficiently condense pDNA to form polyplexes. They displayed high transfection efficiency, which increased with the N/P ratio. All polyplexes containing α-CD grafted with linear oligoethylenimine exhibited higher transfection efficiency and lower cytotoxicity than did those with branched oligoethylenimine [114]. β-CD was threaded on a triblock copolymer chain: poly(ethylene glycol)–poly(propylene glycol)–poly(ethylene glycol) [113,116]. Here, too, the cationic polyrotaxanes obtained condensed pDNA; they showed low cytotoxicity and high transfection efficiency.

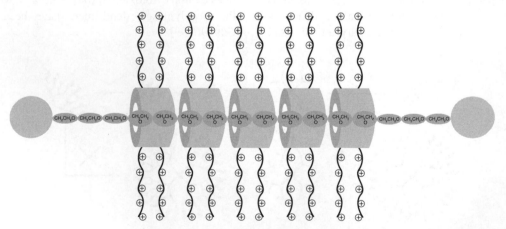

Figure 8. Structure of cationic polyrotaxane with multiple oligoethylenimine-grafted β-cyclodextrin rings. (From [113], with permission of Wiley-VCH Verlag GMbH & Co KGaA, and the authors.)

4. CONCLUSIONS

After this review on different uses of CDs in nanoparticle systems, it appears that CDs represent a prominent tool in the design of new drug delivery systems. Not only can they be added in the preparation process of already well-known polymeric nanoparticles to improve their loading capacity, but additional M-β-CD can significantly decrease the P-glycoprotein expression. Cyclodextrin can also be linked chemically to the nanoparticle polymer, to from a co-polymer with high loading capacity, but also with the possibility of encapsulating proteins.

Very encouraging is the use of CDs in the preparation of polyplexes, whatever their structure (linear, dendrimers, stars, polyrotaxanes). In this use, CDs showed their potential to decrease the cytotoxicity of the parent polyplexes and to increase their transfection efficiency. Due to their strong affinity for adamantane, CDs have demonstrated their abilty to transform polyplexes in stealthy, targeted, or immobilized systems.

The interesting role of CDs in drug delivery systems is not limited to their role in nanoparticle formation and characteristics—they can also be used in supramolecular structures; for example, two CDs most often β-CD, or their derivatives can form complexes with fullerenes C_{60} [117–121], making them water-soluble products and thus easier to use for biological applications, among which are anti-HIV or DNA cleavage activities [122–124].

Studied intensively for the past 30 years, with more and more sophisticated applications in the drug delivery sciences, CDs will continue to astonish us in the future with new and not yet predictable uses.

REFERENCES

1. Loftsson, T., Duchêne, D. (2007). Cyclodextrins and their pharmaceutical applications (historical perspectives). *Int. J. Pharm.*, *329*, 1–11.

2. Duchêne, D., Ponchel, G., Wouessidjewe, D. (1999). Cyclodextrins in targeting, applications to nanoparticles. *Adv. Drug Deliv. Rev.*, *36*, 29–40.

3. Duchêne, D., Wouessidjewe, D., Ponchel, G. (1999). Cyclodextrins and carrier systems. *J. Controll. Release*, *62*, 263–268.

4. Monza da Silveira, A., Ponchel, G., Puisieux, F., Duchêne, D. (1998). Combined poly(isobutylcyanoacrylate) and cyclodextrins for enhancing the encapsulation of lipophilic drugs. *Pharm. Res.*, *15*, 1051–1055.

5. Duchêne, D., Ponchel, G., Boudad, H., Monza da Silveira, A. (2001). Emploi des cyclodextrins dans la formulation de nanoparticules de poly(cyanoacrylate d'alkyle) chargées en divers principes actifs. *Ann. Pharm. Fr.*, *59*, 384–391.

6. Monza da Silveira, M., Duchêne, D., Ponchel, G. (2004). Drug release characteristics from combined poly(isobutylcyanoacrylate) and cyclodextrin nanoparticles loaded with progesterone. *Polym. Sci. A*, *46*, 1937–1944.

7. Monza da Silveira, A., Duchêne, D., Ponchel, G. (2000). Influence of solubility and partition coefficient on the loading of combined poly(isobutylcyanoacrylate) and hydroxypropyl-β-cyclodextrin nanoparticles by steroids. *STP Pharma Sci.*, *10*, 309–314.

8. Boudad, H., Legrand, P., Lebas, G., Cheron, M., Duchêne, D., Ponchel, G. (2001). Combined hydroxypropyl-β-cyclodextrin and poly(alkylcyanoacrylate) nanoparticles intended for oral administration of saquinavir. *Int. J. Pharm.*, *218*, 113–124.

9. Boudad, H., Legrand, P., Appel, M., Coconnier, M.-H., Ponchel, G. (2001). Formulation of combined cyclodextrin poly(alkylcyanoacrylate) nanoparticles on Caco-2 cells monolayers intended for oral administration of saquinavir. *STP Pharma Sci.*, *11*, 369–375.

10. Duchêne, D., Ponchel, G. (2002). Combined poly(alkyl cyanoacrylate)/cyclodextrin nanoparticles. *J. Inclusion Phenom. Macrocyclic Chem.*, *14*, 15–16.

11. Boudad, H., Legrand, P., Besnard, M., Duchêne, D., Ponchel, G. (2002). Drug release characteristics from combined poly(alkylcyanoacrylate) and cyclodextrin nanoparticles loaded with saquinavir. *Proceedings of the 4th World Meeting on Pharmaceutics, Biopharmaceutics, Pharmaceutical Technology,* Florence, Italy, pp. 807–808.

12. Boudad, H., Legrand, P., Coconnier, M.-H., Appel, M., Duchêne, D., Ponchel, G. (2002). Absorption of saquinavir from combined hydroxypropyl-β-cyclodextrin and poly (alkylcyanoacrylate) nanoparticles through Caco-2 cells monolayers. *Proceedings of the 4th World Meeting on Pharmaceutics, Biopharmaceutics, Pharmaceutical Technology,* Florence, Italy, pp. 1385–1386.

13. Maestrelli, F., Garcia-Fuentes, M., Mura, P., Alonso, M.J. (2006). A new drug nanocarrier consisting of chitosan and hydroxypropylcyclodextrin. *Eur. J. Pharm. Biopharm.*, *63*, 79–86.

14. Calvo, P., Remuñán-López, C., Vila-Jato, J.L., Alonso, M.J. (1997). Novel hydrophilic chitosan–polyethylene oxide nanoparticles as protein carriers. *J. Appl. Polym. Sci.*, *63*, 125–132.

15. Krauland, A.H., Alonso, M.J. (2007). Chitosan/cyclodextrin nanoparticles as macromolecular drug delivery system. *Int. J. Pharm.*, *340*, 134–142.

16. Trapani, A., Garcia-Fuentes, M., Alonso, M.J. (2008). Novel drug nanocarriers combining hydrophilic cyclodextrins and chitosan. *Nanotechnology*, *19*. 185101/1–185101/10.

17. Teijeiro-Osorio, D., Remuñán-López, C., Alonso, M.J. (2009). Chitosan/cyclodextrin nanoparticles can efficiently transfect the airway epithelium in vitro. *Eur. J. Pharm. Biopharm.*, *71*, 257–263.

18. Teijeiro-Osorio, D., Remuñán-López, C., Alonso, M.J. (2009). New generation of hybrid poly/oligosaccharide nanoparticles as carriers for the nasal delivery of macromolecules. *Biomacromolecules*, *10*, 243–249.

19. Ieva, E., Trapani, A., Cioffi, N., Ditaranto, N., Monopoli, A., Sabbatini, L. (2009). Analytical characterization of chitosan

nanoparticles for peptide drug delivery applications. *Anal. Bioanal. Chem.*, *393*, 207–215.

20. Chen, Y., Siddalingappa, B., Chan, P.H.H., Benson, H.A.E. (2008). Development of a chitosan-based nanoparticle formulation for delivery of a hydrophilic hexapeptide, dalargin. *Biopolym. (Pept. Sci.)* *90*, 663–670.

21. Chen, Y., Mohanraj, V., Parkin, J. (2003). Chitosan-dextran sulfate nanoparticles for delivery of an anti-angiogenesis peptide. *Int. J. Peptide Res. Ther.*, *10*, 621–629.

22. Fernández-Urrusuno, R., Romani, D., Calvo, P. Vila-Jato, J.L., Alonso, M.J. (1999). Development of freeze-dried formulation of insulin-loaded chitosan nanoparticles intended for nasal administration. *STP Pharma Sci.*, *9*, 429–436.

23. Arbós, P., Wirth, M., Arangoa, M.A., Gabor, F., Irache, J.M. (2002). Gantrez® AN as a new polymer for the preparation of ligand-nanoparticle conjugates. *J. Controll. Release*, *83*, 321–330.

24. Salman, H.H., Gamazo, C., Campanero, M.A., Irache, J.M. (2005). *Samonella*-like bioadhesive nanoparticles. *J. Controll. Release*, *106*, 1–13.

25. Salman, H.H., Gamazo, C., Campanero, M.A., Irache, J.M. (2006). Bioadhesive mannosylated nanoparticles for oral delivery. *J. Nanosci. Nanotechnol.*, *6*, 3203–3209.

26. Salman, H.H., Gamazo, C., Agüeros, M., Irache, J.M. (2007). Bioadhesive capacity and immunoadjuvant properties of thiamine-coated nanoparticles. *Vaccines*, *25*, 8123–8132.

27. Salman, H.H., Gamazo, C., De Smidt, P.C., Russell-Jones, G., Irache, J.M. (2008). Evaluation of bioadhesive capacity and immunoadjuvant properties of vitamin B12–Gantrez nanoparticles. *Pharm. Res.*, *25*, 2859–2868.

28. Arbós, P., Campanero, M.A., Arangoa, M.A., Irache, J.M. (2004). Nanoparticles with specific bioadhesive properties to circumvent the pre-systemic degradation of fluorinated pyrimidines. *J. Controll. Release*, *96*, 55–65.

29. Agüeros, M., Areses, P., Campanero, M.A., Salman, H., Quincoces, G., Peñuelas, I., Irache, J.M. (2009). Bioadhesive properties and biodistribution of cyclodextrin–poly(anhydride) nanoparticles. *Eur. J. Pharm. Sci.* *37*, 231–240.

30. Gref, R., Amiel, C., Molinard, K., Daoud-Mahammed, S., Sébille, B., Gillet, B., Beloeil, J.-C., Ringard, C., Rosilio, V., Poupaert, J., Couvreur, P. (2006). New self-assembled nanogels based on host–guest interactions: characterization and drug loading. *J. Controll. Release*, *111*, 316–324.

31. Daoud-Mahammed, S., Couvreur, P., Gref, R. (2007). Novel self-assembling nanogels: stability and lyophilisation studies. *Int. J. Pharm.*, *332*, 185–191.

32. Renard, E., Deratani, A., Volet, G., Sébille, B. (1997). Characterization of water-soluble high molecular weight β-CD–epichlorhydrin polymers. *Eur. Polym. J.*, *33*, 49–57.

33. Arranz, F., Sanchez-Chaves, M. (1988). ^{13}C nuclear magnetic resonnance spectral study on the distribution of substituents in relation to the preparation method of partially acetylated dextrans. *Polymers*, *29*, 507–512.

34. Amiel, C., Moine, L., Sandier, A., Brown, W., David, C., Hauss, F., Renard, E., Gosselet, M., Sébille, B. (2001). Macromolecular assempblies generated by inclusion complexes between amphipatic polymers and β-cyclodextrin polymers in aqueous media. In: McCormick, C.L., Ed., *Stimuli-Responsive Water Soluble and Amphiphilic Polymers.* American Chemical Society, Washington, DC, pp. 58–81.

35. Harada, A., Adachi, H., Kawaguchi, Y., Kamachi, M. (1997). Recognition of alkyl groups on a polymer chain by cyclodextrins. *Macromolecules*, *30*, 5181–5182.

36. Gelb, R.I., Schwartz, L.M. (1989). Complexation of carboxylic acids and anions by α and β cyclodextrins. *J. Inclusion Phenom.*, *7*, 465–476.

37. Daoud-Mahammed, S., Ringard-Lefebvre, C., Razzouq, N., Rosilio, V., Gillet, B., Couvreur, P., Amiel, C., Gref, R. (2007). Spontaneous association of hydrophibized dextran and poly-β-cyclodextrin into nanoassemblies: formation and interaction with a hydrophobic drug. *J. Colloid Interface Sci.*, *307*, 83–93.

38. Wintgens, V., Daoud-Mahammed, S., Gref, R., Bouteiller, L., Amiel, C. (2008). Aqueous polysaccharide associations mediated by β-cyclodextrin polymers. *Biomacromolecules*, *9*, 1434–1442.

39. Crini, G., Cosentino, C., Bertini, S., Naggi, A., Torri, G., Vecchi, C., Janus, L., Morcellet, M. (1998). Solid state NMR spectroscopy study of molecular motion in cyclomaltoheptaose (β-cyclodextrin) crosslinked with epichlorohydrin. *Carbohydr. Res.*, *308*, 37–45.

40. Daoud-Mahammed, S., Couvreur, P., Bouchemal, K., Chéron, M., Lebas, G., Amiel, C., Gref, R. (2009). Cyclodextrin and polysaccharide-based nanogels: entapment of two hydrophobic molecules, benzophenone and tamoxifen. *Biomacromolecules*, *10*, 547–554.

41. Harada, A. (2006). Supramolecular polymers based on cyclodextrins. *J. Polym. Sci. A*, *44*, 5113–5119.

42. Aime, S., Botta, I.M., Fedeli, F., Gianolio, E., Terreno, E., Anelli, P. (2001). High-relaxivity contrast agents for magnetic resonance imaging based on multisite interactions between β-cyclodextrin oligomers and suitably functionalized Gd^{III} chelates. *Chem. Eur. J.*, *7*, 5261–5269.

43. Aime, S., Botta, M., Fasano, M., Geninatti Crich, S., Terreno, E. (1999). ^{1}H and ^{17}O-NMR relaxometric investigations of paramagnetic contrast agentsfor MRI: clues for higher relaxivities. *Coord. Chem. Rev.*, *185–186*: 321–333.

44. Battistini, E., Gianolio, E., Gref, R., Couvreur, P., Fuzerova, S., Othman, M., Aime, S., Badet, B., Durand, P. (2008). High-relaxivity magnetic resonance imaging (MRI) contrast agent based on supramolecular assembly between a gadolinium chelate, a modified dextran, and poly-β-cyclodextrin. *Chem. Eur. J.*, *14*, 4551–4561.

45. Brannon-Peppas, L. (1995). Recent advances on the use of biodegradable microparticles and nanoparticles in controlled drug delivery. *Int. J. Pharm.*, *116*, 1–9.

46. Anderson, J.M., Shive, M.S. (1997). Biodegradation and biocompatibility of PLA and PLGA microspheres. *Adv. Drug Deliv. Rev.*, *28*, 5–24.

47. Gao, H., Wang, Y.N., Fan, Y.G., Ma, J.B. (2005). Synthesis of a biodegradable tadpole-shapped polymer via the coupling

reaction of polylactide onto mono(6-(2-aminoethyl)amino-6-deoxy)-β-cyclodextrin and its properties as the new carrier of protein delivery system. *J. Controll. Release, 107,* 158–173.

48. Gao, H., Yang, Y.-W., Fan, Y.G., Ma, J.B. (2006). Conjugates of poly(DL-lactic acid) with ethylenediamino or diethylenetriamino bridged bis(β-cyclodextrin)s and their nanoparticles as protein delivery systems. *J. Controll. Release, 112,* 301–311.

49. Gao, H., Wang, Y.-N., Fan, Y.-G., Ma, J.-B. (2007). Conjugates of poly(DL-lactide-*co*-glycolide) on amino cyclodextrins and their nanoparticles as protein delivery system. *J. Biomed. Mater. Res. A, 80,* 111–122.

50. Peter, R.C., Salek, J.S., Sikorski, C.T., Kumaravel, G., Lin, F. T. (1990). Cooperative binding by aggregated mono-6-(alkylamino)-β-cyclodextrins. *J. Am. Chem. Soc., 112,* 3860–3868.

51. Liu, Y., Yang, Y.W., Cao, R., Song, S.H., Zhang, H.Y., Wanf, L. H. (2003). Thermodynamic origin of molecular selective binding of bile salts by aminated β-cyclodextrins. *J. Phys. Chem. B, 107,* 14130–14139.

52. Gref, R., Rodrigues, J., Couvreur, P. (2002). Polysaccharides grafted with polyesters: novel amphiphilic copolymers for biomedical applications. *Macromolecules, 35,* 9861–9867.

53. Rodrigues, J.S., Santos-Magalhaes, N.S., Coelho, L.C.B.B., Couvreur, P., Ponchel, G., Gref, R. (2003). Novel core (polyester)–shell (polysaccharide) nanoparticles: protein loading and surface modifiaction with lectins. *J. Controll. Release, 92,* 103–112.

54. Govender, T., Stolnik, S., Garnett, M.C., Illum, L., Davis, S.S. (1999). PLGA nanoparticles prepared by nanoprecipitation: drug loading and release studies of a water soluble drug. *J. Controll. Release, 57,* 171–185.

55. Gao, H., Wang, Y.N., Fan, Y.G., Ma, J.B. (2006). Spectroscopic studies on the interactions of modified mono-bis-β-CDs with bovine serum albumin. *Bioorg. Med. Chem., 14,* 131–137.

56. Bencini, M., Ranucci, E., Ferruti, P., Manfredi, A., Trotta, F., Cavalli, R. (2008). Poly(4-acryloylmorpholine) oligomers carrying a β-cyclodextrin residue at one terminus. *J. Polym. Sci. A, 46,* 1607–1617.

57. Cavalli, R., Donalisio, M., Civra, A., Ferruti, P., Ranucci, E., Trotta, F., Lembo, D. (2009). Enhanced antiviral activity of acyclovir loaded into β-cyclodextrin–poly(4-acryloylmorpholine) conjugate nanoparticles. *J. Controll. Release, 137,* 116–122.

58. Schubert, M.A., Müller-Goymann, C.C. (2003). Solvent injection as a new approach for manufacturing lipid nanoparticles: evaluation of the method and process parameters. *Eur. J. Pharm. Biopharm., 55,* 125–131.

59. Eguchi, M., Du, Y.-Z., Taira, S., Kodaka, M. (2005). Functional nanoparticle based on β-cyclodextrin: preparation and properties. *NanoBiotechnology, 1,* 165–169.

60. Eguchi, M., Du, Y.-Z., Ogawa, Y., Okada, T., Yumoto, N., Kodaka, M. (2006). Effects of conditions for preparing nanoparticles composed of aminoethylcarbamoyl-β-cyclodextrin and ethylene glycol diglycidyl ether on trap efficiency of a guest molecule. *Int. J. Pharm., 311,* 215–222.

61. Yamazaki, N., Du, Y.-Z., Nagai, M., Omi, S. (2003). Preparation of polyepoxide microcapsule via interfacial polyaddition reaction in w/o and o/w emulsion systems. *Colloids Surf. B, 29,* 159–169.

62. Martínez Barbosa, M.E., Montembault, V., Cammas-Marion, S., Ponchel, G., Fontaine, L. (2007). Synthesis and characterization of novel poly(γ-benzyl-L-glutamate) derivatives tailored for the preparation of nanoparticles of pharmaceutical interest. *Polym. Int., 56,* 317–324.

63. Martínez Barbosa, M.E., Bouteiller, L., Cammas-Marion, S., Montembault, V., Fontaine, L., Ponchel, G. (2007). Synthesis and ITC characterization of novel nanoparticles constituted by poly(γ-benzyl L-glutamate)-β-cyclodextrin. *J. Mol. Recognit., 21,* 169–178.

64. Fontaine, L., Menard, L., Brosse, J.C., Sennyey, G., Senet, J.P. (2001). New polyurethanes derived from amino acids: synthesis and characterization of α,ω-diaminooligopeptides by ring-opening polymerization of glutamate N-carboxyanhydrides. *React. Funct. Polym., 47,* 11–21.

65. Thioune, O., Fessi, H., Devissaguet, J.P., Puisieux, F. (1997). Preparation of pseudolatex by nanoprecipitation: influence of the solvent nature on intrinsic viscosity and interaction constant. *Int. J. Pharm., 146,* 233–238.

66. Felgner, P.L., Barenholz, Y., Behr, J.P., Cheng, S.H., Cullis, P., Huang, L., Jessee, J.A., Seymour, L., Szoka, F., Thierry, A.R., Wagner, E., Wu, G. (1997). Nomenclature for synthetic gene delivery systems. *Hum. Gene Ther., 8,* 511–512.

67. Gonzalez, H., Hwang, S.J., Davis, M.E. (1999). New class of polymers for the delivery of macromolecular therapeutics. *Bioconjugate Chem., 10,* 1068–1074.

68. Hwang, S.J., Bellocq, N., Davis, M.E. (2001). Effects of structure of β-cyclodextrin-containing polymers on gene delivery. *Bioconjugate Chem., 12,* 280–290.

69. Pun, S.H., Bellocq, N.C., Liu, A., Jensen, G., Machemer, T., Quijano, E., Schluep, T., Wen, S., Engler, H., Heidel, J., Davis, M.E. (2004). Cyclodextrin-modified polyethylenimine polymers for gene delivery. *Bioconjugate Chem., 15,* 831–840.

70. Davis, M.E., Pun, S.H., Bellocq, N.C., Reineke, T.M., Popielarski, S.R., Mishra, S., Heidel, J.D. (2004). Self-assembling nucleic acid delivery vehicles via linear, water-slouble, cyclodextrin-containing polymers. *Curr. Med. Chem., 11,* 179–197.

71. Park, I.K., von Recum, H.A., Jiang, S., Pun, S.H. (2006). Supramolecular assembly of cyclodextrin-based nanoparticles on solid surfaces for gene delivery. *Langmuir, 22,* 8478–8484.

72. Forrest, M.L., Gabrielson, N., Pack, D.W. (2004). Cyclodextrin–polyethylenimine conjugates for targeted in vitro gene delivery. *Biotechnol. Bioeng., 89,* 416–423.

73. Reineke, T.M., Davis, M.E. (2003). Structural effects of carbohydrate-containing polycations on gene delivery: 1. Carbohydrate size and its distance from charge centers. *Bioconjugate Chem., 14,* 247–254.

74. Reineke, T.M., Davis, M.E. (2003). Structural effects of carbohydrate-containing polycations on gene delivery: 2. Charge center type. *Bioconjugate Chem., 14,* 255–261.

75. Reineke, T.M., Davis, M.E. (2003). Structural effects of carbohydrate-containing polycations on gene delivery: 3. Cyclodextrin type and functionalization. *Bioconjugate Chem.*, *14*, 672–678.

76. Amiel, C., Sébille, B. (1999). Association between amphiphilic poly(ethylene oxide) and β-cyclodextrin polymers: aggregation and phase separation. *Adv. Colloid Interface Sci.* *79*, 105–122.

77. Sandier, A., Brown, W., Mays, H. (2000). Interaction between an adamantane end-capped poly(ethylene oxide) and a β-cyclodextrin polymer. *Langmuir*, *16*, 1634–1642.

78. Hwang Pun, S., Davis M.E. (2002). Development of a non viral gene delivery vehicle for systemic application. *Bioconjugate Chem.*, *13*, 630–639.

79. Bellocq, N.C., Pun, S.H., Jensen, G.S., Davis, M.E. (2003). Transferrin-containing, cyclodextrin polymer-based particles for tumor-targeted gene delivery. *Bioconjugate Chem.*, *14*, 1122–1132.

80. Bartlett, D.W., Davis, M.E. (2007). Physicochemical and biological characterization of targeted, nucleic acid-containing nanoparticles. *Bioconjugate Chem.*, *18*, 456–468.

81. Pun, S.H., Tack, F., Bellocq, N.C., Cheng, J., Grubbs, B.H., Jensen, G.S., Davis, M.E., Brewster, M., Janicot, M., Janssens, B., Floren, W., Bakker, A. (2004). Targeted delivery of RNA-cleaving DNA enzyme (DNAzyme) to tumor tissue by transferrin-modified, cyclodextrin based particles. *Cancer Biol. Ther.*, *3*, 641–650.

82. Gref, R., Lück, M., Quellec, P., Marchand, M., Delacherie, E., Harnisch, S., Blunk, T., Müller, R. (2000). "Stealth" corona-core nanoparticles surface modified by poly(ethylene glycol) (PEG): influence of the corona (PEG chain length and surface density) and of the core composition on phagocytic uptake and plasma protein adsorption. *Colloids Surf., B*, *18*, 301–313.

83. Roy, R., Hernández-Mateo, F., Santoyo-González, F. (2000). Synthesis of persialylated β-cyclodextrins. *J. Org. Chem.*, *65*, 8743–8746.

84. Vargas-Berenguel, A., Ortega-Caballero, F., Santoyo-González, F., García-López, J.J., Giménez-Martínez, J.J., García-Fuentes, L., Ortiz-Salmerón, E. (2002). Dendritic galactosides based on a β-cyclodextrin core for the construction of site-specific molecular delivery systems: synthesis and molecular recognition studies. *Chem. Eur. J.*, *8*, 812–827.

85. García-López, J.J., Hernández-Mateo, F., Isac-García, J., Kim, J.M., Roy, R., Santoyo-González, F., Vargas-Berenguel, A. (1999). Synthesis of per-glycosylated β-cyclodextrins having enhanced lectin binding affinity. *J. Org. Chem.*, *64*, 522–531.

86. García-Barrientos, A., García-López, J.J., Isac-García, J., Ortega-Caballero, F., Uriel, C., Vargas-Berenguel, A., Santoyo-González, F. (2001). Synthesis of β-cyclodextin, per-*O*-glycosylated through an ethylene glycol spacer arm. *Synthesis*, 1057–1064.

87. Ortega-Caballero, F., Giménez-Martínez, J.J., García-Fuentes, L., Ortiz-Salmerón, E., Santoyo-González, F., Vargas-Berenguel, A. (2001). Binding affinity properties of dendritic glycosides based on a β-cyclodextrin core toward guest molecules and concanavalin A. *J. Org. Chem.*, *66*, 7786–7795.

88. Ortiz Mellet, C., Defaye, J., García Fernández, J.M. (2002). Multivalent cyclooligosaccharides: versatile carbohydrate clusters with dual role as molecular receptors and lectin ligands. *Chem. Eur. J.*, *8*, 1983–1990.

89. Adeli, M., Zarnegar, Z., Kabiri, R. (2008). Amphiphilic star copolymer containing cyclodextrin core and their application as nanocarrier. *Eur. Polym. J.*, *44*, 1921–1930.

90. Bryson, J.M., Chu, W.-J., Lee, J.-H., Reineke, T.M. (2008). A β-cyclodextrin "click cluster" decorated with seven paramagnetic chelates containing two water exchange sites. *Bioconjugate Chem.*, *19*, 1505–1509.

91. Yang, C., Li, H., Goh, S.H., Li, J. (2007). Cationic star polymers consisting of α-cyclodextrin core and oligoethylenimine arms as nonviral gene delivery vectors. *Biomaterials*, *28*, 3245–3254.

92. Srinivasachari, S., Fichter, K.M., Reineke, T.M. (2008). Polycationic β-cyclodextrin "click clusters": monodisperse and versatile schaffolds for nucleic acid delivery. *J. Am. Chem. Soc.*, *130*, 4618–4627.

93. Xu, F.J., Zhang, Z.X., Ping, Y., Li, J., Kang, E.T., Neoh, K.G. (2009). Star-shaped cationic polymers by atom transfer radical polymerization from β-cyclodextrin cores for nonviral gene delivery. *Biomacromolecules*, *10*, 285–293.

94. Roessler, B., Bielinska, A.U., Janczak, K., Lee, I., Baker, J.R., Jr. (2001). Substituted β-cyclodextrins interact with PAMAM dendrimer–DNA complexes and modify transfection efficiency. *Biochem. Biophys. Res. Commun.*, *283*, 124–129.

95. Turnbull, W.B., Stoddart, J.F. (2002). Design and synthesis of glycodendrimers. *Rev. Mol. Biotechnol.*, *90*, 231–255.

96. Arima, H., Kihara, F., Hirayama, F., Uekama, K. (2001). Enhancement of gene expression by polyamidoamine dendrimer conjugates with α-, β-, and γ-cyclodextrins. *Bioconjugate Chem.*, *12*, 476–484.

97. Kihara, F., Arima, H., Tsutsumi, T., Hirayama, F., Uekama, K. (2002). Effects of structure of polyamidoamine dendrimer on gene transfer efficiency of the dendrimer conjugate with α-cyclodextrin. *Bioconjugate Chem.*, *13*, 1211–1219.

98. Kihara, F., Arima, H., Tsutsumi, T., Hirayama, F., Uekama, K. (2003). In vitro and in vivo gene transfer by an optimized α-cyclodextrin conjugate with polyamidoamine dendrimer. *Bioconjugate Chem.*, *14*, 342–350.

99. Zhang, W., Chen, Z., Song, X., Si, J., Tang, G. (2008). Low generation polypropylenimine dendrimer graft β-cyclodextrin: an efficient vector for gene delivery system. *Technol. Cancer Res. Treat.*, *7*, 103–108.

100. Nijhuis, C.A., Sinha, J.K., Wittstock, G., Huskens, J., Ravoo, B.J., Reinhoudt, D.N. (2006). Controlling the supramolecular assembly of redox-active dendrimers at molecular printboards by scanning electrochemical microscopy. *Langmuir*, *22*, 9770–9775.

101. Nijhuis, C.A., Dolatowska, K.A., Ravoo, B.J., Huskens, J., Reinhoudt, D.N. (2007). Redox-controlled interaction of biferrocenyl-terminated dendrimers with β-cyclodextrin molecular printboards. *Chem., Eur. J.*, *13*, 69–80.

102. Bielinska, A., Kukowska-Latallo, J.F., Johnson, J., Tolalia, D.A., Baker, J.R., Jr. (1998). Regulation of in vitro gene

expression using antisense oligonucleotides or antisense expression plasmids transfected using starburst PAMAM dendrimers. *Nucleic Acid Res.*, *24*, 2176–2182.

103. Qiun, L., Pahud, D.R., Ding, Y., Bielinska, A.U., Kukowska-Latallo, J.F., Baker, J.R., Jr., Bromberg, J.S. (1998). Efficient transfer of genes into murine cardiac grafts by starburst polyamidoamine dendrimers. *Hum. Gene Ther.*, *9*, 553–560.

104. Kukowska-Latallo, J.F., Bielinska A.U., Johnson, J., Spindler, R., Tomalia, D.A., Baker, J.R., Jr., (1996). Efficient transfer of genetic material into mammalian cells using starburst polyamidoamine dendrimers. *Proc. Nat. Acad. Sci. USA*, *93*, 4897–4902.

105. Wenz, G. (1991). Cyclodextrins as building blocks for supramolecular structures and functional units. *Angew. Chem. Int. Ed. Engl.*, *33*, 803–822.

106. Wenz, G., Han, B.-H., Müller, A. (2006). Cyclodextrins rotaxanes and polyrotaxanes. *Chem. Rev.*, *106*, 782–817.

107. Li, J., Loh, X.J. (2008). Cyclodextrin-based supramolecular architectures: synthese, structures and applications for drug and gene delivery. *Adv. Drug Deliv. Rev.*, *60*, 1000–1017.

108. Nelson, A., Stoddart, J.F. (2003). Dynamic multivalent lactosides displayed on cyclodextrin beads dangling from polymer strings. *Org. Lett.*, *5*, 3783–3786.

109. Nelson, A., Belitsky, J.M., Vidal, S., Joiner, C.S., Baum, L.G., Stoddart, J.F. (2004). A self-assembled multivalent pseudo-polyrotoxane for binding galectin-1. *J. Am. Chem. Soc.*, *126*, 11914–11922.

110. Belitsky, J.M., Nelson, A., Stoddart, J.F. (2006). Monitoring cyclodextrin–polyviologen pseudopolyrotaxanes with the Bradford assay. *Org. Biomol. Chem.*, *4*, 250–256.

111. Belitsky, J.M., Nelson, A., Hernandez, J.D., Baum, L.G., Stoddart, J.M.F. (2007). Multivalent interactions between lectins and supramolecular complexes: galectin-1 and self-assembled pseudoplyrotaxanes. *Chem. Biol.*, *14*, 1140–1151.

112. Ooya, T., Choi, H.S., Yamashita, A., Yui, Y., Sugaya, Y., Kano, A., Maruyama, A., Akita, H., Ito, R., Kogure, K., Harashima, H. (2006). Biocleavable polyrotaxane–plasmid DNA polyplex for enhanced gene delivery. *J. Am. Chem. Soc.*, *128*, 3852–3853.

113. Li, J., Yang, C., Li, H., Wang, X., Goh, S.H., Ding, J.L., Wang, D.Y., Leong, K.W. (2006). Cationic supramolecules composed of multiple oligoethylenimine-grafted β-cyclodextrins threated on a polymer chain for efficient gene delivery. *Adv. Mater.*, *18*, 2969–2974.

114. Yang, C., Wang, X., Li, H., Goh, S.H., Li, J. (2007). Synthesis and characterization of polyrotaxanes consisting of cationic α-cyclodextrins threaded on poly(ethylene oxide)-*ran*-(propylene oxide) as gene carriers. *Biomacromolecules*, *8*, 3365–3374.

115. Yang, G., Li, H., Wang, X., Li, J. (2009). Cationic supramolecules consisting of oligoethylenimine-grafted α-cyclodextrins threated on poly(ethylene oxide) for gene delivery. *J. Biomed. Mater. Res. A*, *89*, 13–23.

116. Yang, C., Wang, X., Li, H., Tan, E., Lim, C.T., Li, J. (2009). Cationic polyrotaxanes as gene carriers: physicochemical properties and real-time observation of DNA complexation, and gene transfection in cancer cells. *J. Phys. Chem.*, *113*, 7903–7911.

117. Hu, H.-C., Liu, Y., Zhang, D.-D., Wang, L.-F. (1999). Studies on water-soluble α-, β- or γ-cyclodextrin prepolymer inclusion complexes with C_{60}. *J. Inclusion Phenom. Macrocyclic Chem.*, *33*, 295–305.

118. Komatsu, K., Fujiwara, K., Murata, Y., Braun, T. (1999). Aqueous solubilization of crystalline fullerenes by supramolecular complexation with γ-CD and sulfocalix [8] arene under mechanochemical high-speed vibration milling. *J. Chem. Soc., Perkin Trans. 1*, 2963–2966.

119. Filippone, S., Heimann, F., Rassat, A. (2002). A highly water-soluble 2 : 1 β-cyclodextrin–fullerene conjugate. *Chem. Commun.*, *14*, 1508–1509.

120. Murthy, C.N., Choi, S.J., Geckeler, K.E. (2002). Nanoencapsulation of [60]fullerene by a novel sugar-based polymer. *J. Nanosci. Nanotechnol.*, *2*, 129–132.

121. Fukami, T., Mugishima, A., Suzuki, T., Hidaka, S., Endo, T., Ueda, H., Tomono, K. (2004). Enhancement of water solubility of fullerene by cogrinding with mixture of cycloamyloses, novel cyclic α-1,4-glucans, via solid–solid mechanochemical reaction. *Chem. Pharm. Bull.*, *52*, 961–964.

122. Jensen, A.W., Wilson, S.R., Schuster, D.I. (1996). Biological applications of fullerenes. *Bioorg. Med. Chem.*, *4*, 767–779.

123. Samal, S., Geckeler, K.E. (2001). DNA-cleavage by fullerene-based synzymes. *Macromol. Biosci.*, *1*, 329–331.

124. Bosi, S., Da Ros, T., Spalluta, G., Prato, M. (2003). Fullerene derivatives: an attractive tool for biological applications. *Eur. J. Med. Chem.*, *38*, 913–923.

INDEX

Cyclodextrins in Pharmaceutics, Cosmetics, and Biomedicine: Current and Future Industrial Applications, First Edition. Edited by Erem Bilensoy.
© 2011 John Wiley & Sons, Inc. Published 2011 by John Wiley & Sons, inc.